Chern-Simons Gauge Theory: 20 Years After

AMS/IP
Studies in Advanced Mathematics

Volume 50

Chern-Simons Gauge Theory: 20 Years After

Jørgen E. Andersen
Hans U. Boden
Atle Hahn
Benjamin Himpel
Editors

American Mathematical Society • International Press

Shing-Tung Yau, General Editor

2000 *Mathematics Subject Classification.* Primary 11F23, 14E20, 16S10, 19L10, 20C20, 20F99, 20L05, 30F60, 32G15, 46E25, 53C50, 53D20, 53D99, 54C40, 55R80, 57M25, 57M27, 57M60, 57M50, 57N05, 57N10, 57R56, 58D27, 58D30, 58E09, 58J28, 58J30, 58J52, 58Z05, 70S15, 81T08, 81T13, 81T25, 81T30, 81T45, 83C80.

Library of Congress Cataloging-in-Publication Data

Chern-Simons gauge theory: 20 years after / Jørgen E. Andersen ... [et al.], editors.
p. cm. (AMS/IP studies in advanced mathematics ; v. 50)
Proceedings of a workshop held at the Max Planck Institute for Mathematics in Bonn, Aug. 3–7, 2009.
Includes bibliographical references.
ISBN 978-0-8218-5353-5 (alk. paper)
1. Number theory—Congresses. 2. Algebraic topology—Congresses. 3. Associative rings—Congresses. 4. K-theory—Congresses. 5. Group theory—Congresses. I. Andersen, Jørgen E. 1965–

QA241.C6355 2010
514′.74—dc22

2011012166

∞ The paper used in this book is acid-free and falls within the guidelines
established to ensure permanence and durability.
Visit the AMS home page at http://www.ams.org/
Visit the International Press home page at http://www.intlpress.com/

10 9 8 7 6 5 4 3 2 1   16 15 14 13 12 11

Contents

Photo courtesy of Michael Bohn.

Participants of Chern-Simons Gauge Theory: 20 years after.

Preface

The workshop *Chern-Simons gauge theory: 20 years after* was held at the Max-Planck-Institute for Mathematics in Bonn from August 3 until August 7, 2009. It brought together mathematicians and physicists from many countries, including Austria, Brazil, Canada, Denmark, France, Germany, Japan, Korea, Luxembourg, Netherlands, Poland, Portugal, Romania, Spain, Sweden, Switzerland, United Kingdom and the United States of America, to discuss their research programs in geometric topology, stochastic analysis and mathematical physics. What was shared by this diverse group was a strong interest in Chern-Simons theory, and this is reflected in the talks of the conference as well as in the papers in this volume.

Witten introduced Quantum Chern-Simons theory to knot theory in 1989, when he described the Jones polynomial of a link in S^3 as a certain (non-rigorous) Feynman path integral, more specifically as the quantum field theory whose action is a multiple of the Chern-Simons function. He interpreted these invariants using the axioms of topological quantum field theory (TQFT) as well as via an asymptotic expansion.

Since then, there have been several advances in understanding both approaches to Chern-Simons theory, starting with the first rigorous construction of the TQFT version by Reshetikhin and Turaev. This gave rise to quantum topology, a vibrant field which is currently moving in a number of directions including:

- categorification,
- modular forms, making contact with number theory,
- Gromov-Witten theory via Large N duality, and
- asymptotic analysis making contact with finite type invariants and hyperbolic geometry, for example via the volume conjecture and the AJ-conjecture.

Many of these topics were touched upon in the talks delivered at the conference and are discussed further in the papers of this volume.

One of the outstanding open problems in Chern-Simons theory is to provide a mathematically rigorous treatment for the Feynman path integral, and the difficulty has to do with the problem of constructing a suitable gauge-invariant measure on the infinite dimensional space of all connections. A new approach to this difficult problem is provided by stochastical analysis, and this approach was also discussed at the conference and is the focus for several papers in this volume.

Another important aspect of Chern-Simons gauge theory that was featured at the conference has to do with Floer homology theory and related invariants for 3-manifolds. The starting point is Taubes' result from 1989 which allows one to interpret the Casson invariant as the Euler characteristic of the instanton homology defined by Floer, and some new directions of research in Floer homology and Casson invariants are presented in the papers of this volume.

In addition to its many scientific activities, the conference featured a number of pleasant social activities, including a boat ride on the Rhine, a hike up the Drachenfels, a visit to the Biergarten in the Rheinau, and a fun wine and pretzel party at the Hausdorff Institute with musical performances in classical, jazz, and rock music by a number of talented individuals, including Dana Fine, Chris Herald, Ben Himpel, Diane and Louis Kauffman, Paul Kirk, Markus Land, Cristina Martinez, Justin Roberts, Roland van der Veen and Lucy Zhang. We are especially grateful to Matthias Kreck for making the Hausdorff Institute available to us for this memorable evening.

There were many people who helped to make the conference a success, and we would like to take this opportunity to recognize their contributions. To begin, the conference received funding through the Hausdorff Center for Mathematics in Bonn and the ITGP network of the European Science Foundation, and we would like to thank all the staff at the Hausdorff Institute and the Max-Planck-Institute in Bonn for their friendly assistance during the conference. The editors would also like to thank Arthur Greenspoon for his sage advice and Debbie Iscoe for her expertise in preparing the final versions of the manuscripts. Finally, we would like to thank all the authors who contributed their work to this volume, and especially Edward Witten, who delivered an inspirational lecture at the conference and contributed another monumental paper that is sure to stimulate further activity in this and related fields.

Jørgen E. Andersen, Hans U. Boden, Atle Hahn, and Benjamin Himpel
November 2010

List of Speakers

Dave Auckly: *Gauge-string duality and the structure of large rank Chern-Simons invariants*

Dror Bar-Natan: *Convolutions on Lie groups and Lie algebras and ribbon 2-knots*

Dave Auckly: *Gauge-string duality and the structure of large rank Chern-Simons invariants*

Dror Bar-Natan: *Convolutions on Lie groups and Lie algebras and ribbon 2-knots*

Chris Beasley: *Localization for Wilson loops in Chern-Simons theory*

Dana Fine: *A geometric alternative to gauge fixing in Chern-Simons theory on* $S^1 \times \Sigma$

Stavros Garoufalidis: *Chern-Simons theory and arithmetic*

Enore Guadagnini: *Functional integration and abelian link invariants*

Sergei Gukov: *Exact results for perturbative Chern-Simons theory with complex gauge group*

Christopher M. Herald: *An* $SU(3)$ *Casson invariant for rational homology spheres*

Kazuhiro Hikami: *WRT invariants and modular forms*

Rinat M. Kashaev: *On rings associated with ideal triangulations of knot complements*

Louis H. Kauffman: *Khovanov homology and the Potts model*

Mikhail Khovanov: *Categorification of quantum groups*

Paul Kirk: *Non-abelian representations, homology 3-spheres, and knot concordance*

Albrecht Klemm: *Chern-Simons theory and topological string theory on non-compact Calabi-Yau manifolds*

Christine Lescop: *On the cube of the equivariant linking pairing for closed 3-manifolds of rank one*

Marcos Mariño: *Chern-Simons theory, the* $1/N$ *expansion, and string theory*

Gregor Masbaum: *Integral structures in TQFT and the mapping class group*

Catherine Meusburger: *Getting physics from 3d gravity: What does an observer in 3d gravity see?*

Aleksandar Miković: *Invariants of spin networks embedded in 3-manifolds*

HITOSHI MURAKAMI:	*$SL(2,\mathbb{C})$-representations and asymptotic behaviors of the colored Jones polynomial of a knot*
ROBERT C. PENNER:	*Fatgraphs and finite type invariants*
ALBERT SCHWARZ:	*Generalizations of Chern-Simons theory*
AMBAR N. SENGUPTA:	*Functional integrals in low-dimensional gauge theories*
GEORGE THOMPSON:	*Chern-Simons theory on Seifert 3-manifolds*
JONATHAN WEITSMAN:	*Fermionization and convergent perturbation expansions in Chern-Simons gauge theory*
EDWARD WITTEN:	*Branes and quantization*

List of Participants

Sergio Albeverio, University of Bonn
Jørgen E. Andersen, University of Aarhus
Dave Auckly, Kansas State University
Dror Bar-Natan, University of Toronto
Stefan Bauer, University of Bielefeld
Chris Beasley, Stony Brook University
Stefan Behrens, University of Bonn
Adara Blaga, West University of Timisoara
Hans U. Boden, McMaster University
Francesco Costantino, IRMA, University of Strasbourg
Tien Cuong Dinh, Professor at Paris 6 University
Jérome Dubois, Université Paris 7, Denis Diderot
Magnus Engenhorst, Institut für Mathematik, Universität Augsburg
João Faria Martins, Centro de Matemática da Universidade do Porto
Alexander Felshtyn, University of Szczecin and MPIM, Bonn
Dana Fine, University of Massachusetts
Stavros Garoufalidis, Georgia Institute of Technology
Masha Gordina, University of Connecticut/University of Bielefeld
Enore Guadagnini, University of Pisa
Sergei Gukov, California Institute of Technology
Atle Hahn, Universidade de Lisboa
Luiz Hartmann, USP - Universidade de São Paulo
Andriy Haydys, University of Bielefeld
Chris Herald, University of Nevada
Michael Heusener, Université Blaise Pascal Clermont II
Kazuhiro Hikami, University of Tokyo
Benjamin Himpel, University of Bonn
Iulia Elena Hirica, University of Bucharest
Fuji Hiroyuki, Nagoya University
Lotte Hollands, University of Amsterdam
Saeid Jafari, College of Vestsjaelland South (VUC)
Franck Jedrzejewski, CEA (French Atomic Commission)
Alexander Kahle, University of Göttingen
Uwe Kaiser, Boise State University
Rinat Kashaev, University of Geneva
Louis H. Kauffman, Math Dept, University of Illinois at Chicago
Gerald Kelnhofer, University of Wien
Mikhail Khovanov, Columbia University
Hoil Kim, Kyungpook National University
Paul Kirk, Indiana University

Albrecht Klemm, University of Bonn
Valentin Krasontovitsch, University of Bonn (Math student)
Markus Land, University of Bonn
Jonatan Lenells, Cambridge University
Christine Lescop, University of Grenoble I
Poon Chuan Adrian Lim, University of Luxembourg
Christoph Luedeling, University of Bonn
Marcos Mariño, University of Geneva
Cristina Martinez, Autonoma University of Barcelona
Gregor Masbaum, Université Paris 7
Gwenael Massuyeau, CNRS, University of Strasbourg
Andrew McIntyre, Bennington College
Catherine Meusburger, University of Hamburg
Jouko Mickelsson, University of Helsinki and Royal Inst. of Technology
Aleksandar Mikovic, Universidade Lusofona, Lisbon
Hitoshi Murakami, Tokyo Institute of Technology
Chiara Nappi, Institute for Advanced Study
Robert Penner, University of California, Los Angeles
Arturo Prat-Waldron, University of California, Berkeley and MPIM Bonn
Justin Roberts, University of California, San Diego
Nobuya Sato, Rikkyo University
Albert Schwarz, UCDavis/MPIM
Ambar Sengupta, Louisiana State University
Jan Swoboda, MPIM, Bonn
George Thompson, International Centre for Theoretical Physics ICTP
Giorgio Trentinaglia, Georg-August-Universität Göttingen
Alessandro Valentino, University of Göttingen
Roland van der Veen, University of Amsterdam
Jonathan Weitsman, Northeastern University
Edward Witten, Institute for Advanced Study
Don Zagier, Max-Planck-Institute for Mathematics
José Miguel Zapata Rolón, University of Bonn
Lucy Liuxuan Zhang, University of Toronto

AMS/IP Studies in Advanced Mathematics
Volume 50, 2011

Remarks on Wilson Loops and Seifert Loops in Chern-Simons Theory

Chris Beasley

ABSTRACT. As noted long ago by Atiyah and Bott, the classical Yang-Mills action on a Riemann surface admits a beautiful symplectic interpretation as the norm-square of a moment map associated to the Hamiltonian action by gauge transformations on the affine space of connections. Here I will explain how certain Wilson loop observables in Chern-Simons gauge theory on a Seifert three-manifold can be given an analogous symplectic description. Among other results, this symplectic description implies that the stationary-phase approximation to the Wilson loop path integral is exact for torus knots, an empirical observation made previously by Lawrence and Rozansky. This article reviews selected material from the larger work "Localization for Wilson Loops in Chern-Simons Theory," arXiv:0911.2687.

1. Introduction

This brief article is intended as an introduction to the study of Chern-Simons gauge theory via non-abelian localization [**4, 5**]. I will not attempt to give here a comprehensive overview of the subject. Instead, my goal is to highlight two very beautiful ideas, one old and one new, which enter the story of non-abelian localization in an essential way and which may have applications elsewhere. See also [**6, 15**] for alternative approaches to path integral localization in Chern-Simons theory.

To set the stage, we consider Chern-Simons theory on a compact, oriented three-manifold M with gauge group G. By assumption G will be a compact, connected, simply-connected, and simple Lie group. For instance, G could be $SU(N)$ for any $N > 1$. These assumptions on G ensure that any principal G-bundle P over M is topologically trivial, a technical convenience. The gauge field A of Chern-Simons theory is then a connection on P.

Let me now introduce the Wilson loop operators in Chern-Simons theory. Quite generally, a Wilson loop operator $W_R(C)$ in any gauge theory on a manifold M is described by the data of an oriented, closed curve C which is smoothly[1] embedded

2010 *Mathematics Subject Classification.* Primary 81T45; Secondary 53D20 57M27.

[1]The condition that C be smoothly embedded in M is merely for convenience and is not strictly required to define $W_R(C)$ as a sensible operator in gauge theory. Indeed, the Wilson loop expectation value in Chern-Simons theory can be computed exactly even for the case that C is an arbitrary closed graph [**25**] in M.

in M and which is decorated by an irreducible representation R of the gauge group G. As a classical functional of the connection A, the Wilson loop operator $W_R(C)$ is then given simply by the trace in R of the holonomy of A around C,

$$\begin{aligned} W_R(C) &= \mathrm{Tr}_R \mathrm{Hol}_C(A)\,, \\ &= \mathrm{Tr}_R P\exp\Bigl(-\oint_C A\Bigr)\,. \end{aligned} \tag{1.1}$$

Following the standard practice in physics, we describe the holonomy $\mathrm{Hol}_C(A)$ in the second line of (1.1) in terms of a path-ordered exponential $P\exp(-\oint_C A)$, which describes solutions to the first-order differential equation for parallel transport[2] along C.

With the Wilson loop operator in hand, we finally introduce the absolutely-normalized Wilson loop path integral

$$Z(k; C, R) = \frac{1}{\mathrm{Vol}(\mathcal{G})} \int_{\mathcal{A}} \mathcal{D}A\; W_R(C)\, \exp\Bigl[i\, \frac{k}{4\pi}\, \mathrm{CS}(A)\Bigr]\,. \tag{1.2}$$

Here $\mathrm{CS}(A)$ is the Chern-Simons action,

$$\mathrm{CS}(A) = \int_M \mathrm{Tr}\Bigl(A\wedge dA + \frac{2}{3} A\wedge A\wedge A\Bigr)\,, \tag{1.3}$$

and 'Tr' denotes a suitably-normalized, negative-definite, invariant quadratic form on the Lie algebra $\mathfrak{g}$ of G. For instance, if G is $SU(N)$, the quadratic form 'Tr' is given concretely by the trace in the fundamental N-dimensional representation of $SU(N)$.

For later use, let me recall three essential properties of the Chern-Simons action. First, the Chern-Simons action is purely topological, insofar as it depends only on the choice of an orientation, not a metric, on M. Second, the critical points of the Chern-Simons action are precisely the flat connections on M, for which

$$F_A = dA + A\wedge A = 0. \tag{1.4}$$

Finally, even though the Chern-Simons action is not manifestly gauge-invariant — and indeed the Chern-Simons action is *not* gauge-invariant — the Chern-Simons action is almost gauge-invariant, in the sense that it is invariant under homotopically trivial gauge transformations. Under homotopically nontrivial gauge transformations, the functional $\mathrm{CS}(A)$ shifts by integral multiples of $8\pi^2$, where the relevant integer is determined by the homotopy class of the gauge transformation as a map from M to G. So long as the Chern-Simons level $k \in \mathbb{Z}$ is quantized to be an integer, the exponential of the Chern-Simons action in (1.2) is then honestly gauge-invariant.

Otherwise, we present $Z(k; C, R)$ as an integral over the infinite-dimensional affine space $\mathcal{A}$ of connections on M. As usual in gauge theory, we divide that integral by the volume of the infinite-dimensional group $\mathcal{G}$ of gauge transformations acting on $\mathcal{A}$.

Before proceeding, let me make one elementary remark. When R is the trivial representation, the Wilson loop operator $W_R(C)$ is the identity operator, and the

[2]Because we work in conventions for which $d_A = d + A$ is the covariant derivative, a minus sign appears in the argument of the path-ordered exponential.

absolutely-normalized Wilson loop path integral reduces immediately to the path integral which describes the Chern-Simons partition function $Z(k)$ on M,

$$Z(k) = \frac{1}{\mathrm{Vol}(\mathcal{G})} \int_{\mathcal{A}} \mathcal{D}A \, \exp\left[i \frac{k}{4\pi} \, \mathrm{CS}(A)\right]. \tag{1.5}$$

In terms of $Z(k)$ and $Z(k; C, R)$, the Wilson loop expectation value $\langle W_R(C) \rangle$ is then given by the ratio

$$\langle W_R(C) \rangle = \frac{Z(k; C, R)}{Z(k)}. \tag{1.6}$$

Though the Wilson loop expectation value is very convenient to consider for certain purposes, we work exclusively with $Z(k; C, R)$ today.

As it stands, the Wilson loop path integral in (1.2) is a purely formal expression. Nonetheless, over twenty years ago Witten [**24**] gave a completely precise prescription to define the observable $Z(k; C, R)$, based upon the canonical quantization of Chern-Simons theory in the Hamiltonian formalism. This formalism leads to a well-known algebraic description [**23**] of $Z(k; C, R)$ in terms of a presentation of M via surgery on S^3, combined with data about certain modular representations associated to two-dimensional rational conformal field theories.

Beyond providing a rigorous means to define the observable $Z(k; C, R)$, the Hamiltonian formalism is also very powerful, insofar as it can be used to perform explicit, exact calculations in Chern-Simons theory. For instance, among the classic results in [**24**], Witten computed the unknot observable $Z\big(k; \bigcirc, \boldsymbol{j}\big)$ for Chern-Simons theory on S^3 with gauge group $SU(2)$,

$$Z\big(k; \bigcirc, \boldsymbol{j}\big) = \sqrt{\frac{2}{k+2}} \, \sin\left(\frac{\pi \, j}{k+2}\right), \qquad j = 1, \ldots, k+1\,. \tag{1.7}$$

Here $\boldsymbol{j}$ is the irreducible representation of $SU(2)$ with dimension j, and as indicated, $\boldsymbol{j}$ runs without loss over the finite set of representations which are integrable in the affine Lie algebra for $SU(2)$ at level k.

Yet despite its computability, the algebraic definition of $Z(k; C, R)$ in the Hamiltonian formalism obscures many features which are manifest in the preceding path integral (1.2) and which one would like to understand more deeply. As a simple example, in the semi-classical limit that k is large, a naive stationary-phase approximation can be applied to the path integral describing the Chern-Simons partition function, and this approximation implies asymptotic behavior for $Z(k)$ that is far from evident in the complicated, exact expressions that arise from conformal field theory. Nonetheless, the predicted asymptotic behavior can be checked in examples, as was done early on by Freed and Gompf [**10**], Jeffrey [**13**], and Garoufalidis [**11**]. See for instance §7 of [**20**] for a survey of continuing work in this area.

This article concerns a very special and very beautiful situation in which the stationary-phase approximation to the Wilson loop path integral is actually exact. Namely, the three-manifold M is a Seifert manifold, equipped with a distinguished locally-free $U(1)$ action, and the curve C is a Seifert fiber of M. By definition, a locally-free $U(1)$ action is one for which the generating vector field is nowhere vanishing. Equivalently, all stabilizers under a locally-free $U(1)$ action are proper, necessarily discrete, subgroups of $U(1)$.

Let me introduce a bit of terminology. The existence of a locally-free $U(1)$ action implies that the Seifert manifold M decomposes geometrically as the total space of a nontrivial circle bundle over a Riemann surface Σ,

$$\begin{array}{ccc} S^1 & \longrightarrow & M \\ & & \downarrow \pi \\ & & \Sigma \end{array} \quad . \tag{1.8}$$

Here Σ is allowed to have orbifold points, and the circle bundle is allowed to be a corresponding orbifold bundle, so long as M itself is smooth.

Each Seifert manifold carries a distinguished set of Wilson loop operators which respect the $U(1)$ action by rotations in the fiber of (1.8). For these Wilson loops, the curve C is an orbit of the given $U(1)$ action, and C appears geometrically in M as the S^1 fiber over a basepoint $\sigma \in \Sigma$. Assuming that σ is a smooth (non-orbifold) point of Σ, the topology of C in M does not depend upon the choice of σ, so such Wilson loops are determined entirely by the choice of the representation R. Henceforth, we refer to these special Wilson loop operators which wrap the generic Seifert fibers of M as "Seifert loop operators" to distinguish them from general Wilson loops in M, about which we will also have some things to say.

As a concrete example, S^3 admits countably-many locally-free $U(1)$ actions, each associated to a distinct Seifert presentation. According to a basic result of Moser [**19**], the corresponding knots which can be realized as Seifert fibers in S^3 are precisely the torus knots. So the Seifert loop operators in S^3 are just the Wilson loop operators which wrap torus knots.

Some Experimental Evidence

The exactness of the stationary-phase approximation to the Seifert loop path integral was discovered by Lawrence and Rozansky [**16**] on an empirical basis, through a detailed analysis of the explicitly-known formulae for $Z(k; C, R)$ in the case $G = SU(2)$.[3] Specifically, after a rather involved series of algebraic manipulations, Lawrence and Rozansky were able to rewrite $Z(k; C, R)$ very compactly as a finite sum of analytic expressions, each being either a contour integral or the residue of a meromorphic function. These summands in $Z(k; C, R)$ could then be associated in a one-to-one fashion with the connected components in the moduli space of flat connections on M. Since the flat connections on M are the critical points of the Chern-Simons action, such a form for $Z(k; C, R)$ strongly suggests that the stationary-phase approximation to the Seifert loop path integral is exact.

By way of illustration, Lawrence and Rozansky would rewrite the exact formula for the unknot observable $Z\big(k; \bigcirc, \boldsymbol{j}\big)$ in (1.7) as

$$\begin{aligned} &Z\big(k; \bigcirc, \boldsymbol{j}\big) = \\ &\frac{1}{2\pi i} \mathrm{e}^{-\frac{i\pi(1+j^2)}{2(k+2)}} \int_{-\infty}^{+\infty} dx \, \mathrm{ch}_{\boldsymbol{j}}\Big(\mathrm{e}^{\frac{i\pi}{4}} \frac{x}{2}\Big) \sinh^2\Big(\mathrm{e}^{\frac{i\pi}{4}} \frac{x}{2}\Big) \exp\Big(-\frac{(k+2)}{8\pi} x^2\Big) . \end{aligned} \tag{1.9}$$

Here $\mathrm{ch}_{\boldsymbol{j}}$ is the character of $SU(2)$ associated to the representation $\boldsymbol{j}$,

$$\mathrm{ch}_{\boldsymbol{j}}(y) = \frac{\sinh(j\, y)}{\sinh(y)} = \mathrm{e}^{(j-1)y} + \mathrm{e}^{(j-3)y} + \cdots + \mathrm{e}^{-(j-3)y} + \mathrm{e}^{-(j-1)y} , \tag{1.10}$$

and the equality between the expressions in (1.7) and (1.9) follows by evaluating (1.9) as a sum of elementary Gaussian integrals.

[3]See [**12, 17**] for the generalization to other gauge groups G.

Of course, the only flat connection on S^3 is the trivial connection. As a result, the aforementioned sum in $Z(k;\bigcirc,\boldsymbol{j})$ contains only a single term, given by the integral over the real variable x in (1.9). According to Lawrence and Rozansky, this integral is to be interpreted as the stationary-phase contribution from the trivial connection to the full Wilson loop path integral in (1.2).

One of the main results in [**4, 5**] is to make the semi-classical interpretation of formulae such as (1.9) completely precise. Briefly, the contour integral over x arises geometrically as an integral over the Cartan subalgebra of $SU(2)$, regarded as the group of constant gauge transformations on S^3. The constant gauge transformations are the stabilizer of the trivial connection in the group $\mathcal{G}$ of all gauge transformations, and the presence of this stabilizer group plays an important role in the semi-classical analysis of the Wilson loop path integral. Moreover, all dependence on the $SU(2)$ representation $\boldsymbol{j}$ enters the integrand of (1.9) through the character $\mathrm{ch}_{\boldsymbol{j}}$. As a result, the character $\mathrm{ch}_{\boldsymbol{j}}$ can be naturally interpreted the avatar of the unknot Wilson loop operator itself when the path integral in (1.2) is reduced to the contour integral in (1.9).

As an aside, let me mention two interesting generalizations of the semi-classical formula for $Z\big(k;\bigcirc,\boldsymbol{j}\big)$ in (1.9). First, this formula extends directly to gauge groups G other than $SU(2)$, in which case the unknot Wilson loop operator for any irreducible representation R of G reduces naturally to the corresponding character ch_R. Moreover, the semi-classical formula in (1.9) generalizes in a surprisingly simple way to the arbitrary $(\mathbf{p},\mathbf{q})$-torus knot $K_{\mathbf{p},\mathbf{q}}$ in S^3,

(1.11)

$$Z\big(k;K_{\mathbf{p},\mathbf{q}},\boldsymbol{j}\big) = \frac{1}{2\pi i}\frac{1}{\sqrt{\mathbf{pq}}}\exp\left[-\frac{i\pi}{2(k+2)}\left(\frac{\mathbf{p}}{\mathbf{q}}+\frac{\mathbf{q}}{\mathbf{p}}+\mathbf{pq}\,(j^2-1)\right)\right]\times$$
$$\times\int_{-\infty}^{+\infty} dx\;\mathrm{ch}_{\boldsymbol{j}}\left(\mathrm{e}^{\frac{i\pi}{4}}\frac{x}{2}\right)\sinh\left(\mathrm{e}^{\frac{i\pi}{4}}\frac{x}{2\mathbf{p}}\right)\sinh\left(\mathrm{e}^{\frac{i\pi}{4}}\frac{x}{2\mathbf{q}}\right)\exp\left[-\frac{(k+2)}{8\pi}\left(\frac{x^2}{\mathbf{pq}}\right)\right].$$

Modulo a rather subtle overall phase, the integrand in (1.11) merely acquires denominators proportional to $\mathbf{p}$ and $\mathbf{q}$. From the semi-classical perspective, these denominators are associated to the $U(1)$ stabilizers of the exceptional Seifert fibers in S^3. I refer the interested reader to §7.2 in [**4**] for a detailed discussion of (1.11).

Non-Abelian Localization

My goal here is to explain a second, more conceptual way to understand the exactness of the stationary-phase approximation to the Seifert loop path integral. In this approach, we apply non-abelian localization, as introduced by Witten in [**26**], to study Chern-Simons theory on a Seifert manifold.

Very briefly, non-abelian localization provides a cohomological interpretation for a special class of symplectic integrals which are intimately related to symmetries. These integrals take the canonical form

$$Z(\epsilon) = \int_X \exp\left[\Omega - \frac{1}{2\epsilon}\,(\mu,\mu)\right]. \tag{1.12}$$

Here X is an arbitrary symplectic manifold with symplectic form Ω. We assume that a Lie group H acts on X in a Hamiltonian fashion with moment map $\mu : X \to \mathfrak{h}^*$, where $\mathfrak{h}^*$ is the dual of the Lie algebra $\mathfrak{h}$ of H. We also introduce an invariant quadratic form $(\cdot\,,\,\cdot)$ on $\mathfrak{h}$ and dually on $\mathfrak{h}^*$ to define the function $S = \frac{1}{2}(\mu, \mu)$ appearing in the integrand of $Z(\epsilon)$. Finally, $\epsilon \in \mathbb{R}$ is a coupling parameter.[4]

The symplectic integral in (1.12) has a number of important properties, which for sake of brevity I merely state. See [**21, 26**] for proofs of the following statements. First, the integrand of (1.12) admits an interpretation in terms of the H-equivariant cohomology ring of X. Using this interpretation, one can then show that the symplectic integral itself localizes onto the critical points of the invariant function $S = \frac{1}{2}(\mu, \mu)$ on X. Moreover, a non-abelian localization formula, roughly analogous to the Duistermaat-Heckman formula [**9**], exists to describe the contributions from the critical locus of S. In a smooth situation, these contributions are given by the integrals of certain de Rham cohomology classes over the critical loci. See §6 of [**4**] for a precise statement of the non-abelian localization formula applicable to Chern-Simons theory.

Example: Two-Dimensional Yang-Mills Theory

Given the special form of $Z(\epsilon)$, one should not be surprised that this integral has special properties. But why consider such an integral in the first place? One answer, following Witten [**26**], is that the path integral of two-dimensional Yang-Mills theory assumes precisely the canonical symplectic form in (1.12).

To explain the latter observation, I will simply exhibit the counterparts of X, Ω, H, and μ relevant to describe Yang-Mills theory on a Riemann surface Σ. The Yang-Mills path integral is formally an integral over the affine space $\mathcal{A}$ of connections on a fixed principal G-bundle P over Σ, so clearly we must set

$$X = \mathcal{A} \,. \tag{1.13}$$

The affine space $\mathcal{A}$ also possesses a natural symplectic form Ω given by the intersection pairing on Σ,

$$\Omega = -\int_\Sigma \mathrm{Tr}\big(\delta A \wedge \delta A\big) \,. \tag{1.14}$$

Here δ denotes the exterior derivative acting on $\mathcal{A}$. Since A serves as a coordinate on $\mathcal{A}$, δA is a one-form on $\mathcal{A}$, and Ω is a two-form on $\mathcal{A}$ which is manifestly non-degenerate and closed. Of course on Σ itself, δA transforms as a section of the bundle $\Omega^1_\Sigma \otimes \mathrm{ad}(P)$ of adjoint-valued one-forms.

The obvious group which acts on $\mathcal{A}$ is the group $\mathcal{G}$ of gauge transformations. As shown long ago by Atiyah and Bott [**2**], the action of $\mathcal{G}$ on $\mathcal{A}$ is Hamiltonian with moment map given by the curvature $F_A = dA + A\wedge A$. That is, since elements in the Lie algebra of $\mathcal{G}$ appear on Σ as sections of the adjoint bundle $\mathrm{ad}(P)$, the curvature F_A, as a section of $\Omega^2_\Sigma \otimes \mathrm{ad}(P)$, can naturally be considered as a function on $\mathcal{A}$ taking values in the dual of the Lie algebra of $\mathcal{G}$. Thus,

$$H = \mathcal{G} \,, \qquad\qquad \mu = F_A \,. \tag{1.15}$$

Finally, the Lie algebra of $\mathcal{G}$ admits an invariant form given by

$$(\phi, \phi) = -\int_\Sigma \mathrm{Tr}\big(\phi \wedge \star \phi\big) \,. \tag{1.16}$$

---

[4]To make sense of the measure on X in (1.12), we expand the exponential $\exp(\Omega)$ in its Taylor series and pick out the term $\frac{1}{n!}\Omega^n$ of proper degree to integrate over X. Hence $\exp(\Omega)$ conveniently describes the symplectic measure on X.

Here ϕ is an element of the Lie algebra of $\mathcal{G}$, transforming on Σ as a section of $\mathrm{ad}(P)$. With the quadratic form in (1.16), the invariant function $S = \frac{1}{2}(\mu, \mu)$ appearing in the canonical symplectic integral over $\mathcal{A}$ immediately becomes the Yang-Mills action,

$$S = \frac{1}{2}(\mu, \mu) = -\frac{1}{2}\int_\Sigma \mathrm{Tr}\big(F_A\wedge\star F_A\big)\,. \tag{1.17}$$

The metric on Lie($\mathcal{G}$) in (1.16) is defined using a duality operator $\star$ on Σ. For two-dimensional Yang-Mills theory, $\star$ relates zero-forms to two-forms, and to obtain such an operator, we only require a symplectic structure, as opposed to a metric, on Σ. Given a symplectic form ω on Σ, we define $\star$ by the condition $\star 1 = \omega$. The symplectic form ω is invariant under all area-preserving diffeomorphisms of Σ, and this large group acts as a symmetry of two-dimensional Yang-Mills theory. As a result, two-dimensional Yang-Mills theory is essentially a topological gauge theory.

In the remainder of this article, I want to explain how to recast the Seifert loop path integral (1.2) as a symplectic integral of the canonical form (1.12). Once this step is accomplished, the general arguments in [**26**] imply that the Seifert loop path integral localizes onto critical points of the classical action $S = \frac{1}{2}(\mu, \mu)$. Furthermore, using the non-abelian localization formula, one can perform exact computations of the Seifert loop path integral and thus obtain a cohomological description for the Seifert loop operator itself. Specifically, as demonstrated in §7.3 of [**4**], the Seifert loop operator reduces naturally to the Chern character of an associated universal bundle over the moduli space of flat connections on M.

Two key ideas are required to obtain a symplectic description of the Seifert loop path integral. The first idea, which appears in §3 of [**5**], pertains to the basic Chern-Simons path integral in (1.5) and really has nothing to do with the Wilson loop operator. In contrast, the second idea concerns the Wilson loop operator itself and really has nothing to do with Chern-Simons theory. Nonetheless, both of these ideas fit together in a very elegant way.

2. The symplectic geometry of Chern-Simons theory

The path integral which describes the partition function of two-dimensional Yang-Mills theory automatically assumes the canonical symplectic form in (1.12). As a special case of our general study of $Z(k; C, R)$, I now want to review how the path integral (1.5) which describes the partition function $Z(k)$ of Chern-Simons theory on a Seifert manifold M can also be cast as such a symplectic integral.

In order to obtain a symplectic interpretation of the two-dimensional Yang-Mills path integral, we found it necessary to introduce a symplectic structure on Σ. To obtain a corresponding symplectic interpretation for the Chern-Simons path integral, we must introduce the analogous geometric structure on the three-manifold M – namely, a contact structure.

Globally, a contact structure on M is described by an ordinary one-form κ, a section of Ω^1_M, which at each point of M satisfies the contact condition

$$\kappa\wedge d\kappa \neq 0\,. \tag{2.1}$$

By a classic theorem of Martinet [**18**], any compact, orientable[5] three-manifold admits a contact structure, so we do not necessarily assume M to be Seifert at

[5] Any three-manifold admitting a contact structure must be orientable, since the nowhere vanishing three-form $\kappa\wedge d\kappa$ defines an orientation.

this stage. However, if M is a Seifert manifold, then we certainly want the contact form κ to respect the $U(1)$ action on M. Such a contact form can be exhibited as follows.

We recall that the Seifert manifold M is the total space of an S^1-bundle of degree n over Σ,

$$\begin{array}{ccc} S^1 & \xrightarrow{n} & M \\ & & \downarrow \pi \\ & & \Sigma \end{array}, \tag{2.2}$$

or an orbifold version thereof. For simplicity, I will phrase the following construction of κ in the language of smooth manifolds, but the orbifold generalization is immediate.

Regarding M as the total space of a principal $U(1)$-bundle, we take κ to be a $U(1)$-connection on this bundle which satisfies

$$d\kappa \;=\; n\,\pi^*(\omega)\,. \tag{2.3}$$

Here ω is any unit-area symplectic form on Σ, and we recall that a $U(1)$-connection on Σ appears upstairs on M as an ordinary one-form. Because $d\kappa$ represents the Euler class of the S^1-bundle over Σ, the degree n necessarily appears in (2.3). As an abelian connection, κ is automatically invariant under the $U(1)$-action on M. Also, since the pullback of κ to each S^1 fiber is non-vanishing, the contact condition in (2.1) is satisfied so long as $n \neq 0$ and the bundle is non-trivial, as we assume.

Chern-Simons theory is often considered to be an intrinsically three-dimensional gauge theory. However, one of the more interesting results in [**5, 4**] is to show that Chern-Simons theory is *not quite* a three-dimensional gauge theory, since one of the three components of A can be completely decoupled from all topological observables.

In order to decouple one component of A from the Chern-Simons path integral, we introduce a new, infinite-dimensional "shift" symmetry $\mathcal{S}$ which acts on A as

$$\delta A \;=\; \sigma\,\kappa\,. \tag{2.4}$$

Here σ is an arbitrary adjoint-valued scalar, a section of $\Omega^1_M \otimes \mathfrak{g}$, that parametrizes the action of $\mathcal{S}$ on $\mathcal{A}$.

Of course, the Chern-Simons action $\mathrm{CS}(\,\cdot\,)$ does not respect the shift of A in (2.4), so we must play a little path integral trick, of the sort familiar from path integral derivations of T-duality or abelian S-duality. See §8 in [**27**] for a nice review of the path integral derivations of these dualities.

We first introduce a new field Φ which transforms like σ as an adjoint-valued scalar, a section of $\Omega^1_M \otimes \mathfrak{g}$, and which is completely gauge-trivial under $\mathcal{S}$. Thus $\mathcal{S}$ acts on Φ as

$$\delta\Phi \;=\; \sigma\,. \tag{2.5}$$

Next, we consider a new, shift-invariant action $S(A,\Phi)$ incorporating both A and Φ such that, if Φ is set identically to zero via (2.5), then $S(A,\Phi)$ reduces to the Chern-Simons action for A. This condition fixes $S(A,\Phi)$ to be

$$\begin{aligned} S(A,\Phi) \;&=\; \mathrm{CS}(A - \kappa\,\Phi)\,,\\ &=\; \mathrm{CS}(A) - \int_M \Big[2\kappa\wedge\mathrm{Tr}(\Phi F_A) - \kappa\wedge d\kappa\,\mathrm{Tr}(\Phi^2)\Big]\,. \end{aligned} \tag{2.6}$$

Finally, using (2.6) we introduce an *a priori* new path integral[6] over both A and Φ,

$$\widetilde{Z}(k) = \int \mathcal{D}A\,\mathcal{D}\Phi\, \exp\left[i\frac{k}{4\pi}\,S(A,\Phi)\right]. \tag{2.7}$$

On one hand, if we set $\Phi \equiv 0$ using the shift-symmetry in (2.5), $\widetilde{Z}(k)$ immediately reduces[7] to the usual Chern-Simons partition function $Z(k)$. Hence,

$$\widetilde{Z}(k) = Z(k). \tag{2.8}$$

On the other hand, due to the elementary fact that $\kappa\wedge\kappa = 0$, the field Φ appears only quadratically in $S(A,\Phi)$. So we can simply perform the Gaussian integral over Φ in (2.7) to obtain a new path integral description of the Chern-Simons partition function,

$$Z(k) = \int \mathcal{D}A\, \exp\left[i\frac{k}{4\pi}\,S(A)\right], \tag{2.9}$$

where

$$S(A) = \mathrm{CS}(A) - \int_M \frac{1}{\kappa\wedge d\kappa}\mathrm{Tr}\left[(\kappa\wedge F_A)^2\right]. \tag{2.10}$$

In performing the Gaussian integral over Φ, we use the contact condition on κ in (2.1), since this condition ensures that quadratic term for Φ in (2.6) is everywhere non-degenerate on M. In particular, the inverse "$1/\kappa\wedge d\kappa$" appearing in (2.10) is defined as follows. Because $\kappa\wedge d\kappa$ is everywhere non-vanishing, we can always write $\kappa\wedge F_A = \varphi\,\kappa\wedge d\kappa$ for some section φ of $\Omega^0_M \otimes \mathfrak{g}$. Thus, we set $\kappa\wedge F_A/\kappa\wedge d\kappa = \varphi$, and the second term in $S(A)$ becomes $\int_M \kappa\wedge\mathrm{Tr}(F_A\varphi)$.

By construction, $S(A)$ is invariant under the shift of A in (2.4), as can be checked directly. Thus, we have obtained a new description of the Chern-Simons partition function for which one component of A completely decouples from the path integral. Further, we have yet to use the condition that M be a Seifert manifold, so the reformulation of $Z(k)$ via the shift-invariant action in (2.10) holds for any three-manifold M endowed with a contact structure. See [**14**] for a detailed analysis of the shift-invariant reformulation of $Z(k)$ in the special case that the gauge group is $U(1)$.

Symplectic Data

If M is a Seifert manifold, an additional miracle occurs, and the path integral in (2.9) becomes an integral of the canonical symplectic form to which non-abelian localization applies. For sake of time, let me merely summarize the symplectic data associated to (2.9).

First, the space over which we integrate in (2.9) and which must play the role of X is the quotient of the affine space $\mathcal{A}$ by the group $\mathcal{S}$,

$$X = \mathcal{A}/\mathcal{S}. \tag{2.11}$$

In three dimensions, the affine space $\mathcal{A}$ is not symplectic. However, once we take the quotient by $\mathcal{S}$ in (2.11), the space $\mathcal{A}/\mathcal{S}$ carries a natural symplectic form Ω

[6]I will not be very careful about the overall normalizations for the path integrals that appear here, but see §3 of [**5**] for a detailed accounting of formal normalization factors.

[7]We note that the Jacobian associated to the gauge-fixing condition $\Phi \equiv 0$ for $\mathcal{S}$ is trivial.

given by

$$\Omega = -\int_M \kappa\wedge\mathrm{Tr}(\delta A\wedge\delta A)\ . \tag{2.12}$$

Clearly Ω is both gauge-invariant and shift-invariant, and Ω descends to a non-degenerate symplectic form on $\mathcal{A}/\mathcal{S}$.

We must now find a Hamiltonian group acting on $\mathcal{A}/\mathcal{S}$ such that the shift-invariant action $S(A)$ is the square of the corresponding moment map. As an initial guess, motivated by the example of two-dimensional Yang-Mills theory, one might consider the group $\mathcal{G}$ of gauge transformations. However, this guess cannot be correct. By construction, the square of the moment map for the action of $\mathcal{G}$ on $\mathcal{A}/\mathcal{S}$ would be invariant under $\mathcal{G}$. However, $S(A)$ is plainly not invariant under $\mathcal{G}$, since the Chern-Simons term appearing in (2.10) is not invariant under the large gauge transformations in $\mathcal{G}$ (while the remaining term in (2.10) is manifestly invariant under $\mathcal{G}$).

As a second guess, one might replace $\mathcal{G}$ with its identity component $\mathcal{G}_0$, which does preserve the shift-invariant action $S(A)$. However, one can show that $\mathcal{G}_0$ is obstructed from acting in a Hamiltonian fashion on $\mathcal{A}/\mathcal{S}$ by a non-trivial Lie algebra cocycle c,

$$c(\phi,\psi) = -\int_M d\kappa\wedge\mathrm{Tr}(\phi\, d\psi)\ . \tag{2.13}$$

Here ϕ and ψ are elements of the Lie algebra of $\mathcal{G}_0$, transforming as sections of $\Omega^0_M\otimes\mathfrak{g}$ on M. Parenthetically, this cocycle is closely related to a cocycle that appears in the theory of loop groups [**22**], which also provides useful background for the identification of H below.

To remedy the situation, we consider the central extension[8] $\widetilde{\mathcal{G}}_0$ of $\mathcal{G}_0$ determined by $c(\phi,\psi)$,

$$U(1)_{\mathrm{Z}} \xrightarrow{c} \widetilde{\mathcal{G}}_0 \longrightarrow \mathcal{G}_0\,. \tag{2.14}$$

Here we use the subscript 'Z' to emphasize that $U(1)_{\mathrm{Z}}$ is central in $\widetilde{\mathcal{G}}_0$. The natural action of $\mathcal{G}_0$ on $\mathcal{A}/\mathcal{S}$ extends to an action by $\widetilde{\mathcal{G}}_0$, for which $U(1)_{\mathrm{Z}}$ acts trivially. By construction, the action of $\widetilde{\mathcal{G}}_0$ on $\mathcal{A}/\mathcal{S}$ is then Hamiltonian.

However, $\widetilde{\mathcal{G}}_0$ is still not the group which is to play the role of the Hamiltonian group H! In order to define the canonical symplectic integral in (1.12), the Lie algebra of H must carry a non-degenerate, invariant quadratic form $(\cdot\,,\cdot)$. But the Lie algebra of $\widetilde{\mathcal{G}}_0$ does not admit such a form, essentially because we have no generator to pair with the generator of the central $U(1)_{\mathrm{Z}}$.

However, we have also yet to apply the Seifert condition on M. We do so now. To avoid confusion, let me denote the Seifert $U(1)$ acting on M by $U(1)_{\mathrm{R}}$, to distinguish it from the central $U(1)_{\mathrm{Z}}$. The action by $U(1)_{\mathrm{R}}$ on M induces a corresponding action on both $\widetilde{\mathcal{G}}_0$ and $\mathcal{A}/\mathcal{S}$, and we finally take H to be the semi-direct product

$$H = U(1)_{\mathrm{R}} \ltimes \widetilde{\mathcal{G}}_0\,. \tag{2.15}$$

[8]A Lie algebra two-cocycle always determines a central extension of algebras. Provided that the cocycle is properly quantized, as is the cocycle in (2.13), the central extension of algebras lifts to a corresponding central extension of groups.

The Lie algebra of H does admit a non-degenerate, invariant quadratic form $(\cdot\,,\cdot)$, under which the generators of $U(1)_{\mathrm{R}}$ and $U(1)_{\mathrm{Z}}$ are paired. Furthermore, the action of H on $\mathcal{A}/\mathcal{S}$ is Hamiltonian with moment map μ (for which a completely explicit though perhaps not so illuminating formula exists), and the corresponding invariant function $S = \frac{1}{2}(\mu, \mu)$ on $\mathcal{A}/\mathcal{S}$ is precisely $S(A)$. Given the amount of symmetry respected by both $\frac{1}{2}(\mu, \mu)$ and $S(A)$, the latter result could hardly have been otherwise.

The presentation here is a regrettably quick sketch of a fairly miraculous result, and I refer the reader to §3 of [**5**] for a complete discussion.

3. Inclusion of Wilson loop operators

I now want to explain how the prior statements concerning the Chern-Simons partition function can be generalized to allow for insertions of Wilson loop operators. (See §4 of [**4**] for an expanded version of the material here.) As it happens, only one new idea is required.

We clearly need a new idea, because a naive attempt to reapply the previous path integral manipulations to the Wilson loop path integral in (1.2) runs immediately aground. To illustrate the difficulty with the direct approach, let us consider the obvious way to rewrite the Wilson loop path integral in a shift-invariant form,

$$Z(k; C, R) = \int \mathcal{D}\!A\, \mathcal{D}\Phi\; \mathcal{W}_R(C)\, \exp\left[i\frac{k}{4\pi}\mathrm{CS}(A - \kappa\,\Phi)\right]. \tag{3.1}$$

Here $\mathcal{W}_R(C)$ denotes the generalized Wilson loop operator defined not using A but using the shift-invariant combination $A - \kappa\,\Phi$, so that

$$\mathcal{W}_R(C) = \mathrm{Tr}_R\, P\exp\left[-\oint_C (A - \kappa\,\Phi)\right]. \tag{3.2}$$

Exactly as for our discussion of (2.7), we can use the shift symmetry to fix $\Phi \equiv 0$, after which the path integral in (3.1) reduces trivially to the standard Wilson loop path integral in (1.2).

However, to learn something useful from (3.1) we must perform the path integral over Φ, and as it stands, this integral is not easy to do. Because the generalized Wilson loop operator $\mathcal{W}_R(C)$ is expressed in (3.2) as a complicated, non-local functional of Φ, the path integral over Φ in (3.1) is not a Gaussian integral that we can trivially evaluate as we did for (2.7).

A more fundamental perspective on our problem is the following. Let us return to the description of the ordinary Wilson loop operator $W_R(C)$ as the trace in the representation R of the holonomy of A around C,

$$W_R(C) = \mathrm{Tr}_R\, P\exp\left(-\oint_C A\right). \tag{3.3}$$

As observed by Witten in one of the small gems of [**24**], this description of $W_R(C)$ should be regarded as intrinsically quantum mechanical, for the simple reason that $W_R(C)$ can be naturally interpreted in (3.3) as the partition function of an auxiliary quantum system attached to the curve C. Briefly, the representation R is to be identified with the Hilbert space of this system, the holonomy of A is to be identified with the time-evolution operator around C, and the trace over R is the usual trace over the Hilbert space that defines the partition function in the Hamiltonian formalism.

Because the notion of tracing over a Hilbert space is inherently quantum mechanical, any attempts to perform essentially classical path integral manipulations involving the expressions in (3.2) or (3.3) are misguided at best. Rather, if we hope to generalize the semi-classical path integral manipulations which we used to study the Chern-Simons partition function, we need to use an alternative, semi-classical description for the Wilson loop operator itself.

More precisely, we want to replace the quantum mechanical trace over R in (3.3) by a path integral over an auxiliary bosonic field U which is attached to the curve C and coupled to the connection A as a background field, so that schematically

$$W_R(C) = \int \mathcal{D}U \, \exp\Big[i\, \mathrm{cs}_\alpha\big(U; A|_C\big)\Big]. \tag{3.4}$$

Here $\mathrm{cs}_\alpha\big(U; A|_C\big)$ is an action, depending upon the representation R through its highest weight α, which is a local, gauge-invariant, and indeed topological functional of the defect field U and the restriction of A to C. Not surprisingly, this semi-classical description (3.4) of $W_R(C)$ turns out to be the key ingredient required to reformulate the Wilson loop path integral in a shift-invariant fashion.

The idea of representing the Wilson loop operator by a path integral as in (3.4) is a very old and very general piece of gauge theory lore. In the context of four-dimensional Yang-Mills theory, this idea goes back (at least) to work of Balachandran, Borchardt, and Stern [**3**] in the 1970's. See also [**8**] and §7.7 in [**27**] for other appearances of the path integral in (3.4).

The basic idea behind the path integral description (3.4) of the Wilson loop operator is very simple. We interpret the closed curve C as a periodic "time" for the field U, and we apply the Hamiltonian formalism to rewrite the path integral over U axiomatically as the quantum mechanical trace of the corresponding time-evolution operator around C,

$$W_R(C) = \mathrm{Tr}_{\mathscr{H}}\, P\exp\left(-i\oint_C \mathbf{H}\right). \tag{3.5}$$

Here $\mathscr{H}$ is the Hilbert space obtained by quantizing U, and $\mathbf{H}$ is the Hamiltonian which acts on $\mathscr{H}$ to generate infinitesimal translations along C.

Comparing the conventional description of the Wilson loop operator in (3.3) to the axiomatic expression in (3.5), we see that the two agree if we identify[9]

$$\begin{gathered} R \longleftrightarrow \mathscr{H}\,, \\ P\exp\left(-\oint_C A\right) \longleftrightarrow P\exp\left(-i\oint_C \mathbf{H}\right). \end{gathered} \tag{3.6}$$

Hence to make the Wilson loop path integral in (3.4) precise, we need only exhibit a classical theory on C, for which the gauge group G acts as a symmetry, such that upon quantization we obtain a Hilbert space $\mathscr{H}$ isomorphic to R and for which the time-evolution operator around C is given by the holonomy of A, acting as an element of G on R.

[9]We follow the standard physical definition according to which $\mathbf{H}$ is a hermitian operator, accounting for the '$-i$' in (3.5). We also recall that the gauge field A is valued in the Lie algebra $\mathfrak{g}$, so A is anti-hermitian and no 'i' appears in the holonomy.

A Semi-Classical Description of the Wilson Loop Operator

Let me now tell you what classical theory to place on C to realize the quantum identifications in (3.6).

Of the two identifications in (3.6), the more fundamental by far is the identification of the irreducible representation R with a Hilbert space $\mathscr{H}$, obtained by quantizing some classical phase space upon which G acts as a symmetry. So before we even consider what classical theory must live on C to describe the Wilson loop operator, we can ask the simpler and more basic question — what classical phase space must we quantize to obtain R as a Hilbert space?

As well-known, this question is beautifully answered by the Borel-Weil-Bott theorem [**7**]. In order to recall this theorem, let me first fix a maximal torus $T \subset G$, for which $\mathfrak{t} \subset \mathfrak{g}$ is the associated Cartan subalgebra. Given the irreducible representation R and some choice of positive roots for G, I also introduce the associated highest weight α. Canonically, the weight α lies in the dual $\mathfrak{t}^*$ of $\mathfrak{t}$, but given the invariant form 'Tr' on $\mathfrak{g}$, we are free to identify $\mathfrak{t}^* \cong \mathfrak{t}$ and hence to regard α as an element of $\mathfrak{t}$,

$$\alpha \in \mathfrak{t}^* \cong \mathfrak{t}\,. \tag{3.7}$$

Though mathematically unnatural, the convention in (3.7) proves to be convenient later.

The Borel-Weil-Bott theorem concerns the geometry of the orbit $\mathcal{O}_\alpha \subset \mathfrak{g}$ which passes through α under the adjoint action of G. Equivalently, the adjoint orbit $\mathcal{O}_\alpha$ can be realized as a quotient G/G_α, where G_α is the stabilizer of α under the adjoint action of G. Explicitly, the identification between G/G_α and $\mathcal{O}_\alpha$ is given by the map

$$g\,G_\alpha \longmapsto g\,\alpha\,g^{-1}\,, \qquad g \in G\,. \tag{3.8}$$

As will be essential in a moment, $\mathcal{O}_\alpha$ is a compact complex manifold which admits a natural Kähler structure invariant under G. For instance, if $G = SU(2)$ and α is any non-zero[10] weight, then $\mathcal{O}_\alpha = SU(2)/U(1)$ can be identified as $\mathbb{CP}^1$ endowed with the round, Fubini-Study metric.

In a nutshell, the Borel-Weil-Bott theorem states that the irreducible representation R can be realized geometrically as the space of holomorphic sections of a certain unitary line bundle $\mathfrak{L}(\alpha)$ over $\mathcal{O}_\alpha$. That is,

$$R \cong H^0_{\bar\partial}\big(\mathcal{O}_\alpha,\, \mathfrak{L}(\alpha)\big)\,, \tag{3.9}$$

where the action of G on the sections of $\mathfrak{L}(\alpha)$ is induced from its action on $\mathcal{O}_\alpha$.

As a unitary line bundle over a Kähler manifold, $\mathfrak{L}(\alpha)$ carries a natural unitary connection Θ_α which is also invariant under G. The connection Θ_α enters the path integral description of the Wilson loop operator, so let me exhibit it explicitly. When pulled back to G, the line bundle $\mathfrak{L}(\alpha)$ trivializes, and the connection Θ_α appears as the following left-invariant one-form on G,

$$\Theta_\alpha = \mathrm{Tr}\big(\alpha \cdot g^{-1}dg\big)\,. \tag{3.10}$$

I have introduced the connection Θ_α because its curvature $\nu_\alpha = d\Theta_\alpha$ is precisely the Kähler form on $\mathcal{O}_\alpha$. As a result, $\mathfrak{L}(\alpha)$ is a prequantum line bundle over $\mathcal{O}_\alpha$, and the Borel-Weil-Bott isomorphism in (3.9) identifies R as the Hilbert space obtained by Kähler quantization of $\mathcal{O}_\alpha$.

---

[10]Of course, if $\alpha = 0$, then $\mathcal{O}_\alpha$ is merely a point.

Perhaps more physically, the Borel-Weil-Bott theorem can be interpreted as identifying the space of groundstates for a charged particle moving on $\mathcal{O}_\alpha$ in the presence of a background magnetic field given by ν_α. Briefly, because of the non-zero magnetic field, the wavefunctions which describe this particle transform on $\mathcal{O}_\alpha$ not as functions but as sections of the line bundle $\mathfrak{L}(\alpha)$. As standard, the Hamiltonian which describes free propagation on $\mathcal{O}_\alpha$ is proportional to the Laplacian $\triangle$ acting on sections of $\mathfrak{L}(\alpha)$, and by Hodge theory, the kernel of $\triangle$ can be identified as the space of holomorphic sections of $\mathfrak{L}(\alpha)$. Hence the role of (3.9) is to realize the representation R in terms of groundstates on $\mathcal{O}_\alpha$.

Given the previous quantum mechanical interpretation for R, the corresponding path integral description (3.4) for the Wilson loop operator follows immediately. Ignoring the coupling to A for a moment, if we simply wish to describe the low-energy effective dynamics of an electron moving on $\mathcal{O}_\alpha$ in the background magnetic field ν_α, we consider a one-dimensional sigma model of maps

$$U : C \longrightarrow \mathcal{O}_\alpha\,, \tag{3.11}$$

with sigma model action

$$\mathrm{cs}_\alpha(U) \,=\, \oint_C U^*(\Theta_\alpha) \,=\, \oint_C \mathrm{Tr}(\alpha\cdot g^{-1}dg)\,. \tag{3.12}$$

Here $U^*(\Theta_\alpha)$ denotes the pullback of Θ_α to a connection over C. If U is lifted as a map to $\mathcal{O}_\alpha = G/G_\alpha$ by a corresponding map

$$g : C \longrightarrow G\,, \tag{3.13}$$

then the pullback of Θ_α appears explicitly as in (3.12). As a word of warning, I will freely switch between writing formulae in terms of U or g as convenient.

From a physical perspective, the first-order action cs_α simply describes the minimal coupling of the charged particle on $\mathcal{O}_\alpha$ to the background magnetic field specified by Θ_α, and we have omitted the second-order kinetic terms for U as being irrelevant at low energies. From a more geometric perspective, cs_α is a one-dimensional Chern-Simons action for the abelian connection $U^*(\Theta_\alpha)$ over C. As such, the quantization of the parameter $\alpha \in \mathfrak{t}$ as a weight of G follows just as for the quantization of the Chern-Simons level k. More physically, the quantization of α follows from the quantization of magnetic flux on a compact space.

This sigma model on C clearly respects the action of G on $\mathcal{O}_\alpha$ as a global symmetry. To couple the sigma model to the restriction of the bulk gauge field A, we simply promote the global action of G on $\mathcal{O}_\alpha$ to a gauge symmetry. That is, we consider the covariant version of (3.12),

$$\begin{aligned} \mathrm{cs}_\alpha(U; A|_C) \,&=\, \oint_C U^*(\Theta_\alpha(A)) \,=\, \oint_C \mathrm{Tr}(\alpha\cdot g^{-1}d_A g)\,,\\ d_A g \,&=\, dg + A|_C \cdot g\,. \end{aligned} \tag{3.14}$$

In the second line, we indicate the action of the covariant derivative d_A on g. The action by d_A on g descends to a corresponding covariant action by d_A on U as well.

I now claim that the quantization of the gauged sigma model on C with action $\mathrm{cs}_\alpha(U; A|_C)$ leads to the identifications in (3.6) required to describe the Wilson loop operator. First, the classical equation of motion for U simply asserts that U is covariantly constant,

$$d_A U \,=\, 0\,. \tag{3.15}$$

As a result, the classical phase space for U can be identified with the orbit $\mathcal{O}_\alpha$, and by the Borel-Weil-Bott isomorphism in (3.9), the corresponding Hilbert space $\mathscr{H}$ for U is identified as the representation R. Similarly, since the classical time-evolution for U is given by parallel transport, the quantum time-evolution operator around C is immediately given by the holonomy of A, acting as an element of G on R.

The Shift-Invariant Wilson Loop in Chern-Simons Theory

Because A only enters as a background field in (3.14), the path integral description (3.4) of $W_R(C)$ is completely general and applies to any gauge theory in any dimension. Nonetheless, this description of the Wilson loop operator is precisely what we need to obtain a shift-invariant formulation of the Wilson loop path integral in Chern-Simons theory.

Let us first apply (3.4) to rewrite the Wilson loop path integral in (1.2) as a path integral over both A and U,

$$Z(k; C, R) = \frac{1}{\mathrm{Vol}(\mathcal{G})} \int_{\mathcal{A}\times L\mathcal{O}_\alpha} \mathcal{D}\!A\, \mathcal{D}U\, \exp\left[i\frac{k}{4\pi}\mathrm{CS}(A) + i\,\mathrm{cs}_\alpha(U; A|_C)\right]. \tag{3.16}$$

Here we introduce the free loopspace $L\mathcal{O}_\alpha$ of $\mathcal{O}_\alpha$ to parametrize configurations of U.

Once we introduce the defect field U coupling to A in (3.16), the classical equation of motion for A becomes

$$\mathcal{F}_A \stackrel{\text{def}}{=} F_A + \frac{2\pi}{k} U \cdot \delta_C = 0\,. \tag{3.17}$$

Here δ_C is a two-form with delta-function support on C which represents the Poincaré dual of the curve. Using δ_C, we rewrite $\mathrm{cs}_\alpha(U; A|_C)$ as a bulk integral over M,

$$\mathrm{cs}_\alpha(U; A|_C) = \oint_C \mathrm{Tr}(\alpha \cdot g^{-1} d_A g) = \int_M \delta_C \wedge \mathrm{Tr}(\alpha \cdot g^{-1} d_A g)\,, \tag{3.18}$$

from which (3.17) follows.

As we see from (3.17), in the presence of the operator $W_R(C)$, classical configurations for A are given by connections which are flat on the knot complement

$$M^o = M - C\,, \tag{3.19}$$

and otherwise have delta-function curvature along C. The singularity in A along C manifests itself on M^o as a non-trivial monodromy of the connection around a transverse circle linking C. Though I will not have time to say more, the moduli space of such flat connections with monodromies is the space onto which the Seifert loop path integral localizes. This space is directly related to the moduli space of representations of the knot group of C in G and, in suitable circumstances, fibers over the moduli space of (non-singular) flat connections on M.

At the cost of introducing defect degrees-of-freedom along C, we have managed to describe $W_R(C)$ in terms of a completely local — and indeed linear — functional of A. Consequently, the same path integral trick that we used to decouple one component of A from the Chern-Simons partition function applies immediately to (3.16). We simply replace A in (3.16) by the shift-invariant[11] combination $A - \kappa\,\Phi$,

[11] The shift-symmetry $\mathcal{S}$ acts trivially on U.

and we then perform the Gaussian integral over Φ. In the process, the only new ingredient is that we obtain an extra term linear in Φ from the coupling of A to U.

Without discussing any more details, let me present the resulting shift-invariant formulation for the Wilson loop path integral,

$$Z(k;C,R) = \frac{1}{\mathrm{Vol}(\mathcal{G})} \int_{\mathcal{A}/\mathcal{S}\times L\mathcal{O}_\alpha} \mathcal{D}A\,\mathcal{D}U\,\exp\left[i\frac{k}{4\pi}S(A,U)\right], \tag{3.20}$$

where

$$S(A,U) = \mathrm{CS}(A) + \frac{4\pi}{k}\mathrm{cs}_\alpha(U;A|_C) - \int_M \frac{1}{\kappa\wedge d\kappa}\mathrm{Tr}\left[(\kappa\wedge\mathcal{F}_A)^2\right]. \tag{3.21}$$

The shift-invariant action $S(A,U)$ appears much as the shift-invariant action (2.10) for A alone, with the replacement therein of F_A by the generalized curvature $\mathcal{F}_A$. Thus for an arbitrary Wilson loop (or link) operator in Chern-Simons theory, the path integral can be rewritten such that one component of A completely decouples.

Moreover, if M is a Seifert manifold and C is a Seifert fiber of M, the shift-invariant Seifert loop path integral is again an integral of the canonical symplectic form in (1.12). The relevant symplectic space X is just the product

$$X = \mathcal{A}/\mathcal{S} \times L\mathcal{O}_\alpha, \tag{3.22}$$

where the loopspace $L\mathcal{O}_\alpha$ carries a natural symplectic (and indeed Kähler) form inherited from the Kähler form ν_α on $\mathcal{O}_\alpha$. The Hamiltonian group H which acts on X is the same group that appears in (2.15). In fact, the loopspace $L\mathcal{O}_\alpha$ can be interpreted formally as an infinite-dimensional coadjoint orbit of H. Finally, the square of the moment map for the diagonal action of H on X is precisely the shift-invariant action $S(A,U)$ appearing in (3.21).

For a last bit of furious hand-waving, let me remark that the description of the Seifert loop operator as a character follows quite naturally from the appearance of the loopspace $L\mathcal{O}_\alpha$ in (3.20). In essence, non-abelian localization on $L\mathcal{O}_\alpha$ is related to index theory on $\mathcal{O}_\alpha$, and index theory on $\mathcal{O}_\alpha$ provides a classic derivation [**1**] of the famous Weyl character formula. See §7.2 in [**4**] for a complete discussion.

Acknowledgments. I take pleasure in thanking the organizers and participants of the Bonn workshop "Chern-Simons Gauge Theory: 20 Years After," where this article was presented. I especially thank Edward Witten, both for our prior collaboration on the subject and for posing the question which sparked this work.

References

[1] M. Atiyah and R. Bott, *A Lefschetz fixed point formula for elliptic complexes, II*, Ann. Math. **88** (1968), 451–491.

[2] ———, *Yang-Mills equations over Riemann surfaces*, Phil. Trans. R. Soc. Lond. **A308** (1982), 523–615.

[3] A. Balachandran, S. Borchardt, and A. Stern, *Lagrangian and Hamiltonian descriptions of Yang-Mills particles*, Phys. Rev. D **17** (1978), 3247–3256.

[4] C. Beasley, *Localization for Wilson loops in Chern-Simons theory*, `arXiv:0911.2687 [hep-th]`.

[5] C. Beasley and E. Witten, *Non-abelian localization for Chern-Simons theory*, J. Diff. Geom. **70** (2005), 183–323, `hep-th/0503126`.

[6] M. Blau and G. Thompson, *Chern-Simons theory on S^1-bundles: Abelianisation and q-deformed Yang-Mills theory*, JHEP **0605** (2006), 003, `arXiv:hep-th/0601068`.

[7] R. Bott, *Homogeneous vector bundles*, Ann. of Math. **66** (1957), 203–248.

[8] D. Diakonov and V. Y. Petrov, *A formula for the Wilson loop*, Phys. Lett. B **224** (1989), 131–135.

[9] J. J. Duistermaat and G. J. Heckman, *On the variation in the cohomology of the symplectic form of the reduced phase space*, Invent. Math. **69** (1982), 259–268; Addendum, Invent. Math. **72** (1983), 153–158.

[10] D. Freed and R. Gompf, *Computer calculation of Witten's 3-manifold invariant*, Commun. Math. Phys. **141** (1991), 79–117.

[11] S. Garoufalidis, *Relations among 3-manifold invariants*, Ph.D. thesis, University of Chicago, 1992.

[12] S. Hansen and T. Takata, *Reshetikhin-Turaev invariants of Seifert three-manifolds for classical simple Lie algebras and their asymptotic expansions*, J. Knot Theory Ramifications **13** (2004), 617–668, `math.GT/0209403`.

[13] L. Jeffrey, *Chern-Simons-Witten invariants of lens spaces and torus bundles, and the semi-classical approximation*, Commun. Math. Phys. **147** (1992), 563–604.

[14] L. Jeffrey and B. McLellan, *Eta-invariants and anomalies in $U(1)$ Chern-Simons theory*, `arXiv:1004.2913 [math.SG]`.

[15] A. Kapustin, B. Willett, and I. Yaakov, *Exact results for Wilson loops in superconformal Chern-Simons theories with matter*, `arXiv:0909.4559 [hep-th]`.

[16] R. Lawrence and L. Rozansky, *Witten-Reshetikhin-Turaev invariants of Seifert manifolds*, Commun. Math. Phys. **205** (1999), 287–314.

[17] M. Mariño, *Chern-Simons theory, matrix integrals, and perturbative three-manifold invariants*, Commun. Math. Phys. **253** (2004), 25–49, `hep-th/0207096`.

[18] J. Martinet, *Formes de contact sur les varietétés de dimension 3*, Springer Lecture Notes in Math **209** (1971), 142–163.

[19] L. Moser, *Elementary surgery along a torus knot*, Pacific J. Math **38** (1971), 737–745.

[20] *Problems on Invariants of Knots and 3-Manifolds*, Ed. by T. Ohtsuki with an introduction by J. Roberts, in *Invariants of Knots and 3-Manifolds (Kyoto, 2001)*, pp. 377–572, Geom. Topol. Monogr. 4, Geom. Topol. Publ., Coventry, 2002, `math.GT/0406190`.

[21] P.-E. Paradan, *The moment map and equivariant cohomology with generalized coefficients*, Topology **39** (2000), 401–444.

[22] A. Pressley and G. Segal, *Loop Groups*, Clarendon Press, Oxford, 1986.

[23] N. Reshetikhin and V. Turaev, *Invariants of 3-manifolds via link polynomials and quantum groups*, Invent. Math. **103** (1991), 547–597.

[24] E. Witten, *Quantum field theory and the Jones polynomial*, Commun. Math. Phys. **121** (1989), 351–399.

[25] ———, *Gauge theories and integrable lattice models*, Nucl. Phys. B **322** (1989), 629–697.

[26] ———, "Two-dimensional Gauge Theories Revisited," J. Geom. Phys. **9** (1992), 303–368, `hep-th/9204083`.

[27] ———, *Dynamics of quantum field theory*, in *Quantum Fields and Strings: A Course for Mathematicians, Vol. 2*, Ed. by P. Deligne et al., American Mathematical Society, Providence, Rhode Island, 1999.

Simons Center for Geometry and Physics, Stony Brook University, Stony Brook, New York 11794-3636

E-mail address: `cbeasley@insti.physics.sunysb.edu`

AMS/IP Studies in Advanced Mathematics
Volume 50, 2011

Quantum Field Theory and the Volume Conjecture

Tudor Dimofte and Sergei Gukov

ABSTRACT. The volume conjecture states that for a hyperbolic knot K in the three-sphere S^3 the asymptotic growth of the colored Jones polynomial of K is governed by the hyperbolic volume of the knot complement $S^3\backslash K$. The conjecture relates two topological invariants, one combinatorial and one geometric, in a very nonobvious, nontrivial manner. The goal of the present lectures is to review the original statement of the volume conjecture and its recent extensions and generalizations, and to show how, in the most general context, the conjecture can be understood in terms of topological quantum field theory. In particular, we consider: *a*) generalization of the volume conjecture to families of incomplete hyperbolic metrics; *b*) generalization that involves not only the leading (volume) term, but the entire asymptotic expansion in $1/N$; *c*) generalization to quantum group invariants for groups of higher rank; and *d*) generalization to arbitrary links in arbitrary three-manifolds.

1. Preliminaries

Let K be an oriented knot (or link) in the three-sphere S^3. The original volume conjecture [**24, 29**] relates the N-colored Jones polynomial of K to the hyperbolic volume of the knot complement $S^3\backslash K$:

$$\begin{array}{ccc} N\text{-colored Jones poly of } K & \longleftrightarrow & \text{hyperbolic volume of } S^3\backslash K \\ \text{(combinatorial, rep. theory)} & & \text{(geometric)}\,. \end{array} \tag{1.1}$$

We begin by reviewing some of the definitions and ingredients that enter on the two sides here in order to make this statement more precise, and to serve as a precursor for its subsequent generalization.

Jones polynomials. The (non-colored) Jones polynomial $J(K;q)$ of a knot or link can be defined combinatorially via the skein relation

$$q\,J(\nwarrow\!\!\!\!\nearrow) - q^{-1}J(\nearrow\!\!\!\!\nwarrow) = (q^{\frac{1}{2}} - q^{-\frac{1}{2}})\,J(\,)\,(\,)\,, \tag{1.2}$$

along with the normalization[1]

$$J(\bigcirc) = q^{\frac{1}{2}} + q^{-\frac{1}{2}} \qquad \text{for } \bigcirc = \text{unknot}\,, \tag{1.3}$$

2010 *Mathematics Subject Classification.* Primary 58J28, 81T45; Secondary 58Z05, 57M27.
The work of SG is supported in part by DOE Grant DE-FG03-92-ER40701, in part by NSF Grant PHY-0757647, and in part by the Alfred P. Sloan Foundation. Opinions and conclusions expressed here are those of the authors and do not necessarily reflect the views of funding agencies.

[1] The most common normalization for the unknot seen in the mathematics literature is $J(\bigcirc) = 1$. For the connection with topological quantum field theory, however, (1.3) is much more natural.

and the rule

$$J(K_1 \sqcup K_2) = J(K_1)\,J(K_2) \tag{1.4}$$

for any disjoint union of knots or links. Thus, for example, the (right-handed) trefoil and figure-eight knots have Jones polynomials

$$J(\mathbf{3_1}) = q^{-\frac{1}{2}} + q^{-\frac{3}{2}} + q^{-\frac{5}{2}} - q^{-\frac{9}{2}}\,,$$
$$J(\mathbf{4_1}) = q^{\frac{5}{2}} + q^{-\frac{5}{2}}\,.$$

In general, $J(K,q)$ is a Laurent polynomial, $J(K,q) \in \mathbb{Z}[q^{\frac{1}{2}}, q^{-\frac{1}{2}}]$.

The combinatorial construction of the Jones polynomial is intimately related to representation theory of $SU(2)$ — or the closely related representation theories of the quantum group $U_q(su(2))$ or the affine Lie algebra $\widehat{su(2)}$. In particular, the classical Jones polynomial above is obtained by "coloring" the knot (or link) K in S^3 with the 2-dimensional representation of $SU(2)$. More generally, such a knot or link can be colored with any finite-dimensional representation R of $SU(2)$, leading to a colored Jones polynomial $J_R(K,q)$. The *N-colored* Jones polynomial $J_N(K,q)$ takes R to be the irreducible N-dimensional representation [**40, 36, 25**]. The colored Jones polynomial can again be computed in a purely algebraic/combinatorial manner, by using the two rules

$$J_{\oplus_i R_i}(K;q) = \sum_i J_{R_i}(K;q) \tag{1.5a}$$

and

$$J_R(K^n;q) = J_{R^{\otimes n}}(K;q)\,, \tag{1.5b}$$

together with

$$J_R(K_1 \sqcup K_2) = J_R(K_1) J_R(K_2)\,, \tag{1.6}$$

and the fact that $J_1(K;q) = J_{R=\oslash}(K;q) \equiv 1$. The first rule says that if R is reducible, then J_R splits as a sum over irreducible components. The second rule says that the R-colored Jones polynomial for the n-cabling of a knot (formed by taking n copies of the knot or link, slightly displaced away from one another[2]) is equal to the colored Jones polynomial of the original knot but in representation $R^{\otimes n}$.

For example, from (1.5a-b) and the fact that $J_{N=2}(K,q) = J(K,q)$, it is easy to see that

$$J_N(\bigcirc) = \frac{q^{\frac{N}{2}} - q^{-\frac{N}{2}}}{q^{\frac{1}{2}} - q^{-\frac{1}{2}}}\,. \tag{1.7}$$

More generally, for any knot K, relations (1.5) can be used to reduce $J_N(K;q)$ to ordinary Jones polynomials of K and its cablings. We have

$$J_1(K;q) = 1\,,$$
$$J_2(K;q) = J(K;q)\,,$$
$$J_3(K;q) = J(K^2;q) - 1\,,$$
$$J_4(K;q) = J(K^3;q) - 2J(K;q)\,,$$
$$\ldots,$$

where the expressions for J_3, J_4, etc. follow from the rules for decomposing representations of $SU(2)$: $\mathbf{2}^{\otimes 2} = \mathbf{1} \oplus \mathbf{3}\,,\ \mathbf{2}^{\otimes 3} = (\mathbf{1} \oplus \mathbf{3}) \otimes \mathbf{2} = \mathbf{2} \oplus \mathbf{2} \oplus \mathbf{4}$, etc. Since $J(K;q) \in \mathbb{Z}[q^{\frac{1}{2}}, q^{-\frac{1}{2}}]$ for any K, it is clear that the colored Jones polynomials $J_N(K;q)$ will also be elements of $\mathbb{Z}[q^{\frac{1}{2}}, q^{-\frac{1}{2}}]$.

---

[2]This displacement must be done in a way that produces zero linking number between the various copies.

We have explained the left side of (1.1), completely, if somewhat tersely, in terms of algebra and combinatorics. The right side has a very different interpretation.

Hyperbolic volumes. It was conjectured by Thurston [**37**] (and is now proved [**34**]) that every three-manifold may be decomposed into pieces that admit exactly one of eight different geometric structures. The most common structure by far is hyperbolic. Indeed, in the case of knot complements in S^3 this statement can be made exact: a knot complement has a hyperbolic structure if and only if it is not a torus or satellite knot [**37**]. By definition, a "hyperbolic structure" refers to a geodesically-complete metric of constant curvature -1. If a hyperbolic structure does exist on a manifold M, then it is unique, and the corresponding hyperbolic volume $\mathrm{Vol}(M)$ is a well-defined topological invariant.

In fact, there also exists a natural complexification of the hyperbolic volume of a three-manifold M, obtained as

$$\mathrm{Vol}(M) + i\mathrm{CS}(M)\,, \tag{1.8}$$

where $\mathrm{CS}(M)$ is the so-called Chern-Simons invariant of M. To understand this, suppose that M allows a spin structure (as all knot/link complements in S^3 do) and consider flat $SL(2,\mathbb{C})$ connections on M in place of hyperbolic metrics.[3] There exists a flat connection $\mathcal{A}$ whose real and imaginary parts can be interpreted, respectively, as the vielbein and spin connection of the hyperbolic metric. The real part of the quantity

$$\frac{i}{2} I_{CS}(\mathcal{A}) = \frac{i}{2}\int_M \mathrm{Tr}\left(\mathcal{A}\wedge d\mathcal{A} + \frac{2}{3}\mathcal{A}\wedge\mathcal{A}\wedge\mathcal{A}\right) \tag{1.9}$$

then reproduces $\mathrm{Vol}(M)$, while the imaginary part defines $\mathrm{CS}(M)$. The expression $I_{CS}(\mathcal{A})$ is the so-called Chern-Simons functional of $\mathcal{A}$. Further details can be found *e.g.* in [**38, 39**] or [**19, 10**]. Under gauge transformations acting on $\mathcal{A}$, the functional $I_{CS}(\mathcal{A})$ is only well-defined up to shifts of $8\pi^2$, leading to an ambiguity of $4\pi^2$ in the definition of $\mathrm{CS}(M)$. Because of this, it is often convenient to exponentiate the complexified volume (1.8), writing it in the unambiguous form

$$Z(M) = e^{\frac{i}{4\pi} I_{CS}(\mathcal{A})} = e^{\frac{1}{2\pi}\left(\mathrm{Vol}(M)+i\mathrm{CS}(M)\right)}\,. \tag{1.10}$$

For hyperbolic knot complements, the full complexified volume $Z(M)$ can be efficiently computed in terms of ideal hyperbolic triangulations, *cf.* [**8, 32, 44**].

The Volume Conjecture. We have not said much yet about the variable q appearing in the Jones polynomials. Strictly speaking, this variable should be a root of unity[4]

$$q = e^{\frac{2\pi i}{k}}\,, \qquad k \in \mathbb{Z}_+\,. \tag{1.11}$$

At the special value $k = N$, all Jones polynomials $J_N(K;q)$ vanish, but the ratio

$$V_N(K,q) = \frac{J_N(K;q)}{J_N(\bigcirc;q)} \tag{1.12}$$

remains finite. The original volume conjecture [**24, 29**] then states that

$$\lim_{N\to\infty} \frac{2\pi \log\left|V_N(K;q=e^{\frac{2\pi i}{N}})\right|}{N} = \mathrm{Vol}(M)\,. \tag{1.13}$$

---

[3] Recall that a "G-connection" on a principal G-bundle $E \to M$ can be written locally as a $\mathfrak{g}$-valued one-form $\mathcal{A}$. The bundle E is typically taken to be trivial in the present context, $E = G\times M$. A gauge transformation (a change of coordinates on E) induced by an element $g\in\Gamma(E)$ acts locally on the connection as $\mathcal{A}\mapsto g^{-1}\mathcal{A}g + g^{-1}dg$.

[4] In terms of representation theory, the integer k is identified as the level of the affine Lie algebra $\widehat{su(2)}_k$. The representation theory of the quantum group $U_q(su(2))$ also (crucially) simplifies greatly when q is a root of unity, becoming essentially equivalent to the representation theory of $\widehat{su(2)}_k$. See also Sections 3.3.1-3.3.2.

It is also possible to remove the absolute value and exponentiate to obtain the complexified generalization (*cf.* [**30**])

$$\boxed{V_N(K; q = e^{\frac{2\pi i}{N}}) \overset{N\to\infty}{\sim} Z(M)^N = e^{\frac{N}{2\pi}\left(\mathrm{Vol}(M) + i\mathrm{CS}(M)\right)}} \,. \tag{1.14}$$

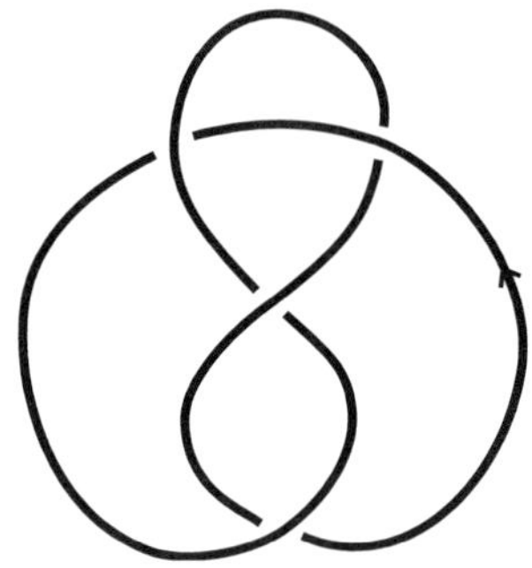

FIGURE 1. The figure-eight knot, $\mathbf{4_1}$.

As an example, consider the figure-eight knot (Figure 1), the simplest hyperbolic knot. The colored Jones polynomial (see *e.g.* [**24**] or [**22**]) is

$$V_N(\mathbf{4_1}; q = e^{\frac{2\pi i}{N}}) = \sum_{m=0}^{N-1} (q)_m (q^{-1})_m \,, \qquad (x)_m := (1-x)(1-x^2)\cdots(1-x^m)\,. \tag{1.15}$$

The hyperbolic volume of the figure-eight knot complement is

$$\mathrm{Vol}(S^3\backslash\mathbf{4_1}) = 2\,\mathrm{Vol}(\Delta) = 2.02988\ldots\,, \tag{1.16}$$

where $\mathrm{Vol}(\Delta) = \mathrm{Im}\,\mathrm{Li}_2(e^{i\frac{\pi}{3}})$ denotes the volume of a regular hyperbolic ideal tetrahedron. The Chern-Simons invariant $\mathrm{CS}(S^3\backslash\mathbf{4_1})$ vanishes. It is fairly straightforward (and an informative exercise[5]) to show that in the limit $N\to\infty$ one has, as expected,

$$\lim_{N\to\infty} \frac{2\pi \log V_N(\mathbf{4_1}; e^{\frac{2\pi i}{N}})}{N} = \mathrm{Vol}(S^3\backslash\mathbf{4_1})\,. \tag{1.17}$$

2. The many dimensions of the volume conjecture

There are several natural ways in which one might try to generalize the basic volume conjecture (1.14). One possibility is to consider not just $k = N$ (or $q = e^{\frac{2\pi i}{N}}$), but arbitrary values of k (or q). Another option would be to ask what happens to subleading terms in the asymptotic expansion of $V_N(K;q)$ as $N\to\infty$. It might also be interesting to consider not just hyperbolic knots in S^3 but arbitrary links in more complicated three-manifolds. It turns out that all these generalizations make sense, and can be nicely combined and interpreted in terms of Chern-Simons theory with complex gauge group [**19**]. In this section, we detail each of them (and one additional generalization) in turn, and begin to explain what kind of new objects one should expect on the right-hand-side of (1.14). Then, in section 3, our goal will be to explain where such generalizations come from.

---

[5]One method involves analytically continuing the summand as a ratio of quantum dilogarithm functions (*cf.* [**13, 10**]), approximating the sum by an integral, and evaluating it at its saddle point.

2.1. Parametrized VC. The original volume conjecture only held for a special root of unity $q = e^{\frac{2\pi i}{N}}$. In order to generalize to arbitrary $q = e^{\frac{2\pi i}{k}}$, the appropriate limit to consider is

$$k \to \infty\,, \qquad N \to \infty\,, \qquad u := i\pi \frac{N}{k} \quad \text{fixed} \tag{2.1}$$

(or $q \to 1$, $q^N = e^{2u}$ fixed). The question, then, is how to understand

$$\lim_{k,N\to\infty} J_N(K;q)^{1/k} \quad ? \tag{2.2}$$

The answer, described in [**19**], uses the fact that in correspondence with the "deformation" in the colored Jones polynomial, there exists a one-parameter deformation of the hyperbolic structure on a knot complement $S^3\backslash K$. To understand this, let μ be a small loop linking the excised knot K, as in Figure 2a. In terms of flat $SL(2,\mathbb{C})$ connections, the geodesically complete hyperbolic metric has a parabolic $SL(2,\mathbb{C})$ holonomy around μ,

$$\mathrm{Hol}(\mu, \text{complete}) = \pm \begin{pmatrix} 1 & 1 \\ 0 & 1 \end{pmatrix}, \tag{2.3}$$

whereas the incomplete, u-deformed hyperbolic metric/$SL(2,\mathbb{C})$ connection is defined to have a holonomy conjugate to

$$\mathrm{Hol}(\mu, u) = \begin{pmatrix} e^u & 1 \\ 0 & e^{-u} \end{pmatrix}. \tag{2.4}$$

(As long as $e^u \neq e^{-u}$, this deformed holonomy is also conjugate to the purely diagonal matrix $\mathrm{diag}(e^u, e^{-u})$.) The resulting metric is not complete. For example, when u is purely imaginary, the u-deformed metric has a conical cusp of angle $2\mathrm{Im}(u)$ at the knot K.

The complexified hyperbolic volume for this one-parameter family of metrics can again be defined in terms of the Chern-Simons functional $I_{CS}(\mathcal{A})$ appearing in (1.9). Now, however, $\mathcal{A} = \mathcal{A}(u)$ should be a flat $SL(2,\mathbb{C})$ connection with prescribed holonomy (2.4). The "parametrized" volume conjecture then takes the form [**19**]

$$\boxed{J_N(K;q) \overset{k,N\to\infty}{\sim} e^{-\frac{k}{4\pi i} I_{CS}(\mathcal{A}(u))}}\,. \tag{2.5}$$

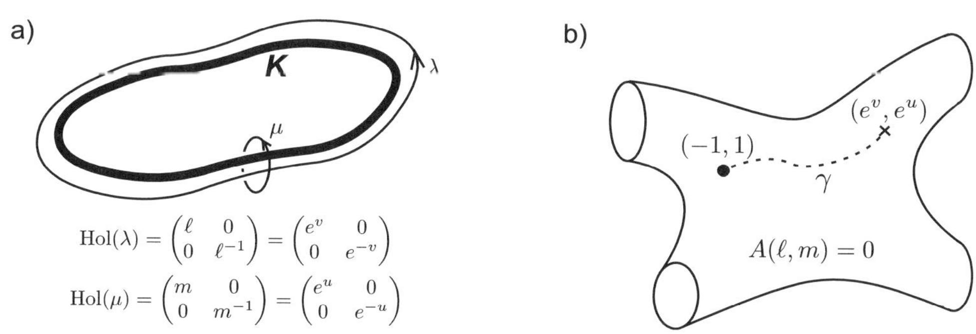

FIGURE 2. a) The "longitude" λ and "meridian" μ holonomy paths in the knot complement $S^3\backslash K$. b) Integration on the A-polynomial curve to find the deformed complex volume.

The quantity $I_{CS}(\mathcal{A}(u))$ can be described very explicitly. Indeed, suppose that we require a hyperbolic metric (expressed in terms of a flat $SL(2,\mathbb{C})$ connection) to have holonomies conjugate to $\mathrm{diag}(e^u, e^{-u})$ and $\mathrm{diag}(e^v, e^{-v})$, respectively, along the meridian and longitude loops depicted in Figure 2a. Such a metric exists if and only if the so-called A-polynomial of K vanishes [**7**],

$$A(\ell, m) = 0 \qquad \text{for} \qquad \ell = e^v\,, \quad m = e^u\,. \tag{2.6}$$

Given a fixed $e^u \in \mathbb{C}^*$, exactly one of the solutions $v = v^{\text{hyp}}(u)$ of this equation corresponds to the u-deformed hyperbolic metric. The Chern-Simons functional evaluated at the flat connection $\mathcal{A}(u)$ can then be written as

$$I_{CS}(\mathcal{A}(u)) = I_{CS}(\mathcal{A}(i\pi)) + 4\int_\gamma \theta\,, \tag{2.7}$$

where $\mathcal{A}(i\pi)$ is the non-deformed hyperbolic flat connection,

$$\theta = -(v' + i\pi)\, du' \tag{2.8}$$

is a one-form on the curve $A(e^{v'}, e^{u'}) = 0$, and γ is a path on this curve that connects the complete hyperbolic structure at $(e^{v'}, e^{u'}) = (-1, \pm 1)$ to the u-deformed metric at $(e^{v'}, e^{u'}) = (e^{v^{\text{hyp}}(u)}, e^u)$, as in Figure 2b.[6]

As our recurrent example, consider again the figure-eight knot. The complete colored Jones polynomial, *cf.* [**22**], is

$$J_N(\mathbf{4_1}; q) = \frac{q^{\frac{N}{2}} - q^{-\frac{N}{2}}}{q^{\frac{1}{2}} - q^{-\frac{1}{2}}} \sum_{j=0}^{N-1} q^{Nj} \prod_{k=1}^{j} (1 - q^{k-N})(1 - q^{-k-N})\,. \tag{2.10}$$

The A-polynomial of the figure-eight knot is

$$A(\ell, m) = (\ell - 1)(m^4\ell^2 - (1 - m^2 - 2m^4 - m^6 + m^8)\ell + m^4\ell^2)\,, \tag{2.11}$$

and from (2.7) and (1.16), it results (after some algebra) that the Chern-Simons functional can be written as

$$I_{CS}(\mathcal{A}(u)) = 2\mathrm{Li}_2(e^{-p-u}) - 2\mathrm{Li}_2(e^{p-u}) + 8(p - i\pi)(u - i\pi)\,, \tag{2.12}$$

where $x = e^p$ is the solution to $m^3x^2 + (1 - m^2 - m^4)x + m^3 = 0$ with smallest negative imaginary part. For irrational $u/i\pi$ in a neighborhood of $u = i\pi$ it can then be shown (*cf.* [**27, 31**]) that the proposed asymptotics (2.5) indeed hold.

The necessity for taking $u/i\pi$ irrational here may appear a little strange at first glance. It stems fundamentally from the fact that the Jones polynomials $J_N(K; q = e^{\frac{2\pi i}{k}})$ are really only defined for $N, k \in \mathbb{Z}$. A subtle analytic continuation in either N or k is necessary to achieve $u/i\pi = N/k \notin \mathbb{Q}$. As anticipated in [**19**] and explained recently in [**41**], it is this continuation that causes the growth of the colored Jones polynomial to be exponential. We will remark on this further in Section 3.3.2.

In light of this argument, one might ask now why the original volume conjecture at the rational value $k = N$ or $u = i\pi$ held in the first place. Recall that $J_N(K; q)$ actually vanished at $k = N$, so it was necessary to divide by $J_N(\bigcirc; q)$ to obtain the non-vanishing ratio $V_N(K; q)$. Examining $V_N(K; q)$ at $u \to 0$ is equivalent to considering the *derivative* of $J_N(K; q)$ at $u = i\pi$, which of course knows about analytic continuation.[7]

---

[6]The actual complexified volume that appears in the literature on hyperbolic geometry (*cf.* [**33, 43, 23**]) is related to $I_{CS}(\mathcal{A}(u))$ as

$$\mathrm{Vol}(S^3\backslash K; u) + i\mathrm{CS}(S^3\backslash K; u) = \frac{i}{2} I_{CS}(\mathcal{A}(u)) + 2iv(u)\mathrm{Re}(u) - 2\pi u + 2\pi^2 i\,. \tag{2.9}$$

Note that $I_{CS}(\mathcal{A}(u))$ is analytic in u, whereas $\mathrm{Vol}(u) + i\mathrm{CS}(u)$ is not.

[7]We thank E. Witten for useful observations on this subject.

2.2. Quantum VC. The second option for generalizing the volume conjecture (1.14) is to ask for higher-order terms in the asymptotic expansion of the colored Jones polynomial. Let us define a new "quantum" parameter $\hbar$ as

$$\hbar = \frac{i\pi}{k}, \tag{2.13}$$

so that

$$q = e^{2\hbar}. \tag{2.14}$$

The two parameters N and k of the colored Jones polynomial can be traded for $\hbar$ and u, and the limit (2.1) is simply $\hbar \to 0$. At $u = i\pi$, higher-order asymptotics are then predicted [**19**] to have the form

(2.15)
$$V_N(K; q = e^{\frac{2\pi i}{N}}) \overset{N\to\infty}{\sim} \exp\left(\frac{1}{2\hbar}(\mathrm{Vol} + i\mathrm{CS}) - \frac{3}{2}\log\hbar + \frac{1}{2}\log\frac{-i\pi T_K}{4} + \sum_{n=2}^{\infty} \tilde{S}_n \hbar^{n-1} \right).$$

Here, for example, T_K is the Ray-Singer torsion of the knot complement $S^3\backslash K$. It can be defined after putting any background metric on $S^3\backslash K$ [**35**] as

$$T(M) = \exp\left(-\frac{1}{2}\sum_{n=0}^{3} n(-1)^n \log\det{}'\Delta_n \right) = \frac{(\det{}'\Delta_0)^{\frac{3}{2}}}{(\det{}'\Delta_1)^{\frac{1}{2}}}, \tag{2.16}$$

where Δ_n is the Laplacian acting on n-forms.

It is fairly straightforward to combine the present quantum deformation with the parametrization of the volume conjecture in u. The expectation is that

(2.17)
$$\boxed{J_N(K;q) \overset{N,k\to\infty}{\sim} \exp\left(-\frac{1}{4\hbar} I_{CS}(\mathcal{A}(u)) - \frac{3}{2}\log\hbar + \frac{1}{2}\log\frac{iT_K(u)}{4\pi} + \sum_{n=2}^{\infty} S_n(u)\hbar^{n-1} \right)}.$$

Here, $T_K(u)$ is a u-deformed torsion, and is related to the Alexander polynomial of K [**16, 28**]. The higher-order coefficients in (2.15) are related to those in (2.17) as

$$\sum_{n\geq 2} \tilde{S}_n \hbar^{n-1} = \sum_{n\geq 2} S_n(i\pi)\hbar^{n-1} - \log\frac{\sinh\hbar}{\hbar}. \tag{2.18}$$

For the figure-eight knot, the quantum volume conjecture (2.17) was tested to first subleading order in [**20**] (see also [**1**]), using the Ray-Singer torsion

$$T_{\mathbf{4_1}}(u) = \frac{4\pi^2}{\sqrt{-m^{-4} + 2m^{-2} + 1 + 2m^2 - m^4}}. \tag{2.19}$$

Higher-order coefficients $S_n(u)$ can also be computed [**10**]. For example,

$$S_2(u) = \frac{-i(T_{\mathbf{4_1}})^3}{12(4\pi^2)^3 m^6}\left(1 - m^2 - 2m^4 + 15m^6 - 2m^8 - m^{10} + m^{12}\right), \tag{2.20}$$

$$S_3(u) = \frac{-2(T_{\mathbf{4_1}})^6}{(4\pi^2)^6 m^6}\left(1 - m^2 - 2m^4 + 5m^6 - 2m^8 - m^{10} + m^{12}\right) - \frac{1}{6}. \tag{2.21}$$

These expressions appear to be new, unexplored knot invariants with distinctive number-theoretic properties. Needless to say, it would be interesting to test the quantum volume conjecture (2.17) for other hyperbolic knots and/or to higher order in the $\hbar$-expansion.

Just as the generalization of the volume conjecture to $u \neq 0$ was interpreted in terms of the $SL(2,\mathbb{C})$ Chern-Simons functional, there is also a Chern-Simons interpretation of the quantum volume conjecture. One must consider how the functional $I_{CS}(\mathcal{A})$ behaves when the connection $\mathcal{A}$ undergoes "quantum fluctuations" away from the flat connection $\mathcal{A}(u)$. This is accomplished in physics via perturbative quantum field theory. Symbolically, we can write $\mathcal{A} = \mathcal{A}(u) + \mathcal{A}'$, where $\mathcal{A}'$ contains the fluctuations away from flatness, and define a perturbative "partition function" via the path integral

$$Z(S^3\backslash K; u; \hbar)_{\text{pert}} = \int \mathcal{D}\mathcal{A}' \, e^{-\frac{1}{4\hbar} I_{CS}(\mathcal{A}(u)+\mathcal{A}')} . \tag{2.22}$$

The exponent in the integrand has a critical point at $\mathcal{A}' = 0$, and a saddle point expansion around this point yields the right-hand-side of (2.17). (To be very precise, $J_N(K;q) \sim Z(S^3\backslash K; \hbar; u)/Z(S^3;\hbar)$, where $Z(S^3;\hbar) = \sqrt{2/k}\sin(\pi/k)$ is the partition function of the three-sphere S^3.)

2.3. Groups and representations. So far, we have considered two continuous deformations of the volume conjecture, in u and $\hbar$, as drawn schematically in Figure 3. In addition, there are two discrete generalizations that we can make.

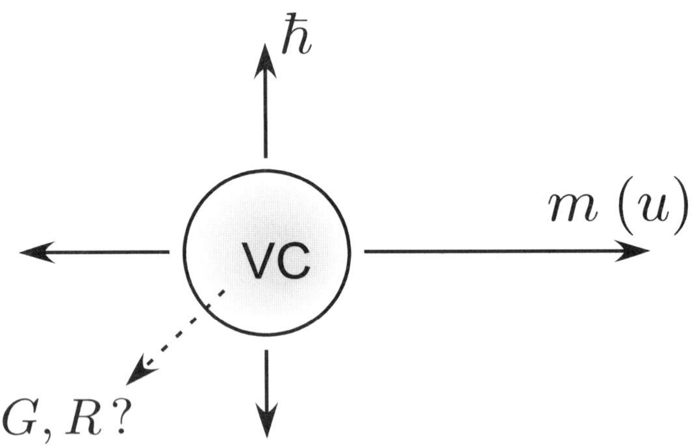

FIGURE 3. Continuous and discrete generalizations of the volume conjecture.

The first such generalization involves the "gauge groups" and representations that define colored Jones polynomials. Recall from Section 1 that the N-colored Jones polynomial is a quantum $SU(2)$ invariant that corresponds to coloring a knot with the N-dimensional representation of $SU(2)$. More generally, one can consider "quantum $SU(n)$ invariants," or in fact invariants for any compact Lie group G. Knots or links should then be colored by finite-dimensional representations R of G. For semisimple G and irreducible R, the representation can be labelled by a highest weight λ in the weight lattice $\Lambda_{wt} \subset \mathfrak{g}^*$, where $\mathfrak{g} = \text{Lie}(G)$. The resulting quantum polynomial invariant of a knot in S^3 may be denoted

$$P^G_{R_\lambda}(K;q) . \tag{2.23}$$

Just like the colored Jones polynomial, $P^G_R(K;q)$ depends on a root of unity $q = e^{\frac{2\pi i}{k}}$. Also like the colored Jones, these invariants satisfy

$$P^G_{\oplus_i R_i}(K;q) = \sum_i P^G_{R_i}(K;q) , \qquad \text{and} \qquad P^G_{R^{\otimes n}}(K;q) = P^G_R(K^n;q) . \tag{2.24}$$

More general tensor products can also be produced by cabling a knot or link and coloring each component of the cable with a different representation. When $G = SU(n)$ and R is the fundamental representation (or any of its conjugates), the polynomial $P^G_R(K;q)$ satisfies a skein relation similar to (1.2).

Using the positive nondegenerate trace form $-\mathrm{Tr} : \mathfrak{g} \times \mathfrak{g} \to \mathbb{R}$, the weight λ can be identified with its dual element λ^* in $\mathfrak{t}$, the Cartan subalgebra of $\mathfrak{g}$. Let us also define ρ to be half the sum of positive roots, and $\rho^* \in \mathfrak{t} \subset \mathfrak{g}$ its dual. Then the interesting limit to consider for $P^G_{R_\lambda}(K, q)$ is

$$k \to \infty\,, \qquad \lambda^* \to \infty\,, \qquad u := \frac{i\pi}{k}(\lambda^* + \rho^*)\ \text{fixed}\,, \tag{2.25}$$

or

$$q = e^{2\hbar} = e^{\frac{2\pi i}{k}} \to 1 \quad (\hbar \to 0)\,, \qquad q^{\lambda^*+\rho^*} = e^{2u}\ \text{fixed}\,. \tag{2.26}$$

The parameter u has now become a diagonal matrix, an element of $\mathfrak{t}_{\mathbb{C}}$. Coming back to the case of $SU(2)$ and an N-dimensional representation, in this notation we have

$$\lambda^* = \begin{pmatrix} N-1 & 0 \\ 0 & -(N-1) \end{pmatrix}, \qquad \rho^* = \begin{pmatrix} 1 & 0 \\ 0 & -1 \end{pmatrix}, \qquad u = i\pi \begin{pmatrix} \frac{N}{k} & 0 \\ 0 & -\frac{N}{k} \end{pmatrix}. \tag{2.27}$$

The asymptotics of the invariant $P^G_R(K; q)$ should look very similar to those of the colored Jones polynomial, namely

$$\boxed{P^G_{R_\lambda}(K; q) \overset{\hbar \to 0}{\sim} \exp\left(-\frac{1}{4\hbar} I_{CS}(u) - \frac{\delta}{2} \log \hbar + \frac{1}{2} \log \frac{iT(u)}{4\pi} + \sum_{n=2}^{\infty} S_n(u) \hbar^{n-1} \right)}\,. \tag{2.28}$$

The leading term $I_{CS}(u)$ is now the Chern-Simons functional (1.9) evaluated at a flat $G_{\mathbb{C}}$ connection $\mathcal{A}(u)$ — in other words, a connection taking values in the complexified Lie algebra $\mathfrak{g}_{\mathbb{C}}$ — whose holonomy around the meridian of the knot as in Figure 2a is

$$\mathrm{Hol}(\mu) = m = e^u\,. \tag{2.29}$$

For generic u, this holonomy is an element of the complexified maximal torus $T_{\mathbb{C}} \subset G_{\mathbb{C}}$. Again, $I_{CS}(u)$ can be expressed as

$$I_{CS}(u) = \text{const.} + 4 \int_{\gamma(u)} \theta\,, \tag{2.30}$$

where $\theta \sim -\sum_{i=1}^r v_i\, du_i$ + exact is a differential on an r-dimensional complex variety cut out by r equations $A_j(e^v, e^u) = 0$, with $r = \mathrm{rank}(G)$. The equations $A_j(e^v, e^u) = 0$ describe the moduli space of flat $G_{\mathbb{C}}$ connections on $S^3 \backslash K$.

Subleading terms on the right side of (2.28) also have a geometric interpretation. The function $T(u)$ is the Ray–Singer torsion of the knot complement twisted by the flat connection $\mathcal{A}(u)$, and the number δ is a fixed integer which can be computed in terms of cohomology of $S^3 \backslash K$ with coefficients in the associated flat bundle, with structure group $G_{\mathbb{C}}$ and connection $\mathcal{A}(u)$ (*cf.* [**4, 10**]). More generally, the full asymptotic expansion can be written as a perturbative path integral just like (2.22), which takes into account the quantum fluctuations of a flat $G_{\mathbb{C}}$ connection.

2.4. Links and 3-manifolds. The final generalization of the volume conjecture that we consider is to arbitrary links in arbitrary three-manifolds. Here we really begin to require a true TQFT description of the "quantum G-invariants" of knots and links. This was supplied by quantum Chern-Simons theory with compact gauge group G in [**40**], and reinterpreted via quantum groups and R-matrices in [**36**]. Using either of these approaches, one may define a quantum partition function

$$Z^G(M, L; \{R_a\}; \hbar) \tag{2.31}$$

for a link L in any three-manifold M, where each component of the link is colored with a different representation R_a. The "polynomial" P_R^G is obtained from this after normalizing by the partition function of an empty manifold (a manifold with no link),

$$P^G_{\{R_a\}}(M,L;q) = \frac{Z^G(M,L;\{R_a\};\hbar)}{Z^G(M;\hbar)}\,, \qquad (q=e^{2\hbar})\,. \tag{2.32}$$

Thus, in the case of the colored Jones polynomial,

$$J_N(K;q) = \frac{Z^{SU(2)}(S^3,K;R_N;\hbar)}{Z^{SU(2)}(S^3;\hbar)}\,. \tag{2.33}$$

The integer k (appearing in $q = e^{2\hbar} = e^{\frac{2\pi i}{k}}$) is identified with the "level" or coupling constant of the compact Chern-Simons theory.[8]

The partition function (2.31) supplies the left-hand-side of the volume conjecture. We then want to understand the asymptotics of $Z^G(M,L;\{R_a\};\hbar)$ in the limit $\hbar \to 0$, with a parameter $u_a = \hbar(\lambda_a^* + \rho^*)$ held fixed for each separate link component. The answer should be given by perturbative, quantum Chern-Simons theory with *complex* gauge group $G_{\mathbb{C}}$, evaluated on the link complement $M\backslash L$, in the background of a flat connection with fixed holonomy

$$m_a = \exp(u_a) \tag{2.34}$$

around the meridian of each excised link component. Denoting this perturbative Chern-Simons partition function by

$$Z^{G_{\mathbb{C}}}_{\text{pert}}(M\backslash L;\{u_a\};\hbar) = \exp\left(-\frac{1}{4\hbar}I_{CS}(\{u_a\}) - \frac{\delta}{2}\log\hbar + \ldots\right)\,, \tag{2.35}$$

we expect that

$$\boxed{Z^G(M,L;\{R_a\};\hbar) \overset{\hbar\to 0}{\sim} Z^{G_{\mathbb{C}}}_{\text{pert}}(M\backslash L;\{u_a\};\hbar)}\,. \tag{2.36}$$

This discussion can also be rephrased in a somewhat more symmetric manner, using link complements on both sides of the volume conjecture. It turns out that in compact Chern-Simons theory the partition function of a knot (or link) $K \subset M$ colored by representation R_λ is equivalent to the partition function of the knot complement $M\backslash K$ with fixed meridian holonomy

$$m = \exp\left(i\pi\frac{\lambda^*+\rho^*}{k}\right) = \exp\left(\hbar(\lambda^*+\rho^*)\right) = \exp(u)\,. \tag{2.37}$$

For the compact G theory to make sense, the eigenvalues of the matrix $u/i\pi$ must clearly be rational. However, interesting asymptotics — potentially with exponential growth as in (2.36) — occur when u is *analytically continued* away from such rational values. This process of analytic continuation naturally lands one in the regime of Chern-Simons theory with complex gauge group $G_{\mathbb{C}}$ [**10**].

After so many generalizations, it may be unclear that the volume conjecture has anything to do with volumes anymore. Indeed, for higher-rank gauge groups G, "volume" should not be a hyperbolic volume but rather the "volume" of a holonomy representation

$$\varrho : \pi_1(M\backslash K) \to G_{\mathbb{C}}\,. \tag{2.38}$$

Even in the case of $G = SU(2)$ and knots in the three-sphere, one may run across cases of non-hyperbolic knot complements. It was clear from the initial days of the volume

[8] To be completely precise, the integer k used throughout these lectures is the sum of the Chern-Simons level and the dual Coxeter number of G.

conjecture [**29**] that even in these cases the asymptotics of $J_N(K;q)$ could still be given by an appropriate flat (but non-hyperbolic/non-metric) $SL(2,\mathbb{C})$ structure.

3. TQFT

We have just seen that the volume conjecture admits a multitude of generalizations, all of which seem to be related to Chern-Simons quantum field theory. The most complete statement of the volume conjecture (2.36) involves Chern-Simons theory with compact gauge group G on the left-hand side and Chern-Simons theory with complex gauge group $G_{\mathbb{C}}$ on the right:

$$\begin{array}{ccc} \underline{\textit{combinatorics/rep. theory}} & & \underline{\textit{geometry}} \\ \text{quantum } G\text{-invariants} & & \text{volumes of representations} \\ J_N(K;q)\,,\ P^G_{R_\lambda}(K;q)\,, & \longleftrightarrow & \varrho:\pi_1(M\backslash K)\to G_{\mathbb{C}}\,, \\ Z^G(M,K;u;\hbar)\,,\ \text{etc.} & & Z^{G_{\mathbb{C}}}_{\text{pert}}(M\backslash K;u;\hbar)\,,\ \text{etc.} \end{array} \tag{3.1}$$

$$q=e^{\frac{2\pi i}{k}}=e^{2\hbar}\,,\qquad u=i\pi\frac{\lambda^*+\rho^*}{k}\,.$$

Chern-Simons theory is a topological quantum field theory (TQFT). In addition to the basic implication that partition functions such as $Z^G(M,K;u;\hbar)$ or $Z^{G_{\mathbb{C}}}_{\text{pert}}(M\backslash K;u;\hbar)$ are topological invariants of colored knots and links in three-manifolds, the structure of TQFT provides powerful methods for actually computing them in multiple ways. It also shows why a general correspondence like (3.1) might be expected to hold.

3.1. Cutting and gluing. In its more mathematical incarnation, a 3-dimensional TQFT can be thought of as a functor Z that assigns

$$\begin{array}{ccc} \text{closed 3-manifold } M & \rightsquigarrow & \text{number } Z(M) \\ \text{closed 2-manifold } \Sigma & \rightsquigarrow & \text{vector space } Z(\Sigma) \\ \text{closed 1-manifold } S^1 & \rightsquigarrow & \text{category } Z(S^1) \\ \text{point } p & \rightsquigarrow & \text{2-category } Z(p)\,. \end{array} \tag{3.2}$$

For our applications to Chern-Simons theory, we will really only need the top two levels $Z(M)$ and $Z(\Sigma)$. The finer structure of categories and 2-categories has recently been explored in *e.g.* [**14**].

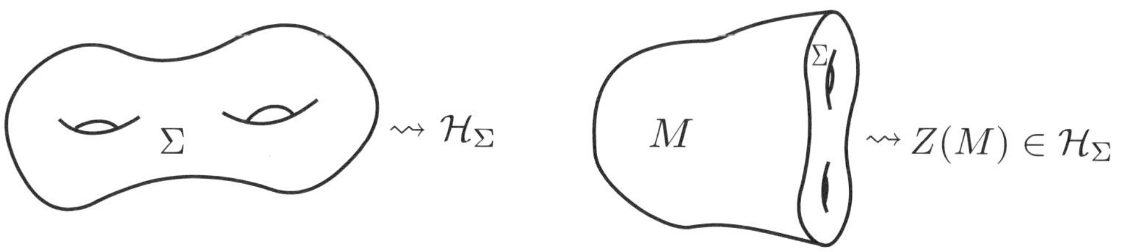

FIGURE 4. Hilbert space assigned to a surface Σ and partition function assigned to a three-manifold M in TQFT.

If a 3-manifold M has a boundary $\Sigma=\partial M$, the object $Z(M)$ is no longer a number, but an element of the vector space $Z(\Sigma)$ assigned to the boundary, as shown in Figure 4. This vector space is in fact a Hilbert space, so let us denote it as $\mathcal{H}_\Sigma=Z(\Sigma)$. At the top two levels, the TQFT must then satisfy the following axioms of Atiyah and Segal (*cf.* [**2**]).

(1) A change of orientation $\Sigma\to-\Sigma$ dualizes the Hilbert space, $\mathcal{H}_{-\Sigma}=\mathcal{H}^*_\Sigma$.
(2) For a boundary consisting of multiplet disjoint components, $\mathcal{H}_{\Sigma_1\sqcup\Sigma_2}=\mathcal{H}_{\Sigma_1}\otimes\mathcal{H}_{\Sigma_2}$.

(3) Using the first two axioms, we see that for a manifold M with $\partial M = (-\Sigma_1) \sqcup \Sigma_2$ one obtains a map $Z(M) : \mathcal{H}_{\Sigma_1} \to \mathcal{H}_{\Sigma_2}$. Then, given a 3-manifold N that can be written as $N = M_1 \cup_{\Sigma_2} M_2$, with $\partial M_1 = (-\Sigma_1) \sqcup \Sigma_2$ and $\partial M_2 = (-\Sigma_2) \sqcup \Sigma_3$ as illustrated below, the functoriality property $Z(N) = Z(M_2) \circ Z(M_1)$ must hold.

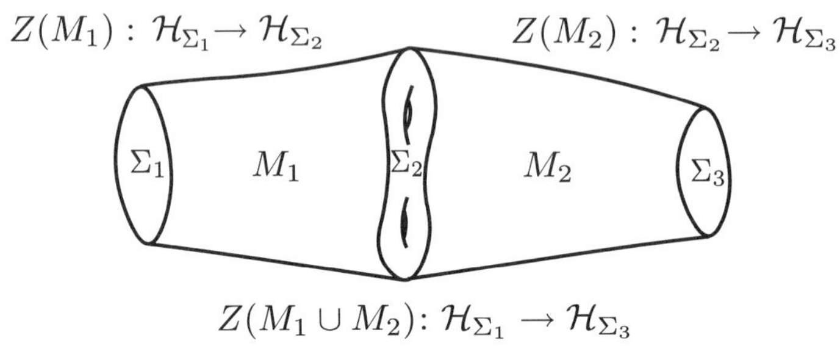

(4) For the empty boundary, $\mathcal{H}_{\Sigma=\varnothing} = \mathbb{C}$.

(5) For $M = \Sigma \times I$, the map $Z(M) : \mathcal{H}_\Sigma \overset{\text{id}}{\to} \mathcal{H}_\Sigma$ is just the identity.

Using these axioms, the partition function $Z(M)$ of any three-manifold, with or without boundary, may be constructed by cutting the manifold into pieces and taking inner products in boundary Hilbert spaces to glue the pieces back together. For this purpose, it is often convenient to know how the mapping class group of a surface Σ acts on $\mathcal{H}_\Sigma$, in order to properly identify the Hilbert spaces on two sides of a gluing.

There are many examples of three-dimensional TQFT, differing essentially in the definitions of the boundary Hilbert spaces $\mathcal{H}(\Sigma)$, as well as the action of the mapping class groups on these spaces. In the case of Chern-Simons theory with gauge group G (whether compact or complex), $\mathcal{H}_\Sigma$ is a quantization of the space $\mathcal{M}_{\text{flat}}(G;\Sigma)$ of flat G-connections on Σ:

$$\mathcal{M}_{\text{flat}}(G;\Sigma) = \left\{ \begin{array}{r} \text{connections } \mathcal{A} \text{ on principal} \\ G\text{-bundle over } \Sigma \end{array} \middle|\; F_{\mathcal{A}} = 0 \right\} \Big/ \text{gauge equivalence}\,. \tag{3.3}$$

(Recall that a connection is flat if the curvature $F_{\mathcal{A}} = d\mathcal{A} + \mathcal{A} \wedge \mathcal{A}$ vanishes.) The precise meaning of the quantization used to obtain $\mathcal{H}_\Sigma$ from $\mathcal{M}_{\text{flat}}(G;\Sigma)$ will be the subject of Section 3.2. It depends on the level $k = i\pi\hbar^{-1}$ (or coupling constant) of Chern-Simons theory, the only adjustable parameter in the TQFT.

In Chern-Simons theory, one is also interested in colored knots or links embedded in 3-manifolds. Suppose for the moment that we have compact Chern-Simons theory with gauge group G and level $k \in \mathbb{Z}$. The intersection of a knot and a boundary surface Σ shows up as a puncture on Σ and TQFT would assign the boundary S^1 surrounding this puncture in Σ the category of representations of the affine Lie algebra $\widehat{\mathfrak{g}}_k$,

$$Z(S^1) \sim \text{reps of } \widehat{\mathfrak{g}}_k\,. \tag{3.4}$$

The definition of the Hilbert space $\mathcal{H}(\Sigma)$ of a multiple-punctured Σ would then have to be altered to include the space of homomorphisms between such representations. For our purposes, however, the complication of knots can be conveniently avoided by *excising the knots and trading representations that color the knots for boundary conditions on knot complements.*

This trick was already mentioned in Section 2.4. In the language of TQFT, it can be described the following way. Suppose that we have a knot K colored by representation R_λ inside the closed manifold M. We cut out a tubular neighborhood N_K of the knot, so that

$$M = (M \backslash N_K) \cup_{T^2} N_K\,, \qquad N_K \simeq D^2 \times S^1\,. \tag{3.5}$$

Of course, $M\backslash N_K \simeq M\backslash K$ is just the knot complement, and N is topologically a 2-disk times S^1 that contains the knot running through its center. The partition functions $Z(M\backslash K; u; \hbar)$ and $Z(N_K; R_\lambda; \hbar)$ are both vectors in the boundary Hilbert space $\mathcal{H}_{T^2}$; therefore, by TQFT,

$$Z(K \subset M; R_\lambda; \hbar) = \langle\, Z(M\backslash K; u; \hbar)\, ,\, Z(N_K; R_\lambda; \hbar)\,\rangle_{\mathcal{H}_{T^2}} \in \mathbb{C}\,. \tag{3.6}$$

As we will see in the next section, the Hilbert space $\mathcal{H}_{T^2}$ can be understood as a space of functions of the variable u that describes the holonomy of flat connections around the meridian of T^2 (as in Figure 2a). The crucial fact, then, is that the vector $Z(N; R_\lambda; \hbar) \in \mathcal{H}_{T^2}$ is only supported on the part of this space with

$$e^u = \mathrm{Hol}(\mu) = \exp\left(i\pi \frac{\lambda^* + \rho^*}{k}\right). \tag{3.7}$$

In other words, $Z(N; R_\lambda; \hbar)$ acts like a delta-function $\delta(u - i\pi\frac{\lambda^*+\rho^*}{k})$. Therefore, coloring by R_λ is equivalent to restricting $Z(M\backslash K)$ to an appropriate one-dimensional subspace of $\mathcal{H}_{T^2}$:

$$Z(K \subset M; R_\lambda; \hbar) = Z(M\backslash K; u; \hbar)\big|_{u = i\pi\frac{\lambda^*+\rho^*}{k}} \in \mathbb{C}\,. \tag{3.8}$$

Our plan now is to give a complete description of $\mathcal{H}_{T^2}$ and to explain how the elements $Z(M\backslash K) \in \mathcal{H}_{T^2}$ may be calculated for knot complements, in the case of Chern-Simons theory with both compact and complex gauge groups. (The extension to links is straightforward and will not be mentioned explicitly hereafter.) This will first require a brief discussion of quantization.

3.2. Quantization. The basic problem of quantization begins with a pair $(\mathcal{M}, \omega)$, where $\mathcal{M}$ is a manifold with symplectic structure ω, called a classical "phase space." Quantization takes this pair and constructs a quantum Hilbert space $\mathcal{H}$. Moreover, quantization should map the algebra of functions on $\mathcal{M}$ to an algebra $\mathcal{A}_\hbar$ of operators on $\mathcal{H}$:

$$\begin{array}{ccc} (\mathcal{M}, \omega) & \rightsquigarrow & \mathcal{H}\ (= \text{Hilbert space}) \\ & & \circlearrowleft \\ \text{alg. of functions on } \mathcal{M} & \rightsquigarrow & \text{alg. of operators on } \mathcal{H} \\ f & \mapsto & \mathcal{O}_f : \mathcal{H} \to \mathcal{H}\,. \end{array} \tag{3.9}$$

The functions on $\mathcal{M}$ form a Poisson algebra with respect to the usual pointwise multiplication of functions and a Lie algebra structure $\{\bullet, \bullet\}$ induced by the symplectic structure. Quantization must map this algebra to an associative but noncommutative algebra $\mathcal{A}_\hbar$, such that

$$[\mathcal{O}_f, \mathcal{O}_g] = -i\hbar\, \mathcal{O}_{\{f,g\}} + \dots\,, \tag{3.10}$$

where $[\bullet, \bullet]$ is the commutator of operators. Here $\hbar$ is a parameter that is involved in the determination of $\mathcal{H}$ itself as well as the algebra of operators.

Very roughly, the Hilbert space $\mathcal{H}$ consists of L^2 sections of a complex line bundle over $\mathcal{M}$ with curvature $\frac{1}{\hbar}\omega$. Locally, these sections are only allowed to depend on *half* of the coordinates of $\mathcal{M}$. In a standard physical setup, $\mathcal{M}$ can be thought of as the space of all possible positions x_i and momenta p_i of particles; thus the elements of $\mathcal{H}$ are functions ("wavefunctions") that depend on either positions or moments, but not both.

In addition to the construction of $\mathcal{H}$, the process of quantization must also explain how classical motions or trajectories of a physical system are associated to quantum states in $\mathcal{H}$. A classical trajectory (or "semiclassical state") is described by a Lagrangian submanifold $\mathcal{L} \subset \mathcal{M}$. Being Lagrangian means that $\mathcal{L}$ is middle-dimensional and $\omega|_{\mathcal{L}} = 0$. Let θ be

1-form (called a Liouville 1-form[9]) that satisfies $\omega = d\theta$. Notice that $\theta|_{\mathcal{L}}$ is closed. Then the Lagrangian $\mathcal{L}$ is called quantizable if

$$\oint_{\gamma} \theta \in 2\pi\hbar\mathbb{Z} \tag{3.11}$$

for any closed cycle $\gamma \subset \mathcal{L}$. The vector (or wavefunction) $Z \in \mathcal{H}$ corresponding to $\mathcal{L}$ can be written as

$$Z = Z(x) = \exp\left(\frac{i}{\hbar}S_0(x) + \dots\right), \tag{3.12}$$

with

$$S_0(x) = \int_{x_0}^{x} \theta \tag{3.13}$$

for some fixed x_0 and varying $x \in \mathcal{L}$. Due to the quantization (3.11), the expression (3.12) is completely well-defined.

Expression (3.12) only defines Z to leading order in $\hbar$. To find subleading corrections, it is useful to employ a complementary approach. Suppose that the Lagrangian submanifold $\mathcal{L}$ is cut out by certain equations $f_i = 0$ on $\mathcal{M}$. Quantization promotes these functions to operators $\mathcal{O}_{f_i}$ acting on $\mathcal{H}$, and the vector Z can also be defined as a solution to the equations

$$\mathcal{O}_{f_i} \cdot Z = 0 \qquad \forall\, i\,. \tag{3.14}$$

If the $\mathcal{O}_{f_i}$ are properly quantized, then the solution to these equations will be the exact wavefunction.

3.2.1. *Methods.* The problem of quantization can be approached in many different ways. Each approach has its advantages and disadvantages, but in the end all methods are expected to yield the same result. The classic approach of *geometric quantization* (*cf.* [**42**]) starts by defining a prequantum line bundle $L \to \mathcal{M}$ with a unitary connection of curvature $\frac{1}{\hbar}\omega$. Note that such a line bundle only exists for

$$\frac{1}{2\pi\hbar}\omega \in H^2(\mathcal{M};\mathbb{Z})\,, \tag{3.15}$$

which can lead to a quantization of $\hbar^{-1}$ (*i.e.* a restriction of $\hbar$ to a discrete set of values in $\mathbb{C}^*$). The local choice of "position" versus "momentum" coordinates is encoded in the choice of a set of $\frac{1}{2}\dim_{\mathbb{R}}\mathcal{M}$ vector fields $\mathcal{P}_j$, called a polarization, and the Hilbert space $\mathcal{H}$ is then defined as the set of square-integrable, $\mathcal{P}_j$-invariant sections of $\mathcal{L}$. This gives a very concrete definition of $\mathcal{H}$, although it can be very hard to show that the construction is independent of the choice of polarization. (The problem becomes more manageable if $\mathcal{M}$ is Kähler.) Moreover, it is often difficult in geometric quantization to find the full quantum expressions for operators $\mathcal{O}_{f_i}$.

An alternative, *deformation quantization* [**5**] partially solves this latter problem. It describes a formal $\hbar$-deformation of the ring of functions on $\mathcal{M}$, using a noncommutative

[9]There is an ambiguity in choosing θ, directly related to the choice of coordinates of $\mathcal{M}$ (positions versus momenta) that elements of $\mathcal{H}$ are to depend on.

product of the type

$$
\begin{aligned}
f \star_\hbar g \;=\;& fg + \hbar \sum_{i,j} \alpha^{ij} \partial_i(f)\partial_j(g) + \frac{\hbar^2}{2} \sum_{i,j,k,l} \alpha^{ij}\alpha^{kl}\partial_i\partial_k(f)\partial_j\partial_l(g) \\
& + \frac{\hbar^2}{3}\left(\sum_{i,j,k,l} \alpha^{ij}\partial_j(\alpha^{kl})(\partial_i\partial_k(f)\partial_l(g) - \partial_k(f)\partial_i\partial_l(g)) \right) + \ldots, \qquad (3.16)
\end{aligned}
$$

where $\alpha = \omega^{-1}$ is the Poisson structure corresponding to the symplectic form ω. In local coordinates $\{f, g\} = \alpha^{ij}\partial_i(f)\partial_j(g)$. One important advantage of deformation quantization is that it is completely canonical and does not require any auxiliary choices. In particular, there is an explicit formula for the $\star_\hbar$-product (3.16) due to Kontsevich [**26**], that allows one to express it as a sum over *admissible graphs*,

$$
f \star_\hbar g := \sum_{n=0}^{\infty} \hbar^n \sum_{\substack{\text{graphs } \Gamma \\ \text{of order } n}} w(\Gamma) B_\Gamma(f,g)\,, \qquad (3.17)
$$

where $w(\Gamma)$ is a weight (number) assigned to Γ, and $B_\Gamma(f,g)$ are bilinear differential operators whose coefficients are differential polynomials, homogeneous of degree n in the components of the bivector field α on $\mathcal{M}$. By definition, an admissible graph of order n is an ordered pair of maps $i, j : \{1, \ldots, n\} \to \{1, \ldots, n, L, R\}$ where neither map has fixed points and both maps are distinct at every point. There are $n^n(n+1)^n$ such graphs.

For example, the graph of order 2 corresponding to the first term in the second line of eq. (3.16) has 4 vertices and 4 edges:

(3.18) $\Gamma =$ 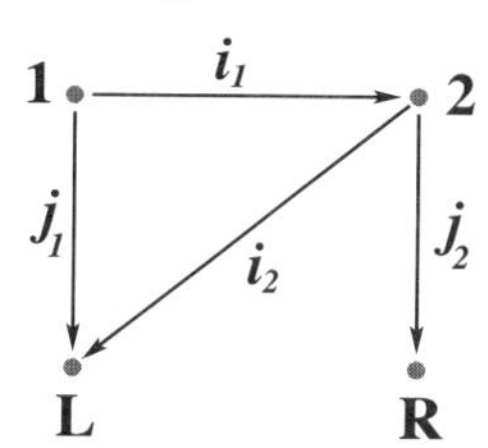

$$
i_1 = (1,2), \quad j_1 = (1,L), \quad i_2 = (2,L), \quad j_2 = (2,R)
$$

An example of a more complicated admissible graph (of order 4) is shown on Figure 5. The corresponding bidifferential operator is

$$
B_\Gamma(f,g) = \sum \alpha^{i_4 j_4}(\partial_{i_3}\alpha^{i_1 j_1})(\partial_{j_1}\partial_{j_4}\alpha^{i_2 j_2})(\partial_{i_2}\partial_{i_4}\alpha^{i_3 j_3})(\partial_{i_1}\partial_{j_3} f)(\partial_{j_2} g)\,. \qquad (3.19)
$$

When the Poisson structure is flat, a graph with an edge ending in a vertex other than L or R will have zero contribution to the sum (3.17), since it will involve derivatives of α. In this case the $\star_\hbar$-product (3.17) becomes the usual Moyal product

$$
f \star_\hbar g(x) = \exp\left(\hbar \alpha^{ij} \frac{\partial}{\partial x^i} \frac{\partial}{\partial y^j} \right) f(x) g(y)|_{y=x} \qquad (3.20)
$$

Deformation quantization is a powerful method for finding the operators $\mathcal{O}_{f_i}$. It is important to stress, however, that, by itself, it does *not* explain how to construct the space $\mathcal{H}$ (it is not an honest quantization), and can not capture the fact that $\hbar^{-1}$ should ever be discretized.

A third option, *brane quantization* [**21**], is a marriage of geometric and deformation quantizations in a physical context. It approaches the problem of quantization by complexifying $\mathcal{M}$ and ω, and constructing a certain (secondary) topological quantum field theory on the resulting space $\mathcal{M}_\mathbb{C}$. It has the advantage of easily characterizing the various choices that one must make in quantization, and provides simple geometric criteria that describe

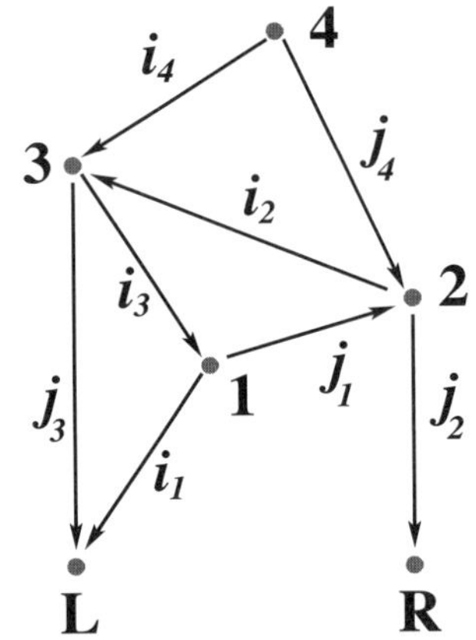

FIGURE 5. An example of an admissible graph of order 4.

quantizable $(\mathcal{M},\omega;\hbar)$. In this approach, the Hilbert space $\mathcal{H}$ is constructed as the space of morphisms (space of open strings),

$$\mathcal{H} = \mathrm{Hom}(\mathcal{B}_{cc},\mathcal{B}')\,, \tag{3.21}$$

where $\mathcal{B}_{cc}$ and $\mathcal{B}'$ are objects (branes) of a certain category associated to the symplectic manifold $\mathcal{M}_{\mathbb{C}}$. Moreover, in this approach, independence of $\mathcal{H}$ on various choices can be reformulated as a problem of constructing a flat connection on the space of such choices, which identifies the space of ground states in the secondary TQFT. In a closely related context, this problem has been studied in the mathematical physics literature [**6, 11**], and leads to a beautiful story that involves integrable systems and tt^* equations.

3.2.2. *Simple examples.* Let us now adapt the general statements here to some specific examples.

Harmonic oscillator. The quintessential simplest nontrivial problem of quantization is the harmonic oscillator. Consider a classical system that consists of a particle moving on a line (with coordinate $x=x(t)$) with a potential energy $V=\frac{1}{2}x^2$. This is depicted in Figure 6. The total (potential + kinetic) energy of the particle at any moment of time is given by the Hamiltonian

$$H=\frac{1}{2}x^2+\frac{1}{2}p^2\,, \tag{3.22}$$

where classically $p=\dot{x}=\frac{dx}{dt}$ is the momentum. This total energy H is conserved. The classical phase space $\mathcal{M}$ is just $\mathbb{R}^2=\{(x,p)\}$, endowed with a symplectic structure $\omega = dp\wedge dx$. A classical trajectory with energy $H=E$ is just a circle of radius $\sqrt{2E}$ in phase space. This defines a Lagrangian submanifold $\mathcal{L}_{(E)}\simeq S^1$.

Now let us quantize the system. Since $H^2(\mathcal{M};\mathbb{Z})=0$, there is no restriction or quantization of $\hbar^{-1}$. On the other hand, there *is* a restriction on $\mathcal{L}$ which quantizes the energy. Namely, according to (3.11), for a Liouville 1-form θ such that $\omega=d\theta$, the integral

$$\oint_{\mathcal{L}}\theta=\oint_{S^1}\theta=2\pi E \tag{3.23}$$

must be an element of $2\pi\hbar\mathbb{Z}$, implying that $E=n\hbar$ for positive $n\in\mathbb{Z}$. In fact, this equation is corrected by quantum effects — a Maslov correction in geometric quantization — to

$$E=\hbar\Big(n+\frac{1}{2}\Big)\,. \tag{3.24}$$

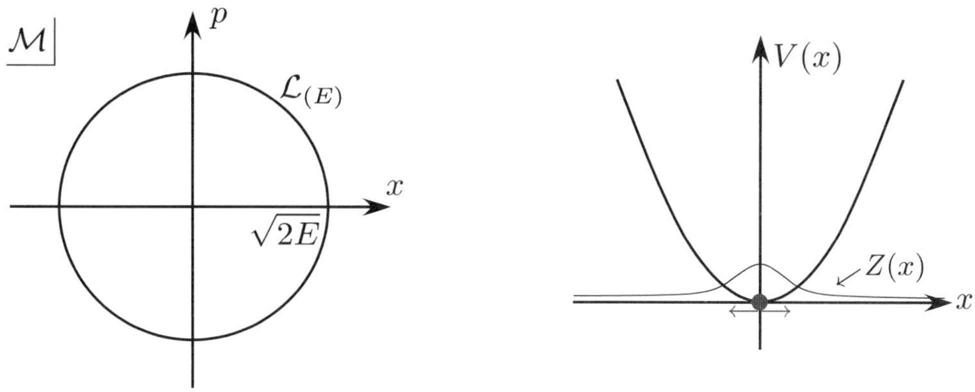

FIGURE 6. The harmonic oscillator: potential $V(x) = \frac{1}{2}x^2$ in physical space, phase space $\mathcal{M}$, a classical trajectory $\mathcal{L}_{(E)}$ in phase space, and the ground state quantum wavefunction $Z(x)$.

This leads to the famous result that the lowest possible energy of a quantum harmonic oscillator (at $n = 0$) is nonzero.

Suppose we choose a polarization $\partial/\partial p = 0$, and a corresponding Liouville 1-form $\theta = p\,dx$. The Hilbert space $\mathcal{H}$ can simply be identified as $L^2(\mathbb{R}) \sim \{\text{functions of } x\}$, on which the functions x and p act as operators

$$\hat{x} := \mathcal{O}_x = x\,, \qquad \hat{p} := \mathcal{O}_p = -i\hbar\frac{d}{dx}\,. \tag{3.25}$$

In this case, the exact quantum expression for the Hamiltonian is

$$\mathcal{O}_H = \frac{1}{2}(\hat{x}^2 + \hat{p}^2)\,. \tag{3.26}$$

It is then easy to find the quantum wavefunctions corresponding to classical states $\mathcal{L}_{(E)}$. From (3.12), we find a leading contribution

$$Z(x) \sim \exp\left(\frac{i}{\hbar}\int_0^x \theta\right) = \exp\left(\frac{i}{\hbar}\int_0^x \sqrt{2E - x^2}\,dx\right) \simeq \exp\left(-\frac{1}{2\hbar}x^2 + \dots\right)\,. \tag{3.27}$$

Since the Lagrangian $\mathcal{L}_{(E)}$ is defined classically by $H - E = 0$, the complete expression for $Z(x)$ can be obtained by solving the operator equation $(\mathcal{O}_H - E)Z = 0$. This eigenvalue equation has square-integrable solutions only for the quantized energies (3.24); for example, at the ground state energy $E = \hbar/2$, the exact solution is $Z(x) = \exp\left(-\frac{1}{2\hbar}x^2\right)$.

Representations of Lie groups. Another famous application of quantization is the construction of unitary representations of Lie groups by quantization of coadjoint orbits. A basic premise of this approach, also known as the *orbit method*, is that coadjoint orbits come equipped with a natural symplectic structure (the Kostant-Kirillov-Souriau symplectic structure), therefore providing interesting examples for quantization.

Continuing with our default notations in these notes, we use G for a compact Lie group (that we usually assume to be simple), $G_{\mathbb{C}}$ for its complexification, and $G_{\mathbb{R}}$ for some real form of the complex group $G_{\mathbb{C}}$ (that may be equal to G). We denote by $\mathfrak{g}_{\mathbb{R}}$ the Lie algebra of $G_{\mathbb{R}}$, and similarly for G and $G_{\mathbb{C}}$. Given an element $\lambda \in \mathfrak{g}_{\mathbb{R}}^*$ (the highest weight of the desired unitary representation R_λ) one constructs $\mathcal{M} = G_{\mathbb{R}} \cdot \lambda$ as the coadjoint orbit of $G_{\mathbb{R}}$ in $\mathfrak{g}_{\mathbb{R}}^*$ passing through λ.

In the case of compact groups, the phase space $\mathcal{M}$ is compact and its quantization leads to a finite-dimensional Hilbert space $\mathcal{H}$ as the space of the unitary representation R_λ. This is the statement of the Borel-Bott-Weil theorem. Moreover, the condition $\frac{1}{2\pi\hbar}\omega \in H^2(\mathcal{M};\mathbb{Z})$

that ensures the existence of a prequantum line bundle becomes equivalent to the condition that λ be an element of the weight lattice $\Lambda_w \subset \mathfrak{g}^*$.

As a very simple illustration, consider the group $SU(2)$. In this case, a non-trivial coadjoint orbit is topologically equivalent to the flag manifold

$$SU(2)/U(1) \simeq \mathbb{P}^1 \,. \tag{3.28}$$

Letting ω be the unit volume form on $\mathbb{P}^1$, we see that $(\mathcal{M},\omega)$ is quantizable for

$$\hbar^{-1} = 2\pi\lambda\,, \qquad \lambda \in \mathbb{Z}_{(+)}\,. \tag{3.29}$$

The prequantum line bundle with curvature $\hbar^{-1}\omega$ is simply $\mathcal{O}(\lambda) \to \mathbb{P}^1$. Choosing a holomorphic polarization, so that $\mathcal{H}$ is defined as the space of holomorphic sections of $\mathcal{O}(\lambda)$, we see that $\dim \mathcal{H} = \lambda + 1$. The Hilbert space is precisely the space of the $(\lambda+1)$-dimensional representation of $SU(2)$.

Similarly, some infinite-dimensional representations, such as unitary principal series representations of $SL(n,\mathbb{C})$ or $SL(n,\mathbb{R})$, can be described as quantized orbits. Nevertheless, there remain some outstanding puzzles: there exist unitary representations that don't appear to correspond to orbits, and, conversely, there are real orbits that don't seem to correspond to unitary representations. An example of first kind occurs even in the basic case of the real group $G_{\mathbb{R}} = SL(2,\mathbb{R})$ and the complementary series representations. To illustrate the second phenomenon, one can take $G_{\mathbb{R}}$ to be a real group of Cartan type B_N, *i.e.* $G_{\mathbb{R}} = SO(p,q)$ with $p+q = 2N+1$. The minimal orbit $\mathcal{O}_{\min}$ of B_N is a nice symplectic manifold of (real) dimension $4N-4$, for any values of p and q. On the other hand, the minimal representation of $SO(p,q)$ exists only if $p \le 3$ or $q \le 3$ [**9**]. Both of these issues can be resolved in the brane quantization approach [**21**], at the cost of replacing classical geometric objects (namely, coadjoint orbits) with their quantum or "stringy" analogs (branes). In particular, in the case of B_N one finds that, while the minimal orbit exists for any values of p and q, the corresponding brane exists only if $p \le 3$ or $q \le 3$. (In general, the condition is that the second Stieffel-Whitney class $w_2(\mathcal{M}) \in H^2(\mathcal{M};\mathbb{Z}_2)$ must be a mod 2 reduction of a torsion class in the integral cohomology of $\mathcal{M}$.)

3.3. Chern-Simons theory. Finally, we arrive at our goal, Chern-Simons theory. Let us recall for a second why we began discussing quantization in the first place. In Section 3.1, we reviewed how partition functions in TQFT could be obtained by cutting and gluing three-manifolds. We explained that the partition function of a manifold with a knot is equivalent to the partition function of the corresponding knot complement, projected onto appropriate boundary conditions in $\mathcal{H}(T^2)$ as in (3.8). To make complete sense of this, however, and to actually calculate partition functions, we must understand what $\mathcal{H}(T^2)$ really is. Using Section 3.2 we are finally in a position to do so.

3.3.1. *Quantization of Chern-Simons theory.* Consider Chern-Simons theory with gauge group G — either compact or complex — on a knot complement $M = \widetilde{M}\backslash K$, with $\partial M = T^2$. The phase space $\mathcal{M}$ associated to T^2 is simply the space of flat G-connections on T^2, modulo gauge equivalence. Since a flat connection is completely determined by the conjugacy classes of its holonomies, we have

$$\mathcal{M} = \mathcal{M}_{\text{flat}}(G;T^2) \tag{3.30}$$

$$= \big\{\text{representations}: \ \pi_1(T^2) \to G\big\}/\text{conjugation}\,. \tag{3.31}$$

The fundamental group $\pi_1(T^2) \simeq \mathbb{Z}\oplus\mathbb{Z}$ is abelian, generated by the meridian and longitude of the torus. The holonomies along these loops can therefore be simultaneously diagonalized[10] into the maximal torus $\mathbf{T} \subset G$. Coordinates on $\mathcal{M}$ are then given by the $2r$ independent eigenvalues $(m_1, ..., m_r)$ and $(\ell_1, ..., \ell_r)$ of the meridian and longitude holonomies, where r is the rank of G. We must also divide by the Weyl group $\mathcal{W}$ of G, which simultaneously permutes both sets of eigenvalues, to obtain

$$\mathcal{M} \simeq (\mathbf{T}^r \times \mathbf{T}^r)/\mathcal{W} = \mathbf{T}^{2r}/\mathcal{W}\,. \tag{3.32}$$

For example, for a compact group $G = SU(n)$ the phase space is $\mathcal{M} = (S^1)^{2(n-1)}/S_n$, where S_n is the symmetric group on n elements. Similarly, for its complexification $G_{\mathbb{C}} = SL(n,\mathbb{C})$, the phase space is $\mathcal{M} = (\mathbb{C}^*)^{2(n-1)}/S_n$. In general, ignoring subtleties in high codimension that are not pertinent to quantization, the relation between compact and complex phase spaces can be described as

$$\mathcal{M}_{\text{flat}}(G_{\mathbb{C}};\Sigma) = [\mathcal{M}_{\text{flat}}(G;\Sigma)]_{\mathbb{C}} \simeq T^*\mathcal{M}_{\text{flat}}(G;\Sigma)\,. \tag{3.33}$$

(In particular, the last relation is only a birational equivalence.)

Compact theory. In order to quantize $\mathcal{M}$, we need a symplectic structure. In *compact* Chern-Simons theory, it is given by

$$\omega = \frac{1}{4}\int_{T^2} \mathrm{Tr}\big[\delta\mathcal{A}\wedge\delta\mathcal{A}\big]\,. \tag{3.34}$$

This can be expressed more concretely in coordinates $\{m_i, \ell_i\} = \{e^{u_i}, e^{v_i}\}$ as

$$\omega = \sum_i d\log m_i \wedge d\log \ell_i = \sum_i du_i \wedge dv_i\,. \tag{3.35}$$

The holonomy variables u_i and v_i function as "positions" and "momenta," respectively. Now, the parameter $\hbar = i\pi/k$ that appeared naturally in the discussion of the volume conjecture in Section 2 is rescaled from the standard geometric quantization parameter $\hbar$ of Section 3.2 by a factor of i. In terms of k, the quantization condition (3.15) simply takes the form $k \in \mathbb{Z}$. The integer k is identified as the Chern-Simons level, modulo the shift mentioned in Footnote 8.

The last ingredient we need to describe the Hilbert space $\mathcal{H}$ is a choice of polarization. For clarity, let us take $G = SU(2)$ to be of rank one, and let us choose the polarization $\partial/\partial v = 0$, so that $\mathcal{H}_{T^2}$ essentially consists of periodic and Weyl-invariant functions of u, $f(u) = f(u+2\pi i) = f(-u)$. Being somewhat more careful, and thinking of these not as functions but as sections of the line bundle with curvature $\frac{k}{\pi}\omega$, one finds that the simultaneous periodicity in the momentum v and the position u restricts u to take values in $\frac{i\pi}{k}\mathbb{Z}$. Therefore, a function $f(u)$ only takes nonzero values at $k+1$ distinct points $u = 0, \frac{i\pi}{k}, \frac{2i\pi}{k}..., i\pi$, and the space $\mathcal{H}_{T^2}$ is finite-dimensional. For general compact semi-simple G, the Hilbert space $\mathcal{H}$ takes the form [**12, 3**]

$$\mathcal{H}_{T^2} \simeq \frac{\Lambda_w}{\mathcal{W}\ltimes k\Lambda_r}\,. \tag{3.36}$$

where Λ_w, Λ_r are the weight and root lattices of G. In other words, $\mathcal{H}_{T^2}$ is the set of weights (hence representations) in a level-k affine Weyl chamber.

Given a "wavefunction" $Z(M;u;\hbar) \in \mathcal{H}_{T^2}$ associated to the knot complement $M = \widetilde{M}\backslash K$ (with $\partial M = T^2$), the partition function $Z^G(\widetilde{M},K;R_\lambda;\hbar)$ for $K \in \widetilde{M}$ colored by representation R_λ is simply given by evaluating $Z(M;u;\hbar)$ at $u = i\pi\frac{\lambda^*+\rho^*}{k}$ as in (3.8). For example, in the case of $SU(2)$ theory, we evaluate $Z(M;u;\hbar)$ at $u = i\pi N/k$ (and normalize by the partition function of S^3) to find the colored Jones polynomial $J_N(K;q)$. The single

[10]If G is not compact, there may be elements that are not so diagonalizable, but they form lower-dimensional components of $\mathcal{M}$ which should not be considered in the quantization.

wavefunction $Z(M;u;\hbar)$ in $\mathcal{H}_{T^2}$ comprises the entire family of colored Jones polynomials $J_N(K;q)$, $N \in \mathbb{Z}$.

How is such a wavefunction obtained in the compact theory? For any three-manifold M, there is a Lagrangian submanifold $\mathcal{L} \subset \mathcal{M}$ corresponding to the semi-classical "state" M. This manifold $\mathcal{L}$ is simply defined as the set of flat connections on T^2 that can extend to a flat connection on all of M. It is the so-called G-character variety of M and can be described by a set of polynomial equations in the eigenvalues ℓ_i and m_i:

$$\mathcal{L}: \quad A_j(\ell, m) = 0\,. \tag{3.37}$$

Depending on whether we restrict to $\ell, m \in S^1$ or $\ell, m \in \mathbb{C}^*$, these same equations describe flat G or $G_{\mathbb{C}}$ connections. In the rank-one case, there is just a single equation, the A-polynomial of the knot complement. Upon quantization, the functions A_j get mapped to quantum operators

$$\widehat{A}_j(\widehat{\ell}, \widehat{m}, q = e^{\frac{2\pi i}{\hbar}}) := \mathcal{O}_{A_j}\,, \tag{3.38}$$

where $\widehat{\ell}_i := \mathcal{O}_{\ell_i} = e^{\widehat{v}_i}$ and $\widehat{m}_i := \mathcal{O}_{m_i} = e^{\widehat{u}_i}$ act on $\mathcal{H}_{T^2}$ as

$$\widehat{\ell}_i\, Z(u) = Z(u_i + \hbar) \quad (\text{shifting only } u_i)\,, \qquad \widehat{m}_i\, Z(u) = e^{u_i} Z(u)\,. \tag{3.39}$$

In terms of the colored Jones polynomial $J_N(K,q)$, this means $\widehat{\ell} J_N(K,q) = J_{N+1}(K,q)$ and $\widehat{m} J_N(K,q) = q^{N/2} J_N(K,q)$. The wavefunction $Z(M;u)$ must satisfy [**19, 10**]

$$\widehat{A}_j\, Z(M;u) = 0 \qquad \forall j\,, \tag{3.40}$$

which leads to a set of recursion relations on polynomial invariants of the knot K. In the mathematical literature, such a recursion relation for the colored Jones polynomial (*i.e.* in the case of $G = SU(2)$) is known as the *AJ conjecture* [**15, 17**] (also *cf.* [**18**]).

Complex theory. Now, we would like to relate partition functions in Chern-Simons theory with compact gauge group G to Chern-Simons theory with complex gauge group $G_{\mathbb{C}}$. In the case of complex gauge group, the phase space is $\mathcal{M} = \mathcal{M}_{\text{flat}}(G_{\mathbb{C}}; T^2) = \big((\mathbb{C}^*)^r \times (\mathbb{C}^*)^r\big)/\mathcal{W}$, and the full symplectic structure induced by Chern-Simons theory is

$$\omega = \frac{\tau}{2}\,\omega_0 + \frac{\tilde{\tau}}{2}\,\overline{\omega_0}\,, \tag{3.41}$$

with $\omega_0 = \frac{1}{4}\int_{T^2} \mathrm{Tr}(\delta\mathcal{A} \wedge \delta\mathcal{A})$ as in (3.34). The connection $\mathcal{A}$ now takes values in $\mathfrak{g}_{\mathbb{C}}$, and a priori there are two independent coupling constants τ and $\tilde{\tau}$. These are the analog of the level k in the compact theory; we include them here in the definition of ω. Since $\mathcal{M}$ is noncompact, the quantization condition (3.15) is less restrictive, only fixing $\tau + \tilde{\tau} \in \mathbb{Z}$.

The noncompactness of $\mathcal{M}$ changes the nature of the Hilbert space $\mathcal{H}$ — as in the case of the harmonic oscillator, it is no longer finite-dimensional. Choosing a polarization $\partial/\partial v = 0$, we can effectively take $\mathcal{H}$ to consist of Weyl-invariant square-integrable functions $f(u,\bar{u}) \in L^2((\mathbb{C}^*)^r)$. However, the fact that (3.41) is a simple sum of holomorphic and antiholomorphic pieces means that at a perturbative level any wavefunction $Z(M;u) \in \mathcal{H}$ will *factorize* into holomorphic and antiholomorphic components. Put more concretely, the exact wavefunction $Z(M;u,\bar{u};\hbar = \frac{2\pi i}{\tau}, \tilde{\hbar} = \frac{2\pi i}{\tilde{\tau}})$ corresponding to complex Chern-Simons theory on the knot complement M can be written as [**10, 41**]

$$Z(M;u,\bar{u};\hbar,\tilde{\hbar}) = \sum_{\alpha,\bar{\alpha}} n_{\alpha,\bar{\alpha}}\, Z^{\alpha}_{\text{pert}}(M;u;\hbar)\, \overline{Z}^{\bar{\alpha}}_{\text{pert}}(M;\bar{u};\tilde{\hbar})\,, \tag{3.42}$$

for some coefficients $n_{\alpha,\bar{\alpha}}$, where, as $\hbar \to 0$, each component $Z^{\alpha}_{\text{pert}}(M;u)$ can be expressed as a perturbative series

$$Z^{\alpha}_{\text{pert}}(M;u) = \exp\left(-\frac{1}{\hbar} S_0(u) + \frac{\delta}{2}\log\hbar + S_1(u) + \hbar\, S_1(u) + \ldots\right)\,, \qquad \hbar = \frac{2\pi i}{\tau}\,. \tag{3.43}$$

Each partition function $Z^{\alpha}_{\text{pert}}(M;u;\hbar)$ corresponds to complex Chern-Simons theory in the background of a fixed flat connection on M that has meridian holonomy eigenvalues $m = e^u$. The set of such connections, labelled by α, is nothing but the (finite) set of solutions $\{v^{\alpha}(u)\}$ (mod $2\pi i$) to the equations

$$A_j(\ell, m) = 0 \tag{3.44}$$

at fixed $m = e^u$. In the case of $SL(2,\mathbb{C})$ theory, one of these flat connections is the geometric one, corresponding to a hyperbolic metric on M.

Since the complex phase space $\mathcal{M}$ is just the complexification of the phase space of the compact theory, the quantization of the functions $A_j(\ell, m)$ is formally identical to the quantization in the compact case. In other words, the operators $\mathcal{O}_{A_j} = \widehat{A}_j(\hat{\ell}, \hat{m}, q = e^{2\hbar})$ are identical to those of the compact theory. Every component $Z^{\alpha}_{\text{pert}}(M;u;\hbar)$ must therefore satisfy [**19, 10**]

$$\widehat{A}_j(\widehat{\ell}, \widehat{m}, e^{2\hbar})\, Z^{\alpha}_{\text{pert}}(M;u;\hbar) = 0 \qquad \forall\, j, \alpha\,, \tag{3.45}$$

with $\widehat{\ell}_i Z(M;u;\hbar) = Z(M; u_i + \hbar; \hbar)$ (in other words $\widehat{v}_i = \hbar\partial_{u_i}$) and $\widehat{m}_i Z(M;u;\hbar) = u^{u_i} Z(M;u;\hbar)$. In particular, at leading order in $\hbar$, we can write

$$Z^{\alpha}_{\text{pert}}(M;u;\hbar) = \exp\left(-\frac{1}{\hbar}\int_{\gamma_\alpha} \theta + \ldots\right), \tag{3.46}$$

where $\theta \sim -\sum_i v_i\, du_i$ is a Liouville 1-form and γ_α is a path on the complex variety $\mathcal{L} = \{A_j = 0\}$ ending at the point $(e^{v^{\alpha}(u)}, e^u)$, as in Figure 2b. Now that u is a continuous parameter in the complex theory, this integral expression makes complete sense.

3.3.2. *Synthesis.* It is fairly clear from the above discussion of quantization that there should be a relation between the partition function for Chern-Simons theory with compact gauge group G and the partition function for Chern-Simons theory with complex gauge group $G_{\mathbb{C}}$. Essentially the same equations (3.40) and (3.45) define the two partition functions — though in one case they are difference equations and in the other they are differential equations. This relation was developed in [**19, 10**], and was recently explained very concretely in [**41**] in terms of analytic continuation.

Algebraically, there may be several solutions to the difference equations (3.40) of the compact theory. Let us label them as $Z^{\alpha}_G(M;u)$. The exact partition function of the compact theory (*i.e.* the colored Jones polynomial for $G = SU(2)$) is given as a linear combination

$$Z^G(M;u;\hbar) = \sum_{\alpha} n_\alpha Z^{\alpha}_G(M;u;\hbar)\,. \tag{3.47}$$

The $\hbar \to 0$ asymptotics of each component in this sum are then governed by the corresponding solution $Z^{\alpha}_{\text{pert}}(M;u)$ to the differential equation (3.45), written in the form (3.43). These are holomorphic pieces of the $G_{\mathbb{C}}$ partition function. The physical statement of the volume conjecture for $SU(2)$ is that the component of the sum (3.47) with the *dominant* leading asymptotics corresponds to the $SL(2,\mathbb{C})$ partition function $Z^{\alpha=\text{hyp}}_{\text{pert}}(M;u)$ around the hyperbolic $SL(2,\mathbb{C})$ flat connection. Of all the flat $SL(2,\mathbb{C})$ connections, this has the largest volume in a neigborhood of the complete hyperbolic point $u = i\pi$. Therefore, *if* the solution $Z^{\text{hyp}}_G(M;u)$ of the difference equations contributes to the colored Jones polynomial in (3.47), it will have the dominant asymptotic. One must simply assure that

$$\text{Physical volume conjecture:} \qquad n_{\text{hyp}} \neq 0\,. \tag{3.48}$$

For higher-rank groups, it is again clear that the overall asymptotics of $Z^G(M;u)$ will be controlled by the flat $G_{\mathbb{C}}$ connection with the largest volume that makes a corresponding contribution to (3.47). One may expect by comparison to $SU(2)$ theory that the connection with the largest overall volume (the analog of the hyperbolic flat connection) in fact contributes and dominates. This has yet to be explored.

The expansions (3.47) and (3.42) for compact and complex Chern-Simons theory, and the relation between them, were explained in [**41**] using analytic continuation of the Chern-Simons path integral. The path integral provides yet another method for quantizing a topological quantum field theory, with its own inherent advantages. Let us finish by saying a few words about this.

The path integral for compact Chern-Simons theory takes the form

$$Z^G(M;u;\hbar) = \int \mathcal{D}\mathcal{A}_{(u)} \exp\left(\frac{ik}{4\pi} I_{CS}(\mathcal{A})\right) = \int \mathcal{D}\mathcal{A}_{(u)} \exp\left(-\frac{1}{4\hbar} I_{CS}(\mathcal{A})\right), \tag{3.49}$$

where $I_{CS}(\mathcal{A}) = \int_M \mathrm{Tr}\big(\mathcal{A}d\mathcal{A} + \frac{2}{3}\mathcal{A}^3\big)$ is the Chern-Simons action as in (1.9) and $k \in \mathbb{Z}$. The integral is over all G-connections on $M = \widetilde{M}\backslash K$, modulo gauge equivalence, with fixed holonomy eigenvalues e^u at the meridian of K. (In order to obtain a nonzero answer, $u/\hbar \sim \lambda^* + \rho^*$ must be integral.) For $\mathfrak{g}$-values connections $\mathcal{A}$, the action $I_{CS}(\mathcal{A})$ is real. Therefore, for $k \in \mathbb{Z}$, the integral (3.49) is oscillatory and can be calculated by appropriately regulating the oscillations as $\mathcal{A} \to \infty$. In [**41**], however, the problem was posed of analytically continuing to $k \in \mathbb{C}$. Roughly speaking, to accomplish this one must also complexify the gauge connection $\mathcal{A}$ so that it is $\mathfrak{g}_{\mathbb{C}}$-valued. When $k \in \mathbb{R}$, the integral (3.49) is then interpreted as a holomorphic contour integral along the real subspace in the space of complex connections. As k is pushed away from the real line, this integration contour must also move. In general, the appropriate integration contour for $k \in \mathbb{C}$ is a sum of contours going through the various saddle points of the complexified action $I_{CS}(\mathcal{A})$. Each saddle point is a flat $G_{\mathbb{C}}$ connection, and an expression of the form (3.47) results.

For complex Chern-Simons theory, the procedure is quite similar. The path integral is

$$Z^{G_{\mathbb{C}}}(M;u;\hbar,\tilde{\hbar}) = \int \mathcal{D}\mathcal{A}_{(u)}\mathcal{D}\overline{\mathcal{A}}_{(\bar{u})} \exp\left(\frac{i\tau}{8\pi} I_{CS}(\mathcal{A}) + \frac{i\tilde{\tau}}{8\pi} I_{CS}(\overline{\mathcal{A}})\right) = \int \mathcal{D}\mathcal{A}_{(u)}\mathcal{D}\overline{\mathcal{A}}_{(\bar{u})} \exp\left(-\frac{1}{4\hbar} I_{CS}(\mathcal{A}) - \frac{1}{4\tilde{\hbar}} I_{CS}(\overline{\mathcal{A}})\right), \tag{3.50}$$

for a $\mathfrak{g}_{\mathbb{C}}$-valued $G_{\mathbb{C}}$-connection $\mathcal{A}$. The integrand is initially well-defined when $\tau + \tilde{\tau} \in \mathbb{Z}$, and leads to a convergent oscillatory integral when the exponent is imaginary — *i.e.* for $\tilde{\tau} = \bar{\tau}$. In order to analytically continue to independent $\tau, \tilde{\tau} \in \mathbb{C}$, one must treat $\mathcal{A}$ and $\overline{\mathcal{A}}$ as independent connections and again complexify *each* of them. One then deforms the contour of integration away from the "real" subspace when $\tilde{\tau} \neq \bar{\tau}$, and writes the resulting contour as a sum over *pairs* of saddle points for $\mathcal{A}$ and $\overline{\mathcal{A}}$. Since $(\mathfrak{g}_{\mathbb{C}})_{\mathbb{C}} \simeq \mathfrak{g}_{\mathbb{C}} \times \mathfrak{g}_{\mathbb{C}}$, however, these are just pairs of saddle points of flat $G_{\mathbb{C}}$-connections. An expression of the form (3.42) results:

$$Z^{G_{\mathbb{C}}}(M;u;\hbar,\tilde{\hbar}) = \sum_{\alpha,\bar{\alpha}} n_{\alpha,\bar{\alpha}} Z^{\alpha}_{G_{\mathbb{C}}}(M;u;\hbar)\overline{Z^{\alpha}_{G_{\mathbb{C}}}}(M;\bar{u};\tilde{\hbar})\,. \tag{3.51}$$

The functions $Z^{\alpha}_{G_{\mathbb{C}}}(M;u;\hbar)$ and $Z^{\alpha}_{G}(M;u;\hbar)$ here and in (3.47) should be *identical*, since they both correspond to $G_{\mathbb{C}}$ connections.

In [**41**], it is explained how the coefficients n_α and $n_{\alpha,\bar{\alpha}}$ may be calculated for specific examples, like the trefoil and figure-eight knot complements. As expected, the coefficient of the hyperbolic component "$\alpha = \mathrm{hyp}$" of the $SU(2)$ partition function is nonzero, leading to another demonstration of the volume conjecture.

The careful reader may still be wondering why it is only the growth of the colored Jones polynomial at *non*rational N/k that shows exponential behavior. The answer comes from a final subtlety in the analytic continuation of the path integral: for $k \notin \mathbb{Z}$, the sum (3.47) can have *multiple* contributions from the same flat connection, differing by a multiplicative factor $e^{2\pi i k}$. (If analytically continuing in N as well, factors of $e^{\frac{2\pi i u}{\hbar}}$ may also arise.) This behavior originates from the fact that $\exp\left(\frac{ik}{4\pi} I_{CS}\right)$ is not completely gauge-invariant when $k \notin \mathbb{Z}$. For example, in the case of the figure-eight knot, the actual hyperbolic contribution to (3.47) goes like

$$(e^{i\pi k} - e^{-i\pi k}) Z^{\text{hyp}}_{G_{\mathbb{C}}}(M; u; \hbar)\,, \tag{3.52}$$

which vanishes at $k \in \mathbb{Z}$, leading to polynomial rather than exponential growth of $J_N(K; q)$ for $u/i\pi \sim N/k \in \mathbb{Q}$. It is expected that this feature is fairly generic for hyperbolic knots.

Acknowledgments. We would like to thank Edward Witten, Don Zagier, and Jonatan Lenells for enlightening discussions on subjects considered in these notes. We would also like to thank the organizers of the workshop *Chern-Simons Gauge Theory: 20 years after* and the Hausdorff Center for Mathematics, University of Bonn, for their generous support, accommodations, and collaborative working environment.

References

[1] J.E. Andersen and S.K. Hansen, *Asymptotics of the quantum invariants for surgeries on the figure 8 knot*, J. Knot Theory Ramifications **15** (2006), no. 4, 479–548.
[2] M. Atiyah, *Geometry and physics of knots*, Cambridge Univ. Press (1990).
[3] S. Axelrod, S. Della Pietra, and E. Witten, *Geometric quantization of Chern-Simons gauge theory*, J. Diff. Geom. **33** (1991), 787–902.
[4] D. Bar-Natan and E. Witten, *Perturbative expansion of Chern-Simons theory with non-compact gauge group*, Comm. Math. Phys. **141** (1991), no. 2, 423–440.
[5] F. Bayen, M. Flato, C. Fronsdal, A. Lichnerowicz, and D. Sternheimer, *Deformation theory and quantization. I. Deformations of symplectic structures*, Ann. Physics **111** (1978), no. 1, 61–110.
[6] S. Cecotti and C. Vafa, *Topological antitopological fusion*, Nucl. Phys. **B367** (1991), 359–461.
[7] D. Cooper, M. Culler, H. Gillet, and D. Long, *Plane curves associated to character varieties of 3-manifolds*, Invent. Math. **118** (1994), no. 1, 47–84.
[8] D. Coulson, O.A. Goodman, C.D. Hodgson, and W.D. Neumann, *Computing arithmetic invariants of 3-manifolds*, Experimental Mathematics **9** (2000), no. 1, 127–152.
[9] Jr. D.A. Vogan, *Singular unitary representations*, Noncommutative harmonic analysis and Lie groups (Marseille, 1980) (J. Carmona and M. Vergne, eds.), Lecture Notes in Math., vol. 880, Springer, Berlin, 1981, pp. 506–535. MR MR644845 (83k:22036)
[10] T. Dimofte, S. Gukov, J. Lenells, and D. Zagier, *Exact results for perturbative Chern-Simons theory with complex gauge group*, Comm. Num. Thy. and Phys. **3** (2009), no. 2, 363–443.
[11] B. Dubrovin, *Integrable systems in topological field theory*, Nucl. Phys. **B379** (1992), 627–689.
[12] S. Elitzur, G. Moore, A. Schwimmer, and N. Seiberg, *Remarks on the canonical quantization of the Chern-Simons-Witten theory*, Nucl. Phys. **B326** (1989), 108.
[13] L.D. Faddeev and R.M. Kashaev, *Quantum dilogarithm*, Modern Phys. Lett. **A9** (1994), no. 5, 427–434.
[14] D.S. Freed, M.J. Hopkins, J. Lurie, and C. Teleman, *Topological quantum field theories from compact Lie groups*, "A Celebration of the Mathematical Legacy of Raoul Bott" (2009).
[15] S. Garoufalidis, *On the characteristic and deformation varieties of a knot*, Geometry and Topology Monographs **7** (2004), 291–304.
[16] S. Garoufalidis and T.T.Q. Le, *Asymptotics of the colored Jones function of a knot*, (2005).
[17] ———, *The colored Jones function is q-holonomic*, Geom. Topol. **9** (2005), 1253–1293.
[18] S. Garoufalidis and X. Sun, *The non-commutative a-polynomial of twist knots*, (2008).
[19] S. Gukov, *Three-dimensional quantum gravity, Chern-Simons theory, and the A-polynomial*, Commun. Math. Phys. **255** (2005), 577–627.
[20] S. Gukov and H. Murakami, *SL(2,C) Chern-Simons theory and the asymptotic behavior of the colored Jones polynomial*, (2006).
[21] S. Gukov and E. Witten, *Branes and quantization*, (2008).
[22] K. Habiro, *On the colored Jones polynomials of some simple links*, S = urikaisekikenky = usho K = oky = uroku **1172** (2000), 34–43.
[23] H.M. Hilden, M.T. Lozano, and J.M. Montesinos-Amilibia, *On volumes and Chern-Simons invariants of geometric 3-manifolds*, J. Math. Sci. Univ. Tokyo **1996** (3), no. 3, 723–744.

[24] R.M. Kashaev, *The hyperbolic volume of knots from quantum dilogarithm*, Lett. Math. Phys. **39** (1997), 269–265.
[25] R. Kirby and P. Melvin, *The 3-manifold invariants of Witten and Reshetikhin-Turaev for sl(2,C)*, Invent. Math. **105** (1991), no. 1, 1432–1297.
[26] M. Kontsevich, *Deformation quantization of Poisson manifolds*, Lett. Math. Phys. **66** (2003), no. 3, 157–216.
[27] H. Murakami, *Some limits of the colored Jones polynomials of the figure-eight knot*, Kyungpook Math. J. **44** (2004), no. 3, 369–383.
[28] ———, *The colored Jones polynomials and the Alexander polynomial of the figure-eight knot*, J. Geom. Topol. **7** (2007), no. 2, 249–269.
[29] H. Murakami and J. Murakami, *The colored Jones polynomials and the simplicial volume of a knot*, Acta Math. **186** (2001), no. 1, 85–104.
[30] H. Murakami, J. Murakami, M. Okamoto, T. Takata, and Y. Yokota, *Kashaev's conjecture and the Chern-Simons invariants of knots and links*, Experiment. Math. **11** (2002), no. 3, 427–435.
[31] H. Murakami and Y. Yokota, *The colored Jones polynomials of the figure-eight knot and its Dehn surgery spaces*, J. Reine Angew. Math **607** (2007), 47–68.
[32] W.D. Neumann, *Extended Bloch group and the Cheeger-Chern-Simons class*, Geom. Topol. **8** (2004), 413–474.
[33] W.D. Neumann and D. Zagier, *Volumes of hyperbolic three-manifolds*, Topology **24** (1985), no. 3, 307–332.
[34] G. Perelman, *Ricci flow with surgery on three-manifolds*, (2003).
[35] J. Porti, *Torsion de reidemesiter poir les variétés hyperboliques*, Mem. Amer. Math. Soc. **128** (1997), no. 612.
[36] N. Reshetikhin and V. Turaev, *Invariants of 3-manifolds via link polynomials and quantum groups*, Invent. Math. **103** (1991), no. 3, 547–597.
[37] W.P. Thurston, *Three dimensional manifolds, Kleinian groups, and hyperbolic geometry*, Bull. AMS **6** (1982), no. 3, 357–381.
[38] E. Witten, *2+1 dimensional gravity as an exactly soluble system*, Nucl. Phys. **B311** (1988), 46.
[39] ———, *Quantization of Chern-Simons gauge theory with complex gauge group*, Commun. Math. Phys. **137** (1991), 29.
[40] ———, *Quantum field theory and the Jones polynomial*, Adv. Ser. Math. Phys. **17** (1994), 361–451.
[41] ———, *Analytic continuation of Chern-Simons theory*, (2010).
[42] N. Woodhouse, *Geometric quantization*, New York: Oxford Univ. Press (1992).
[43] T. Yoshida, *The eta-invariant of hyperbolic 3-manifolds*, Invent. Math. **81** (1985), 473–514.
[44] C.K. Zickert, *The Chern-Simons invariant of a representation*, (2007).

Department of Physics, California Institute of Technology 452-48, Pasadena, CA 91125
E-mail address: tdd@theory.caltech.edu

Department of Physics, California Institute of Technology 452-48, Pasadena, CA 91125
E-mail address: gukov@theory.caltech.edu

AMS/IP Studies in Advanced Mathematics
Volume 50, 2011

Computational Aspects in Reidemeister Torsion and Chern–Simons Theories

Jérôme Dubois

ABSTRACT. This paper reviews some computational, and practical, aspects in Reidemeister torsion and Chern–Simons theories which has appeared over the last 20 years for most of them. The paper focuses on the non–abelian Reidemeister torsion on the $SL_2(\mathbb{C})$-character variety, on L^2-torsion and on the Chern–Simons invariant of knots. Moreover, many examples of explicit computations are presented in it.

1. Introduction

The aim of this article is to present some computational aspects in Reidemeister torsion and Chern–Simons theories, and applications concerning the asymptotic expansion of the colored Jones polynomial related to the volume conjecture in the case of torus knots (which is "trivial" and well known in that case, see [**18**]).

Reidemeister torsions and the Chern–Simons invariants can both be considered as functionals on the $SL_2(\mathbb{C})$-character variety of the fundamental group of a 3–manifold (with or without boundary). Abelian Reidemeister torsions have been extensively studied for a long time: initiated by the interpretation of the Alexander polynomial in terms of torsion given by Milnor [**29**] in the sixties, the theory of abelian Reidemeister torsion was later developed by Turaev in [**36, 37**]). E. Witten [**41**] constructed a non-abelian torsion, viewed as a 1-form on the SU(2)-character variety of a Riemannian surface, and proved a remarkable formula between this torsion–form and the symplectic volume of the SU(2)-character variety of the Riemannian surface. Later, J. Porti [**34**] studied a non-abelien Reidemeister torsion (twisted by the adjoint representation associated to a generic irreducible representation) for hyperbolic 3–manifold with or without torus boundary. Using Porti's method and Turaev's refined torsion, I have constructed a non–abelian Reidemeister torsion for knot exteriors with sign in [**6**], and provided computations it in the case of fibered knots [**7**], and in the case of twist knots in [**8**].

Much of the recent research has been oriented toward twisted Alexander invariants and their relationship to twisted Reidemeister torsion. We refer to the

2010 *Mathematics Subject Classification.* Primary 57M25; Secondary 57M27.
Key words and phrases. Reidemeister torsion; L^2-torsion; Twisted Alexander polynomial; Chern–Simons invariant; Character variety; Hyperbolic 3–manifold; Torus knot; Fibered knot.

extensive survey of S. Friedl and S. Vidussi [**13**], which gives an excellent overview of these developments.

E. Witten [**40**] gave a spectacular interpretation of the Jones polynomial in terms of Chern–Simons theory by using a path integral. This path integral is formally a topological invariant, which can be defined rigorously and it is sometimes called the Witten–Reshetikin–Turaev invariant. Moreover, Reidemeister torsion (non–abelian) and Chern–Simons invariant seem to appear as subleading terms in the expansion of this path integral. This is suggested for example by some results of J. Andersen and S. Hansen [**1**] in the case of surgeries on the figure 8 knot, K. Hikami and H. Murakami [**16**], and [**9**] in the case of torus knots.

Organization

The paper is organized as follows. Section 2 gives a quick presentation of the structure of the $\mathrm{SL}_2(\mathbb{C})$-character variety of a knot group. The special case of torus knots, which will be the main example of the present paper, is completely treated in Theorem 2.3. In Section 3, we construct the non–abelian Reidemeister torsion for knot exteriors twisted by the adjoint representation associated to certain irreducible $\mathrm{SL}_2(\mathbb{C})$-representations (the so–called *regular* representations). We discuss some methods of computations and give an explicit formula for the non–abelian Reidemeister torsion in the case of torus knots (see Proposition 3.7) and for fibered knots (see Corollary 3.9). Section 4 is a report on joint work with C. Wegner and deals with L^2-torsion, more precisely a L^2-version of Alexander invariants. After a quick review of the notion of L^2-torsion and of the definition of the L^2-Alexander invariant under consideration, we explain how to compute it for trefoil knots, and more generally for torus knots of type $(2, q)$ using an explicit Wirtinger presentation of the knot group. Section 5 gives a method, due to Kirk and Klassen, to explicitly compute the Chern-Simons invariant of a knot exterior. The difficulty consists in the fact that knot exteriors have a boundary component, thus the Chern-Simons invariant consists in a functional with values in a certain bundle. The last section explains how the Reidemeister torsion and Chern-Simons invariants appear as subleading terms in the asymptotic expansion of the colored Jones polynomial for torus knots. These results are interesting because of their connection to the volume conjecture of Kashaev and Murakami–Murakami, which is known to be trivially true in the case of torus knots.

Acknowledgments

I would like to warmly thank the organizers of the conference *"Chern–Simons Gauge Theory: 20 years after"*: Jørgen Andersen, Hans Boden, Atle Hahn and Benjamin Himpel for their invitation and for giving me the opportunity to write this paper. I also want to warmly thank Christian Wegner and the referee of this paper for their careful reading of the manuscript and for many helpful comments.

The author is partially supported by ANR, project n°ANR-08-JCJC-0114-01.

2. Review on $\mathrm{SL}_2(\mathbb{C})$–character variety

Given a finitely generated group π we let

$$R(\pi) = R(\pi; \mathrm{SL}_2(\mathbb{C})) = \mathrm{Hom}(\pi; \mathrm{SL}_2(\mathbb{C}))$$

denote the space of $\mathrm{SL}_2(\mathbb{C})$-representation of π. This space is endowed with the compact-open topology.

A representation $\rho\colon \pi \to \mathrm{SL}_2(\mathbb{C})$ is called *abelian* if $\rho(\pi)$ is an abelian subgroup of $\mathrm{SL}_2(\mathbb{C})$. A representation ρ is called *reducible* if there exists a proper subspace $U \subset \mathbb{C}^2$ such that $\rho(g)(U) \subset U$, for all $g \in \pi$. Of course any abelian representation is reducible. A non reducible representation is called *irreducible*.

The group $\mathrm{SL}_2(\mathbb{C})$ acts on the representation space $R(\pi)$ by conjugation, but the naive quotient $R(\pi)/\mathrm{SL}_2(\mathbb{C})$ is not Hausdorff in general. Following [**5**], we will focus on the *character variety* $X(\pi) = X(\pi; \mathrm{SL}_2(\mathbb{C}))$ which is the set of *characters* of π. Associated to $\rho \in R(\pi)$ is its character $\chi_\rho\colon \pi \to \mathbb{C}$, defined by $\chi_\rho(g) = \mathrm{tr}(\rho(g))$, where tr denotes the trace of matrices.

In some sense $X(\pi)$ is the "algebraic quotient" of $R(\pi)$ by the action by conjugation of $\mathrm{PSL}_2(\mathbb{C})$. It is well known that $R(\pi)$ and $X(\pi)$ have the structure of complex algebraic affine sets (see [**5**]).

Let $R^{\mathrm{irr}}(\pi)$ denote the subset of irreducible representations of π in $\mathrm{SL}_2(\mathbb{C})$, and let $X^{\mathrm{irr}}(\pi)$ denote its image under the map $R(\pi) \to X(\pi)$. Moreover, one can observe that two irreducible representations of π in $\mathrm{SL}_2(\mathbb{C})$ are conjugate if and only if their characters are equal: the invariance of the trace by conjugation trivially implies that two conjugate representations (not necessarily irreducible) have the same character, the converse is true for irreducible representations by [**5**, Proposition 1.5.2].

For a knot K in S^3, let Π_K denote its group. Let Π'_K be the subgroup generated by the commutators of Π_K. It is well known that $\Pi_K/\Pi'_K \cong H_1(S^3 \setminus K; \mathbb{Z}) \cong \mathbb{Z}$ is generated by the meridian μ of K. As a consequence, each abelian representation of Π_K in $\mathrm{SL}_2(\mathbb{C})$ is conjugate either to $\varphi_z\colon \Pi_K \to \mathrm{SL}_2(\mathbb{C})$ defined by $\varphi_z(\mu) = \begin{pmatrix} e^z & 0 \\ 0 & e^{-z} \end{pmatrix}$, with $z \in \mathbb{C}$, or to a representation ρ with $\rho(\mu) = \pm \begin{pmatrix} 1 & 1 \\ 0 & 1 \end{pmatrix}$.

Let $R^{\mathrm{ab}}(\Pi_K)$ denote the space of all abelian representations. We write $X^{\mathrm{ab}}(\Pi_K)$ for the image of $R^{\mathrm{ab}}(\Pi_K)$ under the map $R(\Pi_K) \to X(\Pi_K)$. Similarly, we write $X^{\mathrm{nab}}(\Pi_K)$ for the image of $R(\Pi_K) \setminus R^{\mathrm{ab}}(\Pi_K)$ under $R(\Pi_K) \to X(\Pi_K)$. Observe that $X^{\mathrm{irr}}(\Pi_K) \subset X^{\mathrm{nab}}(\Pi_K)$ but $X^{\mathrm{nab}}(\Pi_K) \not\subset X^{\mathrm{irr}}(\Pi_K)$ in general.

If M is a 3–manifold, then we shall write $X^{\mathrm{ab}}(M) = X^{\mathrm{ab}}(\pi_1(M))$, $X^{\mathrm{nab}}(M) = X^{\mathrm{nab}}(\pi_1(M))$, $X^{\mathrm{irr}}(M) = X^{\mathrm{irr}}(\pi_1(M))$ etc.

EXAMPLE 2.1. For any hyperbolic 3–manifold M, there exist a discrete and faithful representation $\rho\colon \pi_1(M) \to \mathrm{PSL}_2(\mathbb{C})$ which lift to $\mathrm{SL}_2(\mathbb{C})$ defining an irreducible representation called the *holonomy* representation.

EXAMPLE 2.2 (Character variety for torus knots). Let $T(p,q)$ denote the torus knot of type (p,q) where $p,q > 1$ are coprime integers and set $M(p,q) = S^3 \setminus V(T(p,q))$, where $V(T(p,q))$ denotes an open tubular neighborhood of $T(p,q)$. The group of $T(p,q)$ admits the following well known presentation:

$$\Pi(p,q) = \langle a, b \mid a^p = b^q \rangle.$$

Let $r, s \in \mathbb{Z}$ be such that $ps - qr = 1$. The meridian of $T(p,q)$ is represented, into $\Pi(p,q)$, by the word $\mu = a^{-r}b^s$ and the longitude by $\lambda = a^p\mu^{-pq} = b^q\mu^{-pq}$.

THEOREM 2.3. *The non abelian part $X^{\mathrm{nab}}(M(p,q))$ of the character variety $X(M(p,q))$ is the disjoint union of $N_{p,q} = (p-1)(q-1)/2$ copies of $\mathbb{C}$ indexed by* *the pair* (α, β) *which satisfies the following conditions:*

$$1 \leqslant \alpha \leqslant p-1,\ 1 \leqslant \beta \leqslant q-1,\ \alpha \equiv \beta \,(\mathrm{mod}\, 2). \tag{2.1}$$

On the component indexed by (α, β)*, we have* $\chi(a) = 2\cos\left(\frac{\alpha\pi}{p}\right)$, $\chi(b) = 2\cos\left(\frac{\beta\pi}{q}\right)$.

Moreover, we have $X^{\mathrm{irr}}(M(p,q)) = X^{\mathrm{nab}}(M(p,q)) \setminus X^{\mathrm{ab}}(M(p,q))$ *and the component of* $X^{\mathrm{nab}}(M(p,q))$ *indexed by* (α, β) *is attached to* $X^{\mathrm{ab}}(M(p,q))$ *at two abelian representations* $\varphi_{\frac{k-}{pq}\pi i}$ *and* $\varphi_{\frac{k+}{pq}\pi i}$ *where* k^- *and* k^+ *satisfy the following conditions:*

$$0 < k^- < k^+ < pq, \tag{2.2}$$

$$k^- = \varepsilon^-(\beta ps - \varepsilon\alpha qr) \,(\mathrm{mod}\, 2pq), \tag{2.3}$$

$$k^+ = \varepsilon^+(\beta ps + \varepsilon\alpha qr) \,(\mathrm{mod}\, 2pq). \tag{2.4}$$

Here $\varepsilon, \varepsilon^-, \varepsilon^+ \in \{\pm 1\}$.

The first part of Theorem 2.3 is due to Klassen [**24**] and Le [**25**], the second one is given in [**9**]. Further notice the following remarks.

REMARK 2.4. The integers k^- and k^+ satisfying conditions (2.3) and (2.4) also satisfy the following properties:

(1) $k^\pm$ is not divisible neither by p nor by q;
(2) $k^+ \pm k^- \in 2\mathbb{Z}$;
(3) either p divides $k^+ + k^-$, and q divides $k^+ - k^-$; or q divides $k^+ + k^-$, and p divides $k^+ - k^-$.

REMARK 2.5. If we define positive integers $m = \frac{k^+ - k^-}{2}$ and $l = \frac{(k^+)^2 - (k^-)^2}{4pq}$, one has the following properties:

(1) $m < \frac{pq}{2}$,
(2) either p divides m or q divides m,
(3) m divides pql,
(4) $\frac{m^2}{pq} < l < m - \frac{m^2}{pq}$.

REMARK 2.6. In the case of $(2,q)$ torus knots one has a very simple explicit description of the character variety $X(M(2,q))$, see [**25**] for details. We use the fact that $T(2,q)$ is a two-bridge knot and its group admits the following Wirtinger presentation:

$$\Pi(2,q) = \langle a, b \mid wa = bw \rangle \text{ where } w = (ab)^{(q-1)/2}.$$

We write $x = \operatorname{tr}\rho(a)$, $y = \operatorname{tr}\rho(ab)$. The character variety is parametrized by x and y. Specifically, its abelian part $X^{\mathrm{ab}}(M(p,q))$ is described by the equation $y - x^2 + 2 = 0$.

The non abelian part $X^{\mathrm{nab}}(M(2,q))$ is described by the polynomial equation $\Phi(2,q) = 0$, where $\Phi(2,q) \in \mathbb{Z}[x,y]$ satisfying the linear recurrence:

$$\Phi(2,q) = y\Phi(2,q-2) - \Phi(2,q-4) \text{ and } \Phi(2,1) = 1, \Phi(2,3) = y - 1. \tag{2.5}$$

The non abelian part $X^{\mathrm{nab}}(M(2,q))$ of the character variety is the disjoint union of $(q-1)/2$ copies of $\mathbb{C}$ parametrized by x, indexed by $\ell \in \{1, \ldots, (q-1)/2\}$ and attached to $X^{\mathrm{ab}}(M(2,q))$ at the $(q-1)$ abelian representations φ_z for

$$z \in \left\{ i\pi - i\pi\frac{2\ell - 1}{2q}, i\pi\frac{2\ell - 1}{2q} \right\}.$$

As a consequence, the $(q-1)/2$ pairs of integers (k^-, k^+) are given by the following formulae:

$$k_\ell^- = 2\ell - 1 \text{ and } k_\ell^+ = 2q - 2\ell + 1,$$

where $\ell \in \{1, \ldots, (q-1)/2\}$.

3. The non-abelian Reidemeister torsion

This section deals with the non-abelian Reidemeister torsion for knots exteriors. We essentially focus on the notion of the Reidemeister torsion for knot exteriors twisted by the adjoint representation associated to an irreducible $\mathrm{SL}_2(\mathbb{C})$-representation of the knot group and give some methods of computation.

3.1. Preliminaries: sign-determined torsion of a CW-complex. We review the basic notions and results about the sign-determined Reidemeister torsion introduced by Turaev which are needed in this paper. Further details can be found in Milnor's survey [**30**] or in Turaev's monograph [**37**].

Torsion of a chain complex. Let $C_* = (0 \longrightarrow C_n \xrightarrow{d_n} C_{n-1} \xrightarrow{d_{n-1}} \cdots \xrightarrow{d_1} C_0 \longrightarrow 0)$ be a chain complex of finite dimensional vector spaces over $\mathbb{C}$. Choose a basis $\mathbf{c}^i$ for C_i and a basis $\mathbf{h}^i$ for the i-th homology group $H_i = \ker(d_i \colon C_i \to C_{i-1})/\operatorname{im}(d_{i+1} \colon C_{i+1} \to C_i)$. The torsion of C_* with respect to these choice of bases is defined as follows.

Let $\mathbf{b}^i$ be a sequence of vectors in C_i such that $d_i(\mathbf{b}^i)$ is a basis of $B_{i-1} = \operatorname{im}(d_i \colon C_i \to C_{i-1})$ and let $\widetilde{\mathbf{h}}^i$ denote a lift of $\mathbf{h}^i$ in $Z_i = \ker(d_i \colon C_i \to C_{i-1})$. The set of vectors $d_{i+1}(\mathbf{b}^{i+1})\widetilde{\mathbf{h}}^i\mathbf{b}^i$ is a basis of C_i. Let $[d_{i+1}(\mathbf{b}^{i+1})\widetilde{\mathbf{h}}^i\mathbf{b}^i/\mathbf{c}^i] \in \mathbb{C}^*$ denote the determinant of the transition matrix between those bases (the entries of this matrix are coordinates of vectors in $d_{i+1}(\mathbf{b}^{i+1})\widetilde{\mathbf{h}}^i\mathbf{b}^i$ with respect to $\mathbf{c}^i$). The *sign-determined Reidemeister torsion* of C_* (with respect to the bases $\mathbf{c}^*$ and $\mathbf{h}^*$) is the following alternating product (see [**36**, Definition 3.1]):

$$\text{(3.1)} \qquad \operatorname{Tor}(C_*, \mathbf{c}^*, \mathbf{h}^*) = (-1)^{|C_*|} \cdot \prod_{i=0}^{n} [d_{i+1}(\mathbf{b}^{i+1})\widetilde{\mathbf{h}}^i\mathbf{b}^i/\mathbf{c}^i]^{(-1)^{i+1}} \in \mathbb{C}^*.$$

Here $|C_*| = \sum_{k \geqslant 0} \alpha_k(C_*)\beta_k(C_*)$, where $\alpha_i(C_*) = \sum_{k=0}^{i} \dim C_k$ and $\beta_i(C_*) = \sum_{k=0}^{i} \dim H_k$.

The torsion $\operatorname{Tor}(C_*, \mathbf{c}^*, \mathbf{h}^*)$ does not depend on the choices of $\mathbf{b}^i$ and $\widetilde{\mathbf{h}}^i$. Further observe that if C_* is acyclic (i.e. if $H_i = 0$ for all i), then $|C_*| = 0$.

Torsion of a CW-complex. Let W be a finite CW-complex and consider a representation $\rho \in \operatorname{Hom}(\pi_1(W); \mathrm{SL}_2(\mathbb{C}))$. We define the $\mathfrak{sl}_2(\mathbb{C})_\rho$-twisted cochain complex of W to be

$$C^*(W; \mathfrak{sl}_2(\mathbb{C})_\rho) = \operatorname{Hom}_{\pi_1(X)}(C_*(\widetilde{W}; \mathbb{Z}); \mathfrak{sl}_2(\mathbb{C})_\rho).$$

Here $C_*(\widetilde{W}; \mathbb{Z})$ is the complex of the universal covering with integer coefficients which is in fact a $\mathbb{Z}[\pi_1(W)]$-module (via the action of $\pi_1(W)$ on $\widetilde{W}$ as covering group), and $\mathfrak{sl}_2(\mathbb{C})_\rho$ denotes the $\mathbb{Z}[\pi_1(W)]$-module via the composition $Ad \circ \rho$, where $Ad \colon \mathrm{SL}_2(\mathbb{C}) \to \operatorname{Aut}(\mathfrak{sl}_2(\mathbb{C})), A \mapsto Ad_A$ is the adjoint representation. This cochain complex $C^*(W; \mathfrak{sl}_2(\mathbb{C})_\rho)$ computes the $\mathfrak{sl}_2(\mathbb{C})_\rho$-twisted cohomology of W which we denote as $H^*(W; \mathfrak{sl}_2(\mathbb{C})_\rho)$.

Let $\{e_1^{(i)}, \ldots, e_{n_i}^{(i)}\}$ be the set of i-dimensional cells of W. We lift them to the universal covering and we choose an arbitrary order and an arbitrary orientation for the cells $\left\{\tilde{e}_1^{(i)}, \ldots, \tilde{e}_{n_i}^{(i)}\right\}$. If $\mathcal{B} = \{\mathbf{a}, \mathbf{b}, \mathbf{c}\}$ is an orthonormal basis of $\mathfrak{sl}_2(\mathbb{C})$, then we consider the corresponding "dual" basis

$$\mathbf{c}^i_{\mathcal{B}} = \left\{\tilde{e}^{(i)}_{1,\mathbf{a}}, \tilde{e}^{(i)}_{1,\mathbf{b}}, \tilde{e}^{(i)}_{1,\mathbf{c}}, \ldots, \tilde{e}^{(i)}_{n_i,\mathbf{a}}, \tilde{e}^{(i)}_{n_i,\mathbf{b}}, \tilde{e}^{(i)}_{n_i,\mathbf{c}}\right\}$$

of $C^i(W; \mathfrak{sl}(2)_\rho) = \mathrm{Hom}_{\pi_1(X)}(C_*(\widetilde{W}; \mathbb{Z}); \mathfrak{sl}_2(\mathbb{C})_\rho)$. Now choosing for each i a basis $\mathbf{h}^i$ for the twisted cohomology $H^i_\rho(W)$, we can compute using Equation (3.1)

$$\mathrm{Tor}(C^*(W; \mathfrak{sl}_2(\mathbb{C})_\rho), \mathbf{c}^*_{\mathcal{B}}, \mathbf{h}^*).$$

The cells $\{\tilde{e}_j^{(i)}\}_{0 \leqslant i \leqslant \dim W, 1 \leqslant j \leqslant n_i}$ are in one-to-one correspondence with the cells of W and their order and orientation induce an order and an orientation for the cells $\{e_j^{(i)}\}_{0 \leqslant i \leqslant \dim W, 1 \leqslant j \leqslant n_i}$. Again, corresponding to these choice, we get a basis c^i over $\mathbb{R}$ for $C^i(W; \mathbb{R})$.

Choose a *cohomology orientation* of W, i.e. an orientation of the real vector space $H^*(W; \mathbb{R}) = \bigoplus_{i \geqslant 0} H^i(W; \mathbb{R})$. Let $\mathfrak{o}$ denote this chosen orientation. Provide each vector space $H^i(W; \mathbb{R})$ with a reference basis h^i such that the basis $\{h^0, \ldots, h^{\dim W}\}$ of $H^*(W; \mathbb{R})$ is positively oriented with respect to $\mathfrak{o}$. Compute the sign-determined Reidemeister torsion $\mathrm{Tor}(C^*(W; \mathbb{R}), c^*, h^*) \in \mathbb{R}^*$ of the resulting based and cohomology based chain complex and consider its sign

$$\tau_0 = \mathrm{sgn}\,(\mathrm{Tor}(C^*(W; \mathbb{R}), c^*, h^*)) \in \{\pm 1\}. \tag{3.2}$$

We define the *sign-determined* $Ad \circ \rho$*-twisted Reidemeister torsion* of W to be

$$\mathrm{TOR}(W; Ad \circ \rho, \mathbf{h}^*, \mathfrak{o}) = \tau_0 \cdot \mathrm{Tor}(C^*(W; \mathfrak{sl}_2(\mathbb{C})_\rho), \mathbf{c}^*_{\mathcal{B}}, \mathbf{h}^*) \in \mathbb{C}^*. \tag{3.3}$$

This definition only depends on the combinatorial class of W, the conjugacy class of ρ, the choice of $\mathbf{h}^*$ and the cohomology orientation $\mathfrak{o}$. It is independent of the orthonormal[1] basis $\mathcal{B}$ of $\mathfrak{sl}_2(\mathbb{C})$, of the choice of the lifts $\tilde{e}_j^{(i)}$, and of the choice of the positively oriented basis of $H^*(W; \mathbb{R})$. Moreover, it is independent of the order and the orientation of the cells (because they appear twice).

One can prove that TOR is invariant under cellular subdivision, homeomorphism and simple homotopy equivalences (see Turaev' Monographs [**36, 37**] for proofs). In fact, it is precisely the sign $(-1)^{|C_*|}$ in (3.3) which ensures all these important invariance properties to hold.

3.2. Abelian Reidemeister torsion. In the sixties, J. Milnor proved that the abelian Reidemeister torsion of a knot exterior is related to the Alexander polynomial of the knot. Later, in the seventies, V. Turaev undertook an extensive study of abelian Reidemeister torsions.

For a 3–manifold M with torus boundary components and first Betti number equal to 1, the (classical) Alexander polynomial $\Delta_M(t)$ is defined by using the infinite abelian covering. The next result was proved by J. Milnor and describes the relaionship between this invariant and the torsion.

[1]If the Euler characteristic of W is equal to 0, as for a knot exterior, then the definition is independent of the choice of *any* basis of $\mathfrak{sl}_2(\mathbb{C})$.

PROPOSITION 3.1 ([**29**]). *The torsion of* $C_*(M;\mathbb{C}(t))$ *is given by*

$$\mathrm{Tor}(C_*(M;\mathbb{C}(t))) = \frac{\Delta_M(t)}{(t-1)}. \tag{3.4}$$

3.3. The Reidemeister torsion twisted by the adjoint representation associated to a $\mathrm{SL}_2(\mathbb{C})$-representation. The Reidemeister torsion twisted by the adjoint representation was considered by Witten in [**41**] for Riemannian surfaces. Porti [**34**] studied the case of hyperbolic closed 3–manifolds and hyperbolic 3–manifolds with torus boundary. In [**6, 7**], I proposed a sign–refined version of this torsion.

3.3.1. *Regularity for representations.* In this subsection we briefly review two notions of regularity (see [**34, 7**]). In the sequel $K \subset S^3$ denotes an oriented knot. We let $M_K = S^3 \setminus V(K)$ denote the knot exterior, where $V(K)$ is an open tubular neighborhood of K. Observe that M_K is a compact 3–manifold whose boundary consists in a 2–torus. Furthermore, let $\Pi_K = \pi_1(M_K)$ denote the knot group. The meridian μ of K is supposed to be oriented according to the rule $\ell k(K,\mu) = +1$, while the longitude λ is oriented according to the condition $\mathrm{int}(\mu,\lambda) = +1$. Here $\mathrm{int}(\cdot,\cdot)$ denotes the intersection form on ∂M_K.

We say that $\rho \in R^{\mathrm{irr}}(\Pi_K)$ is *regular* if $\dim H^1(M_K;\mathfrak{sl}_2(\mathbb{C})_\rho) = 1$. This notion is invariant under conjugation and thus it is well defined for irreducible characters.

EXAMPLE 3.2. For the torus knot $T(p,q)$, one can prove that each irreducible representation of $\Pi(p,q)$ in $\mathrm{SL}_2(\mathbb{C})$ is regular.

Observe (see [**34, 7**] for a proof) that for a regular representation ρ, we have

$$\dim H^1(M_K;\mathfrak{sl}_2(\mathbb{C})_\rho) = 1, \ \dim H^2(M_K;\mathfrak{sl}_2(\mathbb{C})_\rho) = 1$$

and $H^j(M_K;\mathfrak{sl}_2(\mathbb{C})_\rho) = 0$ for all $j \neq 1, 2$.

Let γ be a simple closed unoriented curve in ∂M_K. Among irreducible representations we focus on the γ-regular ones. We say that $\rho \in R^{\mathrm{irr}}(\Pi_K)$ is *γ-regular*, if (see [**34**, Definition 3.21]):

(1) the inclusion $\alpha\colon \gamma \hookrightarrow M_K$ induces an *injective* map

$$\alpha^*\colon H^1(M_K;\mathfrak{sl}_2(\mathbb{C})_\rho) \to H^1(\gamma;\mathfrak{sl}_2(\mathbb{C})_\rho),$$

(2) if $\mathrm{tr}(\rho(\pi_1(\partial M_K))) \subset \{\pm 2\}$, then $\rho(\gamma) \neq \pm\mathbf{1}$.

Notice that γ-regularity is invariant under conjugation and hence is well–defined for irreducible characters $\chi \in X^{\mathrm{irr}}(M_K)$.

EXAMPLE 3.3. For the torus knot $T(p,q)$, one can prove that each irreducible representation of $\Pi(p,q)$ in $\mathrm{SL}_2(\mathbb{C})$ is μ-regular and also λ-regular.

Here is an alternative formulation, see [**7**, Proposition 3]. Fix a generator P^ρ of $H^0(\partial M_K;\mathfrak{sl}_2(\mathbb{C})_\rho) \simeq \mathbb{C}$. The inclusion $\alpha\colon \gamma \hookrightarrow M_K$ and the cup product induce the linear form $f^\rho_\gamma\colon H^1(M_K;\mathfrak{sl}_2(\mathbb{C})_\rho) \to \mathbb{C}$ defined by

$$f^\rho_\gamma(v) = B_{\mathfrak{sl}_2(\mathbb{C})}\left(P^\rho, v(\gamma)\right), \text{ for all } v \in H^1(M_K;\mathfrak{sl}_2(\mathbb{C})_\rho),$$

where $B_{\mathfrak{sl}_2(\mathbb{C})}$ is the Killing form of the Lie algebra $\mathfrak{sl}_2(\mathbb{C})$.

PROPOSITION 3.4 (Proposition 3 of [**7**]). *An irreducible representation* $\rho \in R^{\mathrm{irr}}(\Pi_K)$ *is γ-regular if and only if the linear form* $f^\rho_\gamma\colon H^1(M_K;\mathfrak{sl}_2(\mathbb{C})_\rho) \to \mathbb{C}$ *is an isomorphism.*

3.4. Review on Reidemeister torsion for knot exteriors. This subsection gives a review of the constructions made in [**6**, § 6]. In particular, we shall explain how to construct distinguished bases for the twisted cohomology of knot exteriors.

Canonical orientation of knot exteriors. We equip the exterior of K with its *canonical cohomology orientation* defined as follows (see [**37**, Section V.3]). We have

$$H^*(M_K;\mathbb{R}) = H^0(M_K;\mathbb{R}) \oplus H^1(M_K;\mathbb{R})$$

and we base this $\mathbb{R}$-vector space with $\{[\![pt]\!], m^*\}$. Here $[\![pt]\!]$ is the cohomology class of a point, and $\mu^* \colon \mu \mapsto 1$ is the dual of the meridian μ of K. This reference basis of $H^*(M_K;\mathbb{R})$ induces the so-called canonical cohomology orientation of M_K. In the sequel, we let $\mathfrak{o}$ denote the canonical cohomology orientation of M_K.

How to construct natural bases for the twisted cohomology. Let ρ be a regular representation of Π_K. One has a distinguished isomorphism induced by the cup product and the killing form (which explicitly depends on the invariant vector P^ρ), see [**6**, Lemmas 5.1 & 5.2]:

$$\phi_{P^\rho} \colon H^2(M_K;\mathfrak{sl}_2(\mathbb{C})_\rho) \to H^2(M_K;\mathbb{Z}) \otimes \mathbb{C}.$$

Let c be the generator of $H^2(\partial M_K;\mathbb{Z}) = \mathrm{Hom}(H_2(\partial M_K;\mathbb{Z}),\mathbb{Z})$ corresponding to the fundamental class $[\![\partial M_K]\!] \in H_2(\partial M_K;\mathbb{Z})$ induced by the orientation of ∂M_K. The *reference generator* of $H^2(M_K;\mathfrak{sl}_2(\mathbb{C})_\rho)$ is defined by

$$h^{(2)}_\rho = \phi^{-1}_{P^\rho}(c). \tag{3.5}$$

Let ρ be a λ-regular representation of Π_K. The *reference generator* of the first twisted cohomology group $H^1(M_K;\mathfrak{sl}_2(\mathbb{C})_\rho)$ is defined by

$$h^{(1)}_\rho(\lambda) = (f^\rho_\lambda)^{-1}(1). \tag{3.6}$$

The Reidemeister torsion for knot exteriors. Let $\rho \colon \Pi_K \to \mathrm{SL}_2(\mathbb{C})$ be a λ-regular representation. The *Reidemeister torsion* $\mathbb{T}^K_\lambda$ at ρ is defined to be

$$\mathbb{T}^K_\lambda(\rho) = \mathrm{TOR}\left(M_K; Ad\circ\rho, \{h^{(1)}_\rho(\lambda), h^{(2)}_\rho\}, \mathfrak{o}\right) \in \mathbb{C}^*.$$

It is an invariant of knots (of course independent of the choice of the generator P^ρ of $H^0(\partial M_K;\mathfrak{sl}_2(\mathbb{C})_\rho)$). Moreover, if ρ_1 and ρ_2 are two λ-regular representations which have the same character then $\mathbb{T}^K_\lambda(\rho_1) = \mathbb{T}^K_\lambda(\rho_2)$. Thus $\mathbb{T}^K_\lambda$ defines a map on the set $X^{\mathrm{irr}}_\lambda(M_K) = \{\chi \in X^{\mathrm{irr}}(M_K) \mid \chi \text{ is } \lambda\text{-regular}\}$.

Twisted Alexander invariant for knot exteriors. As a generalization of the Reidemeister-Milnor-Turaev torsion, or generalization of the homological definition of the usual Alexander polynomial, one can consider the *twisted Alexander polynomial.* Twisted Alexander polynomials were originally introduced by X-S. Lin in the nineties and later generalized and studied by M. Wada [**38**], T. Kitano [**23**], P. Kirk–C. Livingston [**22**], S. Friedl–T. Kim [**12**], etc... In this paper, we only "twist" by the (special) representation

$$\varphi \otimes Ad\circ\rho \colon \Pi_K \to Aut(\mathbb{C}(t) \otimes \mathfrak{sl}_2(\mathbb{C})) = Aut(\mathfrak{sl}_2(\mathbb{C}(t))),$$

where $\varphi \colon \Pi_K \to \mathbb{Z} = \langle t \rangle$ is the usual abelianization and $\rho \colon \Pi_K \to \mathrm{SL}_2(\mathbb{C})$ is a "generic" irreducible representation. For a more general theory of twisted Alexander polynomial, one can refer to Friedl and Vidussi's survey [**13**].

Let $\rho\colon \Pi_K \to \mathrm{SL}_2(\mathbb{C})$ be a λ-regular representation. In [**10**, Proposition 3], it is proved that the associated twisted complex

$$C_*(M_K; \varphi \otimes Ad \circ \rho) = \mathfrak{sl}_2(\mathbb{C}(t)) \otimes_{\mathbb{Z}[\Pi_K]} C_*(\widetilde{M_K}; \mathbb{Z})$$

is acyclic, i.e. $H_*(M_K; \varphi \otimes Ad \circ \rho) = 0$, thus the Reidemeister torsion is well–defined:

$$\Delta_K^{\varphi\otimes Ad\circ\rho}(t) = \tau_0 \cdot \mathrm{Tor}(C_*(M_K; \varphi \otimes Ad \circ \rho); \emptyset; \mathfrak{o}) \in \mathbb{C}(t) \setminus \{0\}.$$

3.5. Computation of the non-abelian Reidemeister torsion from the twisted Alexander polynomial. In this section, we explain how to compute the non-abelian Reidemeister torsion (a non-acyclic torsion) as a special value of the twisted Alexander polynomial (an acyclic torsion).

3.5.1. *The result.* Consider $\varphi\colon \Pi_K \to \mathbb{Z} = \langle t \rangle$ the usual abelianization and ρ a λ-regular representation of Π_K into $\mathrm{SL}_2(\mathbb{C})$. One has:

THEOREM 3.5. *The following equality holds:*

$$-\lim_{t\to 1} \frac{\Delta_K^{\varphi\otimes Ad\circ\rho}(t)}{(t-1)} = \mathbb{T}_\lambda^K(\rho)^{-1}. \tag{3.7}$$

Equation (3.7) was first proved by Yamaguchi in [**42**] (see also [**10**] for a more general result). It is a powerful method to compute the non-abelian Reidemeister torsion because the twisted Alexander polynomial can be computed using the twisted Fox matrix. For example, such a result has been used in [**8**] to explicitly compute the non-abelian Reidemeister torsion for the family of twist knots (a special family of two–bridge knots).

REMARK 3.6. The inverse of non-abelian Reidemeister torsion appears in Equation (3.7) because we use twisted cohomology groups whereas Yamaguchi used twisted homology groups in [**42**].

Now we explain, following a method initiated by Kitano [**23**], how to compute the twisted Alexander polynomial $\Delta_K^{\varphi\otimes Ad\circ\rho}(t)$ from Fox–calculus. Take a Wirtinger presentation of Π_K as follows

$$\Gamma = \langle g_1, \ldots, g_k \,|\, r_1, \ldots, r_{k-1} \rangle. \tag{3.8}$$

Let W_K be the 2-dimensional CW–complex constructed from the presentation (3.8) in the usual way. The 0-skeleton of W_K consists of a single 0-cell pt, the 1-skeleton is a wedge of k oriented 1-cells corresponding to the generators $g_1, \ldots g_k$ and the 2-skeleton consists in $(k-1)$ 2-cells $D_1, \ldots, D_{k-1}$ with attaching maps given by the relations $r_1, \ldots, r_{k-1}$ of the presentation (3.8).

F. Waldhausen [**39**] proved that the Whitehead group of a knot group is trivial. As a result, W_K has the same simple homotopy type as M_K. So, the CW–complex W_K can be used to compute $\Delta_K^{\varphi\otimes Ad\circ\rho}(t)$. Therefore it is enough to consider Reidemeister torsion of the $\mathfrak{sl}_2(\mathbb{C}(t))$-twisted chain complex $C_*(W_K; \varphi \otimes Ad \circ \rho)$.

The twisted complex $C_*(W_K; \varphi \otimes Ad \circ \rho)$ thus becomes:

$$0 \longrightarrow (\mathbb{C}(t) \otimes_{\mathbb{C}} \mathfrak{sl}_2(\mathbb{C}))^{k-1} \xrightarrow{\partial_2} (\mathbb{C}(t) \otimes_{\mathbb{C}} \mathfrak{sl}_2(\mathbb{C}))^{k} \xrightarrow{\partial_1} \mathbb{C}(t) \otimes_{\mathbb{C}} \mathfrak{sl}_2(\mathbb{C}) \longrightarrow 0. \tag{3.9}$$

Here we briefly denote the l-times direct sum of $\mathbb{C}(t) \otimes_{\mathbb{C}} \mathfrak{sl}_2(\mathbb{C})$ by $(\mathbb{C}(t) \otimes_{\mathbb{C}} \mathfrak{sl}_2(\mathbb{C}))^l$. In the complex of Equation (3.9), we have

$$\partial_1 = (\Phi(x_1 - 1), \Phi(x_2 - 1), \ldots, \Phi(x_k - 1)).$$

and ∂_2 is expressed using the Fox differential calculus and the action given by $\Phi = \varphi \otimes (Ad \circ \rho)$:

$$\partial_2 = \begin{pmatrix} \Phi(\frac{\partial r_1}{\partial x_1}) & \ldots & \Phi(\frac{\partial r_{k-1}}{\partial x_1}) \\ \vdots & \ddots & \vdots \\ \Phi(\frac{\partial r_1}{\partial x_k}) & \ldots & \Phi(\frac{\partial r_{k-1}}{\partial x_k}) \end{pmatrix} \tag{3.10}$$

Let $A^i_{K,\Phi}$ denote the $3(k-1) \times 3(k-1)$–matrix obtained from the matrix of Equation (3.10) of ∂_2 by deleting its i-th row. The twisted Alexander polynomial $\Delta_K^{\varphi\otimes Ad\circ\rho}(t)$ can be described, up to a factor t^m ($m \in \mathbb{Z}$), as follows (for more details see [**23**]):

$$\Delta_K^{\varphi\otimes Ad\circ\rho}(t) = \tau_0 \cdot \frac{\det A^i_{K,\Phi}}{\det(\Phi(x_i - 1))}, \tag{3.11}$$

where τ_0 is the sign defined in Equation (3.2). Equation (3.11) is in fact independent of i; this property comes from a careful study of the rows of the Fox–matrix $A_{K,\Phi}$ and using the well–known identity:

$$\sum_{j=1}^{k} \frac{\partial u}{\partial x_j}(1 - x_j) = 1 - u,$$

for any u in the free group generated by $x_1, \ldots, x_k$.

Using Theorem 3.5, the non-abelian Reidemeister torsion $\mathbb{T}^K_\lambda(\rho)$ can be thus expressed as

$$\mathbb{T}^K_\lambda(\rho)^{-1} = -\lim_{t\to 1} \frac{\Delta_K^{\varphi\otimes Ad\circ\rho}(t)}{(t-1)} = -\lim_{t\to 1}\left(\tau_0 \cdot \frac{\det A^i_{K,\Phi}}{(t-1)\det(\Phi(x_i - 1))}\right). \tag{3.12}$$

3.5.2. *Example: the case of torus knots.* As an illustration of Equation (3.12), we write down the non-abelian Reidemeister torsion for torus knots (with respect to the longitude).

PROPOSITION 3.7 ([**7**]). *If $\chi \in X^{\mathrm{irr}}(M_{p,q})$ lies on the component of the character variety indexed by the pair (α, β) which satisfies conditions (2.1), then*

$$\mathbb{T}^{T_{p,q}}_\lambda(\chi) = \frac{16}{p^2q^2}\sin^2\left(\frac{\pi\alpha}{p}\right)\sin^2\left(\frac{\pi\beta}{q}\right). \tag{3.13}$$

The very special form of the non–abelian Reidemeister torsion for torus knots is due to the fact that a torus knot exterior is a Seifert fibered manifold with two exceptional fibers and a regular one.

3.6. Non abelian Reidemeister torsion for fibered 3–manifolds. A 3–manifold M is called fibered if it has the structure of a fibered manifold over the circle S^1. It is known from a long time that the (usual) Alexander polynomial is the characteristic polynomial of the (usual) monodromy (see for example [**2**]). The non-abelian case is similar using the so–called *twisted monodromy*.

3.6.1. *General statements.* Consider a 3–manifold M fibered over the circle. There exists a surface F, called the fiber, and a map $\phi\colon F \to F$, called the monodromy, such that M is homeomorphic to $(F\times[0,1])/\left((x,0)\sim(\phi(x),1)\right)$. The monodromy induces an endomorphism of $H_*(F)$. The so–called twisted monodromy is the induced homomorphism on $H_1(F;\varphi\otimes Ad\circ\rho)$. A knot K is fibered if its exterior M_K is a fibered 3–manifold. It is well–known that the longitude λ is the boundary of the fiber F.

THEOREM 3.8 ([**10**]). *We let $\varphi\colon \Pi_K \to \pi_1(S^1) = \mathbb{Z}$ be the homomorphism induced by the projection $M_K \to S^1$, and let $\rho\colon \Pi_K \to \mathrm{SL}_2(\mathbb{C})$ be a λ-regular representation such that the restriction $\rho|_{\pi_1(S)}$ is non–abelian. Then $\Delta_{M_K}^{\varphi\otimes Ad\circ\rho}(t)$ is (up to a sign) the characteristic polynomial of the twisted monodromy.*

As a consequence, we have:

COROLLARY 3.9 ([**7**]). *Let K be a fibered knot in S^3 and suppose that its fiber F is of genus g. Let ε_0 be the sign of the isomorphism $\mathrm{Id}-\phi^*\colon H^1(F;\mathbb{R}) \to H^1(F;\mathbb{R})$ induced by the monodromy $\phi\colon F \to F$. If ρ is a λ-regular representation such that $\rho_{|\pi_1(F)}$ is non–abelian, then the tangent map at $\rho_{|\pi_1(F)}$ to $R(\phi)\colon X^{\mathrm{irr}}(F) \to X^{\mathrm{irr}}(F)$ admits 1 as simple eigenvalue. If we let $e_1,\dots,e_{6g-4}$ denote its other eigenvalues, then the non-abelian Reidemeister torsion $\mathbb{T}_\lambda^K(\chi)$ is equal to*

$$\mathbb{T}_\lambda^K(\chi_\rho) = -\varepsilon_0\cdot\prod_{i=1}^{6g-4}(1-e_i).$$

3.6.2. *Example: the case of the figure eight knot.* We let K denote the figure eight knot. It is well–known that K is a fibered knot, so we let F denote its fiber. It is well–known that the fiber F is genus one. As a fibered knot, the group Π_K of the figure eight knot K admits the following presentation:

$$\Pi_K = \langle a,b,s \mid s^{-1}as = ab,\, s^{-1}bs = bab\rangle.$$

The monodromy $\phi : F \to F$ induces an endomorphism $\phi_1 : H_1(F;\mathbb{Z}) \to H_1(F;\mathbb{Z})$ such that the matrix of ϕ_1 in the basis $\{a_*,b_*\}$ is $\begin{pmatrix}1&1\\1&2\end{pmatrix}$. As a consequence, $\det(\mathbf{1}-\phi_1) = -1$.

Set $x_1 = I_a$, $x_2 = I_b$ and $x_3 = I_{ab}$, where $I_\gamma\colon X(M_K) \to \mathbb{C}$ is given by $I_\gamma(\chi) = \chi(\gamma)$. We know that the character variety $X(F)$ is parametrized by x_1, x_2, x_3 (see [**34**, § 4.5]). The action of the monodromy is given by

$$P = (x_3,\, x_2x_3 - x_1,\, x_2x_3^2 - x_1x_3 - x_2) \in \mathbb{Z}[x_1,x_2,x_3]^3.$$

Further notice that for an irreducible representation $\rho\colon \Pi_K \to \mathrm{SL}_2(\mathbb{C})$ we have

$$x_1 = x_3 \text{ and } x_1 + x_2 = x_1x_2. \tag{3.14}$$

All but two irreducible representations $\rho\colon \Pi_K \to \mathrm{SL}_2(\mathbb{C})$ are λ-regular, up to conjugation. From now on ρ is supposed to be λ-regular.

Using Theorem 3.8, the polynomial torsion $\Delta_{M_K}^{\varphi\otimes Ad\circ\rho}$ is the characteristic polynomial of the twisted monodromy $\phi_*^\rho\colon H_1(F;\mathfrak{sl}_2(\mathbb{C})_\rho) \to H_1(F;\mathfrak{sl}_2(\mathbb{C})_\rho)$. Observe that 1 is a simple eigenvalue of ϕ_*^ρ because the boundary of the fiber is invariant by the monodromy and ρ is supposed to be λ-regular. The two others are inverse of each other. Let ℓ, ℓ^{-1} be the eigenvalues of ϕ_*^ρ different from 1. It is easy to see that

$$\ell + \ell^{-1} = \mathrm{tr}\,\phi_*^\rho - 1 = 2(x_1+x_2) - 1.$$

Moreover $I_\mu^2 = 1 + x_1 + x_2$, thus $\ell + \ell^{-1} = 2I_\mu^2 - 3$. As a result the polynomial torsion is:

$$\Delta_{M_K}^{\varphi\otimes Ad\circ\rho}(t) = -(t-1)\left(t^2 - (2I_\mu^2(\rho) - 3)t + 1\right).$$

4. L^2-torsion

This section is a report on joint work in progress with C. Wegner [**11**] and outlines an approach for computing the L^2-torsion for certain torus knots.

The Reidemeister torsion is defined using matrices, i.e. operators with finite spectrum, and using the usual notion of determinant. The analytic Ray–Singer torsion is defined for operators with (infinite) discrete spectrum, and it is well–known that for closed 3–manifolds analytic and Reidemeister torsions are equal by the celebrated theorem of Cheeger–Müller (see [**4, 31**]). For certain operators whose spectrum is no more discrete but continuous, the notion of L^2-torsion have been introduced around 15 years ago by Carey–Mathai, Lott, Lück–Rothenberg, Novikov–Shubin (see in particular Lück's monograph [**28**] for a complete history of the story). One of the most significant results in this area, is due to W. Lück and says that the L^2-torsion of a hyperbolic 3–manifold is proportional to the hyperbolic volume of M up to a universal constant (which is in fact equal to $-\frac{1}{6\pi}$).

L^2-Alexander invariants for knots were first introduced by Li and Zhang [**26**] as a sort of Alexander type invariant but using Fuglede–Kadison determinant instead of the usual one as we do for twisted Alexander invariant. Futhermore, observe that the L^2-Alexander invariant of a knot evaluated at $t = 1$ is precisely the L^2-torsion of the knot complement. In general, L^2-torsions are very difficult to compute explicitly, and only few results are known on the values of the L^2-torsion. Here we compute the L^2-Alexander invariant of torus knots of type $(2, q)$ (the general case will appear in [**11**]).

4.1. Preliminaries. Consider the Hilbert space $l^2(\pi)$ defined as follows. We define an inner product on the complex group ring $\mathbb{C}\pi$ by

$$\langle \sum_{g\in\pi} c_g \cdot g\,,\, \sum_{g\in\pi} d_g \cdot g\rangle = \sum_{g\in\pi} c_g \cdot \overline{d_g}.$$

The Hilbert space completion with respect to the inner product $\langle\,,\,\rangle$ is denoted as $l^2(\pi)$.

There are several equivalent definitions of the von Neumann algebra $\mathcal{N}(\pi)$. Let us take the following one: $\mathcal{N}(\pi)$ is the algebra of all bounded linear endomorphisms of $l^2(\pi)$ that commute with the left $\mathbb{C}\pi$-action. The *trace* of an element $\phi \in \mathcal{N}(\pi)$ is defined by $\mathrm{tr}_{\mathcal{N}(\pi)}(\phi) = \langle \phi(e)\,,\, e\rangle$ where $e \in \mathbb{C}\pi \subset l^2(\pi)$ denotes the unit element. We can extend this trace to $n \times n$-matrices over $\mathcal{N}(\pi)$ by considering the sum of the traces of the entries on the diagonal.

A finitely generated Hilbert $\mathcal{N}(\pi)$-module V is a Hilbert space V with a linear left π-action such that there exists a $\mathbb{C}\pi$-linear embedding of V into an orthogonal direct sum of a finite number of copies of $l^2(\pi)$.

The *von Neumann dimension* of a finitely generated Hilbert $\mathcal{N}(\pi)$-module V is defined by $\dim_{\mathcal{N}(\pi)}(V) = \mathrm{tr}_{\mathcal{N}(\pi)}(\mathrm{pr}_V) \in \mathbb{R}^{>0}$. Here

$$\mathrm{pr}_V : \bigoplus_{i=1}^{k} l^2(\pi) \to \bigoplus_{i=1}^{k} l^2(\pi)$$

denotes the orthogonal projection onto V. The von Neumann dimension does not depend on the choice of the embedding of V into a finite number of copies of $l^2(\pi)$. We list three fundamental properties of the von Neumann dimension:

(1) $\dim_{\mathcal{N}(\pi)}(V) = 0$ if and only if $V = 0$.
(2) $\dim_{\mathcal{N}(\pi)}(l^2(\pi)) = 1$
(3) If π is finite then $\dim_{\mathcal{N}(\pi)}(V) = \dim_{\mathbb{C}}(V)/|\pi|$.

Let $f\colon U \to V$ be a map of finitely generated Hilbert $\mathcal{N}(\pi)$-modules. The spectral density function of f is the function $F(f)(\lambda)$ defined as follows:

$$F(f)(\lambda) = \dim_{\mathcal{N}(\pi)}\left(\operatorname{im}(E^{f^*f}_{\lambda^2})\right).$$

Here $\{E^{f^*f}_{\lambda}\colon U \to U \mid \lambda \in \mathbb{R}\}$ is the family of spectral projections of the positive endomorphism $f^*f\colon U \to U$. Observe that $F(t)(\lambda)$ is monotone and upper-semi-continuous. It defines a measure on the Borel σ-algebra on $\mathbb{R}$ which is uniquely determined by $dF(f)((a,b]) = F(f)(b) - F(f)(a)$ for $a < b$.

Let $f\colon U \to V$ be a map of finitely generated Hilbert $\mathcal{N}(\pi)$-modules. We define the *Fuglede-Kadison determinant* of f by

$$\det_{\mathcal{N}(\pi)}(f) = \exp\Big(\int_{0^+}^{\infty} \ln(\lambda)\, dF(f)(\lambda)\Big)$$

if $\int_{0^+}^{\infty} \ln(\lambda)\, dF(f)(\lambda) > -\infty$ and by $\det_{\mathcal{N}(\pi)}(f) = 0$ otherwise.

4.2. Definition. In this section, we recall the definition of the L^2-Alexander invariants for knots. Let $K \subset S^3$ be a knot and fix a Wirtinger presentation Γ of the knot group Π_K

$$\Gamma = \langle g_1, \dots, g_k \,|\, r_1, \dots r_{k-1}\rangle.$$

Let $\phi\colon \Pi_K \to \mathbb{Z}$ be the homomorphism of the knot group given by $g_i \mapsto 1$. For $t \in \mathbb{C}^*$ we obtain a ring homomorphism

$$\psi_t\colon \mathbb{C}\Pi_K \to \mathbb{C}\Pi_K,\ \sum_{g\in\Pi_K} c_g \cdot g \mapsto \sum_{g\in\Pi_K} c_g \cdot t^{\phi(g)} \cdot g.$$

Let F_j $(1 \le j \le k)$ be the matrix obtained from the Fox matrix $F = (\partial r_i/\partial g_j)$ by removing its jth column. We obtain a matrix $\psi_t(F_j) \in \mathcal{M}((k-1)\times(k-1); \mathbb{C}\Pi_K)$ by applying ψ_t entry-wise to the matrix F_j.

The L^2-Alexander invariant $\Delta^{(2)}_{K,\Gamma}(t)$ of the knot K with respect to the Wirtinger presentation Γ of Π_K is defined as the Fuglede–Kadison determinant of the map

$$r^{(2)}_{\psi_t(F_1)}\colon l^2(\Pi_K)^{k-1} \to l^2(\Pi_K)^{k-1}$$

given by right multiplication with the matrix $\psi_t(F_1)$.

$$\Delta^{(2)}_{K,\Gamma}(t) = \det_{\mathcal{N}(\Pi_K)}\big(r^{(2)}_{\psi_t(F_1)}\colon l^2(\Pi_K)^{k-1} \to l^2(\Pi_K)^{k-1}\big) \in [0,\infty)$$

where $\mathcal{N}(\Pi_K)$ is the Von–Neumann algebra.

To define the L^2-Alexander invariant, we suppose that one (and thus all) Wirtinger presentation Γ of the knot group Π_K of K has the following property: for all $t \in \mathbb{C}^*$, the map $r^{(2)}_{\psi_t(F_1)}$ is injective with $\det_{\mathcal{N}(\Pi_K)}(r^{(2)}_{\psi_t(F_1)}) > 0$. With such a hypothesis Li and Zhang proved [**26**, Proposition 3.4], that the L^2-torsion $\Delta^{(2)}_{K,\Gamma}(t)$ does not depend, up to a power of $|t|$, on the choice of the Wirtinger presentation, and thus defines a knot invariant.

4.3. L^2-torsion of the trefoil knot.

PROPOSITION 4.1. *Consider the trefoil knot K with the Wirtinger presentation*

$$\Gamma = \langle a, b \,|\, aba = bab \rangle$$

of the associated knot group Π_K. We obtain

$$\Delta^{(2)}_{K,\Gamma}(t) = \max(|t|, 1)^2.$$

PROOF. We have $\psi_t(F_2) = (1 - tb + t^2 ab)$ and hence $\Delta^{(2)}_{K,\Gamma}(t) = \det_\Gamma(r^{(2)}_{1-tb+t^2ab})$. For $n \in \mathbb{Z}$ we set

$$d_n = \sum_{k=0}^{\lfloor n/3 \rfloor} (-1)^k \cdot b^{n-3k} \cdot (bab)^k \in \mathbb{C}\Gamma.$$

Notice that $d_n - b \cdot d_{n-1}$ is non-zero if and only if n is a non-negative multiple of 3. In this case we have

$$d_n - b \cdot d_{n-1} = (-1)^{n/3} \cdot (bab)^{n/3}. \tag{4.1}$$

For $|t| < 1$ the element

$$c_t = \sum_{n=0}^{\infty} \big(d_n - ab \cdot d_{n-2}\big) \cdot t^n$$

lies in $l^1(\Pi_K)$ because of

$$\begin{aligned} \Big\| \sum_{n=0}^{\infty} \big(d_n - ab \cdot d_{n-2}\big) \cdot t^n \Big\|_{l^1} &\leq \sum_{n=0}^{\infty} \big(\lfloor \frac{n+3}{3} \rfloor + \lfloor \frac{n+1}{3} \rfloor\big) \cdot |t|^n \\ &\leq \sum_{n=0}^{\infty} n \cdot |t|^n = \frac{|t|}{(1-|t|)^2} < \infty. \end{aligned}$$

An easy calculation shows $(1 - tb + t^2 ab) \cdot c_t = 1$ (using equation (4.1).) We conclude that $r^{(2)}_{1-tb+t^2ab}$ is invertible in $\mathcal{N}(\Pi_K)$ with inverse $r^{(2)}_{c_t}$. We define

$$f_{t,u} = r^{(2)}_{1-(tu)b+(tu)^2 ab} \text{ for } 0 \leq u \leq 1.$$

Futhermore, [**3**, Theorem 1.10(e)] tells us

$$\begin{aligned} \Delta^{(2)}_{K,\Gamma}(t) &= \det_{\mathcal{N}(\Pi_K)}(f_{t,1}) \\ &= \exp\Big(- \operatorname{Re}\Big(\int_0^1 \operatorname{tr}_{\mathcal{N}(\Pi_K)} \big(\sum_{k=0}^{\infty} r^{(2)}_{c_{tu}} \circ r^{(2)}_{-tb+2t^2uab} \big)\, du \Big)\Big). \end{aligned}$$

We have

$$\begin{aligned} r^{(2)}_{c_{tu}} \circ r^{(2)}_{-tb+2t^2uab} &= r^{(2)}_{(-tb+2t^2uab)\cdot c_{tu}} \\ &= \sum_{n=0}^{\infty} r^{(2)}_{(-tb+2t^2uab)\cdot \sum_{n=0}^{\infty}(d_n - ab\cdot d_{n-2})\cdot (tu)^n}. \end{aligned}$$

Under the group homomorphism $\Pi_K \to \mathbb{Z}, a, b \mapsto 1$, every element in the support of $d_n - ab \cdot d_{n-2}$ is mapped to $n \in \mathbb{N}^*$. This shows that every element in the support of $(-tb + 2t^2 uab) \cdot \sum_{n=0}^{\infty}(d_n - ab \cdot d_{n-2}) \cdot (tu)^n$ is mapped to a positive integer; in particular, the unit element $e \in \Pi_K$ does not lie in the support. Hence

$$\operatorname{tr}_{\mathcal{N}(\Pi_K)} \Big(r^{(2)}_{(-tb+2t^2uab)\cdot \sum_{n=0}^{\infty}(d_n - ab\cdot d_{n-2})\cdot (tu)^n} \Big) = 0.$$

This shows $\Delta^{(2)}_{K,\Gamma}(t) = \exp(0) = 1$ for all $0 < |t| < 1$.
In the case $|t| > 1$ we can argue analogously. We obtain

$$\det_{\mathcal{N}(\Pi_K)}\big(r^{(2)}_{1-t^{-1}a^{-1}+t^{-2}b^{-1}a^{-1}}\big) = 1$$

which implies

$$\begin{aligned}\Delta^{(2)}_{K,\Gamma}(t) &= \det_{\mathcal{N}(\Pi_K)}(r^{(2)}_{1-tb+t^2ab}) \\ &= \det_{\mathcal{N}(\Pi_K)}\big(r^{(2)}_{1-t^{-1}a^{-1}+t^{-2}b^{-1}a^{-1}}\big)\cdot \det_{\mathcal{N}(\Pi_K)}\big(r^{(2)}_{t^2ab}\big) = |t|^2.\end{aligned}$$

For $|t| = 1$ we have $\Delta^{(2)}_{K,\Gamma}(t) = \exp\big(-\rho^{(2)}(K)\big)$. It is a well–known fact that $\rho^{(2)}(K) = 0$. □

4.4. L^2-torsion of the torus knots. The same technique as for the trefoil knot can be used for torus knots of type (2, q). Here we outline the main steps without all details (further details could be find in [**11**]).

Let $q = 2k + 1$ be an odd integer and consider the $(2, q)$-torus knot $T(2, q)$. The fundamental group of the exterior of $T(2, q)$ admits the following well–known Wirtinger presentation:

$$\Gamma_{2,q} = \langle a, b \mid a(ba)^k b^{-1}(ba)^{-k}\rangle.$$

The matrix F_2 obtained by removing the second column of the Fox matrix F is reduced to a single element of $\Gamma_{2,q}$ and is equal to:

$$F_2 = \frac{\partial}{\partial a}\big(a(ba)^k b^{-1}(ba)^{-k}\big) = 1 - b + ab - bab + \cdots + (ab)^k.$$

Thus,

$$\psi_t(F_2) = 1 - tb + t^2ab - t^3bab + \cdots + t^{2k}(ab)^k$$

can be written as follows

$$\psi_t(F_2) = 1 + \sum_{j=0}^{k-1} -t^{2j+1}b(ab)^j + t^{2j+2}(ab)^{j+1} = 1 + (-tb + t^2ab)\sum_{j=0}^{k-1} t^{2j}(ab)^j.$$

Set

$$d_n = \sum_{j=0}^{\lfloor n/q\rfloor} (-1)^j b^{n-qj}(b(ab)^k)^j.$$

In the same way as for the trefoil knot, we have by a direct computation the following claim.

CLAIM 4.2. *We have*

$$d_n - bd_{n-1} = (-1)^{\lfloor n/q\rfloor}(bab)^{\lfloor n/q\rfloor}. \tag{4.2}$$

For $|t| < 1$, set

$$c_t = \sum_{n=0}^{\infty} \big(d_n - ab\cdot d_{n-2}\big)\cdot t^n.$$

In the same way as for the trefoil knot, $c_t \in l^1(\Pi(T_{2,q}))$.

CLAIM 4.3. *We have*

$$\psi_t(F_2)\cdot c_t = 1 \tag{4.3}$$

PROOF. Using Eq. (4.2), some telescopic sequences appear, thus Eq. (4.3) follows. □

Thus we obtain

PROPOSITION 4.4. *We have*

$$\Delta^{(2)}_{T(2,q),\Gamma_{2,q}}(t) = \max(|t|, 1)^{q-1}.$$

5. Chern–Simons invariant

5.1. Preliminaries. In the late 1980s, E. Witten considered a quantum field theory whose Lagrangian is the Chern–Simons functional. He argued that the Chern–Simons path integral on a (closed) 3–manifold with an embedded link gives a (formal) three–dimensional interpretation of the Jones polynomial of links. Unfortunately the path integral is not yet well–defined mathematically. But, the perturbative expansion method can be used and gives mathematical definition of knot invariants. Here we use another approach.

Let M be a 3–manifold. Let $\mathcal{A}$ denote the space of $\mathfrak{sl}_2(\mathbb{C})$-valued 1-forms on M. The Chern–Simons functional $CS\colon \mathcal{A} \to \mathbb{C}$ is defined by

$$CS(A) = \frac{1}{8\pi^2} \int_M \operatorname{tr}(A \wedge dA + \frac{2}{3} A \wedge A \wedge A). \tag{5.1}$$

The gauge group $\mathcal{G} = C^\infty(M; \mathrm{SL}_2(\mathbb{C}))$ is the space of smooth functions on M with values in $\mathrm{SL}_2(\mathbb{C})$, one can prove that CS is up to an integer gauge–invariant for closed manifolds. The orbit space $\mathcal{F}/\mathcal{G}$ of gauge–equivalence classes of flat connections is well–known to be homeomorphic to the character variety $X(M)$ of M (see for example Kirk's survey [**19**, § 2.2]). As a consequence, for us the Chern–Simons invariant of a closed manifold M is a functional on $X(M)$ with values into $\mathbb{C}/\mathbb{Z}$.

In the case of a manifold with boundary, the integral of Equation (5.1) is *not* gauge invariant even to an integer. In what follow, we will recall the definition of the Chern–Simons invariant for knot exteriors (and more generally for 3–manifolds whose boundary consists of a 2–torus), which is a section of a certain bundle over $X(T^2)$.

5.2. How to compute the Chern–Simons invariant? Here we review two papers [**20, 21**] by P. Kirk and E. Klassen which give us a method to explicitly compute the Chern-Simons invariant. First, we focus on 3–manifold without boundary and give a surgery formula for computing the Chern–Simons invariant (viewed as a functional on the character variety with values in $\mathbb{C}/\mathbb{Z}$) of a manifold obtained by surgery; second, we consider 3–manifold with a 2-torus boundary, typically a knot exterior, and give a method to compute its Chern–Simons invariant (viewed as a functional on the character variety in a certain bundle).

5.2.1. *Manifold without boundary obtained by surgery.* The main result to compute the Chern–Simons invariant of a manifold obtained by surgery on a knot is a theorem proved by Kirk–Klassen in 1990. This theorem tells us that the difference of Chern–Simons invariants can be computed entirely using the algebraic data contained in the restriction map $X(M_K) \to X(\partial M_K)$ induced by the usual inclusion $\partial M_K \hookrightarrow M_K$.

THEOREM 5.1. *Let $K \subset M$ be a knot in a closed 3–manifold. Consider two representations $\rho_0, \rho_1 \in R(M)$ and suppose there exists a path of representations*

$\rho_t \colon I \to R(M_K)$ from $\rho_{0|\Pi_K}$ to $\rho_{1|\Pi_K}$. Let $(\alpha(t), \beta(t))$, $t \in I$, be a path in $\mathbb{C}^2$ such that

$$\rho_t(\mu) = \begin{pmatrix} e^{2\pi i \alpha(t)} & 0 \\ 0 & e^{-2\pi i \alpha(t)} \end{pmatrix}, \quad \rho_t(\lambda) = \begin{pmatrix} e^{2\pi i \beta(t)} & 0 \\ 0 & e^{-2\pi i \beta(t)} \end{pmatrix}.$$

The Chern–Simons invariant CS_M satisfies the following equation:

$$CS_M(\rho_1) - CS_M(\rho_0) = -2\int_0^1 \beta(t)\alpha'(t)\,dt \in \mathbb{C}/\mathbb{Z}.$$

EXAMPLE 5.2 (The case of Lens spaces). The Lens space $L(p,q)$ is obtained by surgery of slope p/q on the trivial knot $\bigcirc$. The fundamental group of $L(p,q)$ is known to be the cyclic group of finite order p: $\pi_1(L(p,q)) = \langle z \mid z^p = 1\rangle$, and the character variety is a finite set:

$$X(L(p,q)) = \left\{\rho_n \mid n = 0, \ldots, \left[\frac{p}{2}\right]\right\} \text{ with } \rho_n(z) = \begin{pmatrix} e^{2\pi i n/p} & 0 \\ 0 & e^{-2\pi i n/p} \end{pmatrix}.$$

The exterior of the trivial knot is the solid torus $M_\bigcirc = S^1 \times D^2$ with $x = S^1 \times 1$ and $y = 1 \times S^1$ and $\Pi(\bigcirc) = \mathbb{Z} = \langle x \rangle$. The Lens space $L(p,q)$ is obtained from $M_\bigcirc$ by attaching another solid torus V such that the meridian μ of $\bigcirc$ is $\mu = px + qy$ and bounds a disc in V. Let $r, s \in \mathbb{Z}$ be such that $ps - qr = 1$ and set $q^* = -r$. The longitude of $\bigcirc$ is given by $\lambda = rx + sy$. Let $\rho_t \colon \Pi_\bigcirc \to \mathrm{SL}_2(\mathbb{C})$ be such that $\rho_t(x) = \begin{pmatrix} e^{2\pi i t} & 0 \\ 0 & e^{-2\pi i t} \end{pmatrix}$. Observe that $\rho_t \in X(L(p,q))$ if, and only if, $pt \in \mathbb{Z}$. Associated to ρ_t one can choose $(\alpha(t), \beta(t)) \in \mathbb{R}^2$ as follows:

$$\alpha(t) = pt \text{ and } \beta(t) = -q^* t.$$

Applying Theorem 5.1 with t between 0 and n/p where $n \in \{0, 1, 2, \ldots, [p/2]\}$, we obtain

$$CS_{L(p,q)}(\rho_n) = -2\int_0^{n/p} \beta(t)\alpha'(t)\,dt = pq^*\left(\frac{n}{p}\right)^2 = q^*\frac{n^2}{p}.$$

5.2.2. *Manifold with torus boundary: the case of knot exteriors.* Let M denote a 3–manifold whose boundary is non-empty and consists of a single 2–torus $T = \partial M$. Following ideas from Ramadas, Singer and Weitsman [**35**], the Chern–Simons invariant of M is considered in [**21**] as a lift of the restriction map induced by the inclusion $T \hookrightarrow M$:

$$\begin{array}{ccc} & & \mathcal{B}_T \\ & \overset{c_M}{\nearrow} & \downarrow \\ X(M) & \longrightarrow & X(T) \end{array}$$

Here $\mathcal{B}_T$ is a $\mathbb{C}^*$-bundle over $X(T)$ defined as follows. Let (μ, λ) be an oriented basis for $\pi_1(T)$. The map $t \colon X(T) \to \mathbb{C}^3$ given by $\rho \mapsto (\operatorname{tr}\rho(\mu), \operatorname{tr}\rho(\lambda), \operatorname{tr}\rho(\mu\lambda))$ is an algebraic embedding of the character variety $X(T)$. Let $V(T)$ be the 2-dimensional vector space (over $\mathbb{C}$):

$$V(T) = \mathrm{Hom}(\pi_1(T); \mathbb{C}).$$

The map $V(T) \to X(T)$ defined by $v \mapsto (x \mapsto 2\cos(2\pi i v(x)))$ is a branched covering. The covering group G is isomorphic to a semi–direct product of $\mathbb{Z} \oplus \mathbb{Z}$ and $\mathbb{Z}/2$ with the presentation

$$G = \langle x, y, b \mid xyx^{-1}y^{-1} = bxbx = byby = b^2 = 1\rangle.$$

Via the isomorphism $V(T) \to \mathbb{C}^2$ defined by
$$v \mapsto (v(\mu), v(\lambda))$$
the action of G on $V(T) \cong \mathbb{C}^2$ is as follows
$$x(\alpha, \beta) = (\alpha + 1, \beta), \quad y(\alpha, \beta) = (\alpha, \beta + 1), \quad b(\alpha, \beta) = (-\alpha, -\beta).$$
Now we extend the action of G to the the product $V(T) \times \mathbb{C}^*$ by the formulas:
$$x(\alpha, \beta; z) = (\alpha + 1, \beta; ze^{2\pi i\beta}) \tag{5.2}$$
$$y(\alpha, \beta; z) = (\alpha, \beta + 1; ze^{-2\pi i\alpha}) \tag{5.3}$$
$$b(\alpha, \beta; z) = (-\alpha, -\beta; z). \tag{5.4}$$
Thus, the quotient $\mathbb{C}^*$-bundle $\mathcal{B}_T$ over $X(T)$ is defined by the formula
$$\mathcal{B}_T = V(T) \times \mathbb{C}^*/G.$$
We use the following notation for points in $\mathcal{B}_T$. Since $\mathcal{B}_T$ is a quotient of $\mathbb{C}^2 \times \mathbb{C}^*$ we write $[\alpha, \beta; z]$ for equivalence classes, so for example
$$[\alpha, \beta; z] = [\alpha + m, \beta + n; ze^{2\pi i(m\beta - n\alpha)}].$$

The map $C_M : \rho \mapsto C_M(\rho) = [\gamma_\mu, \gamma_\lambda, e^{2\pi i CS_M(\rho)}]$ defines a lift of the restriction map induced by $T \hookrightarrow M$, see [**21**, Theorems 2.1 and 3.2]. Here $(\gamma_\mu, \gamma_\lambda)$ denotes a lift in $\mathbb{C}^2$ of the restriction $\rho_{|\pi_1(T)}$.

One can compute C_M using the following Kirk–Klassen formula.

THEOREM 5.3 ([**21**], Theorem 2.7, Corollary 2.6 and Theorem 3.2). *Let M denote an oriented 3–manifold whose boundary $\partial M = T$ consist of a 2–torus. Let (μ, λ) denote an oriented basis for $\pi_1(T)$.*

(1) *Let $\rho(t)\colon \pi_1(M) \to \mathrm{SL}_2(\mathbb{C})$, $t \in [0, 1]$, be a path of representations. Let $(\gamma_\mu(t), \gamma_\lambda(t))$ denote a lift of $\rho(t)_{|\pi_1(T)}$ to $\mathbb{C}^2$. Suppose*
$$C_M(\rho(t)) = [\gamma_\mu(t), \gamma_\lambda(t); z(t)]$$
for all t. Then
$$z(1) \cdot z(0)^{-1} = \exp\left(2\pi i \int_0^1 \gamma_\mu(t)\gamma'_\lambda(t) - \gamma'_\mu(t)\gamma_\lambda(t)\, dt\right). \tag{5.5}$$
Furthermore, if $\rho(1)$ is the trivial representation, then $z(1) = 1$.

(2) *There is an inner product*
$$\langle \cdot, \cdot \rangle \colon \mathcal{B}_T \times \mathcal{B}_{-T} \to \mathbb{C}^* \tag{5.6}$$
given by taking the pair $([\gamma_\mu, \gamma_\lambda, z], [\gamma_\mu, \gamma_\lambda, w])$ to $z/w \in \mathbb{C}^*$.

5.3. The case of torus knots. We can combine Theorem 5.3 to the explicit description of the character variety of torus knots given in Theorem 2.3 to obtain the following result.

PROPOSITION 5.4. *Recall that $M(p, q)$ denotes the exterior of a torus knot of type (p,q), where p, q are coprime integers. Let $\rho : \Pi(p, q) \to \mathrm{SL}_2(\mathbb{C})$ be a non–abelian representation whose character lies in the component of $X(M(p, q))$ parametrized by (α, β) (see Theorem 2.3). If the matrix $\rho(\mu)$ is conjugate to*
$$\rho(\mu) \simeq \begin{pmatrix} e^{2\pi i\gamma_\mu} & * \\ 0 & e^{-2\pi i\gamma_\mu} \end{pmatrix}$$

then

$$C_{M(p,q)}(\rho) = \left[\gamma_\mu, \frac{1}{2} - pq\gamma_\mu; \exp\left(2\pi i \left(\frac{(\beta ps + \varepsilon\alpha qr)^2}{4pq} - \frac{\gamma_\mu}{2}\right)\right)\right]. \tag{5.7}$$

Here $ps - qr = 1$ *and the result does not depend on the choice of* $\varepsilon \in \{\pm 1\}$.

The detailed proof can be found in [**9**].

6. An application in the asymptotic expansion of the colored Jones polynomial for torus knots related to the volume conjecture

This last part will be the more informal one. We will show that the non–abelian Reidemeister torsion and the Chern-Simons invariant for torus knots appear as subleading terms in the asymptotic expansion of the colored Jones polynomial for torus knots.

The volume conjecture of Kashaev [**17**] and Murakami-Murakami [**32**] relates some special values of the colored Jones polynomial to the simplicial volume.

To be more precise, let $\|\cdot\|$ be the simplicial or Gromov norm of a 3–manifold defined in [**14**] as follows. For a 3–manifold M, let $[M]$ denote its fundamental class in $H_3(M;\mathbb{R})$, the Gromov norm of M is the infimum of the L_1-norm on the singular cycles representing the fundamental class $[M]$:

$$\|M\| = \inf_{[\sum_i r_i\sigma_i]=[M]} \sum_i |r_i|.$$

We also consider

$$\langle K\rangle_N = \lim_{h\to 2\pi i/N} J'_N(K;h), \quad \forall N \in \mathbb{Z}_{>1},$$

where

$$J'_N(K;h) = \frac{J_N(K;h)}{J_N(\bigcirc;h)}$$

denotes the normalized N-th colored Jones polynomial. Here $J_N(K;h)$ is the N-th colored Jones polynomial and $\bigcirc$ stands for the unknot. One can observed that $J'_2(K;h)$ is equal to the (usual) Jones polynomial. In the standard normalization one has

$$J_N(\bigcirc;h) = \frac{\sinh(Nh/2)}{\sinh(h/2)}$$

and it is known that in this normalization the N-th colored Jones polynomial vanishes at the point $h = 2\pi i/N$ for any knot or link. Thus, the quantity $\langle K\rangle_N$ is a well defined invariant. The "volume conjecture" of [**17, 32**] states that for a knot K we have

$$2\pi \lim_{N\to\infty} \frac{\log|\langle K\rangle_N|}{N} = v_3\|S^3 \setminus K\| \tag{6.1}$$

where v_3 denotes the hyperbolic volume of a regular ideal tetrahedron in $\mathbb{H}^3$.

For any torus knot K, the volume conjecture appears to be trivially true in the sense that (see [**18**])

$$2\pi \lim_{N\to\infty} \frac{\log|\langle K\rangle_N|}{N} = 0. \tag{6.2}$$

The non–abelian Reidemeister torsion and the Chern–Simons invariant appear as subleading terms in the asymptotic expansion of $\langle K\rangle_N$. Below is the precise form of this result.

Let K be the torus knot of type (p, q). The non–abelian part of its character variety, $X^{\text{nab}}(M_K)$, consists of $N_{p,q} = (p-1)(q-1)/2$ connected components, and each component intersects the abelian part in exactly two bifurcation points (see Theorem 2.3). The ℓ-th non abelian component is characterized by a unique pair of distinct positive integers $0 < k_\ell^- < k_\ell^+ < pq$ satisfying certain conditions (see Theorem 2.3 for details). We consider the euclidean picture in Fig. 1 where $O = (0,0)$, $O' = (\frac{1}{2}, 0)$, $P^\pm = (\frac{k_\ell^\pm}{2pq}, 0)$, and two parallel segments passing through the points $P^\pm$ with slope $-pq$. In fact, this picture essentially describes the real slice of ℓ-th component of the character variety. Let $A_\ell^\diamond$ and $A_\ell^\triangleright$ be the double areas

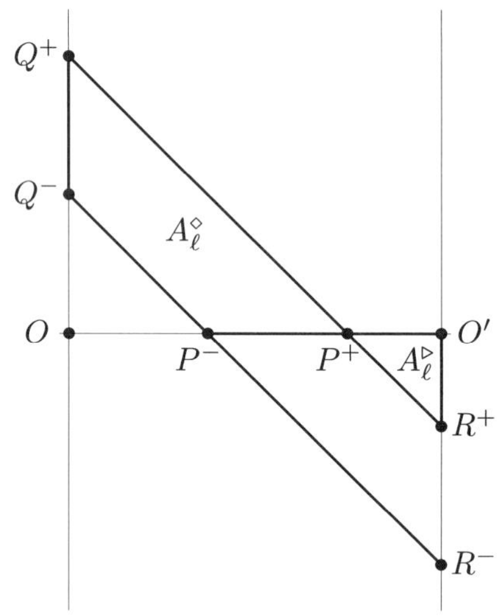

FIGURE 1. Euclidean picture associated to the ℓ-th component of $X^{\text{nab}}(M_K)$.

of the trapezoid $P^-P^+Q^+Q^-$ and the triangle P^+R^+O', respectively (a priori these are positive real numbers, but $A_\ell^\diamond$ is in fact an integer). We note that the set

$$\left\{A_\ell^\triangleright = \exp\left(2\pi i\left(\frac{(\beta ps + \varepsilon\alpha qr)^2}{4pq}\right)\right) (\operatorname{mod} \mathbb{Z}) \mid 1 \leqslant \ell \leqslant (p-1)(q-1)/2\right\}$$

is essentially the set of values of the Chern–Simons invariant of the knot K (see Proposition 5.4). Define the quantity

$$Z_N(K) = \sum_{\ell=1}^{N_{p,q}} \varepsilon_\ell \sqrt{\left|\mathbb{T}_\lambda^K(\chi_\ell)\right|}\, A_\ell^\diamond\, e^{-2\pi i N A_\ell^\triangleright}, \quad \varepsilon_\ell = (-1)^{[k_\ell^+/p]+[k_\ell^+/q]} \tag{6.3}$$

where $N_{p,q} = (p-1)(q-1)/2$ is the number of connected components of $X(M_K)$, χ_ℓ is any character in the ℓ-th component (as the torsion is locally constant), and the positive value of the square root is assumed.

THEOREM 6.1 ([**9**]). *For the (p, q) torus knot K the following asymptotic equality holds:*

$$4\pi e^{\frac{i\pi}{2N}(\frac{p}{q}+\frac{q}{p})} \cdot \langle K\rangle_N = N^{3/2} Z_N(K) + \pi i^{pqN} \sum_{n=1}^{\infty} \frac{a_n(K)}{n!}\left(\frac{i\pi}{2pqN}\right)^{n-1} \tag{6.4}$$

where $a_n(K)$ are finite type invariants.

Equation (6.4) gives a geometrical interpretation in terms of classical invariants of knots for each contribution in the asymptotic expansion of $\langle K\rangle_N$. We can note some consistency with the *Asymptotic Expansion Conjecture* (see [**33**, § 7.2]), namely the appearance of Chern–Simons invariant and Reidemeister torsion, but we can see also some peculiarity of our formula for example the presence of the multiplication factor $A_\ell^\diamond$ and ε_ℓ. In [**1**], J. Andersen and S. Hansen explore the Asymptotic Expansion Conjecture for surgeries on the figure–eight knot. In [**15**], K. Hikami proposes some identities concerning an asymptotic expansion of q-series which reduces to the invariant $\langle K\rangle_N$ with q being the N-th root of unity.

References

[1] J. Andersen and S. Hansen, *Asymptotics of the quantum invariants for surgeries on the figure 8 knot*, J. of Knot Theory and Its Ramifications 15 (2006), 479–548.

[2] G. Burde and H. Zieschang, *Knots*, de Gruyter Studies in Mathematics, no. 5, Walter de Gruyter, 2003.

[3] A. Carey, M. Farber and V. Mathai *Determinant lines, von Neumann algebras and L^2-torsion* J. Reine Angew. Math. **484** (1997), 153–181.

[4] J. Cheeger, *Analytic torsion and the heat equation*, Ann. Math **109** (1979), 259–322.

[5] M. Culler and P. Shalen, *Varieties of group representations and splittings of 3-manifolds*, Ann. of Math. **117** (1983), 109–146.

[6] J. Dubois, *Non abelian Reidemeister torsion and volume form on the* SU(2)*-representation space of knot groups*, Ann. Institut Fourier, **55** (2005), 1685–1734.

[7] J. Dubois, *Non abelian twisted Reidemeister torsion for fibered knots*, Canad. Math. Bull. **49** (2006), 55–71.

[8] J. Dubois, V. Huynh, and Y. Yamaguchi, *Non-abelian Reidemeister torsion for twist knots*, J. Knot Theory Ramications **18** (2009), 303–341.

[9] J. Dubois and R. Kashaev, *On the asymptotic expansion of the colored Jones polynomial for torus knots*, Math. Ann. **339** (2007), 757–782.

[10] J. Dubois and Y. Yamaguchi, *Multivariable twisted Alexander polynomial for hyperbolic three-manifolds with boundary*, preprint 2009.

[11] J. Dubois and C. Wegner, *L^2-Alexander invariant for knots*, in preparation.

[12] S. Friedl and T. Kim, *Thurston norm, fibered manifolds and twisted Alexander polynomials* GT/0505594, to be published by Topology, 28 pages (2005).

[13] S. Friedl and S. Vidussi, *A survey of twisted Alexander polynomials* preprint arXiv : GT/0905.0591v2 (2009).

[14] M. Gromov, *Volume and bounded cohomology*, Inst. Hautes Etudes Sci. Publ. Math. **56** (1982), 5–99

[15] K. Hikami, *Volume Conjecture and asymptotic expansion of q-series*, Experiment. Math., **12** (2003), 319–337.

[16] K. Hikami and H. Murakami, *Representations and the colored Jones polynomial of a torus knot*, preprint arXiv : GT/1001.2680 (2010).

[17] R. Kashaev, *A link invariant from dilogarithm*, Mod. Phys. Lett. A **10** (1995), 1409–1418.

[18] R. Kashaev and O. Tirkkonen, *Proof of the volume conjecture for torus knots*, J. Math. Sci. New York 115 (2003), 2033–2036.

[19] P. Kirk, SU(2)*-Representation varieties of* 3*-manifolds, gauge theory invariants, and surgery on knots*, Proceedings of the GARC workshop on Geometry and Topology'93 ed. H.J. Kim, Lecture note series number **18**, Seoul National University (1993), 137–176.

[20] P. Kirk and E. Klassen, *Chern-Simons invariants of 3–manifolds and representation spaces of knot groups*, Math. Annalen **287** (1990), 343–367.

[21] ———, *Chern–Simons invariants of 3–manifolds decomposed along tori and the circle bundle over the representation space of T^2*, Commun. Math. Phys. 153 (1993) 521–557.

[22] P. Kirk and C. Livingston, *Twisted Alexander invariants, Reidemeister torsion, and Casson-Gordon invariants*, Topology **38** (1999), 635–661.

[23] T. Kitano, *Twisted Alexander polynomial and Reidemeister torsion*, Pacific J. Math. **174** (1996), 431–442.
[24] E. Klassen, *Representations of knot groups in* SU(2), Trans. Amer. Math. Soc. **326** (1991), 795–828.
[25] Le, Thang, *Varieties of representations and their subvarieties of cohomology jumps for certain knot groups*, Russian Acad. Sci. Sb. Math.**78** (1994), 187–209.
[26] W. Li and W. Zhang, *An L^2-Alexander invariant for knots*, Commun. Contemp. Math. **8** (2006), no. 2, 167–187.
[27] ———, *An L^2-Alexander-Conway invariant for knots and the volume conjecture*, Differential geometry and physics, 303–312, Nankai Tracts Math., 10, World Sci. Publ., Hackensack, NJ, 2006.
[28] W. Lück, *L^2-invariants: theory and applications to geometry and K-theory*, Ergebnisse der Mathematik und ihrer Grenzgebiete. 3. Folge. A Series of Modern Surveys in Mathematics, 44. Springer-Verlag, Berlin, 2002.
[29] J. Milnor, *A Duality theorem for Reidemeister torsion*, Ann. of Math. **76** (1962), 134–147.
[30] ———, *Whitehead torsion*, Bull. Amer. Math. Soc. **72** (1966), 358–426.
[31] W. Müller, *Analytic torsion and R-torsion of Riemannian manifolds*, Adv. in Math. **28** (1978), 233–305.
[32] H. Murakami and J. Murakami, *The colored Jones polynomials and the simplicial volume of a knot*, Acta Math. **186** (2001), 85–104.
[33] T. Ohtsuki, Problems on invariants of knots and 3-manifolds, *Geom. Topol. Monogr.* **4** (2002), 377–572.
[34] J. Porti, *Torsion de Reidemeister pour les variétés hyperboliques*, vol. 128, Mem. Amer. Math. Soc., no. 612, AMS, 1997.
[35] T. Ramadas, I. Singer, and J. Weitsman, *Some comments on Chern–Simons gauge theory*, Comm. Math. Phys. **126** (1989), 409–420.
[36] V. Turaev, *Introduction to Combinatorial Torsions*, Birkhäuser 2001.
[37] ———, *Torsions of* 3*-Dimensional Manifolds*, Birkhäuser 2002.
[38] M. Wada, *Twisted Alexander polynomial for finitely presentable groups*, Topology **33** (1994), 241–256.
[39] F. Waldhausen, *Algebraic K-theory of generalized free products I, II.*, Ann. of Math. **108** (1978), 135–204.
[40] E. Witten, *Quantum field theory and the Jones polynomial*, Commun. Math. Phys. **121** (1989), 351–399.
[41] ———, *On quantum gauge theories in two dimensions*, Commun. Math. Phys. **141** (1991), 153–209.
[42] Y. Yamaguchi, *A relationship between the non-acyclic Reidemeister torsion and a zero of the acyclic Reidemeister torsion*, Ann. Inst. Fourier (Grenoble) **58** (2008), 337–362.

Institut de Mathématiques de Jussieu, Université Paris Diderot–Paris 7, UFR de Mathématiques, Case 7012, Bâtiment Chevaleret, 2, place Jussieu, 75205 Paris Cedex 13 France

E-mail address: dubois@math.jussieu.fr

AMS/IP Studies in Advanced Mathematics
Volume 50, 2011

Functional Integration and Abelian Link Invariants

Enore Guadagnini

ABSTRACT. The functional integral computation of the various topological invariants, which are associated with the Chern-Simons field theory, is considered. The standard perturbative setting in quantum field theory is rewieved and new developments in the path-integral approach, based on the Deligne-Beilinson cohomology, are described in the case of the abelian $U(1)$ Chern-Simons field theory formulated in $S^1 \times S^2$.

1. Introduction

The main subject of my talk concerns the use of the so-called path-integral —or functional integration— in the definition and in the computation of the various topological invariants which are associated with the quantum Chern-Simons field theory. This argument has already been introduced in several talks at this conference, so I shall skip the preliminaries and I will concentrate on the following question, which has indirectly been posed yesterday by one of the speakers.

We all agree that the path-integral has not a precise meaning. In particular, if S_{CS} *denotes the Chern-Simons action, the functional integral*

$$I(M) = \int_M DA \, e^{iS_{CS}[A]} \, , \tag{1.1}$$

which should correspond to an invariant of the 3-manifold M*, is not well defined. So, what is the meaning of expression* (1.1) ?

I shall try to present the answer to this question in simple but rather complete terms. Some basic features of the use of the functional integration in quantum field theory will be described in section 2 and section 3. In section 4 some new developments in the path-integral computation of observables in Chern-Simons field theory will be presented. I will show how to bypass the difficulties of standard perturbation theory in the case of the abelian $U(1)$ Chern-Simons theory formulated in a closed non-trivial 3-manifold M; the example $M = S^1 \times S^2$ will be discussed in detail. A few observations on the 3-manifold invariants associated with the Chern-Simons theory are contained in section 5 and section 6.

2010 *Mathematics Subject Classification.* Primary 57R56, 57M27; Secondary 58D30, 81T45.
Key words and phrases. Topological quantum field theory, Invariants of knots and 3-manifolds.

2. Perturbative quantum field theory

Functional integration can be used in perturbative quantum field theory. Given a set of fields —denoted by $\phi(x)$— and an action functional $S[\phi]$, in physics one is usually interested in functional integrations with a quite peculiar "measure" which naively corresponds to the following product

$$D\phi \, e^{iS[\phi]} = (\text{const.}) \left(\prod_x d\phi(x) \right) e^{iS[\phi]} \; . \tag{2.1}$$

Alternatively, one can introduce a complete set of orthonormal functions $\{\psi_n(x)\}$, with $n \in \mathbb{N}$, so that each classical configuration $\phi(x)$ can be written as a linear combination of these functions, $\phi(x) = \sum_n c_n \psi_n(x)$. By varying the (real) coefficients $\{c_n\}$ one gets an infinitesimal variation $d\phi(x)$ of the fields, $d\phi(x) = \sum_n dc_n \psi_n(x)$. Then, one can replace expression (2.1) with

$$D\phi \, e^{iS[\phi]} = (\text{const.}) \left(\prod_{n=1}^{\infty} dc_n \right) e^{iS[\phi]} \; . \tag{2.2}$$

It is well known that both expressions (2.1) and (2.2) are not well defined. With an infinite number of integration variables, the result of an integration looks like the product of an infinite number of coefficients that, apart from very exceptional cases, is not well defined. So, independently of the choice of the value of the multiplicative constant which appears in equation (2.1) or in equation (2.2), both expressions (2.1) and (2.2) do not represent a well defined integration measure. In facts, the integral of the measure $I = \int D\phi \, e^{iS}$, which can be imagined to represent some kind of "partition function" —precisely like expression (1.1)— is in general not well defined.

2.1. Observables. On the other hand, it is a fact that perturbative quantum field theory provides a rather accurate description of physical phenomena. For instance, in the $SU(3)_c \times SU(2)_L \times U(1)_Y$ Standard Model, by means of the functional integration method one can compute the value of the magnetic moment of the electron (or of the other charged leptons). The prediction of the Standard Model can be compared with the observed experimental value of the magnetic moment of the electron. The actual agreement [**1**] between the computed value and the real observed value is within less then one part in a million. In addition to the remarkable correspondence of the experiments with the predictions of quantum field theory, the puzzling question is:

How is it possible, by using a ill-defined functional integration, to make a prediction ?

For, in order to make a prediction, no ambiguities must occur, all the steps of the computation have to be well defined and, independently of any choice of notations or conventions, the final expression/value of the prediction must be unique.

The point is that all the predictions of quantum field theory that can be compared with experiments —quantities of this type are called "observables"— never take the form of the partition function $I = \int D\phi \, e^{iS}$. In standard quantum field theory, the observables are strictly connected with expectation values of the type

$$\langle F[\phi] \rangle = \frac{\int D\phi \, e^{iS[\phi]} \, F[\phi]}{\int D\phi \, e^{iS[\phi]}} \; , \tag{2.3}$$

where $F[\phi]$ is a functional of the fields. At first sight, expression (2.3) also appears to be not well defined because, in order to compute $\langle F[\phi] \rangle$, one could imagine to evaluate separately the numerator and the denominator and then to take the ratio. However, this is not what physicists do in order to compute the expectation values.

In standard perturbative quantum field theory equation (2.3) means: choose some kind of "regularization" in order to give a meaning to the numerator and to the denominator simultaneoulsy. Then, for the regularized ratio, consider the limit in which the regularization is removed. This limit exists, or it exists at least for the quantum field theory models that appear to be somehow related to the description of physical phenomena. I shall present one example in a while. The regularization of the path-integral measure must necessarily be expressed in terms of a finite number of integration variables. Thus, for example, one could regularize expression (2.2) by considering the finite product

$$D\phi \, e^{iS[\phi]} \Big|_{\text{reg}} = D\phi_{N} \; e^{iS[\phi]} = (\text{const.}) \left(\prod_{n=1}^{N} dc_n \right) \; e^{iS[\phi]} \; . \tag{2.4}$$

In this case, expression (2.3) should be interpreted as

$$\langle F[\phi] \rangle = \lim_{N \to \infty} \frac{\int D\phi_{N} \; e^{iS[\phi]} \, F[\phi]}{\int D\phi_{N} \; e^{iS[\phi]}} \; . \tag{2.5}$$

After this first logical settlement, the next step consists in disentangling the difficulties in the computation of $\langle F[\phi] \rangle$ that are related to the form of the action $S[\phi]$ from those that are connected with the structure of the functional $F[\phi]$ itself. To this end, it is convenient to consider the so-called correlation functions

$$\langle \phi(x_1) \cdots \phi(x_n) \rangle = \frac{\int D\phi \; e^{iS[\phi]} \, \phi(x_1) \cdots \phi(x_n)}{\int D\phi \; e^{iS[\phi]}} \; . \tag{2.6}$$

(From now on it is understood that the meaning of a ratio of two functional integrals, like expression (2.6), is specified by a prescription of the type illustrated in equation (2.5).) The correlation functions are determined by the form of the action $S[\phi]$ and, provided they are well defined, one can then consider the problem of expressing $\langle F[\phi] \rangle$ in terms of the correlation functions. In several field theory applications in particle physics, this second task is rather trivial because from the correlation functions one can obtain directly the transition amplitudes for the various particles processes. However, in the Chern-Simons field theory, the solution of this second problem presents peculiar non-trivial aspects. In the computation of the expectation values of the holonomies associated with oriented links, one has to introduce a framing procedure in order to eliminate the ambiguities —which occur in the product of two A fields at the same point— and preserve the ambient isotopy invariance.

In order to control the whole set of correlation functions, it is useful to introduce the generating functional

$$Z[J] = \frac{\int D\phi \; e^{iS[\phi]} \; e^{i \int J\phi}}{\int D\phi \; e^{iS[\phi]}} \; , \tag{2.7}$$

where $J = J(x)$ is a classical "source" variable. Perturbative quantum field theory is based on the existence of the so-called "free" fields. Hence, let us illustrate the meaning of expression (2.7) in the simple case in which the action is a quadratic functional $S = S_0$ of the fields

(2.8) $$S_0[\phi] = \tfrac{1}{2} \int dx \, \phi(x) \nabla \phi(x) \, ,$$

where ∇ is an appropriate differential operator[1]. The following identity, which is not related to the path-integral at all, plays an important role. One has

(2.9)
$$\begin{aligned} e^{iS_0[\phi]} \, e^{i\int J\phi} &= \exp\left\{ \tfrac{i}{2} \int dx \, [\phi(x)\nabla\phi(x) + 2J(x)\phi(x)] \right\} \\ &= \exp\left\{ \tfrac{i}{2} \int dx \, \widetilde{\phi}(x) \nabla \widetilde{\phi}(x) \right\} \, \exp\left\{ -\tfrac{i}{2} \int dx \, dy \, J(x) \nabla^{-1}(x,y) J(y) \right\} \\ &= e^{iS_0[\widetilde{\phi}]} \, e^{-\frac{i}{2}\int J\nabla^{-1}J} \, , \end{aligned}$$

where

(2.10) $$\widetilde{\phi}(x) = \phi(x) + \int dy \, \nabla^{-1}(x,y) J(y) \equiv \phi(x) + \text{const.} \quad ,$$

and $\nabla^{-1}(x,y)$ is a Green function for the ∇ operator

(2.11) $$\nabla \cdot \nabla^{-1}(x,y) = \delta(x-y) \, .$$

In general, $\nabla^{-1}(x,y)$ satisfies certain analytic properties[2] which, in particle physics, must be consistent with some observed properties of the experimental data like causality and energy positivity. For the moment, let us assume that $\nabla^{-1}(x,y)$ exists, I shall return to this point later. The identity (2.9) implies that the generating functional $Z_0[J]$ for free fields can be written as

(2.12) $$Z_0[J] = e^{-\frac{i}{2}\int J\nabla^{-1}J} \times \frac{\int D\phi \; e^{iS_0[\phi + \text{const.}]}}{\int D\phi \; e^{iS_0[\phi]}} \, .$$

Now one finds the crucial point. In order to compute the ratio of the two functional integrations shown in equation (2.12), one must consider the limit (in which the regularization is removed) of regularized functional integrations. With a finite number of integration variables, the result of the integral is invariant under translation of these variables, thus

(2.13) $$\frac{\int D\phi \; e^{iS_0[\phi + \text{const.}]}}{\int D\phi \; e^{iS_0[\phi]}} \equiv \lim_{N\to\infty} \frac{\int D\phi_N \; e^{iS_0[\phi + \text{const.}]}}{\int D\phi_N \; e^{iS_0[\phi]}} = 1 \, ,$$

and the generating functional is then

(2.14) $$Z_0[J] = \exp\left\{ -\tfrac{i}{2} \int dx \, dy \, J(x) \nabla^{-1}(x,y) J(y) \right\} \, .$$

Note that, even if the computation of $Z_0[J]$ that has been presented here is somehow based on the functional integration method, in the whole argument no ill-defined functional integration has been really computed. $Z_0[J]$ is well defined and determines the value of all the observables of the free theory. For instance, the poles

---

[1] For instance, when the field model is used to describe one kind of free spinless particles, the operator ∇ is given by $\nabla = -\eta^{\mu\nu}\partial^2/\partial x^\mu \partial x^\nu - m^2$, where $\eta^{\mu\nu}$ denotes the Minkowski metric.

[2] The Green function $i\nabla^{-1}(x,y)$ is usually called the Feynman propagator and, in the case of free spinless particles, it is given by $i\nabla^{-1}(x,y) = i\int [d^4p/(2\pi)^4] \, e^{-ip(x-y)} \, (p^2 - m^2 + i\epsilon)^{-1}$.

in the Fourier transform of $\nabla^{-1}(x,y)$ fix the value of the particles mass. So, all the observables of the free theory do not depend at all on the value that one could imagine to give to the partition function $I_0 = \int D\phi\, e^{iS_0}$. This remains true also in the case of interacting fields models, where the action contains cubic or quartic terms in powers of the fields. In fact, perturbative quantum field theories can also be formulated [**2, 3**] without the introduction of functional integration.

REMARK 2.1. Equation (2.13) can also be interpreted as a *defining relation*, because equality (2.13) is precisely the only property of the functional integration that is used in standard perturbative quantum field theory.

To sum up, in the path-integral formulation of perturbative quantum field theories there is really no need of computing the partition function $I = \int D\phi\, e^{iS}$, any functional integration of this type is not well defined[3] and all the observables do not depend on it. Clearly, the fact that the functional integration $I = \int D\phi\, e^{iS}$ is not well defined is not connected with the possible existence of an analytic continuation of the field model in the euclidean region (this subject is related to the analytic properties of the Feynman propagator). Also, as far as the observables of quantum field theory are concerned, the question whether, in the functional integration, one has to sum over smooth or singular configurations for the field variables is a completely irrelevant issue.

Finally, in analogy with the result of a gaussian integral with a finite number of integration variables, sometimes in literature one finds the relation

$$I_0 = \int D\phi\, e^{iS_0} = (\text{const.})\, \text{Det}^{-1/2}\, (-i\nabla)\ . \tag{2.15}$$

Equation (2.15) is not a definition of the value of the partition function because the expression appearing on the r.h.s. (for the differential operators ∇ that normally enter the action functional) is not well defined. Expression (2.15) can be used as a guess-suggesting reminder for the properties of the regularized functional integral. For instance, when ∇ smoothly depends on a parameter (or on a set of parameters) λ, from equation (2.15) one can guess the expression for the logarithmic variation of I_0 with respect to λ,

$$I_0^{-1}\, \frac{\partial I_0}{\partial \lambda} = -\tfrac{1}{2}\, \text{Tr}\left(\nabla^{-1}\, \frac{\partial \nabla}{\partial \lambda} \right)\ . \tag{2.16}$$

Differently from equation (2.15), expression (2.16) is well defined and, in facts, its structure is similar to the structure of the correlation functions —or of many of the observables— in quantum field theory [**4**].

In a "free fields" model, the correlations functions can be derived from expression (2.14),

$$\langle \phi(x_1)\cdots\phi(x_n)\rangle_0 = \frac{\int D\phi\ e^{iS_0[\phi]}\, \phi(x_1)\cdots\phi(x_n)}{\int D\phi\ e^{iS_0[\phi]}} = \left. \frac{(-i)^n\, \delta^n\, Z_0[J]}{\delta J(x_1)\cdots\delta J(x_n)} \right|_{J=0} . \tag{2.17}$$

[3]Something similar also happens in statistical mechanics where the partition function Z, which takes the form $Z = \text{Tr}\, e^{-H/kT}$, is not an observable. The observables are combinations of the normalized mean values or can be derived from the thermodynamic potentials in the thermodynamic limit. For example, in order to determine the free energy of the system, one only needs to consider the leading term of the expansion of $\ln Z$ in powers of the volume (for instance) in the thermodynamic limit. As a result, if one modifies the partition function and multiplies it by, say, a factor five, $Z \to Z' = 5Z$, the expression of the free energy does not change. Thus, Z cannot be —and in facts it is not— an observable, whereas the free energy is.

If the functional $F[\phi]$ can be written as a smooth linear combination of the correlation functions, one can then evaluate the observable $\langle F[\phi]\rangle$. Since the correlations functions are really distributions, the computation of $\langle F[\phi]\rangle$ may present ambiguities when, for instance, a correlation function is integrated with a function that is not a test function or when, in a correlation function, one needs to consider the limit of two (or more) coincident points. This problem, which is also present in an interacting fields model, is related (in part) to the so-called composite operators problem.

2.2. Interactions and renormalization. In the case of an interacting fields model, the action $S[\phi]$ is written as the sum of two terms, $S[\phi] = S_0[\phi] + S_I[\phi]$, where S_0 denotes the "free" action and S_I contains the interaction terms. The generating functional $Z[J]$ of equation (2.7) can be written as

$$Z[J] = e^{iZ^c[J]} = \frac{\int D\phi\, e^{iS_0}\, e^{iS_I}\, e^{i\int J\phi}}{\int D\phi\, e^{iS_0}} \Bigg/ \frac{\int D\phi\, e^{iS_0}\, e^{iS_I}}{\int D\phi\, e^{iS_0}}\ , \tag{2.18}$$

where the numerator

$$\langle e^{iS_I}\, e^{i\int J\phi}\rangle_0 = \frac{\int D\phi\, e^{iS_0}\, e^{iS_I}\, e^{i\int J\phi}}{\int D\phi\, e^{iS_0}} \tag{2.19}$$

denotes the sum of all the Feynman diagrams (which are constructed with the Feynman propagator $i\nabla^{-1}$, the interaction vertices of S_I, and in which each external leg corresponds to $-\int dy\, \nabla^{-1}(x,y)J(y)$); and the denominator

$$\langle e^{iS_I}\rangle_0 = \frac{\int D\phi\, e^{iS_0}\, e^{iS_I}}{\int D\phi\, e^{iS_0}} \tag{2.20}$$

just corresponds to the sum of the vacuum-to-vacuum diagrams, i.e. diagrams with no external legs. Actually, the sum of the vacuum-to-vacuum diagrams factorizes in the numerator and cancels out with the denominator. So, there is no need of computing the vacuum-to-vacuum diagrams (which remain divergent even after the standard regularization/renormalization procedure has been introduced). In conclusion, in the derivation of the correlation functions and of the observables of an interacting field theory, one never has to compute the value of the partition function $I = \int D\phi\, e^{iS}$.

A few remarks on the meaning of the renormalization in field theory are in order. By means of a Legendre transformation of the functional $Z^c[J]$ of the connected correlation functions, one can introduce the effective action $\Gamma[\varphi]$,

$$\varphi(x) = \frac{\delta Z^c[J]}{\delta J(x)}\quad , \quad \Gamma[\varphi] = Z^c[J] - \int dx J(x)\varphi(x)\ , \tag{2.21}$$

which is the sum of the one-particle-irreducible Feynman diagrams in which the external legs are represented by the classical variable φ. The perturbative expansion of a generic (nontrivial) correlation function —of the field theory defined by the action $S[\phi]$— is equal to the perturbative expansion containing diagrams at the *tree level* only of a new field model defined by the action that coincides with the functional $\Gamma[\varphi]$. Diagrams at the tree level contain no loops: so, they present no ultraviolet divergences and maintain all the symmetries of the action. This means that, provided $\Gamma[\varphi]$ is well defined, all the correlations functions are well defined. Since the effective action establishes how the symmetries of the theory are realized and determines the values of the observables[4], $\Gamma[\varphi]$ is the fundamental functional that must be considered in the renormalization task.

Quite often, some of the diagrams contributing to Γ are not well defined and present ambiguities. The root of these ambiguities is usually due to the presence of divergences which can be eliminated (in agreement with the action principle) by local conterterms, i.e. by terms which have the form of integrals of polynomials of the field variables (and their derivatives) defined in the same point with divergent coefficients. Generally, in the intermediate steps of the renormalization, one makes use of an arbitrary regularization and, after the introduction of appropriate local counterterms (which also depend on the choice of the regularization), one takes the limit in which the regularization is removed. The whole renormalization procedure consists of:

- introduction of local counterterms (with divergent and finite coefficients) in the diagrams computations in order to make the effective action Γ finite and maintain the maximum number of symmetries (Lorentz symmetry, internal and gauge symmetries,...);
- introduction of the normalization conditions, which determine the meaning of the finite physical parameters on which the renormalized Γ eventually depends.

In renormalizable models, only a finite number of parameters need to be fixed (coupling constants, particle masses, fields or wave functions normalizations). The finite values of the coupling constants and of the particle masses —that must agree with the experimental values— are also called the renormalized or physical parameters. The coupling constants usually correspond to the values of certain transition amplitudes in particular kinematic conditions (for instance, in the limit of vanishing momenta), and the particle masses correspond to the poles in the energy variable of the dressed propagators. This means that, in the case of an interacting field theory, a few specific terms of Γ (and not of the action S) determine the values of the renormalized or physical parameters. The renormalized parameters are observables, whereas the so-called bare parameters —which enter the action functional $S[\phi]$— are not observables.

The abelian Chern-Simons field theory in $\mathbb{R}^3$ is a "free fields" theory because the action is a quadratic functional of the fields $S_{CS} = 2\pi k \int A \wedge dA$; the renormalization is trivial in this case because the effective action coincides with the action.

The Chern-Simons field theory in $\mathbb{R}^3$ with a simple non-abelian gauge group is an interacting theory and a non-trivial renormalization is required because some of the contributions to the effective action have ambiguities. For instance, the one-loop correction to the two point function for the connection field A is the sum of two terms: each term is divergent but, in their sum, the two divergent parts tend to cancel. Consequently, this sum is not well defined ($\infty - \infty$ is not well defined) and

[4]The magnetic moment of the electron, for instance, is described by the three-point proper vertex containing two spinor electron fields and one vector electromagnetic field. Namely, the renormalized effective action Γ of the Standard Model admits an expansion in powers of the fields. Consider now the term $B \in \Gamma$ given by $B = \int dxdydz\, \overline{\psi}_\alpha(x)\Lambda^\mu_{\alpha\beta}(x,y,z)\psi_\beta(y)A_\mu(z)$, where $\psi(x)$ is the spinor field associated with the electron and $A_\mu(x)$ denotes the 4-vector potential of electromagnetism. The function $\Lambda^\mu_{\alpha\beta}(x,y,z)$ (three-point proper vertex) describes how the electron interacts with the electromagnetic field and contains, in particular, the required information on the magnetic moment of the electron.

has ambiguities. In agreement with the behaviour of all perturbative quantum field theories, these ambiguities take the form of a finite local term, namely $b \int A^a \wedge dA^a$, where b is an arbitrary finite parameter. Since this term has the same structure of a lagrangian term, the ambiguity in the value of b is totally irrelevant and produces no physical observable effects. (This kind of ambiguities, which is well known in quantum field theory, concerns the finite terms arbitrariness in the renormalization process.) In facts, any choice of the finite value of b does not modify the structure of the proper vertices contained in Γ (it simply changes the name of some bare unobservable parameter); consequently, the perturbative expansion of a generic correlation function —which is equal to the perturbative expansion made of tree-level diagrams only, constructed with the functional Γ— is not modified by a change of the value of b.

Thus, in the non-abelian Chern-Simons field theory in $\mathbb{R}^3$ with compact gauge group, the non-trivial aspect of the renormalization is concentrated in the normalization conditions, which state how to identify the coupling constant. In particular, consider the complete (i.e. the sum of the contributions to all orders of perturbation theory) two-point proper vertex for the A field contained in Γ; its structure is fixed by the symmetries of the theory and takes the form

$$\Gamma \Big|_{AA} = \alpha \int A^a \wedge dA^a \, , \tag{2.22}$$

where the nonvanishing real coefficient α admits a power expansion in terms of the bare parameters. The dependence of α on the bare parameters is not unique and can be arbitrarily modified by changing the regularization; but how α depends on the bare parameters is not observable and then it is totally irrelevant. When the renormalized Γ preserves the BRS symmetry, the normalization condition states that the coupling constant k of the Chern-Simons theory is given by

$$\alpha = k/8\pi \, . \tag{2.23}$$

Like in any renormalizable field theory, all observables depend unambiguosly[5] on the renormalized parameters (in our case, on k). In particular, a second order computation in perturbation theory of the expectation values of the holonomies shows that, in terms of the coupling constant k, the deformation parameter q turns out to be

$$q = e^{-2\pi i/k} \, . \tag{2.24}$$

The coupling constant of the non-abelian Chern-Simons field theory in three dimensions should not be confused with the so-called "level" parameter which appears in two-dimensional conformal field theories.

3. Perturbative Chern-Simons field theory

Let us now concentrate on the Chern-Simons field theory [**5**] and on the path-integral computation of the topological invariants. It is important to distinguish the cases in which the topological model is defined in $\mathbb{R}^3$ or in a closed 3-manifold M. Also, it is significant to distinguish the expectation values $\langle W[A] \rangle$, of a gauge

[5]For example, in the Standard Model the magnetic moment of the electron depends unambiguosly [**1**] on the value of the renormalized electromagnetic coupling constant $\alpha_{em} \simeq 1/137$. Whereas the dependence of α_{em} on the "bare" coupling constant is not unique and is not observable; in facts, can you imagine how to measure it by means of an experiment?

invariant functional $W[A]$ of the A fields, from (a possible variant of) the partition function $I = \int DA\, e^{iS_{CS}}$.

Let us firstly recall the results obtained in perturbation theory in $\mathbb{R}^3$. The non-abelian Chern-Simons field theory formulated in $\mathbb{R}^3$ with compact gauge group is perturbatively renormalizable. Several features of the perturbative expansion have been explored with the covariant gauge-fixing of the so-called Landau gauge. As far as the correlation functions, the BRS transformations and the structure of the renormalized proper vertices in the effective action are concerned, explicit computations [**6**] have been produced up to two (Feynman) loops. The primitive ultraviolet divergences associated with the single diagrams —constructed with the connections and ghosts fields— tend to cancel in the construction of the proper vertices. Actually, the theory is finite [**7**] to all orders of perturbation theory. Therefore, in the renormalization process, only the finite local conterterms freedom remains to be fixed by means of the normalization conditions. These conditions determine how to identify the coupling constant in the effective action.

Since renormalizability is determined by the short-distance behaviour of the model, the ultraviolet properties of the Chern-Simons theory formulated in $\mathbb{R}^3$ or formulated in a generic 3-manifold M are obviously the same. This does not imply that the use of perturbation theory extends trivially from $\mathbb{R}^3$ to M. There are in fact obstacles —for both the abelian and the non-abelian Chern-Simons theories formulated in a closed 3-manifold M— to produce real perturbative computations of the observables, but these difficulties are not related to the ultraviolet divergences.

For the abelian Chern-Simons theory in $\mathbb{R}^3$, the expectation values of the holonomies, which are associated with coloured oriented and framed links, can explicitly be computed in closed form by means of standard perturbation theory.

In the non-abelian case with compact gauge group, the explicit computation of the expectation values of the Wilson line operators in $\mathbb{R}^3$ has been produced at the third non-trivial order of perturbation theory [**8**]. The results of perturbation theory are in agreement with what is expected on the basis of general arguments. In facts, by taking into account the relevant symmetry properties of the expectation values in quantum field theory —namely, ambient isotopy invariance, validity of satellite relations, structure of the representation ring of the gauge group, covariance of the expectation values under a modification of the framing— one finds that the expression of the expectation values (of the trace of the holonomies) is unique.

These invariants of framed and coloured links take the form of generalized Jones polynomials [**9**]; the algebraic structure of these polynomials, which is determined by the characters of simple Lie groups, is very general. In facts, these link invariants can also be obtained or defined by means of skein relations [**10**], quantum group Hopf algebra methods [**11**], statistical state models [**12**]. For each simple Lie algebra, the corresponding braid group representations entering the construction of these link polynomials have a universal and canonical structure (the classification of these braid group representations is somewhat similar to the classification of the irreducible representations of simple Lie algebras). These braid group representations also appear, for example, as monodromy representations in conformal bidimensional models [**13, 14, 15**].

The perturbative setting of the Chern-Simons field theory can be imagined to be extended from $\mathbb{R}^3$ to a generic closed 3-manifold M. Indeed, provided the fields propagator is well defined, the whole perturbative construction based on the Wick contractions trivially follows. However, as a matter of facts, no explicit example of a real functional integral computation —in standard perturbation theory— of an observable in a closed 3-manifold $M \neq S^3$ has been produced in the last twenty years.

4. Functional integral and Deligne-Beilinson cohomology

When the Chern-Simons field theory is formulated in a nontrivial closed oriented manifold $M \neq S^3$, the explicit computation of the observables by means of the standard perturbation theory within the path-integral method presents technical difficulties, which are related to the gauge-fixing procedure and the definition of the fields propagator. Let me show how it is possible to overcome these problems in the abelian case [**16**]. I will consider the example $M = S^1 \times S^2$ because standard perturbation theory cannot be used in this case. On the other hand, when M is a homology sphere, different approaches can be used to determine the expectation values of the Wilson line operators for an abelian gauge group; for instance, the Referee has pointed out the articles [**17, 18**].

4.1. Basic definitions. The abelian Chern-Simons theory [**19, 20, 21**] with gauge group $U(1)$ is defined by means of a $U(1)$-connection A in a closed oriented 3-manifold M. The holonomy associated with an oriented knot $C \subset M$ is given by the integral $\int_C A$; this integral is invariant under ordinary $U(1)$ gauge transformations acting on A. In the standard field theory formulation of abelian gauge theories, the configuration space locally coincides with the set of 1-forms modulo exact forms, $A \sim A + d\Lambda$. But if one assumes [**5, 22**] that a complete set of observables is given by the exponential of the holonomies $\{\exp[2\pi i \int_C A]\}$ which are associated with oriented knots C in M, the invariance group of the observables is actually larger than the standard gauge group; in facts, the observables are locally defined on the classes of 1-forms modulo forms $\widehat{A}$ with integer periods, $A \sim A + \widehat{A}$, $\int_C \widehat{A} = n \in \mathbb{Z}$. More precisely, the configuration space is defined in terms of the Deligne-Beilinson (DB) cohomology classes [**22, 23, 24, 25, 26**]. While in normal quantum electrodynamics the symmetry group choice between the gauge group $U(1)$ and its noncompact covering is not fixed a priori, the fact that, in the Chern-Simons theory context, a complete set of observables is given by the expectation values of the Wilson operators necessarily implies —in order to ensure consistency— that the DB formulation has to be adopted.

The DB class associated with the connection A will be denoted by the same symbol $A \in H^1_D(M)$, where $H^1_D(M)$ represents the DB cohomology group of M of degree 1. Let $H^3_D(M)$ be the space of the DB classes of degree 3; the *-product of two classes is a pairing of the DB cohomology groups that defines a natural mapping [**27**]

$$H^1_D(M) \otimes H^1_D(M) \longrightarrow H^3_D(M) \; ;$$

the *-product of A with A just corresponds to the abelian Chern-Simons lagrangian

$$A * A \longrightarrow A \wedge dA \; . \tag{4.1}$$

Like the integral of any element of $H^3_D(M)$, the Chern-Simons action

$$S_{CS}[A] = \int_M A * A \longrightarrow \int_M A \wedge dA$$

is defined modulo integers; consequently, the path-integral phase factor

$$\exp\Big\{2\pi i k S_{CS}[A]\Big\} = \exp\Big\{2\pi i k \int_M A * A\Big\}$$

is well defined when the coupling constant k takes integer values, $k \in \mathbb{Z}$, $(k \neq 0)$. Let us now consider a framed, oriented and coloured link $L \subset M$ with N components $\{C_1, C_2, ..., C_N\}$. The colour of each component C_j, with $j = 1, 2, ..., N$, is represented by an integer charge $q_j \in \mathbb{Z}$. The classical expression $W(L)$ of the Wilson line is given by

$$W(L) = \prod_{j=1}^{N} \exp\Big\{2\pi i q_j \int_{C_j} A\Big\} = \exp\Big\{2\pi i \sum_j q_j \int_{C_j} A\Big\} , \tag{4.2}$$

and the observables of the Chern-Simons gauge theory in M are given by the expectation values

$$\langle W(L)\rangle\Big|_M = \frac{\int_M DA\, e^{2\pi i k S_{CS}[A]}\, W(L)}{\int_M DA\, e^{2\pi i k S_{CS}[A]}} , \tag{4.3}$$

where the path integral should be defined on the DB classes which belong to $H_D^1(M)$.

Remark 4.1. One usually assumes that expression (4.3) is well defined. But one should keep in mind that, for certain manifolds M and for certain values of the coupling constant k, expression (4.3) could not be well defined.

The structure of the functional space admits a natural description in terms of the homology groups of M, as indicated by the following exact sequence [**28, 29**]

$$0 \longrightarrow \Omega^1(M)/\Omega^1_{\mathbb{Z}}(M) \longrightarrow H_D^1(M) \longrightarrow H^2(M) \longrightarrow 0 , \tag{4.4}$$

where $\Omega^1(M)$ is the space of 1-forms on M, $\Omega^1_{\mathbb{Z}}(M)$ is the space of closed 1-forms with integer periods on M and $H^p(M)$ is the $(p)^{th}$ integral cohomology group of M. Thus, $H_D^1(M)$ can be understood as an affine bundle over $H^2(M)$, whose fibres have a typical underlying (infinite dimensional) vector space structure given by $\Omega^1(M)/\Omega^1_{\mathbb{Z}}(M)$.

4.2. Distributional forms. Now, the crucial observation is that, in order to compute the observables (4.3), the introduction of a gauge-fixing and of the fields propagator is not essential. In order to illustrate this point, let us consider first the case $M = S^3$.

The integral of a one-form A along an oriented knot $C \subset S^3$ can be written as the integral on the whole S^3 of the external product $A \wedge J_C$, where the current J_C is a distributional 2-form with support on the knot C; that is, one can write $\int_C A = \int_{S^3} A \wedge J_C$. Since C can be understood as the boundary of a Seifert surface Σ_C in S^3, one has $J_C = d\eta_C$ for some 1-form η_C with support on Σ_C. One then finds $\int_C A = \int_{S^3} A \wedge d\eta_C$. For example, consider the unknot C in S^3 shown in Figure 1, with a simple disc as Seifert surface. Inside the open domain depicted in Figure 1, the oriented knot is described —in local coordinates (x, y, z)— by a piece of the y-axis and the corresponding distributional forms J_C and η_C are given by

$$J_C = \delta(z)\,\delta(x)\, dz \wedge dx \qquad , \qquad \eta_C = \delta(z)\,\theta(-x)\, dz \, .$$

In terms of DB classes, one has

$$\exp\Big\{2\pi i\int_C A\Big\} = \exp\Big\{2\pi i\int_{S^3} A*\eta_C\Big\}\,, \tag{4.5}$$

where η_C denotes the DB class —associated with the knot C— that is locally represented by the distributional form η_C.

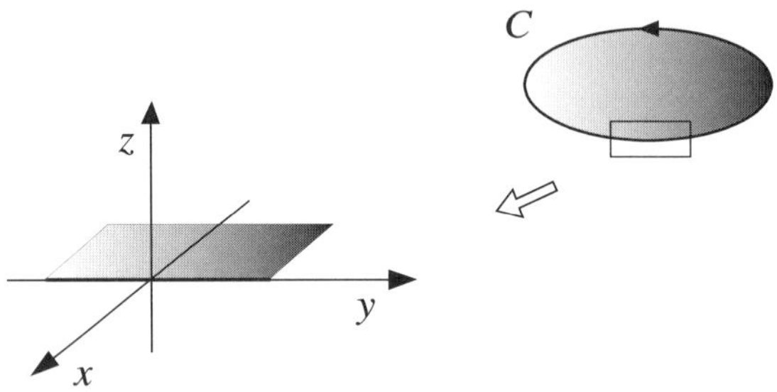

FIGURE 1. In a open domain with local coordinates (x, y, z), a piece of the knot C can be identified with the y axis, and the disc that it bounds can be identified with a portion of the half plane $(x < 0, y, z = 0)$.

For the coloured link $L \subset S^3$, one can write

$$W(L) = \exp\Big\{2\pi i \sum_j q_j \int_{C_j} A\Big\} = \exp\Big\{2\pi i\int_{S^3} A*\eta_L\Big\}\,, \tag{4.6}$$

where $\eta_L = \sum_j q_j \eta_j$ denotes the DB class associated with the link L. Since $H^2(S^3)$ is trivial, sequence (4.4) implies $H^1_D(S^3) \simeq \Omega^1(S^3)/\Omega^1_{\mathbb{Z}}(S^3)$. The analogue of equation (2.9) now takes the form

$$\begin{aligned} e^{2\pi i k S_{CS}[A]}\, W(L) &= \exp\left\{2\pi i k\int_{S^3} A*A + 2\pi i\int_{S^3} A*\eta_L\right\} \\ &= \exp\left\{2\pi i k\int_{S^3} \widetilde{A}*\widetilde{A}\right\} \exp\left\{-(2\pi i/4k)\int_{S^3}\eta_L*\eta_L\right\} \\ &= e^{2\pi i k S_{CS}[\widetilde{A}]}\exp\left\{-(2\pi i/4k)\int_{S^3}\eta_L*\eta_L\right\}\,, \end{aligned} \tag{4.7}$$

where $\widetilde{A} = A-(1/2k)\,\eta_L$. The ambiguities in $\int_{S^3}\eta_L*\eta_L$ that are related to the self-linking number can be fixed in the standard way by the introduction of a framing for the link L. The integral $\int\eta_L*\eta_L$ simply counts the number of intersections (with signs) of the framing components of the link with a Seifert surface associated with L (more details can be found in Ref.[**16**]). At this point, assuming invariance under translation of the functional integral (as shown in equation (2.13)), one finally gets

$$\langle W(L)\rangle\Big|_{S^3} = \exp\left\{-(2\pi i/4k)\int_{S^3}\eta_L*\eta_L\right\} = \exp\Big\{-(2i\pi/4k)\sum_{ij} q_i\mathbb{L}_{ij}q_j\Big\}\,, \tag{4.8}$$

where $\mathbb{L}_{ij}$ are the matrix elements of the linking matrix associated with the link L. Equation (4.8) coincides with the result of standard perturbation theory but, in the derivation of expression (4.8), both gauge-fixing and Feynman propagator have not been introduced.

The DB formalism turns out to be particularly useful for the path-integral computation of the observables in a generic 3-manifold M because sequence (4.4) describes the non-trivial structure of the functional space and equation (4.5) remains valid also in the case of a knot C in a generic 3-manifold M. Indeed, the class $\eta_C \in H^1_D(M)$ which is canonically associated with a knot $C \subset M$ is well defined for arbitrary 3-manifold M.

4.3. Observables in non-trivial manifolds. As an example of functional integration in a non-trivial manifold, let us consider the Deligne-Beilinson formalism when the Chern-Simons theory is formulated in the manifold $M = S^1 \times S^2$, which can be represented by the region of $\mathbb{R}^3$ which is delimited by two concentric 2-spheres, with the convention that the points on the two surfaces with the same angular coordinates are identified. Since $H^2(S^1 \times S^2) = \mathbb{Z}$, from relation (4.4) it follows that, as shown in Figure 2, $H^1_D(S^1 \times S^2)$ can be understood as an affine bundle over $\mathbb{Z}$ in which each fibre has a linear space structure isomorphic to $\Omega^1(S^1 \times S^2)/\Omega^1_{\mathbb{Z}}(S^1 \times S^2)$.

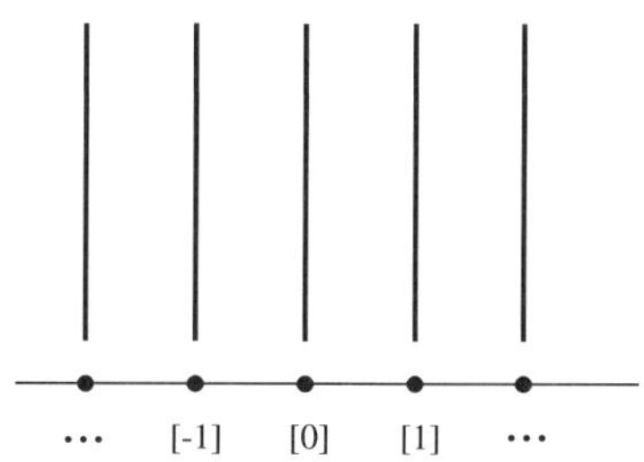

FIGURE 2. Presentation of the Deligne-Beilinson affine bundle $H^1_D(S^1 \times S^2)$.

In order to fix an origin in $H^1_D(S^1 \times S^2)$, we introduce the "diagonal" section s,

$$\begin{aligned} s : \mathbb{Z} &\rightarrow H^1_D(S^1 \times S^2) \\ n &\mapsto s(n) \equiv n\,\gamma_0\,, \end{aligned} \tag{4.9}$$

where $\gamma_0 \in H^1_D(S^1 \times S^2)$ denotes the DB class which is canonically associated with the knot G_0, shown in Figure 3, that can be taken as a generator of $H_1(S^1 \times S^2) = \mathbb{Z}$.

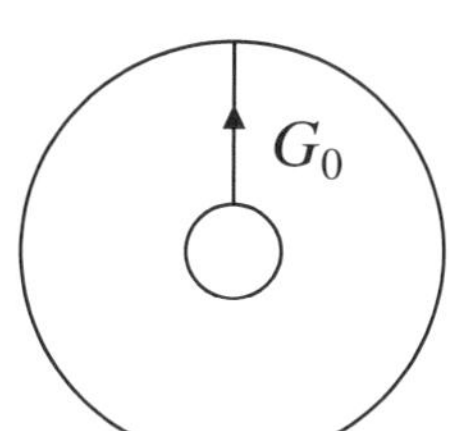

FIGURE 3. In the region of $\mathbb{R}^3$ that provides a description of $S^1 \times S^2$, the oriented loop $G_0 \subset S^1 \times S^2$ —generator of $H_1(S^1 \times S^2)$— is represented.

Each element $A \in H^1_D\left(S^1 \times S^2\right)$ can then be written as

(4.10) $$A = n\,\gamma_0 + \alpha\,,$$

for some integer n and $\alpha \in \Omega^1\left(S^1 \times S^2\right)/\Omega^1_{\mathbb{Z}}\left(S^1 \times S^2\right)$. The (not regularized) functional measure takes the form

(4.11) $$\begin{aligned} DA\, e^{2\pi i k S_{CS}[A]} &= \sum_{n=-\infty}^{+\infty} D\alpha \exp\left\{2\pi i k \int_{S^1\times S^2} (n\gamma_0 + \alpha) * (n\gamma_0 + \alpha)\right\} \\ &= \sum_{n=-\infty}^{+\infty} D\alpha \exp\left\{2\pi i k \int_{S^1\times S^2} \left[\alpha * \alpha + 2n\alpha * \gamma_0\right]\right\}\,. \end{aligned}$$

Because of the non-trivial homology of the manifold $S^1 \times S^2$, the functional measure has a physical (not related to the gauge invariance) zero mode $\beta_0 \in H^1_D(S^1 \times S^2)$. More precisely, let us represent a generator of $H_2(S^1 \times S^2)$ by a oriented 2-dimensional sphere S_0; S_0 is isotopic with the component S^2 of $S^1 \times S^2$ and, if one represents $S^1 \times S^2$ by the region of $\mathbb{R}^3$ which is delimited by two concentric spheres, S_0 can just be represented by a third concentric sphere, as shown in Figure 4.

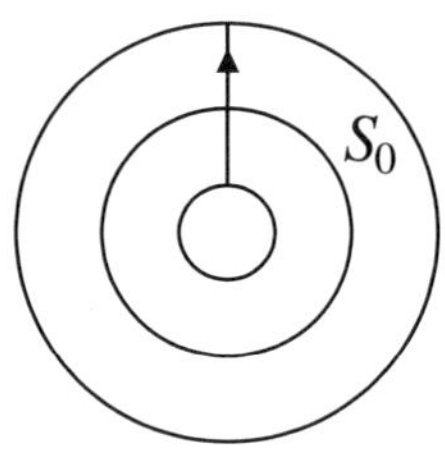

FIGURE 4. The sphere $S_0 \subset S^1 \times S^2$.

Let β_0 be the distributional 1-form which is globally defined in $S^1 \times S^2$ and has support on S_0, and let $\beta_0 \in H^1_D(S^1 \times S^2)$ be the class which is locally described by the distributional 1-form β_0. The overall sign of β_0 is fixed by the orientation of S_0 so that

(4.12) $$\int_{G_0} \beta_0 = 1\,.$$

Since the boundary of the closed surface S_0 is trivial, one has $d\beta_0 = 0$. From equation (4.12) and $d\beta_0 = 0$ it follows that, for any integer m,

(4.13) $$\exp\left\{2\pi i k S_{CS}[A]\right\} = \exp\left\{2\pi i k S[A + (m/2k)\beta_0]\right\}\,,$$

that implies

(4.14) $$\exp\left\{2\pi i k S_{CS}[A]\right\} = \frac{1}{2k}\sum_{m=0}^{2k-1} \exp\left\{2\pi i k S[A + (m/2k)\beta_0]\right\}\,.$$

Consider now the observable (4.3) associated with the link $L \subset S^1 \times S^2$ for fixed integer k. The integral

(4.15) $$[L] = \int_L \beta_0$$

takes integer values; the value of $[L]$ is equal to the sum of the intersection numbers (weighted with the values of the colour of the link components) of the link L with the surface S_0. Property (4.14) implies that $\langle W(L)\rangle\big|_{S^1\times S^2}$ vanishes unless $[L]\equiv 0 \mod 2k$ and, in that case, one can use again an identity of the type shown in equation (4.7) to get [**16**]

$$\langle W(L)\rangle\Big|_{S^1\times S^2} = \begin{cases} 0 \text{ if } [L]\not\equiv 0 \mod 2k\,, \\ \exp\Big\{-(2i\pi/4k)\sum_{ij} q_i \mathbb{L}_{ij} q_j\Big\} \text{ if } [L]\equiv 0 \mod 2k\,. \end{cases} \tag{4.16}$$

Note that, when M is not a homology sphere, the value of the self linking number of a framed knot $C\subset M$ is in general not unique, unless C is homologically trivial. Similarly, precisely because $[L]\equiv 0 \mod 2k$, the exponential of the sum of the linking matrix elements appearing in equation (4.16), namely the quantity $\exp\Big\{-(2i\pi/4k)\sum_{ij} q_i\mathbb{L}_{ij}q_j\Big\}$, is well defined; a discussion on this point can be found in Ref.[**16**]. Expression (4.16) is the result of a real (non perturbative) functional integration computation in a non-trivial manifold. Equation (4.16) is in agreement with the prediction of the Reshetikhin-Turaev surgery rules, that will be discussed in the next section.

The model considered in this section is an example of topological field theory in which standard perturbation theory cannot be used. In the quantum Chern-Simons field theory formulated in $S^1\times S^2$, the standard Feynman propagator for the A field does not exist because of the normalizable zero mode that corresponds to the class $\beta_0\in H^1_D(S^1\times S^2)$. Among the field configurations, one can find a globally defined 1-form A_0 such that $dA_0=0$, but $A_0\neq d\phi$ for some well defined ϕ. The Deligne-Beilinson formalism has been also applied [**16**] to the torsion-free manifolds $M=S^1\times\Sigma_g$ where Σ_g is a closed 2-surface of genus $g\geq 1$; and one example of manifold with torsion, $M=RP^3$, has been studied by Thuillier in [**30**].

5. Surgery invariants

In order to compute —by means of the quantum groups modular algebra— topological invariants in a generic 3-manifold M, Reshetikhin and Turaev have produced appropriate surgery rules [**31**] that are in agreement with the suggestions contained in [**5**]. These rules have been developed also by Kohno [**32**], by Lickorish [**33**] and by Morton and Strickland [**34**].

According to the Reshetikhin-Turaev surgery rules (adapted to the case of the abelian Chern-Simons field theory), for fixed integer k, the expectation value of the Wilson line $W(L)$ associated with a link L in the 3-manifold $M=M_{\mathcal{L}}$ can be written as a ratio

$$\text{surgery rules:}\qquad \langle W(L)\rangle\Big|_{M_{\mathcal{L}}} = \langle W(L)\,W(\mathcal{L})\rangle\Big|_{S^3}\Big/\,\langle W(\mathcal{L})\rangle\Big|_{S^3}\,. \tag{5.1}$$

In equation (5.1), $\mathcal{L}$ is the surgery link that codifies a Dehn surgery [**35**] presentation of M in S^3; the integer surgery coefficients are determined by the framings of the components of $\mathcal{L}$. In the computation of the expectation values in S^3, one has to sum over the values $q=0,1,2,...,2k-1$ of the colours which are associated with the components of $\mathcal{L}$. The structure of equation (5.1) is somehow similar to the structure of the corresponding functional integral expression

(5.2) functional integral: $$\langle W(L)\rangle\Big|_M = \frac{\int_M DA\, e^{2\pi i k S_{CS}[A]}\, W(L)}{\int_M DA\, e^{2\pi i k S_{CS}[A]}} \; .$$

But equation (5.1) is not based on a path-integral in the manifold M; the numerator and the denominator (which appear on the r.h.s of the equation) —that must be computed in S^3— are separately well defined and, when the denominator is not vanishing, the ratio (5.1) is well defined. Expression (5.1) refers to a particular surgery presentation described by the surgery link $\mathcal{L}$ but, since the ratio (5.1) is invariant under Kirby moves [**36**], expression (5.1) is an isotopy invariant of the link L in the oriented 3-manifold M.

In all the examples considered so far, the computations of the expectation values $\langle W(L)\rangle\big|_M$ of equation (5.2), which have been obtained by means of the non-perturbative path-integral formalism based on the Deligne-Beilinson cohomology, are in agreement with expression (5.1).

Assuming that equations (5.1) and (5.2) are equivalent, one finds that for certain manifolds and for particular integer values of the coupling constant k, the path-integral formula (5.2) is not well defined. For instance, in the case $M = RP^3$ the denominator of the ratio (5.1) is vanishing when k is an odd integer and then expression (5.1) is not well defined. In the functional integral approach, the constraint on the values of k when $M = RP^3$ has been discussed in [**30**].

Equation (5.1) implies that [**37**] :

- the set of expectation values of the $U(1)$ Chern-Simons theory in S^3 and in any homology sphere M_0 coincide

(5.3) $$\left\{ \langle W(L)\rangle\Big|_{M_0} \right\} = \left\{ \langle W(L)\rangle\Big|_{S^3} \right\} .$$

- For any coloured, oriented and framed link $L \subset M$, one can introduce a new link $\widetilde{L} \subset M$ (called the simplicial satellite of L) which is a satellite of L and which is obtained from L by replacing each link component of colour q with $|q|$ parallel copies of the same component, each copy with unitary colour. If the simplicial satellite $\widetilde{L}$ of the link L in a generic manifold M is homologically trivial mod $2k$, then there exists a link L' in S^3 such that

(5.4) $$\langle W(L)\rangle\Big|_M = \langle W(L')\rangle\Big|_{S^3} .$$

By means of the Reshetikhin-Turaev surgery rules, for fixed integer k, one can also define a 3-manifold invariant

(5.5) $$I_k(M) = I_k(M_{\mathcal{L}}) = (2k)^{-N_{\mathcal{L}}/2}\; e^{i\pi\sigma(\mathcal{L})/4} \langle W(\mathcal{L})\rangle\Big|_{S^3} ,$$

where $N_{\mathcal{L}}$ denotes the number of components of $\mathcal{L}$ and $\sigma(\mathcal{L})$ represents the so-called signature of the linking matrix associated with $\mathcal{L}$, i.e. $\sigma(\mathcal{L}) = n_+ - n_-$ where $n_\pm$ is the number of positive/negative eigenvalues of the linking matrix which is defined by the framed link $\mathcal{L}$. Some properties of $I_k(M_{\mathcal{L}})$ (and of its generalizations) have been studied, for instance, in [**38, 39, 40**].

If M_0 is a homology 3-sphere, then [**37**] one finds $I_k(M_0) = 1$. One could suspect that the invariant $I_k(M)$ only depends on the homology group $H_1(M)$; the following counterexamples show that this is not the case. The lens spaces $L(5,1)$

and $L(5,2)$ are not homeomorphic but they have the same homology group $\mathbb{Z}_5$; equation (5.5) gives

$$I_2(L(5,1)) = -1 \quad , \quad I_2(L(5,2)) = 1 \; . \tag{5.6}$$

Similarly, the manifolds $L(9,1)$ and $L(9,2)$ are not homeomorphic; they have the same homology group $\mathbb{Z}_9$ and are of the same homotopy type. One finds

$$I_3(L(9,1)) = i\sqrt{3} \quad , \quad I_3(L(9,2)) = -i\sqrt{3} \; . \tag{5.7}$$

6. Partition function

In order to complete the answer to the question formulated in the Introduction, let us recall that, in quantum field theory, any well defined functional integration takes really the form of a ratio of functional integrations. Therefore one can imagine that the "suitably normalized" partition function of the Chern-Simons theory formulated in the closed 3-manifold M,

$$R_{N_0}(M) = \frac{\int_M DA \; e^{iS_{CS}}}{N_0} \; , \tag{6.1}$$

should correspond to the Reshetikhin-Turaev surgery invariant for the manifold M. Equation (6.1) should be interpreted as the result of some limit prescription for the ratio of two regularized functional integrals, as indicated in equation (2.5). So, N_0 stands for an appropriate path-integral that introduces a reference point for the integration. Presumably, N_0 is not unique; the specific choice of N_0 is precisely the crucial point that will make expression (6.1) well defined.

Several variants of the Reshetikhin-Turaev surgery invariant have been introduced in literature and have been used to obtain well defined results; but all these combinatorial invariants are not really based on a functional integration. We hope that, in the future, a true functional integral derivation of an explicit and well defined expression of a 3-manifold invariant will be produced.

Acknowledgments. I wish to thank F. Thuillier for useful discussions.

References

[1] V. W. Hughes and T. Kinoshita, Rev. of Mod. Phys. **71** (1999), S133; B. Odom, D. Hanneke, B. D'Urso and G. Gabrielse, Phys. Rev. Lett. **97**, 030801 (2006); G. Gabrielse, D. Hanneke, T. Kinoscita, M. Nio and B. Odom, Phys. Rev. Lett. **97**, 030802 (2006).

[2] N. N. Bogolioubov and D. V. Chirkov, *Introduction a la théorie quantique des champs*, Dunod Editeur (Paris, 1960).

[3] C. Itzykson and J-B. Zuber, *Quantum Field Theory*, McGraw-Hill (Singapore, 1980).

[4] J. Schwinger, Phys. Rev. **82** (1951), 664.

[5] E. Witten, Commun. Math. Phys. **121** (1989), 351.

[6] E. Guadagnini, M. Martellini, and M. Mintchev, Phys. Lett. **B227** (1989), 111.

[7] F. Delduc, O. Piguet, C. Lucchesi, and S.P. Sorella, Nucl. Phys. B **346** (1990), 313.

[8] A. C. Hirshfeld and U. Sassenberg, Journal of Knot Theory and its Ramifications, **5** (1996), 805.

[9] V. F. R. Jones, Ann. Math. **126** (1987), 335.

[10] P. Freyd, D. Yetter, J. Hoste, W. B. R. Lickorish, K. Millett and A. Ocneanu, Bull. Amer. Math. Soc. **12** (1985), 239; J. H. Przytycki and P. Traczyk, Kobe J. Math. **4** (1987), 115.

[11] V. G. Turaev, Invent. Math. **92** (1988), 527.

[12] L. H. Kauffman, Topology **26** (1987), 395.

[13] V. G. Knizhnik and A. B. Zamolodchikov, Nucl. Phys. B **247** (1984), 83.

[14] T. Kohno, Ann. Inst. Fourier, Grenoble, **37** (1987), 139.

[15] A. Tsuchiya and Y. Canie, Adv. Stud. Pure Math. **16** (1988), 297.

[16] E. Guadagnini and F. Thuillier, SIGMA **4** (2008), 078, arXiv:0801.1445.
[17] D. H. Adams, Phys. Rev. Lett. **78** (1997), 4155.
[18] S. Albeverio and J. Schäfer, J. Math. Phys. **36** (1995), 2157.
[19] A. S. Schwarz, Lett. Math. Phys. **2** (1978), 247.
[20] ———, Commun. Math. Phys. **67** (1979), 1.
[21] C. R. Hagen, Ann. Phys. **157** (1984), 342.
[22] M. Bauer, G. Girardi, R. Stora, and F. Thuillier, JHEP **0508** (2005), 027.
[23] P. Deligne, *Théorie de Hodge II*, Publ. Math. I.H.E.S. **40** (1971), 5.
[24] A. A. Beilinson, J. Soviet Math. **30** (1985), 2036.
[25] K. Gomi, J. Math. Sci. Univ. Tokyo **8** (2001), 223242.
[26] R. Dijkgraaf and E. Witten, Commun. Math. Phys. **129** (1990), 393.
[27] J. Cheeger and J. Simons, *Differential characters and geometric invariants*, Stony Brook Preprint 1973; reprinted in Lecture Notes in Mathematics 1167, Geometry and Topology Proc. 1983-84, Eds J. Alexander and J. Harer, Springer 1985.
[28] J. L. Brylinski, *Loop spaces, characteristic classes and geometric quantization*, Progress in Mathematics 107 (Boston, 1993).
[29] R. Harvey, B. Lawson, and J. Zweck, Amer. J. Math. **125** (2003), 791.
[30] F. Thuillier, J. Math. Phys. **50** (2009), 122301; arXiv:0901.2485.
[31] N. Y. Reshetikhin and and V. G. Turaev, Commun. Math. Phys. **127** (1990), 1; and Invent. Math. **103** (1991), 547.
[32] T. Kohno, Topology **31** (1992), 203.
[33] W. B. R. Lickorish, Pacific J. Math. **149** (1991), 337.
[34] H. R. Morton and P. M. Strickland, *Satellites and Surgery Invariants*, in Knots 90, ed. A. Kawauchi, Walter de Gruyter, Berlin, 1992.
[35] D. Rolfsen, *Knots and Links*, AMS Chelsea Publishing, Providence, Rhode Island, 2003.
[36] R. Kirby, Invent. Math. **45** (1978), 35.
[37] E. Guadagnini and F. Mancarella, J. Math. Phys. **51** (2010), 06230; arXiv:1004.5211v1.
[38] H. Murakami, T. Ohtsuki, and M. Okada, J. Math. **29** (1992), 545.
[39] F. Deloup, C. R. Acad. Sci. Paris, t. 326, Série i (1998), 69.
[40] S. K. Hansen, J. K. Slingerland, and P. R. Turner, Adv. Theor. Math. Phys. **9** (2005), 321.

PHYSICS DEPARTMENT OF PISA UNIVERSITY, AND INFN SEZIONE DI PISA
Current address: Dipartimento di Fisica, Largo B. Pontecorvo, 3, 56127 Pisa, Italy
E-mail address: guada@df.unipi.it

AMS/IP Studies in Advanced Mathematics
Volume 50, 2011

Chern-Simons Invariants, SO(3) Instantons, and $\mathbb{Z}/2$ Homology Cobordism

Matthew Hedden and Paul Kirk

Abstract. We review the $SO(3)$ instanton gauge theory of Fintushel and Stern and recast it in the context of 4-manifolds with cylindrical ends. Applications to the $\mathbb{Z}/2$ homology cobordism group of $\mathbb{Z}/2$ homology 3-spheres are given.

1. Introduction

Among the many results of instanton gauge theory, one, due to Furuta, stands out because no alternative proof has been found despite enormous progress in gauge theory made through the Seiberg-Witten and Ozsváth-Szabó theories.

Theorem (**Furuta** [**16**], see also [**14**]). *The Seifert fibered integer homology 3-spheres* $\Sigma(p,q,pqk-1), k=1,2,\cdots\infty$ *are linearly independent in the homology cobordism group,* Θ^3_H.

Furuta proved this theorem using the machinery of instantons on pseudofree orbifolds, as developed by Fintushel and Stern [**13, 14**]. In light of the (perhaps imprecise) expectation that the geometric information contained in Donaldson theory should coincide with that of the more recent approaches to gauge theory, the difficulty in proving this theorem in the context of Seiberg-Witten or Ozsváth-Szabó theory is quite mysterious, and motivates the present work.

The aim of this article is to revisit the technique used in the proof of Furuta's result and to recast it in the light of advances in the theory of instantons on manifolds with cylindrical ends. It is our hope that understanding the result in this context will increase its power as a tool for studying smooth cobordism or knot concordance, and will perhaps shed light on how one could extract similar information from the modern invariants.

Before describing some of the nuances involved in the cylindrical end reformulation, we present the following application afforded by the present approach (see Theorem 3.3 for a stronger statement and Section 4 for further applications).

2010 *Mathematics Subject Classification.* Primary 57M25.

Key words and phrases. Chern-Simons invariants, instantons, rational homology spheres.

This work was supported in part by the National Science Foundation under grants 0604310, 0706979, 0906258, and 1007196.

Theorem 1. *Suppose that p, q, d are relatively prime positive odd integers. For k large enough, the rational homology spheres obtained by $\frac{d}{d^k-1}$ surgeries on the right-handed (p, q) torus knot are linearly independent in the $\mathbb{Z}/2$ homology cobordism group $\Theta^3_{\mathbb{Z}/2}$.*

As mentioned, the mechanism underlying Furuta's argument relied on the use of orbifold connections and the machinery of instantons on manifolds with cylindrical ends, introduced in Taubes's ground-breaking article [**33**]. Here, we reformulate the entire argument in terms of gauge theory on 4-manifolds with cylindrical ends in the manner pioneered by Taubes [**33**] and Floer [**15**]. This yields additional flexibility, in particular providing cobordism obstructions outside of the realm of pseudofree orbifolds. In addition, however, our treatment allows us to carefully address several technical points. In the present context we are forced to consider the L^2 moduli space of instantons when addressing the issue of compactness, and hence must appeal to the results of [**28**]. Here, the Chern-Simons invariants of flat connections provide a lower bound on the quanta of energy that can escape in a non-convergent sequence of instantons.

It is perhaps implicit (if not explicitly stated) in [**16**] and [**14**] that the boundary 3-manifolds should not admit degenerate flat $SO(3)$ connections, and in any case this is automatic when the boundary manifolds are lens spaces or Seifert fibered homology spheres $\Sigma(p, q, r)$ with *three* singular fibers. But for more general 3-manifolds extra care with hypotheses is needed, and we hope our exposition provides this. In addition, new wrinkles arise in the enumeration of reducible instantons (see Theorem 2.16). These and other points are presented in a context that eschews orbifolds in favor of manifolds with cylindrical ends. This material can also be considered as a generalization of the results of Ruberman and Matić, [**27, 32**], which correspond to the case $p_1(E, \alpha) = 0$.

In a companion article [**18**], we will use this machinery to show that a certain infinite family of untwisted, positive-clasped Whitehead doubles of torus knots is linearly independent in the smooth concordance group. We anticipate many more applications to the study of concordance and cobordism groups.

Finally, we remark that the general argument is reminiscent of techniques that use Casson-Gordon invariants to establish linear independence in the *topological* knot concordance group [**19**], and this intriguing similarity provides motivation for further study. Roughly speaking, in the context of Casson-Gordon invariants arguments focus on one prime p at a time and corresponding D_{2p} (dihedral) representations. Knots whose branched covers have no p-torsion in their homology contribute nothing to the obstruction to linear independence. In the Furuta/Fintushel-Stern argument, one focuses on the smallest Chern-Simons invariant of flat $SO(3)$ connections. Homology spheres whose smallest Chern-Simons invariant exceed this minimum contribute nothing to the obstruction to linear independence.

Acknowledgments. It is our pleasure to thank Tom Mrowka, Charles Livingston, and Ron Fintushel for many interesting conversations. The second author thanks the organizers of the inspiring *Chern-Simons Gauge Theory: 20 years after* conference.

2. Adapted bundles and instantons

2.1. Adapted bundles. We begin by reviewing Donaldson's description [**9**] of gauge theory on adapted bundles over 4-manifolds with cylindrical ends, in the context of $SO(3)$ bundles. Given a flat $SO(3)$ connection α on a 3-manifold Y, denote by $H^*_\alpha(Y)$ the cohomology of Y with coefficients in the corresponding flat $\mathbb{R}^3$ bundle. The flat connection α is called *non-degenerate* if $H^1_\alpha(Y) = 0$. If Y is a rational homology sphere, then the trivial connection is non-degenerate.

Consider a compact oriented 4-manifold X with boundary $\partial X = Y = \sqcup_{i=1}^c Y_i$, a disjoint union of closed 3-manifolds. Endow X with a Riemannian metric which is isometric to $[-1, 0] \times Y$ in a collar neighborhood of the boundary. Throughout the rest of this section, we make the following assumption.

The 4-manifold X is path connected, satisfies $H^1(X; \mathbb{Z}/2) = 0$, and has non-empty boundary $Y = \sqcup_{i=1}^c Y_i$ which is a disjoint union of rational homology 3-spheres.

Form the non-compact manifold $X_\infty = X \cup_Y \big([0,\infty)\times Y\big)$ by adding an infinitely long collar to each boundary component. Choose a Riemannian metric on X_∞ whose restriction to each cylinder is the product metric. The submanifold $[0,\infty)\times Y_i$ will be called an *end* of X_∞.

Definition 2.1. An *adapted bundle* (E, α) *over* X is an $SO(3)$ vector bundle $E \to X_\infty$, together with a fixed flat connection α_i on each end $[0,\infty) \times Y_i$. We use α as shorthand for the set $\{\alpha_i\}$. Two adapted bundles (E, α) and (E', α') are called *equivalent* if there is a bundle isomorphism from E to E' identifying the flat connections α and α' over $[r, \infty) \times Y$ for some $r \geq 0$.

Any adapted bundle (E, α) is equivalent to one in which the flat connection α_i on each end $[0, \infty) \times Y_i$ is in *cylindrical form*, that is $\alpha_i = \pi^*(\tilde\alpha_i)$ where $\tilde\alpha_i$ is a flat connection on Y_i and $\pi : [0,\infty) \times Y_i \to Y_i$ is the projection to the second factor. Indeed, since α is flat, there always exists a gauge transformation $g : E \to E$ which equals the identity on the interior of X so that $g^*(\alpha)$ has this form, obtained by parallel transport along rays $[0,\infty) \times \{y\}$. We will tacitly assume that each α_i is in cylindrical form whenever convenient, and use the same notation α_i for the restriction of this flat connection to Y_i.

Since X has non-empty boundary, $SO(3)$ vector bundles over X_∞ are isomorphic if and only if their second Stiefel-Whitney classes are equal. In particular, if (E, α) and (E', α') are equivalent adapted bundles over X_∞, then $w_2(E) = w_2(E')$.

2.2. The Pontryagin charge, instantons, and Chern-Simons invariants. Given any $SO(3)$ connection A on a bundle E over a (not necessarily compact) 4–manifold, Z, let $F(A)$ denote its curvature 2-form. Define the *Pontryagin charge of* A to be the real number

$$p_1(A) = -\frac{1}{8\pi^2}\int_Z \mathrm{Tr}(F(A) \wedge F(A)), \tag{2.1}$$

provided this integral converges. When Z is closed, $p_1(A) = \langle p_1(E), [Z]\rangle \in \mathbb{Z}$.

Definition 2.2. An $SO(3)$ connection A on a Riemannian manifold Z is called an *instanton* if its curvature form satisfies the equation

$$F(A) = - * F(A).$$

If A is an instanton, then

$$p_1(A) = \frac{1}{8\pi^2}\int_Z \mathrm{Tr}(F(A)\wedge *F(A)) = \frac{1}{8\pi^2}\|F(A)\|^2_{L^2}$$

whence $p_1(A)$ is defined and non-negative for any instanton A whose curvature has finite L^2 norm.

Given an adapted bundle (E,α) over X_∞, choose an $SO(3)$ connection A_0 on E which extends the given flat connection α on the union of the ends. Then its curvature $F(A_0)$ is supported on the compact submanifold X, and so $p_1(A_0)\in\mathbb{R}$ is defined. Given any other extension A_1 of α, then Chern-Weil theory shows that $\mathrm{Tr}(F(A_1)\wedge F(A_1)) = \mathrm{Tr}(F(A_0)\wedge F(A_0)) + d\gamma$ where

$$\gamma = -\int_0^1 \mathrm{Tr}\left((A_0 - A_1)\wedge 2F(tA_0 + (1-t)A_1)\right)dt.$$

See, for example, [**36**, Lemma 3.3]. Since $A_0 - A_1$ vanishes on the collar, Stokes' theorem implies that $p_1(A_0) = p_1(A_1)$, so we denote this quantity by $p_1(E,\alpha)$.

DEFINITION 2.3. Given a pair α_0, α_1 of $SO(3)$ connections on the same bundle E over a closed 3-manifold Y, define the *relative Chern-Simons invariant*, $\mathrm{cs}(Y,\alpha_0,\alpha_1)$
$\in\mathbb{R}$, by

$$\mathrm{cs}(Y,\alpha_0,\alpha_1) = p_1(\mathbb{R}\times E,\alpha),$$

where α equals α_0 on $(-\infty,0]\times Y$ and α_1 on $[1,\infty)\times Y$.

DEFINITION 2.4. Given a connection α on an $SO(3)$ bundle E over a closed 3-manifold Y, the bundle extends to a bundle with connection A over some 4-manifold Z with boundary Y, since $H_3(BSO(3)) = 0$. Define the *Chern-Simons invariant of α modulo $\mathbb{Z}$*, $\mathrm{cs}(Y,\alpha)\in\mathbb{R}/\mathbb{Z}$ by

$$\mathrm{cs}(Y,\alpha) = p_1(A) \bmod \mathbb{Z}.$$

Since the integral of the first Pontryagin form of a connection over a *closed* 4-manifold is an integer, it follows that $\mathrm{cs}(Y,\alpha)$ is well defined in $\mathbb{R}/\mathbb{Z}$.

These two invariants are related by

$$\mathrm{cs}(Y,\alpha_0,\alpha_1) = \mathrm{cs}(Y,\alpha_1) - \mathrm{cs}(Y,\alpha_0) \bmod \mathbb{Z}$$

whenever the left side is defined.

If α extends to a *flat* connection A over some 4-manifold then $p_1(A) = 0$, and hence the Chern-Simons invariant modulo $\mathbb{Z}$ is a flat cobordism invariant. In particular, if $E\to Y$ is the trivial bundle and α a flat connection on E, then $\mathrm{cs}(Y,\alpha)$ has a canonical lift to $\mathbb{R}$, namely $\mathrm{cs}(Y,\theta,\alpha)$, where θ denotes the trivial connection with respect to the given trivialization of E.

Bundle automorphisms $h: E\to E$ covering the identity which preserve the $SO(3)$ structure group are called *gauge transformations.* The group $\mathcal{G}(E)$ of gauge transformations acts on the space of $SO(3)$ connections $\mathcal{A}(E)$ by pullback.

If $g: E\to E$ is a gauge transformation, then $\mathrm{cs}(Y,\alpha,g^*(\alpha))$ equals $p_1(\tilde{E})$, where $\tilde{E}$ is the bundle over $Y\times S^1$ obtained by taking the mapping torus of g. Since $Y\times S^1$ is closed, Chern-Weil theory implies that $p_1(\tilde{E})$ is an integer, and so we conclude that $\mathrm{cs}(Y,\alpha,g^*(\alpha))\in\mathbb{Z}$. Notice that if g is homotopic to the identity gauge transformation, then the bundle $\tilde{E}$ is isomorphic to the pullback of E over Y via the projection $\pi: Y\times S^1\to Y$, and hence $\mathrm{cs}(Y,\alpha,g^*(\alpha)) = p_1(\pi^*(E)) = 0$.

A similar argument shows that the integer $\mathrm{cs}(Y,\alpha,g^*(\alpha))$ depends only on the path component of g in $\mathcal{G}(E)$. It follows easily that the reduction modulo $\mathbb{Z}$ of $\mathrm{cs}(Y,\alpha_0,\alpha_1)$ depends only on the orbits of α_0 and α_1 under the action of $\mathcal{G}(E)$. Similarly, $\mathrm{cs}(Y,\alpha)\in\mathbb{R}/\mathbb{Z}$ depends only on the orbit of α.

Dold and Whitney proved that $SO(3)$ bundles over 4-complexes are determined up to isomorphism by their second Stiefel-Whitney class w_2 and their first Pontryagin class p_1 [**7**]. Moreover, these are related by $p_1\equiv\mathcal{P}(w_2)$ mod 4, where $\mathcal{P}:H^2(-;\mathbb{Z}/2)\to H^4(-;\mathbb{Z}/4)$ denotes the Pontryagin square, a cohomology operation which lifts the cup square $H^2(-;\mathbb{Z}/2)\to H^4(-;\mathbb{Z}/2)$, $y\mapsto y^2$.

As a consequence, if Y is a $\mathbb{Z}/2$ homology 3-sphere, then $H^2(Y\times S^1;\mathbb{Z}/2)=0$, and so $\mathrm{cs}(Y,\alpha,g^*(\alpha))\in 4\mathbb{Z}$ for any gauge transformation g.

2.3. The moduli space. Let (E,α) be an adapted bundle over X_∞. The elliptic operators $*d_{\alpha_i}-d_{\alpha_i}*$ acting on even degree $E|_{Y_i}$–valued forms on Y_i are self-adjoint with discrete spectrum. Choose (and fix for the remainder of this section) a $\delta>0$ smaller than the absolute value of any non-zero eigenvalue of any of the $*d_{\alpha_i}-d_{\alpha_i}*$. Let $f:X_\infty\to[0,\infty)$ be a smooth function which equals 0 on X and $t\delta$ on $\{t\}\times Y_i$ for $t>1$. Then define the weighted Sobolev spaces $L^{p,\delta}_n$ of sections of bundles over X_∞ as the completion of the space of compactly supported sections with respect to the norm $\|\phi\|_{L^{p,\delta}_n}:=\|e^f\phi\|_{L^p_n}$.

Fix an extension A_0 of α to E. Identify E with $so(E)$ via the identification $\mathbb{R}^3=so(3)$ of $SO(3)$ representations, and let $SO(E)$ denote the $SO(3)$ bundle automorphisms. Connections on E take the form A_0+a for a 1-form a with values in E. Following [**9**, Chapter 4], define

$$\mathcal{A}^\delta(E,\alpha)=\{A_0+a\mid a\in L^{3,\delta}_1(\Omega^1_{X_\infty}(E))\}$$

and

$$\mathcal{G}^\delta(E,\alpha)=\{g\in SO(E)\mid \nabla_{A_0}g\in L^{3,\delta}_1(T^*X_\infty\otimes\mathrm{End}(E))\}$$

(where ∇_{A_0} denotes the covariant derivative associated to A_0). Then pulling back connections extends to an action of the completions $\mathcal{G}^\delta(E,\alpha)$ on $\mathcal{A}^\delta(E,\alpha)$. The *moduli space* $\mathcal{M}(E,\alpha)$ *of instantons on* (E,α) is defined to be

$$\mathcal{M}(E,\alpha)=\{A\in\mathcal{A}^\delta(E,\alpha)\mid F(A)=-*F(A)\}/\mathcal{G}^\delta(E,\alpha).$$

The idea underlying these definitions is that $\mathcal{A}^\delta(E,\alpha)$ consists of connections that limit exponentially along the ends of X_∞ to α, and $\mathcal{G}^\delta(E,\alpha)$ consists of gauge transformations that limit exponentially along the ends of X_∞ to a gauge transformation that stabilizes α.

DEFINITION 2.5. A *reducible connection* on (E,α) is a connection A in $\mathcal{A}^\delta(E,\alpha)$ whose stabilizer $\Gamma_A\subset\mathcal{G}^\delta$ is non-trivial. Let $\mathcal{M}(E,\alpha)_{red}\subset\mathcal{M}(E,\alpha)$ denote the subset of gauge equivalence classes of reducible instantons. Its complement $\mathcal{M}(E,\alpha)\setminus\mathcal{M}(E,\alpha)_{red}$ is denoted $\mathcal{M}(E,\alpha)^*$.

Restricting a gauge transformation to the fiber of E over a base point in X embeds the stabilizer Γ_A in $SO(3)$ and identifies it with the centralizer of the holonomy group of A at the base point. In particular, the subgroup $\mathcal{G}^\delta_0(E,\alpha)\subset\mathcal{G}^\delta(E,\alpha)$ of gauge transformations which restrict to the identity in the fiber of E over the base point of X acts freely on $\mathcal{A}^\delta(E,\alpha)$, and the quotient $SO(3)\cong\mathcal{G}^\delta(E,\alpha)/\mathcal{G}^\delta_0(E,\alpha)$ acts on $\mathcal{A}^\delta(E,\alpha)/\mathcal{G}^\delta_0(E,\alpha)$ with stabilizers Γ_A. The restriction of this action to the irreducible instantons defines a principal $SO(3)$ bundle $\mathcal{E} \to \mathcal{M}(E,\alpha)^*$ called the *base point fibration* [**13**, Section 9].

2.4. Calculation of the index. Define $\mathrm{Ind}^+(E,\alpha)$ to be the Fredholm index of the operator

$$S^\delta_{A_0} = d^*_{A_0} \oplus d^+_{A_0} : L^{3,\delta}_1(\Omega^1_{X_\infty}(E)) \to L^{3,\delta}(\Omega^0_{X_\infty}(E) \oplus \Omega^+_{X_\infty}(E))$$

where Ω^+ denotes the self-dual 2-forms: $\omega = *\omega$. The Atiyah-Patodi-Singer theorem can be used to calculate this index. Let $\eta(Y,\alpha)$ denote the Atiyah-Patodi-Singer spectral invariant $\eta(B_\alpha, 0)$ of the operator on Y:

$$B_\alpha : \Omega^0_Y(E|_Y) \oplus \Omega^1_Y(E|_Y) \to \Omega^0_Y(E|_Y) \oplus \Omega^1_Y(E|_Y),$$
$$B_\alpha(\phi_0,\phi_1) = (-d^*_\alpha\phi_1, d_\alpha\phi_0 + *d_\alpha\phi_1),$$

and let

$$\rho(Y,\alpha) = \eta(Y,\alpha) - 3\eta(Y),$$

denote the Atiyah-Patodi-Singer ρ invariant of (Y,α). (In general, for a connection γ on a bundle $V \to Y$, $\rho(Y,\gamma) = \eta(Y,\gamma) - \dim(V)\cdot\eta(Y,\theta)$, where θ is the trivial connection on the trivial 1-dimensional line bundle.) The real number $\rho(Y,\alpha)$ depends only on the gauge equivalence class of α.

In general, the kernel of B_{α_i} equals the direct sum $\mathcal{H}^0_{\alpha_i}(Y_i) \oplus \mathcal{H}^1_{\alpha_i}(Y_i)$ of the d_{α_i}-harmonic 0- and 1-forms on Y_i with coefficients in the flat bundle supporting α_i (see [**9**, Section 2.5.4]). The assumption that each flat connection α_i is non-degenerate and the Hodge theorem imply that $\ker B_\alpha \cong \oplus_{i=1}^c H^0_{\alpha_i}(Y_i)$. Setting $h_\alpha = \dim H^0_\alpha(Y)$ and $h_{\alpha_i} = \dim H^0_\alpha(Y_i)$ we conclude that $h_\alpha = \sum h_{\alpha_i}$. Note that $\rho(Y_i,\alpha_i) = 0$ and $h_{\alpha_i} = 3$ when α_i has trivial holonomy, as such a connection is gauge equivalent to the trivial connection.

The discussion following the proof of Proposition 3.15 of [**2**] shows that the index $\mathrm{Ind}^+(E,\alpha)$ is equal to the Atiyah-Patodi-Singer index $\mathrm{Ind}\, S^{APS}_{A_0}$ of the corresponding operator on (the compact manifold) X with Atiyah-Patodi-Singer boundary conditions. This uses the non-degeneracy of the α_i to conclude that L^2 solutions of $S_{A_0}\phi = 0$ decay along the ends faster than $e^{-t\delta}$. Thus $\mathrm{Ind}^+(E,\alpha)$ can be computed using the Atiyah-Patodi-Singer theorem. For the convenience of the reader we outline the calculation; similar calculations can be found in the literature; for example it is a consequence of (the more general) [**28**, Proposition 8.4.1]. Let $b^+(X)$ denote the dimension of a maximal positive definite subspace of $H^2(X;\mathbb{R})$ with respect to the intersection form (see Section 2.6 below for the definition of the intersection form in this context).

PROPOSITION 2.6. *Let (E,α) be an adapted bundle over a path connected 4-manifold X with $c \geq 1$ boundary components. Assume further that the flat connection α_i on each Y_i is non-degenerate, and that $H^1(X;\mathbb{Q}) = 0 = H^1(Y;\mathbb{Q})$. Then*

$$\mathrm{Ind}^+(E,\alpha) = 2p_1(E,\alpha) - 3(1+b^+(X)) + \tfrac{1}{2}\sum_{\alpha_i\,\text{nontrivial}} (3 - h_{\alpha_i} - \rho(Y_i,\alpha_i)).$$

PROOF. We use the Atiyah-Patodi-Singer index theorem [**2**] twice, once on the bundle E with operator

$$S^{APS}_{A_0} = d^*_{A_0} \oplus d^+_{A_0} : L^2_1(\Omega^1_X(E); P^+_\alpha) \to L^2(\Omega^0_X(E) \oplus \Omega^+_X(E))$$

with index

$$\operatorname{Ind} S^{APS}_{A_0} = \int_X f_{X,E}(R_X, F(A_0)) - \tfrac{1}{2}(h_\alpha + \eta(Y,\alpha)),$$

and again on the trivial 1-dimensional $\mathbb{R}$ bundle with trivial connection

$$S^{APS} = d^* \oplus d^+ : L^2_1(\Omega^1_X; P^+) \to L^2(\Omega^0_X \oplus \Omega^+_X)$$

with index

$$\operatorname{Ind} S^{APS} = \int_X f_{X,\epsilon}(R_X, F(\theta)) - \tfrac{1}{2}(h + \eta(Y)).$$

In these formulas P^+_α denotes the spectral projection onto the non-negative eigenspace of B_α, and $\eta(Y_i, \alpha_i)$ and h_α, are as above. The corresponding unadorned symbols refer to their analogues on the trivial, untwisted $\mathbb{R}$ bundle ϵ. Since Y_i is a rational homology sphere, $h = c$.

The (inhomogeneous) differential form $f_{X,E}(R_X, F(A))$ is a function of the Riemannian curvature R_X and the curvature $F(A)$ of a connection A. It equals

$$\big(\dim E - \tfrac{1}{8\pi^2}\operatorname{Tr}(F(A)\wedge F(A))\big)\big(2 + \tfrac{1}{2}(\chi(R_X) + \sigma(R_X))\big),$$

where $\chi(R_X)$ is the Euler form and $\sigma(R_X)$ is the Hirzebruch signature form. Hence

$$\int_X f_{X,E}(R_X, F(A_0)) = 2p_1(A_0) + 3\int_X f_{X,\epsilon}(R_X, F(\theta)).$$

By convention, the integral of an inhomogeneous form over an n-manifold is defined to be the integral of its n-dimensional component.

Since $\operatorname{Ind}^+(E,\alpha)$ is equal to $\operatorname{Ind} S^{APS}_{A_0}$, subtracting $3\operatorname{Ind} S^{APS}$ from $\operatorname{Ind} S^{APS}_{A_0}$ and simplifying yields

$$\operatorname{Ind}^+(E,\alpha) = 2p_1(E,\alpha) + 3\operatorname{Ind} S^{APS} - \tfrac{1}{2}\rho(Y,\alpha) + \tfrac{1}{2}(3c - h_\alpha). \tag{2.2}$$

To obtain the desired formula, we must show $\operatorname{Ind} S^{APS} = -1 - b^+(X)$. The kernel of S^{APS} consists of L^2 harmonic 1-forms on X_∞. Proposition 4.9 of [**2**] identifies this space with the image $H^1(X,Y;\mathbb{R}) \to H^1(X;\mathbb{R})$. The hypothesis that $H^1(X;\mathbb{Q}) = 0$ (which holds, in any event, by our standing assumption that $H^1(X;\mathbb{Z}/2) = 0$) implies that $H^1(X;\mathbb{R}) = 0$, and so $\ker S^{APS} = 0$.

We now identify the cokernel of S^{APS}. The cokernel is isomorphic to the kernel of the adjoint which, in turn, is isomorphic to the extended L^2 solutions to $S^*\phi = 0$, i.e., the space of pairs $\phi = (\phi_0, \phi_+) \in \Omega^0_{X_\infty} \oplus \Omega^+_{X_\infty}$ satisfying $d\phi_0 + d^*\phi_+ = 0$ and for which $(\phi_0, \phi_+)|_{\{t\}\times Y} - \gamma$ decays exponentially on the collar for some harmonic form $\gamma = (\gamma_0, \gamma_1) \in \ker B$ on Y [**2**]. Since Y is a rational homology sphere, $\ker B \cong H^0(Y)$. It follows that $\gamma = (\gamma_0, 0)$ where γ_0 is a harmonic 0-form, that is, a locally constant function on Y. Thus $\phi_0|_{\{t\}\times Y} - \gamma_0$ and $\phi_+|_{\{t\}\times Y}$ are exponentially decaying forms.

Let $X(t) = X \cup ([0,t]\times Y)$. Then

$$\langle d\phi_0, d^*\phi_+\rangle_{L^2(X(t))} = \pm\langle d\phi_0|_{\{t\}\times Y}, \phi_+|_{\{t\}\times Y}\rangle_{L^2(\{t\}\times Y)} \xrightarrow{t\to\infty} 0$$

This implies that $d\phi_0$ and $d^*\phi_+$ are L^2-orthogonal. Therefore, $d\phi_0 + d^*\phi_+ = 0$ implies that ϕ_0 is a harmonic 0–form (that is, a constant function) and ϕ_+ is an exponentially decaying harmonic self-dual 2-form. Since X is connected we conclude $\operatorname{coker} S^{APS} = 1 + b^+(X)$, and hence $\operatorname{Ind} S^{APS} = -1 - b^+(X)$.

If α_i has trivial holonomy, then $\rho(Y, \alpha_i) = 0$ and $h_{\alpha_i} = \dim H^0_{\alpha_i}(Y) = 3$. Therefore,

$$(2.3) \quad -\rho(Y,\alpha) + 3c - h_\alpha = \sum_i 3 - h_{\alpha_i} - \rho(Y_i, \alpha_i) = \sum_{\alpha_i \text{nontrivial}} 3 - h_{\alpha_i} - \rho(Y_i, \alpha_i)$$

Combining Equations (2.2) and (2.3) yields the result. □

Suppose that one of the flat connections, say α_1, on the boundary component Y_1 is trivial with respect to some trivialization of E over Y_1. Suppose further that Z is a compact, negative definite 4–manifold with boundary $-Y_1 \sqcup_j W_j$, with the trivialized bundle $Z \times \mathbb{R}^3 \to Z$ over it. Then glue X to Z along Y_1 to produce the manifold $X' = X \cup_{Y_1} Z$. Let E' be the bundle over X' obtained by gluing E to the trivial bundle using the given trivializations over Y_1. Extending the connection A_0 over X' by the trivial connection over Z gives a new adapted bundle (E', α') over X'_∞. Since $p_1(E, \alpha) = p_1(E', \alpha')$, Proposition 2.6 then shows that $\operatorname{Ind}^+(E, \alpha) = \operatorname{Ind}^+(E', \alpha')$.

If $\operatorname{Ind}^+(E, \alpha) \geq 0$, $p_1(E, \alpha) > 0$, and each α_i is non-degenerate, then after varying the Riemannian metric inside X if necessary, $\mathcal{M}(E, \alpha)^*$ is a smooth orientable manifold of dimension $\operatorname{Ind}^+(E, \alpha)$. In the present context this is a consequence of Lemma 8.8.4, Theorem 9.0.1 and Remark 9.2.7 of [**28**]. When consulting [**28**], it is useful to note that because we assume the flat connections α_i are non-degenerate, the appropriate path components of the L^2 moduli spaces and thickened moduli spaces of [**28**] coincide with $\mathcal{M}(E, \alpha)$.

When $p_1(E, \alpha) = 0$, then $\mathcal{M}(E, \alpha)$ is the moduli space of flat connections and hence is unchanged by varying the Riemannian metric on X_∞. In this situation one can instead perturb the equation $F(A) + *F(A) = 0$ to make $\mathcal{M}(E, \alpha)^*$ a smooth orientable manifold of dimension $\operatorname{Ind}^+(E, \alpha)$; see [**17, 9**].

2.5. Compactness. A critical question about $\mathcal{M}(E, \alpha)$ is whether it is compact. To understand this, we define an invariant $\hat{\tau}(Y, \alpha) \in (0, 4]$, essentially the minimal relative $SO(3)$ Chern-Simons invariant over the path components of Y. This invariant provides a sufficient condition to guarantee compactness of $\mathcal{M}(E, \alpha)$ (Proposition 2.9 below).

We first introduce some notation. For any path connected space Z and compact Lie group G, denote by $\chi(Z, G)$ the space

$$\chi(Z, G) = \operatorname{Hom}(\pi_1(Z), G)/\text{conjugation}.$$

This space is a compact real algebraic variety and is analytically isomorphic to the space of gauge equivalence classes of flat G connections [**12**]. To a flat G connection α we associate its holonomy $\operatorname{hol}_\alpha \in \chi(Z, G)$. Notice that $\chi(Z, G)$ is partitioned into disjoint compact subspaces corresponding to the isomorphism classes of G-bundles over Z which support the various flat connections. When Z is not path connected then take $\chi(Z, G)$ to be the product of the $\chi(Z_i, G)$ over the path components Z_i of Z.

We can now define the invariant $\hat{\tau}(Y, \alpha) \in (0, 4]$, which is an adaptation of invariants found in [**16, 14**] and [**28**, Definition 6.3.5] to our context.

For each boundary component Y_i of X, the bundle $E|_{Y_i}$ carries the flat connection α_i and is determined up to isomorphism by the gauge equivalence class of α_i. Hence α_i determines its second Stiefel-Whitney class. Thus the holonomy gives a decomposition into a disjoint union

$$\chi(Y_i, SO(3)) = \bigsqcup_{w \in H^2(Y_i;\mathbb{Z}/2)} \chi(Y_i, SO(3))_w \tag{2.4}$$

If γ is any flat connection on $E|_{Y_i}$ and g any gauge transformation of $E|_{Y_i}$, then $\mathrm{cs}(Y_i, \alpha_i, \gamma) - \mathrm{cs}(Y_i, \alpha_i, g^*(\gamma))$ is an integer, as explained in Section 2.2. This implies that $\mathrm{cs}(Y_i, \alpha_i, -)$ descends to a well-defined $\mathbb{R}/\mathbb{Z}$ valued function on the compact subspace $\chi(Y_i, SO(3))_{w_2(E|_{Y_i})} \subset \chi(Y_i, SO(3))$. This function is locally constant, see e.g. [**21**], and hence takes finitely many values in $\mathbb{R}/\mathbb{Z}$. This in turn implies that the set

$$S(\alpha_i) = \{\mathrm{cs}(Y_i, \alpha_i, \gamma) \bmod 4 \mid \gamma \text{ a flat connection on } E|_{Y_i}\} \subset \mathbb{R}/4\mathbb{Z}$$

is finite, since it is contained in the preimage of a finite set under $\mathbb{R}/4\mathbb{Z} \to \mathbb{R}/\mathbb{Z}$.

At the moment, taking Chern-Simons invariants mod 4 might seem unmotivated, but it occurs for two reasons. First, in the compactness result below, energy can only bubble off at interior points in multiples of 4. Second, if Y_i is a $\mathbb{Z}/2$ homology sphere, then $\mathrm{cs}(Y_i, \alpha_i, \gamma) - \mathrm{cs}(Y_i, \alpha_i, g^*(\gamma)) = \mathrm{cs}(Y_i, g^*(\gamma), \gamma)$ is in fact always four times an integer, as explained in Section 2.2.

DEFINITION 2.7. Let $b : \mathbb{R}/4\mathbb{Z} \to (0, 4]$ be the obvious bijection. Define $\tau(Y_i, \alpha_i) \in (0, 4]$ to be the minimal value which b attains on the set $S(\alpha_i)$. Informally, $\tau(Y_i, \alpha_i)$ is the minimal relative Chern-Simons invariant $\mathrm{cs}(Y_i, \alpha_i, \gamma)$ for $\gamma \in \chi(Y_i, SO(3))_{w_2(E|_{Y_i})}$, taken modulo 4.

Then define

$$\hat{\tau}(Y, \alpha) = \min_{Y_i \subset \partial X} \{\tau(Y_i, \alpha_i)\} \in (0, 4] \tag{2.5}$$

Note that $\hat{\tau}(Y, \alpha)$ depends only on $Y = \sqcup_i Y_i$ and the gauge equivalence classes of the flat connections α_i.

The following lemma will prove useful for estimating $\hat{\tau}(Y, \alpha)$. We omit the proof, which is a simple consequence of the remarks in the paragraph following Definition 2.3.

LEMMA 2.8. *Let $E \to Y$ be an $SO(3)$ vector bundle over a closed 3-manifold and α, γ two flat connections on E. Choose any pair (W_α, E_α) where W_α is a 4-manifold with boundary Y, $E_\alpha \to W_\alpha$ is a vector bundle extending E. Similarly choose (W_γ, E_γ).*

Then $\mathrm{cs}(Y, \alpha, \gamma)$ mod 4, taken in $(0, 4]$, is greater than or equal to the fractional part of $p_1(E_\gamma, \gamma) - p_1(E_\alpha, \alpha)$, taken in $(0, 1]$. In particular, if $p_1(E_\gamma, \gamma)$ is rational with denominator dividing $k \in \mathbb{Z}_{>0}$ for all flat connections γ on E, then $\tau(Y, \alpha) \geq \frac{1}{k}$ for all α. □

As mentioned, $\hat{\tau}(Y, \alpha)$ provides a sufficient condition for compactness of the moduli space.

PROPOSITION 2.9. *Suppose that (E, α) is an adapted bundle with each α_i nondegenerate and $0 \leq p_1(E, \alpha) < \hat{\tau}(Y, \alpha)$. Then $\mathcal{M}(E, \alpha)$ is compact.*

PROOF. This follows from the *convergence with no loss of energy* theorem of Morgan-Mrowka-Ruberman [**28**, Theorem 6.3.3] which says that a sequence of gauge equivalence classes of finite energy instantons on (E, α) have a geometric limit which is the union of an idealized instanton on X_∞ and a union of finite energy instantons on tubes (see also [**33**] and [**15**] for earlier versions of this result). We give an outline and refer the reader to [**28**] for details.

This result states that any sequence of gauge equivalence classes of instantons $\{[A_i]\}_{i=1}^\infty$ in $\mathcal{M}(E,\alpha)$ has a subsequence which converges in the following sense. The limit is a sequence of connections $(A_\infty, B_1, \cdots, B_k)$, where A_∞ is an idealized instanton on an adapted bundle (E', α') over X_∞, and B_j, $j = 1, \cdots, k$ are instantons on the bundle $\mathbb{R} \times E|_{Y_{i_j}}$ over $\mathbb{R} \times Y_{i_j}$. Here $i_j \in \{1, 2, \cdots, c\}$.

More precisely, A_∞ is a smooth instanton away from a finite set of points $\{x_m\}$ in X_∞, and the curvature density $\|F(A_\infty)\|^2$ is the sum of a non-negative smooth function and a Dirac measure supported on these points with weight $32\pi^2 n_m$ for some non-negative integers n_m. The instantons B_j on the cylinders have finite, positive energy: $0 < \|F(B_j)\|^2 < \infty$. In particular none of the B_j are flat.

The connections A_∞ and B_j have compatible boundary values, which means that the formal sums of gauge equivalence classes of flat connections on $Y = \partial X$ are equal

$$\alpha_1' + \cdots + \alpha_c' = \alpha_1 + \cdots + \alpha_c + \sum_{j=1}^{k} \big(\lim_{r\to-\infty} B_j(r) - \lim_{r\to\infty} B_j(r) \big),$$

where $B_j(r) = B_j|_{\{r\}\times Y_{i_j}}$. Finally, no energy is lost, i.e. there is an equality of non-negative numbers

$$p_1(E,\alpha) = p_1(E',\alpha') + \sum_{\{x_m\}} 4n_m + \sum_{j=1}^{k} p_1(B_j). \tag{2.6}$$

Since $p_1(E,\alpha) < 4$ and every term in (2.6) is non-negative, the idealized instanton A_∞ is necessarily an (honest) instanton, that is, each n_m is zero, the set $\{x_m\}$ is empty, the bundle E' equals E, and A_∞ is a smooth instanton on E. Informally, no energy can bubble off at interior points.

Next, suppose that $k > 0$. Then the identification of the limiting flat connections implies that one of the instantons B_j on $\mathbb{R} \times E|_{Y_{i_j}}$ over $\mathbb{R} \times Y_{i_j}$ has left handed limit one of the α_i, that is, $\lim_{r\to-\infty} B_j(r) = \alpha_{i_j}$. Denote by γ the limit $\lim_{r\to\infty} B_j(r)$. Then $p_1(B_j) \equiv cs(Y_{i_j}, \alpha_{i_j}, \gamma) \bmod 4$, and since B_j is not flat, $p_1(B_j) > 0$. Therefore

$$p_1(B_j) \geq \tau(Y_{i_j}, \alpha_{i_j}) \geq \hat{\tau}(Y,\alpha) > p_1(E,\alpha),$$

which is impossible, since every term in (2.6) is non-negative. Therefore $k = 0$. Informally, no energy can escape down the ends of X_∞. Thus $A_\infty \in \mathcal{M}(E,\alpha)$, as desired. □

2.6. Reducible connections. We now discuss reducible adapted bundles and reducible connections. We first remind the reader of the extended intersection form of a 4-manifold whose boundary is a union of rational homology spheres. Start with the pairing

$$H^2(X;\mathbb{Z}) \times H^2(X,Y;\mathbb{Z}) \to \mathbb{Z}, \qquad x \cdot y = (x \cup y) \cap [X,Y].$$

Let d be any positive integer so that $d \cdot H^2(Y;\mathbb{Z}) = 0$. Then this produces a well-defined pairing

$$H^2(X;\mathbb{Z}) \times H^2(X;\mathbb{Z}) \to \tfrac{1}{d}\mathbb{Z} \tag{2.7}$$

by

$$x \cdot y := \tfrac{1}{d}(x \cdot z), \text{ where } z \in H^2(X,Y;\mathbb{Z}) \text{ satisfies } i^*(z) = d \cdot y.$$

This can also be described as the restriction of the composite

$$H^2(X;\mathbb{Z}) \times H^2(X;\mathbb{Z}) \to H^2(X;\mathbb{R}) \times H^2(X;\mathbb{R}) \cong H^2(X,Y;\mathbb{R}) \times H^2(X,Y;\mathbb{R}) \xrightarrow{\cup} H^4(X,Y;\mathbb{R}) \cong \mathbb{R}.$$

Note that $x \cdot y = 0$ if x or y is a torsion class. Thus we say X *has a negative definite intersection form* or more briefly X *is negative definite* if $x \cdot x < 0$ whenever x is not a torsion class. One defines positive definite analogously. Call X *indefinite* if it is neither positive nor negative definite.

Let $b^+(X)$ denote the maximal dimension of any subspace of $H^2(X;\mathbb{R})$ on which the intersection form is positive definite, and similarly define $b^-(X)$. The assumption that X has boundary a union of rational homology 3-spheres implies that $\dim(H^2(X;\mathbb{R})) = b^+(X) + b^-(X)$. Therefore $b^+(X) = 0$ if and only if X is negative definite.

If X is indefinite, then for a generic Riemannian metric the moduli space $\mathcal{M}(E,\alpha)$ is a smooth manifold. However, in contrast to some arguments in gauge theory, the argument of [**14, 16**] makes use of the fact that reducibles cannot be perturbed away when X is negative definite. This technique originates in Donaldson's proof of his celebrated theorem that definite intersection forms of closed smooth 4-manifolds are diagonalizable [**8**]. Therefore we make the following assumption for the remainder of this section:

The intersection form of X is negative definite; that is, $x \cdot x < 0$ unless x is a torsion class.

Near the orbit of a reducible instanton A, an application of the slice theorem shows that $\mathcal{M}(E,\alpha)$ has the structure V/Γ_A for a vector space V of dimension $\operatorname{Ind}^+(E,\alpha) + \dim \Gamma_A$ [**9**, Theorem 4.13]. This is understood by considering the base point fibration (described above) $SO(3) \to \mathcal{A}^\delta(E,\alpha)/\mathcal{G}_0^\delta(E,\alpha) \to \mathcal{A}^\delta(E,\alpha)/\mathcal{G}^\delta(E,\alpha)$. Restricting to $\mathcal{M}(E,\alpha)$ gives an $SO(3)$ action on a smooth manifold of dimension $\operatorname{Ind}^+(E,\alpha) + 3$ with quotient $\mathcal{M}(E,\alpha)$ and stabilizer Γ_A over the instanton A.

Subgroups of $SO(3)$ have several possible centralizers, and this can lead to different types of singularities in the instanton moduli space. For our purposes, it is sufficient to deal solely with orbit types of reducibles that have $SO(2)$-stabilizers. In light of our assumption that $H_1(X)$ has no 2-torsion, we can ensure this with the following (which, hereafter, will be assumed unless otherwise stated):

The adapted bundle (E,α) is non-trivial. Equivalently, α does not extend to a connection on E with trivial holonomy.

In typical applications either the bundle E is non-trivial, or else one of the α_i has non-trivial holonomy, and so the assumption holds. With this assumption, the singularities of $\mathcal{M}(E,\alpha)$ are cones on $\mathbb{C}P^n$, where $2n+1 = \operatorname{Ind}^+(E,\alpha)$. If $n > 0$ the second Stiefel-Whitney class of the base point fibration $\mathcal{E} \to \mathcal{M}(E,\alpha)^*$ restricts on the link of each reducible instanton to the generator of $H^2(\mathbb{C}P^n;\mathbb{Z}/2) = \mathbb{Z}/2$ [**13**, Section 9].

Suppose there exists a reducible connection $A \in \mathcal{A}^\delta(E,\alpha)$. Since Γ_A centralizes the holonomy group of A, the assumptions that (E,α) is non-trivial and

$H^1(X;\mathbb{Z}/2)=0$ imply that Γ_A is conjugate to $SO(2)\subset SO(3)$. Moreover, for each boundary component Y_i, α_i is a reducible flat connection on Y_i whose stabilizer contains a maximal torus.

The action of Γ_A on the fiber E_x of E over a point $x\in X_\infty$ gives a splitting $E_x = L_x\oplus\epsilon_x$, where ϵ_x is the 1-eigenspace of Γ_A and L_x is its 2-dimensional orthogonal complement on which Γ_A acts non-trivially. Parallel transport using A then gives a decomposition $E=L_A\oplus\epsilon_A$ into the sum of an orthogonal plane bundle L_A and a real line bundle ϵ_A. The bundle ϵ_A is trivial since $H^1(X;\mathbb{Z}/2)=0$. A choice of trivialization of ϵ_A together with the orientation of E determines an orientation (and hence the structure of an $SO(2)$ vector bundle) on L_A. The connection A splits correspondingly as the direct sum of connections $A=a\oplus\theta$, where a is an $SO(2)$ connection on L_A and θ is the trivial connection on the trivial bundle. In particular, the connection α_i on each end splits as

$$\alpha_i=\beta_i\oplus\theta \text{ on } L_A|_{Y_i}\oplus\epsilon,$$

where $\beta_i=\lim_{r\to\infty} a|_{\{r\}\times Y_i}$. The flat connection β_i is essentially uniquely determined by the subbundle $L_A|_Y\subset E|_Y$, as the following lemma shows.

LEMMA 2.10. *Every $SO(2)$ vector bundle L over a rational homology sphere Z admits a flat $SO(2)$ connection. Moreover, given any two flat $SO(2)$ connections β, β' on L, there exists a path of $SO(2)$ gauge transformations g_t with g_0 the identity and $g_1^*(\beta')=\beta$.*

PROOF. For simplicity identify $SO(2)$ with $U(1)$. Choose any $U(1)$ connection b on L. Since $H^2(Z;\mathbb{R})=0$, the real form $\frac{i}{2\pi}F(b)$ is exact. Adding a 1-form $i\omega$ to b changes $F(b)$ to $F(b)+id\omega$, and so there exists an ω for which $F(b+i\omega)=0$, i.e. $\beta=b+i\omega$ is flat.

The group of gauge transformations of the bundle L, $\mathcal{G}(L)=C^\infty(Z,U(1))$, has a single path component, since $[Z,U(1)]=H^1(Z;\mathbb{Z})=0$. A flat connection β on L has a holonomy representation

$$\mathrm{hol}_\beta\in\chi(\pi_1 Z,U(1))=\mathrm{Hom}(\pi_1 Z,U(1))=H^1(Z;\mathbb{R}/\mathbb{Z}).$$

The Bockstein in the coefficient exact sequence $H^1(Z;\mathbb{R}/\mathbb{Z})\to H^2(Z;\mathbb{Z})$ takes hol_β to $c_1(L)$ (see below) and is an isomorphism since Z is a rational homology sphere. Since the holonomy determines the gauge equivalence class of a flat connection, it follows that any two flat connections β,β' on L are gauge equivalent, and that there is a path g_t of gauge transformations with $g_0=Id$ and $g_1^*(\beta')=\beta$. □

The relationship between the holonomy $U(1)$ representation and the first Chern class used in the proof of Lemma 2.10 can be explained in the following way. Let $U(1)_d$ denote $U(1)$ with the discrete topology. The identity map $i:U(1)_d\to U(1)$ induces a map $Bi: BU(1)_d\to BU(1)$ on classifying spaces. Since $BU(1)_d=K(\mathbb{R}/\mathbb{Z},1)$ and $BU(1)=K(\mathbb{Z},2)$, we have a sequence of (natural) isomorphisms for any finite abelian group A

$$\mathrm{Hom}(A,U(1))=H^1(K(A,1);\mathbb{R}/\mathbb{Z})=[K(A,1),BU(1)_d]$$
$$\xrightarrow{Bi_*}[K(A,1),BU(1)]=H^2(K(A,1);\mathbb{Z}).$$

The identity map $BU(1)\to BU(1)$, viewed as a class in $H^2(BU(1);\mathbb{Z})$, is the first Chern class of the universal line bundle, and so naturality implies that a homomorphism $h:A\to U(1)$ is sent to $c_1(L)$, where L is the pull back of the universal bundle via $K(A,1) \xrightarrow{Bh} K(U(1)_d,1) \xrightarrow{Bi} BU(1)$. Moreover, the map Bi_* can be identified with the Bockstein.

Given a manifold W equipped with a homomorphism $h : \pi_1(W) \to U(1)$ with finite image, view h as an element of $H^1(W;\mathbb{R}/\mathbb{Z})$. Then h is sent by the Bockstein to the first Chern class of the $U(1)$ bundle that supports a flat connection with holonomy h.

In the case of a rational homology 3-sphere Z, we have a sequence of isomorphisms

$$\mathrm{Hom}(\pi_1(Z),\mathbb{Q}/\mathbb{Z}) = \mathrm{Hom}(H_1(Z),\mathbb{Q}/\mathbb{Z}) = H^1(Z;\mathbb{Q}/\mathbb{Z}) \xrightarrow{B} H^2(Z;\mathbb{Z}) \xrightarrow{\cap[Z]} H_1(Z;\mathbb{Z})$$

with B the Bockstein and the last isomorphism Poincaré duality. The inverse of this composite is given by the linking form, $m \mapsto \mathrm{lk}(m,-) : H_1(Z;\mathbb{Z}) \to \mathbb{Q}/\mathbb{Z}$. Thus we have shown the following.

LEMMA 2.11. *Suppose Z is a rational homology 3-sphere and $c_1 \in H^2(Z;\mathbb{Z})$ a class with Poincaré dual $m \in H_1(Z;\mathbb{Z})$. Then the holonomy of the flat $U(1)$ bundle with first Chern class c_1 is given by* $\mathrm{hol}(\gamma) = \mathrm{lk}(m,\gamma)$, *where*

$$\mathrm{lk} : H_1(Z;\mathbb{Z}) \times H_1(Z;\mathbb{Z}) \to \mathbb{Q}/\mathbb{Z} \xrightarrow{\exp(2\pi i -)} U(1)$$

is the linking form. □

Identifying $U(1)$ and $SO(2)$ via $\mathbb{C} = \mathbb{R}^2$ allows us to make the analogous statement, substituting the Euler class for c_1.

Suppose that (E,α) is an adapted bundle which admits a reducible connection A. Then $(E,\alpha) = (L \oplus \epsilon, \beta \oplus \theta)$, where β is unique in the sense of Lemma 2.10. Conversely, if $\epsilon \subset E$ is a trivializable real line bundle and L the orthogonal 2-plane bundle, then orient ϵ (and hence L). Let β denote the flat $SO(2)$ connection carried by L over the end and set $\alpha = \beta \oplus \theta$. Extend β to a connection a on L. Then $A = a \oplus \theta$ is a reducible connection in $\mathcal{A}(E,\alpha)$. This establishes the first assertion in the following lemma.

LEMMA 2.12. *An adapted bundle (E,α) supports a reducible connection if and only if there exists a nowhere zero section s of E, and hence a splitting $E = L \oplus \epsilon$ with L the orthogonal complement of $\epsilon = \mathrm{span}\, s$, and, on the ends, $\alpha = \beta \oplus \theta$ where β is the (unique up to homotopy) flat connection on L.*

If an adapted bundle (E,α) supports a reducible connection inducing a splitting $E = L \oplus \epsilon$, it supports a reducible instanton in $\mathcal{M}(E,\alpha)$ inducing the same splitting.

PROOF. The first part was justified in the preceding paragraph. Suppose that $(E,\alpha) = (L, \beta \oplus \theta)$. Since Y is a union of rational homology spheres, Proposition 4.9 of [**2**] shows that the space of L^2 harmonic 2-forms $\{\omega \in L^2(\Omega^2_{X_\infty}) | d\omega = 0 = d^*\omega\}$ maps isomorphically to $H^2(X;\mathbb{R})$. Thus there exists an L^2 harmonic 2-form ω on X_∞ representing the class $e(L) \in H^2(X;\mathbb{R})$. In light of the facts that X is negative definite, the Hodge $*$-operator preserves L^2 harmonic 2-forms, and $*^2 = 1$, it follows that $*\omega = -\omega$. Choose a connection a on L. Then $\frac{1}{2\pi}\mathrm{Pf}(F(a))$ and ω are cohomologous by Chern-Weil Theory, where Pf denotes the pfaffian of an $so(2)$ matrix. By adding an $so(2)$ valued 1-form to a one can change $\mathrm{Pf}(F(a))$ arbitrarily within its cohomology class. Thus we may choose a so that $\frac{1}{2\pi}\mathrm{Pf}(F(a)) = \omega$.

The connection a limits to a flat connection on L along the ends. Thus Lemma 2.10 implies a limits to $h^{-1}(\beta)$ for some gauge transformation $h : L|_Y \to L|_Y$ which can be connected to the identity by a path of gauge transformations. Extend h to a gauge transformation $\tilde{h}$ of all of L by the identity on X, the path from the identity to h on $[0,1] \times Y$, and h on $[1,\infty) \times Y$.

Replacing a by $\tilde{h}(a)$ does not change $F(a)$ since $SO(2)$ is abelian, and $h(a)$ limits to β. Let $A = \tilde{h}(a) \oplus \theta$ on the $SO(3)$ vector bundle $L \oplus \epsilon$. Then A is a reducible instanton on $(E,\alpha) = (L\oplus\epsilon, \beta\oplus\theta)$ with L^2-curvature. Since we assumed that the flat connection α on Y is non-degenerate, it follows from [**9**, Theorem 4.2] that $A \in L_1^{p,\delta}$ and so $A \in \mathcal{M}(E,\alpha)$. □

The following proposition determines the Pontryagin charge of a reducible instanton in terms of the intersection form.

PROPOSITION 2.13. *Let $e \in H^2(X;\mathbb{Z})$. Let $L \to X_\infty$ be the $SO(2)$ vector bundle with Euler class e. Let β be the flat connection on the restriction $L|_Y$, and $(E,\alpha) = (L\oplus\epsilon, \beta\oplus\theta)$ the corresponding adapted bundle.*

Then

$$p_1(E,\alpha) = -e\cdot e \in \tfrac{1}{d}\mathbb{Z} \text{ and } w_2(E) \equiv e \bmod 2.$$

PROOF. Choose (arbitrarily) a connection a_0 on L which extends the flat connection β on the ends, and let $A_0 = a_0 \oplus \theta$ be the corresponding connection on $E = L \oplus \epsilon$; this connection agrees with $\alpha = \beta \oplus \theta$ over the ends, and hence lies in $\mathcal{A}^\delta(E,\alpha)$.

Let $\hat{e}$ denote the differential 2-form on X_∞

$$\hat{e} = \tfrac{1}{2\pi}\mathrm{Pf}(F(a_0)).$$

A straightforward calculation starting with the formula for the inclusion of Lie algebras $so(2) \subset so(3)$ gives

$$\begin{aligned}
-\frac{1}{8\pi^2}\,\mathrm{Tr}(F(A_0)\wedge F(A_0)) &= -\tfrac{1}{2}\,\mathrm{Tr}(\tfrac{1}{2\pi}F(A_0)\wedge \tfrac{1}{2\pi}F(A_0)) \\
&= -\tfrac{1}{2\pi}\mathrm{Pf}(F(a_0))\wedge \tfrac{1}{2\pi}\mathrm{Pf}(F(a_0)) \\
&= -\hat{e}\wedge\hat{e}.
\end{aligned}$$

Hence

$$p_1(A_0) = p_1(E,\alpha) = -\int_{X_\infty} \hat{e}\wedge\hat{e}.$$

Chern-Weil theory implies that $\hat{e}$ is a closed 2-form. It vanishes on the ends since a_0 is flat on the ends. On a *closed* 4-manifold Z Chern-Weil theory implies that $\hat{e}$ represents (in DeRham cohomology) the image of the Euler class e under the coefficient homomorphism $H^2(Z;\mathbb{Z}) \to H^2(Z;\mathbb{R})$. Since $H^2(X;\mathbb{R}) = \mathrm{Hom}(H_2(X;\mathbb{Z}),\mathbb{R})$ and every class in $H_2(X;\mathbb{Z})$ is represented by a closed surface, it follows that $\hat{e}$ represents the image of e in $H^2(X;\mathbb{R})$.

For $x, y \in H^2(X,Y;\mathbb{R})$ represented by closed forms that vanish on the boundary, $x\cdot y = \int_X x\wedge y$. Since $H^2(X,Y;\mathbb{R}) \to H^2(X;\mathbb{R})$ is an isomorphism, it follows that

$$p_1(E,\alpha) = -\int_X \hat{e}\wedge\hat{e} = -e\cdot e.$$

The congruence $w_2(E) \equiv e \bmod 2$ follows from the general congruence $w_2(L\oplus\epsilon) \equiv e(L) \bmod 2$, which is valid for any $SO(2)$ bundle L over any space. □

2.7. Enumeration of reducible instantons. We next address the problem of enumerating the reducible instantons in $\mathcal{M}(E,\alpha)$. The results of the previous section allow us to reduce this problem to the enumeration of bundle reductions. These in turn determine certain cohomology classes and motivate the following definition.

DEFINITION 2.14. Let $(E,\alpha) = (L \oplus \epsilon, \beta \oplus \theta)$ with L non-trivial. Fix an orientation of ϵ, orienting L. Let $e = e(L) \in H^2(X;\mathbb{Z})$ denote the Euler class of L. Define

$$C(e) = \{e' \in H^2(X;\mathbb{Z}) \mid e' \cdot e' = e \cdot e,\ e' \equiv e \bmod 2,\ i_j^*(e') = \pm i_j^*(e),\ j = 1, \cdots, c\}/\pm 1$$

where $i_j : Y_j \subset Y \subset X$ denotes the inclusion of the jth boundary component. Our standing assumption that X is negative definite implies that $C(e)$ is a finite set.

Suppose that L, L' are two orthogonal plane subbundles of E and that $(E,\alpha) = (L \oplus \epsilon, \beta \oplus \theta) = (L' \oplus \epsilon', \beta' \oplus \theta)$. Orient L and L' arbitrarily and let $e = e(L)$, $e' = e(L')$. Then $e \equiv e' \bmod 2$ and $e \cdot e = e' \cdot e'$ by Proposition 2.13. Fix a boundary component Y_j of X. If α_j is non-trivial, then since $\beta_j \oplus \theta = \alpha_j = \beta_j' \oplus \theta$, the *unoriented subbundles* $L|_{Y_j}$ and $L'|_{Y_j}$ are equal, since their orthogonal complements coincide. But if α_j is the trivial connection, $L|_{Y_i}$ and $L'|_{Y_i}$ need not coincide, although they are both trivial bundles. In either case, $i_j^*(e) = \pm i_j^*(e')$. Changing the orientation of L' changes the sign of e'. Since X is connected, it follows that the function taking a reducible connection $A \in \mathcal{A}^\delta(L \oplus \epsilon, \beta \oplus \theta)_{red}$ to $e(L_A) \in C(e)$ is well-defined.

PROPOSITION 2.15. *The function $\Phi : \mathcal{M}(E,\alpha)_{red} \to C(e)$ defined by $\Phi([A]) = e(L_A)/\pm 1$ is well-defined and injective. If $C(e)$ consists of a single point then Φ is a bijection.*

PROOF. We first show that Φ is well-defined on gauge equivalence classes. Suppose that A is a reducible instanton on (E,α) and $g \in \mathcal{G}^\delta(E,\alpha)$ a gauge transformation. Then the decomposition $E = L_A \oplus \epsilon$ is sent to $E = g(L_A) \oplus g(\epsilon)$. Hence L_A and $g(L_A)$ are isomorphic bundles over X_∞, and so $\Phi(A) = \Phi(g(A))$. Thus Φ is well-defined on gauge equivalence classes.

Suppose that A and B are reducible instantons on (E,α) which satisfy $\Phi(A) = \Phi(B)$. Then, after reorienting L_B and ϵ_B if needed, $e(L_A) = e(L_B)$. Since $SO(2)$ bundles are determined up to isomorphism by their Euler class, there exists a bundle isomorphism $h : L_A \to L_B$. Extending by the unique orthogonal orientation-preserving isomorphism $I : \epsilon_A \to \epsilon_B$ produces a bundle isomorphism $g = h \oplus I : E = L_A \oplus \epsilon_A \to L_B \oplus \epsilon_B = E$. Since the decompositions agree on the ends with the given decomposition (E,α), $\lim_{r\to\infty} g|_{\{r\}\times Y_j}$ exists and lies in the stabilizer of α_j, so that $g \in \mathcal{G}^\delta(E,\alpha)$. Therefore, A and B represent the same point in $\mathcal{M}(E,\alpha)$.

Lemma 2.12 shows that $\mathcal{M}(L \oplus \epsilon, \beta \oplus \theta)$ is non-empty, and so if $C(e) = \{e\}$, Φ is surjective. □

Determining the image of Φ can be tricky when Y is not a union of $\mathbb{Z}/2$ homology spheres. Lemma 2.12 shows that $e = e(L)$ itself is always in the image, and hence, as noted in Proposition 2.15, if $C(e)$ consists of a single point, then Φ is a bijection.

THEOREM 2.16. *Let $(E, \alpha) = (L \oplus \epsilon, \beta \oplus \theta)$ be a reducible adapted bundle with Euler class $e = e(L)$ and let $\Phi : \mathcal{M}(E, \alpha)_{red} \to C(e)$ be the injective function defined above.*

(1) *If $H^2(X; \mathbb{Z})$ splits orthogonally with respect to the intersection form as $H^2(X; \mathbb{Z}) = \operatorname{span}(e) \oplus W$ with $\operatorname{span}(e) \cong \mathbb{Z}$ and W free abelian, then $C(e)$ consists of a single point and so Φ is a bijection.*
(2) *To each $e' \in C(e)$ one can assign an obstruction in $H^1(Y; \mathbb{Z}/2)$ whose vanishing implies that $e' = \Phi([A])$ has a solution $[A] \in \mathcal{M}(E, \alpha)_{red}$.*
(3) *If each component of Y is a $\mathbb{Z}/2$ homology sphere, then Φ is a bijection.*

PROOF. Suppose first that $H^2(X; \mathbb{Z})$ splits orthogonally with respect to the intersection form as $H^2(X; \mathbb{Z}) = \operatorname{span}(e) \oplus W$. Then any class e' satisfying $e' \equiv e$ mod 2 takes the form $e' = (1 + 2k)e + 2w$, with $w \in W$. Hence $e' \cdot e' = (1 + 4k + 4k^2)e \cdot e + 4w \cdot w$. If $e' \in C(e)$ then $e' \cdot e' = e \cdot e$, so that $4(k^2 + k)e \cdot e + 4w \cdot w = 0$. Since $e \cdot e < 0$ and $w \cdot w \leq 0$, this is possible if and only if $k = 0$ or -1 and $w = 0$, so that $e' = \pm e$, as claimed.

We turn now to the second assertion, which clearly implies the third assertion. Given a class $e' \in C(e)$, let $L' \to X$ be an $SO(2)$ vector bundle with $e' = e(L')$. Then since $e' \equiv e$ mod 2, the bundles $E = L \oplus \epsilon$ and $L' \oplus \epsilon'$ over X are isomorphic. Fix an identification $E = L \oplus \epsilon = L' \oplus \epsilon'$. Let β_j be the unique flat connection on $L|_{Y_j}$, β'_j be the unique flat connection on $L'|_{Y_j}$. Then $\alpha_j = \beta_j \oplus \theta$ and $\alpha'_j = \beta'_j \oplus \theta$ are two flat connections on $E|_{Y_j}$.

If there exists a gauge transformation $g : E \to E$ so that $g|_{Y_j}^*(\alpha'_j) = \alpha_j$, then Lemma 2.12 implies that there exists an instanton $A' \in \mathcal{M}(E, \alpha)$ inducing the splitting $E = g(L') \oplus g(\epsilon')$, and hence $e' = \Phi(A')$ has the solution A'. So we seek to construct such a gauge transformation g. We construct g first on the boundary and then try to extend over the interior.

For each j, choose an isomorphism $h_j : L|_{Y_j} \to L'|_{Y_j}$ which preserves or reverses orientation according to whether $i_j^*(e') = i_j^*(e)$ or $i_j^*(e') = -i_j^*(e)$ (if $2i_j^*(e) = 0$ choose h_j to preserve orientation, for definiteness). Then $h_j^*(\beta'_j)$ and β_j are flat connections on $L|_{Y_j}$ and so we may assume, by modifying β'_j if needed, that $h_j^*(\beta'_j) = \beta_j$. Let $g_j : E|_{Y_j} = L|_{Y_j} \oplus \epsilon \to E = L'|_{Y_j} \oplus \epsilon'$ be the unique orientation-preserving orthogonal extension (so that on the trivial $\mathbb{R}$ factors, the orientation is preserved or reversed according to the sign). By construction, $g_j^*(\alpha'_j) = \alpha_j$. Let $g = \sqcup_j g_j : E|_Y \to E|_Y$.

The stabilizer of α_j is connected except for those j for which $i_j^*(e)$ has order 2, in which case the stabilizer is isomorphic to $O(2)$. Thus the desired gauge transformation exists if and only if g can be extended to a gauge transformation over X, after perhaps composing g_j with an appropriate gauge transformation k_j for those path components such that $i_j^*(e)$ has order 2.

Let P be the principal $SO(3)$ bundle associated to $E \to X$, so that gauge transformations are sections of $P \times_{Ad} SO(3)$. Obstruction theory identifies the primary obstruction to extending g to all of X as a class $\psi \in H^2(X, Y; \pi_1(SO(3))) = H^2(X, Y; \mathbb{Z}/2)$. Naturality of primary obstructions, together with the fact that $P \times_{Ad} SO(3)$ admits sections (e.g. the identity transformation) shows that ψ is in $\ker H^2(X, Y; \mathbb{Z}/2) \to H^2(X; \mathbb{Z}/2)$, hence ψ lies in the image of the injection $H^1(Y; \mathbb{Z}/2) \to H^2(X, Y; \mathbb{Z}/2)$. Let $\vartheta \in H^2(Y; \mathbb{Z}/2)$ denote the unique element mapped to ψ.

When $\vartheta = 0$, then g extends over the 2-skeleton of X. It further extends further to the 3-skeleton, because $\pi_2(SO(3)) = 0$. The last obstruction to extending g over X lies in $H^4(X, Y; \pi_3(SO(3))) = H^4(X, Y; \mathbb{Z})$; the coefficients are untwisted since our standing assumption that $H^1(X; \mathbb{Z}/2) = 0$ implies that $\pi_1(X)$ admits no non-trivial homomorphisms to $\mathrm{Aut}(\mathbb{Z}) = \mathbb{Z}/2$. This class is the obstruction to extending g over the top cell of X/Y.

Suppose that $D \subset X$ is a fixed 4-ball, and that $g : E|_{X\setminus D} \to E|_{X\setminus D}$ is the gauge transformation which we wish to extend to all of E. Choose a connection $A \in \mathcal{A}^\delta(L \oplus \epsilon, \beta_L \oplus \theta)$. The connection $g^*(A|_{X\setminus D})$ extends to a connection $B \in \mathcal{A}^\delta(L' \oplus \epsilon', \beta_{L'} \oplus \theta')$.

Since $\mathrm{Tr}(F(g^*(A)) \wedge F(g^*(A))) = \mathrm{Tr}(F(A) \wedge F(A))$ pointwise (on $X \setminus D$), it follows that

$$p_1(A) - p_1(B) = \tfrac{1}{8\pi^2} \int_D \mathrm{Tr}(F(A) \wedge F(A)) - \tfrac{1}{8\pi^2} \int_D \mathrm{Tr}(F(B) \wedge F(B)), \tag{2.8}$$

The condition $e' \cdot e' = e \cdot e$ implies that $p_1(A) = p_1(B)$, and hence the left side of Equation (2.8) is zero.

But Chern-Weil theory implies that the right side of Equation (2.8) equals $p_1(V)$, where V is the bundle over S^4 obtained by clutching two copies of $E|_D$ using $g|_{\partial D}$. Since $p_1(V) = 0$, the bundle V is trivial, and thus the restriction of g to ∂D extends over D, giving the desired gauge transformation g.

Thus we have seen that there is a class $\vartheta \in H^1(Y; \mathbb{Z}/2)$ whose vanishing guarantees that g extends over X. □

2.8. The argument of Furuta. With all the necessary machinery in place, we can now describe our variant of the argument of Furuta. Theorem 2.17 provides the mechanism to establish that certain definite 4-manifolds cannot exist. Furuta used this argument to prove the linear independence of the Seifert fibered homology spheres $\{\Sigma(p, q, pqk - 1)\}_{k=1}^\infty$ in Θ^3_H. We will give further applications below.

Recall from Section 2.6 that if X is a negative definite four-manifold satisfying $H^1(X; \mathbb{Z}/2) = 0$, and ∂X consists of a union of rational homology spheres, then any $SO(2)$ bundle L determines a unique adapted $SO(3)$ bundle $(E, \alpha) = (L \oplus \epsilon, \beta \oplus \theta)$. Letting $e(L)$ denote the Euler class of L, Proposition 2.13 shows that $p_1(E, \alpha) = -e(L) \cdot e(L) \in \frac{1}{d}\mathbb{Z}$, where d is an integer which annihilates $H_1(Y_i; \mathbb{Z})$ for each i. Finally, recall the invariant $\hat{\tau}(Y, \alpha) \in (0, 4]$ of Definition 2.7.

THEOREM 2.17. *Let X be a negative definite 4-manifold with $H^1(X; \mathbb{Z}/2) = 0$ whose boundary consists of a union of rational homology spheres. Let L be an $SO(2)$ bundle over X_∞ with Euler class $e = e(L)$, and let $(E, \alpha) = (L \oplus \epsilon, \beta \oplus \theta)$ be the corresponding adapted bundle. Assume that*

(1) *Each $\alpha_i = \beta_i \oplus \theta$ is non-degenerate,*
(2) $\mathrm{Ind}^+(E, \alpha) \geq 0$, *where*

$$\mathrm{Ind}^+(E, \alpha) = -2e \cdot e - 3 + \tfrac{1}{2} \sum_{\beta_i \text{nontrivial}} (3 - h_{\alpha_i} - \rho(Y_i, \alpha_i)),$$

(3) $0 < -e \cdot e < \hat{\tau}(Y, \alpha) \leq 4$.

Then the number of gauge equivalence classes of reducible instantons on (E, α) is even, greater than 1, and $\mathrm{Ind}^+(E, \alpha)$ is odd.

PROOF. The irreducible moduli space $\mathcal{M}(E,\alpha)^*$ is a smooth oriented manifold of dimension $\text{Ind}^+(E,\alpha)$. Let k denote the number of singular points in the full moduli space. Proposition 2.15 identifies k with the image of $\Phi : \mathcal{M}(E,\alpha)_{red} \to C(L)$, and shows that $k \geq 1$. Each singular point $A \in \mathcal{M}(E,\alpha)_{red}$ has a neighborhood homeomorphic to a cone on $\mathbb{C}P^n$ for some n, and so $2n+1 = \text{Ind}^+(E,\alpha)$. Proposition 2.9 implies that $\mathcal{M}(E,\alpha)$ is compact.

Removing cone neighborhoods of the k singular points yields a $2n+1$ dimensional compact manifold $\mathcal{M}^0$, whose boundary consists of a disjoint union of k copies of $\mathbb{C}P^n$ (perhaps with differing orientations). This manifold is equipped with an $SO(3)$ bundle $\mathcal{E} \to \mathcal{M}^0$ that restricts to a bundle on each $\mathbb{C}P^n$ with nontrivial second Stiefel-Whitney class. Since $H^*(\mathbb{C}P^n;\mathbb{Z}/2)$ is a polynomial ring on w_2, $0 \neq w_2(\mathcal{E}|_{\mathbb{C}P^n})^n$ for each boundary component. But since $(\sqcup_{i=1}^k \mathbb{C}P^n, \mathcal{E})$ extends to $(\mathcal{M}^0, \mathcal{E})$, it follows that k is even, as this is the only way for the Stiefel-Whitney numbers of $(\sqcup_{i=1}^k \mathbb{C}P^n, \mathcal{E})$ to vanish. □

2.9. Computing Atiyah-Patodi-Singer ρ invariants and Chern-Simons invariants of flat SO(2) connections on rational homology spheres. Before we turn to applications and examples we discuss the use of flat cobordisms to compute the ρ invariants and Chern-Simons invariants which appear in the expression for $\text{Ind}^+(E,\alpha)$ and the definition of $\hat{\tau}(Y,\alpha)$.

The Atiyah-Patodi-Singer ρ invariants of flat $SO(3)$ connections on 3-manifolds can be difficult to compute. However, in Theorem 2.17 we require Y to be a union of rational homology spheres and the flat connection α to be reducible; in fact, to take the form $\alpha = \beta \oplus \theta$ for θ the trivial connection. Thus, what is required in applications of this theorem is a calculation of ρ invariants of such reducible flat $SO(3)$ connections on rational homology spheres. This is an easier task, and can often be accomplished as follows.

Since $\rho(Y,\alpha)$ depends only on the gauge equivalence class of α, and taking the holonomy induces a 1-1 correspondence from gauge equivalence classes of flat connections to conjugacy classes of homomorphisms $\pi_1(Y) \to SO(3)$, we henceforth use the notation α for both a flat connection and its holonomy.

Then $\alpha : \pi_1(Y) \to SO(3)$ is reducible if it is conjugate to a representation $\beta \oplus 1 : \pi_1(Y) \to SO(3)$ with image in

$$\begin{pmatrix} SO(2) & 0 \\ 0 & 1 \end{pmatrix} \subset SO(3).$$

Complexification preserves eigenvalues, and hence $\rho(Y,\alpha)$ is unchanged by including $SO(3)$ in $SU(3)$. The complexification of $\beta \oplus 1$ is conjugate to the diagonal representation $\beta_{\mathbb{C}} \oplus \bar{\beta}_{\mathbb{C}} \oplus 1$ where $\beta_{\mathbb{C}} : \pi_1(Y) \to U(1)$ is just β viewed via the identification $SO(2) = U(1)$. Because $\rho(Y,\theta) = \eta(Y,\theta) - \eta(Y,\theta) = 0$,

$$\rho(Y,\alpha) = \rho(Y,\beta_{\mathbb{C}}) + \rho(Y,\bar{\beta}_{\mathbb{C}}). \tag{2.9}$$

Therefore the task is reduced to computing $\rho(Y,\gamma)$ for 1-dimensional unitary representations γ.

A successful strategy to compute $\rho(Y,\gamma)$ for $\gamma : \pi_1(Y) \to U(1)$ is to find a cobordism W from Y to a lens space (or union of lens spaces) $L(Y)$ over which γ extends. Finding such a cobordism is a familiar problem; see for example [**5, 23**]. The key idea is that finding such a W is equivalent to showing (Y,γ) vanishes in an appropriate oriented bordism group over $\mathbb{R}/\mathbb{Z}$ or $\mathbb{Z}$. This can be addressed with straightforward methods.

Such a W and extension of γ provides a *flat* cobordism from Y to $L(Y)$. Applying the Atiyah-Patodi-Singer signature theorem [**2, 3**] one obtains:

$$\rho(Y,\gamma)-\rho(L(Y),\gamma)=\mathrm{Sign}_\gamma(W)-\mathrm{Sign}(W).$$

By this approach, computing the ρ invariants which arise in Theorem 2.17 reduces to computing $U(1)$ signatures of W and computing $U(1)$ ρ invariants of lens spaces.

The calculation of $\rho(L(a,b),\gamma)$ for a lens space $L(a,b)$ and a $U(1)$ representation $\gamma:\pi_1(L(a,b))\to U(1)$ has been carried out in [**10, 3**]. We state the result in the form we need. Let a be a positive integer and $0<b<a$ an integer relatively prime to a. Denote by $L(a,b)$ the oriented lens space obtained by $-a/b$ surgery on the unknot U. The meridian μ of the unknot generates $\pi_1(S^3\setminus U)$, and hence also $\pi_1(L(a,b))=H_1(L(a,b))=\mathbb{Z}/a$.

PROPOSITION 2.18. *Let $\beta:\pi_1(L(a,b))\to SO(2)$ be the representation sending the generator μ to the rotation of angle $\frac{2\pi\ell}{a}$ for some integer ℓ. Then*

$$\rho(L(a,b),\beta)=\tfrac{4}{a}\sum_{k=1}^{a-1}\cot(\tfrac{\pi k}{a})\cot(\tfrac{\pi kb}{a})\sin^2(\tfrac{\pi k\ell}{a}).$$

If $\alpha=\beta\oplus\theta:\pi_1(L(a,b))\to SO(3)$, then $\rho(L(a,b),\alpha)=\rho(L(a,b),\beta)$.

PROOF. Equation (2.9) shows that

$$\rho(L(a,b),\beta)=\rho(L(a,b),\beta_{\mathbb{C}})+\rho(L(a,b),\bar\beta_{\mathbb{C}})$$

where $\beta_{\mathbb{C}}:\pi_1(L(a,b))\to U(1)$ is given by $\beta_{\mathbb{C}}(\mu)=e^{2\pi i\ell/a}$.

The lens space $L(a,b)$ is orientation-preserving diffeomorphic to the quotient of S^3 by the free action $g\cdot(z,w)=(e^{2\pi i/a}z,e^{2\pi ib/a}w)$ for $(z,w)\in S^3\subset\mathbb{C}^2$. This diffeomorphism identifies the homotopy class of μ with the generator g of the deck transformations. Proposition 4.1 of [**10**] calculates the corresponding η invariant; the formula is

$$\eta(L(a,b),\beta_{\mathbb{C}})=-\tfrac{1}{a}\sum_{k=1}^{a-1}\cot(\tfrac{\pi k}{a})\cot(\tfrac{\pi kb}{a})e^{2\pi ik\ell/a}.$$

Since $\rho(L(a,b),\beta_{\mathbb{C}})+\rho(L(a,b),\bar\beta_{\mathbb{C}})=\eta(L(a,b),\beta_{\mathbb{C}})+\eta(L(a,b),\bar\beta_{\mathbb{C}})-2\eta(L(a,b),\theta)$,

$$\begin{aligned}\rho(L(a,b),\beta)&=-\tfrac{1}{a}\sum_{k=1}^{a-1}\cot(\tfrac{\pi k}{a})\cot(\tfrac{\pi kb}{a})(e^{2\pi ik\ell/a}+e^{-2\pi ik\ell/a}-2)\\&=-\tfrac{2}{a}\sum_{k=1}^{a-1}\cot(\tfrac{\pi k}{a})\cot(\tfrac{\pi kb}{a})(\cos(\tfrac{2\pi k\ell}{a})-1)\\&=\tfrac{4}{a}\sum_{k=1}^{a-1}\cot(\tfrac{\pi k}{a})\cot(\tfrac{\pi kb}{a})\sin^2(\tfrac{\pi k\ell}{a}).\end{aligned}$$

The last assertion follows from the facts that ρ invariants add with respect to the direct sum of representations and that $\rho(Z,\theta)=\eta(Z,\theta)-\eta(Z,\theta)=0$ for any Z. □

The formula of Neumann-Zagier [**31**]

$$\frac{2}{a}\sum_{k=1}^{a-1}\cot(\tfrac{\pi kc}{a})\cot(\tfrac{\pi k}{a})\sin^2(\tfrac{\pi k}{a}) = \tfrac{2c^*}{a} - 1 \tag{2.10}$$

when $(a,c)=1$ and $0<c^*<a$ satisfies $cc^* \equiv -1 \mod a$,

gives an explicit method for computing the ρ invariants which appear in Proposition 2.18 in the case when $\ell = b$ (by reindexing the sum). In the general case the extensions of this calculation due to Lawson [**24, 25**] are useful.

The strategy of using a flat cobordism also plays an important role in computing Chern-Simons invariants and hence the invariant $\hat{\tau}(Y,\alpha)$. This is because if W is a flat cobordism from (Y,α) to (Y',α'), then $\mathrm{cs}(Y,\alpha)=\mathrm{cs}(Y',\alpha') \mod \mathbb{Z}$.

For *reducible* α the flat cobordisms as described above permits one to express Chern-Simons invariants in terms of Chern-Simons invariants of lens spaces. Indeed, one can quickly obtain estimates for Chern-Simons invariants of lens spaces (or, more generally, spherical space forms) as follows. First note that $L(a,b)$ has universal cover S^3, and every flat $SO(3)$ connection on S^3 is gauge equivalent to the trivial connection, which has Chern-Simons invariant a multiple of 4. Integrating the Chern-Simons form associated to the pullback of any flat connection over a fundamental domain shows that

$$\tau(L(a,b),\alpha) = \frac{4k}{a} \geq \frac{4}{a}.$$

Computing or estimating Chern-Simons invariants for irreducible flat $SO(3)$ connections, however, requires more sophisticated techniques, such as those introduced in [**21, 22, 23, 4**].

3. Seifert fibered examples

Propositions 2.6 and 2.18 allow us to identify $\mathrm{Ind}^+(E,\alpha)$ with the invariant $R(a_1,\cdots,a_n)$ of Fintushel-Stern [**14**] for X the truncated mapping cylinder of the Seifert fibration of the Seifert fibered homology sphere $\Sigma(a_1,\cdots,a_n)$. We explain this and give their proof that when $R(a_1,\cdots,a_n)>0$, $\Sigma(a_1,\cdots,a_n)$ does not bound a positive definite 4-manifold.

Fix an n-tuple $(a_1,b_1),\cdots,(a_n,b_n)$ of relatively prime pairs of integers satisfying $(a_1\cdots a_n)\sum_{i=1}^n \frac{b_i}{a_i}=1$ with the a_i positive. These uniquely define an oriented Seifert fibered homology sphere $\Sigma(a_1,\cdots,a_n)$ whose Kirby diagram is illustrated in Figure 1.

Starting with the mapping cylinder C of the Seifert fibration $\Sigma \to S^2$, remove neighborhoods of the n singular points of C leaving a 4-manifold X. These neighborhoods are homeomorphic to cones on lens spaces $L(a_i,b_i)$. Thus X is a manifold with boundary, and this boundary consists of the disjoint union of Σ and the lens spaces $L(a_i,b_i)$. For convenience we denote $Q=\sqcup_i L(a_i,b_i)$, so that $\partial X = Q \sqcup \Sigma$.

The 4-dimensional Kirby diagram for X is also given by Figure 1. More precisely, this diagram can be thought of as a cobordism from a connected sum of lens spaces $-L(a_i,b_i)$ to Σ, obtained by adding a single 0-framed 2-handle. Then X is obtained by gluing this cobordism to a cobordism from the disjoint union of lens spaces to the connected sum of lens spaces obtained by attaching $(n-1)$ 1-handles to form the connected sum. Thus X is obtained from $I\times(-Q)$ by adding $(n-1)$ 1-handles and one 2-handle; alternatively X is obtained from $I\times\Sigma$ by adding

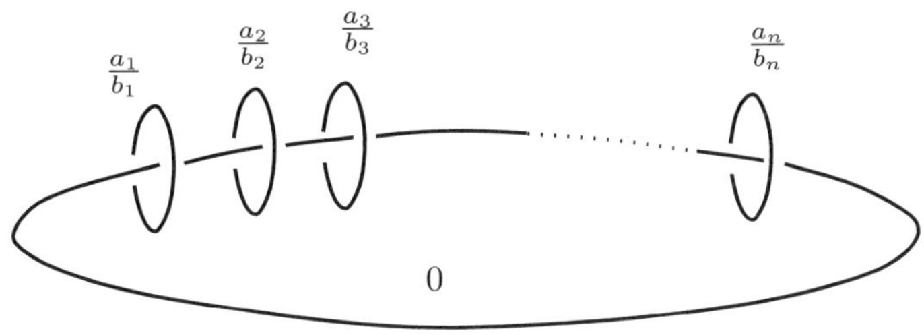

FIGURE 1

one 2-handle (a 0-framed meridian to the 0-framed circle indicated in Figure 1) and $(n-1)$ 3-handles along 2-spheres separating the components. Therefore $H_1(X) = 0$ and $H_2(X) \cong \mathbb{Z}$.

Consider the surface obtained from the union of the punctured disk bounded by the 0-framed component together with the core of the 2-handle. Pushing the interior of this surface slightly into the interior of X gives a properly embedded n-punctured sphere $F \subset X$ representing a generator $[F, \partial F] \in H_2(X, Q) = H_2(X, \partial X) \cong \mathbb{Z}$. Orient X so that it is negative definite. With this orientation, $\partial X = \Sigma \sqcup_i L(a_i, b_i)$ (recall that as an oriented 3-manifold $L(p,q)$ is defined to be $-\frac{p}{q}$ Dehn surgery on the unknot).

Let $L \to X$ be the $SO(2)$ bundle whose Euler class $e(L) \subset H^2(X)$ equals the Poincaré dual to F, and $(E, \alpha) = (L \oplus \epsilon, \beta \oplus \theta)$ the corresponding adapted bundle. We compute the terms that appear in the formula for $\mathrm{Ind}^+(E, \alpha)$ given by Proposition 2.6.

The product $a = a_1 \cdots a_n$ annihilates $H_1(\partial X)$. The boundary components of F are meridian curves representing the generators $\mu_i \in H_1(L(a_i, b_i))$. A straightforward geometric argument using Figure 1 shows that

$$F \cdot aF = -(a_1 \cdots a_n) \sum_{i=1}^{n} \tfrac{b_i}{a_i} = -1$$

Thus

$$p_1(E, \alpha) = -e(L) \cdot e(L) = -\tfrac{1}{a}(F \cdot aF) = \tfrac{1}{a}.$$

Next, note that the flat $SO(2)$ connection on Σ is trivial since Σ is a integer homology sphere, or, equivalently, F misses Σ.

This leaves the contributions from the $L(a_i, b_i)$. We use Proposition 2.18 to compute the terms $\rho(L(a_i, b_i), \alpha_i)$. The representation $\alpha_i = \beta_i \oplus \theta$ sends μ_i to $e^{2\pi i \ell_i / a_i}$ where, according to Lemma 2.11, the integers ℓ_i are determined by $\ell_i / a_i = \mathrm{lk}(\mu_i, \mu_i)$. It follows that $\ell_i = -b_i$ and so β_i is non-trivial. Hence $H^0(L(a_i, b_i); \mathbb{R}^2_{\beta_i}) = 0 = H^1(L(a_i, b_i); \mathbb{R}^2_{\beta_i})$ and so

$$h_{\alpha_i} = h^0_{\alpha_i} + h^1_{\alpha_i} = h^0_{\beta_i} + h^0_{\theta} + h^1_{\beta_i} + h^1_{\theta} = 0 + 1 + 0 + 0 = 1.$$

Therefore,

$$\rho(L(a_i, b_i), \alpha_i) = \tfrac{4}{a_i} \sum_{k=1}^{a_i - 1} \cot(\tfrac{\pi k}{a_i}) \cot(\tfrac{\pi k(-b_i)}{a_i}) \sin^2(\tfrac{\pi k(-b_i)}{a_i}).$$

Thus Proposition 2.6 implies that

$$\operatorname{Ind}^+(E,\alpha) = \tfrac{2}{a} - 3 + n + \sum_{i=1}^{n} \tfrac{2}{a_i} \sum_{k=1}^{a_i-1} \cot(\tfrac{\pi k}{a_i})\cot(\tfrac{\pi k b_i}{a_i})\sin^2(\tfrac{\pi k b_i}{a_i}). \tag{3.1}$$

This agrees with the formula of Fintushel-Stern [**13**] for their invariant $R(a_1,\cdots,a_n)$, an equivariant index computed using the Atiyah-Segal G-index theorem [**1**] rather than the Atiyah-Patodi-Singer theorem as done here. Indeed, their formula is obtained from this one by making the summation index substitution $r = kb_i$ in the inner sum, and using the fact that $a\sum_{j=1}^{n}\frac{b_j}{a_j} = 1$ to express the inverse of b_i mod a_i as $b_i^{-1} = \frac{a}{a_i}$. The Neumann-Zagier formula (2.10) can then be applied to the summands in (3.1) to explicitly compute $\operatorname{Ind}^+(E,\alpha)$ in terms of the Seifert invariants $a_1,\cdots,a_n$. After some simplification one obtains

$$\operatorname{Ind}^+(E,\alpha) = 2n - 3 - 2\sum_{i=1}^{n} K_i$$

where K_i are integers satisfying $0 < b_i + K_i a_i < a_i$.

Assume that for the choice of L whose Euler class $e = e(L)$ is Poincaré dual to F, $\operatorname{Ind}^+(L\oplus\epsilon,\alpha) > 0$. Assume further that $\Sigma = \Sigma(a_1,\cdots,a_n)$ bounds a positive definite manifold B with $H_1(B;\mathbb{Z}/2) = 0$.

Extend E over the union of $N = X\cup_\Sigma(-B)$ by taking the trivial bundle over $-B$. In this way we obtain a bundle over N which we denote E_N. The remark after Proposition 2.6 shows that $\operatorname{Ind}^+(E_N,\alpha) = \operatorname{Ind}^+(L\oplus\epsilon,\alpha) > 0$. Since $\tau(L(a_i,b_i),\alpha) \geq \frac{1}{a_i}$, it follows that $\hat\tau(\partial N) \geq \min\{\frac{1}{a_i}\} > \frac{1}{a} = p_1(E_N,\alpha)$, and hence $\mathcal{M}(E_N,\alpha)$ is compact.

Since Σ is a homology sphere, $H^2(N)$ splits orthogonally with respect to the intersection form as $H^2(X)\oplus H^2(-B) = \operatorname{span}(e)\oplus W$ with W free abelian. Thus Theorem 2.16 implies that $C(e)$ contains a single point, contradicting Theorem 2.17.

This shows that if $\operatorname{Ind}^+(E,\alpha) > 0$, then $\Sigma(a_1,\cdots,a_n)$ cannot bound any positive definite manifold B with $H_1(B;\mathbb{Z}/2) = 0$. This recovers the original result of Fintushel-Stern [**13**] using cylindrical end moduli spaces in place of orbifold moduli spaces.

Notice that the argument did not use any information about the 0-framed component in Figure 1. In particular, we can replace this unknot by *any knot* (and the punctured disk which makes up F by a punctured Seifert surface for K) to get the following slight generalization of the Fintushel-Stern result:

THEOREM 3.1. *Let $a_1,\cdots,a_n$ be a set of relatively prime integers which satisfy $R(a_1,\cdots,a_n) > 0$.*

Let K be any knot in S^3, and let Σ be the homology 3-sphere obtained from $\frac{a_i}{b_i}$ Dehn surgeries on n parallel copies of the meridian of K and 0-surgery on K, where the b_i are integers chosen so that $(a_1\cdots a_n)\sum_i\frac{b_i}{a_i} = 1$. Then Σ does not bound a positive definite 4-manifold B with $H_1(B;\mathbb{Z}/2) = 0$. □

To be explicit, for any relatively prime positive integers p, q and positive integer k, the hypothesis of Theorem 3.1 holds with $(a_1,a_2,a_3) = (p,q,pqk-1)$. This is easily checked using the Neumann-Zagier formula.

We next prove Theorem 1, a generalization to Seifert fibered rational homology spheres of the theorem of Furuta stated in the introduction. The main difficulty not present in the original theorem concerns the count of reducibles: in contrast to the homology sphere case, $C(e)$ may have more than one class, and so the task will be to establish that the order of $C(e)$ is odd, in order to use Theorem 2.17 to reach a contradiction. Our purpose was not to find the most general theorem, but rather to illustrate the new complexities that arise when leaving the comfortable domain of integer homology spheres.

We mention the related results in the articles of Endo [**11**], B. Yu [**37**], and Mukawa [**30**]. Endo considers 2-fold branched covers of a certain infinite family of pretzel knots. These branched covers are Seifert fibered integer homology spheres, and Endo proves the knots are linearly independent in the smooth knot concordance group by using Furuta's theorem to rule out the existence of certain $\mathbb{Z}/2$-homology cobordisms which would exist if the knots satisfied a linear relation. Yu proves that certain Montesinos knots are not slice by considering their 2-fold covers, which are Seifert fibered rational homology spheres. Mukawa proves that certain Seifert fibered rational homology spheres have infinite order in the rational homology cobordism group. The emphasis in Mukawa's article is on removing the $\mathbb{Z}/2$ constraints on the homology of the bounding 4-manifolds. In a forthcoming article, [**18**], we will use this machinery to establish that the positive clasped untwisted Whitehead doubles of the $(2, 2^k - 1)$ torus knots are linearly independent in the smooth knot concordance group.

We begin with some generalities on Seifert fibered rational homology spheres. Any Seifert fibered 3-manifold which fibers over S^2 has the Kirby diagram given in Figure 1 for some choices of relatively prime pairs of integers (a_i, b_i) with $a_i > 0$. Let $a = a_1 \cdots a_n$ and $d = a\sum_i \frac{b_i}{a_i}$.

As before, let X be the 4-manifold indicated in Figure 1 (together with 3-handles) so that the boundary of X is the union of lens spaces $Q = \sqcup_i L(a_i, b_i)$ and the Seifert fibered 3-manifold $S = S(0; (a_1, b_1), \cdots, (a_n, b_n))$. Then S is a rational homology sphere if and only if d is non-zero; in fact, the order of $H_1(S)$ is $|d|$. Reversing orientation changes the sign of d. Assume, then that S is a rational homology sphere and orient S so that $d > 0$.

There is an exact sequence

$$0 \to H_2(X) \to H_2(X, Q) \xrightarrow{\partial} H_1(Q) \to H_1(X) \to 0. \tag{3.2}$$

From the handle description one computes this to be

$$0 \to \mathbb{Z} \to \mathbb{Z} \to \oplus_i \mathbb{Z}/a_i \to H_1(X) \to 0.$$

The surface F constructed from the punctured disk bounded by the 0-framed component and the core of the 0-handle has boundary meeting the $L(a_i, b_i)$ in circles μ_i, and is disjoint from S. The circles μ_i generate $H_1(L(a_i, b_i))$. The class $[F, \partial F]$ generates $H_2(X, Q) = \mathbb{Z}$, and $\partial([F, \partial F]) = [\partial F] = \mu_1 + \cdots + \mu_n \in H_1(Q)$. In particular, $H_1(X) \cong \big(\oplus_i \mathbb{Z}/a_i\big)/\langle \mu_1 + \cdots + \mu_n\rangle$. Our construction requires $H_1(X; \mathbb{Z}/2) = 0$; this happens if and only if at most one a_i is even.

As before,

$$F \cdot aF = -(a_1 \cdots a_n) \sum_{i=1}^{n} \tfrac{b_i}{a_i} = -d.$$

Taking $E = L \oplus \epsilon$ with $e(L) \in H^2(X)$ Poincaré dual to F, one calculates $\mathrm{Ind}^+(E, \alpha)$ just as in (3.1) and obtains
(3.3)

$$\mathrm{Ind}^+(E,\alpha) = \tfrac{2d}{a} - 3 + n + \sum_{i=1}^{n} \tfrac{2}{a_i} \sum_{k=1}^{a_i - 1} \cot(\tfrac{\pi k}{a_i}) \cot(\tfrac{\pi k b_i}{a_i}) \sin^2(\tfrac{\pi k b_i}{a_i}) = 2n - 3 - 2\sum_{i=1}^{n} K_i.$$

where K_i are integers satisfying $0 < b_i + K_i a_i < a_i$. The second equality follows from (2.10).

Chern-Simons invariants of flat $SO(3)$ connections on Seifert fibered rational homology spheres can be computed just as for their integer homology sphere counterparts. For our purposes the the following result will suffice.

LEMMA 3.2. *Let S be the rational homology sphere $S = S(0; (a_1, b_1), \cdots, (a_n, b_n))$ with $d = a \sum_i \frac{b_i}{a_i}$ odd and positive, where the $a_i > 0$ are relatively prime and $a = a_1 \cdots a_n$.*

(1) *If γ is a flat irreducible $SO(3)$ connection on the trivial bundle $E = S \times \mathbb{R}^3$, $\mathrm{cs}(S, \gamma) = \frac{q}{a} \bmod \mathbb{Z}$ for some integer q.*
(2) *If γ is a flat reducible $SO(3)$ connection on the trivial bundle, then $\mathrm{cs}(S, \gamma) = \frac{4q}{d} \bmod \mathbb{Z}$ for some integer q.*

PROOF. The fundamental group $\pi = \pi_1(S)$ has the presentation

$$\pi = \langle x_1, \cdots, x_n, h \mid h \text{ central}, \ x_i^{a_i} h^{b_i}, x_1 \cdots x_n \rangle.$$

Since d is odd, $H^1(S; \mathbb{Z}/2) = 0$, and so any representation $\pi_1(S) \to O(2)$ takes values in $SO(2)$. It follows that an irreducible $SO(3)$ representation necessarily takes h to $1 \in SO(3)$, since the centralizer of any non-abelian subgroup of $SO(3)$ not conjugate into $O(2)$ is trivial. The fundamental group of X is the quotient of π by h, and hence any irreducible representation of π extends to $\pi_1(X)$. Thus X provides a flat $SO(3)$ cobordism from S to a disjoint union of lens spaces.

The Chern-Simons invariants of the lens space $L(a, b)$ are fractions with denominator a and since Chern-Simons invariants modulo $\mathbb{Z}$ are flat cobordism invariants, the Chern-Simons invariant of an irreducible representation of π is a sum of fractions with denominators $a_1, \cdots, a_n$. Thus every irreducible flat $SO(3)$ connection γ on S satisfies $\mathrm{cs}(S, \gamma) = \frac{q}{a}$ for some integer q.

A reducible $SO(3)$ representation factors through $SO(2)$, hence has image in $\mathbb{Z}/d \subset SO(2)$. Integrating over a fundamental domain in the d-fold cover and using the fact that the pullback flat connection has trivial holonomy and hence has Chern-Simons invariant a multiple of 4 shows that $\mathrm{cs}(S, \gamma) = \frac{4q}{d}$ for some integer q. □

Let $\Theta^3_{\mathbb{Z}/2}$ denote the group of oriented $\mathbb{Z}/2$ homology spheres modulo $\mathbb{Z}/2$ homology bordism. We give the following variant of the result of Furuta, which corresponds to the case of $d = 1$. This theorem implies Theorem 1 of the introduction.

THEOREM 3.3. *Suppose that p, q, d are pairwise relatively prime positive odd integers. Let $n_1, n_2, \cdots$ be a strictly increasing sequence of even positive integers satisfying*

$$\gcd(d, n_k) = 1,$$
$$n_k > dn_i - d(d-1)/pq \textit{ for all } k > i,$$

and

$$n_1 > \frac{d}{pq}\max\{1+\frac{d}{pq}, 1+\frac{1}{p}, 1+\frac{1}{q}\}.$$

For example, when $d>1$ one can take $n_k = d^{k+s}-1$ for some fixed $s \gg 0$.

Then the rational homology spheres obtained by $-\frac{d}{n_k}$ surgeries on the left-handed (p,q) torus knot $k \geq 1$ are linearly independent in the $\mathbb{Z}/2$ homology cobordism group $\Theta^3_{\mathbb{Z}/2}$, and hence generate an infinitely generated free abelian subgroup.

PROOF. Choose integers r, s so that $ps + rq = -1$. Given an even integer $n \geq n_1$ satisfying $(d,n) = 1$, denote by S_n the Seifert fibered rational homology sphere

$$S_n = S(0; (p,r), (q,s), (pqn-d, n))$$

(the hypotheses imply that $pqn - d > n$). This manifold is diffeomorphic to $-\frac{d}{n}$ surgery on the $(p,-q)$ torus knot. One way to see this is to observe from Figure 1 that S_n is obtained from the Seifert fibration of $S^3 = S(0;(p,r),(q,s))$ by performing Dehn surgery on a regular fiber, which is the left handed (p,q) torus knot. For details see [**29**].

Let X_n denote the corresponding negative definite 4-manifold described in Figure 1, with boundary $S_n \sqcup Q$, where $Q = L(p,r) \sqcup L(q,s) \sqcup L(pqn-d,n)$. With the constructions given above, $H_1(S_n)$ is cyclic of order d and $H_1(X_n) = 0$.

We argue by contradiction. Suppose there exists an oriented $\mathbb{Z}/2$ homology punctured 4-ball B with boundary

$$\partial B = -\sum_{i=1}^{N} a_{n_i} S_{n_i}$$

with $a_{n_N} \neq 0$. We may assume, by changing orientation, that $a_{n_N} > 0$.

Let Z_{n_N} be a negative definite 4-manifold with boundary S_{n_N}. For example, a simply connected Z_{n_N} can be constructed by gluing negative definite manifolds to X_{n_N} along the lens spaces. Note that this is possible since the lens space $L(a,b)$, equipped with either orientation, bounds a negative definite 4-manifold obtained by plumbing disk bundles over S^2 according to the continued fraction expansion of $\frac{a}{b}$ (or, for $-L(a,b)$, the expansion of $\frac{a}{a-b}$).

Thus, we construct a negative definite 4-manifold N with boundary

$$\partial N = L(p,r) \sqcup L(q,s) \sqcup L(pqn_N - d, n_N) \sqcup_{i=1}^{N-1} a_{n_i} S_{n_i},$$

as an identification space from B, X_{n_N}, and $a_{n_N} - 1$ copies of Z_{n_N}. Here, N is obtained by gluing B to X_{n_N} along one S_{n_N}, and gluing the copies of Z_{n_N} to B along the remaining copies of S_{n_N}.

For the convenience of the reader we simplify notation and illustrate it in Figure 2. Let

- $Q = L(p,r) \sqcup L(q,s) \sqcup L(pqn_N - d, n_N)$
- S denote the first copy of S_{n_N},
- $X = X_{n_N}$,
- Z denote the disjoint union of $a_{n_N} - 1$ copies of Z_{n_N},
- $T = \sqcup_{i=1}^{N-1} a_{n_i} S_{n_i}$.

Thus $N = X \cup B \cup Z$, the boundary is given by $\partial N = Q \sqcup T$, and S separates N into X and $B \cup Z$.

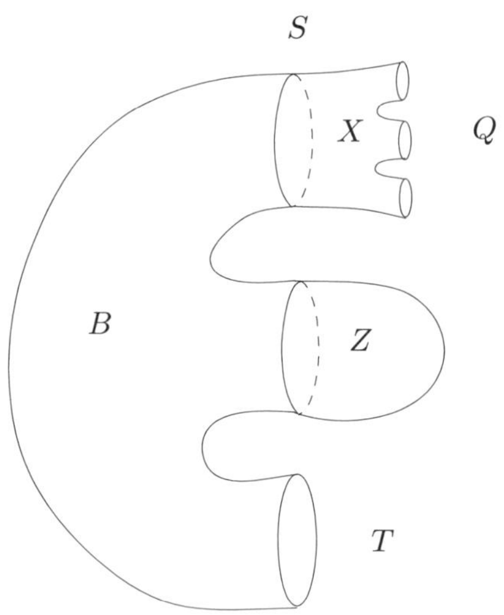

FIGURE 2.

Let $e \in H^2(N,T)$ denote the extension (by zero) of the class in $H^2(X,S)$ which is Lefschetz dual to the generator $[F,\partial F] \in H_2(X,Q) = \mathbb{Z}$. Let $L \to N$ denote the corresponding $SO(2)$ vector bundle with Euler class e, which is trivial over $B \cup Z$. Let $E = L \oplus \epsilon$, and (E,α) the corresponding adapted $SO(3)$ bundle.

Then

$$\begin{aligned} p_1(E,\alpha) &= -F\cdot F = \tfrac{1}{pq(pqn_N-d)}\big(rq(pqn_N-d)+ps(pqn_N-d)+pqn_N\big) \\ &= \frac{d}{pq(pqn_N-d)}. \end{aligned}$$

Suppose that K_1 and K_2 are integers satisfying $0 < r+K_1p < p$ and $0 < s+K_2q < q$. Then

$$-1 = ps+qr = p(s+K_2q)+q(r+K_1p)-pq(K_1+K_2) \leq p(q-1)+q(p-1)-pq(K_1+K_2)$$

so that

$$p+q-1 \leq pq(2-(K_1+K_2))$$

and hence $K_1+K_2 \leq 1$. Similarly

$$-1 \geq p+q-pq(K_1+K_2),$$

which implies $K_1+K_2 \geq 1$. Thus $K_1+K_2 = 1$.

Taking $(a_1,b_1) = (p,r), (a_2,b_2) = (q,s), (a_3,b_3) = (pqn_N-d, n_N)$ and $a = a_1a_2a_3$ in (3.3) we obtain (since $0 < n_N < pqn_N - d$ we take $K_3 = 0$):

(3.4)

$$\mathrm{Ind}^+(E,\alpha) = \tfrac{2d}{a} + \sum_{i=1}^{3} \tfrac{2}{a_i} \sum_{k=1}^{a_i-1} \cot(\tfrac{\pi k}{a_i})\cot(\tfrac{\pi k b_i}{a_i})\sin^2(\tfrac{\pi k b_i}{a_i}) = 3-2(K_1+K_2) = 1.$$

Each component S_{n_i} of T is a Seifert fibered rational homology sphere with singular fibers of multiplicities p, q and $pqn_i - d$. The bundle E is trivial over T. Lemma 3.2 implies that each irreducible flat connection on the trivial $SO(3)$ bundle over S_{n_i} has Chern-Simons invariant a rational number with denominator

$pq(pqn_i - d)$ and reducible flat connections have Chern-Simons invariant a rational number with denominator d. The (relative) Chern-Simons invariants of the lens spaces that make up Q are fractions with denominators p, q, and $pqn_N - d$ respectively. Lemma 2.8 then implies that

$$\min\{\tfrac{1}{pq(pqn_i-d)}, \tfrac{1}{d}, \tfrac{1}{p}, \tfrac{1}{q}, \tfrac{1}{pqn_N-d}\} \leq \hat{\tau}(Q \sqcup T, \alpha).$$

The hypothesis $n_N \geq n_1 > \frac{d}{pq}(1 + \frac{d}{pq})$ implies that

$$\tfrac{d}{pq(pqn_N-d)} < \tfrac{1}{d}.$$

Similarly, $n_N > \frac{d}{pq}(1 + \frac{1}{p})$ implies

$$\tfrac{d}{pq(pqn_N-d)} < \tfrac{1}{q}$$

and $n_N > \frac{d}{pq}(1 + \frac{1}{q})$ implies

$$\tfrac{d}{pq(pqn_N-d)} < \tfrac{1}{p}.$$

Finally, the hypothesis $n_N > dn_i - d(d-1)/pq$ implies that

$$\tfrac{d}{pq(pqn_N-d)} < \tfrac{1}{pq(pqn_i-d)}.$$

Taking the minimum of the right side for $i = 1, 2, \cdots, N-1$ we conclude

$$\tfrac{d}{pq(pqn_N-d)} = p_1(E, \alpha) < \hat{\tau}(Q \sqcup T, \alpha). \tag{3.5}$$

Since the path components of Q are lens spaces, the restriction of α to each component of Q is non-degenerate. Since T is a union of rational homology spheres and α restricts to the trivial connection on T, the restriction of α to each component of T is non-degenerate. With Equations (3.4) and (3.5), this shows that the hypotheses of Theorem 2.17 hold.

We now turn to the count of singular points (i.e. boundary points) of $\mathcal{M}(E, \alpha)$. Since each component of T has first homology $\mathbb{Z}/d$ with d odd and since $p, q, pqn_N - d$ are odd, every component of ∂N is a $\mathbb{Z}/2$ homology sphere. Theorem 2.16 then implies that $\mathcal{M}(E, \alpha)$ has $|C(e)|$ boundary components, where

$$C(e) = \{e' \in H^2(N) \mid e' \cdot e' = -\tfrac{d}{a}, e' \equiv e \bmod 2, e'|_T = 0, e'|_{Q_i} = \pm m_i, i = 1, 2, 3\}/\pm 1.$$

Here, Q_i are the three lens space components of Q and $m_i \in H^2(Q_i)$ is the Poincaré dual to the meridional generator of $H_1(Q_i)$.

Since p, q, and d are pairwise relatively prime, $H_1(Q) = \mathbb{Z}/p \oplus \mathbb{Z}/q \oplus \mathbb{Z}/(pqn_N - d) = \mathbb{Z}/a$, generated by $m_1 + m_2 + m_3$. Suppose that $e' \in C(e)$. Choose the unique integer k with $0 \leq k < a$ so that the restriction of e' to $H^2(Q)$ equals $k(m_1 + m_2 + m_3)$. Since $e' \in C(e)$, e' vanishes on T. Note that the sign of e' is ambiguously defined. It can be uniquely specified by requiring e' to restrict to m_1 in $H^2(L(p, r))$ (since p is odd), and so we assume this. This implies that $k \equiv 1 \bmod p$. Since $e' \in C(e)$, $k \equiv \pm 1 \bmod q$ and $k \equiv \pm 1 \bmod (pqn_N - d)$.

The class de' vanishes on S, and hence can be expressed as $de' = e_1 + e_2$ where $e_1 \in H^2(B \cup Z, \partial(B \cup Z))$ and $e_2 \in H^2(X, S)$. Since $B \cup Z$ is negative definite, $e_1 \cdot e_1 = -\ell_1$ for some non-negative integer ℓ_1. The restriction of e_2 to $H^2(Q)$ equals $dk(m_1 + m_2 + m_3)$, and since $H^2(X, S) \cong \mathbb{Z}\langle e \rangle$, we have $e_2 = (dk + a\ell_2)e$ for some $\ell_2 \in \mathbb{Z}$, and so $e_2 \cdot e_2 = -(dk + a\ell_2)^2 d/a$. Note $e_1 \cdot e_2 = 0$.

Then

$$-\frac{d}{a} = e' \cdot e' = \frac{1}{d^2}(e_1 + e_2) \cdot (e_1 + e_2) = -\frac{\ell_1}{d^2} - \frac{(dk + a\ell_2)^2}{da}.$$

We rewrite this as $d^3 = \ell_1 a + (dk + a\ell_2)^2 d$. Since d and a are relatively prime, $\ell_1 = d\ell_3$ for some integer $\ell_3 \geq 0$, and so

$$d^2 = \ell_3 a + (dk + a\ell_2)^2.$$

The fact that $\frac{d}{pq(pqn_N - d)} < \frac{1}{d}$ implies $a > d^2$, from which it follows that $\ell_3 = 0$. We therefore have $d = \pm(dk + a\ell_2)$. Reducing this last equation mod p yields $d \equiv \pm d \bmod p$. Since p is odd and $(d,p) = 1$, it follows that $d = dk + a\ell_2$. Reducing this equation mod q and mod $pqn_N - d$ shows that $k \equiv 1 \bmod q$ and $k \equiv 1 \bmod (pqn_N - d)$, and so $k = 1$. This implies that $\ell_2 = 0$, and that the restriction of e' to Q *equals* the restriction of e to Q.

Hence $e' - e$ vanishes on ∂N and

$$(e'-e)\cdot(e'-e) = -\tfrac{2d}{a} - 2e'\cdot e = -\tfrac{2d}{a} - \tfrac{2}{d}(de')\cdot e = -\tfrac{2d}{a} - \tfrac{2}{d}e_2\cdot e = -\tfrac{2d}{a} - \tfrac{2}{d}(de)\cdot e = 0.$$

Thus $e' - e \in H^2(N, \partial N)$ is a torsion class. Since $H^2(N, \partial N)$ has only odd torsion, it follows that $e' = e + 2x$ for a torsion class $x \in H^2(N, \partial N)$.

Conversely, every class of the form $e' + 2x$ for $x \in H^2(N, \partial N)$ torsion is in $C(e)$, and hence we have established a bijection between $C(e)$ and $\mathrm{Torsion}(H^2(N, \partial N))$. In particular, $C(e)$ is odd.

This contradicts the conclusion of Theorem 2.17, and hence the putative manifold B cannot exist. $\square$

For an explicit example of Theorem 3.3, consider $(p,q,d) = (3,5,7)$. Then the homology lens spaces obtained from $-\frac{7}{7^k - 1}$ surgery on the $(3,5)$ torus knot for $k = 1, 2, \cdots \infty$ are linearly independent in $\Theta^3_{\mathbb{Z}/2}$.

4. Example: non Seifert fibered homology 3-spheres

Using various techniques of the type discussed in Section 2.9 and topological constructions, one can greatly extend the range of applications of our method. For example, in [**18**] we prove that the untwisted Whitehead doubles of the $(2, 2^k - 1)$ torus knots are linearly independent in the smooth concordance group. We content ourselves for now with one further application, this time in a context which is homologically identical but geometrically distinct from the case of Seifert fibered homology spheres. The motivation here is to illustrate some of the new issues which arise when one moves beyond the study of Seifert fibered 3-manifolds.

Suppose that $L = K_1 \sqcup \cdots \sqcup K_n \subset S^3$ is a split link and $K_0 \subset S^3 \setminus L$ an additional component which links each K_i once. Perform Dehn surgery on $L \sqcup K_0$ with surgery coefficient 0 on K_0 and $\frac{a_i}{b_i}$ on K_i, where $(a_1 \cdots a_n) \sum \frac{b_i}{a_i} = 1$. This configuration defines a negative definite 4-manifold X, whose boundary is a union of homology lens spaces $Y_i = -S^3_{a_i/b_i}(K_i)$ and a homology sphere $Y_0 = \Sigma$ (the construction is homologically the same as the construction in Section 3). A Seifert surface for the 0-framed component can be chosen that meets each K_i transversely once. Let F denote the corresponding punctured Seifert surface together with the core of the 2-handle, pushed slightly into the interior of X. Let $e \in H^2(X;\mathbb{Z})$ be Poincaré dual to the relative homology class represented by F.

The analysis of Section 3 can be extended to this setting. Indeed, the homological data is identical to that of the Fintushel-Stern truncated mapping cylinder, and in the same way determines an adapted bundle $(E, \alpha) = (L_e \oplus \epsilon, \beta \oplus \theta)$. Therefore the part of the data that depends only on the homology, namely $C(e)$,

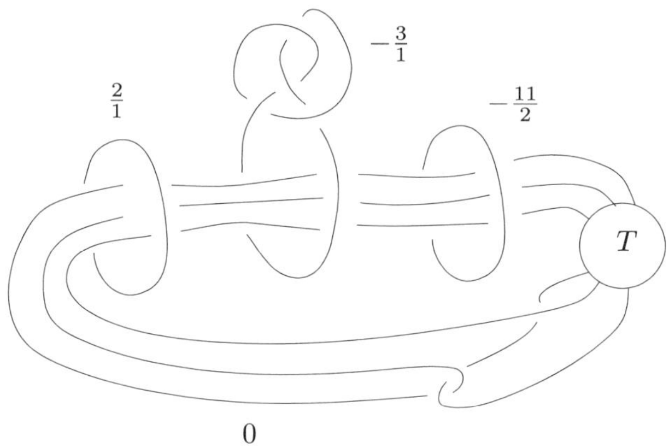

FIGURE 3.

$p_1(E,\alpha) = -e \cdot e$, and the holonomy of the flat connection β_i on Y_i, is the same as for the corresponding Seifert fibered example.

However, some of the invariants that appear in the formula for $\mathrm{Ind}^+(E,\alpha)$ and $\hat{\tau}(Y,\alpha)$ depend on more than the homological data. Since the Y_i are no longer lens spaces, to apply Theorem 2.17 one one needs to check non-degeneracy of the α_i on Y_i. One also must compute ρ invariants of the reducible flat connections α_i on Y_i. Finally, one needs to estimate Chern-Simons invariants of flat connections on Σ and $Y_i, i \geq 1$.

The flat connection α_0 is trivial on the homology sphere Y_0 and hence it is non-degenerate. However, the flat connections $\alpha_i = \beta_i \oplus \theta$ for $i \geq 1$ need not be non-degenerate for arbitrary choices of K_i and $\frac{a_i}{b_i}$. In fact α_i is non-degenerate if and only if $\Delta_{K_i}(\exp(2\pi i b_i/a_i)) \neq 0$ [**20**], where $\Delta_K(t)$ denotes the Alexander polynomial of K.

We show how this works with one example, illustrated in Figure 3. In this figure T can be any 3-stranded tangle so that the resulting K_0 is connected. The components K_1 and K_3 are unknotted and K_2 is the figure 8 knot. By adding two 3-handles one obtains a negative definite 4-manifold X with boundary the union of the homology sphere $Y_0 = \Sigma$ given by this surgery diagram and the disjoint union of $Y_1 = L(2,1), Y_3 = L(11,-2)$ and the homology lens space $Y_2 = -S^3_{-3/1}(K_2) = S^3_{3/1}(K_2)$, where the latter equality follows from the fact that K_2 is the amphicheiral figure 8 knot.

Since the homological data in this example is the same as that for $\Sigma(2,3,11)$, we see that $C(e)$ consists of a single point, $p_1(E,\alpha) = \frac{1}{66}$, and the holonomy of the flat connection β_i takes μ_i to $\exp(2\pi i(-b_i)/a_i)$ for $i = 1,2,3$, and is trivial for $i = 0$.

The Alexander polynomial of the figure 8 knot (resp. unknot) has no roots on the unit circle and so α_2 (resp. α_1, α_3) is non-degenerate. Also, $h_{\alpha_i} = 1$, and so applying Proposition 2.6 we obtain

$$\mathrm{Ind}^+(E,\alpha) = \tfrac{2}{66} - \tfrac{1}{2}\sum_{k=1}^{3} \rho(Y_i,\alpha_i).$$

The ρ invariants of $Y_1 = L(2,1)$ (resp. $Y_3 = L(11,-1)$) with respect to α_1 (resp. α_3) were computed above. We compute $\rho(Y_2, \alpha_2)$ using the Atiyah-Patodi-Singer theorem, as described in Section 2.9.

For any knot $K \subset S^3$ there exists a cobordism (rel boundary) Z_K from $S^3 \backslash n(K)$ to $S^3 \setminus n(U)$ (here U is the unknot), together with a homomorphism $H_1(Z_K) \to \mathbb{Z}$ which extends the abelianization map on the knot complement. One construction of such a Z_K is to push a Seifert surface for K into D^4 and perform surgery along circles to turn the Seifert surface into a 2–disk. The resulting 4-manifold contains a 2–disk whose complement is Z_K. For more details on this construction, we refer the reader to [**23**]. An alternate construction for such a Z_K can be found in [**18**].

One can then glue $I \times (S^1 \times D^2)$ to Z_K to obtain a cobordism W_K from $-\frac{a}{b}$ surgery on K to $L(a,b)$ over which β (and hence α) extends. Since $\alpha = \beta \oplus 1$ which complexifies to $\beta_{2\mathbb{C}} \oplus \bar{\beta}_{2\mathbb{C}} \oplus 1$, we conclude that

$$\rho(Y_2, \alpha_2) = \rho(L(3,1), \alpha_2) + \operatorname{Sign}_{\beta_{2\mathbb{C}}}(W_{K_2}) + \operatorname{Sign}_{\bar{\beta}_{2\mathbb{C}}}(W_{K_2}) - 2\operatorname{Sign}(W_{K_2}).$$

The quantity $\operatorname{Sign}_{\beta_{2\mathbb{C}}}(W_{K_2}) - \operatorname{Sign}(W_{K_2})$ equals the Levine-Tristram signature ([**26, 34**]) of the knot K_i at the $U(1)$-representation $\beta_{2\mathbb{C}}$ which sends the meridian of K_i to $e^{2\pi i(-b_2)/a_2}$. This can be proven using Wall non-additivity [**35**], and the details are carried out in [**23**]. Since K_2 is the figure 8 knot, all the Levine-Tristram signatures of K_2 are zero. This follows, for instance, from the fact that the Alexander polynomial of the figure 8 knot has no roots on the unit circle. Hence $\rho(Y_2, \alpha_2) = \rho(L(3,1), \alpha_2)$ and so

$$\operatorname{Ind}^+(E, \alpha) = R(2,3,11) = 1. \tag{4.1}$$

To apply Theorem 2.17 requires an understanding of $\hat{\tau}(Y, \alpha)$. This is the point where the method requires a deeper analysis of flat $SO(3)$ connections on Y_i. In the Seifert fibered case each K_i is an unknot, and Y_i is a lens space. Thus all $SO(3)$ representations are abelian, hence reducible, and their Chern-Simons invariants are easy to compute.

For the example of Figure 3,

$$p_1(E, \alpha) = \tfrac{1}{66}, \ \tau(Y_1, \alpha_1) \geq \tfrac{4}{2}, \text{ and } \tau(Y_3, \alpha_3) \geq \tfrac{4}{11}, \tag{4.2}$$

since $Y_1 = L(2,1)$ and $Y_3 = L(11,-2)$. The Chern-Simons is a flat cobordism invariant modulo $\mathbb{Z}$, and so, using the cobordism described above when computing ρ invariants of Y_2, we conclude that the Chern-Simons invariants of the reducible flat connections over Y_2 agree modulo $\mathbb{Z}$ with those of $L(3,1)$. Hence

$$\{\operatorname{cs}(Y_2, \alpha_2, \gamma)\} \geq \tfrac{1}{3},$$

where $\{x\}$ denotes the fractional part of of a real number x.

However, Y_2 admits irreducible flat $SO(3)$ connections, whose Chern-Simons invariants may contribute to $\hat{\tau}$. A method for computing Chern-Simons invariants of surgeries on the figure 8 knot is described in [**21**]. In particular, [**21**] calculates Chern-Simons invariants of -3 surgery on the figure 8 knot. There are two gauge equivalence classes of irreducible $SO(3)$ flat connections on Y_2, and both have Chern-Simons invariants rational numbers with denominator 24. Hence for γ an irreducible flat connection on Y_2, $\operatorname{cs}(Y_2, \alpha_2, \gamma)$ is a difference of fractions with denominators 3 and 24. Therefore

$$\{\operatorname{cs}(Y_2, \alpha_2, \gamma))\} \geq \tfrac{1}{24}.$$

Lemma 2.8 implies that $\tau(Y_2, \alpha_2) \geq \frac{1}{24}$. Together with Equation (4.2) this shows that $\hat{\tau}(Y, \alpha) > p_1(E, \alpha)$.

Theorem 2.17 then implies that Y_0 does not bound a positive definite 4-manifold. Moreover, a simple adaptation of the argument shows that Y_0 is of infinite order in $\Theta^3_{\mathbb{Z}/2}$. For this case, one must obstruct the existence of a punctured $\mathbb{Z}/2$ homology ball whose boundary consists of a disjoint union of manifolds, each of which are orientation-preserving diffeomorphic to $-Y_0$. For this, we cap off all of the $-Y_0$ with the negative definite 4–manifold X from the previous argument. Consider the adapted bundle (E, α) over the resulting manifold which is the stabilization of the $SO(2)$ bundle whose Euler class is Poincaré dual to the generator of $H^2(X)$ (for one copy of X). The estimates of the Chern-Simons invariant for flat connections on lens spaces and Y_2 given above show that the punctured $\mathbb{Z}/2$-homology ball does not exist, proving that Y_0 has infinite order. Note that to apply Theorem 2.17 here one must bound the relative Chern-Simons invariants $\mathrm{cs}(Y_i, \theta_i, \gamma)) > \frac{1}{66}$, where θ_i is the trivial connection on the trivial bundle (corresponding to the remaining copies of X on which the $SO(2)$ bundle restricts trivially). This estimate, however, follows exactly as in the discussion giving the bounds $\mathrm{cs}(Y_i, \alpha_i, \gamma)) > \frac{1}{66}$.

References

[1] M. F. Atiyah and G. B. Segal, *The index of elliptic operators. II,* Ann. Math. **87** (1968), 531–545.

[2] M. F. Atiyah, V. K. Patodi, and I. M. Singer, *Spectral asymmetry and Riemannian geometry. I,* Math. Proc. Cambridge Philos. Soc. **77** (1975), 43–69.

[3] ———— *Spectral asymmetry and Riemannian geometry. II,* Math. Proc. Cambridge Philos. Soc. **78** (1975), 405–432.

[4] D. R. Auckly, *Topological methods to compute Chern-Simons invariants,* Math. Proc. Cambridge Philos. Soc. **115** (1994), 229–251.

[5] A. J. Casson and C. McA. Gordon, *On slice knots in dimension three,* Algebraic and geometric topology (Proc. Sympos. Pure Math., Stanford Univ., Stanford, CA, 1976), Part 2, pp. 39–53, Proc. Sympos. Pure Math., XXXII, Amer. Math. Soc., Providence, RI, 1978.

[6] T. D. Cochran and W. B. R. Lickorish, *Unknotting information from 4-manifolds,* Trans. Amer. Math. Soc. **297** (1986), 125–142.

[7] A. Dold and H. Whitney, *Classification of oriented sphere bundles over a 4-complex,* Ann. Math. **69** (1959), 667–677.

[8] S. K. Donaldson, *An application of gauge theory to four-dimensional topology,* J. Differential Geom. **18** (1983), 279–315.

[9] ———— *Floer homology groups in Yang-Mills theory.* With the assistance of M. Furuta and D. Kotschick. Cambridge Tracts in Mathematics, 147. Cambridge University Press, Cambridge, 2002. viii+236 pp.

[10] H. Donnelly, *Eta invariants for G-spaces,* Indiana Univ. Math. J. **27** (1978), 889–918.

[11] H. Endo, *Linear independence of topologically slice knots in the smooth cobordism group,* Topology Appl. **63** (1995), 257–262.

[12] B. Fine, P. Kirk, and E. Klassen, E. *A local analytic splitting of the holonomy map on flat connections,* Math. Ann. **299** (1994), 171–189.

[13] R. Fintushel and R. J. Stern, *Pseudofree orbifolds,* Ann. Math. **122** (1985), 335–364.

[14] ———— *Instanton homology of Seifert fibred homology three spheres,* Proc. London Math. Soc. **61** (1990), 109–137.

[15] A. Floer, *An instanton-invariant for 3-manifolds,* Comm. Math. Phys. **118** (1988), 215–240.

[16] M. Furuta, *Homology cobordism group of homology 3-spheres,* Invent. Math. **100** (1990), 339–355.

[17] ———— *Perturbation of moduli spaces of self-dual connections,* J. Fac. Sci. Univ. Tokyo Sect. IA Math. **34** (1987), 275–297.

[18] M. Hedden and P. Kirk, *Instantons, Concordance and Whitehead doubling,* preprint (2010).

[19] B. J. Jiang, *A simple proof that the concordance group of algebraically slice knots is infinitely generated,* Proc. Amer. Math. Soc. **83** (1981), 189–192.
[20] E. P. Klassen, *Representations of knot groups in* SU(2), Trans. Amer. Math. Soc. **326** (1991), 795–828.
[21] P. A. Kirk and E. P. Klassen, *Chern-Simons invariants of* 3*-manifolds and representation spaces of knot groups,* Math. Ann. **287** (1990), 343–367.
[22] ——— *Chern-Simons invariants of* 3*-manifolds decomposed along tori and the circle bundle over the representation space of* T^2, Comm. Math. Phys. **153** (1993), 521–557.
[23] P. Kirk, E. P. Klassen, and D. Ruberman, *Splitting the spectral flow and the Alexander matrix,* Comment. Math. Helv. **69** (1994), 375–416.
[24] T. Lawson, *Invariants for families of Brieskorn varieties,* Proc. Amer. Math. Soc. **99** (1987), 187–192.
[25] ——— *A note on trigonometric sums arising in gauge theory,* Manuscripta Math. **80** (1993), 265–272.
[26] J. Levine, *Invariants of knot cobordism,* Invent. Math. **8** (1969), 98–110.
[27] G. Matić, SO(3)*-connections and rational homology cobordisms,* J. Differential Geom. **28** (1988), 277–307.
[28] J. W. Morgan, T. Mrowka, and D. Ruberman, *The* L^2*-moduli space and a vanishing theorem for Donaldson polynomial invariants,* Monographs in Geometry and Topology, II. International Press, Cambridge, MA, 1994. ii+222 pp.
[29] L. Moser, *Elementary surgery along a torus knot,* Pacific J. Math. **38** (1971), 737–745.
[30] T. Mukawa, *Rational homology cobordisms of Seifert fibred rational homology three spheres,* J. Math. Kyoto Univ. **42** (2002), 551–577.
[31] W. D. Neumann and D. Zagier, *A note on an invariant of Fintushel and Stern,* Geometry and topology (College Park, Md., 1983/84), 241–244, Lecture Notes in Math., 1167, Springer, Berlin, 1985.
[32] D. Ruberman, *Rational homology cobordisms of rational space forms,* Topology **27** (1988), 401–414.
[33] C. H. Taubes, *Gauge theory on asymptotically periodic* 4*-manifolds,* J. Differential Geom. **25** (1987), 363–430.
[34] A. Tristram, *Some cobordism invariants for links,* Proc. Camb. Phil. Soc. **66** (1969), 251–264.
[35] C. T. C. Wall, *Non-additivity of the signature,* Invent. Math. **7** (1969), 269–274.
[36] R. O. Wells, Jr. *Differential analysis on complex manifolds,* Third edition. With a new appendix by Oscar Garcia-Prada. Graduate Texts in Mathematics, 65. Springer, New York, 2008. xiv+299 pp.
[37] B. Z. Yu, *A note on an invariant of Fintushel and Stern,* Topology Appl. **38** (1991), 137–145.

Department of Mathematics, Michigan State University, East Lansing, MI 48824
E-mail address: mhedden@math.msu.edu

Department of Mathematics, Indiana University, Bloomington, IN 47405
E-mail address: pkirk@indiana.edu

AMS/IP Studies in Advanced Mathematics
Volume 50, 2011

Extending the $SU(3)$ Casson Invariant to Rational Homology 3-Spheres

Christopher M. Herald

ABSTRACT. An extension of the integer valued $SU(3)$ Casson invariant of integral homology 3-spheres in [**4**] to rational homology 3-spheres is described. This invariant counts irreducible flat connections modulo gauge, and incorporates correction terms involving flat reducible connections to arrive at a quantity that is independent of perturbation.

Introduction

These notes summarize the development in [**3**] of a gauge theoretic invariant of rational homology 3-spheres ($\mathbb{Q}$HSs) which generalizes the Casson invariant. Specifically, it is an extension of the $SU(3)$ invariant of integral homology spheres ($\mathbb{Z}$HSs) developed in [**4**] to the larger class of 3-manifolds. The approach is roughly to view the Chern-Simons function as a gauge invariant Bott-Morse function on the space of $SU(3)$ connections, and use it to define an infinite dimensional analog of the Euler characteristic.

This work, as well as several earlier parts of the program to study generalizations of the Casson invariant, has been the result of an enjoyable collaboration with Hans Boden. It is also a pleasure to thank the organizers of the "Chern-Simons Gauge Theory: 20 Years Later" workshop, where this work was presented. Finally, thanks to the referee for suggesting several improvements in the paper.

1. Historical overview

In 1985, Casson defined an invariant of oriented integral homology 3-spheres, i.e., closed, oriented manifolds X^3 with the same homology as S^3 (see [**1**]). Roughly speaking, this invariant is (one half of) a signed count of representations $\rho : \pi_1 X \to SU(2)$ modulo conjugation. This invariant reduces modulo 2 to the Rochlin invariant. Casson provided a Dehn surgery formula that described the effect on the invariant of surgery on a knot, in terms of the Alexander invariant of the knot.

Casson's definition of the invariant begins with a Heegaard decomposition $X = X_1 \cup_F X_2$. Consider the diagram of maps between the groups $\pi_1 F$, $\pi_1 X_1$,

---

2010 *Mathematics Subject Classification.* Primary 57M27; Secondary 58J30, 58E09, 58D27.
The author was partially supported by NSF Grant #0709625.

$\pi_1 X_2$ and $\pi_1 X$, determined by Seifert Van Kampen Theorem. Applying the functor $Hom(\cdot, SU(2))$/conjugation gives a description of the representations of $\pi_1 X$ as "representations of the surface group which extend over both handlebodies," which may be viewed as intersections of the representation varieties of X_1 and X_2 in the representation variety of F. By giving these three representation varieties orientations, Casson defined the invariant as an oriented intersection number. Although there are singularities in the representation varieties, the intersections in the smooth portion form a compact set and hence there is a well-defined intersection number of the smooth strata.

The Casson invariant was reinterpreted in terms of gauge theory by Taubes in 1992 (see [**12**]). He showed that it could be viewed as an infinite dimensional analog of the Euler characteristic, using the Chern-Simons function in place of a Morse function. An immediate consequence of Taubes' formulation was that the Casson invariant could be seen as the Euler characteristic of Floer's $\mathbb{Z}_8$-graded homology groups, known as instanton Floer homology (see [**9, 8**]).

Casson's invariant was extended to $\mathbb{Q}$HSs by Walker in 1992 in [**13**]. For $\mathbb{Q}$HSs, Casson's count of intersections of the smooth strata of the representation varieties is not independent of the perturbations necessary to make the intersections transverse, so Walker devised a correction term, using a sum over the intersections in the singular stratum (corresponding to abelian representations of $\pi_1 X$ to $SU(2)$). This correction term is another perturbation dependent quantity which cancels out the perturbation dependence of the first count of intersections.

Walker's extension to $\mathbb{Q}$HSs admits a purely combinatorial formulation, one that would not have have been possible without the extension beyond $\mathbb{Z}$HSs. This combinatorial description was used by Lescop in 1995 [**11**] to extend the invariant to all 3-manifolds. Lescop's extension has so far not been described in terms of representations of $\pi_1 X$.

Since Casson's invariant involves representations of $\pi_1 X$ into $SU(2)$, an obvious question is whether one can define analogous invariants for higher rank Lie groups. In 1992, Cappell, Lee and Miller proposed a program to extend Casson's intersection definition of the invariant to representation varieties associated to larger Lie groups [**6**]. Difficulties related to the singularities in the representation varieties remain an obstacle to completing this program, at least from this approach.

A different approach was taken in [**2**], extending Taubes' approach to the Casson invariant as an infinite dimensional Morse theoretic Euler characteristic to $SU(3)$, leading to an $SU(3)$ invariant of integral homology spheres. In this approach, the challenge of getting well-defined intersections for singular representation varieties is replaced by the challenge that the space of connections modulo gauge has singularities at the orbits of reducible connections. A signed count of critical points in the smooth stratum is not perturbation independent because critical points can disappear into the reducible stratum. Understanding this process requires some transversality results in the context of parameterized equivariant Morse functions.

After this initial work, the transversality results proven there were used to define a simpler, integer-valued $SU(3)$ invariant of integral homology spheres in [**4**]. In addition, Cappell, Lee and Miller adopted a similar gauge theory approach to define another $SU(3)$ invariant in [**7**]. All three of these $SU(3)$ invariants count irreducible flat connections modulo gauge with sign, but they differ in the nature of the correction term involving the singular stratum (i.e., the flat reducible connections). The invariants of [**2**] and [**4**], in particular, differ by a "weighted $SU(2)$ Casson invariant," a notion described in [**4**].

The work in [**3**], described in these notes, is aimed at extending the invariant of [**4**] beyond $\mathbb{Z}$HSs to $\mathbb{Q}$HSs. One motivation for doing so is that calculations in [**4**] and [**5**] for many families of Seifert fibered $\mathbb{Z}$HSs, such as all integral surgeries on torus knots, turned up some intriguing patterns. Extending to $\mathbb{Q}$HSs will allow more overlapping families of surgeries where similar methods for calculations can be applied.

We finish this overview with a philosophical note. The $SU(2)$ Casson invariant has been shown to be expressible in a number of ways: as the Euler characteristic of the Floer instanton homology, as a finite type invariant, or as the restriction to homology spheres of a global invariant defined via a surgery formula. From these different points of view, extending to higher rank might mean different things or might not make any sense at all. For example, when rank is larger than two, ordinary Floer homology is not well-defined, independent of perturbation; perhaps one could make sense of equivariant Floer homology, but this is necessarily infinitely generated, in which case the Euler characteristic is not well-defined. In this paper, we focus on extension in one particular direction. Namely, in the context of higher rank connections, we focus on finding a way to count irreducible connections modulo gauge with sign, in analogy with Taubes' approach, and providing a suitable correction involving reducibles, so that the result is perturbation invariant. (It is in this sense that the three invariants of [**2**], [**4**], and [**7**] all qualify as $SU(3)$ Casson invariants. One should think of them as the same invariant normalized differently by addition of different $SU(2)$-related invariants.)

Given that Taubes' formulation of the Casson invariant is an analog of the Euler characteristic in finite dimensions, obtained from a Morse function by counting critical points with sign, one can ask what is the finite dimensional analog of the generalized invariant we seek? The answer to this is the following. The $SU(3)$ generalization for integral homology spheres in [**4**] may be viewed as an analog of the relative Euler characteristic $\chi(M/S^1, F/S^1)$, where M is a manifold with semifree S^1 action, and F is the fixed point stratum. Consider an invariant Bott-Morse function f with isolated critical orbits with the property that, for any critical point p in F, the Morse index of p with respect to the function $f|_F$ equals the Morse index of p with respect to $f : M \to \mathbb{R}$. In other words, the function increases in all directions normal to F. For such a function, the relative Euler characteristic is just a signed count of the critical points in the smooth stratum (and the correction term is zero). The finite dimensional analog of [**4**] allows us to calculate this relative Euler characteristic from the critical set and Morse index data of *any* invariant Bott-Morse function with isolated critical orbits, whether or not it satisfies the hypothesis above concerning the normal directions to F.

In this paper, we will discuss an extension of this invariant to rational homology spheres. Once again, there is a finite dimensional analog. Let $M \subset SU(3)^k$ be a submanifold which is invariant under the adjoint $SU(3)$ action $g(g_1, \ldots, g_k) = (gg_1g^{-1}, \ldots, gg_kg^{-1})$, and which misses the bottom orbit type stratum $\left(Z_{SU(3)}\right)^k$, where $Z_{SU(3)}$ denotes the center of $SU(3)$. Then, an invariant Bott-Morse function with isolated critical orbits can be chosen so that, for every critical point that is not in the top stratum, the overall Morse index equals the Morse index of the function restricted to the stratum. In the analog of the Casson invariant formula we derive here, the correction term is zero and the formula reduces to a signed count of critical points in the top stratum. Thus it is a Morse theoretic formula for $\chi(M/SU(3), S)$, where S denotes the singular set of the quotient space. Once again, the correction term allows one to use the formula to calculate this invariant for a broader family of Bott-Morse functions.

2. Taubes' gauge theoretic approach to the Casson invariant

The reinterpretation of the Casson invariant as an infinite dimensional Morse theoretic Euler characteristic by Taubes is based on the following observations. Consider the set $\mathcal{A}$ of $SU(2)$ connections on the trivial bundle $E = X \times \mathbb{C}^2$, where X is a closed, oriented 3-manifold. The Chern-Simons function $cs : \mathcal{A} \to \mathbb{R}$ has the property that its critical points are exactly the flat connections, i.e., those with curvature zero. Once a basepoint x_0 is chosen and a basis is chosen for E_{x_0}, each flat connection determines a holonomy representation from $\pi_1 X$ to $SU(2)$.

The gauge group, $\mathcal{G} = Map(X, SU(2))$, consisting of automorphisms of E, acts on the space of connections, and the Chern-Simons function descends to a function $cs : \mathcal{A}/\mathcal{G} \to \mathbb{R}/\mathbb{Z}$. Here the critical points are the gauge orbits of flat connections, which correspond bijectively to the representations $\rho : \pi_1 X \to SU(2)$ up to conjugation. In this way, we view the Casson invariant as a count of critical points of cs on $\mathcal{A}/\mathcal{G}$.

Several points bear further explanation. $\mathcal{A}/\mathcal{G}$ is not a smooth manifold everywhere; it has quotient singularities at the orbits of reducible connections. Rank two reducible connections have the form $A = A_1 \oplus A_2$ on a splitting $E = E_1 \oplus E_2$. (The fact that $A_1 \oplus A_2$ is an $SU(2)$ connection forces $A_2 = \overline{A}_1$, but this is unimportant for the discussion of stabilizers.) The stabilizer of such a connection is isomorphic to either a 1-torus or $SU(2)$, depending on whether A_1 is gauge equivalent for A_2. By comparison, irreducible connections have stabilizer isomorphic to $Z_{SU(2)} \cong \mathbb{Z}_2$. Let $\mathcal{A}^*$ denote the set of irreducible connections.

If X^3 is a $\mathbb{Z}HS$, there is little interaction of the critical set of cs with the reducible connections. Specifically, this homology condition means that the abelianization of $\pi_1 X$ is trivial, so the only reducible flat connection up to gauge is the trivial connection. The linearization of the flatness equation on a slice to the $\mathcal{G}$ action on $\mathcal{A}$ can be related to cohomology of X with twisted $su(2)$ coefficients, and cohomological considerations show that the trivial connection orbit is isolated. Since the moduli space of flat connections is compact, this implies that the critical set in the smooth part $\mathcal{A}^*/\mathcal{G}$ of $\mathcal{A}/\mathcal{G}$ is compact.

As with Casson's intersection approach to the invariant, where the representation varieties of the two handlebodies may need to be perturbed before they intersect transversely in finitely many points, Taubes' approach also requires perturbation. Taubes and Floer developed a class of functions $h : \mathcal{A}/\mathcal{G} \to \mathbb{R}$ based on holonomies around curves (averaged over a tubular neighborhood, for analytical reasons), which can be added to the Chern-Simons function; the critical set $\mathcal{M}_h$ of $cs + h$ is called the *perturbed moduli space* associated to h. For generic choices of perturbation, the moduli space consists of finitely many orbits, cut out transversely by the perturbed flatness equation.

The cohomology condition at the trivial connection implies that any two generic small perturbations can be connected by a path such that the parameterized moduli space is a compact cobordism between the irreducible parts of the two perturbed moduli spaces. With a suitable orientation on the cobordism, related to the orientations on points in the moduli space, this shows the signed count of points in the moduli space is perturbation independent. (Taubes had to do more work to show that the invariant was equal to that defined by Casson, but we highlight this cobordism because it is central to the discussion of correction terms later in the paper.)

Recall that given a compact, finite dimensional manifold M and a Morse function $f : M \to \mathbb{R}$, we can express the Euler characteristic in terms of the Morse index $\mu : \text{crit}(f) \to \mathbb{Z}$ with the formula

$$\chi(M) = \sum_{c \in \text{crit}(f)} (-1)^{\mu(c)}.$$

We described Taubes' sign convention earlier as an infinite dimensional analog of this Morse theoretic invariant. We now explain the analogy in more detail.

The quotient space $\mathcal{A}/\mathcal{G}$ is infinite dimensional and, at any critical point, the Hessian of cs has infinitely many positive and negative eigenvalues. Thus the Morse index is infinite, and the Euler characteristic formula does not immediately extend to this situation. Instead, we replace the Morse index with a relative Morse index, defined as the spectral flow of (the negative of) the Hessian of $cs : \mathcal{A}^*/\mathcal{G} \to \mathbb{R}/\mathbb{Z}$ from some fixed basepoint to each critical point. Here spectral flow means the number of eigenvalues that cross zero, counted with sign, along a path from the base point to the critical point.

Taubes observed that one could instead use the spectral flow of the twisted signature operator

$$K_A = \begin{bmatrix} 0 & d_A^* \\ d_A & -\operatorname{Hess} cs(A) \end{bmatrix} : \Omega^0(X) \oplus \Omega^1(X) \to \Omega^0(X) \oplus \Omega^1(X),$$

indexed by the connection A, where all differential forms have $su(2)$ coefficients. This has the advantage that it makes sense between any critical points in $\mathcal{A}/\mathcal{G}$, not just irreducible ones, so Taubes used the trivial connection θ as the basepoint.

As mentioned above, it may be necessary to perturb the flatness equation in order to get a nondegenerate, zero dimensional critical set (and hence, by compactness, a finite set of perturbed flat gauge orbits). To get signs for perturbed flat connections, we extend the family K_A to a family parameterized by both a connection A and a perturbation h, namely

$$K_{A,h} = \begin{bmatrix} 0 & d_A^* \\ d_A & -\operatorname{Hess}(cs + h)(A) \end{bmatrix}.$$

Then, if h is a perturbation used to make the moduli space nondegenerate, and A is a perturbed flat connection, $[A]$ is counted with sign $(-1)^{\mathrm{SF}(K_{\theta,0}, K_{A,h})}$. We will, however, abuse the notation and leave the reference to the perturbation h out of the notation for the spectral flow.

Several comments about spectral flow are in order. First, it is a homotopy invariant of paths, relative to the endpoints. This is why we have not specified a path along which to take the spectral flow. Secondly, while spectral flow does depend on the gauge representative of the endpoint orbits, its parity, and hence the sign, is well-defined on gauge orbits.

THEOREM 2.1 ([**12**]). *For a $\mathbb{Z}$HS X, the quantity*

$$\sum_{[A]\in\mathcal{M}_h^*(X)} (-1)^{\mathrm{SF}(K_{\theta,0},K_{A,h})}$$

is independent of small, generic perturbation h, and equals twice the Casson invariant of X.

3. Reducible connections

The subject of this paper, an $SU(3)$ invariant for rational homology spheres, differs from the setting considered by Taubes and Casson in two ways. First, with the rank increased to three, there are now more partitions of the vector bundle dimension, and hence more types of reducible connections are possible. Secondly, relaxing the assumption on $H_1(X;\mathbb{Z})$ allows nontrivial gauge orbits of abelian flat connections. The $\mathbb{Q}$HS condition still forces the abelian flat moduli space to be a finite collection of points, but it does not require them to be isolated (i.e., these points may be in the closure of the other strata of the flat moduli space).

In the gauge theory framework, either of these changes means that there are more types of singularities in $\mathcal{A}/\mathcal{G}$ that may interfere with the Morse theoretic approach of Taubes. From Casson's point of view, the representation varieties have a more complicated singular structure than before. Either way, the reducibles must be dealt with, because a signed count of irreducible points is not an invariant. The generalization to structure group $SU(3)$ for $\mathbb{Z}$HSs was considered in [**2**], [**4**] and [**7**]. The $SU(2)$ extension to $\mathbb{Q}$HSs was performed by Walker. In both situations, there is only one stratum of reducibles with which $\mathcal{M}^*$ interacts, so each necessitates a single correction term. Presently, we will consider structure group $SU(3)$ and $\mathbb{Q}$HSs, so there are multiple singular strata to deal with.

We will now detail the types of reductions in the present case. First, note that there are three ways to partition the rank three into a sum of nonzero dimensions, which we denote 3, $2\oplus 1$, and $1\oplus 1\oplus 1$. Correspondingly, a connection on a $\mathbb{C}^3$ bundle can be irreducible, reducible into a direct sum of connections of ranks two and one (but no further), or reducible into a direct sum of three rank one connections. In the latter case, the connection has abelian holonomy, but there are actually three orbit types for abelian rank three connections. Specifically, the orbit type depends on whether any of the rank one summand connections are isomorphic to one another. To reflect the different multiplicities (up to gauge), we thus denote the possible orbit types: 3 (irreducible), $2\oplus 1$ (reducible, but nonabelian), $1\oplus 1\oplus 1$ (abelian with distinct summands), $1\oplus 1^2$, and 1^3 (central). We depart from this notation slightly by continuing to denote the irreducible type with a $*$ superscript.

In the previous discussion, the statements about orbit type do not depend on whether we are considering $SU(3)$ or $U(3)$ connections, although in each case the size of the stabilizer depends on whether we take the $SU(3)$ or $U(3)$ gauge group acting on the connections.

4. Bifurcations in the moduli space

In this section, we summarize key transversality results concerning changes in the topology of the moduli space under perturbation. These results are used to show that the $SU(3)$ Casson invariant formula is independent of perturbation. The space of holonomy perturbations can be given the structure of a Banach space (see [**10**], for example). Let B_δ denote the open ball of radius δ about zero in this space.

Given a path $\gamma = \{h_t \mid 0 \leq t \leq 1\}$ of perturbations, we define the parameterized moduli space to be

$$W_\gamma = \{([A], t) \mid [A] \in \mathcal{M}_{h_t}, 0 \leq t \leq 1\}.$$

We call a point in W_γ irreducible, if the connection component is irreducible, and similarly with the other orbit type descriptions. In particular, we can decompose W_γ into orbit type strata,

$$W_\gamma = W_\gamma^* \cup W_\gamma^{2\oplus 1} \cup W_\gamma^{1\oplus 1\oplus 1} \cup W_\gamma^{1\oplus 1^2} \cup W_\gamma^{1^3}.$$

THEOREM 4.1 (Theorem 65, [**10**]). *There exists a $\delta > 0$ such that for h in a residual subset of B_δ, the perturbed flat moduli space $\mathcal{M}_h(X)$ is a finite set of gauge orbits $[A]$ that are cut out transversely. In other words, $cs + h$ is Bott-Morse with finitely many isolated critical orbits.*

If δ is sufficiently small, then for any two such perturbations h_0 and h_1 in B_δ, generic paths γ in B_δ from h_0 to h_1 have the following properties: W_γ^σ is a union of arcs for each $\sigma \in \{*, 2\oplus 1, 1\oplus 1\oplus 1, 1\oplus 1^2, 1\}$. The individual arcs are compact except for noncompact ends that limit to distinct points in a more reducible stratum, as follows.

- *ends of W_γ^* may limit to points in $W_\gamma^{2\oplus 1}$,*
- *ends of $W_\gamma^{2\oplus 1}$ may limit to points in $W_\gamma^{1\oplus 1\oplus 1} \cup W_\gamma^{1\oplus 1^2}$, and*
- *$W_\gamma^{1^3}$, $W_\gamma^{1\oplus 1^2}$ and $W_\gamma^{1\oplus 1\oplus 1}$ are compact product cobordisms.*

Figure 1 illustrates the structure of the parameterized moduli space, as described in the theorem. The parameterized moduli space connecting any two nondegenerate perturbed moduli spaces is a 1-dimensional cobordism, with constant orbit type on each arc, except for three types of singularities. Each type of singularities involves an arc of one type limiting to a point on a more reducible arc, forming a T-intersection, illustrated in Figure 2; the types that can interact this way are listed in the theorem.

We make the following comments on terminology. A *T-intersection* describes the topology of the parameterized moduli space in a neighborhood of one of the singularities. From a different point of view, a *bifurcation* is the change that occurs in the moduli space, or equivalently in the t-slice $\mathcal{M}_{h_t} \times \{t\}$ of the parameterized moduli space, as t is varied and a T-intersection is passed. (As indicated in Figure 1, there may also be other changes in the topology which are simply standard Morse births and deaths of pairs of points within one stratum; these do not reflect singularities in the cobordism.)

Note that, as in the case of structure group $SU(2)$ on $\mathbb{Z}$HSs considered by Taubes, the central stratum does not interact with the others, so it can be safely ignored. For this reason, this stratum is left out of Figure 1.

5. Correction term for $2 \oplus 1$ stratum

In this section, we discuss the construction of a correction term for the $2 \oplus 1$ points. Specifically, when a bifurcation occurs involving the disappearance of an irreducible point into $\mathcal{M}^{2\oplus 1}$, the count of irreducibles changes by ± 1, so we seek

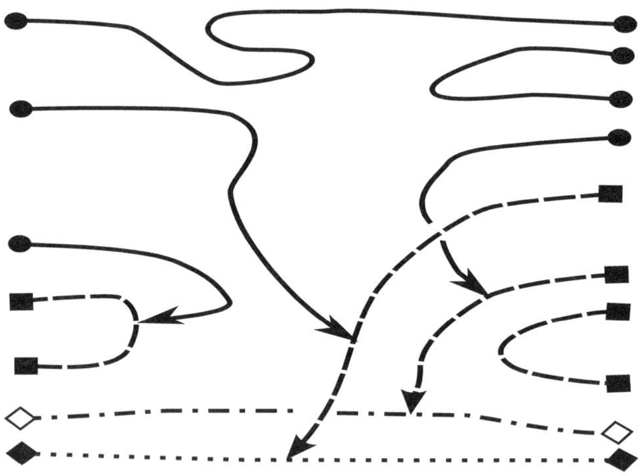

FIGURE 1. W_γ is made up of cobordisms with T-intersections. W_γ^* is represented by solid arcs, $W_\gamma^{2\oplus 1}$ by dashed arcs, and the $W_\gamma^{1\oplus 1\oplus 1}$ and $W_\gamma^{1\oplus 1^2}$ strata by the two arcs across the bottom. Arrows indicate a noncompact end of an arc limiting to a different stratum. Endpoints on the left and right represent points in $\mathcal{M}_{h_0}$ and $\mathcal{M}_{h_1}$, respectively.

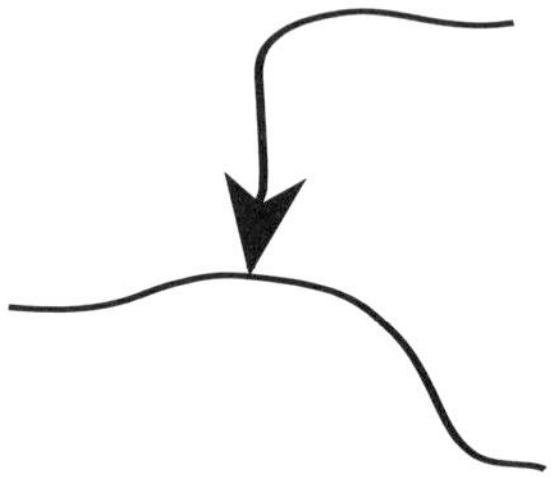

FIGURE 2. Each non-manifold point in W_γ has a neighborhood homeomorphic to a T-intersection. There are three orbit type combinations that can interact this way, listed in Theorem 4.1.

a formula involving spectral flows for points in $\mathcal{M}^{2\oplus 1}$ that changes in just such a way that it cancels out this change in the algebraic count of the points in $\mathcal{M}^*$.

To motivate the construction, we begin with a discussion of a simple model, illustrating an analogous bifurcation and a suitable correction term. Let $U(1)$ act on $M = \mathbb{C}$ by the usual action, and let z denote the usual complex coordinate on $\mathbb{C}$. Then $f_t(z) = \frac{1}{4}|z|^4 - \frac{t}{2}|z|^2$, $-1 \le t \le 1$, is a family of $U(1)$ invariant functions, parameterized by t. When $t < 0$, the only critical point is $z = 0$. But when $t > 0$, there is, in addition to $z = 0$, a whole circle of other critical points, defined by $|z|^2 = t$. In either the $t < 0$ or $t > 0$ cases, the critical set is a finite collection of nondegenerate critical orbits, the best one can hope for with invariant functions. One can show that this bifurcation behavior is codimension one in the space of invariant functions, or codimension zero in the space of paths of functions, so it cannot be eliminated by perturbing the path of functions.

If we divide out by the group action, $M/U(1)$ is a ray with an endpoint (which is a singularity in the quotient space), and we can view f_t as a path of functions on it. For $t < 0$, its critical set on the smooth stratum is empty; for $t > 0$, it consists of one nondegenerate critical point. It is instructive to consider the parameterized critical set, modulo the group action, in this case. It consists of a 1-manifold $\{([0], t) \mid -1 \leq t \leq 1\}$, as well as a noncompact 1-manifold $\{([z], t) \mid |z|^2 = t,\ t > 0\}$ which limits to the point $([0], 0)$ in the previous 1-manifold. The two pieces together, in the quotient topology of $M/U(1)$, fit together to form a T-intersection as in Figure 2.

Clearly, no signed count of critical points in the smooth stratum can be independent of perturbation. What's more, no signed count of orbits of critical points in both strata can be perturbation invariant, because the parity changes. In the context of this simple model, in order to get an "Euler characteristic" type of invariant that counts critical points in the smooth stratum of $M/U(1)$, even controlling for the limiting behavior of f_t at infinity (which has nothing to do with the problem illustrated above), we must add a correction term involving the critical point fixed by the group action. A suitable correction term is $-\frac{\mu(0)}{2}$, one half the Morse index of the fixed point critical point $z = 0$.

More generally, if M is a manifold with semifree $U(1)$ action, F is the submanifold of fixed points, and $M^* = M \setminus F$, then a similar behavior takes place: circle orbits of critical points in M^* can pop out of fixed point critical points in F, as the $U(1)$ invariant function is varied. At any critical point p in F for an invariant function, the Hessian is $U(1)$ equivariant, and hence it decomposes into a direct sum of symmetric bilinear forms on T_pF and N_pF. Let $\mu_N(p)$ denote the normal Morse index, i.e., the number of negative eigenvalues in the normal summand.

The quantity

$$\sum_{[c] \in \text{crit}(f) \cap (M^*/U(1))} (-1)^{\mu([c])} - \sum_{b \in \text{crit}(f) \cap F} (-1)^{\mu(b)} \frac{\mu_N(b)}{2}$$

expresses the relative Euler characteristic $\chi(M/U(1), F/U(1))$, for any $U(1)$ invariant Bott-Morse function f with isolated critical orbits.

We now return to the $SU(3)$ gauge theory setting. From the existence of T-intersections where irreducible points disappear into the $2 \oplus 1$ stratum, it is clear that a signed count of irreducible points in the moduli space alone will not be independent of perturbation. It is also clear that adding a signed count of $2 \oplus 1$ points will not fix the problem; rather, we must use a correction term involving $2 \oplus 1$ points that changes by ± 1 each time such a bifurcation occurs.

Consider first a bifurcation wherein a $2 \oplus 1$ point gives birth to a new irreducible point. Specifically, suppose $[B_t] \in \mathcal{M}_{h_t}^{2\oplus 1}$, for $t \in [0,1]$, and $[A_t] \in \mathcal{M}_{h_t}^*$ for $t > \frac{1}{2}$ represents the irreducible point created in the bifurcation. In analogy with the $M/U(1)$ discussion above, we need to count the $2 \oplus 1$ points with some weight $F(B_t)$ such that $F(B_1) - F(B_0) = (-1)^{\mathrm{SF}(\theta, A_1)}$. Such bifurcations on the $2 \oplus 1$

stratum coincide with spectral flow of K_{B_t,h_t}. Specifically, an eigenvalue of K_{B_t,h_t} with 2-dimensional eigenspace normal to the $2\oplus 1$ stratum crosses zero. There is a key relationship between $\mathrm{SF}(\theta, A_1)$, the normal spectral flow $\mathrm{SF}_N(B_0, B_1)$, and the overall spectral flow $\mathrm{SF}(\theta, B_0)$ (see Proposition 7.1 for details).

A strategy to generalize the $M/U(1)$ correction term above to the present gauge theory setting is the following. Fix a $2\oplus 1$ connection A_0 to use as a basepoint. Then, for each perturbed flat $[A] \in \mathcal{M}_h^{2\oplus 1}$, use the normal spectral flow from A_0 to A, in place of the normal Morse index, and place a factor of $(-1)^{\mathrm{SF}(\theta,A)}$ in front of the normal spectral flow to take care of the signs. What we mean here by normal spectral flow is the following. For a path A_t of $2\oplus 1$ connections, the tangent bundle to $\mathcal{A}$ at A_t decomposes into the tangent bundle to the reducibles and the normal bundle. There is a corresponding decomposition of the operator K_{A_t,h_t} into tangential and normal components, and normal spectral flow is the spectral flow of the latter component.

An immediate problem with this strategy is that we need the correction term to be well-defined on the gauge orbit $[A]$, but $\mathrm{SF}_N(A_0, gA)$ and $\mathrm{SF}_N(A_0, A)$ differ by $4 \deg g$, where $\deg : \mathcal{G} \to \mathbb{Z}$ is a surjective map. In other words, the quotient $\mathcal{A}/\mathcal{G}$ has nontrivial fundamental group, and spectral flow is a nontrivial homomorphism from this group to $\mathbb{Z}$.

Assume for the moment that X is a $\mathbb{Z}$HS. In this case, the only T-intersections involve ends of W_γ^* running into $W_\gamma^{2\oplus 1}$. We now recall the method used in [**4**] to create a well-defined correction term in this setting. While $\mathcal{M}^*$ need not be compact, $\mathcal{M}^{2\oplus 1}$ is compact, and so there are finitely many path components $C_1, \ldots, C_n$ of $\mathcal{M}^{2\oplus 1}$. Suppose that, for each component C_i, we choose a basepoint in $[\widehat{A}_i] \in C_i$. For small perturbations, the perturbed flat connections each remain close to some well-defined unperturbed flat moduli space component, and can be connected by short paths in $\mathcal{A}^{2\oplus 1}/\mathcal{G}$, along which spectral flow is well-defined. Thus, we fix a collection of disjoint neighborhoods U_i of the components C_i in $\mathcal{A}^{2\oplus 1}/\mathcal{G}$, and then we insist on using sufficiently small perturbations that the moduli space remains in the union of the neighborhoods, $\mathcal{M}_h^{2\oplus 1} = \bigcup_{i=1}^n \left(\mathcal{M}_h^{2\oplus 1} \cap U_i\right)$. A suitable correction term would then be

$$-\sum_{i=1}^{n} \sum_{[A]\in\mathcal{M}_h^{2\oplus 1}\cap U_i} (-1)^{\mathrm{SF}(\theta,A)} \frac{\mathrm{SF}_N(\widehat{A}_i, A)}{2},$$

where we insist that the normal spectral flow be computed using a path of connections that stays in the neighborhood.

There remains the question of making a canonical choice of basepoint $[\widehat{A}_i]$ in each component C_i. Given that the parity of $\mathrm{SF}(\theta, A)$ to an irreducible is unchanged under change of orientation on X, it would be nice to make the basepoint choice orientation independent, also. For this reason, we select the two extremes, choosing two connections $\widehat{A}_i^{\pm}$ such that the normal spectral flow along paths between them in C_i is maximized, and then average the normal spectral flow from the two to get the correction term. (Heuristically, one can think of this as picking a basepoint in the "middle" of C_i, in terms of normal spectral flow.)

We will write $\mathrm{SF}_N(\widehat{A}_i^{\mathrm{avg}}, A) = \frac{1}{2}\left(\mathrm{SF}_N(\widehat{A}_i^{+}, A) + \mathrm{SF}_N(\widehat{A}_i^{-}, A)\right)$, even though we haven't defined a connection $\widehat{A}_i^{\mathrm{avg}}$, to simplify the notation.

THEOREM 5.1 ([4]). *The quantity*

$$\sum_{[A]\in\mathcal{M}_h^*}(-1)^{\mathrm{SF}(\theta,A)}-\sum_{i=1}^{n}\sum_{[A]\in\mathcal{M}_h^{2\oplus 1}\cap U_i}(-1)^{\mathrm{SF}(\theta,A)}\frac{\mathrm{SF}_N(\widehat{A}_i^{\mathrm{avg}},A)}{2}$$

is independent of nondegenerate small perturbation and defines an invariant of $\mathbb{Z}$HSs.

We now consider $\mathbb{Q}$HSs once again. We begin by noting that isomorphism classes of one-dimensional subbundles $L\subset E=X\times\mathbb{C}^3$ correspond bijectively to elements $H^2(X;\mathbb{Z})$, via the map $L\mapsto c_1(L)$. For $\omega\in H^2(X;\mathbb{Z})$, let L_ω denote the line bundle with this Chern class. Similarly, rank two subbundles are also classified by c_1; in particular, each rank two subbundle $M\subset E$ is isomorphic to $M_\omega=L_\omega\oplus L_0$ where $c_1(M)=\omega$. In summary, each splitting into a rank two plus a rank one subbundle is isomorphic (as a direct sum) to $M_{-\omega}\oplus L_\omega$ for a unique $\omega\in H^2(X;\mathbb{Z})$. Note that each line bundle L_ω admits a unique gauge orbit of flat connections, so there are flat connections respecting every $2\oplus 1$, $1\oplus 1\oplus 1$, or $1\oplus 1^2$ bundle splitting.

In the prescription for choosing basepoints above, the compactness of $\mathcal{M}^{2\oplus 1}$, and hence of its path components, is essential, for it guarantees the existence of extrema $\widehat{A}_i^{\pm}$. On $\mathbb{Q}$HSs, $\mathcal{M}^{2\oplus 1}$ need not be compact; its closure may contain points in the abelian moduli space. This means that if we want to use the "average of the extremes" approach above for choosing basepoints, we must consider abelian connections which are limits of $2\oplus 1$ connections as candidates for basepoints, from which to measure normal spectral flow relative to points in $\mathcal{M}_h^{2\oplus 1}$.

On the other hand, an abelian connection on $L_{\omega_1}\oplus L_{\omega_2}\oplus L_{\omega_3}$ may be a limit point of $2\oplus 1$ flat connections three different bundle splittings: $M_{-\omega_1}\oplus L_{\omega_1}$, $M_{-\omega_2}\oplus L_{\omega_2}$, and $M_{-\omega_3}\oplus L_{\omega_3}$. In this case, a single component of the closure of $\mathcal{M}^{2\oplus 1}$ contains reducible connections on nonisomorphic splittings. Since $2\oplus 1$ connections in this component cannot be connected by a path of connections respecting a consistent splitting into ranks two and one, normal spectral flow on this component does not make sense.

For $\mathbb{Q}$HSs, we modify the definition of the $2\oplus 1$ correction term to be a sum of correction terms indexed by ω, the Chern class of the line bundle summand. Let $\mathcal{M}_\omega$ denote the flat moduli space of connections respecting a splitting isomorphic to $M_{-\omega}\oplus L_\omega$. Each component of $\mathcal{M}_\omega$ is compact, perhaps containing some abelian points. But on each component of $\mathcal{M}_\omega$, there is a consistent way of viewing all gauge orbits as $2\oplus 1$ type, so we can make sense of normal spectral flow throughout the component.

In order to extend the method described above, we also need neighborhoods of these components where we can make sense of normal spectral flow. We set

$$\mathcal{A}_\omega=\{A\in\mathcal{A}\mid A\cong A_1\oplus A_2 \text{ on } M_{-\omega}\oplus L_\omega\}.$$

Note that $\mathcal{A}_\omega$ is a gauge invariant set, containing not only $2\oplus 1$ connections (with the indicated line bundle) but also abelian ones. If $C\subset\mathcal{M}_\omega$ is a path component, then we can choose a path connected neighborhood of C in $\mathcal{A}_\omega/\mathcal{G}$ where normal spectral flow (with respect to the $M_{-\omega}\oplus L_\omega$ splitting) is well-defined on gauge orbits.

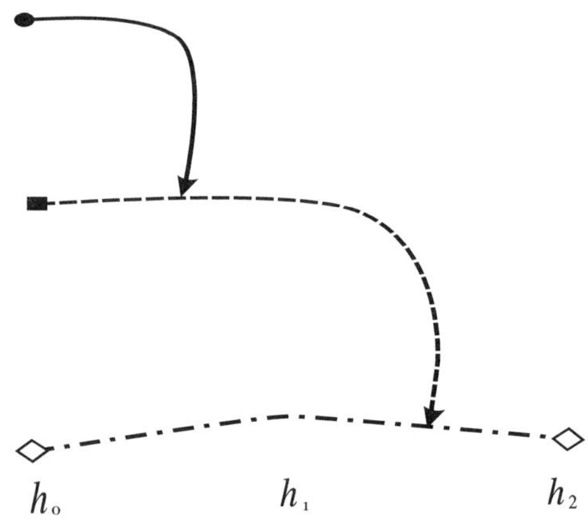

FIGURE 3. A scenario illustrating the need for abelian correction terms. $\mathcal{M}_{h_0}$ contains an irreducible point, a $2 \oplus 1$ point, and one abelian point. $\mathcal{M}_{h_1}$ contains only a $2 \oplus 1$ and an abelian point. $\mathcal{M}_{h_2}$ contains only an abelian point.

This allows us to extend the $2 \oplus 1$ correction term to this setting with a formula of the form

$$c_{2\oplus 1}(X, h, \omega) = -\sum_{i=1}^{n} \sum_{[A]\in \mathcal{M}^{2\oplus 1}_{\omega,h} \cap U_i} (-1)^{\mathrm{SF}(\Theta, A)} \frac{\left(\mathrm{SF}_N(\widehat{A}_i^{\mathrm{avg}}, A)\right)}{2}. \tag{5.1}$$

Specifically, this correction term changes by ± 1 in such a way that it compensates for the loss or gain of one irreducible point whenever one dies into or is born out of the $2 \oplus 1$ stratum, as in the top two T-intersections in Figure 1.

We must also deal with a second complication in the $\mathbb{Q}$HS setting, however, namely the noncompactness of the $2 \oplus 1$ stratum of the parameterized moduli space. Consider, for example, a path of perturbations for which the parameterized moduli space has the form indicated in Figure 3. As h_0 is varied to h_1, one irreducible point disappears into the $2 \oplus 1$ stratum; the change in the count of irreducibles would be counterbalanced by a change in the $2 \oplus 1$ correction term. On the other hand, if h_1 is varied further, over to h_2, then the $2 \oplus 1$ point in the moduli space disappears into the abelian stratum. When such a bifurcation occurs, although the signed count of irreducibles is not affected, the $2 \oplus 1$ correction term changes. Notice that this problem arises if the latter bifurcation involves either the $1 \oplus 1 \oplus 1$ or the $1 \oplus 1^2$ stratum.

In the scenario illustrated in Figure 3, the formula for the invariant will give different answers for h_0 and h_2 unless there is also a correction term involving abelian points which compensates for loss or gain of $2 \oplus 1$ points. Furthermore, the abelian correction terms must compensate with more than a ± 1, since the contribution of the $2 \oplus 1$ point that is gained or lost is more complicated.

In conclusion, we need $1\oplus 1\oplus 1$ and $1\oplus 1^2$ correction terms, in addition to the $2\oplus 1$ correction term described above, in order to get an invariant of $\mathbb{Q}$HSs. The next section describes these correction terms associated to the abelian strata, after defining the relevant notions of spectral flow that go into their definitions.

6. Correction terms for the abelian strata

Before discussing abelian correction terms in detail, we note a weakness in our notation regarding the $2\oplus 1$ correction term. Ultimately, we'll be writing down two more correction terms, one for each abelian stratum, and each involving the normal spectral flow with respect to that stratum. Our notation SF_N does not indicate the stratum with respect to which we are taking the "normal component." We now discuss the normal bundle structures in the different cases, and switch to more specific notation for the different normal spectral flow quantities.

Corresponding to each reduction type is a block form subgroup of $SU(3)$, which is the maximal holonomy subgroup for a connection of this type. Keeping in mind that the tangent space to $\mathcal{A}$ corresponds to the space of $su(3)$ valued 1-forms, we therefore look at the following Lie subalgebras of $su(3)$:

$$\begin{aligned}\mathfrak{s} &= Lie(S(U(2)\times U(1))),\\ \mathfrak{t} &= Lie(S(U(1)\times U(1)\times U(1))),\\ \mathfrak{u} &= Lie\left(\{\operatorname{diag}(\lambda^{-2},\lambda,\lambda)\mid \lambda\in U(1))\}\right).\end{aligned}$$

Let $\mathfrak{s}^\perp$, $\mathfrak{t}^\perp$, and $\mathfrak{u}^\perp$ denote their orthogonal complements in $su(3)$. Since $\mathfrak{u}\subset\mathfrak{t}\subset\mathfrak{s}$, we have $\mathfrak{s}^\perp\subset\mathfrak{t}^\perp\subset\mathfrak{u}$. For a path of $2\oplus 1$ connections, K_{A_t} respects the decomposition of $su(3)$ forms into $\mathfrak{s}$ and $\mathfrak{s}^\perp$ components, and these two components reflect the tangential and normal components. Similarly, $\mathfrak{t}$ and $\mathfrak{t}^\perp$ valued forms reflect the decompositition of T_A into $T_A\mathcal{A}^{1\oplus1\oplus1}\oplus N_A\mathcal{A}^{1\oplus1\oplus1}$, and similarly with $\mathfrak{u}$ and $\mathfrak{u}^\perp$ at the $1\oplus 1^2$ connections.

The diagonal maximal torus $S(U(1)\times U(1)\times U(1))$ of $SU(3)$ is its own centralizer, and under the action of this group $\mathfrak{t}^\perp\cong\mathbb{C}_{12}\oplus\mathbb{C}_{13}\oplus\mathbb{C}_{23}$ decomposes into complex lines with different weights. Similarly, for a path of $1\oplus 1\oplus 1$ connections, the normal spectral flow breaks up into three corresponding components, under the action of the stabilizer torus. For fixed ω, when considering $1\oplus 1\oplus 1$ connections on $L_{\omega'}\oplus L_{\omega''}\oplus L_\omega$ as a potential source of births and deaths of $2\oplus 1$ connections on $M_{-\omega}\oplus L_\omega$, however, different roles are played by the different components and it is helpful to write

$$\mathfrak{t}^\perp=(\mathfrak{t}^\perp\cap\mathfrak{s})\oplus\mathfrak{s}^\perp. \tag{6.1}$$

To understand the normal bundle to the $1\oplus 1^2$ connections, note that the stabilizer is isomorphic to $U(2)$, embedded in $SU(3)$ by the map $U\in U(2)\mapsto U\oplus\det(U^{-1})\in SU(3)$. Under the action of $U(2)$, we have a decomposition of $\mathfrak{u}^\perp$ into two irreducible representations, $\mathfrak{u}^\perp\cong su(2)\oplus\mathbb{C}^2$, and we have a similar decomposition of the normal spectral flow for $1\oplus 1^2$ connections. Homology considerations (for small perturbations) insure trivial $su(2)$ component spectral flow. This corresponds to the fact that Theorem 4.1 disallows $1\oplus1\oplus1$ arcs limiting to the $1\oplus 1^2$ arcs in W_γ. Thus the $\mathfrak{u}^\perp$ spectral flow is really just the $\mathbb{C}^2$ component, which comes in 4-dimensional eigenspaces due to the $U(2)$ symmetry. When considering births of $2\oplus 1$ connections on the splitting $M_{-\omega}\oplus L_\omega$ from $1\oplus 1^2$ connections on $L_{-2\omega}\oplus L_\omega\oplus L_\omega$, there is a decomposition $\mathfrak{u}^\perp=(\mathfrak{u}^\perp\cap\mathfrak{s})\oplus\mathfrak{s}^\perp$ analogous to (6.1), but due to the symmetry we can express these components in terms of the overall $\mathfrak{u}^\perp$ spectral flow.

Since the 2⊕1 correction term is a sum over splittings indexed by $\omega \in H^2(X;\mathbb{Z})$, the abelian correction terms, which account for the $2 \oplus 1$ points running into the abelian strata, will also be described as sums over those splittings.

Fix $\omega \in H^2(X;\mathbb{Z})$, and set $\mathcal{M}_\omega^{1\oplus1\oplus1} = \mathcal{M}_\omega \cap \mathcal{M}^{1\oplus1\oplus1}$, the moduli space of $1 \oplus 1 \oplus 1$ connections where one summand is on L_ω. Set $\mathcal{M}_\omega^{1\oplus1^2}$ denote the moduli space of $1 \oplus 1^2$ flat connections on $L_{-2\omega} \oplus L_\omega \oplus L_\omega$.

Choose small, path-connected neighborhoods $U_i \subset \mathcal{A}_\omega/\mathcal{G}$ of the path components C_i of $\mathcal{M}_\omega$, and choose basepoints $[\widehat{A}_i^\pm] \in C_i$, as indicated in the previous section. We will again abbreviate the average normal spectral flow by $\mathrm{SF}_{\mathfrak{s}^\perp}(\widehat{A}_i^{\mathrm{avg}}, A)$.

For each splitting of E into line bundles, there is a unique gauge orbit of abelian flat connections, and it varies continuously under small perturbation. For any $[A] \in \mathcal{M}_h^{1\oplus1\oplus1} \cup \mathcal{M}_h^{1\oplus1^2}$, we denote by $[\widehat{A}]$ the abelian flat connection near $[A]$. In particular, the notation $\mathrm{SF}_{\mathfrak{t}^\perp}(\widehat{A}, A)$ or $\mathrm{SF}_{\mathfrak{u}^\perp}(\widehat{A}, A)$ always mean "normal spectral flow between appropriately chosen gauge representatives of the nearby point in $\mathcal{M}^{1\oplus1\oplus1} \cup \mathcal{M}^{1\oplus1^2}$ and $[A]$."

We are now ready to describe the contributions to the abelian correction terms for the splitting determined by ω. Make the following definitions.

$$c_{1\oplus1\oplus1}(X,h,\omega) = -\frac{1}{4}\sum_{i=1}^{n} \sum_{[A]\in\mathcal{M}_{\omega,h}^{1\oplus1\oplus1}\cap U_i} (-1)^{\mathrm{SF}(\Theta,A)}\,\mathrm{SF}_{\mathfrak{s}\cap\mathfrak{t}^\perp}(\widehat{A},A)\left(\mathrm{SF}_{\mathfrak{s}^\perp}(\widehat{A}_i^{\mathrm{avg}},A)\right),$$

and

$$\begin{aligned} c_{1\oplus1^2}(X,h,\omega) &= -\frac{1}{32}\sum_{i=1}^{n} \sum_{[A]\in\mathcal{M}_{\omega,h}^{1\oplus1^2}\cap U_i} (-1)^{\mathrm{SF}(\Theta,A)}\,\mathrm{SF}_{\mathfrak{u}^\perp}(\widehat{A},A) \\ &\quad\times\left(\mathrm{SF}_{\mathfrak{u}^\perp}(\widehat{A},A) - 4 + 4\,\mathrm{SF}_{\mathfrak{s}^\perp}(\widehat{A}_i^{\mathrm{avg}},\widehat{A})\right). \end{aligned}$$

THEOREM 6.1 ([**3**]). *The quantity*

$$\sum_{\mathcal{M}_h^*}(-1)^{\mathrm{SF}(\theta,A)} - \sum_{\omega\in H^2(X;\mathbb{Z})}\left(c_{2\oplus1}(X,h,\omega) + c_{1\oplus1\oplus1}(X,h,\omega) + c_{1\oplus1^2}(X,h,\omega)\right)$$

is independent of (small) perturbation, and hence defines an invariant of X.

In these notes, we have been imprecise about the convention for defining spectral flow between connections where the operator K_A has zero modes. Depending on that convention, the orientation independent correction term formula may look slightly different, as in [**4**].

Also, note that if all points in the reducible and abelian flat moduli space are already nondegenerate, even before perturbation, then the normal spectral flows are all zero and the correction terms vanish. In this case, the invariant is simply a signed count of the irreducible flat orbits.

7. Idea of the proof

We finish with a brief sketch of the proof. Choose two nondegenerate perturbations, and fix a path between them. According to Theorem 4.1, we can assume the parameterized moduli space consists of a cobordism (on each stratum) with only the T-intersection singularities listed there.

Among the changes in topology that occur as the perturbation is varied are standard births and deaths of pairs of points within a stratum. These do not involve T-intersections, and hence there is no normal spectral flow between the points that cancel each other out. As a consequence, the leading signs insure that the quantity given by the formula does not change under these births and deaths.

It remains, then, to show that, given a T-intersection as in Figure 2, the contribution from the left endpoint equals the sum of the two contributions from the right endpoints (for either of the three orbit type combinations). This depends on a technical calculation, which we now summarize.

Let $([A_t], t)$, $t \in [t_-, t_+]$, parameterize a small arc of one reducible stratum W^σ, $\sigma \in \{1 \oplus 1^2, 1 \oplus 1 \oplus 1, 2 \oplus 1\}$. Assume that it only contains one T-intersection point $([A_{t_0}], t_0)$, $t_- < t_0 < t_+$. Let $W^{\sigma'}$ denote the stratum of the less reducible stratum that terminates at this bifurcation point. Assume that $([B_v], \bar{t}(v))$, parameterizes the less reducible arc, where $v \in (0, v_+]$. Assume, also, that $\lim_{v \to 0} \big(B_v, \bar{t}(v)\big) = (A_{t_0}, t_0)$ and $\bar{t}(v_+) = t_+$. We set $B_0 = A_{t_0}$, so that we may view $([B_v], \bar{t}(v))$, $v \in [0, v_+]$, as a deformation of B_0 into a less reducible stratum. According to the proof of Theorem 4.1, one eigenvalue of K_{A_t, h_t}, crosses zero transversely at t_0, and this eigenvalue has multipicity $m = 4$ if $\sigma = 1 \oplus 1^2$ or $m = 2$ if $\sigma = 2 \oplus 1$ or $\sigma = 1 \oplus 1 \oplus 1$. Let $\delta \in \{-1, +1\}$ denote the direction of the spectral flow.

PROPOSITION 7.1 ([**3**]). *Under all the assumptions described above,*

$$\mathrm{SF}(A_{t_-}, B_{v_+}) = \frac{(m-1)(\delta - 1)}{2}.$$

Proposition 7.1 allows one to calculate difference in contributions to the various sums in the Casson invariant formula from one side of a T-intersection to the other. In this way, one can see that each time a bifurcation causes a change in one term, it also causes a change in another term in such a way that the formula overall is invariant.

References

[1] S. Akbulut and J. McCarthy, *Casson's invariant for oriented homology 3-spheres, an exposition*, Math. Notes, No. 36, Princeton Univ. Press, Princeton, NJ, 1990.

[2] H. U. Boden and C. M. Herald, *The SU(3) Casson invariant for integral homology 3-spheres*, J. Differential Geom. **50** (1998), 147–206.

[3] ———, *The SU(3) Casson invariant for rational homology 3-spheres,* in preparation.

[4] H. U. Boden, C. M. Herald, and P. A. Kirk, *An integer valued SU(3) Casson invariant*, Math. Res. Lett. **8** (2001), 589–603.

[5] ———, *The integer valued SU(3) Casson invariant for Brieskorn spheres,* J. Differential Geom. **71** (2005), 23–83.

[6] S. E. Cappell, R. Lee, and E. Y. Miller, *A symplectic geometry approach to generalized Casson's invariants of* 3-*manifolds*, Bull. Amer. Math. Soc. (N.S.) **22** (1990), 269–275.

[7] ———, *A perturbative* SU(3) *Casson invariant, Comment. Math. Helv.* **77** (2002), 491–523.

[8] S. K. Donaldson, *Floer Homology Groups in Yang-Mills Theory*, Cambridge Tracts in Math. 147, Cambridge University Press, Cambridge, 2002.

[9] A. Floer, *An instanton invariant for 3-manifolds,* Comm. Math. Phys. **118** (1989), 215–240.

[10] C. Herald, *Transversality for equivariant exact 1-forms and gauge theory on 3-manifolds*, Adv. Math. **200** (2006), 245–302.

[11] C. Lescop, *Sur l'invariant de Casson-Walker: formule de chirurgie globale et généralisation aux variétés de dimension* 3 *fermées orientées,* C. R. Acad. Sci. Paris Sér. I Math. **315** (1992), 437–440.

[12] C. Taubes, *Casson's invariant and gauge theory,* J. Differential Geom. **31** (1990), 547–599.
[13] K. Walker, *An extension of Casson's invariant*, Ann. of Math. Stud. 126, Princeton University Press, Princeton, NJ, 1992.

Department of Mathematics and Statistics/MS 0084, University of Nevada, Reno, Reno, NV 89557-0084
E-mail address: `herald@unr.edu`

AMS/IP Studies in Advanced Mathematics
Volume 50, 2011

Decomposition of Witten–Reshetikhin–Turaev Invariant: Linking Pairing and Modular Forms

Kazuhiro Hikami

ABSTRACT. We study the $SU(2)$ Witten–Reshetikhin–Turaev invariants for Seifert manifolds associated with the Arnold 14 unimodal singularities. We show that the invariants are decomposed based on a value of linking pairings. Discussed also is a relationship with modular forms with weight-3/2.

1. Introduction

There has been a great progress in quantum topology during 20 years since Witten constructed quantum Chern–Simons invariant for 3-manifolds M [**37**],

$$Z_k(M) = \int e^{iS(A)} \mathcal{D}A, \tag{1.1}$$

where A is a G-gauge connection on the trivial bundle over M, and $S(A)$ is the Chern–Simons action

$$S(A) = \frac{k}{4\pi} \int_M \operatorname{Tr}\left(A \wedge dA + \frac{2}{3} A \wedge A \wedge A \right), \tag{1.2}$$

with $k \in \mathbb{Z}$. Throughout this article, we study $SU(2)$ gauge group. Due to that the critical points of $S(A)$ are the flat connection on M, it is expected [**37, 8, 7**] that in the large k limit we have

$$Z_k(M) \sim \frac{1}{2} e^{-\frac{3}{4}\pi i} \sum_\alpha \sqrt{T_\alpha}\, e^{2\pi i (k+2)\, \mathrm{CS}(A_\alpha)}\, e^{-\frac{2\pi i}{4} I_\alpha}. \tag{1.3}$$

Here a sum of α denotes a gauge equivalent class of flat connections, and $\mathrm{CS}(A_\alpha)$, T_α, and I_α denote the classical Chern–Simons invariant, the Reidemeister torsion, and the spectral flow, respectively.

Asymptotic behavior (1.3) has been computed exactly by use of explicit form of the Witten–Reshetikhin–Turaev (WRT) invariants [**32**] for Seifert manifolds (see, *e.g.*, [**34, 25**]). Based on these results, Lawrence and Zagier [**26**] found a remarkable structure of quantum invariants. It was shown that the WRT invariant for the Poincaré homology sphere is a limiting value of the Eichler integral of vector-valued modular form with weight-3/2. By use of nearly modular properties of the

2010 *Mathematics Subject Classification.* Primary 57M27, 58J28; Secondary 57R56, 11F23.

Eichler integral, not only contributions from critical points (1.3) but also perturbative invariants arising from an asymptotic expansion of the trivial connection contribution can be computed. From the viewpoint of modular forms, perturbative invariants are from period function while the Chern–Simons invariant correspond to the Eichler integral at integers. This result was generalized to Seifert homology spheres [**13, 14, 17, 18**], the Seifert manifolds associated with the ADE singularities [**15**], and torus knots [**20, 11, 38**]. It was also realized that the WRT invariant for Seifert manifolds is closely related to the Ramanujan mock theta functions [**26, 12**].

Results of [**15**] indicate that the WRT invariant for rational homology sphere M is decomposed as

$$\tau_N(M) = \sum_{\boldsymbol{\ell} \in \operatorname{Tors} H_1(M;\mathbb{Z})} \mathrm{e}^{2\pi \mathrm{i} \lambda_M(\boldsymbol{\ell},\boldsymbol{\ell}) N} \, \tau_N^{(\lambda_M(\boldsymbol{\ell},\boldsymbol{\ell}))}(M), \tag{1.4}$$

where λ_M is the linking pairing, and the decomposed WRT invariant $\tau_N^{(\lambda_M(\boldsymbol{\ell},\boldsymbol{\ell}))}(M)$ is a limiting value of certain q-series in $\tau \searrow \frac{1}{N}$. In case that M is the spherical Seifert manifold, each decomposed WRT invariant was identified with the Eichler integral, and the perturbative invariants can also be decomposed for each value of the linking pairings.

Purpose of this article is to study the WRT invariants for Seifert manifolds associated with the Arnold unimodal singularities [**2**]. We shall show that the WRT invariants can be written in terms of the Eichler integrals, and that the decomposition (1.4) is fulfilled. One sees that the Arnold strange duality can be seen as $\lambda_M = -\lambda_{M^*}$ from the viewpoint of (1.4) as studied in [**19**].

This article is organized as follows. In Section 2 we introduce the Seifert manifolds as a spherical neighborhood of isolated singularities, and study their linking pairings. Here we pay attention to ADE and unimodal singularities. In Section 3 we define the Eichler integral of certain weight-3/2 vector modular form, and recall that the WRT invariants for E_7 and E_8 can be written as a linear combination of these Eichler integrals. We briefly study their unified WRT invariants [**10**]. In Section 4 we show results on the WRT invariants for Seifert manifolds associated with the unimodal singularities. We study asymptotic expansions by use of the Eichler integrals. Given are the Chern–Simons invariants and generating functions of perturbative invariants.

2. Singularities and Seifert manifolds

2.1. Seifert manifolds associated with singularities. We set $f(x, y, z)$ as the weighted homogeneous polynomials of weight (d_1, d_2, d_3) and degree d. We have a natural $\mathbb{C}^*$-action on V as

$$f\left(t^{d_1} x, t^{d_2} y, t^{d_3} z\right) = t^d f(x, y, z), \tag{2.1}$$

for $t \in \mathbb{C}^*$. We assume that the variety

$$V = \{f(x, y, z) = 0\}, \tag{2.2}$$

has an isolated singularity at the origin, and we introduce a closed oriented 3-manifold M as a spherical neighborhood of the singularity,

$$M = V \cap S^5. \tag{2.3}$$

Here S^5 is a sufficiently small sphere centered at the origin. With respect to the S^1-action induced from the $\mathbb{C}^*$-action (2.1) on V, M is the Seifert manifold. The Seifert invariant $(g; b; (p_1, q_1), (p_2, q_2), (p_3, q_3))$ of M, where g is a genus of orientable base, and (p_j, q_j) are coprime integers, is fixed explicitly from the weighted polynomial (2.1) [**31**].

In the following, we study the case of $g = 0$. The Seifert manifold $M = M(0; b; (p_1, q_1), (p_2, q_2), (p_3, q_3))$ has a surgery description depicted in Fig. 1. The fundamental group of M has a presentation [**36**]

$$\pi_1(M) = \left\langle x_1, x_2, x_3, h \;\middle|\; \begin{array}{l} h \text{ is center}, \\ {x_i}^{p_i} = h^{-q_i} \text{ for } i = 1, 2, 3, \\ x_1\, x_2\, x_3 = h^b \end{array} \right\rangle . \tag{2.4}$$

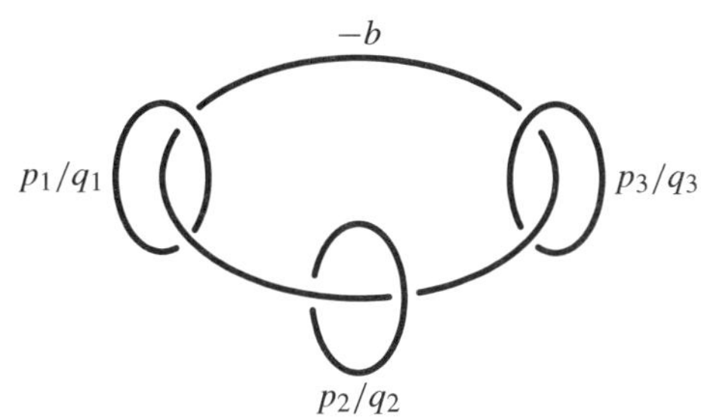

FIGURE 1. Depicted is a surgery description of the Seifert manifold $M(0; b; (p_1, q_1), (p_2, q_2), (p_3, q_3))$.

2.2. Classifications of singularities. Classifications of the isolated singularities have been widely studied (see [**2**] and references therein). Amongst others we pay attention to two classes of the singularities hereafter; the ADE singularities (Table 1) and the Arnold 14 unimodal singularities (Table 2).

- the ADE singularities

  When the weighted homogeneous polynomial $f(x, y, z)$ is of the ADE singularities, the Seifert invariant, b and (p_j, q_j), is given in Table 1 [**31**]. We have $\frac{1}{p_1} + \frac{1}{p_2} + \frac{1}{p_3} > 1$, and the ADE singularities are the quotient singularities associated with spherical triangle $\Delta(p_1, p_2, p_3)$ with angles π/p_1, π/p_2, and π/p_3. Correspondingly the Seifert manifold (2.3) is $SU(2)/\Gamma$ where Γ is a discrete subgroup of SU(2).
- the Arnold 14 unimodal singularities

  The weighted homogeneous polynomial $f(x, y, z)$ given in Table 2 has the exceptional unimodal singularity at the origin. In that Table, the Seifert invariant is also given. We have the hyperbolic triangle $\Delta(p_1, p_2, p_3)$ satisfying $\frac{1}{p_1} + \frac{1}{p_2} + \frac{1}{p_3} < 1$.

All Seifert manifolds associated with these singularities are of $M(g = 0; b = -1; (p_1, 1), (p_2, 1), (p_3, 1))$. It is known that, among the Arnold 14 unimodal singularities, there is a strange duality between the Dolgachev number and the Gabrielov number of singularity X; there exists a unimodal singularity X^* whose Gabrielov number (b_1^*, b_2^*, b_3^*) coincides with the Dolgachev number (p_1, p_2, p_3) of X. This duality was interpreted from various viewpoints, such as the weight system [**35**], the polar duality [**24**], and the mirror symmetry [**33**].

TABLE 1. ADE singularities.

	$f(x,y,z)$	(d_1,d_2,d_3)	d	(p_1,p_2,p_3)	$H_1(M;\mathbb{Z})$	λ_M	CS
D_{4K}	$x^{4K-1}+x\,y^2+z^2$	$(2,4\,K-2,4\,K-1)$	$8\,K-2$	$(2,2,4K-2)$	$\mathbb{Z}_2\oplus\mathbb{Z}_2$	$\begin{pmatrix}0 & \frac{1}{2}\\ \frac{1}{2} & 0\end{pmatrix}$	$\left\{-\frac{(2m+1)^2}{16K-8}\right\}$ for $0\leq m\leq 2K-2$
D_{4K+2}	$x^{4K+1}+x\,y^2+z^2$	$(2,4\,K,4\,K+1)$	$8\,K+2$	$(2,2,4K,1)$	$\mathbb{Z}_2\oplus\mathbb{Z}_2$	$\left(\frac{1}{2}\right)\oplus\left(\frac{1}{2}\right)$	$\left\{-\frac{(2m+1)^2}{16K}\right\}$ for $0\leq m\leq 2K-1$
D_{4K+1}	$x^{4K}+x\,y^2+z^2$	$(2,4\,K-1,4\,K)$	$8\,K$	$(2,2,4K-1)$	$\mathbb{Z}_4$	$\left(\frac{3}{4}\right)$	$\left\{-\frac{(2m+1)^2}{16K-4}\right\}$ for $0\leq m\leq 2K-2$
D_{4K+3}	$x^{4K+2}+x\,y^2+z^2$	$(2,4\,K+1,4\,K+2)$	$8\,K+4$	$(2,2,4K+1)$	$\mathbb{Z}_4$	$\left(\frac{1}{4}\right)$	$\left\{-\frac{(2m+1)^2}{16K+4}\right\}$ for $0\leq m\leq 2K-1$
E_6	$x^4+y^3+z^2$	$(3,4,6)$	12	$(2,3,3)$	$\mathbb{Z}_3$	$\left(\frac{2}{3}\right)$	$\left\{-\frac{1}{24}\right\}$
E_7	$x^3\,y+y^3+z^2$	$(4,6,9)$	18	$(2,3,4)$	$\mathbb{Z}_2$	$\left(\frac{1}{2}\right)$	$\left\{-\frac{1}{48},-\frac{25}{48}\right\}$
E_8	$x^5+y^3+z^2$	$(6,10,15)$	30	$(2,3,5)$	0	$\varnothing$	$\left\{-\frac{1}{120},-\frac{49}{120}\right\}$

TABLE 2. Unimodal singularities.

	$f(x,y,z)$	(d_1,d_2,d_3)	d	Dolgachev (p_1,p_2,p_3)	$H_1(M;\mathbb{Z})$	Gabrielov (b_1,b_2,b_3)	λ_M	CS
E_{12}	$x^7+y^3+z^2$	$(6,14,21)$	42	$(2,3,7)$	0	$(2,3,7)$	$\varnothing$	$\left\{\frac{25}{168},-\frac{47}{168}\right\}$
E_{13}	$x^5\,y+y^3+z^2$	$(4,10,15)$	30	$(2,4,5)$	$\mathbb{Z}_2$	$(2,3,8)$	$\left(\frac{1}{2}\right)$	$\left\{\frac{9}{80},\frac{49}{80}\right\}$
E_{14}	$x^8+y^3+z^2$	$(3,8,12)$	24	$(3,3,4)$	$\mathbb{Z}_3$	$(2,3,9)$	$\left(\frac{2}{3}\right)$	$\left\{\frac{1}{12},\frac{25}{48}\right\}$
Z_{11}	$x^5+x\,y^3+z^2$	$(6,8,15)$	30	$(2,3,8)$	$\mathbb{Z}_2$	$(2,4,5)$	$\left(\frac{1}{2}\right)$	$\left\{\frac{25}{96},\frac{73}{96}\right\}$
Z_{12}	$x^4\,y+x\,y^3+z^2$	$(4,6,11)$	22	$(2,4,6)$	$\mathbb{Z}_2\oplus\mathbb{Z}_2$	$(2,4,6)$	$\left(\frac{1}{2}\right)\oplus\left(\frac{1}{2}\right)$	$\left\{\frac{3}{16},\frac{11}{16}\right\}$
Z_{13}	$x^6+x\,y^3+z^2$	$(3,5,9)$	18	$(3,3,5)$	$\mathbb{Z}_2\oplus\mathbb{Z}_3$	$(2,4,7)$	$\left(\frac{1}{2}\right)\oplus\left(\frac{2}{3}\right)$	$\left\{\frac{2}{15},\frac{19}{30}\right\}$
W_{12}	$x^5+y^4+z^2$	$(4,5,10)$	20	$(2,5,5)$	$\mathbb{Z}_5$	$(2,5,5)$	$\left(\frac{3}{5}\right)$	$\left\{\frac{5}{8},\frac{9}{40}\right\}$
W_{13}	$x^4\,y+y^4+z^2$	$(3,4,8)$	16	$(3,4,4)$	$\mathbb{Z}_8$	$(2,5,6)$	$\left(\frac{5}{8}\right)$	$\left\{\frac{1}{6},\frac{13}{24}\right\}$
Q_{10}	$x^4+y^3+x\,z^2$	$(6,8,9)$	24	$(2,3,9)$	$\mathbb{Z}_3$	$(3,3,4)$	$\left(\frac{1}{3}\right)$	$\left\{\frac{49}{72},\frac{25}{72},-\frac{5}{24}\right\}$
Q_{11}	$x^3\,y+y^3+x\,z^2$	$(4,6,7)$	18	$(2,4,7)$	$\mathbb{Z}_2\oplus\mathbb{Z}_3$	$(3,3,5)$	$\left(\frac{1}{2}\right)\oplus\left(\frac{1}{3}\right)$	$\left\{\frac{19}{112},\frac{75}{112},-\frac{29}{112},\frac{27}{112}\right\}$
Q_{12}	$x^5+y^3+x\,z^2$	$(3,5,6)$	15	$(3,3,6)$	$\mathbb{Z}_3\oplus\mathbb{Z}_3$	$(3,3,6)$	$\left(\frac{1}{3}\right)\oplus\left(\frac{2}{3}\right)$	$\left\{\frac{5}{24},\frac{17}{24},\frac{1}{6},\frac{2}{3}\right\}$
S_{11}	$x^4+y^2\,z+x\,z^2$	$(4,5,6)$	16	$(2,5,6)$	$\mathbb{Z}_8$	$(3,4,4)$	$\left(\frac{3}{8}\right)$	$\left\{\frac{7}{10},\frac{19}{30},\frac{3}{10}\right\}$
S_{12}	$x^3\,y+y^2\,z+x\,z^2$	$(3,4,5)$	13	$(3,4,5)$	$\mathbb{Z}_{13}$	$(3,4,5)$	$\left(\frac{5}{13}\right)$	$\left\{\frac{13}{60},\frac{37}{60},\frac{133}{240},\frac{157}{240}\right\}$
U_{12}	$x^4+y^3+z^3$	$(3,4,4)$	12	$(4,4,4)$	$\mathbb{Z}_4\oplus\mathbb{Z}_4$	$(4,4,4)$	$\begin{pmatrix}\frac{1}{2} & \frac{1}{4}\\ \frac{1}{4} & \frac{1}{2}\end{pmatrix}$	$\left\{\frac{9}{16},\frac{1}{4}\right\}$

2.3. Linking pairing. For closed oriented 3-manifolds M, we have the linking pairing λ_M,

$$\lambda_M : \operatorname{Tors} H_1(M;\mathbb{Z}) \otimes \operatorname{Tors} H_1(M;\mathbb{Z}) \to \mathbb{Q}/\mathbb{Z},$$

where $\operatorname{Tors} H_1(M;\mathbb{Z})$ denotes the torsion part of $H_1(M;\mathbb{Z})$. For $\boldsymbol{a}, \boldsymbol{a}' \in \operatorname{Tors} H_1(M;\mathbb{Z})$, we define λ_M as follows. We choose $s \in \mathbb{Z}_{\neq 0}$ such that $s\,\boldsymbol{a} = 0 \in H_1(M;\mathbb{Z})$, and set a 2-chain $\boldsymbol{B}$ which is bounded as $\partial \boldsymbol{B} = s\,\boldsymbol{a}$. The linking pairing is given from the intersection number as

$$\lambda_M(\boldsymbol{a}, \boldsymbol{a}') = \frac{\#(\boldsymbol{B}\cdot\boldsymbol{a}')}{s} \mod \mathbb{Z}. \tag{2.5}$$

See [**23**] for the classification of linking pairing on 3-manifolds.

For $M = M(0; b; (p_1,q_1),(p_2,q_2),(p_3,q_3))$, abelianization of $\pi_1(M)$ (2.4) gives

$$\begin{aligned} H_1(M;\mathbb{Z}) &\cong \operatorname{span}_{\mathbb{Z}}\{\boldsymbol{x}_1,\boldsymbol{x}_2,\boldsymbol{x}_3,\boldsymbol{h}\} \Big/ \left\{ \begin{matrix} p_i\,\boldsymbol{x}_i + q_i\,\boldsymbol{h} \text{ for } i=1,2,3, \\ \boldsymbol{x}_1+\boldsymbol{x}_2+\boldsymbol{x}_3+\boldsymbol{h} \end{matrix} \right\} \\ &\cong \mathbb{Z}^4/\mathbf{A}_M\,\mathbb{Z}^4, \end{aligned}$$

where $\mathbf{A}_M$ is the linking matrix defined by

$$\mathbf{A}_M = \begin{pmatrix} p_1 & & & q_1 \\ & p_2 & & q_2 \\ & & p_3 & q_3 \\ 1 & 1 & 1 & -b \end{pmatrix}. \tag{2.6}$$

We have $|H_1(M;\mathbb{Z})| = |\det \mathbf{A}_M| = |p_1\,p_2\,p_3\,E_M|$, where E_M is the Euler number

$$E_M = -b - \sum_{j=1}^{3} \frac{q_j}{p_j}. \tag{2.7}$$

We assume $\det \mathbf{A}_M \neq 0$, and the torsion part can be identified with $\mathbb{Z}^4/\mathbf{A}_M\,\mathbb{Z}^4$. Thus for $\boldsymbol{\ell} = (\ell_1,\ell_2,\ell_3,\ell_4)^{\mathrm{T}} \in \mathbb{Z}^4$ representing a generator $\ell_1\,\boldsymbol{x}_1 + \ell_2\,\boldsymbol{x}_2 + \ell_3\,\boldsymbol{x}_3 + \ell_4\,\boldsymbol{h}$, we may set $\boldsymbol{m} = (m_1,m_2,m_3,m_4)^{\mathrm{T}} \in \mathbb{Z}^4$ such that $s\,\boldsymbol{\ell} = \mathbf{A}_M\,\boldsymbol{m}$. As m_i is the number of meridian disks of the i-th solid torus, the intersection number in (2.5) gives

$$\lambda_M(\boldsymbol{\ell},\boldsymbol{\ell}') = \frac{1}{s}\,\boldsymbol{m}^{\mathrm{T}}\,\operatorname{diag}(q_1,q_2,q_3,1)\,\boldsymbol{\ell}' = \boldsymbol{\ell}^{\mathrm{T}}\,\left(\mathbf{A}'_M\right)^{-1}\,\boldsymbol{\ell}',$$

where $\mathbf{A}'_M$ is given by

$$\mathbf{A}'_M = \begin{pmatrix} \frac{p_1}{q_1} & & & 1 \\ & \frac{p_2}{q_2} & & 1 \\ & & \frac{p_3}{q_3} & 1 \\ 1 & 1 & 1 & -b \end{pmatrix}.$$

By use of an integral unimodular matrix $\mathbf{P}$, the matrix $\left(\mathbf{A}'_M\right)^{-1}$ is block-diagonalized as

$$\mathbf{P}^{\mathrm{T}}\,\left(\mathbf{A}'_M\right)^{-1}\,\mathbf{P} = \mathbf{\Lambda}_M \oplus (\pm 1) \oplus \cdots \oplus (\pm 1) \mod \mathbb{Z}.$$

As a result, with $\boldsymbol{a}, \boldsymbol{a}' \in \operatorname{Tors} H_1(M;\mathbb{Z})$ the linking pairing is written as

$$\lambda_M(\boldsymbol{a},\boldsymbol{a}') = \boldsymbol{a}^{\mathrm{T}}\,\mathbf{\Lambda}_M\,\boldsymbol{a}'. \tag{2.8}$$

See Tables 1 and 2 for explicit forms of λ_M for Seifert manifolds related to the ADE and unimodal singularities.

3. WRT invariant for Seifert manifolds

3.1. SU(2) WRT invariant. Mathematically rigorous definition of the Witten invariant (1.1) was given by Reshetikhin and Turaev [**32**]. The SU(2) Reshetikhin–Turaev invariant $\tau_N(M)$ for 3-manifold M is related to the Witten invariant (1.1) with SU(2) gauge group by

$$Z_k(M) = \frac{\tau_{k+2}(M)}{\tau_{k+2}\left(S^2 \times S^1\right)}, \tag{3.1}$$

where we have $\tau_N\left(S^2 \times S^1\right) = \sqrt{\frac{N}{2}}\frac{1}{\sin(\pi/N)}$. Explicitly when M is constructed by rational surgery p_j/q_j on the j-th component of the n-component link L, we have

$$\begin{aligned}\tau_N(M) = {} & \mathrm{e}^{\frac{\pi \mathrm{i}}{4}\frac{N-2}{N}\left(\sum_{j=1}^{n}\Phi(\mathbf{U}^{(p_j,q_j)})-3\,\mathrm{sign}(\mathbf{L})\right)} \\ & \times \sum_{k_1,\dots,k_n=1}^{N-1} J_L(k_1,\dots,k_n)\prod_{j=1}^{n}\left[\rho(\mathbf{U}^{(p_j,q_j)})\right]_{k_j,1}.\end{aligned} \tag{3.2}$$

Here $J_L(k_1,\dots,k_n)$ is the colored Jones polynomial with color-k_j on the j-th component of L, and $\mathbf{L}$ is the linking matrix of L. The Rademacher function $\Phi(\mathbf{U}^{(p,q)})$ for $\mathbf{U}^{(p,q)} = \left(\begin{smallmatrix} p & r \\ q & s\end{smallmatrix}\right) \in SL(2;\mathbb{Z})$ is

$$\Phi\left(\begin{pmatrix} p & r \\ q & s\end{pmatrix}\right) = \begin{cases} \frac{p+s}{q} - 12\,s(q,p), & \text{for } q \neq 0, \\ 0, & \text{for } q = 0.\end{cases}$$

The function $s(b,a)$ is the Dedekind sum

$$s(b,a) = \mathrm{sign}(a)\sum_{k=1}^{|a|-1}\left(\!\left(\frac{k}{a}\right)\!\right)\left(\!\left(\frac{k\,b}{a}\right)\!\right),$$

$$(\!(x)\!) = \begin{cases} x - \lfloor x \rfloor - \frac{1}{2}, & \text{if } x \notin \mathbb{Z}, \\ 0, & \text{if } x \in \mathbb{Z}.\end{cases}$$

$\rho(\mathbf{U}^{(p,q)})$ is given by

$$\begin{aligned}\left[\rho(\mathbf{U}^{(p,q)})\right]_{a,b} = {} & -\mathrm{i}\frac{\mathrm{sign}(q)}{\sqrt{2\,N\,|q|}}\,\mathrm{e}^{-\frac{\pi \mathrm{i}}{4}\Phi(U^{(p,q)})+\frac{\pi \mathrm{i}}{2Nq}sb^2} \\ & \times \sum_{\substack{\gamma \mod 2Nq \\ \gamma = a \mod 2N}} \mathrm{e}^{\frac{\pi \mathrm{i}}{2Nq}p\gamma^2}\left(\mathrm{e}^{\frac{\pi \mathrm{i}}{Nq}\gamma b} - \mathrm{e}^{-\frac{\pi \mathrm{i}}{Nq}\gamma b}\right).\end{aligned} \tag{3.3}$$

The WRT invariant for Seifert manifolds $M(0;b;(p_1,q_1),(p_2,q_2),(p_3,q_3))$ is computable from (3.2) via Fig. 1.

3.2. Unified WRT invariants. For integral homology sphere M, Habiro constructed the unified WRT invariant $I_q(M)$ [**10**]. The invariant $I_q(M)$ takes values in $\varprojlim_n \mathbb{Z}[q]/(q)_n$, and is written as an infinite sum

$$I_q(M) = \sum_{n=0}^{\infty} f_n(q)\cdot(q)_n, \tag{3.4}$$

with $f_n(q) \in \mathbb{Z}[q]$. Here and hereafter we use a standard notation of q-analysis, $(x)_n = (x;q)_n = (1-x)(1-x\,q)\cdots(1-x\,q^{n-1})$. The WRT invariant is given by evaluation at the root of unity,

$$\tau_N(M) = \mathrm{ev}_{q=\mathrm{e}^{2\pi\mathrm{i}/N}}\left[I_q(M)\right], \tag{3.5}$$

in which the infinite series (3.4) terminates at a finite sum. The expression (3.4) follows from a fact that the colored Jones polynomial for knot K, which is normalized to be $J_{\mathrm{unknot}}(N) = 1$ and $J_K(1) = 1$, has a cyclotomic expansion [**9**]

$$J_K(N) = \sum_{n=0}^{\infty} C_K(n)\cdot (q^{1+N})_n\,(q^{1-N})_n, \tag{3.6}$$

where $C_K(n) \in \mathbb{Z}[q,q^{-1}]$. For example, the trefoil has

$$J_{\mathrm{trefoil}}(N) = \sum_{n=0}^{\infty} q^n\,(q^{1+N})_n\,(q^{1-N})_n. \tag{3.7}$$

We note that an inverse of (3.6) is

$$C_K(n) = q^n \sum_{k=0}^{n} (-1)^k\, q^{\frac{1}{2}k(k-1)}\, \frac{(1-q^{k+1})\,(1-q^{2k+2})}{(q)_{n-k}\,(q)_{n+k+2}}\, J_K(k+1).$$

A case of the rational homology sphere was studied in [**3, 4**] by use of the surgery formula [**5**]. For instance, when M_s is s-surgery of knot K whose colored Jones polynomial is (3.6), we have

$$(1-q)\,I_q(M_{+1}) = \sum_{n=0}^{\infty}(-1)^n\, q^{-\frac{n(n+3)}{2}}\, C_K(n)\,(q^{n+1})_{n+1}, \tag{3.8}$$

$$(1-q)\,I_q(M_{-1}) = \sum_{n=0}^{\infty} C_K(n)\,(q^{n+1})_{n+1}. \tag{3.9}$$

We have $\tau_N(M_{\pm 2}) = 0$ for odd N, and $\tau_N(M_{\pm 2})$ for even N follows from

$$(1-q)\,I_q(M_2) = \sqrt{2}\,q^{\frac{1}{4}} \sum_{n\geq 0}(-1)^n\, q^{-\frac{n}{2}}\, C_K(n)\,(q^{\frac{1}{2}};-q^{\frac{1}{2}})_{2n+1}, \tag{3.10}$$

$$(1-q)\,I_q(M_{-2}) = \sqrt{2}\,q^{\frac{1}{4}} \sum_{n=0}^{\infty} C_K(n)\,(q^{\frac{1}{2}};-q^{\frac{1}{2}})_{2n+1}. \tag{3.11}$$

3.3. WRT invariants and modular form. We use $q = \mathrm{e}^{2\pi\mathrm{i}\tau}$ where τ is in the upper half-plane, $\tau \in \mathbb{H}$. For $P \in \mathbb{Z}_{>0}$ and $a \in \mathbb{Z}$, we define

$$\Psi_P^{(a)}(\tau) = \frac{1}{2}\sum_{n\in\mathbb{Z}} n\,\psi_{2P}^{(a)}(n)\,q^{\frac{n^2}{4P}}, \tag{3.12}$$

where $\psi_{2P}^{(a)}(n)$ is an odd function of period $2\,P$,

$$\psi_{2P}^{(a)}(n) = \begin{cases} \pm 1, & \text{for } n \equiv \pm a \mod 2\,P, \\ 0, & \text{otherwise.} \end{cases} \tag{3.13}$$

The q-series $\Psi_P^{(a)}(\tau)$ is a vector-valued modular form with weight $3/2$ satisfying

$$\frac{\Psi_N^{(a)}}{[\eta]^3}(\gamma(\tau)) = \sum_{b=1}^{N-1} \left[\rho(\gamma)\right]_{a,b}\, \frac{\Psi_N^{(b)}}{[\eta]^3}(\tau),$$

where $\gamma \in SL(2;\mathbb{Z})$, and ρ is given in (3.3). η denotes the Dedekind η-function, $\eta(\tau) = q^{\frac{1}{24}}(q)_\infty$. The Eichler integral [**26, 38**] is defined by

$$\widetilde{\Psi}_P^{(a)}(\tau) = \sum_{n=0}^{\infty} \psi_{2P}^{(a)}(n)\, q^{\frac{n^2}{4P}}, \tag{3.14}$$

which can be regarded as a half-integration of $\Psi_P^{(a)}(\tau)$ with respect to τ. A limiting value of the Eichler integral in $\tau \searrow \mathbb{Q}$ can be computed, *e.g.*,

$$\widetilde{\Psi}_P^{(a)}(1/N) = -\sum_{k=0}^{2PN} \psi_{2P}^{(a)}(k)\, \mathrm{e}^{\frac{k^2}{2PN}\pi \mathrm{i}}\, B_1\left(\frac{k}{2PN}\right), \tag{3.15}$$

$$\widetilde{\Psi}_P^{(a)}(N) = \left(1 - \frac{a}{P}\right) \mathrm{e}^{\frac{a^2}{2P}\pi \mathrm{i} N}, \tag{3.16}$$

where $N \in \mathbb{Z}$, and $B_k(x)$ denotes the k-th Bernoulli polynomial defined by $\frac{t\,\mathrm{e}^{x\,t}}{\mathrm{e}^t - 1} = \sum_{k=0}^{\infty} \frac{B_k(x)}{k!}\, t^k$. These two limiting values of the Eichler integral (3.14) are related to each other by nearly modular property as [**11**]

$$\widetilde{\Psi}_P^{(a)}(1/N) + \sqrt{\frac{N}{\mathrm{i}}} \sum_{b=1}^{P-1} \sqrt{\frac{2}{P}} \sin\left(\frac{a\,b}{P}\pi\right) \widetilde{\Psi}_P^{(b)}(-N) \simeq \sum_{k=0}^{\infty} \frac{L\left(-2\,k, \psi_{2P}^{(a)}\right)}{k!} \left(\frac{\pi\,\mathrm{i}}{2\,P\,N}\right)^k. \tag{3.17}$$

Here $L\left(k, \psi_{2P}^{(a)}\right)$ denotes the Dirichlet L-function, and at negative integers it is written in terms of the Bernoulli polynomial as

$$L\left(-k, \psi_{2P}^{(a)}\right) = -\frac{(2\,P)^k}{k+1} \sum_{n=1}^{2P} \psi_{2P}^{(a)}(n)\, B_{k+1}\left(\frac{n}{2\,P}\right). \tag{3.18}$$

The generating function of the L-function in (3.17) is

$$\frac{\sinh((P-a)\,z)}{\sinh(P\,z)} = \sum_{k=0}^{\infty} \frac{L\left(-2\,k, \psi_{2P}^{(a)}\right)}{(2\,k)!}\, z^{2k}, \tag{3.19}$$

where $0 < a < P$. Note that an infinite series in (3.17) follows from a period integral of $\Psi_P^{(a)}(\tau)$, and that such transformation formula is reminiscent of the Ramanujan mock theta function.

Since the work of Lawrence–Zagier [**26**], the WRT invariants for certain Seifert manifolds are known to be related to limiting values of the Eichler integral (3.15). We show an example. We take M to be the Poincaré homology sphere, E_8 in Table 1. The WRT invariant is computed to be [**26**]

$$\mathrm{e}^{\frac{121}{60N}\pi \mathrm{i}} \left(\mathrm{e}^{\frac{2\pi \mathrm{i}}{N}} - 1\right) \cdot \tau_N(E_8) = \mathrm{e}^{\frac{\pi \mathrm{i}}{60N}} - \frac{1}{2}\, \widetilde{\Psi}_{30}^{(1)+(11)+(19)+(29)}(1/N). \tag{3.20}$$

Here and hereafter we use a notation, $\widetilde{\Psi}_P^{k_a(a)+k_b(b)+\cdots}(\bullet) = k_a\, \widetilde{\Psi}_P^{(a)}(\bullet) + k_b\, \widetilde{\Psi}_P^{(b)}(\bullet) + \cdots$, for brevity. By use of (3.17), we obtain an asymptotic expansion

(3.21)

$$e^{\frac{121}{60N}\pi i}\left(e^{\frac{2\pi i}{N}}-1\right)\cdot\tau_N(E_8)\simeq\sqrt{\frac{N}{i}}\frac{2}{\sqrt{5}}\left(\sin\left(\frac{\pi}{5}\right)e^{-\frac{1}{60}\pi i N}+\sin\left(\frac{2\pi}{5}\right)e^{-\frac{49}{60}\pi i N}\right)$$
$$+e^{\frac{\pi i}{60N}}-\frac{1}{2}\sum_{k=0}^{\infty}\frac{L\left(-2k,\psi_{60}^{(1)+(11)+(19)+(29)}\right)}{k!}\left(\frac{\pi i}{60N}\right)^k.$$

One finds that an infinite series with the term $e^{\frac{\pi i}{60N}}$ generates perturbative invariants [**28, 27**]. Note that the L-function is given by

$$2\frac{\cosh(5z)\cosh(9z)}{\cosh(15z)}=\sum_{k=0}^{\infty}\frac{L\left(-2k,\psi_{60}^{(1)+(11)+(19)+(29)}\right)}{(2k)!}z^{2k}.$$

The unified WRT invariant $I_q(E_8)$ is as follows. Recalling that the Poincaré homology sphere is (-1)-surgery of trefoil, we get from (3.7) and (3.9)

$$1+q(1-q)I_q(E_8)=\sum_{n=0}^{\infty}q^n(q^n)_n. \tag{3.22}$$

$I_q(E_8)$ coincides with $\tau_N(E_8)$ (3.20) at the N-th root of unity. We see that the right hand side in (3.22) is $2-\chi_0(1/q)$ [**12**], where $\chi_0(1/q)$ means an extended value outside the unit circle of the fifth order mock theta function $\chi_0(q)$ defined by

$$\chi_0(q)=\sum_{n=0}^{\infty}\frac{q^n}{(q^{n+1})_n}.$$

Note that the unified WRT invariant generates the L-function in (3.21) when we set $q=e^{-t}$ and take a limit $t\searrow 0$,

$$\sum_{n=0}^{\infty}q^n(q^n)_n\big|_{q=e^{-t}}=\frac{1}{2}e^{\frac{1}{120}t}\sum_{k=0}^{\infty}\frac{L\left(-2k,\psi_{60}^{(1)+(11)+(19)+(29)}\right)}{k!}\left(\frac{-t}{120}\right)^k.$$

This type of identities has been studied in [**38, 16, 21, 1, 29**].

3.4. Decomposition of the WRT invariants. It is conjectured [**15**] that the WRT invariant for rational homology sphere M is decomposed as (1.4). In case that M is a homology sphere, we have $\lambda_M=\varnothing$, and $\tau_N^{(0)}(M)$ is an evaluation of the unified WRT invariant $I_q(M)$ at the N-th root of unity [**9**]. Explicit forms of $\tau_N^{(0)}(M)$ were given in [**13, 14, 17**] for some Seifert homology spheres. When $H_1(M;\mathbb{Z})=\mathbb{Z}_2$ and $\lambda_M=\left(\frac{1}{2}\right)$, we have $\tau_N(M)=0$ for odd N, and the decomposition (1.4) is read as $\tau_N^{(0)}(M)=\tau_N^{(\frac{1}{2})}(M)$. Corresponding to (1.4) the perturbative invariants arising from an asymptotic expansion of the trivial connection contribution can also be decomposed in terms of the linking pairing.

The first verification of (1.4) based on the explicit form of the $SU(2)$ WRT invariant is for the lens space $L(p,q)$, which is constructed by a p/q-surgery on trivial knot. The WRT invariant (3.2) is computed as [**22**]

$$e^{-\frac{6s(q,p)+1}{N}\pi i}\left(e^{\frac{2\pi i}{N}}-1\right)\cdot\tau_N(L(p,q)) \tag{3.23}$$
$$=\sum_{n=1}^{p}e^{2\pi i\frac{q}{p}n^2N}\sum_{\varepsilon=\pm1}\frac{\varepsilon}{\sqrt{p}}e^{2\pi i\frac{q}{p}n+\varepsilon\frac{1+2Nn}{Np}\pi i}.$$

As we have $H_1(L(p,q);\mathbb{Z}) = \mathbb{Z}_p$, and a linking pairing is $\lambda_{L(p,q)} = \left(\frac{q}{p}\right)$, the conjecture (1.4) is true for the lens space.

Our previous results **[15]** on the Seifert manifolds associated with the ADE singularities in Table 1 also support (1.4). Therein the decomposed WRT invariants were explicitly constructed as a linear combination of the Eichler integrals (3.15). For instance, in case of E_7 in Table 1 we have $\lambda_{E_7} = \left(\frac{1}{2}\right)$, and

$$\begin{aligned}(3.24)\quad & \mathrm{e}^{\frac{37}{24N}\pi\mathrm{i}} \left(\mathrm{e}^{\frac{2\pi\mathrm{i}}{N}} - 1\right) \tau_N(E_7) \\ &= \left(1 + (-1)^N\right) \frac{\sqrt{2}}{4} \left(2\,\mathrm{e}^{\frac{\pi\mathrm{i}}{24N}} - \widetilde{\Psi}_{12}^{(1)+(5)+(7)+(11)}(1/N)\right),\end{aligned}$$

which is a sum of $\tau_N^{(0)}(E_7)$ and $\tau_N^{(\frac{1}{2})}(E_7)$. Nearly modularity (3.17) gives an asymptotic expansion as

(3.25)

$$\begin{aligned}\mathrm{e}^{\frac{37}{24N}\pi\mathrm{i}} \left(\mathrm{e}^{\frac{2\pi\mathrm{i}}{N}} - 1\right) \cdot \tau_N\left(E_7\right) \simeq \sqrt{\frac{N}{\mathrm{i}}}\, \frac{1}{\sqrt{2}} \left(\mathrm{e}^{-\frac{1}{24}\pi\mathrm{i}N} + \mathrm{e}^{\frac{23}{24}\pi\mathrm{i}N}\right) + \frac{1 + (-1)^N}{\sqrt{2}}\, \mathrm{e}^{\frac{\pi\mathrm{i}}{24N}} \\ - \frac{\sqrt{2}}{4} \left(1 + (-1)^N\right) \sum_{k=0}^{\infty} \frac{L\left(-2\,k, \psi_{24}^{(1)+(3)+(7)+(11)}\right)}{k!} \left(\frac{\pi\,\mathrm{i}}{24\,N}\right)^k.\end{aligned}$$

The first term is a contribution (1.3) from the critical points in the Chern–Simons path integral. The second term and an infinite series are the perturbative invariants, and a generating function of the L-function is

$$2\,\frac{\cosh(3\,z)\,\cosh(2\,z)}{\cosh(6\,z)} = \sum_{k=0}^{\infty} \frac{L\left(-2\,k, \psi_{24}^{(1)+(5)+(7)+(11)}\right)}{(2\,k)!}\, z^{2k}.$$

In case that N is even, $\tau_N(E_7)$ is an evaluation of the unified invariant at the root of unity, which can be computed from (3.7) and (3.11) as **[6]**

$$(3.26)\qquad (1-q)\, I_q(E_7) = \sqrt{2}\, q^{\frac{1}{4}} \sum_{n=0}^{\infty} q^n\, (q^{\frac{1}{2}}; -q^{\frac{1}{2}})_{2n+1}.$$

We note that $I_q(M)$ generates the L-function in (3.25) as

$$\mathrm{e}^t + \sum_{n=0}^{\infty} q^n\, (q^{\frac{1}{2}}; -q^{\frac{1}{2}})_{2n+1}\Big|_{q=\mathrm{e}^{-t}} = \frac{1}{2}\,\mathrm{e}^{\frac{49}{48}t} \sum_{k=0}^{\infty} \frac{L\left(-2\,k, \psi_{24}^{(1)+(5)+(7)+(11)}\right)}{k!} \left(\frac{-t}{48}\right)^k.$$

4. WRT invariants associated with the unimodal singularities

We shall give an explicit form of the WRT invariant for the Seifert manifolds associated with the Arnold unimodal singularities, and show that they are written as a linear combination of the Eichler integrals. We omit details of computations, as they are tedious but analogous to **[26, 15]**. These results support the decomposition conjecture (1.4), and we list in Table 2 the classical Chern–Simons invariant given from an asymptotic expansion (1.3). It should be noted that the Arnold strange duality can also be interpreted from the linking pairing **[19]**,

$$(4.1)\qquad \lambda_M = -\lambda_{M^*},$$

where M and M^* are dual in Arnold's sense. In the following, we show results separately for self-dual cases (E_{12}, Z_{12}, W_{12}, Q_{12}, S_{12}, U_{12}) and other cases ($E_{13} \leftrightarrow Z_{11}$, $Q_{10} \leftrightarrow E_{14}$, $Q_{11} \leftrightarrow Z_{13}$, $S_{11} \leftrightarrow W_{13}$). Also given are generating functions for linking-pairing decomposed perturbative invariants in terms of hyperbolic functions.

4.1. Self-dual cases.

4.1.1. E_{12}. This is the Brieskorn homology sphere, whose WRT invariant was computed in [**13**]. We have (orientation is opposite to [**13**])

$$\mathrm{e}^{-\frac{1}{84N}\pi\mathrm{i}}\left(\mathrm{e}^{\frac{2\pi\mathrm{i}}{N}}-1\right)\cdot\tau_N(E_{12})=\frac{1}{2}\,\widetilde{\Psi}_{42}^{(1)-(13)-(29)+(41)}(-1/N). \tag{4.2}$$

By use of (3.17), we can compute the asymptotic expansion in $N\to\infty$ as

$$\begin{aligned}
&\mathrm{e}^{-\frac{1}{84N}\pi\mathrm{i}}\left(\mathrm{e}^{\frac{2\pi\mathrm{i}}{N}}-1\right)\tau_N\left(E_{12}\right)\\
&\quad\simeq-\sqrt{\mathrm{i}\,N}\,\frac{2}{\sqrt{7}}\left(-\sin\left(\frac{2\,\pi}{7}\right)\mathrm{e}^{\frac{25}{84}\pi\mathrm{i}N}-\sin\left(\frac{3\,\pi}{7}\right)\mathrm{e}^{-\frac{47}{84}\pi\mathrm{i}N}\right)\\
&\qquad\qquad-\frac{1}{2}\sum_{k=0}^{\infty}\frac{L\left(-2\,k,\psi_{84}^{(1)-(13)-(29)+(41)}\right)}{k!}\left(\frac{\pi}{84\,N\,\mathrm{i}}\right)^k,
\end{aligned}$$

where the perturbative invariants are generated from

$$\frac{2\,\sinh(6\,z)\,\sinh(14\,z)}{\cosh(21\,z)}=\sum_{k=0}^{\infty}\frac{L\left(-2\,k,\psi_{84}^{(1)-(13)-(29)+(41)}\right)}{(2\,k)!}\,z^{2k}.$$

The unified invariant is given from (3.7) and (3.8) as [**18**]

$$(q-1)\,I_q(E_{12})=\sum_{n=0}^{\infty}(-1)^n\,q^{-\frac{n(n+1)}{2}}\,(q^{n+1})_{n+1}, \tag{4.3}$$

which also generates the L-function

$$\begin{aligned}
&\sum_{n=0}^{\infty}(-1)^n\,q^{-\frac{1}{2}n(n+1)}\,(q^{n+1})_{n+1}\Big|_{q=\mathrm{e}^{-t}}\\
&\qquad=\frac{1}{2}\,\mathrm{e}^{-\frac{1}{168}t}\sum_{n=0}^{\infty}\frac{L\left(-2\,k,\psi_{84}^{(1)-(13)-(29)+(41)}\right)}{k!}\left(\frac{t}{168}\right)^k.
\end{aligned}$$

4.1.2. Z_{12}. The WRT invariant is given by

$$\mathrm{e}^{-\frac{1}{24N}\pi\mathrm{i}}\left(\mathrm{e}^{\frac{2\pi\mathrm{i}}{N}}-1\right)\cdot\tau_N(Z_{12})=\frac{1}{2}\,\left(1+(-1)^N\right)\widetilde{\Psi}_{12}^{(1)-(5)-(7)+(11)}(-1/N). \tag{4.4}$$

An asymptotic expansion follows by use of (3.17) as

$$\begin{aligned}
\mathrm{e}^{-\frac{1}{24N}\pi\mathrm{i}}\left(\mathrm{e}^{\frac{2\pi\mathrm{i}}{N}}-1\right)\cdot\tau_N(Z_{12})&\simeq-\sqrt{\mathrm{i}\,N}\cdot\frac{1}{\sqrt{3}}\left(\mathrm{e}^{\frac{3}{8}\pi\mathrm{i}N}+\mathrm{e}^{\frac{11}{8}\pi\mathrm{i}N}\right)\\
&\quad+\frac{1}{2}\left(1+(-1)^N\right)\sum_{k=0}^{\infty}\frac{L\left(-2\,k,\psi_{24}^{(1)-(5)-(7)+(11)}\right)}{k!}\left(\frac{\pi}{24\,N\,\mathrm{i}}\right)^k,
\end{aligned}$$

where the perturbative invariants are generated by

$$\frac{2\sinh(3z)\sinh(2z)}{\cosh(6z)} = \sum_{k=0}^{\infty} \frac{L\left(-2k, \psi_{24}^{(1)-(5)-(7)+(11)}\right)}{(2k)!} z^{2k}.$$

4.1.3. W_{12}. We have

$$\text{(4.5)}\quad \mathrm{e}^{-\frac{1}{20N}\pi\mathrm{i}} \left(\mathrm{e}^{\frac{2\pi\mathrm{i}}{N}} - 1\right) \cdot \tau_N(W_{12}) = \frac{1}{2\sqrt{5}} \left\{ \widetilde{\Psi}_{10}^{(1)-2(5)+(9)}(-1/N) \right.$$
$$\left. + 2\,\mathrm{e}^{\frac{4}{5}\pi\mathrm{i}N}\, \widetilde{\Psi}_{10}^{(1)-2\cos\frac{2\pi}{5}(5)+(9)}(-1/N) + 2\,\mathrm{e}^{-\frac{4}{5}\pi\mathrm{i}N}\, \widetilde{\Psi}_{10}^{(1)-2\cos\frac{4\pi}{5}(5)+(9)}(-1/N) \right\}.$$

Here we have three decomposed WRT invariants, $\tau_N^{(0)}(W_{12})$ and $\tau_N^{(\pm\frac{2}{5})}(W_{12})$. An asymptotic expansion in $N \to \infty$ is written as

$$\mathrm{e}^{-\frac{1}{20N}\pi\mathrm{i}} \left(\mathrm{e}^{\frac{2\pi\mathrm{i}}{N}} - 1\right) \cdot \tau_N(W_{12})$$
$$\simeq -\sqrt{\mathrm{i}N} \cdot \frac{2}{5} \left(4 \sin\left(\frac{\pi}{5}\right) \sin\left(\frac{2\pi}{5}\right) \mathrm{e}^{\frac{5}{4}\pi\mathrm{i}N} - \sqrt{5} \cos\left(\frac{4\pi}{5}\right) \mathrm{e}^{\frac{9}{20}\pi\mathrm{i}N} \right)$$
$$+ \frac{1}{2\sqrt{5}} \sum_{k=0}^{\infty} \frac{1}{k!} \left(\frac{\pi}{20\,\mathrm{i}\,N}\right)^k \left\{ L\left(-2k, \psi_{20}^{(1)-2(5)+(9)}\right) \right.$$
$$\left. + 2\,\mathrm{e}^{\frac{4}{5}\pi\mathrm{i}N}\, L\left(-2k, \psi_{20}^{(1)-2\cos\frac{2\pi}{5}(5)+(9)}\right) + 2\,\mathrm{e}^{-\frac{4}{5}\pi\mathrm{i}N}\, L\left(-2k, \psi_{20}^{(1)-2\cos\frac{4\pi}{5}(5)+(9)}\right) \right\}.$$

The perturbative invariants from decomposed invariants, $\tau_N^{(0)}$ and $\tau_N^{(\pm\frac{2}{5})}$, are generated from $(a = 0, 1, 2)$

$$2\,\frac{\sinh\left(2z + \frac{a}{5}\pi\mathrm{i}\right) \sinh\left(2z - \frac{a}{5}\pi\mathrm{i}\right)}{\cosh(5z)} = \sum_{k=0}^{\infty} \frac{L\left(-2k, \psi_{20}^{(1)-2\cos\frac{2a\pi}{5}(5)+(9)}\right)}{(2k)!} z^{2k}.$$

4.1.4. Q_{12}. We get

$$\text{(4.6)}\quad \mathrm{e}^{-\frac{1}{24N}\pi\mathrm{i}} \left(\mathrm{e}^{\frac{2\pi\mathrm{i}}{N}} - 1\right) \cdot \tau_N(Q_{12}) = \frac{1}{6} \left\{ \widetilde{\Psi}_6^{5(1)+(3)-4(5)}(-1/N) \right.$$
$$\left. + 2\,\mathrm{e}^{\frac{2}{3}\pi\mathrm{i}N}\, \widetilde{\Psi}_6^{(1)-(3)+(5)}(-1/N) + \mathrm{e}^{-\frac{2}{3}\pi\mathrm{i}N}\, \widetilde{\Psi}_6^{2(1)+(3)+2(5)}(-1/N) \right\},$$

which induces an asymptotic expansion as

$$\mathrm{e}^{-\frac{1}{24N}\pi\mathrm{i}} \left(\mathrm{e}^{\frac{2\pi\mathrm{i}}{N}} - 1\right) \cdot \tau_N(Q_{12})$$
$$\simeq -\sqrt{\mathrm{i}N} \cdot \frac{1}{4\sqrt{3}} \left(2\,\mathrm{e}^{\frac{1}{12}\pi\mathrm{i}N} + 4\,\mathrm{e}^{\frac{17}{12}\pi\mathrm{i}N} + 2\sqrt{3}\,\mathrm{e}^{\frac{1}{3}\pi\mathrm{i}N} + \sqrt{3}\,\mathrm{e}^{\frac{4}{3}\pi\mathrm{i}N} \right)$$
$$+ \frac{1}{6} \sum_{k=0}^{\infty} \frac{1}{k!} \left(\frac{\pi}{12\,\mathrm{i}\,N}\right)^k \left\{ L\left(-2k, \psi_{12}^{5(1)+(3)-4(5)}\right) \right.$$
$$\left. + 2\,\mathrm{e}^{\frac{2}{3}\pi\mathrm{i}N}\, L\left(-2k, \psi_{12}^{(1)-(3)+(5)}\right) + \mathrm{e}^{-\frac{2}{3}\pi\mathrm{i}N}\, L\left(-2k, \psi_{12}^{2(1)+(3)+2(5)}\right) \right\}.$$

Generating functions of the decomposed perturbative invariants are

$$2\cosh(z) \left(5\,\frac{\sinh(4z)}{\sinh(6z)} - 4\,\frac{\sinh(2z)}{\sinh(6z)} \right) = \sum_{k=0}^{\infty} \frac{L\left(-2k, \psi_{12}^{5(1)+(3)-4(5)}\right)}{(2k)!} z^{2k},$$

$$\frac{1}{2\cosh(z)} = \sum_{k=0}^{\infty} \frac{L\left(-2\,k, \psi_{12}^{(1)-(3)+(5)}\right)}{(2\,k)!} z^{2k},$$

$$\frac{4\cosh(2\,z)+1}{2\cosh(3\,z)} = \sum_{k=0}^{\infty} \frac{L\left(-2\,k, \psi_{12}^{2(1)+(3)+2(5)}\right)}{(2\,k)!} z^{2k}.$$

4.1.5. U_{12}. We have

$$\text{(4.7)}\quad \mathrm{e}^{-\frac{1}{16N}\pi\mathrm{i}} \left(\mathrm{e}^{\frac{2\pi\mathrm{i}}{N}} - 1\right) \cdot \tau_N(U_{12}) = \frac{1}{2}\left\{\widetilde{\Psi}_4^{(1)-3(3)}(-1/N) + 3\,(-1)^N\, \widetilde{\Psi}_4^{(1)+(3)}(-1/N)\right\},$$

and an asymptotic expansion in $N \to \infty$ is given by

$$\mathrm{e}^{-\frac{1}{16N}\pi\mathrm{i}} \left(\mathrm{e}^{\frac{2\pi\mathrm{i}}{N}} - 1\right) \cdot \tau_N(U_{12}) \simeq -\sqrt{\mathrm{i}\,N}\cdot\frac{1}{\sqrt{2}}\left(\sqrt{2}\,\mathrm{e}^{\frac{9}{8}\pi\mathrm{i}N} + \mathrm{e}^{\frac{1}{2}\pi\mathrm{i}N}\right)$$
$$+\frac{1}{2}\sum_{k=0}^{\infty}\frac{1}{k!}\left(\frac{\pi}{8\,N\,\mathrm{i}}\right)^k \left\{L\left(-2\,k, \psi_8^{(1)-3(3)}\right) + 3\,(-1)^N\, L\left(-2\,k, \psi_8^{(1)+(3)}\right)\right\}.$$

Generating functions of the decomposed perturbative invariants are

$$4\,\frac{\sinh^3(z)}{\sinh(4\,z)} = \sum_{k=0}^{\infty} \frac{L\left(-2\,k, \psi_8^{(1)-3(3)}\right)}{(2\,k)!} z^{2k},$$

$$\frac{\cosh(z)}{\cosh(3\,z)} = \sum_{k=0}^{\infty} \frac{L\left(-2\,k, \psi_8^{(1)+(3)}\right)}{(2\,k)!} z^{2k}.$$

4.1.6. S_{12}. The WRT invariant is written as

$$\text{(4.8)}\quad \mathrm{e}^{\frac{2\pi\mathrm{i}}{N}\frac{287}{240}} \left(\mathrm{e}^{\frac{2\pi\mathrm{i}}{N}} - 1\right) \tau_N(S_{12}) = \frac{1}{2\sqrt{13}} \sum_{c=1}^{780} \widetilde{\Psi}_{780}^{(c)}(-1/N)$$
$$\times\left(1 + 2\sum_{a=1}^{6} \mathrm{e}^{\frac{10}{13}(2a-1)^2\pi\mathrm{i}N} \cos\left(\frac{2\,a\,c}{13}\,\pi\right)\right) \psi_{120}^{(13)-(37)-(43)-(53)}(c),$$

which we do not find a simplified expression. An asymptotic expansion is given by

$$\mathrm{e}^{\frac{2\pi\mathrm{i}}{N}\frac{287}{240}} \left(\mathrm{e}^{\frac{2\pi\mathrm{i}}{N}} - 1\right) \tau_N(S_{12}) \simeq -\sqrt{\mathrm{i}\,N}\cdot\frac{1}{\sqrt{5}}\left(\sqrt{2}\sin\left(\frac{2\,\pi}{5}\right)\mathrm{e}^{\frac{13}{30}\pi\mathrm{i}N}\right.$$
$$\left. + \sin\left(\frac{2\,\pi}{5}\right)\mathrm{e}^{\frac{157}{120}\pi\mathrm{i}N} + \sqrt{2}\sin\left(\frac{\pi}{5}\right)\mathrm{e}^{\frac{37}{30}\pi\mathrm{i}N} + \sin\left(\frac{\pi}{5}\right)\mathrm{e}^{\frac{133}{120}\pi\mathrm{i}N}\right)$$
$$+\frac{1}{2\sqrt{13}}\sum_{k=0}^{\infty}\left(\frac{\pi}{1560\,\mathrm{i}\,N}\right)^k \sum_{c=1}^{780} \psi_{120}^{(13)-(37)-(43)-(53)}(c)$$
$$\times\left(1 + 2\sum_{a=1}^{6} \mathrm{e}^{\frac{10}{13}(2a-1)^2\pi\mathrm{i}N} \cos\left(\frac{2\,a\,c}{13}\,\pi\right)\right) \frac{L(-2\,k, \psi_{1560}^{(c)})}{k!}.$$

4.2. Non self-dual cases.

4.2.1. E_{13}. We have

$$\text{(4.9)}\quad \mathrm{e}^{-\frac{21}{40N}\pi\mathrm{i}}\left(\mathrm{e}^{\frac{2\pi\mathrm{i}}{N}}-1\right)\cdot\tau_N(E_{13})=\frac{1}{2\sqrt{2}}\left(1+(-1)^N\right)\widetilde{\Psi}_{20}^{(1)-(9)-(11)+(19)}(-1/N),$$

which gives an asymptotic expansion in $N\to\infty$ as

$$\mathrm{e}^{-\frac{21}{40N}\pi\mathrm{i}}\left(\mathrm{e}^{\frac{2\pi\mathrm{i}}{N}}-1\right)\cdot\tau_N(E_{13})\simeq-\sqrt{\mathrm{i}\,N}\cdot\sqrt{\frac{2}{5}}\sin\left(\frac{2\pi}{5}\right)\left(\mathrm{e}^{\frac{9}{40}\pi\mathrm{i}N}+\mathrm{e}^{\frac{49}{40}\pi\mathrm{i}N}\right)$$
$$+\frac{1}{2\sqrt{2}}\left(1+(-1)^N\right)\sum_{k=0}^{\infty}\frac{L\left(-2\,k,\psi_{40}^{(1)-(9)-(11)+(19)}\right)}{k!}\left(\frac{\pi}{40\,N\,\mathrm{i}}\right)^k.$$

The perturbative invariants follow from

$$\frac{2\sinh(5\,z)\sinh(4\,z)}{\cosh(10\,z)}=\sum_{k=0}^{\infty}\frac{L\left(-2\,k,\psi_{40}^{(1)-(9)-(11)+(19)}\right)}{(2\,k)!}z^{2k}.$$

4.2.2. Z_{11}. We get

$$\text{(4.10)}\quad \mathrm{e}^{\frac{23}{48N}\pi\mathrm{i}}\left(\mathrm{e}^{\frac{2\pi\mathrm{i}}{N}}-1\right)\cdot\tau_N(Z_{11})=\frac{1}{2\sqrt{2}}\left(1+(-1)^N\right)\widetilde{\Psi}_{24}^{(1)-(7)-(17)+(23)}(-1/N),$$

which in large-N limit behaves like

$$\mathrm{e}^{\frac{23}{48N}\pi\mathrm{i}}\left(\mathrm{e}^{\frac{2\pi\mathrm{i}}{N}}-1\right)\cdot\tau_N(Z_{11})\simeq-\sqrt{\mathrm{i}\,N}\cdot\frac{1}{\sqrt{2}}\sin\left(\frac{3\,\pi}{8}\right)\left(\mathrm{e}^{\frac{25}{48}\pi\mathrm{i}N}+\mathrm{e}^{\frac{73}{48}\pi\mathrm{i}N}\right)$$
$$+\frac{1}{2\sqrt{2}}\left(1+(-1)^N\right)\sum_{k=0}^{\infty}\frac{L\left(-2\,k,\psi_{48}^{(1)-(7)-(17)+(23)}\right)}{k!}\left(\frac{\pi}{48\,N\,\mathrm{i}}\right)^k.$$

Here a generating function of L-functions is

$$\frac{2\sinh(3\,z)\sinh(8\,z)}{\cosh(12\,z)}=\sum_{k=0}^{\infty}\frac{L\left(-2\,k,\psi_{48}^{(1)-(7)-(17)+(23)}\right)}{(2\,k)!}z^{2k}.$$

The unified WRT invariant is computed from (3.7) and (3.11) [**6**] as

$$\text{(4.11)}\qquad (q-1)\,I_q(Z_{11})=\sqrt{2}\,q^{\frac{1}{4}}\sum_{n=0}^{\infty}(-1)^n\,q^{\frac{n}{2}}\,(q^{\frac{1}{2}};-q^{\frac{1}{2}})_{2n+1},$$

which also generates the L-function in $t\searrow 0$

$$\sum_{n=0}^{\infty}(-1)^n\,q^{\frac{n}{2}}\,(q^{\frac{1}{2}};-q^{\frac{1}{2}})_{2n+1}\Big|_{q=\mathrm{e}^{-t}}=\frac{1}{2}\,\mathrm{e}^{\frac{47}{96}t}\sum_{k=0}^{\infty}\frac{L\left(-2\,k,\psi_{48}^{(1)-(7)-(17)+(23)}\right)}{k!}\left(\frac{t}{96}\right)^k.$$

4.2.3. Q_{10}. We have

$$\text{(4.12)}\quad \mathrm{e}^{\frac{35}{36N}\pi\mathrm{i}}\left(\mathrm{e}^{\frac{2\pi\mathrm{i}}{N}}-1\right)\cdot\tau_N(Q_{10})$$
$$=\frac{1}{2\sqrt{3}}\left\{\widetilde{\Psi}_{18}^{(1)-(5)-(13)+(17)}(-1/N)+\mathrm{e}^{-\frac{2}{3}\pi\mathrm{i}N}\,\widetilde{\Psi}_{18}^{2(1)+(5)+(13)+2(17)}(-1/N)\right\}.$$

An asymptotic expansion is then given by

$$\mathrm{e}^{\frac{35}{36N}\pi\mathrm{i}}\left(\mathrm{e}^{\frac{2\pi\mathrm{i}}{N}}-1\right)\cdot\tau_N(Q_{10})$$

$$\simeq -\sqrt{\mathrm{i}\,N}\cdot\frac{2}{3}\left(\sin\left(\frac{2\,\pi}{9}\right)\mathrm{e}^{\frac{49}{36}\pi\mathrm{i}N}+\sin\left(\frac{4\,\pi}{9}\right)\mathrm{e}^{\frac{25}{36}\pi\mathrm{i}N}+\frac{\sqrt{3}}{2}\,\mathrm{e}^{-\frac{5}{12}\pi\mathrm{i}N}\right)$$
$$+\frac{1}{2\sqrt{3}}\sum_{k=0}^{\infty}\frac{1}{k!}\left(\frac{\pi}{36\,N\,\mathrm{i}}\right)^k$$
$$\times\left\{L\left(-2\,k,\psi_{36}^{(1)-(5)-(13)+(17)}\right)+\mathrm{e}^{-\frac{2}{3}\pi\mathrm{i}N}\,L\left(-2\,k,\psi_{36}^{2(1)+(5)+(13)+2(17)}\right)\right\}.$$

The decomposed perturbative invariants are generated as

$$\frac{2\sinh(6\,z)\,\sinh(2\,z)}{\cosh\,(9\,z)}=\sum_{k=0}^{\infty}\frac{L\left(-2\,k,\psi_{36}^{(1)-(5)-(13)+(17)}\right)}{(2\,k)!}\,z^{2k},$$
$$\frac{2\,\cosh(8\,z)+\cosh(4\,z)}{\cosh(9\,z)}=\sum_{k=0}^{\infty}\frac{L\left(-2\,k,\psi_{36}^{2(1)+(5)+(13)+2(17)}\right)}{(2\,k)!}\,z^{2k}.$$

Based on the decomposed WRT invariant $\tau_N^{(0)}(Q_{10})$ constructed from surgery on trefoil, we have a hypergeometric generating function for perturbative invariants as

$$(4.13)\quad \sum_{n=0}^{\infty}q^n\sum_{k=-n-1}^{n}\frac{(q)_{2n+1}}{(q)_{n-k}\,(q)_{n+k+1}}\,(-1)^{k+1}\,q^{\frac{1}{6}k(k+3)}\Bigg|_{q=\mathrm{e}^{-t}}$$
$$=\frac{1}{2}\,\mathrm{e}^{\frac{71}{72}t}\sum_{k=0}^{\infty}\frac{L\left(-2\,k,\psi_{36}^{(1)-(5)-(13)+(17)}\right)}{k!}\left(\frac{t}{72}\right)^k.$$

4.2.4. E_{14}. We obtain

$$(4.14)\quad \mathrm{e}^{-\frac{25}{24N}\pi\mathrm{i}}\left(\mathrm{e}^{\frac{2\pi\mathrm{i}}{N}}-1\right)\cdot\tau_N(E_{14})$$
$$=\frac{1}{2\sqrt{3}}\left\{\widetilde{\Psi}_{12}^{(1)-(7)-2(9)}(-1/N)+2\,\mathrm{e}^{\frac{2}{3}\pi\mathrm{i}N}\,\widetilde{\Psi}_{12}^{(1)-(7)+(9)}(-1/N)\right\}.$$

An asymptotic expansion is given by

$$\mathrm{e}^{-\frac{25}{24N}\pi\mathrm{i}}\left(\mathrm{e}^{\frac{2\pi\mathrm{i}}{N}}-1\right)\cdot\tau_N(E_{14})\simeq-\sqrt{\mathrm{i}\,N}\cdot\frac{1}{2}\left(\sqrt{2}\,\mathrm{e}^{\frac{1}{6}\pi\mathrm{i}N}+\mathrm{e}^{\frac{25}{24}\pi\mathrm{i}N}\right)$$
$$+\frac{1}{2\sqrt{3}}\sum_{k=0}^{\infty}\frac{1}{k!}\left(\frac{\pi}{24\,N\,\mathrm{i}}\right)^k\left\{L\left(-2\,k,\psi_{24}^{(1)-(7)-2(9)}\right)+2\,\mathrm{e}^{\frac{2}{3}\pi\mathrm{i}N}\,L\left(-2\,k,\psi_{24}^{(1)-(7)+(9)}\right)\right\},$$

where generating functions of the decomposed perturbative invariants are

$$\frac{4\,\sinh\,(3\,z)\,\sinh^2\,(4\,z)}{\sinh\,(12\,z)}=\sum_{k=0}^{\infty}\frac{L\left(-2\,k,\psi_{24}^{(1)-(7)-2(9)}\right)}{(2\,k)!}\,z^{2k},$$
$$\sinh\,(3\,z)\,\frac{2\cosh(8\,z)+1}{\sinh\,(12\,z)}=\sum_{k=0}^{\infty}\frac{L\left(-2\,k,\psi_{24}^{(1)-(7)+(9)}\right)}{(2\,k)!}\,z^{2k}.$$

4.2.5. Q_{11}. We have

$$
\mathrm{e}^{\frac{25}{56N}\pi\mathrm{i}}\left(\mathrm{e}^{\frac{2\pi\mathrm{i}}{N}}-1\right)\cdot\tau_N(Q_{11})=\frac{1}{2\sqrt{6}}\left(1+(-1)^N\right)
\tag{4.15}
$$

$$
\times\left\{\widetilde{\Psi}_{84}^{(3)-(11)-(17)+(25)-(31)+(39)+(45)-(53)+(59)-(67)-(73)+(81)}(-1/N)\right.
$$

$$
\left.+\,\mathrm{e}^{-\frac{2}{3}\pi\mathrm{i}N}\,\widetilde{\Psi}_{84}^{2(3)+(11)+(17)-(25)+(31)+2(39)+2(45)+(53)-(59)+(67)+(73)+2(81)}(-1/N)\right\},
$$

which induces an asymptotic expansion as

$$
\mathrm{e}^{\frac{25}{56N}\pi\mathrm{i}}\left(\mathrm{e}^{\frac{2\pi\mathrm{i}}{N}}-1\right)\cdot\tau_N(Q_{11})
$$

$$
\simeq-\sqrt{\mathrm{i}\,N}\cdot\sqrt{\frac{2}{7}}\left(\sin\left(\frac{2\,\pi}{7}\right)\left(\mathrm{e}^{\frac{19}{56}\pi\mathrm{i}N}+\mathrm{e}^{\frac{75}{56}\pi\mathrm{i}N}\right)+\sin\left(\frac{3\,\pi}{7}\right)\left(\mathrm{e}^{-\frac{29}{56}\pi\mathrm{i}N}+\mathrm{e}^{\frac{27}{56}\pi\mathrm{i}N}\right)\right)
$$

$$
+\frac{1+(-1)^N}{2\sqrt{6}}\sum_{k=0}^{\infty}\frac{1}{k!}\left(\frac{\pi}{168\,N\,\mathrm{i}}\right)^k
$$

$$
\times\left\{L\left(-2\,k,\psi_{168}^{(3)-(11)-(17)+(25)-(31)+(39)+(45)-(53)+(59)-(67)-(73)+(81)}\right)\right.
$$

$$
\left.+\mathrm{e}^{-\frac{2}{3}\pi\mathrm{i}N}\,L\left(-2\,k,\psi_{168}^{2(3)+(11)+(17)-(25)+(31)+2(39)+2(45)+(53)-(59)+(67)+(73)+2(81)}\right)\right\}.
$$

Here we have

$$
2\sinh(4\,z)\frac{-\sinh(7\,z)-\sinh(21\,z)+\sinh(35\,z)}{\cosh(42\,z)}
$$

$$
=\sum_{k=0}^{\infty}\frac{L\left(-2\,k,\psi_{168}^{(3)-(11)-(17)+(25)-(31)+(39)+(45)-(53)+(59)-(67)-(73)+(81)}\right)}{(2\,k)!}\,z^{2k},
$$

$$
2\,\cosh(21\,z)\,\frac{2\,\cosh(18\,z)+\cosh(10\,z)}{\cosh(42\,z)}+\frac{2\,\sinh(21\,z)\,\sinh(4\,z)}{\cosh(42\,z)}
$$

$$
=\sum_{k=0}^{\infty}\frac{L\left(-2\,k,\psi_{168}^{2(3)+(11)+(17)-(25)+(31)+2(39)+2(45)+(53)-(59)+(67)+(73)+2(81)}\right)}{(2\,k)!}\,z^{2k}.
$$

4.2.6. Z_{13}. We get

$$
\mathrm{e}^{\frac{17}{30N}\pi\mathrm{i}}\left(\mathrm{e}^{\frac{2\pi\mathrm{i}}{N}}-1\right)\cdot\tau_N(Z_{13})=\frac{1}{2\sqrt{6}}\left(1+(-1)^N\right)
\tag{4.16}
$$

$$
\times\left\{\widetilde{\Psi}_{30}^{(2)-(8)-2(12)+2(18)+(22)-(28)}(-1/N)\right.
$$

$$
\left.+2\,\mathrm{e}^{-\frac{1}{3}\pi\mathrm{i}N}\,\widetilde{\Psi}_{30}^{(2)-(8)+(12)-(18)+(22)-(28)}(-1/N)\right\}.
$$

An asymptotic expansion in $N\to\infty$ is computed to be

$$
\mathrm{e}^{-\frac{17}{30N}\pi\mathrm{i}}\left(\mathrm{e}^{\frac{2\pi\mathrm{i}}{N}}-1\right)\cdot\tau_N(Z_{13})\simeq-\sqrt{\mathrm{i}\,N}\cdot\sqrt{\frac{2}{5}}\,\sin\left(\frac{2\,\pi}{5}\right)\left(\mathrm{e}^{\frac{4}{15}\pi\mathrm{i}N}+\mathrm{e}^{\frac{19}{15}\pi\mathrm{i}N}\right)
$$

$$
+\frac{1}{2\sqrt{6}}\left(1+(-1)^N\right)\sum_{k=0}^{\infty}\frac{1}{k!}\left(\frac{\pi}{15\,N\,\mathrm{i}}\right)^k\left\{L\left(-2\,k,\psi_{30}^{(1)-(4)-2(6)+2(9)+(11)-(14)}\right)\right.
$$

$$
\left.+2\,\mathrm{e}^{-\frac{1}{3}\pi\mathrm{i}N}\,L\left(-2\,k,\psi_{30}^{(1)-(4)+(6)-(9)+(11)-(14)}\right)\right\},
$$

where the decomposed perturbative invariants are generated from

$$4\,\frac{\sinh\left(\frac{3}{2}z\right)\sinh^2\left(\frac{5}{2}z\right)}{\sinh\left(\frac{15}{2}z\right)}=\sum_{k=0}^{\infty}\frac{L\left(-2\,k,\psi_{30}^{(1)-(4)-2(6)+2(9)+(11)-(14)}\right)}{(2\,k)!}\,z^{2k}.$$

$$\frac{\sinh\left(\frac{3}{2}\,z\right)}{\sinh\left(\frac{5}{2}\,z\right)}=\sum_{k=0}^{\infty}\frac{L\left(-2\,k,\psi_{30}^{(1)-(4)+(6)-(9)+(11)-(14)}\right)}{(2\,k)!}\,z^{2k}.$$

4.2.7. S_{11}. We have

(4.17)

$$\mathrm{e}^{\frac{13}{30N}\pi\mathrm{i}}\left(\mathrm{e}^{\frac{2\pi\mathrm{i}}{N}}-1\right)\cdot\tau_N(S_{11})=\frac{1}{2\sqrt{2}}\left\{\widetilde{\Psi}_{30}^{(2)-(7)-(8)+(13)-(17)+(22)+(23)-(28)}(-1/N)\right.$$
$$+(-1)^N\widetilde{\Psi}_{30}^{(2)+(7)-(8)-(13)+(17)+(22)-(23)-(28)}(-1/N)$$
$$\left.+2\,\mathrm{e}^{-\frac{3}{4}\pi\mathrm{i}N}\,\widetilde{\Psi}_{30}^{(2)+(8)+(22)+(28)}(-1/N)\right\}.$$

An asymptotic expansion is given by

$$\mathrm{e}^{\frac{13}{30N}\pi\mathrm{i}}\left(\mathrm{e}^{\frac{2\pi\mathrm{i}}{N}}-1\right)\cdot\tau_N(S_{11})$$
$$\simeq-\sqrt{\mathrm{i}\,N}\cdot\frac{4}{\sqrt{30}}\left(\sin\left(\frac{\pi}{5}\right)\mathrm{e}^{\frac{7}{5}\pi\mathrm{i}N}+\sin\left(\frac{2\,\pi}{5}\right)\left(\mathrm{e}^{\frac{19}{15}\pi\mathrm{i}N}+\mathrm{e}^{\frac{3}{5}\pi\mathrm{i}N}\right)\right)$$
$$+\frac{1}{2\sqrt{2}}\sum_{k=0}^{\infty}\frac{1}{k!}\left(\frac{\pi}{60\,N\,\mathrm{i}}\right)^k\left\{L\left(-2\,k,\psi_{30}^{(2)-(7)-(8)+(13)}\right)\right.$$
$$\left.+(-1)^N\,L\left(-2\,k,\psi_{30}^{(2)+(7)-(8)-(13)}\right)+2\,\mathrm{e}^{-\frac{3}{4}\pi\mathrm{i}N}\,L\left(-2\,k,\psi_{60}^{(2)+(8)+(22)+(28)}\right)\right\},$$

where generating functions of each L-function are

$$2\sinh(3\,z)\,\frac{\cosh(10\,z)\pm\cosh(5\,z)}{\sinh(15\,z)}=\sum_{k=0}^{\infty}\frac{L\left(-2\,k,\psi_{30}^{(2)\pm(7)-(8)\mp(13)}\right)}{(2\,k)!}\,z^{2k},$$

$$\frac{2\,\cosh(10\,z)\,\cosh(3\,z)}{\cosh(15\,z)}=\sum_{k=0}^{\infty}\frac{L\left(-2\,k,\psi_{60}^{(2)+(8)+(22)+(28)}\right)}{(2\,k)!}\,z^{2k}.$$

4.2.8. W_{13}. The WRT invariant is computed as

$$(4.18)\quad \mathrm{e}^{-\frac{7}{12N}\pi\mathrm{i}}\left(\mathrm{e}^{\frac{2\pi\mathrm{i}}{N}}-1\right)\cdot\tau_N(W_{13})=\frac{1}{2\sqrt{2}}\left\{\widetilde{\Psi}_{6}^{(1)-2(4)-(5)}(-1/N)\right.$$
$$\left.+(-1)^N\,\widetilde{\Psi}_{6}^{(1)+2(4)-(5)}(-1/N)+2\,\mathrm{e}^{\frac{3}{4}\pi\mathrm{i}N}\,\widetilde{\Psi}_{6}^{(1)-(5)}(-1/N)\right\},$$

which gives an asymptotic expansion in $N\to\infty$ as

$$\mathrm{e}^{-\frac{7}{12N}\pi\mathrm{i}}\left(\mathrm{e}^{\frac{2\pi\mathrm{i}}{N}}-1\right)\cdot\tau_N(W_{13})\simeq-\sqrt{\mathrm{i}\,N}\cdot\frac{1}{\sqrt{2}}\left(\mathrm{e}^{\frac{1}{3}\pi\mathrm{i}N}+\mathrm{e}^{\frac{13}{12}\pi\mathrm{i}N}\right)$$
$$+\frac{1}{2\sqrt{2}}\sum_{k=0}^{\infty}\frac{1}{k!}\left(\frac{\pi}{12\,\mathrm{i}\,N}\right)^k\left\{L\left(-2\,k,\psi_{12}^{(1)-2(4)-(5)}\right)\right.$$
$$\left.+(-1)^N\,L\left(-2\,k,\psi_{12}^{(1)+2(4)-(5)}\right)+2\,\mathrm{e}^{\frac{3}{4}\pi\mathrm{i}N}\,L\left(-2\,k,\psi_{12}^{(1)-(5)}\right)\right\}.$$

Generating functions of the decomposed perturbative invariants are

$$4\,\frac{\sinh(2\,z)\,\sinh^2(\frac{3}{2}z)}{\sinh(6\,z)} = \sum_{k=0}^{\infty} \frac{L\left(-2\,k, \psi_{12}^{(1)-2(4)-(5)}\right)}{(2\,k)!}\,z^{2k},$$

$$4\,\frac{\sinh(2\,z)\,\cosh^2(\frac{3}{2}z)}{\sinh(6\,z)} = \sum_{k=0}^{\infty} \frac{L\left(-2\,k, \psi_{12}^{(1)+2(4)-(5)}\right)}{(2\,k)!}\,z^{2k},$$

$$2\,\frac{\sinh(2\,z)\,\sinh(3\,z)}{\sinh(6\,z)} = \sum_{k=0}^{\infty} \frac{L\left(-2\,k, \psi_{12}^{(1)-(5)}\right)}{(2\,k)!}\,z^{2k}.$$

5. Concluding remarks

We have shown that the WRT invariant for M can be decomposed as (1.4) by use of the linking pairing when M is the Seifert manifold associated with the Arnold unimodal singularities. The classical Chern–Simons invariants can be given from a nearly modular property of the Eichler integral (3.17). Although the linking pairing has a duality corresponding to the Arnold strange duality [**19**], it is not clear at this stage whether there exists duality for the classical Chern–Simons invariant and the decomposed perturbative invariants.

Variants of unified WRT invariants are constructed for three of the Seifert manifolds, E_{12}, Z_{11}, and Q_{10}, based on an s-surgery formula on trefoil. They have an interesting property from the viewpoint of modular form [**6**]. By computer experiments as a generalization of (4.13), hypergeometric generating function for $s \in \mathbb{Z}_{\neq 0,-6}$ seems to have an expansion in $t \searrow 0$

$$\begin{aligned}(5.1)\quad & \sum_{n=0}^{\infty} q^n \sum_{k=-n-1}^{n} (-1)^{k+1} \frac{(q)_{2n+1}}{(q)_{n-k}\,(q)_{n+k+1}}\, q^{\frac{s-2}{2s}k^2+\frac{1}{2}k}\Bigg|_{q=\mathrm{e}^{-t}} \\ &= \frac{1}{2}\,\mathrm{e}^{\left(1-\frac{s}{24(6+s)}\right)t} \sum_{n=0}^{\infty} \frac{L\left(-2\,n, \psi_{2\ell(s)}\right)}{n!} \left(\frac{t}{4\,\ell(s)}\right)^n,\end{aligned}$$

where $\ell(s) = \operatorname{lcm}(2,3,s,6+s)$, and the L-function follows from

$$2\,\frac{\sinh\left(\frac{\ell(s)}{3}\,z\right)\,\sinh\left(\frac{\ell(s)}{6+s}\,z\right)}{\cosh\left(\frac{\ell(s)}{2}\,z\right)} = \sum_{k=0}^{\infty} \frac{L\left(-2\,k, \psi_{2\ell(s)}\right)}{(2\,k)!}\,z^{2k}.$$

To conclude, one sees that the WRT invariants for rational homology spheres could be treated by use of the Eichler integral, which is intimately related to mock modular form (see [**39, 30**] for recent developments). It is not clear whether the decomposition (1.4) holds when M is a hyperbolic rational homology sphere. Analytic properties of the WRT invariant for hyperbolic manifolds will provide an interesting research objects [**40**].

Acknowledgments. The author would like to thank organizers of "Chern–Simons Theory: 20 years after". This work is supported in part by Grant-in-Aid from the Ministry of Education, Culture, Sports, Science and Technology of Japan.

References

[1] G. E. Andrews, J. Jiménez-Urroz, and K. Ono, *q-series identities and values of certain L-functions*, Duke Math. J. **108** (2001), 395–419.

[2] V. I. Arnold, V. V. Goryunov, O. V. Lyashko, and V. A. Vasil'ev, *Singularity Theory I*, Springer, Berlin, 1998.

[3] A. Beliakova, C. Blanchet, and T. T. Q. Le, *Unified quantum invariants and their refinements for homology 3-spheres with 2-torsion*, Fund. Math. **201** (2008), 217–239 .

[4] A. Beliakova, I. Bühler, and T. T. Q. Le, *A unified quantum SO(3) invariant for rational homology 3-spheres*, preprint (2008), [arXiv:0801.3893].

[5] A. Beliakova and T. T. Q. Le, *Integrality of quantum 3-manifold invariants and a rational surgery formula*, Compos. Math. **143** (2007), 1593–1612.

[6] K. Bringmann, K. Hikami, and J. Lovejoy, *On modularity of the unified WRT invariant for certain Seifert manifolds*, Adv. Appl. Math. (2010), to appear.

[7] D. S. Freed, *Remarks on Chern–Simons theory*, Bull. Amer. Math. Soc. **46** (2009), 221–254.

[8] D. S. Freed and R. E. Gompf, *Computer calculation of Witten's 3-manifold invariant*, Commun. Math. Phys. **141** (1991), 79–117.

[9] K. Habiro, *On the quantum sl_2 invariants of knots and integral homology spheres*, Geometry & Topology Monographs **4** (2002), 55–68.

[10] ———, *A unified Witten–Reshetikhin–Turaev invariant for integral homology sphere*, Invent. Math. **171** (2008), 1–81.

[11] K. Hikami, *Quantum invariant for torus link and modular forms*, Commun. Math. Phys. **246** (2004), 403–426.

[12] ———, *Mock (false) theta functions as quantum invariants*, Regular & Chaotic Dyn. **10** (2005), 509–530.

[13] ———, *On the quantum invariant for the Brieskorn homology spheres*, Int. J. Math. **16** (2005), 661–685.

[14] ———, *Quantum invariant, modular form, and lattice points*, Int. Math. Res. Not. IMRN **2005** (2005), 121–154.

[15] ———, *On the quantum invariant for the spherical Seifert manifold*, Commun. Math. Phys. **268** (2006), 285–319.

[16] ———, *q-series and L-functions related to half-derivatives of the Andrews–Gordon identity*, Ramanujan J. **11** (2006), 175–197.

[17] ———, *Quantum invariants, modular forms, and lattice points II*, J. Math. Phys. **47** (2006), 102301.

[18] ———, *Hecke type formula for unified Witten–Reshetikhin–Turaev invariant as higher order mock theta functions*, Int. Math. Res. Not. IMRN **2007** (2007), 22–32.

[19] ———, *Duality of linking pairing in Arnold's singularity*, Proc. Japan Acad. Ser. A **84** (2008), 81–86.

[20] K. Hikami and A. N. Kirillov, *Torus knot and minimal model*, Phys. Lett. B **575** (2003), 343–348.

[21] ———, *Hypergeometric generating function of L-function, Slater's identities, and quantum knot invariant*, Algebra i Analiz **17** (2005), 190–208.

[22] L. C. Jeffrey, *Chern–Simons–Witten invariants of lens spaces and torus bundles, and the semiclassical approximation*, Commun. Math. Phys. **147** (1992), 563–604.

[23] A. Kawauchi and S. Kojima, *Algebraic classification of linking pairings on 3-manifolds*, Math. Ann. **253** (1980), 29–42.

[24] M. Kobayashi, *Duality of weights, mirror symmetry and Arnold's strange duality*, Tokyo J. Math. **31** (2008), 225–251.

[25] R. Lawrence and L. Rozansky, *Witten–Reshetikhin–Turaev invariants of Seifert manifolds*, Commun. Math. Phys. **205** (1999), 287–314.

[26] R. Lawrence and D. Zagier, *Modular forms and quantum invariants of 3-manifolds*, Asian J. Math. **3** (1999), 93–107.

[27] H. Murakami, *Quantum SO(3) invariants dominate the SU(2) invariant of Casson and Walker*, Math. Proc. Cambridge Philos. Soc. **117** (1995), 237–249.

[28] T. Ohtsuki, *Quantum Invariants — A Study of Knots, 3-Manifolds, and Their Sets*, World Scientific, Singapore, 2002.

[29] K. Ono, *The Web of Modularity: Arithmetic of the Coefficients of Modular Forms and q-series*, Amer. Math. Soc., Providence, 2004.
[30] ———, *Unearthing the visions of a master: harmonic Maass forms and number theory*, in D. Jerison, B. Mazur, T. Mrowka, W. Schmid, R. P. Stanley, and S.-T. Yau, eds., *Current Developments in Mathematics 2008*, pp. 347–454, Intl. Press, Boston, 2009.
[31] P. Orlik and P. Wagreich, *Isolated singularities of algebraic surfaces with C^*-action*, Ann. Math. **93** (1971), 205–228.
[32] N. Yu. Reshetikhin and V. G. Turaev, *Invariants of 3-manifolds via link polynomials and quantum groups*, Invent. Math. **103** (1991), 547–597.
[33] S.-S. Roan, *Mirror symmetry and Arnold's duality*, J. Math. Sci. **94** (1999), 1237–1253.
[34] L. Rozansky, *A large k asymptotics of Witten's invariant of Seifert manifolds*, Commun. Math. Phys. **171** (1995), 279–322.
[35] K. Saito, *Duality for regular systems of weights*, Asian J. Math. **2** (1998), 983–1047.
[36] H. Seifert, *Topology of 3-dimensional fibered spaces*, in *Seifert and Threlfall: A Textbook of Topology*, pp. 359–422, Academic Press, New York, 1980, translated by W. Heil.
[37] E. Witten, *Quantum field theory and the Jones polynomial*, Commun. Math. Phys. **121** (1989), 351–399.
[38] D. Zagier, *Vassiliev invariants and a strange identity related to the Dedekind eta-function*, Topology **40** (2001), 945–960.
[39] ———, *Ramanujan's mock theta functions and their applications [d'après Zwegers and Bringmann–Ono]*, Séminaire Bourbaki **986** (2006–2007).
[40] ———, *Quantum modular forms*, preprint (2009).

Department of Mathematics, Naruto University of Education, Tokushima 772-8502, Japan

E-mail address: KHikami@gmail.com

AMS/IP Studies in Advanced Mathematics
Volume 50, 2011

Representations and the Colored Jones Polynomial of a Torus Knot

Kazuhiro Hikami and Hitoshi Murakami

ABSTRACT. We show that for a torus knot the $SL(2;\mathbb{C})$ Chern–Simons invariants and the $SL(2;\mathbb{C})$ twisted Reidemeister torsions appear in an asymptotic expansion of the colored Jones polynomial. This suggests a generalization of the volume conjecture that relates the asymptotic behavior of the colored Jones polynomial of a knot to the volume of the knot complement.

1. Introduction

In 1985, Jones introduced a knot invariant, the Jones polynomial, by using operator algebra [**15**]. It turns out to be a special case of a more general situation. In fact for any simple Lie algebra $\mathfrak{g}$ and its irreducible representation ρ one can define the quantum $(\mathfrak{g},\rho)$ invariant for knots (see for example [**35**]). Then the Jones polynomial is regarded as the quantum $(\mathfrak{sl}(2;\mathbb{C}),V^2)$ invariant, where V^2 is the two-dimensional irreducible representation.

Then, in 1989, Witten used Chern–Simons theory to describe the Jones polynomial in terms of path integral [**37**] and suggested quantum invariants for three-manifolds.

Suppose that we are given a compact Lie group G with Lie algebra $\mathfrak{g}$. Let K be a knot in the three-sphere S^3 and V an irreducible representation of G. Let $\mathcal{A}$ be the set of all G-connection on the trivial G-bundle over S^3. For a G-connection A, define the Chern–Simons functional $L(A)$ to be

$$L(A) := \frac{1}{4\pi}\int_{S^3}\operatorname{Tr}(A\wedge dA+\frac{2}{3}A\wedge A\wedge A).$$

Then Witten proposed the following Feynman path integral as a definition of the quantum invariant:

$$Z(S^3,K) := \int_{\mathcal{A}} e^{\sqrt{-1}k\,L(A)}W_V(K;A)\mathcal{D}A,$$

2010 *Mathematics Subject Classification.* Primary 57M27, 57M25, 57M50; Secondary 58J28, 57R56.

Key words and phrases. Knot, torus knot, volume conjecture, colored Jones polynomial, Chern–Simons invariant, Reidemeister torsion.

The authors are supported by Grant-in-Aid for Challenging Exploratory Research (21654053).

where $W_V(K;A)$ is the Wilson loop, that is, the trace of the image in V by the representation of the element in G given by the parallel transport along K using the connection A.

If $G = SU(2)$ and V is the N-dimensional irreducible representation, this defines the N-dimensional colored Jones polynomial $J_N\big(K;\exp(2\pi\sqrt{-1}/(k+2))\big)$.

Since then there have been many researches about these quantum invariants for knots and three-manifolds by both mathematicians and physicists.

In 1995, Kashaev defined a yet another knot invariant $\langle K\rangle_N$ by using quantum dilogarithm [**16**], where N is an integer greater than one. Moreover in [**17**] he observed that for a few knots the large N limit of $\log\left(|\langle K\rangle_N|\right)/N$ gives the hyperbolic volume of the knot complement $S^3\setminus K$. He also conjectured this would be true for any hyperbolic knot. Here a hyperbolic knot is a knot whose complement possesses a complete hyperbolic metric with finite volume.

J. Murakami and the second author proved that Kashaev's invariant is indeed a special value of the colored Jones polynomial [**27**]. More precisely, letting $J_N(K;q)$ be the colored Jones polynomial associated with the N-dimensional irreducible representation of the Lie algebra $\mathfrak{sl}(2;\mathbb{C})$, we showed that $J_N\big(K;\exp(2\pi\sqrt{-1}/N)\big)$ is (essentially) equal to Kashaev's invariant. We also generalized Kashaev's conjecture to the following conjecture (Volume Conjecture).

CONJECTURE 1.1 (Volume Conjecture,[**27**]). *For any knot, we have*

$$2\pi\lim_{N\to\infty}\frac{\log\left|J_N\big(K;\exp(2\pi\sqrt{-1}/N)\big)\right|}{N}=\mathrm{Vol}(S^3\setminus K).$$

Here Vol *is the simplicial volume (or the Gromov norm)* [**9**] *that is normalized so that it equals the sum of the hyperbolic volumes of the hyperbolic pieces in the JSJ decomposition* [**13, 14**] *of the knot complement.*

Note that we normalize $J_N(K;q)$ so that $J_N(unknot;q)=1$.

The volume conjecture has been proved to be true for the following knots and links.

- any torus knot by Kashaev and Tirkkonen [**18**],
- the torus link of type $(2,2m)$ by the first author [**10**],
- the figure-eight knot by Ekholm (see for example [**24**]),
- the hyperbolic knot 5_2 by Kashaev and Yokota,
- Whitehead doubles of the torus knot of type $(2,a)$ by Zheng [**39**],
- twisted Whitehead links by Zheng [**39**],
- the Borromean rings by Garoufalidis and Lê [**8**],
- Whitehead chains by van der Veen [**36**].

What happens if we replace the Nth root of unity $\exp(2\pi\sqrt{-1}/N)$ with another complex parameter $\exp(\xi/N)$? Yokota and the second author proved that for the figure-eight knot if ξ is close to $2\pi\sqrt{-1}$, then the limit gives the hyperbolic volume and the Chern–Simons invariant of the three-manifold obtained from S^3 by Dehn surgery along the figure-eight knot with coefficient given by ξ [**28**].

Note that the space of Dehn surgeries along a hyperbolic knot is complex one-dimensional [**33**], and the parameter ξ in the colored Jones polynomial can be regarded as a parameter of Dehn surgeries. For a hyperbolic knot, the complete hyperbolic structure with finite volume is given by an irreducible representation (holonomy representation) of the fundamental group of its complement into the

Lie group $PSL(2;\mathbb{C})$. Therefore it would be possible to use ξ to parameterize representations into $PSL(2;\mathbb{C})$ or $SL(2;\mathbb{C})$.

In this paper we show that for torus knots we can relate the colored Jones polynomial evaluated at $\exp(\xi/N)$ to representations of the fundamental group of a knot complement into $SL(2;\mathbb{C})$. Moreover by considering an asymptotic expansion of the colored Jones polynomial we can obtain the $SL(2;\mathbb{C})$ Chern–Simons invariant and the twisted Reidemeister torsion both associated with the corresponding representation.

The paper is organized as follows. In Section 2 we describe the character variety of a torus knot, which is used to introduce the twisted Reidemeister torsion and the Chern–Simons invariant in Sections 3 and 4, respectively. In Section 5 we calculate an asymptotic behavior of the colored Jones polynomial evaluated at $\exp(\xi/N)$ for $N \to \infty$, and in Section 6 we give topological interpretations of its coefficients. In the last section (Section 7) we give some speculation for general knots giving an observation about the figure-eight knot.

ACKNOWLEDGMENTS. The authors would like to thank J. Dubois, V. Muñoz, E. Witten, and Y. Yamaguchi for helpful comments. Thanks are also due to the organizers of the workshop "Chern–Simons Gauge Theory: 20 years after" held at the Max Planck Institute for Mathematics in August 2009 hosted by the Hausdorff Center for Mathematics.

2. $SL(2,\mathbb{C})$ character variety of a torus knot

Let $T(a,b)$ be the torus knot of type (a,b), where a and b are coprime positive integers. Throughout this paper we assume that b is odd. Let $X\bigl(T(a,b)\bigr)$ be the character variety of $\pi_1\bigl(S^3\setminus T(a,b)\bigr)$ of representations of $\pi_1\bigl(S^3\setminus T(a,b)\bigr)$ into $SL(2,\mathbb{C})$ [**1**]. So two homomorphisms from $\pi_1(S^3\setminus T(a,b))$ to $SL(2,\mathbb{C})$ are regarded as equivalent if and only if they have the same trace.

We will describe $X\bigl(T(a,b)\bigr)$ following [**23**]. Note that this was first described by Klassen [**20**]

Let $\langle x,y \mid x^a = y^b\rangle$ be a presentation of $\pi_1(S^3\setminus T(a,b))$.

There is a unique reducible component, which is homeomorphic to $\mathbb{C}$ by assigning $[\varphi_t]\in X\bigl(T(a,b)\bigr)$ to $t+t^{-1}\in\mathbb{C}$, where φ_t sends x to $\begin{pmatrix} t^b & 0 \\ 0 & t^{-b}\end{pmatrix}$ and y to $\begin{pmatrix} t^a & 0 \\ 0 & t^{-a}\end{pmatrix}$. Here square brackets mean the class of a representation in the character variety.

The irreducible characters decompose into $(a-1)(b-1)/2$ components and each of them is homeomorphic to $\mathbb{C}$. They are indexed by a pair of integers (α,β) such that $1\le\alpha\le a-1$, $1\le\beta\le b-1$, and that $\alpha\equiv\beta \pmod 2$. See also [**20**, Theorem 1], [**5**, Theorem 2]. A representation with index (α,β) sends x to an element with trace $2\cos(\pi\alpha/a)$ and y to one with trace $2\cos(\pi\beta/b)$.

The closure of the component indexed by (α,β) intersects the reducible component in two points $\left[\varphi_{\exp\left(k_1\pi\sqrt{-1}/(ab)\right)}\right]$ and $\left[\varphi_{\exp\left(k_2\pi\sqrt{-1}/(ab)\right)}\right]$, where

$$
\begin{aligned}
k_1 &\equiv \alpha \pmod a, & k_1 &\equiv -\beta \pmod b,\\
k_2 &\equiv \alpha \pmod a, & k_2 &\equiv \beta \pmod b.
\end{aligned}
$$

Note that k_1 and k_2 are uniquely determined by the formulas above.

REMARK 2.1. Our pair (k_1, k_2) is different from Dubois and Kashaev's pair (k_-, k_+) [**5**, Theorem 2]. They choose k_- and k_+ so that $k_- \equiv k_+ \pmod 2$.

Conversely, given a positive integer k that is not a multiple of neither a nor b, we can define a pair (α, β) such that $1 \le \alpha \le a-1$, $1 \le \beta \le b-1$, and $\alpha \equiv \beta \pmod 2$ as follows: Define α to be the integer that is congruent modulo a to k with $1 \le \alpha \le a-1$, β' to be the integer that is congruent modulo b to k with $1 \le \beta' \le b-1$. If $\alpha \equiv \beta' \pmod 2$ then put $\beta := \beta'$, and if $\alpha \not\equiv \beta' \pmod 2$ then put $\beta := b - \beta'$. Note that since we assume that b is odd β always has the same parity as α.

REMARK 2.2. If k defines (α, β) as above, then the pair (k_1, k_2) defined by (α, β) is either $(k, -k)$ or $(-k, k) \pmod{ab}$. So the assignment of $k \in \{n \in \mathbb{Z} \mid 1 \le n \le ab-1, a \nmid n, b \nmid n\}$ to $(\alpha, \beta) \in \{l \in \mathbb{Z} \mid 1 \le l \le a-1\} \times \{m \in \mathbb{Z} \mid 1 \le m \le b-1\}$ is a two-to-one correspondence.

Note that in either case $\sin^2(\alpha\pi/a)\sin^2(\beta\pi/b)$, which appears in the twisted Reidemeister torsion (see §6), does not depend on the definition that we use and equals $\sin^2(k\pi/a)\sin^2(k\pi/b)$.

3. Twisted Reidemeister torsion for a knot

Let K be a knot in S^3 and ρ a representation of $\pi_1(S^3 \setminus K)$ into $SL(2;\mathbb{C})$. Put $C^*(S^3 \setminus K; \rho) := \operatorname{Hom}_{\mathbb{Z}[\pi_1(S^3\setminus K)]}(C_*(\widetilde{S^3 \setminus K}; \mathbb{Z}), \mathfrak{sl}(2;\mathbb{C}))$. Here $\widetilde{S^3 \setminus K}$ is the universal cover of $S^3 \setminus K$, $C_*(\widetilde{S^3 \setminus K}; \mathbb{Z})$ is regarded as a $\mathbb{Z}[\pi_1(S^3\setminus K)]$-module by the action of the deck transformation and $\mathfrak{sl}(2;\mathbb{C})$ is regarded as a $\mathbb{Z}[\pi_1(S^3\setminus K)]$-module via the adjoint representation.

Let $\{0\} \to C^0 \xrightarrow{d^0} C^1 \xrightarrow{d^1} C^2 \xrightarrow{d^3} C^3 \to \{0\}$ be the corresponding cochain complex, where $C^i := C^i(S^3 \setminus K; \rho)$ and d^i is the coboundary map induced by the boundary map of $C_*(\widetilde{S^3 \setminus K}; \mathbb{Z})$. Put $B^i := \operatorname{Image}(d^{i-1}) \subset C^i$, $Z^i := \operatorname{Ker}(d^i) \subset C^i$, and $H^i := Z^i/B^i$.

We choose bases $\mathbf{c}^i$ of C^i and $\mathbf{h}^i$ of H^i. Let $\tilde{\mathbf{h}}^i \subset Z^i$ be a lift of $\mathbf{h}^i$ and $\mathbf{b}^i \subset C^i$ be a set of elements such that $d^i(\mathbf{b}^i)$ forms a basis of B^{i+1}. Since $B^{i+1} \cong C^i/Z^i$ and $H^i \cong Z^i/B^i$, the set $d^{i-1}(\mathbf{b}^{i-1}) \cup \tilde{\mathbf{h}}^i \cup \mathbf{b}^i$ forms a basis of C^i. Define $\left[\left(d^{i-1}(\mathbf{b}^{i-1}) \cup \tilde{\mathbf{h}}^i \cup \mathbf{b}^i\right)/\mathbf{c}^i\right]$ to be the determinant of the change-of-basis matrix from $\mathbf{c}^i$ to $d^{i-1}(\mathbf{b}^{i-1}) \cup \tilde{h}^i \cup \mathbf{b}^i$.

Then the Reidemeister torsion ([**31**], [**6**], [**2**], [**22**], [**34**]) with respect to $\mathbf{c}^i$ and $\mathbf{h}^i$ is defined to be

$$\operatorname{Tor}(C^*, \mathbf{c}^*, \mathbf{h}^*) := \prod_{i=0}^{n} \left[\left(d^{i-1}(\mathbf{b}^{i-1}) \cup \tilde{\mathbf{h}}^i \cup \mathbf{b}^i\right)/\mathbf{c}^i\right]^{(-1)^{i+1}}. \tag{3.1}$$

It is known that $\operatorname{Tor}(C^*, \mathbf{c}^*, \mathbf{h}^*)$ does not depend on the choice of $\mathbf{b}^i$ and $\tilde{\mathbf{h}}^i$. It is also known that up to sign it depends only on the choice of $\mathbf{h}^*$. (We need a cohomological orientation to define the sign but in this paper we do not need it. See [**34**] and [**3**] for details.)

To define a basis $\mathbf{h}^i$ of H^i we need to choose a simple closed curve on ∂E_K, where $E_K := S^3 \setminus \operatorname{Int}(N(K))$ with $N(K)$ the regular neighborhood of K in S^3.

An irreducible representation ρ is called γ-regular [**30, 3**] for a simple closed curve $\gamma \subset \partial E_K$ if the following two conditions are satisfied.

- The homomorphism $i^*\colon H^1(E_K;\rho) \to H^1(\gamma;\rho)$ induced by the inclusion $i\colon \gamma \hookrightarrow E_K$ is injective. Note that $H^*(E_K;\rho)$ is isomorphic to $H^*(S^3 \setminus K;\rho)$.
- If $\operatorname{Tr}(\rho(\pi_1(\partial E_K))) \subset \{\pm 2\}$, then $\rho(\gamma)$ is not $\pm I$, where I is the identity matrix.

If ρ is γ-regular, then $\dim H^1(S^3 \setminus K;\rho) = \dim H^2(S^3 \setminus K;\rho) = 1$ and $\dim H^i(S^3 \setminus K;\rho) = 0$ for $i \neq 1, 2$ [**3**, Lemma 2]. So to define the Reidemeister torsion for a γ-regular representation ρ we only need to choose a non-zero element of $H^1(S^3 \setminus K;\rho)$ and a non-zero element of $H^2(S^3 \setminus K;\rho)$. We use γ to define such an element of $H^1(S^3 \setminus K;\rho) = H^1(E_K;\rho)$ and the fundamental class $[\partial E_K] \in H^2(\partial E_K;\mathbb{Z})$ to define such an element of $H^2(S^3 \setminus K;\rho)$ (for details, see [**5**, § 3] for example).

Therefore given a simple closed curve $\gamma \subset \partial E_K$ such that ρ is γ-regular one can define the Reidemeister torsion ([**30**], [**3**]) by (3.1) up to sign. It is denoted by $\mathbb{T}^K_\gamma(\rho)$ and called the twisted Reidemeister torsion.

It is known that for a torus knot, any irreducible representation is both μ-regular and λ-regular, where μ is the meridian, a loop that goes around the knot so that it bounds a disk in the tubular neighborhood of the knot and that its linking number with the knot is one, and λ is the preferred longitude, a loop that goes along the knot so that its linking number with the knot is zero [**3**, Example 1]. It is also known that for a hyperbolic knot K, then an irreducible representation that defines a hyperbolic Dehn surgery is γ-regular, where γ is the simple closed curve on ∂E_K along which the surgery is performed [**30**].

4. Chern–Simons invariant for a knot

We follow [**19**] to define the $SL(2;\mathbb{C})$ Chern–Simons invariant.

For a closed three-manifold M, one can define the $SL(2;\mathbb{C})$ Chern–Simons function $\mathrm{cs}_M\colon X(M) \to \mathbb{C} \pmod{\mathbb{Z}}$, where $X(M)$ is the $SL(2;\mathbb{C})$ character variety of M. Let A be an $\mathfrak{sl}(2;\mathbb{C})$-valued 1-form on M with $dA + A \wedge A = 0$. Then A defines a flat connection of $M \times SL(2;\mathbb{C})$ and so one can define a representation $\rho\colon \pi_1(M) \to SL(2;\mathbb{C})$ by holonomy. The Chern–Simons function is defined to be

$$\mathrm{cs}_M([\rho]) := \frac{1}{8\pi^2}\int_M \operatorname{Tr}(A \wedge dA + \frac{2}{3} A \wedge A \wedge A) \in \mathbb{C} \pmod{\mathbb{Z}},$$

where $[\rho]$ is the class of ρ in $X(M)$.

Now we assume that M has a boundary which is homeomorphic to a torus. Denote by $X(\partial M)$ the $SL(2;\mathbb{C})$ character variety of the boundary ∂M.

We define $E(\partial M)$ as the quotient space of $\operatorname{Hom}(\pi_1(\partial M), \mathbb{C}) \times \mathbb{C}^*$ by a group G, where

$$G := \langle X, Y, B \mid XYX^{-1}Y^{-1} = XBXB = YBYB = B^2 = 1 \rangle$$

and it acts on $\operatorname{Hom}(\pi_1(\partial M), \mathbb{C}) \times \mathbb{C}^*$ by

$$\begin{aligned} X \cdot (s, t; z) &:= (s + 1, t; z \exp(-8\pi\sqrt{-1}t)), \\ Y \cdot (s, t; z) &:= (s, t + 1; z \exp(8\pi\sqrt{-1}s)), \\ B \cdot (s, t; z) &:= (-s, -t; z). \end{aligned} \tag{4.1}$$

Here a pair (s, t) is identified with the element $s\gamma^* + t\delta^* \in \operatorname{Hom}(\pi_1(\partial M), \mathbb{C})$ with a fixed basis (γ, δ) of $\pi_1(\partial M) \cong \mathbb{Z} \oplus \mathbb{Z}$. Then $E(\partial M)$ becomes a $\mathbb{C}^*$-bundle over

$X(\partial M)$. Note that $X(\partial M)$ is identified with $\mathrm{Hom}(\pi_1(\partial M), \mathbb{C})/G$ via the quotient map $q\colon \mathrm{Hom}(\pi_1(\partial M), \mathbb{C}) \to \mathrm{Hom}(\pi_1(\partial M), SL(2;\mathbb{C}))$ defined by

$$q(\kappa) := \left[\gamma \mapsto \begin{pmatrix} e^{2\pi\sqrt{-1}\kappa(\gamma)} & 0 \\ 0 & e^{-2\pi\sqrt{-1}\kappa(\gamma)} \end{pmatrix}\right]$$

for $\gamma \in \pi_1(\partial M)$.

The Chern–Simons function cs_M in this case is defined to be a map from $X(M)$ to $E(\partial M)$ such that $p \circ \mathrm{cs}_M = i^*$, where $p\colon E(\partial M) \to X(\partial M)$ is the projection and $i^*\colon X(M) \to X(\partial M)$ is induced from the inclusion map $i\colon \partial M \hookrightarrow M$.

$$\begin{array}{ccc} & & E(\partial M) \\ & \overset{\mathrm{cs}_M}{\nearrow} & \big\downarrow p \\ X(M) & \xrightarrow{\ i^*\ } & X(\partial M) \end{array}$$

See [**19**, § 3] for the precise definition.

If we have another three-manifold M' with toral boundary, we can construct a closed three-manifold $M \cup_\partial M'$ by identifying ∂M with $-\partial M'$. Given a representation $\rho\colon M \cup_\partial M' \to SL(2;\mathbb{C})$, the Chern–Simons invariant $\mathrm{cs}_{M\cup_\partial M'}([\rho])$ is given by zz' if $\mathrm{cs}_M\left(\left[\rho\big|_M\right]\right) = [s,t;z]$ and $\mathrm{cs}_{M'}\left(\left[\rho\big|_{M'}\right]\right) = [s,t;z']$, where $\rho\big|_M$ and $\rho\big|_{M'}$ are the restrictions of ρ to M and M' respectively. Note that we use the same basis for $\pi_1(\partial M)$ and $\pi_1(-\partial M')$.

Suppose that M is the complement of the interior of the regular neighborhood of a knot K in S^3. Let ρ be a representation sending the meridian μ and the longitude λ to the elements (up to conjugation) shown below.

$$\rho(\mu) = \begin{pmatrix} \exp(u/2) & * \\ 0 & \exp(-u/2) \end{pmatrix},$$
$$\rho(\lambda) = \begin{pmatrix} \exp(v/2) & * \\ 0 & \exp(-v/2) \end{pmatrix}.$$

We also assume that the elements in $\mathrm{Hom}(\pi_1(\partial M), \mathbb{C})$ sending μ to u and λ to v form a basis. Then we introduce the function $\mathrm{CS}_{u,v}([\rho])$ as follows.

$$\mathrm{cs}_M([\rho]) = \left[\frac{u}{4\pi\sqrt{-1}}, \frac{v}{4\pi\sqrt{-1}}; \exp\left(\frac{2}{\pi\sqrt{-1}}\,\mathrm{CS}_{u,v}([\rho])\right)\right].$$

Note that $\mathrm{CS}_{u,v}([\rho])$ is defined modulo $\pi^2\mathbb{Z}$ and that it depends on lifts (u,v) of $(\exp(u/2), \exp(v/2))$.

REMARK 4.1. Note that we are using the $PSL(2;\mathbb{C})$ normalization described in [**19**, P. 543]. So our $\mathrm{CS}_{u,v}([\rho])$ is -4 times $f(u)$ in [**29, 26**], and Kirk and Klassen's (and so Yoshida's [**38**]) $f(u)$ is $\pi\sqrt{-1}/2 \times \mathrm{CS}_{u,v}(u)$.

5. An asymptotic behavior of the colored Jones polynomial of a torus knot

In this section we give asymptotic expansions of the colored Jones polynomial of a torus knot.

Let $J_N(K;q)$ be the N-dimensional colored Jones polynomial of a knot K. We normalize it so that $J_N(\text{unknot};q) = 1$. So using Witten's formulation $J_N(K;q) = Z(S^3,K)/Z(S^3,\text{unknot})$ with $G = SU(2)$ and V is the N-dimensional irreducible representation. Note that $J_2(K;q) = V_K(q^{-1})$ for any knot K, where $V_K(q)$ is the original Jones polynomial [**15**].

Let $\Delta(K;t)$ be the Alexander polynomial for a knot K. We normalize it so that $\Delta(K;t) = \Delta(K;t^{-1})$ and $\Delta(K;1) = 1$.

Now we consider the torus knot $T(a,b)$. For a complex parameter z, we put

$$\tau_{a,b}(z) := \frac{2\sinh(z)}{\Delta\big(T(a,b);e^{2z}\big)}.$$

Since it is well-known that

$$\Delta\big(T(a,b);t\big) = \frac{\left(t^{ab/2} - t^{-ab/2}\right)\left(t^{1/2} - t^{-1/2}\right)}{\left(t^{a/2} - t^{-a/2}\right)\left(t^{b/2} - t^{-b/2}\right)},$$

we have

$$\tau_{a,b}(z) = \frac{2\sinh(az)\sinh(bz)}{\sinh(abz)}.$$

Note that $(t^{1/2} - t^{-1/2})/\Delta(K;t)$ can be regarded as the (abelian) Reidemeister torsion ([**22**, Theorem 4],[**34**, Theorem 1.1.2]).

Let $\mathcal{P}$ be the set of poles of $\tau_{a,b}(z)$, that is, we put

$$\mathcal{P} := \left\{ \frac{k\pi\sqrt{-1}}{ab} \;\middle|\; k \in \mathbb{Z}, a \nmid k, b \nmid k \right\}.$$

We also put

$$A_k(\xi;N) = \sqrt{-\pi}\exp\left(S_k(\xi)\frac{N}{\xi}\right)\left(\frac{N}{\xi}\right)^{1/2}(T_k)^{1/2},$$

where

$$S_k(\xi) := \frac{-\left(2k\pi\sqrt{-1} - ab\xi\right)^2}{4ab}$$

and

$$T_k := \frac{16\sin^2(k\pi/a)\sin^2(k\pi/b)}{ab}.$$

We would like to know an asymptotic behavior of $J_N\big(T(a,b);\exp(\xi/N)\big)$ for large N.

The case where $\xi = 2\pi\sqrt{-1}$ corresponds to the volume conjecture (Conjecture 1.1). In this case, Kashaev and Tirkkonen [**18**] proved the following asymptotic expansion.

$$\begin{aligned}
&J_N\big(T(a,b);\exp(2\pi\sqrt{-1}/N)\big) \\
\sim& e^{(ab-a/b-b/a)\pi\sqrt{-1}/(2N)} \\
&\times\left(\frac{\pi^{3/2}}{2ab}\left(\frac{N}{\xi}\right)^{3/2}\sum_{k=1}^{ab-1}(-1)^{k+1}k^2\exp\left(S_k(\xi)\frac{N}{\xi}\right)(T_k)^{1/2}\right. \\
&\left.+\frac{1}{4}\sum_{j=1}^{\infty}\frac{a_j}{j!}\left(\frac{\xi}{4abN}\right)^{j-1}\right),
\end{aligned}$$

where a_l is the $2l$-th derivative of $2z\sinh z/\Delta(T(a,b);e^{2z}) = z\tau_{a,b}(z)$ at $z=0$. For a relation to characters of conformal field theory, see [**11**]. See also [**5**] for a topological interpretation of this expansion.

When ξ is not an integer multiple of $2\pi\sqrt{-1}$, we have the following theorem.

THEOREM 5.1. *Let ξ be a complex number that is not an integral multiple of $2\pi\sqrt{-1}$. We also assume that $\operatorname{Im}\xi \ge 0$ for simplicity.*

If $\xi/2 \notin \mathcal{P}$, then we have

$$\begin{aligned}&J_N\bigl(T(a,b);\exp(\xi/N)\bigr)\\ &\sim \frac{e^{(ab-a/b-b/a)\xi/(4N)}}{2\sinh(\xi/2)}\left(\tau_{a,b}(\xi/2)+\sum_{j=1}^{\infty}\frac{\tau_{a,b}^{(2j)}(\xi/2)}{j!}\left(\frac{\xi}{4abN}\right)^j\right)\end{aligned}\tag{5.1}$$

when $\operatorname{Re}\xi > 0$ and

$$\begin{aligned}J_N\bigl(T(a,b);\exp(\xi/N)\bigr) &\sim \frac{e^{(ab-a/b-b/a)\xi/(4N)}}{2\sinh(\xi/2)}\\ \times&\left(\tau_{a,b}(\xi/2)+\sum_{k=1}^{\lfloor ab|\xi|/(2\pi)\rfloor}(-1)^{k+1}A_k(\xi;N)+\sum_{j=1}^{\infty}\frac{\tau_{a,b}^{(2j)}(\xi/2)}{j!}\left(\frac{\xi}{4abN}\right)^j\right)\end{aligned}\tag{5.2}$$

when $\operatorname{Re}\xi \le 0$ as $N\to\infty$, where $\tau_{a,b}^{(2j)}(\xi/2)$ is the $(2j)$th derivative of $\tau_{a,b}(z)$ at $z=\xi/2$ and $\lfloor x\rfloor$ means the largest integer that does not exceed x.

If $\xi/2 \in \mathcal{P}$ (and it is not an integer multiple of $\pi\sqrt{-1}$), then we have

$$\begin{aligned}&J_N\bigl(T(a,b);\exp(\xi/N)\bigr)\\ &\sim \frac{e^{(ab-a/b-b/a)\xi/(4N)}}{2\sinh(\xi/2)}\left(\tau_{a,b}^{(0)}(\xi/2)+\frac{1}{2}(-1)^{ab|\xi|/(2\pi)}A_{ab|\xi|/(2\pi)}(\xi;N)\right.\\ &\qquad\left.+\sum_{k=1}^{ab|\xi|/(2\pi)-1}(-1)^{k+1}A_k(\xi;N)+\sum_{j=1}^{\infty}\frac{\tau_{a,b}^{(2j)}(\xi/2)}{j!}\left(\frac{\xi}{4abN}\right)^j\right)\end{aligned}\tag{5.3}$$

as $N\to\infty$, where $\tau_{a,b}^{(0)}(\xi/2)$ is the constant term of the Laurent expansion of $\tau_{a,b}(z)$ around $z=\xi/2$.

REMARK 5.2. If $\operatorname{Re}\xi > 0$, or $\operatorname{Re}\xi \le 0$ and $|\xi| < 2\pi/(ab)$, then $J_N\bigl(T(a,b);\exp(\xi/N)\bigr)$ converges to $\tau_{a,b}(\xi/2)/(2\sinh(\xi/2)) = 1/\Delta\bigl(T(a,b);\exp\xi\bigr)$. Otherwise it diverges. See Figure 1.

Note that Garoufalidis and Lê proved that for any knot K, $J_N\bigl(K;\exp(\xi/N)\bigr)$ converges to $1/\Delta(K;\exp\xi)$ when $|\xi|$ is small enough [**7**].

PROOF OF THEOREM 5.1 FOR ξ WITH NON-ZERO REAL PART. We first prove Theorem 5.1 where $\operatorname{Re}\xi \ne 0$. Recall that we assume $\operatorname{Im}\xi \ge 0$.

In [**18**], Kashaev and Tirkkonen proved that $J_N\bigl(T(a,b);\exp(\xi/N)\bigr)$ is given by the following integral.

$$J_N\bigl(T(a,b);\exp(\xi/N)\bigr) = \Phi_{a,b,\xi}(N)\int_C e^{abN(-z^2/\xi+z)}\tau_{a,b}(z)\,dz,$$

where

$$\Phi_{a,b,\xi}(N) := \frac{1}{2\sinh(\xi/2)}\sqrt{\frac{abN}{\pi\xi}}\,e^{-abN\xi/4+(ab-a/b-b/a)\xi/(4N)}$$

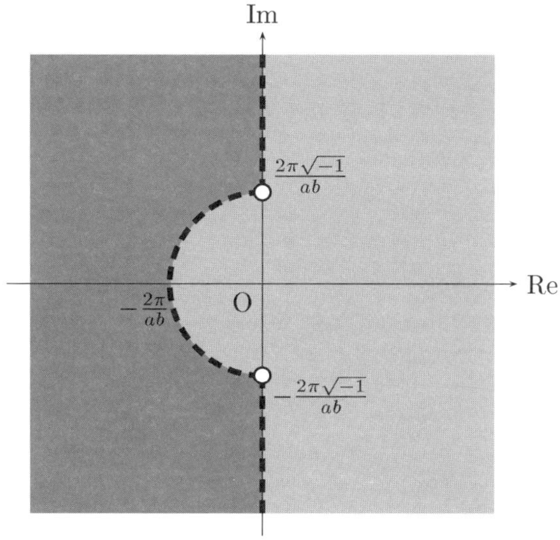

FIGURE 1. The colored Jones polynomial converges in the light gray area, and diverges in the gray area including the dashed lines and semicircle except for $\mathcal{P}$ indicated by the white circles.

and C is the line passing through the origin with slope $\tan(\varphi)$, where φ is chosen so that $(\arg\xi)/2 - \pi/4 < \varphi < (\arg\xi)/2 + \pi/4$. Note that this is to make the integral converges.

Let C_ξ be the line that is parallel to C and passes through $\xi/2$ that is the critical point of the exponent of the integrand. Then we have

$$
\begin{aligned}
&J_N(T(a,b);\exp(\xi/N)) \\
&= \Phi_{a,b,\xi}(N)\left(\int_{C_\xi} e^{abN(-z^2/\xi+z)}\tau_{a,b}(z)\,dz\right. \\
&\qquad \left. + 2\pi\sqrt{-1}\sum_k \operatorname{Res}\left(e^{abN(-z^2/\xi+z)}\tau_{a,b}(z); z = k\pi\sqrt{-1}/(ab)\right)\right) \\
&= \Phi_{a,b,\xi}(N)\left(\int_{C_\xi} e^{abN(-z^2/\xi+z)}\tau_{a,b}(z)\,dz\right. \\
&\qquad \left. + 2\pi\sqrt{-1}\sum_k (-1)^{k+1}\frac{2\sin(k\pi/a)\sinh(k\pi/b)}{ab}\exp\left(N\left(\frac{k^2\pi^2}{ab\xi} + k\pi\sqrt{-1}\right)\right)\right)
\end{aligned}
$$

in a similar way to [**25**], where k runs over integers such that $k\pi\sqrt{-1}/(ab)$ is between C and C_ξ.

First we calculate the asymptotic expansion of the integral. Putting $w := z - \xi/2$, we have

$$
\int_{C_\xi} e^{abN(-z^2/\xi+z)}\tau_{a,b}(z)\,dz
$$

$$\begin{aligned}
&= \int_C e^{abN\left(-(w+\xi/2)^2/\xi+(w+\xi/2)\right)} \tau_{a,b}(w+\xi/2)\,dw \\
&= e^{abN\xi/4} \int_C e^{-abNw^2/\xi} \tau_{a,b}(w+\xi/2)\,dw \\
&= e^{abN\xi/4} e^{\varphi\sqrt{-1}} \int_{-\infty}^{\infty} e^{-abNe^{2\varphi\sqrt{-1}}t^2/\xi} \tau_{a,b}\left(te^{\varphi\sqrt{-1}}+\xi/2\right) dt.
\end{aligned}$$

We use Watson's Theorem, a special case of the steepest descent method, to obtain the asymptotic expansion of this integral.

THEOREM 5.3 ([**21**](Theorem 7.2.7)). *Let $g(z)$ be analytic and bounded on a domain containing the real axis. Set $f(z) := \int_{-\infty}^{\infty} e^{-zy^2/2} g(y)\,dy$ for z real. Then*

$$f(z) \sim \sqrt{\frac{2\pi}{z}} \left(a_0 + \frac{a_2}{z} + \frac{3!!a_4}{z^2} + \frac{5!!a_6}{z^3} + \cdots\right)$$

as $z \to \infty$, $\arg z = 0$, where $g(z) = \sum_{n=0}^{\infty} a_n z^n$ near zero.

Note that this theorem also holds for z with fixed $\arg z$ with $\operatorname{Re} z > 0$.

Since $\arg\left(e^{2\varphi\sqrt{-1}}/\xi\right) = 2\varphi - \arg\xi$, we see $\operatorname{Re}\left(abNe^{2\varphi\sqrt{-1}}/\xi\right) > 0$. Therefore we have the following asymptotic expansion.

$$\begin{aligned}
&\int_{-\infty}^{\infty} e^{-abNe^{2\varphi\sqrt{-1}}t^2/\xi} \tau_{a,b}\left(te^{\varphi\sqrt{-1}}+\xi/2\right) dt \\
\sim&\sqrt{\frac{\pi\xi}{abNe^{2\varphi\sqrt{-1}}}} \left(\sum_{j=0}^{\infty} \frac{(2j-1)!!c_{2j}}{(2abNe^{2\varphi\sqrt{-1}}/\xi)^j}\right)
\end{aligned}$$

as $N \to \infty$, where c_{2j} is the coefficient of t^{2j} in the Taylor expansion of $\tau_{a,b}\left(te^{\varphi\sqrt{-1}}+\xi/2\right)$ around $t=0$. Since $c_{2j} = \dfrac{e^{2j\varphi\sqrt{-1}}}{(2j)!}\tau_{a,b}^{(2j)}(\xi/2)$ with $\tau_{a,b}^{(2j)}(\xi/2)$ the $2j$th derivative of $\tau_{a,b}(z)$ at $\xi/2$, we have

$$\begin{aligned}
&\int_{C_\xi} e^{abN(-z^2/\xi+z)} \tau_{a,b}(z)\,dz \\
\sim&e^{abN\xi/4} e^{\varphi\sqrt{-1}} \sqrt{\frac{\pi\xi}{abNe^{2\varphi\sqrt{-1}}}} \left(\sum_{j=0}^{\infty} \frac{(2j-1)!!\tau_{a,b}^{(2j)}(\xi/2)}{(2j)!(2abN/\xi)^j}\right) \\
=&e^{abN\xi/4} \sqrt{\frac{\pi\xi}{abN}} \left(\sum_{j=0}^{\infty} \frac{\tau_{a,b}^{(2j)}(\xi/2)}{j!} \left(\frac{\xi}{4abN}\right)^j\right).
\end{aligned}$$

Since $\tau_{a,b}^{(0)}(\xi/2) = \tau_{a,b}(\xi/2)$, we finally have the following asymptotic expansion.

$$\begin{aligned}
&J_N\big(T(a,b);\exp(\xi/N)\big) \sim \frac{e^{(ab-a/b-b/a)\xi/(4N)}}{2\sinh(\xi/2)} \\
&\times \quad \left(\tau_{a,b}(\xi/2) + \sum_{j=1}^{\infty} \frac{\tau_{a,b}^{(2j)}(\xi/2)}{j!} \left(\frac{\xi}{4abN}\right)^j\right.
\end{aligned}$$

$$+\sum_k (-1)^{k+1}\frac{4\sin(k\pi/a)\sinh(k\pi/b)\sqrt{-\pi N}}{\sqrt{ab\xi}}$$
$$\times\exp\left(N\left(\frac{k^2\pi^2}{ab\xi}+k\pi\sqrt{-1}-\frac{ab\xi}{4}\right)\right)\Bigg)$$
$$=\frac{e^{(ab-a/b-b/a)\xi/(4N)}}{2\sinh(\xi/2)}$$
$$\times\left(\tau_{a,b}(\xi/2)+\sum_k(-1)^{k+1}A_k(\xi;N)+\sum_{j=1}^{\infty}\frac{\tau_{a,b}^{(2j)}(\xi/2)}{j!}\left(\frac{\xi}{4abN}\right)^j\right).$$

Now we consider the range of k.

We observe that C_ξ crosses the imaginary axis at $\sqrt{-1}(\operatorname{Im}\xi-\operatorname{Re}\xi\tan\varphi)/2$, and when φ increases from $(\arg\xi)/2-\pi/4$ to $(\arg\xi)/2+\pi/4$, the crossing point goes from $\sqrt{-1}\left(\operatorname{Im}\xi-\operatorname{Re}\xi\tan\big((\arg\xi)/2-\pi/4\big)\right)$ to $\sqrt{-1}\left(\operatorname{Im}\xi-\operatorname{Re}\xi\tan\big((\arg\xi)/2+\pi/4\big)\right)$ downwards (upwards, respectively) if $0\le\arg\xi<\pi/2$ (if $\pi/2<\xi\le\pi$, respectively). Note that if $\pi/2<\xi\le\pi$, C_ξ can be parallel to the imaginary axis but we avoid this. Since

$$\begin{aligned}\tan^2((\arg\xi)/2\pm\pi/4)&=\frac{1-\cos(\arg\xi\pm\pi/2)}{1+\cos(\arg\xi\pm\pi/2)}\\&=\frac{1\pm\sin(\arg\xi)}{1\mp\sin(\arg\xi)}\\&=\frac{\big(1\pm\sin(\arg\xi)\big)^2}{\cos^2(\arg\xi)}\\&=\frac{(|\xi|\pm\operatorname{Im}\xi)^2}{(\operatorname{Re}\xi)^2},\end{aligned}$$

we have

$$\tan\big((\arg\xi)/2\pm\pi/4\big)=\frac{\operatorname{Im}\xi\pm|\xi|}{\operatorname{Re}\xi}.$$

and

$$\operatorname{Im}\xi-\operatorname{Re}\xi\tan\big((\arg\xi)/2\pm\pi/4\big)=\mp|\xi|.$$

So if $0\le\arg\xi<\pi/2$, then the crossing point is between $-\sqrt{-1}|\xi|$ and $\sqrt{-1}|\xi|$, and k runs over integers that are not multiples of a or b with $1\le k\le M$ for any integer M satisfying $0<M<ab|\xi|/(2\pi)$. If $\pi/2<\xi\le\pi$, then the crossing point is above $\sqrt{-1}|\xi|$ or below $-\sqrt{-1}|\xi|$, and k runs over all integers that are not multiples of a or b with $1\le k\le M'$ for any integer M' with $M'>ab|\xi|/(2\pi)$.

So when $0\le\arg\xi<\pi/2$, we have

$$J_N\big(T(a,b);\exp(\xi/N)\big)$$
$$=\frac{e^{(ab-a/b-b/a)\xi/(4N)}}{2\sinh(\xi/2)}$$
$$\times\left(A_0(\xi)+\sum_{\substack{1\le k\le M,\\ a\nmid k,b\nmid k}}(-1)^{k+1}A_k(\xi;N)+\sum_{j=1}^{\infty}\frac{\tau_{a,b}^{(2j)}(\xi/2)}{j!}\left(\frac{\xi}{4abN}\right)^j\right)\tag{5.4}$$

for any integer M with $0 < M < ab|\xi|/(2\pi)$. When $\pi/2 < \arg\xi \leq \pi$, we have

$$(5.5)\quad \begin{aligned} &J_N\big(T(a,b);\exp(\xi/N)\big) \\ &\quad = \frac{e^{(ab-a/b-b/a)\xi/(4N)}}{2\sinh(\xi/2)} \\ &\qquad \times \left(A_0(\xi) + \sum_{\substack{1\leq k\leq M',\\ a\nmid k, b\nmid k}} (-1)^{k+1} A_k(\xi;N) + \sum_{j=1}^{\infty} \frac{\tau_{a,b}^{(2j)}(\xi/2)}{j!} \left(\frac{\xi}{4abN}\right)^j \right). \end{aligned}$$

for any integer M' with $M' > ab|\xi|/(2\pi)$.

Note that since the real part of $S_k(\xi)/\xi$ is

$$\left(\frac{k^2\pi^2}{ab|\xi|^2} - \frac{ab}{4}\right) \operatorname{Re}\xi,$$

the real part of the coefficient of N in the exponent in $A_k(\xi;N)$ is positive if and only if $\operatorname{Re}\xi > 0$ and $k > ab|\xi|/(2\pi)$, or $\operatorname{Re}\xi < 0$ and $k < ab|\xi|/(2\pi)$, negative if and only if $\operatorname{Re}\xi > 0$ and $k < ab|\xi|/(2\pi)$, or $\operatorname{Re}\xi < 0$ and $k > ab|\xi|/(2\pi)$, and zero if and only if $k = ab|\xi|/(2\pi)$.

Therefore in (5.4) we can ignore all the k since $A_k(\theta;N)$ decays exponentially, and in (5.5) we can ignore k with $k > ab|\xi|/(2\pi)$. Noting that if a or b divides k then $A_k(\xi;N) = 0$, we finally have

$$\begin{aligned} &J_N\big(T(a,b);\exp(\xi/N)\big) \\ &\qquad \sim \frac{e^{(ab-a/b-b/a)\xi/(4N)}}{2\sinh(\xi/2)} \left(\tau_{a,b}(\xi/2) + \sum_{j=1}^{\infty} \frac{\tau_{a,b}^{(2j)}(\xi/2)}{j!} \left(\frac{\xi}{4abN}\right)^j \right) \end{aligned}$$

when $\operatorname{Re}\xi > 0$ and

$$\begin{aligned} J_N\big(T(a,b);\exp(\xi/N)\big) &\sim \frac{e^{(ab-a/b-b/a)\xi/(4N)}}{2\sinh(\xi/2)} \\ &\times \left(\tau_{a,b}(\xi/2) + \sum_{k=1}^{\lfloor ab|\xi|/(2\pi)\rfloor} (-1)^{k+1} A_k(\xi;N) + \sum_{j=1}^{\infty} \frac{\tau_{a,b}^{(2j)}(\xi/2)}{j!} \left(\frac{\xi}{4abN}\right)^j \right). \end{aligned}$$

when $\operatorname{Re}\xi < 0$. □

REMARK 5.4. When $\operatorname{Re}\xi < 0$ and $|\xi| < 2\pi/(ab)$ there is no $A_k(\xi;N)$ term. When $\operatorname{Re}\xi < 0$ and $|\xi| = 2\pi/(ab)$ the term $A_1(\xi;N)$ oscillates.

PROOF OF THEOREM 5.1 FOR PURELY IMAGINARY ξ. If ξ is purely imaginary, we have already shown the following formulas in [**12**, Proposition 3.2].

If $\operatorname{Re}\xi = 0$ and $\xi/2 \notin \mathcal{P}$, then we have

$$\begin{aligned} J_N\big(T(a,b);\exp(\xi/N)\big) &\sim \frac{e^{(ab-a/b-b/a)\xi/(4N)}}{2\sinh(\xi/2)} \\ &\times \left(\tau_{a,b}(\xi/2) + \sum_{k=1}^{\lfloor ab|\xi|/(2\pi)\rfloor} (-1)^{k+1} A_k(\xi;N) + \sum_{j=1}^{\infty} \frac{\tau_{a,b}^{(2j)}(\xi/2)}{j!} \left(\frac{\xi}{4abN}\right)^j \right). \end{aligned}$$

If $\xi/2$ is in $\mathcal{P}$ but not an integral multiple of $2\pi\sqrt{-1}$, then we have

$$
\begin{aligned}
&J_N\big(T(a,b);\exp(\xi/N)\big)\\
&\sim\frac{e^{(ab-a/b-b/a)\xi/(4N)}}{2\sinh(\xi/2)}\left(\tau_{a,b}^{(0)}(\xi/2)+\frac{1}{2}(-1)^{ab|\xi|/(2\pi)}A_{ab|\xi|/(2\pi)}(\xi;N)\right.\\
&\qquad\left.+\sum_{k=1}^{ab|\xi|/(2\pi)-1}(-1)^{k+1}A_k(\xi;N)+\sum_{j=1}^{\infty}\frac{\tau_{a,b}^{(2j)}(\xi/2)}{j!}\left(\frac{\xi}{4abN}\right)^j\right),
\end{aligned}
$$

where $\tau_{a,b}^{(0)}(\xi/2)$ means the constant term of the Laurent expansion of $\tau_{a,b}(z)$ around $z=\xi/2$. This completes the proof. □

6. A topological interpretation of the asymptotic behavior

In this section we study a topological interpretation of the term $A_k(\xi;N)$ ($k\geq 1$). Given a positive integer k that is not a multiple of a nor b, we associate a pair of integers (α,β) as described in §2.

6.1. A topological interpretation of $S_k(\xi)$. Let $\rho_{\alpha,\beta}$ be an irreducible representation of $\pi_1\big(S^3\setminus T(a,b)\big)$ into $SL(2;\mathbb{C})$ which is in the component of the character variety indexed by (α,β).

The fundamental group of $S^3\setminus T(a,b)$ has a presentation $\pi_1\big(S^3\setminus T(a,b)\big)=\langle x,y\mid x^a=y^b\rangle$. Then the longitude λ can be expressed as $\lambda=x^a\mu^{-ab}$, where μ is the meridian. Up to conjugation, we may assume that the images of μ and λ are as follows.

$$
\rho_{\alpha,\beta}(\mu)=\begin{pmatrix}\mathfrak{m} & *\\ 0 & \mathfrak{m}^{-1}\end{pmatrix},
$$
$$
\rho_{\alpha,\beta}(\lambda)=\begin{pmatrix}\mathfrak{l} & *\\ 0 & \mathfrak{l}^{-1}\end{pmatrix}.
$$

Since $\lambda\mu^{ab}=x^a$ and x^a is in the center of the group $\pi_1\big(S^3\setminus T(a,b)\big)$, we have $\mathfrak{l}=\pm\mathfrak{m}^{-ab}$.

If we put $\mathfrak{m}=\exp(u/2)$ and $\mathfrak{l}=-\exp(v/2)$, we have $\exp(v/2)=\pm\exp(-abu/2)$. Therefore v can be expressed (modulo $2\pi\sqrt{-1}$) in terms of u. We choose $-ab(u+2\pi\sqrt{-1})+2(k-1)\pi\sqrt{-1}$ as such an expression and denote it by $v_k(u)$.

Using the pair $(u,v_k(u))$ with $u:=\xi-2\pi\sqrt{-1}$, we can prove that the function $\mathrm{CS}_{u,v_k(u)}([\rho_{\alpha,\beta}])$ defined in §4 can be expressed in terms of $S_k(\xi)$.

THEOREM 6.1. *Let $\rho_{\alpha,\beta}$ be an irreducible representation such that $[\rho_{\alpha,\beta}]$ is in the component of $X\big(S^3\setminus T(a,b)\big)$ indexed by (α,β). If we put $v_k(u):=-ab(u+2\pi\sqrt{-1})+2(k-1)\pi\sqrt{-1}$, then we have*

$$
\mathrm{CS}_{u,v_k(u)}\left([\rho_{\alpha,\beta}]\right)=S_k(\xi)-\pi\sqrt{-1}u-\frac{uv_k(u)}{4}
$$

with $u:=\xi-2\pi\sqrt{-1}$, that is, the following equality holds.

$$
\mathrm{cs}_{T(a,b)}\big([\rho_{\alpha,\beta}]\big)=\left[\frac{u}{4\pi\sqrt{-1}},\frac{v_k(u)}{4\pi\sqrt{-1}};\exp\left(\frac{2}{\pi\sqrt{-1}}\left(S_k(\xi)-\pi\sqrt{-1}u-\frac{uv_k(u)}{4}\right)\right)\right].
$$

PROOF. From a formula by Dubois and Kashaev [**5**, Proposition 4] we have

$$\begin{aligned}(6.1)\quad &\mathrm{cs}_{T(a,b)}([\rho_{\alpha,\beta}])\\ &= \left[\frac{u}{4\pi\sqrt{-1}}, \frac{1}{2} - \frac{abu}{4\pi\sqrt{-1}}; \exp\left(-8\pi\sqrt{-1}\left(\frac{(\beta ad+\varepsilon\alpha bc)^2}{4ab} - \frac{u}{8\pi\sqrt{-1}}\right)\right)\right],\end{aligned}$$

where integers c and d are chosen so that $ad - bc = 1$, and $\varepsilon = \pm 1$. Note that we are using the $PSL(2;\mathbb{C})$ normalization and so we need to multiply the exponent in the third entry by -4.

Changing the coordinate by using (4.1), we have

$$\begin{aligned}&\mathrm{cs}_{T(a,b)}([\rho_{\alpha,\beta}])\\ &= \left[\frac{u}{4\pi\sqrt{-1}}, \frac{1}{2} - \frac{abu}{4\pi\sqrt{-1}} + \frac{k-ab-2}{2}; \exp\left(\frac{-2k^2\pi\sqrt{-1}}{ab} + u + (k-ab-2)u\right)\right]\\ &= \left[\frac{u}{4\pi\sqrt{-1}}, \frac{v_k(u)}{4\pi\sqrt{-1}}; \exp\left(\frac{-2k^2\pi\sqrt{-1}}{ab} - abu + (k-1)u\right)\right]\\ &= \left[\frac{u}{4\pi\sqrt{-1}}, \frac{v_k(u)}{4\pi\sqrt{-1}}; \exp\left(\frac{2}{\pi\sqrt{-1}}\left(S_k(\xi) - \pi\sqrt{-1}u - \frac{uv_k(u)}{4}\right)\right)\right],\end{aligned}$$

where the first equality follows since $k^2 = (\beta ad + \varepsilon\alpha bc)^2 \pmod{ab}$ and the last equality follows since $u = \xi - 2\pi\sqrt{-1}$. Note that the choice of ε does not matter here. Note also that even if we change the definition of (α,β), the equality still holds (Remark 2.2). □

Since $v_k(u) = 2\frac{d\,S_k(\xi)}{d\,\xi}\Big|_{\xi:=u+2\pi\sqrt{-1}} - 2\pi\sqrt{-1}$, we note that $\mathrm{CS}_{u,v_k(u)}([\rho_{\alpha,\beta}])$ can be determined by $S_k(\xi)$.

6.2. A topological interpretation of T_k. We can show that T_k is the twisted Reidemeister torsion associated with the meridian μ.

LEMMA 6.2. *Let $\rho_{\alpha,\beta}$ be an irreducible representation $\pi_1(S^3\backslash T(a,b)) \to SL(2;\mathbb{C})$ whose character belongs to the component indexed by (α,β) that is determined by k as described in § 2. Then (up to a sign) the Reidemeister torsion $\mathbb{T}_\mu^{T(a,b)}(\rho_{\alpha,\beta})$ associated with the meridian μ is given by*

$$\mathbb{T}_\mu^{T(a,b)}(\rho_{\alpha,\beta}) = \pm\frac{16}{ab}\sin^2\left(\frac{\pi\alpha}{a}\right)\sin^2\left(\frac{\pi\beta}{b}\right).$$

PROOF. If an irreducible representation $\rho_{\alpha,\beta}$ is in the component indexed by (α,β), Dubois [**3**, 6.2] proved that the twisted Reidemeister torsion $\mathbb{T}_\lambda^{T(a,b)}(\rho_{\alpha,\beta})$ associated with the *longitude* λ is given by

$$\mathbb{T}_\lambda^{T(a,b)}(\rho_{\alpha,\beta}) = \frac{16}{a^2b^2}\sin^2\left(\frac{\alpha\pi}{a}\right)\sin^2\left(\frac{\beta\pi}{b}\right). \tag{6.2}$$

Now we use Porti's basis change formula [**30**, Théorème 4.1]. From Remark (ii) to [**30**, Théorème 4.1], we have

$$\mathbb{T}_\mu(\rho_{\alpha,\beta}) = \pm\frac{\partial v}{\partial u}\mathbb{T}_\lambda(\rho_{\alpha,\beta}) \tag{6.3}$$

for an irreducible representation ρ. Here u and v are parameters as described in the previous subsection. Note that we are using cohomological Reidemeister torsion and Porti uses homological one. So our torsion is the inverse of the torsion used in [**30**].

As in the previous subsection $v = -abu + 2n\pi\sqrt{-1}$ for a constant $n \in \mathbb{Z}$. So we have $\partial v/\partial u = -ab$ and the lemma follows. $\square$

REMARK 6.3. In [**30**] Porti uses the twisted homology instead of the twisted cohomology. So the Reidemeister torsion is the inverse of ours. The authors thank J. Dubois for pointing out this.

Since $\sin^2(k\pi/a)\sin^2(k\pi/b) = \sin^2(\alpha\pi/a)\sin^2(\beta\pi/b)$ (Remark 2.2), we have

$$\mathbb{T}_{\mu}^{T(a,b)}(\rho_{\alpha,\beta}) = \pm\frac{16}{ab}\sin^2\left(\frac{k\pi}{a}\right)\sin^2\left(\frac{k\pi}{b}\right).$$

Since T_k is always positive, we have the following theorem.

THEOREM 6.4. *Let $\rho_{\alpha,\beta}$ be an irreducible representation of $\pi_1\bigl(S^3\setminus T(a,b)\bigr)$ into $SL(2;\mathbb{C})$, which is in the component indexed by (α,β) that is associated with an integer k as in §2.*

Then T_k equals the absolute value of the twisted Reidemeister torsion of $\rho_{\alpha,\beta}$ associated with the meridian, that is, we have

$$T_k = \left|\mathbb{T}_{\mu}^{T(a,b)}(\rho_{\alpha,\beta})\right|.$$

6.3. Remaining factor. There remains a strange factor $2\sinh(\xi/2)$ in the asymptotic expression. Recall that we normalize the colored Jones polynomial so that its value for the unknot is one. Another (more natural in physics) normalization is to put the value for the empty link to be one. In this normalization the colored Jones polynomial of the unknot is $[N] = \bigl(q^{N/2} - q^{-N/2}\bigr)/\bigl(q^{1/2} - q^{-1/2}\bigr)$.

Then the factor $2\sinh(\xi/2)$ comes from the following asymptotic expansion at $q = \exp(\xi/N)$ of $\log[N]$.

$$\frac{\sinh(\xi/2)}{\sinh(\xi/(2N))} \sim 2\sinh(\xi/2)\frac{N}{\xi} - \frac{\sinh(\xi/2)}{12}\left(\frac{N}{\xi}\right)^{-1} + \cdots.$$

7. Speculation

Combining the results of Section 6, for a torus knot K and an appropriately chosen parameter ξ we have

$$\lim_{N\to\infty}\left\{J_N(K;\exp(\xi/N))\frac{2\sinh(\xi/2)}{\nu(\xi/N)}\right.$$
$$\left. -\sqrt{-\pi}\sum_{k}(-1)^{k+1}\left(\frac{N}{\xi}\right)^{1/2}\exp\left(S_k(\xi)\frac{N}{\xi}\right)\left(\mathbb{T}_{\lambda}^{K}(\rho_k)\right)^{1/2}\right\} = \frac{2\sinh(\xi/2)}{\Delta(K;\exp\xi)}$$

where $\nu(x)$ is a function that converges to 1 when $x \to 0$, k runs over some irreducible components of the character variety $X(S^3\setminus K)$, ρ_k is an irreducible representation in the component indexed by k, and $S_k(\xi)$ determines the $SL(2;\mathbb{C})$ Chern–Simons invariant $\mathrm{CS}_{u,v_k(u)}([\rho_k])$ as in Theorem 6.1. We expect a similar formula for a general knot.

Here we just give an observation about the figure-eight knot.

In [**28**] Yokota and the second author proved that for the figure-eight knot E, the following holds.

THEOREM 7.1 ([**28**]). *There exists a neighborhood U of 0 in $\mathbb{C}$ such that for any $u \in (U \setminus \pi\sqrt{-1}\mathbb{Q}) \cup \{0\}$, the following limit exists*

$$(u + 2\pi\sqrt{-1}) \lim_{N\to\infty} \frac{\log J_N\bigl(E; \exp((u+2\pi\sqrt{-1})/N)\bigr)}{N}.$$

Moreover if we denote the limit by $H(u)$ and put $v(u) := 2\frac{d\,H(u)}{d\,u} - 2\pi\sqrt{-1}$, then $H(u) - \pi\sqrt{-1}u - uv(u)/4$ coincides with $\mathrm{CS}_{u,v(u)}([\rho])$, where ρ is the representation of $\pi_1(S^3 \setminus E)$ into $SL(2;\mathbb{C})$ sending the meridian to $\begin{pmatrix} \exp(u/2) & * \\ 0 & \exp(-u/2) \end{pmatrix}$ and the longitude to $\begin{pmatrix} -\exp(v(u)/2) & * \\ 0 & -\exp(-v(u)/2) \end{pmatrix}$ up to conjugate.

The $SL(2;\mathbb{C})$ character variety of $S^3 \setminus E$ has two connected components, the abelian one and the non-abelian one.

Non-abelian representations can be calculated explicitly by using the technique described in [**32**] (see also [**26**, §3.1]). Let $\rho_{m\pm}$ be the non-abelian representation of $\pi_1(S^3 \setminus E)$ into $SL(2;\mathbb{C})$ sending the meridian to

$$\begin{pmatrix} m^{1/2} & 1 \\ 0 & m^{-1/2} \end{pmatrix}$$

and the longitude to

$$\begin{pmatrix} \ell(m)^{\pm 1} & (m^{1/2} + m^{-1/2})\sqrt{(m+m^{-1}+1)(m+m^{-1}-3)} \\ 0 & \ell(m)^{\mp 1} \end{pmatrix},$$

where

(7.1)
$$\ell(m) := \frac{m^2 - m - 2 - m^{-1} + m^{-2}}{2} + \frac{m - m^{-1}}{2}\sqrt{(m+m^{-1}+1)(m+m^{-1}-3)}.$$

See [**26**, §3.1] for details. Note that the pair $(m, \ell(m))$ is a zero of the A-polynomial

$$\ell - \bigl(m^2 - m - 2 - m^{-1} + m^{-2}\bigr) + \ell^{-1}. \tag{7.2}$$

REMARK 7.2. Equation (3.8) in [**26**] is mistyped. It should be read as

$$\ell - \bigl(m^2 - m - 2 - m^{-1} + m^{-2}\bigr) + \ell^{-1} = 0.$$

The authors thank E. Witten, who pointed out this.

In [**3**, §6.3] Dubois proves that the twisted Reidemeister torsion $\mathbb{T}^E_\lambda(\rho_{m\pm})$ associated with the longitude λ is given by

$$\mathbb{T}^E_\lambda(\rho_{m\pm}) = \frac{1}{\sqrt{17 + 4\,\mathrm{Tr}(\rho_{m\pm}(\lambda))}} = \frac{1}{2m + 2m^{-1} - 1}.$$

See also [**30**, §4.5] and [**4**].

Therefore from (6.3) the twisted Reidemeister torsion associated with the meridian μ is

$$\mathbb{T}^E_\mu(\rho_{m\pm}) = \pm\frac{\partial\, v}{\partial\, u} \times \frac{1}{2m + 2m^{-1} - 1}.$$

Since in this case $e^{u/2} = m^{1/2}$ and $e^{v/2} = -\ell(m)^{\pm 1}$, we have

$$\begin{aligned}\frac{\partial v}{\partial u} &= \frac{\pm\partial\big(2\log\ell(m)\big)/\partial m}{\partial(\log m)/\partial m}\\ &= \frac{\pm 2m\,d\,\ell(m)/d\,m}{\ell(m)}.\end{aligned}$$

Since the pair $(m, \ell(m))$ is a zero of the A-polynomial, differentiating (7.2) by m we have

$$\frac{d\,\ell(m)}{d\,m} = \frac{2m - 1 + m^{-2} - 2m^{-3}}{1 - \ell(m)^{-2}}.$$

Therefore we finally have

$$\mathbb{T}^E_\mu(\rho_{m\pm}) = \frac{\pm 2}{\sqrt{(m + m^{-1} + 1)(m + m^{-1} - 3)}}.$$

By some computer calculations the following formula seems to hold.

$$\lim_{N\to\infty}\left\{J_N\big(E;\exp(\xi/N)\big)\frac{2\sinh(\xi/2)}{\nu(\xi/N)} - \sqrt{-\pi}\exp\left(H(u)\frac{N}{\xi}\right)\left(\frac{N}{\xi}\right)^{1/2}\sqrt{\mathbb{T}^E_\mu(\rho_{m\pm})}\right\} = \frac{2\sinh(\xi/2)}{\Delta(E;\exp\xi)}$$

where $\nu(x)$ is a function with $\lim_{x\to 0}\nu(x) = 1$.

For a hyperbolic knot K, we expect a similar formula.

$$\lim_{N\to\infty}\left\{J_N\big(K;\exp(\xi/N)\big)\frac{2\sinh(\xi/2)}{\nu(\xi/N)} - \sqrt{-\pi}\exp\left(H(u)\frac{N}{\xi}\right)\left(\frac{N}{\xi}\right)^{1/2}\sqrt{\mathbb{T}^K_\mu(\rho)}\right\} = \frac{2\sinh(\xi/2)}{\Delta(E;\exp\xi)},$$

where $\nu(x)$ is a function with $\lim_{x\to 0}\nu(x) = 1$ and we put $u := \xi - 2\pi\sqrt{-1}$. Moreover ρ, $H(u)$ and $\mathbb{T}^K_\mu(\rho)$ satisfy the following properties. Put $v(u) := 2\frac{d\,H(u)}{d\,u} - 2\pi\sqrt{-1}$.

- $\rho\colon \pi_1(S^3\setminus K) \to SL(2;\mathbb{C})$ sends the meridian to $\begin{pmatrix}\exp(u/2) & * \\ 0 & \exp(-u/2)\end{pmatrix}$ and the longitude to $\begin{pmatrix}-\exp(v(u)/2) & * \\ 0 & -\exp(-v(u)/2)\end{pmatrix}$ up to conjugate.
- $H(u) - \pi\sqrt{-1}u - uv(u)/4$ coincides with $\mathrm{CS}_{u,v(u)}([\rho])$.
- $\mathbb{T}^K_\mu(\rho)$ is the twisted Reidemeister torsion of ρ associated with the meridian.

References

[1] M. Culler and P. B. Shalen, *Varieties of group representations and splittings of 3-manifolds*, Ann. of Math. (2) **117** (1983), no. 1, 109–146. MR MR683804 (84k:57005)

[2] G. de Rham, *Sur les nouveaux invariants topologiques de M. Reidemeister.*, Rec. Math. Moscou, n. Ser. **1** (1936), 737–742 (French).

[3] J. Dubois, *Non abelian twisted Reidemeister torsion for fibered knots*, Canad. Math. Bull. **49** (2006), no. 1, 55–71. MR MR2198719

[4] J. Dubois, V. Huynh, and Y. Yamaguchi, *Non-abelian Reidemeister torsion for twist knots*, J. Knot Theory Ramifications **18** (2009), no. 3, 303–341. MR MR2514847

[5] J. Dubois and R. M. Kashaev, *On the asymptotic expansion of the colored jones polynomial for torus knots*, Math. Ann. **339** (2007), no. 4, 757–782.

[6] W. Franz, *Über die Torsion einer Überdeckung.*, J. Reine Angew. Math. **173** (1935), 245–254 (German).

[7] S. Garoufalidis and T. T. Q. Le, *An analytic version of the Melvin-Morton-Rozansky Conjecture*, arXiv:math.GT/0503641.

[8] ———, *On the volume conjecture for small angles*, arXiv:math.GT/0502163.

[9] M. Gromov, *Volume and bounded cohomology*, Inst. Hautes Études Sci. Publ. Math. (1982), no. 56, 5–99 (1983). MR 84h:53053

[10] K. Hikami, *Quantum invariant for torus link and modular forms*, Comm. Math. Phys. **246** (2004), no. 2, 403–426. MR 2 048 564

[11] K. Hikami and A. N. Kirillov, *Torus knot and minimal model*, Phys. Lett. B **575** (2003), 343–348.

[12] K. Hikami and H. Murakami, *Colored Jones polynomials with polynomial growth*, Commun. Contemp. Math. **10** (2008), no. suppl. 1, 815–834. MR MR2468365

[13] W. H. Jaco and P. B. Shalen, *Seifert fibered spaces in 3-manifolds*, Mem. Amer. Math. Soc. **21** (1979), no. 220, viii+192. MR 81c:57010

[14] K. Johannson, *Homotopy equivalences of 3-manifolds with boundaries*, Lecture Notes in Mathematics, vol. 761, Springer, Berlin, 1979. MR 82c:57005

[15] V. F. R. Jones, *A polynomial invariant for knots via von Neumann algebras*, Bull. Amer. Math. Soc. (N.S.) **12** (1985), no. 1, 103–111. MR 86e:57006

[16] R. M. Kashaev, *A link invariant from quantum dilogarithm*, Modern Phys. Lett. A **10** (1995), no. 19, 1409–1418. MR 96j:81060

[17] ———, *The hyperbolic volume of knots from the quantum dilogarithm*, Lett. Math. Phys. **39** (1997), no. 3, 269–275. MR 98b:57012

[18] R. M. Kashaev and O. Tirkkonen, *A proof of the volume conjecture on torus knots*, Zap. Nauchn. Sem. S.-Peterburg. Otdel. Mat. Inst. Steklov. (POMI) **269** (2000), no. Vopr. Kvant. Teor. Polya i Stat. Fiz. 16, 262–268, 370. MR 1 805 865

[19] P. Kirk and E. Klassen, *Chern-Simons invariants of* 3*-manifolds decomposed along tori and the circle bundle over the representation space of* T^2, Comm. Math. Phys. **153** (1993), no. 3, 521–557. MR 94d:57042

[20] E. P. Klassen, *Representations of knot groups in* SU(2), Trans. Amer. Math. Soc. **326** (1991), no. 2, 795–828. MR MR1008696 (91k:57003)

[21] J. E. Marsden and M. J. Hoffman, *Basic complex analysis*, W. H. Freeman and Company, New York, 1987. MR 88m:30001

[22] J. Milnor, *A duality theorem for Reidemeister torsion*, Ann. of Math. (2) **76** (1962), 137–147. MR MR0141115 (25 #4526)

[23] V. Muñoz, *The* $\mathrm{SL}(2,\mathbb{C})$*-character varieties of torus knots*, Rev. Mat. Complut. **22** (2009), no. 2, 489–497. MR MR2553945

[24] H. Murakami, *The asymptotic behavior of the colored Jones function of a knot and its volume*, Proceedings of 'Art of Low Dimensional Topology VI' (T. Kohno, ed.), January 2000, pp. 87–96.

[25] ———, *Asymptotic behaviors of the colored Jones polynomials of a torus knot*, Internat. J. Math. **15** (2004), no. 6, 547–555. MR MR2078879

[26] ———, *An introduction to the volume conjecture and its generalizations*, Acta Math. Vietnam. **33** (2008), no. 3, 219–253. MR MR2501844

[27] H. Murakami and J. Murakami, *The colored Jones polynomials and the simplicial volume of a knot*, Acta Math. **186** (2001), no. 1, 85–104. MR 2002b:57005

[28] H. Murakami and Y. Yokota, *The colored Jones polynomials of the figure-eight knot and its Dehn surgery spaces*, J. Reine Angew. Math. **607** (2007), 47–68. MR MR2338120

[29] W. D. Neumann and D. Zagier, *Volumes of hyperbolic three-manifolds*, Topology **24** (1985), no. 3, 307–332. MR 87j:57008

[30] J. Porti, *Torsion de Reidemeister pour les variétés hyperboliques*, Mem. Amer. Math. Soc. **128** (1997), no. 612, x+139. MR MR1396960 (98g:57034)

[31] K. Reidemeister, *Homotopieringe und Linsenräume.*, Abh. Math. Semin. Hamb. Univ. **11** (1935), 102–109 (German).

[32] R. Riley, *Nonabelian representations of* 2*-bridge knot groups*, Quart. J. Math. Oxford Ser. (2) **35** (1984), no. 138, 191–208. MR MR745421 (85i:20043)

[33] W. P. Thurston, *The Geometry and Topology of Three-Manifolds*, Electronic version 1.1 - March 2002, http://www.msri.org/publications/books/gt3m/.

[34] V. G. Turaev, *Reidemeister torsion in knot theory*, Uspekhi Mat. Nauk **41** (1986), no. 1(247), 97–147, 240. MR MR832411 (87i:57009)

[35] ———, *Quantum invariants of knots and 3-manifolds*, de Gruyter Studies in Mathematics, vol. 18, Walter de Gruyter & Co., Berlin, 1994. MR MR1292673 (95k:57014)

[36] R. van der Veen, *Proof of the volume conjecture for Whitehead chains*, Acta Math. Vietnam. **33** (2008), no. 3, 421–431. MR MR2501851

[37] E. Witten, *Quantum field theory and the Jones polynomial*, Comm. Math. Phys. **121** (1989), 351–399.

[38] T. Yoshida, *The η-invariant of hyperbolic 3-manifolds*, Invent. Math. **81** (1985), no. 3, 473–514. MR 87f:58153

[39] H. Zheng, *Proof of the volume conjecture for Whitehead doubles of a family of torus knots*, Chin. Ann. Math. Ser. B **28** (2007), no. 4, 375–388. MR MR2348452

Department of Mathematics Education, School of Natural and Living Sciences Education, Naruto University of Education, 748, Nakashima, Takashima, Naruto-cho, Naruto-shi, 772-8502 Japan.

E-mail address: khikami@gmail.com

Department of Mathematics, Tokyo Institute of Technology, Oh-okayama, Meguro, Tokyo 152-8551, Japan

E-mail address: starshea@tky3.3web.ne.jp

AMS/IP Studies in Advanced Mathematics
Volume 50, 2011

Eta-Invariants and Anomalies in $U(1)$ Chern-Simons Theory

Lisa Jeffrey and Brendan McLellan

ABSTRACT. This article studies $U(1)$ Chern-Simons theory and its relation to a construction of Chris Beasley and Edward Witten, [**3**]. The natural geometric setup here is that of a three-manifold with a Seifert structure. We are led to study the stationary phase approximation of the path integral for $U(1)$ Chern-Simons theory after one of the three components of the gauge field is decoupled. This gives an alternative formulation of the partition function for $U(1)$ Chern-Simons theory that is conjecturally equivalent to the usual $U(1)$ Chern-Simons theory, [**12**]. The goal of this article is to establish this conjectural equivalence rigorously using appropriate regularization techniques.

1. Introduction

In [**3**] the authors study the Chern-Simons partition function, [**3**, (3.1)],
(1.1)

$$Z(k) = \frac{1}{\mathrm{Vol}(\mathcal{G})}\left(\frac{k}{4\pi^2}\right)^{\Delta\mathcal{G}} \int_{\mathcal{A}_P} \mathcal{D}A \, \exp\left[i\frac{k}{4\pi}\int_X \mathrm{Tr}\left(A\wedge dA + \frac{2}{3}A\wedge A\wedge A\right)\right],$$

where,

- $A \in \mathcal{A}_P = \{A \in (\Omega^1(P)\otimes\mathfrak{g})^G \mid A(\xi^\sharp) = \xi, \ \forall \xi \in \mathfrak{g}\}$ is a connection on a principal G-bundle $\pi : P \to X$ over a closed three-manifold X. In fact, [**3**] consider only G compact, connected and simple, and for concreteness one may take $G = SU(2)$,
- $\mathfrak{g} = \mathrm{Lie}(G)$ and $\xi^\sharp \in \Gamma(TP)$ is the vector field on P generated by the infinitesimal action of ξ on P,
- $k \in \mathbb{Z}$ (thought of as an element of $H^4(BG,\mathbb{Z})$ that parameterizes the possible Chern-Simons invariants),
- $\mathcal{G} := \{\psi \in (\mathrm{Diff}(P,P))^G \mid \pi\circ\psi = \pi\}$ is the *gauge group*, and
- $\Delta(\mathcal{G})$ is formally defined as the dimension of the gauge group.

In general, the partition function of (1.1) does not admit a general mathematical interpretation in terms of the cohomology of some classical moduli space of

2010 *Mathematics Subject Classification.* Primary 58J28; Secondary 58J52, 54C40.
Key words and phrases. Contact geometry, quantum field theory.
The first author was supported in part by a grant from NSERC.
We would like to thank John Bland, Eckhard Meinrenken, Raphaël Ponge, Edward Witten and especially Frédéric Rochon and Michel Rumin for helpful advice related to this work.

connections, in contrast to Yang-Mills theory for example, [**22**]. The main result of [**3**], however, is that if X is assumed to carry the additional geometric structure of a Seifert manifold, then the partition function of (1.1) *does* admit a more conventional interpretation in terms of the cohomology of some classical moduli space of connections. Using the additional Seifert structure on X, [**3**] decouples one of the components of a gauge field A, and introduces a new partition function [**3**, (3.7)],

$$\begin{aligned} \bar{Z}(k) &= K \cdot \int_{\mathcal{A}'_P} \mathcal{D}A\,\mathcal{D}\Phi\, \exp\left[i\frac{k}{4\pi}\left(CS(A) - \int_X 2\kappa \wedge \mathrm{Tr}(\Phi F_A) \right.\right. \\ &\qquad \left.\left. + \int_X \kappa \wedge d\kappa\, \mathrm{Tr}(\Phi^2) \right)\right], \end{aligned} \tag{1.2}$$

and,

$$CS(A) := \int_X \mathrm{Tr}\left(A \wedge dA + \frac{2}{3} A \wedge A \wedge A \right),$$

where,

- $K := \frac{1}{\mathrm{Vol}(\mathcal{G})} \frac{1}{\mathrm{Vol}(\mathcal{S})} \left(\frac{k}{4\pi^2}\right)^{\Delta\mathcal{G}}$,
- $\kappa \in \Omega^1(X, \mathbb{R})$ is a contact form associated to the Seifert fibration of X [**3**, §3.2],
- $\Phi \in \Omega^0(X, \mathfrak{g})$ is a Lie algebra-valued zero form on X,
- $\mathcal{A}'_P := \mathcal{A}_P \times \mathcal{F}$ where $\mathcal{F}$ is the space of fields $\Phi \in \Omega^0(P, \mathbb{R})$,[1] and $\mathcal{D}\Phi$ is a measure on $\mathcal{F}$,
- $\mathcal{S}$ is the space of local *shift symmetries*[2] that "acts" on the space of connections $\mathcal{A}_P$ and the space of fields Φ, [**3**, §3.1],
- $F_A \in \Omega^2(X, \mathfrak{g})$ is the curvature of A, and
- $CS(A) := \int_X \mathrm{Tr}\left(A \wedge dA + \frac{2}{3} A \wedge A \wedge A\right)$ is the Chern-Simons action.

REMARK 1.3. Note that the partition functions of (1.1) and (1.2) are defined implicitly with respect to the pullback of some trivializing section of the principal G-bundle P. Of course, every principal G-bundle over a three-manifold for G simply connected is trivializable. It is a basic fact that the partition functions of (1.1) and (1.2) are independent of the choice of such trivializations. Note also that the definition of the Chern-Simons partition function in (1.1) is completely heuristic. The measure $\mathcal{D}A$ has not been defined, but only assumed to "exist heuristically," and the volume and dimension of the gauge group, $\mathrm{Vol}(\mathcal{G})$ and $\Delta(\mathcal{G})$, respectively, are at best formally defined.

[**3**] then provides a heuristic argument showing that the partition function computed using the alternative description of (1.2) should be the same as the Chern-Simons partition function of (1.1). In essence, they show, [**3**, §3.1],

$$Z(k) = \bar{Z}(k), \tag{1.4}$$

---

[1]The measure $\mathcal{D}\Phi$ is defined independently of any metric on X and is formally defined by the positive definite quadratic form

$$(\Phi, \Phi) := -\int_X \kappa \wedge d\kappa\, \mathrm{Tr}(\Phi^2),$$

which is invariant under the choice of representative for the contact structure (X, H) on X, i.e. under the scaling $\kappa \mapsto f\kappa$, $\Phi \mapsto f^{-1}\Phi$, for some non-zero function $f \in \Omega^0(X, \mathbb{R})$.

[2]$\mathcal{S}$ may be identified with $\Omega^0(X, \mathfrak{g})$, where the "action" on $\mathcal{A}_P$ is defined as $\delta_\sigma(A) := \sigma\kappa$, and on the space of fields Φ is defined as $\delta_\sigma(\Phi) := \sigma$, for $\sigma \in \Omega^0(X, \mathfrak{g})$.

by gauge fixing $\Phi = 0$ using the shift symmetry. [**3**] then observes that the Φ-dependence in the integral can be eliminated by simply performing the Gaussian integral over Φ in (1.2) directly. [**3**] obtains the alternative formulation:
(1.5)

$$Z(k) = \bar{Z}(k) = K' \cdot \int_{\mathcal{A}_P} \mathcal{D}A \, \exp\left[i\frac{k}{4\pi}\left(CS(A) - \int_X \frac{1}{\kappa \wedge d\kappa} \operatorname{Tr}\left[(\kappa \wedge F_A)^2 \right] \right) \right],$$

where $K' := \frac{1}{\mathrm{Vol}(\mathcal{G})}\frac{1}{\mathrm{Vol}(\mathcal{S})}\left(\frac{-ik}{4\pi^2}\right)^{\Delta\mathcal{G}/2}$. Note that we have abused notation slightly by writing $\frac{1}{\kappa\wedge d\kappa}$. We have done this with the understanding that since $\kappa \wedge d\kappa$ is non-vanishing (since κ is a contact form), then $\kappa \wedge F_A = \phi\, \kappa \wedge d\kappa$ for some function $\phi \in \Omega^0(X, \mathfrak{g})$, and we identify $\frac{\kappa\wedge F_A}{\kappa \wedge d\kappa} := \phi$.

The objective in this article is to study the partition function for $U(1)$ *Chern-Simons theory* using the analogue of (1.5) in this case. Thus, we are also assuming here that X is a Seifert manifold with a "compatible" contact structure, (X, κ) [**3**, §3.2]. Note that any compact, oriented three-manifold possesses a contact structure and one aim of future work is to extend our results to *all* closed three-manifolds using this fact. For now, we restrict ourselves to the case of closed three-manifolds that possess contact compatible Seifert structures (see Definition 3.9 for example). We restrict to the gauge group $U(1)$ so that the action is quadratic and hence the stationary phase approximation is exact. A salient point is that the group $U(1)$ is not simply-connected and there may exist non-trivial principal $U(1)$-bundles P over X. This makes the $U(1)$-theory very different from the $SU(2)$-theory in that one must now incorporate a sum over bundle classes in a definition of the $U(1)$-partition function. As an analogue of (1.1), our basic definition of the partition function for $U(1)$ Chern-Simons theory is now

$$Z_{U(1)}(X,k) = \sum_{p \in \mathrm{Tors}H^2(X;\mathbb{Z})} Z_{U(1)}(X,p,k) \tag{1.6}$$

where

$$Z_{U(1)}(X,p,k) = \frac{1}{\mathrm{Vol}(\mathcal{G}_P)} \int_{\mathcal{A}_P} \mathcal{D}A\, e^{\pi i k S_{X,P}(A)}, \tag{1.7}$$

recalling that the torsion subgroup[3] $\mathrm{Tors}H^2(X;\mathbb{Z}) < H^2(X;\mathbb{Z})$ enumerates the $U(1)$-bundle classes that admit flat connections. Note that the bundle $P \to X$ in (1.7) is taken to be any representative of a bundle class with first Chern class $c_1(P) = p \in \mathrm{Tors}H^2(X;\mathbb{Z})$. Also note that some care must be taken to define the Chern-Simons action, $S_{X,P}(A)$, in the case that $G = U(1)$. We outline this construction in Appendix A.

The main results of this article may be summarized as follows. First, our main objective is the rigorous confirmation of the heuristic result of (1.4) in the case where the gauge group is $U(1)$. This statement is certainly non-trivial and involves some fairly deep facts about the "contact operator" as studied by Michel Rumin, [**17**]. A somewhat surprising observation is that this operator shows up quite naturally in $U(1)$ Chern-Simons theory (see Prop. 4.14 below), and this leads us to make several conjectures motivated by the rigorous confirmation of the heuristic result of (1.4).

---

[3]Recall the definition of the torsion of an abelian group is the collection of those elements which have finite order.

REMARK 1.8. We note that throughout this article quantities involving integrals over the space of connections $\int_{\mathcal{A}_P}$ are generally non-rigorous, and quantities involving integrals over the moduli space of flat connections $\int_{\mathcal{M}_P}$ are rigorously defined. The latter are derived from the former explicitly in this article.

Our main result is the following:

PROPOSITION 1.9. *Let $(X, \phi, \xi, \kappa, g)$ be a closed, quasi-regular K-contact three manifold. If,*

$$\bar{Z}_{U(1)}(X,p,k) = k^{n_X} e^{\pi i k S_{X,P}(A_P)} e^{\frac{\pi i}{4}\left(\eta(\star D) + \frac{1}{512}\int_X R^2\ \kappa\wedge d\kappa\right)} \int_{\mathcal{M}_P} (T_C^d)^{1/2} \tag{1.10}$$

and,

$$Z_{U(1)}(X,p,k) = k^{m_X} e^{\pi i k S_{X,P}(A_P)} e^{\pi i\left(\frac{\eta(\star d)}{4} + \frac{1}{12}\frac{CS(A^g)}{2\pi}\right)} \int_{\mathcal{M}_P} (T_{RS}^d)^{1/2} \tag{1.11}$$

then, after choosing the canonical framing for $TX \oplus TX$ corresponding to a choice of Vielbein, we have

$$Z_{U(1)}(X,k) = \bar{Z}_{U(1)}(X,k)$$

as topological invariants. In fact,

$$\begin{aligned}
\eta(\star d) + \frac{1}{3}\frac{CS(A^g)}{2\pi} &= \eta(\star D) + \frac{1}{512}\int_X R^2\ \kappa\wedge d\kappa \\
&= 1 - \frac{d}{3} + 4\sum_{j=1}^{N} s(\alpha_j, \beta_j),
\end{aligned}$$

where $d = c_1(X) = n + \sum_{j=1}^{N}\frac{\beta_j}{\alpha_j} \in \mathbb{Q}$ and

$$s(\alpha,\beta) := \frac{1}{4\alpha}\sum_{k=1}^{\alpha-1} cot\left(\frac{\pi k}{\alpha}\right) cot\left(\frac{\pi k\beta}{\alpha}\right) \in \mathbb{Q}$$

is the classical Rademacher-Dedekind sum, where $[n; (\alpha_1, \beta_1), \ldots, (\alpha_N, \beta_N)]$ (for $gcd(\alpha_j, \beta_j) = 1$) are the Seifert invariants of X.

Following [**12**], we rigorously define $\bar{Z}_{U(1)}(X,k)$ in §6 using the fact that the stationary phase approximation for our path integral should be exact. This necessitates the introduction of the regularized determinant of D in (7.3), which in turn naturally involves the hypoelliptic Laplacian of (7.5). The rigorous quantity that we obtain for the integrand of (6.2) in §6 is derived in Prop. 7.17. Using an observation from §5 that identifies the volume of the isotropy subgroup of the gauge group $\mathcal{G}_P$, we identify the integrand of (6.2) with the contact analytic torsion T_C^d defined in Def. 9.4. After formally identifying the signature of the contact operator D with the eta-invariant of D in §8, we obtain our fully rigorous definition of $\bar{Z}_{U(1)}(X,k)$ in (9.17) below, which is repeated in (1.10) above.

On the other hand, [**12**] provides a rigorous definition of the partition function $Z_{U(1)}(X,k)$ that does not involve an *a priori* choice of a contact structure on X. The formula for this is recalled in (9.18) below, and is the term $Z_{U(1)}(X,p,k)$ in (1.11) of Prop. 1.9 above.

Our first main step in the proof of Prop. 1.9 is the confirmation of the fact that the Ray-Singer analytic torsion of X, T_{RS}^d, [**16**], is identically equal to the contact analytic torsion T_C^d.[4] We observe that this result follows directly from [**18**, Theorem 4.2].

We also observe in Remark 7.22 that the quantities m_X and n_X that occur in Prop. 1.9 are also equal. This leaves us with the final step in the confirmation of Prop. 1.9 which involves a study of the eta-invariants $\eta(\star d)$, $\eta(\star D)$, that naturally show up in $Z_{U(1)}(X,k)$, $\bar{Z}_{U(1)}(X,k)$, respectively. This analysis is carried out in §10 where we observe that the work of Biquard, Herzlich, and Rumin, [**4**], is our most pertinent reference. Our main observation here is that the quantum anomalies that occur in the computation of $Z_{U(1)}(X,k)$ and $\bar{Z}_{U(1)}(X,k)$ should, in an appropriate sense, be completely equivalent. In our case these quantum anomalies are made manifest precisely in the failure of the eta-invariants to represent topological invariants. As observed by Witten in [**21**], this is deeply connected with the fact that in order to *actually* compute the partition function one needs to make a choice that is tantamount to either a valid gauge choice for representatives of gauge classes of connections, or to some other way of breaking the symmetry of our problem. Such a choice for us is equivalent to a choice of metric, which is encoded in the choice of a quasi-regular K-contact structure on our manifold X. Witten observes in [**21**] that the quantum anomaly that is introduced by our choice of metric may be canceled precisely by adding an appropriate "counter-term" to the eta-invariant, $\eta(\star d)$. This recovers topological invariance and effectively cancels the anomaly.[5] This counter-term is found by appealing to the Atiyah-Patodi-Singer theorem, [**2**], and is in fact identified as the gravitational Chern-Simons term,

$$\mathrm{CS}(A^g) := \frac{1}{4\pi}\int_X \mathrm{Tr}(A^g \wedge dA^g + \frac{2}{3}A^g \wedge A^g \wedge A^g), \tag{1.12}$$

where A^g is the Levi-Civita connection on the spin bundle of X for the metric,

$$g = \kappa \otimes \kappa + d\kappa(\cdot, J\cdot), \tag{1.13}$$

on our quasi-regular K-contact three manifold, $(X, \phi, \xi, \kappa, g)$. In particular, we use the fact that

$$\frac{\eta(\star d)}{4} + \frac{1}{12}\frac{\mathrm{CS}(A^g)}{2\pi}, \tag{1.14}$$

is a topological invariant of X, after choosing the canonical framing. We establish the following in Proposition 10.22,

PROPOSITION 1.15. *Let $(X, \phi, \xi, \kappa, g)$ be a closed, quasi-regular K-contact three-manifold. There exists counter-term, C_T, such that $e^{\frac{\pi i}{4}[\eta(\star_{H} D)+C_T]}$ is a topological invariant that is identically equal to the topological invariant $e^{\pi i\left[\frac{\eta(\star d)}{4}+\frac{1}{12}\frac{CS(A^g)}{2\pi}\right]}$. In fact, after choosing the canonical framing for $TX \oplus TX$ corresponding to a choice of Vielbein, we have*

$$C_T = \frac{1}{512}\int_X R^2\ \kappa \wedge d\kappa,$$

where $R \in C^\infty(X)$ is the Tanaka-Webster scalar curvature of X, and

$$\eta(\star d) + \frac{1}{3}\frac{CS(A^g)}{2\pi} = \eta(\star D) + \frac{1}{512}\int_X R^2\ \kappa \wedge d\kappa.$$

[4]We consider the square roots thereof, viewed as densities on the moduli space of flat connections $\mathcal{M}_X$.

[5]In this case, topological invariance is recovered only up to a choice of two-framing for X. Recall, [**1**], that there is a canonical choice of such framing, and we assume this choice throughout.

This proposition is proven in §10 by appealing to the following result, which is established using a "Kaluza-Klein" dimensional reduction technique for the gravitational Chern-Simons term. This result is modeled after the paper [**10**] and is listed as Proposition 10.19.

PROPOSITION 1.16. [**14**] *Let $(X, \phi, \xi, \kappa, g)$ be a closed, quasi-regular K-contact three-manifold,*

$$\begin{array}{ccc} U(1) & \hookrightarrow & X, \\ & & \downarrow \\ & & \Sigma \end{array}$$

and $g_\epsilon := \epsilon^{-1}\kappa \otimes \kappa + \pi^* h$. After choosing the canonical framing for $TX \oplus TX$, corresponding to a choice of Vielbein,

$$CS(A^{g_\epsilon}) = \left(\frac{\epsilon^{-1}}{2}\right)\int_\Sigma r\,\omega + \left(\frac{\epsilon^{-2}}{2}\right)\int_\Sigma f^2\,\omega \tag{1.17}$$

where $r \in C^\infty_{orb}(\Sigma)$ is the (orbifold) scalar curvature of (Σ, h), $\omega \in \Omega^2_{orb}(\Sigma)$ is the (orbifold) Hodge form of (Σ, h), and $f := \star_h \omega$ (where $\star_h$ denotes the Hodge-$\star$ operator for the metric h). In particular, the adiabatic limit of $CS(A^{g_\epsilon})$ vanishes:

$$\lim_{\epsilon\to\infty} CS(A^{g_\epsilon}) = 0. \tag{1.18}$$

Finally, as a consequence of these investigations, we are able to compute in Proposition 10.23 the $U(1)$ Chern-Simons partition function fairly explicitly.

PROPOSITION 1.19. *Let $(X, \phi, \xi, \kappa, g)$ be a closed, quasi-regular K-contact three-manifold. We have,*

$$\begin{aligned} \eta(\star d) + \frac{1}{3}\frac{CS(A^g)}{2\pi} &= \eta(\star D) + \frac{1}{512}\int_X R^2\,\kappa\wedge d\kappa \\ &= 1 - \frac{d}{3} + 4\sum_{j=1}^{N} s(\alpha_j, \beta_j), \end{aligned}$$

where $d = c_1(X) = n + \sum_{j=1}^N \frac{\beta_j}{\alpha_j} \in \mathbb{Q}$ and

$$s(\alpha, \beta) := \frac{1}{4\alpha}\sum_{k=1}^{\alpha-1} \cot\left(\frac{\pi k}{\alpha}\right)\cot\left(\frac{\pi k\beta}{\alpha}\right) \in \mathbb{Q}$$

is the classical Rademacher-Dedekind sum, where $[n; (\alpha_1, \beta_1), \ldots, (\alpha_N, \beta_N)]$ (for $gcd(\alpha_j, \beta_j) = 1$) are the Seifert invariants of X. In particular, we have computed the $U(1)$ Chern-Simons partition function as:

$$\begin{aligned} Z_{U(1)}(X, p, k) &= k^{n_X} e^{\pi i k S_{X,P}(A_P)} e^{\frac{\pi i}{4}\left(1-\frac{d}{3}+4\sum_{j=1}^N s(\alpha_j,\beta_j)\right)} \int_{\mathcal{M}_P} (T^d_C)^{1/2}, \\ &= k^{m_X} e^{\pi i k S_{X,P}(A_P)} e^{\frac{\pi i}{4}\left(1-\frac{d}{3}+4\sum_{j=1}^N s(\alpha_j,\beta_j)\right)} \int_{\mathcal{M}_P} (T^d_{RS})^{1/2}. \end{aligned}$$

2. Preliminary results

Our starting point is the analogue of (1.5) for the $U(1)$ Chern-Simons partition function:
(2.1)

$$\bar{Z}_{U(1)}(X,p,k) = \frac{e^{\pi i k S_{X,P}(A_P)}}{\mathrm{Vol}(\mathcal{S})\,\mathrm{Vol}(\mathcal{G}_P)} \int_{\mathcal{A}_P} \mathcal{D}A \ \exp\left[\frac{ik}{4\pi}\left(\int_X A\wedge dA - \int_X \frac{(\kappa\wedge dA)^2}{\kappa\wedge d\kappa}\right)\right]$$

where $S_{X,P}(A_P)$ is the Chern-Simons invariant associated to P for A_P a flat connection on P. The derivation of (2.1) can be found in Appendix A. It is obtained by expanding the $U(1)$ analogue of (1.5) around a critical point A_P of the action. Note that the critical points of this action, up to the action of the shift symmetry, are precisely the flat connections, [**3**, (5.3)]. Denote the action in the partition function (2.1) by

$$S(A) := \int_X A\wedge dA - \int_X \frac{(\kappa\wedge dA)^2}{\kappa\wedge d\kappa} \tag{2.2}$$

for $A \in T_{A_P}\mathcal{A}_P$. Also, define

$$\bar{S}(A) := \int_X \frac{(\kappa\wedge dA)^2}{\kappa\wedge d\kappa} \tag{2.3}$$

so that we may write

$$S(A) = CS(A) - \bar{S}(A) \tag{2.4}$$

The primary virtue of (2.1) above is that it is heuristically equal to the original Chern-Simons partition function of (1.7) and yet it is expressed in such a way that the action $S(A)$ is invariant under the shift symmetry. This means that $S(A+\sigma\kappa) = S(A)$ for all tangent vectors $A \in T_{A_P}\mathcal{A}_P \simeq \Omega^1(X)$ and $\sigma \in \Omega^0(X)$. We may naturally view $A \in \Omega^1(H)$, the subset of $\Omega^1(X)$ consisting of the restrictions of the sections of $\Gamma^1(TX)$ to the contact distribution $H \subset TX$. Equivalently, if ξ denotes the Reeb vector field of κ, then $\Omega^1(H) = \{\omega \in \Omega^1(X) \mid \iota_\xi\omega = 0\}$. The remaining contributions to the partition function come from the orbits of $\mathcal{S}$ in $\mathcal{A}_P$, which turn out to give a contributing factor of $\mathrm{Vol}(\mathcal{S})$, [**3**, (3.32)]. We thus reduce our integral to an integral over $\bar{\mathcal{A}}_P := \mathcal{A}_P/\mathcal{S}$ and obtain:

$$\begin{aligned} Z_{U(1)}(X,p,k) &= \frac{e^{\pi i k S_{X,P}(A_P)}}{\mathrm{Vol}(\mathcal{G}_P)} \int_{\bar{\mathcal{A}}_P} \bar{\mathcal{D}}A \ \exp\left[\frac{ik}{4\pi}\left(\int_X A\wedge dA \quad \int_X \frac{(\kappa\wedge dA)^2}{\kappa\wedge d\kappa}\right)\right] \\ &= \frac{e^{\pi i k S_{X,P}(A_P)}}{\mathrm{Vol}(\mathcal{G}_P)} \int_{\bar{\mathcal{A}}_P} \bar{\mathcal{D}}A \ \exp\left[\frac{ik}{4\pi} S(A)\right] \end{aligned}$$

where $\bar{\mathcal{D}}A$ denotes an appropriate quotient measure on $\bar{\mathcal{A}}_P$, and we now view elements $A \in \mathcal{A}_P/\mathcal{S}$ as elements $A \in \Omega^1(H) \simeq T_{A_P}\bar{\mathcal{A}}_P$.

3. Contact structures

At this point, we further restrict the structure on our 3-manifold and assume that the Seifert structure is compatible with a contact metric structure (ϕ, ξ, κ, g) on X. In particular, we restrict to the case of a quasi-regular K-contact manifold. Let us review some standard facts about these structures in the case of dimension three.

REMARK 3.1. Our three manifolds X are assumed to be closed throughout this article.

DEFINITION 3.2. A *K-contact* manifold is a manifold X with a contact metric structure (ϕ, ξ, κ, g) such that the Reeb field ξ is Killing for the associated metric g, i.e. $\mathcal{L}_\xi g = 0$.

Our notation is defined as follows:

- $\kappa \in \Omega^1(X)$ contact form, $\xi =$ Reeb vector field,
- $H := \ker\kappa \subset TX$ denotes the horizontal or contact distribution on (X, κ),
- $\phi \in \mathrm{End}(TX)$, $\phi(Y) = JY$ for $Y \in \Gamma(H)$, $\phi(\xi) = 0$ with complex structure $J \in \mathrm{End}(H)$ on the contact distribution $H \subset TX$,
- $g = \kappa \otimes \kappa + d\kappa(\cdot, \phi\cdot)$.

REMARK 3.3. Note that we will assume that our contact structure is "co-oriented," meaning that the contact form $\kappa \in \Omega^1(X)$ is a global form. Generally, one can take the contact structure to be to be defined only locally by the condition $H := \ker \kappa$, where $\kappa \in \Omega^1(U)$ for open subsets $U \in X$ contained in an open cover of X.

DEFINITION 3.4. The characteristic foliation $\mathcal{F}_\xi$ of a contact manifold (X, κ) is said to be *quasi-regular* if there is a positive integer j and each point has a foliated coordinate chart (U, x) such that each leaf of $\mathcal{F}_\xi$ passes through U at most j times. If $j = 1$ then the foliation is said to be *regular*.

Definitions 9.10 and 3.4 together define a quasi-regular K-contact manifold, $(X, \phi, \xi, \kappa, g)$. Such three-manifolds are necessarily "Seifert" manifolds that fiber over a two dimensional orbifold Σ with with some additional structure. Recall:

DEFINITION 3.5. A *Seifert manifold* is a three manifold X that admits a locally free $U(1)$-action.

Thus, Seifert manifolds are simply $U(1)$-bundles over an orbifold Σ,

$$\begin{array}{ccc} U(1) & \hookrightarrow & X \\ & & \downarrow \\ & & \Sigma \end{array}$$

By [**7**, Theorem 7.5.1] we have the following classification result: X is a quasi-regular K-contact three manifold if and only if X is a $U(1)$-Seifert manifold over a Hodge orbifold surface, or equivalently over a normal projective algebraic variety of real dimension two.

EXAMPLE 3.6. All three-dimensional Lens spaces, $L(p, q)$ and the Hopf fibration $S^1 \hookrightarrow S^3 \to \mathbb{CP}^1$ possess quasi-regular K-contact structures. Note, however, that by [**11**] any trivial $U(1)$-bundle over a Riemann surface Σ_g, $X = U(1) \times \Sigma_g$, possesses *no* K-contact structure and our results do not apply in this case.

REMARK 3.7. Note that in fact our results apply to the class of all closed *Sasakian* three-manifolds. This follows from the observation that every Sasakian three manifold is K-contact, [**6**, Corollary 6.5], and every K-contact manifold possesses a quasi-regular K-contact structure, [**7**, Theorem 7.1.10].

A useful observation for us is that for a quasi-regular K-contact three-manifold, the metric tensor g must take the following form, [**7**, Theorem 6.3.6],

$$g = \kappa \otimes \kappa + \pi^* h \tag{3.8}$$

where $\pi : X \to \Sigma$ is our quotient map, and h represents any (orbifold)Kähler metric on Σ which is normalized so that the corresponding (orbifold)Kähler form, $\omega \in \Omega^2_{orb}(\Sigma, \mathbb{R})$, pulls back to $d\kappa$.

Note that the assumption that the Seifert structure on X comes from a quasi-regular K-contact structure (ϕ, ξ, κ, g) is equivalent to assuming that X is a *CR-Seifert manifold*, [**7**, Prop. 6.4.8]. Recall the following

DEFINITION 3.9. A *CR-Seifert* manifold is a three-dimensional compact manifold endowed with both a strictly pseudoconvex CR structure (H, J) and a Seifert structure, that are compatible in the sense that the circle action $\psi : U(1) \to \mathrm{Diff}(X)$ preserves the CR structure and is generated by a Reeb field ξ. In particular, given a choice of contact form κ, the Reeb field is Killing for the associated metric $g = \kappa \otimes \kappa + d\kappa(\cdot, J\cdot)$.

The assumption that X is CR-Seifert (hence quasi-regular K-contact) is sufficient to ensure that the assumption in [**3**, (3.27)], which states that the $U(1)$-action on X, $\psi : U(1) \to \mathrm{Diff}(X)$, acts by isometries, is satisfied.

We now employ the natural Hodge star operator $\star$, induced by the metric g on X, that acts on $\Omega^\bullet(X)$ taking k forms to $3-k$ forms. As a result of this normalization convention, we have $\star 1 = \kappa \wedge d\kappa$ and $\star\kappa = d\kappa$. Now let

$$\star_H = -\iota_\xi \circ \star \tag{3.10}$$

as in [**3**, (3.30)]. This operator then satisfies

$$\begin{aligned}
\star_H \kappa &= 0 && (3.11)\\
\star_H(\kappa \wedge d\kappa) &= 0 && (3.12)\\
\star_H 1 &= -d\kappa && (3.13)\\
(\star_H)^2 &= -1 && (3.14)
\end{aligned}$$

as is shown in [**3**, Page 20]. We also define a horizontal exterior derivative $d_H : \Omega^q(X; \mathbb{R}) \to \Omega^{q+1}(X; \mathbb{R})$ as the usual exterior derivative d restricted to the space of horizontal forms $\Omega^\bullet(H) := \{\omega \subset \Omega^\bullet(X) \,|\, \iota_\xi \omega = 0\}$. Note that (3.14) is an equality on horizontal forms.

Our key observation is that the action $S(A)$ may now be expressed in terms of these horizontal quantities. Let us start with the term $\bar{S}(A)$. First, the term $\kappa \wedge dA$ in $\bar{S}(A)$ is equal to $\kappa \wedge d_H A$ since the vertical part of dA is annihilated by κ in the wedge product. The term $\frac{\kappa \wedge dA}{\kappa \wedge d\kappa}$ is equivalent to $\star(\kappa \wedge d_H A)$ by the properties of $\star$ above. By the definition of $\star_H$, $\star(\kappa \wedge d_H A) = \star_H d_H A$. We then have,

$$\begin{aligned}
\bar{S}(A) &= \int_X \frac{(\kappa \wedge dA)^2}{\kappa \wedge d\kappa}\\
&= \int_X \star_H(d_H A) \wedge \kappa \wedge d_H A\\
&= \int_X \kappa \wedge [d_H A \wedge \star_H(d_H A)]
\end{aligned}$$

We claim that $\bar{S}(A)$ is now expressed in terms of an inner product on $\Omega^2 H$. More generally, we define an inner product on $\Omega^l(H)$ for $0 \leq l \leq 2$:

DEFINITION 3.15. Define the pairing $\langle\cdot,\cdot\rangle_\kappa^l : \Omega^l H \times \Omega^l H \to \mathbb{R}$ as

$$\langle\alpha,\beta\rangle_\kappa^l := (-1)^l \int_X \kappa \wedge [\alpha \wedge \star_H \beta] \tag{3.16}$$

for any $\alpha, \beta \in \Omega^l H$, $0 \leq l \leq 2$.

PROPOSITION 3.17. *The pairing* $\langle\cdot,\cdot\rangle_\kappa^l$ *is an inner product on* $\Omega^l H$.

PROOF. It can be easily checked that this pairing is just the restriction of the usual L^2-inner product, $\langle\cdot,\cdot\rangle : \Omega^l X \times \Omega^l X \to \mathbb{R}$,

$$\langle\alpha,\beta\rangle := \int_X \alpha \wedge \star\beta \tag{3.18}$$

to horizontal forms. I.e. for any $\beta \in \Omega^l H$, $0 \leq l \leq 2$, we have $\star\beta = \kappa \wedge \star_H \beta$. We then have $\alpha \wedge \star\beta = (-1)^l \kappa \wedge [\alpha \wedge \star_H \beta]$ for any $\alpha, \beta \in \Omega^l H$, $0 \leq l \leq 2$. Thus, $\langle\cdot,\cdot\rangle_\kappa^l = \langle\cdot,\cdot\rangle$ on $\Omega^l H$ and therefore defines an inner product. □

By our definition, we may now write $\bar{S}(A) = \langle d_H A, d_H A\rangle_\kappa^2$. We make the following

DEFINITION 3.19. Define the formal adjoint of d_H, denoted d_H^*, via:

$$\langle d_H^*\gamma, \phi\rangle_\kappa^{l-1} = \langle\gamma, d_H\phi\rangle_\kappa^l$$

for $\gamma \in \Omega^l(H)$, $\phi \in \Omega^{l-1}(H)$ where $l = 1, 2$ and $d_H^*\gamma = 0$ for $\gamma \in \Omega^0(H)$.

PROPOSITION 3.20. $d_H^* = (-1)^l \star_H d_H \star_H : \Omega^l(H) \to \Omega^{l-1}(H)$, $0 \leq l \leq 2$, *where* $\Omega^{-1}(H) := 0$.

PROOF. This just follows from the definition of d^* relative to the ordinary inner product $\langle\cdot,\cdot\rangle$, and the facts that $\langle\cdot,\cdot\rangle_\kappa^{l-1}$ is just this ordinary inner product restricted to horizontal forms and $d^* = (-1)^l \star d\star$. □

Thus, we may now write $\bar{S}(A) = \langle A, d_H^* d_H A\rangle_\kappa^1$ and identify this piece of the action with the second order operator $d_H^* d_H$ on horizontal forms.

Now we turn our attention to the Chern-Simons part of the action $CS(A) = \int_X A \wedge dA$. We would like to reformulate this in terms of horizontal quantities as well. This is straightforward to do; simply observe that $dA = \kappa \wedge \mathcal{L}_\xi A + d_H A$. Thus, we have:

$$\begin{aligned} CS(A) &= \int_X A \wedge dA \\ &= \int_X A \wedge [\kappa \wedge \mathcal{L}_\xi A + d_H A] \\ &= \int_X A \wedge [\kappa \wedge \mathcal{L}_\xi A] + \int_X A \wedge d_H A \\ &= \int_X A \wedge [\kappa \wedge \mathcal{L}_\xi A] \end{aligned}$$

where the last line follows from the fact that $A \wedge d_H A = 0$ since both forms are horizontal. Putting this all together, we may now express the total action $S(A)$ in terms of horizontal quantities as follows:

$$\begin{aligned} S(A) &= CS(A) - \bar{S}(A) \\ &= \int_X A \wedge [\kappa \wedge \mathcal{L}_\xi A] + \int_X A \wedge [\kappa \wedge d_H \star_H d_H A] \\ &= \int_X A \wedge [\kappa \wedge (\mathcal{L}_\xi + d_H \star_H d_H) A] \end{aligned}$$

4. The contact operator D

A surprising observation is that $\kappa \wedge (\mathcal{L}_\xi + d_H \star_H d_H)$ turns out to be well known. It is the second order operator "D" that fits into the complex,

$$C^\infty(X) \xrightarrow{d_H} \Omega^1(H) \xrightarrow{D} \Omega^2(V) \xrightarrow{d_H} \Omega^3(X), \tag{4.1}$$

where,

$$\Omega^2(V) := \{\kappa \wedge \alpha \mid \alpha \in \Omega^2(H)\} = \kappa \wedge \Omega^2(H), \tag{4.2}$$

and for $f \in C^\infty(X)$, $d_H f \in \Omega^1(H)$ stands for the restriction of df to H as usual, while

$$d_H : \Omega^2(V) \to \Omega^3(X), \tag{4.3}$$

is just de Rham's differential restricted to $\Omega^2(V)$ in $\Omega^2(X)$.

DEFINITION 4.4. Denote the complex in (4.1) as $(\mathcal{E}, d_H)$. Thus, we denote $D := d_H$ in the middle degree.[6]

D is defined as follows: since d induces an isomorphism

$$d_0 : \Omega^1(V) \to \Omega^2(H), \text{ with } d_0(f\kappa) = f d\kappa|_{\Lambda^2(H)} \tag{4.5}$$

then any $\alpha \in \Omega^1(H)$ admits a unique extension $l(\alpha)$ in $\Omega^1(X)$ such that $dl(\alpha)$ belongs to $\Omega^2(V)$; i.e. given any initial extension $\bar{\alpha}$ of α, one has

$$l(\alpha) = \bar{\alpha} - d_0^{-1}(d\bar{\alpha})|_{\Lambda^2(H)} \tag{4.6}$$

We then define

$$D\alpha := dl(\alpha) \tag{4.7}$$

We then have [**4**, (39)],

$$D\alpha = \kappa \wedge [\mathcal{L}_\xi + d_H \star_H d_H]\alpha \tag{4.8}$$

for any $\alpha \in \Omega^1(H)$. Thus,

$$\begin{aligned} S(A) &= \int_X A \wedge [\kappa \wedge (\mathcal{L}_\xi + d_H \star_H d_H) A] && (4.9) \\ &= \int_X A \wedge DA && (4.10) \\ &= \langle A, \star DA \rangle && (4.11) \end{aligned}$$

where $\langle \cdot, \cdot \rangle$ is the usual L^2 inner product on $\Omega^1(X)$.

[6] Note that we have abused notation in a few places in this article regarding the operator d_H in middle degree. We will only view $D = d_H$ in terms of the complex $(\mathcal{E}, d_H)$, and otherwise $d_H : \Omega^1(X) \to \Omega^2(H)$ is meant to denote the operator $d : \Omega^1(X) \to \Omega^2(X)$ followed by the projection of $\Omega^2(X)$ onto $\Omega^2(H)$. That is, $d_H : \Omega^1(X) \to \Omega^2(H)$ is defined as $d_H := \bar{\pi} \circ d$ where $\bar{\pi} : \Omega^2(X) \to \Omega^2(H)$ is the projection $\bar{\pi}(\alpha) := \alpha - \kappa \wedge \iota_\xi \alpha$ for $\alpha \in \Omega^q(X)$, $1 \leq q \leq 2$. The notation should also be clear from the context.

Alternatively, we make the following

DEFINITION 4.12. Let $D^1 : \Omega^1(H) \to \Omega^1(X)$ denote the operator

$$D^1 := \mathcal{L}_\xi + d_H \star_H d_H \tag{4.13}$$

and observe that we can also write $S(A) = \langle A, \star_H D^1 A\rangle^1_\kappa$, identifying $S(A)$ with the operator $\star_H D^1$ on $\Omega^1(H)$. Thus, we have proven the following

PROPOSITION 4.14. *The new action, $S(A)$, as defined in (2.2), for the "shifted" partition function of (2.1) can be expressed as a quadratic form on the space of horizontal forms $\Omega^1(H)$ as follows:*

$$S(A) = \langle A, \star DA\rangle \tag{4.15}$$

or equivalently as,

$$S(A) = \langle A, \star_H D^1 A\rangle^1_\kappa \tag{4.16}$$

where D and D^1 are the second order operators defined in (4.8) and (4.13), respectively. $\langle\cdot,\cdot\rangle$ is the usual L^2 inner product on $\Omega^1(X)$, and $\langle\cdot,\cdot\rangle^1_\kappa$ is defined in (3.16).

5. Gauge group and the isotropy subgroup

In order to extract anything mathematically meaningful out of this construction we will need to divide out the action of the gauge group $\mathcal{G}_P$ on $\mathcal{A}_P$. At this point we observe that the gauge group $\mathcal{G}_P \simeq \text{Maps}(X \to U(1))$ naturally descends to a "horizontal" action on $\bar{\mathcal{A}}_P$, which infinitesimally can be written as:

$$\theta \in \text{Lie}(\mathcal{G}_P) : A \mapsto A + d_H\theta \tag{5.1}$$

Following [**20**], we let I_A denote the isotropy subgroup of $\mathcal{G}_P$ at a point $A \in \bar{\mathcal{A}}_P$. Note that I_A can be canonically identified for every $A \in \bar{\mathcal{A}}_P$, and so we simply write I for the isotropy group. The condition for an element of the gauge group $h(x) = e^{i\theta(x)}$ to be in the isotropy group is that $d_H\theta = 0$, given definition 5.1 above. By [**17**, Prop. 12], we see that the condition $d_H\theta = 0$ implies that θ is harmonic, and so $\mathcal{L}_\xi\theta = 0$. Therefore we have $d\theta = 0$ since $d = d_H + \kappa \wedge \mathcal{L}_\xi$. Thus, the group I can be identified with the group of constant maps from X into $U(1)$; hence, is isomorphic to $U(1)$. We let $\text{Vol}(I)$ denote the volume of the isotropy subgroup, computed with respect to the metric induced from $\mathcal{G}_P$, so that

$$\text{Vol}(I) = \left[\int_X \kappa \wedge d\kappa\right]^{1/2} = \left[n + \sum_{j=1}^N \frac{\beta_j}{\alpha_j}\right]^{1/2} \tag{5.2}$$

where $[n; (\alpha_1, \beta_1), \ldots, (\alpha_N, \beta_N)]$ are the Seifert invariants of our Seifert manifold X. The last equality in (5.2) above follows from [**3**, (3.22)].

6. The partition function

We now have

$$
\begin{aligned}
Z_{U(1)}(X,p,k) &= \frac{e^{\pi i k S_{X,P}(A_P)}}{\mathrm{Vol}(\mathcal{G}_P)} \int_{\bar{\mathcal{A}}_P} \bar{\mathcal{D}}A \; e^{\left[\frac{ik}{4\pi} S(A)\right]} \\
&= \frac{\mathrm{Vol}(\mathcal{G}_P)}{\mathrm{Vol}(I)} \frac{e^{\pi i k S_{X,P}(A_P)}}{\mathrm{Vol}(\mathcal{G}_P)} \int_{\bar{\mathcal{A}}_P/\mathcal{G}_P} e^{\left[\frac{ik}{4\pi} S(A)\right]} \left[\det{}'(d_H^* d_H)\right]^{1/2} \mu \\
&= \frac{e^{\pi i k S_{X,P}(A_P)}}{\mathrm{Vol}(I)} \int_{\bar{\mathcal{A}}_P/\mathcal{G}_P} e^{\left[\frac{ik}{4\pi} S(A)\right]} \left[\det{}'(d_H^* d_H)\right]^{1/2} \mu \qquad (6.1)
\end{aligned}
$$

where μ is the induced measure on the quotient space $\bar{\mathcal{A}}_P/\mathcal{G}_P$ and $\det'$ denotes a regularized determinant to be defined later. Since $S(A) = \langle A, \star DA\rangle$ is quadratic in A, we may apply the method of stationary phase, [**19**], [**9**], to evaluate the oscillatory integral of (6.1) exactly. We obtain,

(6.2)

$$
Z_{U(1)}(X,p,k) = \frac{e^{\pi i k S_{X,P}(A_P)}}{\mathrm{Vol}(I)} \int_{\mathcal{M}_P} e^{\frac{\pi i}{4}\ \mathrm{sgn}(\star D)} \frac{\left[\det{}'(d_H^* d_H)\right]^{1/2}}{\left[\det{}'(k\star D)\right]^{1/2}} \nu
$$

where $\mathcal{M}_P$ denotes the moduli space of flat connections modulo the gauge group and ν denotes the induced measure on this space. Note that we have included a factor of k in our regularized determinant since this factor occurs in the exponent multiplying $S(A)$.

7. Zeta function determinants

We will use the following to define the regularized determinant of $k \star D$

PROPOSITION 7.1. [**20**, §2] *Let $\mathcal{H}_0$, $\mathcal{H}_1$ be Hilbert spaces, and $S: \mathcal{H}_1 \to \mathcal{H}_1$ and $T: \mathcal{H}_0 \to \mathcal{H}_1$ such that S^2 and TT^* have well defined zeta functions with discrete spectra and meromorphic extensions to $\mathbb{C}$ that are regular at 0 (with at most simple poles on some discrete subset). If $ST = 0$, and S^2 is self-adjoint, then*

$$\det{}'(S^2 + TT^*) = \det{}'(S^2)\det{}'(TT^*) \qquad (7.2)$$

PROOF. This equality follows from the facts that $S^2TT^* = 0$ and $TT^*S^2 = 0$ (i.e. these operators commute), which both follow from $ST = 0$ and the fact that S^2 and TT^* are both self-adjoint. □

Following the notation of [**20**, §2 (3)–(6)], we set the operators $S = k \star D$ and $T = k d_H d_H^*$ on $\Omega^1(H)$ and observe that $ST = 0$ since (4.1) is a complex. With Prop. 7.1 as *motivation*, we make the formal definition

$$\det{}'(k\star D) := C(k,J)\cdot \frac{[\det{}'(S^2 + TT^*)]^{1/2}}{[\det{}'(TT^*)]^{1/2}} \qquad (7.3)$$

where $S^2 + TT^* = k^2(D^*D + (d_H d_H^*)^2)$, $TT^* = k^2(d_H d_H^*)^2$ and

$$C(k,J) := k^{\left(-\frac{1}{1024}\int_X R^2\,\kappa\wedge d\kappa\right)} \qquad (7.4)$$

is a function of $R \in C^\infty(X)$, the Tanaka-Webster scalar curvature of X, which in turn depends only on a choice of a compatible complex structure $J \in \mathrm{End}(H)$. That is, given a contact form $\kappa \in \Omega^1(X)$, the choice of complex structure $J \in \mathrm{End}(H)$ determines uniquely an associated metric. We have defined $\det'(k\star D)$ in this way to eliminate the metric dependence that would otherwise occur in the k-dependence of this determinant. The motivation for the definition of the factor $C(k, J)$ comes explicitly from Prop. 7.14 below.

The operator

$$\Delta := D^* D + (d_H d_H^*)^2 \tag{7.5}$$

is actually equal to the middle degree Laplacian defined in [**18**, (10)] and has some nice analytic properties.[7] In particular, it is maximally hypoelliptic and invertible in the Heisenberg symbolic calculus, [**17**, Page 300]. We define the regularized determinant of Δ via its zeta function[8]

$$\zeta(\Delta)(s) := \sum_{\lambda \in \mathrm{spec}^*(\Delta)} \lambda^{-s}. \tag{7.6}$$

Note that our definition agrees with [**18**] up to a finite constant term, $\dim H^1(\mathcal{E}, d_H)$.[9] This is a consequence of the following

PROPOSITION 7.7. [**18**, Prop. 2.2] *Let (X, κ) be a contact three-manifold. The contact complex $(\mathcal{E}, d_H)$, defined in Def. 4.4 above, forms a resolution of the constant sheaf $\mathbb{R}$ and its cohomology therefore coincides with the de Rham cohomology of X. Moreover, the natural projection $\pi : \Omega^k(X) \to \mathcal{E}^k$,[10] for $k \leq 1$, and inclusion $i : \mathcal{E}^k \to \Omega^k(X)$, for $k \geq 2$, induce an isomorphism between the two cohomologies.*

Also, $\zeta(\Delta)(s)$ admits a meromorphic extension to $\mathbb{C}$ that is regular at $s = 0$, [**15**, §4].

DEFINITION 7.8. Let $\Delta_0 := (d_H^* d_H)^2$ on $\Omega^0(X)$, $\Delta_1 := \Delta$ on $\Omega^1(H)$ and define $\zeta_i(s) := \zeta(\Delta_i)(s)$. Define the regularized determinant of Δ_i as

$$\det{}'(\Delta_i) := e^{-\zeta'(\Delta_i)(0)}. \tag{7.9}$$

We claim the following

PROPOSITION 7.10. *For any real number $0 < c \in \mathbb{R}$,*

$$\det{}'(c\Delta_i) := c^{\zeta_i(0)} \det{}'(\Delta_i) \tag{7.11}$$

for $i = 0, 1$.

PROOF. To prove this claim, recall that $\zeta_i(s) = \zeta(\Delta_i)(s)$ for $i = 0, 1$, scale as follows:

$$\zeta(c\Delta_i)(s) = c^{-s}\zeta(\Delta_i)(s). \tag{7.12}$$

From here we simply calculate the scaling of the regularized determinants using the definition

$$\det{}'(\Delta_i) := e^{-\zeta'(\Delta_i)(0)} \tag{7.13}$$

and the claim is proven. □

The following will be useful.

---

[7] See (9.3) below for the full definition.

[8] We follow [**18**, Page 10] here.

[9] Recall Def. 4.4 for a definition of the complex $(\mathcal{E}, d_H)$. Also, it is noted in [**18**, Page 11] that $\dim H^1(\mathcal{E}, d_H)$ is finite by hypoellipticity.

[10] Define $\pi(\alpha) := \alpha - \iota_\xi(\alpha)\kappa$, for $\alpha \in \Omega^k(X)$, for $k \leq 1$.

PROPOSITION 7.14. *Let Δ_0, Δ_1, and $\zeta_i(s) := \zeta(\Delta_i)(s)$ be as defined in Def. 7.8. We have*

(7.15)
$$\zeta_0(0) - \zeta_1(0) = \left(-\frac{1}{512}\int_X R^2\, \kappa \wedge d\kappa\right) + \dim \operatorname{Ker} \Delta_1 - \dim \operatorname{Ker} \Delta_0$$

(7.16)
$$= \left(-\frac{1}{512}\int_X R^2\, \kappa \wedge d\kappa\right) + \dim H^1(\mathcal{E}, d_H) - \dim H^0(\mathcal{E}, d_H).$$

where $R \in C^\infty(X)$ is the Tanaka-Webster scalar curvature of X and $\kappa \in \Omega^1(X)$ is our chosen contact form as usual.

PROOF. Let

$$\hat{\zeta_0}(s) := \dim \operatorname{Ker} \Delta_0 + \zeta_0(s)$$
$$\hat{\zeta_1}(s) := \dim \operatorname{Ker} \Delta_1 + \zeta_1(s)$$

denote the zeta functions as defined in [**18**]. From [**18**, Cor. 3.8], one has that

$$\hat{\zeta_1}(0) = 2\hat{\zeta_0}(0)$$

for all three-dimensional contact manifolds. By [**4**, Theorem 8.8], one knows that on CR-Seifert manifolds that

$$\hat{\zeta_0}(0) = \hat{\zeta}(\Delta_0)(0) = \hat{\zeta}(\Delta_H)(0) = \frac{1}{512}\int_X R^2\, \kappa \wedge d\kappa,$$

where $\Delta_H := d_H d_H^* + d_H^* d_H$. Thus,

$$\hat{\zeta_1}(0) = \frac{1}{256}\int_X R^2\, \kappa \wedge d\kappa$$

By our definition of the zeta functions, which differ from that of [**18**] by constant dimensional terms, we therefore have

$$\zeta_0(0) = \frac{1}{512}\int_X R^2\, \kappa \wedge d\kappa - \dim \operatorname{Ker}\ \Delta_0$$
$$\zeta_1(0) = \frac{1}{256}\int_X R^2\, \kappa \wedge d\kappa - \dim \operatorname{Ker}\ \Delta_1$$

Hence,

$$\begin{aligned}\zeta_0(0) - \zeta_1(0) &= \left[\frac{1}{512}\int_X R^2\, \kappa \wedge d\kappa - \dim \operatorname{Ker}\ \Delta_0\right] \\ &\quad - \left[\frac{1}{256}\int_X R^2\, \kappa \wedge d\kappa - \dim \operatorname{Ker}\ \Delta_1\right] \\ &= \left(-\frac{1}{512}\int_X R^2\, \kappa \wedge d\kappa\right) + \dim \operatorname{Ker}\ \Delta_1 - \dim \operatorname{Ker} \Delta_0 \\ &= \left(-\frac{1}{512}\int_X R^2\, \kappa \wedge d\kappa\right) + \dim H^1(\mathcal{E}, d_H) - \dim H^0(\mathcal{E}, d_H).\end{aligned}$$

and the result is proven. □

We now have the following

PROPOSITION 7.17. *The term inside of the integral of* (6.2) *has the following expression in terms of the hypoelliptic Laplacians,* Δ_0 *and* Δ_1*, as defined in Def. 7.8:*

$$\frac{\left[\det'(d_H^* d_H)\right]^{1/2}}{\left[\det'(k \star D)\right]^{1/2}} = k^{n_X} \frac{[\det'(\Delta_0)]^{1/2}}{\left[\det'(\Delta_1)\right]^{1/4}} \tag{7.18}$$

where

$$n_X := \frac{1}{2}(\dim H^1(\mathcal{E}, d_H) - \dim H^0(\mathcal{E}, d_H)). \tag{7.19}$$

PROOF.

(7.20)

$$\frac{\left[\det'(d_H^* d_H)\right]^{1/2}}{\left[\det'(k \star_H D_\kappa^1)\right]^{1/2}} = C(k,J)^{-1} \cdot \frac{\left[\det'(d_H^* d_H)^2\right]^{1/4} \cdot \left[\det' k^2 (d_H d_H^*)^2\right]^{1/4}}{\left[\det'(k^2 \Delta)\right]^{1/4}}$$

$$= C(k,J)^{-1} \cdot \frac{k^{\zeta_0(0)/2}\left[\det'(\Delta_0)\right]^{1/4} \cdot \left[\det'(\Delta_0)\right]^{1/4}}{k^{\zeta_1(0)/2}\left[\det'(\Delta_1)\right]^{1/4}} \tag{7.21}$$

$$= C(k,J)^{-1} \cdot k^{\frac{1}{2}(\zeta_0(0)-\zeta_1(0))} \frac{[\det'(\Delta_0)]^{1/2}}{\left[\det'(\Delta_1)\right]^{1/4}}$$

$$= C(k,J)^{-1} \cdot C(k,J) \cdot k^{n_X} \frac{[\det'(\Delta_0)]^{1/2}}{\left[\det'(\Delta_1)\right]^{1/4}}, \text{ Prop. 7.14,}$$

$$= k^{n_X} \frac{[\det'(\Delta_0)]^{1/2}}{\left[\det'(\Delta_1)\right]^{1/4}}$$

where the second last line comes from (7.16). Also note that $d_H^* d_H$ and $d_H d_H^*$ have the same eigenvalues (by standard arguments), which allows us to proceed to (7.21) from (7.20). □

REMARK 7.22. Note that by Prop. 7.7, the definition of n_X (see (7.19)) here is exactly equal to the quantity $m_X := \frac{1}{2}(\dim H^1(X,d) - \dim H^0(X,d))$ of [**12**, (5.18)]. This shows that our partition function has the same k-dependence as the one in [**12**].

8. The eta invariant

Next we regularize the signature $\operatorname{sgn}(\star D)$ via the eta-invariant and set $\operatorname{sgn}(\star D) = \eta(\star D)(0) := \eta(\star D)$ where

$$\eta(\star D)(s) := \sum_{\lambda \in \operatorname{spec}^*(\star D)} (\operatorname{sgn} \lambda)|\lambda|^{-s} \tag{8.1}$$

Finally, we may now write the result for our partition function

(8.2)

$$Z_{U(1)}(X,p,k) = k^{n_X} e^{\pi i k S_{X,P}(A_P)} e^{\frac{\pi i}{4}\eta(\star D)} \int_{\mathcal{M}_P} \frac{1}{\operatorname{Vol}(I)} \frac{[\det'(\Delta_0)]^{1/2}}{\left[\det'(\Delta_1)\right]^{1/4}}\, \nu$$

where $n_X := \frac{1}{2}(\dim H^1(\mathcal{E}, d_H) - \dim H^0(\mathcal{E}, d_H))$. Note that ν is a measure on $\mathcal{M}_P$ (the moduli space of flat connections modulo the gauge group) relative to the horizontal structure on the tangent space of $\mathcal{M}_P$.

9. Torsion

Now we will study the quantity $\frac{1}{\mathrm{Vol}(I)} \frac{[\det'(\Delta_0)]^{1/2}}{[\det'(\Delta_1)]^{1/4}} \nu$ inside of the integral in (8.2), and in particular how it is related to the analytic contact torsion T_C. First, recall that, [**18**, (16)],

$$T_C := \exp\left(\frac{1}{4}\sum_{q=0}^{3}(-1)^q w(q)\zeta'(\Delta_q)(0)\right), \tag{9.1}$$

where,

$$w(q) = \begin{cases} q & \text{if } q \leq 1, \\ q+1 & \text{if } q > 1, \end{cases} \tag{9.2}$$

in the case where $\dim(X) = 3$. Note that we have chosen a sign convention that leads to the inverse of the definition of T_C in [**18**]. Recall [**18**, (10)],

$$\Delta_q = \begin{cases} (d_H^* d_H + d_H d_H^*)^2, & q = 0, 3, \\ D^*D + (d_H d_H^*)^2, & q = 1, \\ DD^* + (d_H^* d_H)^2, & q = 2. \end{cases} \tag{9.3}$$

We would, however, like to work with torsion when viewed as a density on the determinant line

$$\begin{aligned} |\det H^\bullet(\mathcal{E}, d_H)^*| \quad := \quad & |\det H^0(\mathcal{E}, d_H)| \otimes |\det H^1(\mathcal{E}, d_H)^*| \otimes \\ & \otimes |\det H^2(\mathcal{E}, d_H)| \otimes |\det H^3(\mathcal{E}, d_H)^*| \end{aligned}$$

We follow [**16**] and [**12**] and make the analogous definition.

DEFINITION 9.4. Define the analytic torsion as a density as follows

$$T_C^d := T_C \cdot \delta_{|\det H^\bullet(X,d_H)|}$$

where T_C is as defined in (9.1), and

$$\delta_{|\det H^\bullet(X,d_H)|} := \otimes_{q=0}^{\dim X} |\nu_1^q \wedge \cdots \wedge \nu_{b_q}^q|^{(-1)^q}$$

where $\{\nu_1^q, \cdots, \nu_{b_q}^q\}$ is an orthonormal basis for the space of harmonic contact forms $\mathcal{H}^q(X, d_H)$ with the inner product defined in (3.16). Note that $\mathcal{H}^q(X, d_H)$ is canonically identified with the cohomology space $H^q(X, d_H)$, and $b_q := \dim(H^q(X, d_H))$ is the qth contact Betti number.

Let

$$\nu^{(q)} := \nu_1^q \wedge \cdots \wedge \nu_{b_q}^q$$

and write the analytic torsion of a compact connected Seifert three-manifold X as

$$T_C^d = T_C \times |\nu^{(0)}| \otimes |\nu^{(1)}|^{-1} \otimes |\nu^{(2)}| \otimes |\nu^{(3)}|^{-1}. \tag{9.5}$$

In terms of regularized determinants, we have

$$T_C = \left[(\det'(\Delta_0))^0 \cdot (\det'(\Delta_1))^1 \cdot (\det'(\Delta_2))^{-3} \cdot (\det'(\Delta_3))^4\right]^{1/4} \tag{9.6}$$

where Δ_q, $0 \leq q \leq 3$, denotes the Laplacians on the contact complex as defined in [**18**, (10)] and recalled in (9.3) above. This notation agrees with our notation for

Δ_0, Δ_1 as in (7.11). The Hodge $\star$-operator induces the equivalences $\Delta_q \simeq \Delta_{3-q}$, [**18**, Theorem 3.4], and allows us to write

$$T_C = \left[(\det{}'(\Delta_0))^0 \cdot (\det{}'(\Delta_1))^1 \cdot (\det{}'(\Delta_2))^{-3} \cdot (\det{}'(\Delta_3))^4\right]^{1/4} \tag{9.7}$$

$$= \frac{\det{}'(\Delta_0)}{(\det{}'(\Delta_1))^{1/2}} \tag{9.8}$$

Also, from the isomorphisms $H^q(X,\mathbb{R}) \simeq H^q(X,d_H)$ of [**18**, Prop. 2.2], we have Poincaré duality $H^q(X,d_H) \simeq H^{3-q}(X,d_H)^*$, and therefore

$$T_C^d = T_C \times |\nu^0|^{\otimes 2} \otimes (|\nu^1|^{-1})^{\otimes 2} \tag{9.9}$$

Moreover, by [**17**, Prop. 12], $\mathcal{H}^q(X,d_H) = \mathcal{H}^q(X,\mathbb{R})$, and thus any orthonormal basis $\nu^{(0)}$ of $\mathcal{H}^0(X,d_H) \simeq \mathbb{R}$ is a constant such that

$$|\nu^{(0)}| = \left[\int_X \kappa \wedge d\kappa\right]^{-1/2} \tag{9.10}$$

Also, recall that the tangent space $T_A\mathcal{M}_P \simeq H^1(\mathcal{E},d_H) \simeq H^1(X,\mathbb{R})$, at any point $A \in \mathcal{M}_P$. The measure ν on $\mathcal{M}_P$ that occurs in (8.2) is defined relative to the metric on $H^1(\mathcal{E},d_H) \simeq \mathcal{H}^1(\mathcal{E},d_H)$, which can be identified with the usual L^2-metric on forms. Thus the measure ν may be identified with the inverse of the density $|\nu^{(1)}|$ by dualizing the orthogonal basis $\{\nu_1^1,\ldots,\nu_{b_1}^1\}$ for $\mathcal{H}^1(X,d_H)$; i.e.

$$\nu = |\nu^{(1)}|^{-1} = |\nu_1^1 \wedge \cdots \wedge \nu_{b_1}^1|^{-1} \tag{9.11}$$

Putting together Equations 9.8, 9.10, 9.11 into Equation 9.9, we have

$$T_C^d = T_C \times |\nu^0|^{\otimes 2} \otimes (|\nu^1|^{-1})^{\otimes 2} \tag{9.12}$$

$$= \frac{\det{}'(\Delta_0)}{(\det{}'(\Delta_1))^{1/2}} \cdot \left[\int_X \kappa \wedge d\kappa\right]^{-1} \nu^{\otimes 2} \tag{9.13}$$

$$= \mathrm{Vol}(I)^{-2} \frac{\det{}'(\Delta_0)}{(\det{}'(\Delta_1))^{1/2}} \cdot \nu^{\otimes 2} \tag{9.14}$$

We have thus proven the following,

PROPOSITION 9.15. *The contact analytic torsion, when viewed as a density T_C^d as in definition 9.4, can be identified as follows:*

$$(T_C^d)^{1/2} = \frac{1}{\mathrm{Vol}(I)} \frac{[\det{}'(\Delta_0)]^{1/2}}{[\det{}'(\Delta_1)]^{1/4}}\, \nu \tag{9.16}$$

Our partition function is now

$$\bar{Z}_{U(1)}(X,p,k) = k^{n_X} e^{\pi i k S_{X,P}(A_P)} e^{\frac{\pi i}{4}\eta(\star D)} \int_{\mathcal{M}_P} (T_C^d)^{1/2} \tag{9.17}$$

This partition function should be identical to the partition function defined in [**12**, (7.27)],

$$Z_{U(1)}(X,p,k) = k^{m_X} e^{\pi i k S_{X,P}(A_P)} e^{\frac{\pi i}{4}\eta(\star d)} \int_{\mathcal{M}_P} (T_{RS}^d)^{1/2}. \tag{9.18}$$

Our goal in the remainder is to show that this is indeed the case. Our first observation is that $(T_C^d)^{1/2}$ is equal to the Ray-Singer torsion $(T_{RS}^d)^{1/2}$ that occurs in [**12**, (7.27)]. This follows directly from [**18**, Theorem 4.2]; note that their sign convention makes T_C the inverse of our definition.

10. Regularizing the eta-invariants

Since we have seen that our k-dependence matches that in [**12**] (i.e. $m_X = n_X$; cf. Remark 7.22), the only thing left to do is to reconcile the eta invariants, $\eta(\star D)$ and $\eta(\star d)$. As observed in [**21**], the correct quantity to compare our eta invariant to would be

$$\frac{\eta(\star d)}{4} + \frac{1}{12}\frac{\mathrm{CS}(A^g)}{2\pi}. \tag{10.1}$$

where,

$$\mathrm{CS}(A^g) = \frac{1}{4\pi}\int_X \mathrm{Tr}(A^g \wedge dA^g + \frac{2}{3}A^g \wedge A^g \wedge A^g) \tag{10.2}$$

is the gravitational Chern-Simons term, with A^g the Levi-Civita connection on the spin bundle of X for a given metric g on X. See Appendix B for a short exposition on the regularization of $\eta(\star d)$ in (10.1). It was noticed in [**21**] that in the quasi-classical limit, quantum anomalies can occur that can break topological invariance. Invariance may be restored in this case only after adding a counter-term to the eta invariant. Our job then is to perform a similar analysis for the eta invariant $\eta(\star D)$, which depends on a choice of metric. Of course, our choice of metric is natural in this setting and is adapted to the contact structure. One possible approach is to consider variations over the space of such natural metrics and calculate the corresponding variation of the eta invariant, giving us a local formula for the counter-term that needs to be added. Such a program has already been initiated in [**4**].

Our starting point is the conjectured equivalence that results from the identification of (9.17) and (9.18):

$$e^{\pi i\left[\frac{\eta(\star d)}{4} + \frac{1}{12}\frac{\mathrm{CS}(A^g)}{2\pi}\right]} \text{ “}=\text{” } e^{\frac{\pi i}{4}[\eta(\star D)+C_T]} \tag{10.3}$$

where C_T is some appropriate counter-term that yields an invariant comparable to the left hand of this equation. As noted in Appendix B, the left hand side of this equation depends on a choice of 2-framing on X, and since we have a rule as in (B.9) for how the partition function transforms when the framing is twisted, we basically have a topological invariant. Alternatively, as also noted in Appendix B, one can use the main result of [**1**] and fix the canonical 2-framing on $TX \oplus TX$. We therefore expect the same type of phenomenon for the right hand side of this equation, having at most a $\mathbb{Z}$-dependence on the regularization of our eta invariant, along with a rule that tells us how the partition function changes when our discrete invariants are "twisted," once again yielding a topological invariant.

Our regularization procedure for $\eta(\star D)$ will be quite different than that used for $\eta(\star d)$. Since we are restricted to a class of metrics that are compatible with our contact structure, we are really only concerned with finding appropriate counter-terms for $\eta(\star D)$ that will eliminate our dependence on the choice of contact form κ and complex structure $J \in \mathrm{End}(H)$. In the case of interest, we observe that our regularization may be obtained in one stroke by introducing the *renormalized eta-invariant*, $\eta_0(X,\kappa)$, of X that is discussed in [**4**, §3]. Before giving the definition of $\eta_0(X,\kappa)$, we require the following

LEMMA 10.4. [**4**, Lemma 3.1] *Let (X, J, κ) be a strictly pseudoconvex pseudohermitian three-manifold. The eta-invariants of the family of metrics* $g_\epsilon :=$

$\epsilon^{-1}\kappa \otimes \kappa + d\kappa(\cdot, J\cdot)$ *have a decomposition in homogeneous terms:*

$$\eta(g_\epsilon) = \sum_{i=-2}^{2} \eta_i(X,\kappa)\epsilon^i. \tag{10.5}$$

The terms η_i *for* $i \neq 0$ *are integrals of local pseudohermitian invariants of* (X, κ), *and the* η_i *for* $i > 0$ *vanish when the Tanaka-Webster torsion,* τ, *vanishes.*

We then make the following

DEFINITION 10.6. Let (X, κ) be a compact strictly pseudoconvex pseudohermitian three-dimensional manifold. The *renormalized eta-invariant* $\eta_0(X, \kappa)$ of (X, κ) is the constant term in the expansion of (10.5) for the eta-invariants of the family of metrics $g_\epsilon := \epsilon^{-1}\kappa \otimes \kappa + d\kappa(\cdot, J\cdot)$.

Our assumption that X is K-contact ensures that the Reeb flow preserves the metric. In this situation it is known that the Tanaka-Webster torsion necessarily vanishes, [**4**, §3]. In the case where the torsion of (X, κ) vanishes, the terms $\eta_i(X, \kappa)$ in (10.5) vanish for $i > 0$, so that when $\epsilon \to \infty$, one has

$$\eta_0(X,\kappa) = \lim_{\epsilon\to\infty} \eta(g_\epsilon) := \eta_{ad} \tag{10.7}$$

The limit η_{ad} is known as the *adiabatic limit* and has been studied in [**5**] and [**8**], for example. The adiabatic limit is the case where the limit is taken as ϵ goes to infinity,

$$\eta_{ad} := \lim_{\epsilon\to\infty} \eta(g_\epsilon), \tag{10.8}$$

while the the renormalized eta-invariant, $\eta_0(X, \kappa)$, is naturally interpreted as the constant term in the asymptotic expansion for $\eta(g_\epsilon)$ in powers of ϵ, when ϵ goes to 0. This reverse process of taking ϵ to 0 is also known as the *diabatic* limit. When torsion vanishes (i.e. when the Reeb flow preserves the metric), (10.7) is the statement that the diabatic and adiabatic limits agree. One of the main challenges for our future work will be to extend beyond the case where torsion vanishes. This will naturally involve the study of the diabatic limit. For now, we are restricted to the case of vanishing torsion. In this case, the main result that we will use is the following

THEOREM 10.9. [**4**, Theorem 1.4] *Let* X *be a compact CR-Seifert three-manifold, with* $U(1)$*-action generated by the Reeb field of an* $U(1)$*-invariant contact form* κ. *If* R *is the Tanaka-Webster curvature of* (X, κ) *and* D *is the middle degree operator of the contact complex,*[11] *then*

$$\eta_0(X,\kappa) = \eta(\star D) + \frac{1}{512}\int_X R^2\ \kappa \wedge d\kappa. \tag{10.10}$$

Theorem 10.9 compels us to conjecture that $C_T \sim \left(\frac{1}{512}\int_X R^2\ \kappa \wedge d\kappa\right) \mod 8\mathbb{Z}$. Our motivation for this comes from the fact that $\eta_0(X, \kappa)$ is a topological invariant in our case. We have the following,

[11] See (4.1) and (4.13).

THEOREM 10.11. [**4**, Remark 9.6 and (27)] *If X is a CR-Seifert manifold, then $\eta_0(X,\kappa)$ is a topological invariant and*

$$\eta_0(X,\kappa) = 1 - \frac{d}{3} + 4\sum_{j=1}^{N} s(\alpha_j,\beta_j), \tag{10.12}$$

where $d \in \mathbb{Q}$ is the degree of X as a compact $U(1)$-orbifold bundle and

$$s(\alpha,\beta) := \frac{1}{4\alpha}\sum_{k=1}^{\alpha-1} cot\left(\frac{\pi k}{\alpha}\right) cot\left(\frac{\pi k\beta}{\alpha}\right) \tag{10.13}$$

is the classical Rademacher-Dedekind sum, where $[n;(\alpha_1,\beta_1),\ldots,(\alpha_N,\beta_N)]$ (for $(\alpha_i,\beta_i)=1$ relatively prime) are the Seifert invariants of X.

REMARK 10.14. Note that we have chosen the opposite sign convention of [**4**] for the orbifold first Chern number $d = c_1(X)$ in this article. Thus, d appears with a plus sign in Theorem 10.11 in [**4**].

Thus, we are led to consider the natural topological invariant $e^{\frac{\pi i}{4}[\eta_0(X,\kappa)]}$ and how it compares with the topological invariant $e^{\pi i\left[\frac{\eta(\star d)}{4}+\frac{1}{12}\frac{\mathrm{CS}(A^g)}{2\pi}\right]}$. We consider the limit

$$\lim_{\epsilon\to\infty} e^{\pi i\left[\frac{\eta(\star_\epsilon d)}{4}+\frac{1}{12}\frac{\mathrm{CS}(A^{g_\epsilon})}{2\pi}\right]} \tag{10.15}$$

where $g_\epsilon = \epsilon^{-1}\kappa\otimes\kappa + d\kappa(\cdot, J\cdot)$ is the natural metric associated to X. On the one hand, since this is a topological invariant, and is independent of the metric, we must have

$$\lim_{\epsilon\to\infty} e^{\pi i\left[\frac{\eta(\star_\epsilon d)}{4}+\frac{1}{12}\frac{\mathrm{CS}(A^{g_\epsilon})}{2\pi}\right]} = e^{\pi i\left[\frac{\eta(\star d)}{4}+\frac{1}{12}\frac{\mathrm{CS}(A^{g})}{2\pi}\right]}. \tag{10.16}$$

where we take $g_1 := g$ so that $\star_{g_1} := \star$.

On the other hand, since $\eta(g_\epsilon) = \eta(\star_\epsilon d)$ by definition, and we know that its limit exists as $\epsilon \to \infty$ (in fact $\eta_0(X,\kappa) = \lim_{\epsilon\to\infty}\eta(g_\epsilon)$), we have

$$\lim_{\epsilon\to\infty} e^{\pi i\left[\frac{\eta(\star_\epsilon d)}{4}+\frac{1}{12}\frac{\mathrm{CS}(A^{g_\epsilon})}{2\pi}\right]} = e^{\pi i\left[\frac{\eta_0(X,\kappa)}{4}+\left\{\lim_{\epsilon\to\infty}\frac{1}{12}\frac{\mathrm{CS}(A^{g_\epsilon})}{2\pi}\right\}\right]}. \tag{10.17}$$

Thus, we have

$$e^{\pi i\left[\frac{\eta(\star d)}{4}+\frac{1}{12}\frac{\mathrm{CS}(A^{g})}{2\pi}\right]} = e^{\pi i\left[\frac{\eta_0(X,\kappa)}{4}\right]} e^{\pi i\left\{\lim_{\epsilon\to\infty}\frac{1}{12}\frac{\mathrm{CS}(A^{g_\epsilon})}{2\pi}\right\}}. \tag{10.18}$$

We therefore see that if we can understand the limit $\lim_{\epsilon\to\infty}\frac{1}{12}\frac{\mathrm{CS}(A^{g_\epsilon})}{2\pi}$, we will obtain crucial information for our problem. The following has been established using a "Kaluza-Klein" dimensional reduction technique modeled after the paper [**10**],

PROPOSITION 10.19. [**14**] *Let (X,ϕ,ξ,κ,g) be a closed quasi-regular K-contact three-manifold,*

$$\begin{array}{ccc} U(1) & \hookrightarrow & X \\ & & \downarrow \\ & & \Sigma \end{array}$$

Let $g_\epsilon := \epsilon^{-1}\kappa \otimes \kappa + \pi^* h$. *After choosing a framing for* $TX \oplus TX$, *corresponding to a choice of Vielbein, then,*

$$CS(A^{g_\epsilon}) = \left(\frac{\epsilon^{-1}}{2}\right)\int_\Sigma r\,\omega + \left(\frac{\epsilon^{-2}}{2}\right)\int_\Sigma f^2\,\omega \tag{10.20}$$

where $r \in C^\infty_{orb}(\Sigma)$ *is the (orbifold) scalar curvature of* (Σ, h), $\omega \in \Omega^2_{orb}(\Sigma)$ *is the (orbifold) Hodge form of* (Σ, h), *and* $f := \star_h \omega$. *In particular, the adiabatic limit of* $CS(A^{g_\epsilon})$ *vanishes:*

$$\lim_{\epsilon \to \infty} CS(A^{g_\epsilon}) = 0. \tag{10.21}$$

Proposition 10.19 combined with (10.18) and Theorem 10.9 gives us the following

PROPOSITION 10.22. *Let* $(X, \phi, \xi, \kappa, g)$ *be a closed, quasi-regular K-contact three-manifold. There exists counter-term,* C_T, *such that* $e^{\frac{\pi i}{4}[\eta(\star_H D) + C_T]}$ *is a topological invariant that is identically equal to the topological invariant* $e^{\pi i\left[\frac{\eta(\star d)}{4} + \frac{1}{12}\frac{CS(A^g)}{2\pi}\right]}$. *In fact, after choosing the canonical framing for* $TX \oplus TX$ *corresponding to a choice of Vielbein, we have*

$$C_T = \frac{1}{512}\int_X R^2\, \kappa \wedge d\kappa,$$

where $R \in C^\infty(X)$ *is the Tanaka-Webster scalar curvature of* X, *and*

$$\eta(\star d) + \frac{1}{3}\frac{CS(A^g)}{2\pi} = \eta(\star D) + \frac{1}{512}\int_X R^2\, \kappa \wedge d\kappa.$$

Given Proposition 10.22 and Theorem 10.11, we conclude the following as an immediate consequence,

PROPOSITION 10.23. *Let* $(X, \phi, \xi, \kappa, g)$ *closed, quasi-regular K-contact three-manifold. Then, after choosing the canonical framing for* $TX \oplus TX$ *corresponding to a choice of Vielbein,*

$$\begin{aligned} \eta(\star d) + \frac{1}{3}\frac{CS(A^g)}{2\pi} &= \eta(\star D) + \frac{1}{512}\int_X R^2\, \kappa \wedge d\kappa \\ &= 1 - \frac{d}{3} + 4\sum_{j=1}^N s(\alpha_j, \beta_j), \end{aligned}$$

where $d = c_1(X) = n + \sum_{j=1}^N \frac{\beta_j}{\alpha_j} \in \mathbb{Q}$ *and*

$$s(\alpha, \beta) := \frac{1}{4\alpha}\sum_{k=1}^{\alpha-1} cot\left(\frac{\pi k}{\alpha}\right) cot\left(\frac{\pi k \beta}{\alpha}\right) \in \mathbb{Q}$$

is the classical Rademacher-Dedekind sum, where $[n; (\alpha_1, \beta_1), \ldots, (\alpha_N, \beta_N)]$ *(for* $gcd(\alpha_j, \beta_j) = 1$*) are the Seifert invariants of* X. *In particular, we have computed the* $U(1)$ *Chern-Simons partition function as:*

$$\begin{aligned} Z_{U(1)}(X, p, k) &= k^{n_X} e^{\pi i k S_{X,P}(A_P)} e^{\frac{\pi i}{4}\left(1 - \frac{d}{3} + 4\sum_{j=1}^N s(\alpha_j, \beta_j)\right)} \int_{\mathcal{M}_P} (T^d_C)^{1/2}, \\ &= k^{m_X} e^{\pi i k S_{X,P}(A_P)} e^{\frac{\pi i}{4}\left(1 - \frac{d}{3} + 4\sum_{j=1}^N s(\alpha_j, \beta_j)\right)} \int_{\mathcal{M}_P} (T^d_{RS})^{1/2}. \end{aligned}$$

Appendix A. Basic construction of $U(1)$ Chern-Simons theory

Let X be a closed oriented 3-manifold. For any $U(1)$-connection $A \in \mathcal{A}_P$, [**12**] defines a $SU(2)$-connection $\hat{A}$ on an associated principal $SU(2)$-bundle $\hat{P} = P \times_{U(1)} SU(2)$, via

$$\hat{A}|_{[p,g]} = Ad_{g^{-1}}(\rho_* pr_1^* A|_p) + pr_2^* \vartheta_g$$

where $\rho : U(1) \to SU(2)$ is the diagonal inclusion, $pr_1 : P \times SU(2) \to P$ and $pr_2 : P \times SU(2) \to SU(2)$. Since for any 3-manifold X, $\hat{P}$ is trivializable, let $\hat{s} : X \to \hat{P}$ be a global section. The definition we use for the Chern-Simons action is as follows:

Definition A.1. The Chern-Simons action functional of a $U(1)$-connection $A \in \mathcal{A}_P$ is defined by:

$$S_{X,P}(A) = \int_X \hat{s}^* \alpha(\hat{A}) \ (mod\ \mathbb{Z}) \tag{A.2}$$

where $\alpha(\hat{A}) \in \Omega^3(\hat{P}, \mathbb{R})$ is the Chern-Simons form of the induced $SU(2)$-connection $\hat{A} \in \mathcal{A}_{\hat{P}}$,

$$\alpha(\hat{A}) = \mathrm{Tr}(\hat{A} \wedge F_{\hat{A}}) - \frac{1}{6} \mathrm{Tr}(\hat{A} \wedge [\hat{A}, \hat{A}]) \tag{A.3}$$

We then define the partition function for $U(1)$ Chern-Simons theory as (as in [**12**], [**13**]):

$$Z_{U(1)}(X, k) = \sum_{p \in \mathrm{Tors} H^2(X;\mathbb{Z})} Z_{U(1)}(X, p, k) \tag{A.4}$$

where,

$$Z_{U(1)}(X, p, k) = \frac{1}{\mathrm{Vol}(\mathcal{G}_P)} \int_{\mathcal{A}_P} \mathcal{D}A e^{\pi i k S_{X,P}(A)} \tag{A.5}$$

and

$$S_{X,P}(A) = \int_X \hat{s}^* \alpha(\hat{A}) \tag{A.6}$$

Then for any principal $U(1)$-bundle P we follow [**3**] and define a new action

$$S_{X,P}(A, \Phi) := S_{X,P}(A - \kappa\Phi) \tag{A.7}$$

where we may view $\Phi \in \Omega^0(X)$ and,

$$S_{X,P}(A, \Phi) = \int_X \alpha(\widehat{A - \kappa\Phi}) \tag{A.8}$$

$$= \int_X \alpha(\hat{A} - \kappa\hat{\Phi}) \tag{A.9}$$

$$= S_{X,P}(A) - \int_X [2\kappa \wedge Tr(\hat{\Phi} \wedge F_{\hat{A}}) - \kappa \wedge d\kappa\ Tr(\hat{\Phi}^2)] \tag{A.10}$$

where the second equality follows from the definition of $\hat{A}$ and $\hat{\Phi}$ (where $\hat{\Phi}|_{[p,g]} = Ad_{g^{-1}}(\rho_* pr_1^* \Phi|_p)$) on $\hat{P} = P \times_{U(1)} SU(2)$. The third equality follows from [**3**, (3.6)]. We then define a new partition function

$$\bar{Z}_{U(1)}(X, p, k) := \frac{1}{\mathrm{Vol}(\mathcal{S})} \frac{1}{\mathrm{Vol}(\mathcal{G}_P)} \int_{\mathcal{A}_P} DA\, D\Phi\ e^{\pi i k S_{X,P}(A,\Phi)} \tag{A.11}$$

where $D\Phi$ is defined by the invariant, positive definite quadratic form,

$$(\Phi,\Phi) = -\frac{1}{4\pi^2}\int_X \Phi^2 \kappa\wedge d\kappa \tag{A.12}$$

As observed in [**3**], our new partition function is identically equal to our original partition function defined for $U(1)$ Chern-Simons theory. On the one hand, we can fix $\Phi = 0$ above using the shift symmetry, $\delta\Phi = \sigma$, which will cancel the pre-factor $\mathrm{Vol}(\mathcal{S})$ from the resulting group integral over S and yield exactly our original partition function:

$$Z_{U(1)}(X,p,k) = \frac{1}{\mathrm{Vol}(\mathcal{G}_P)}\int_{\mathcal{A}_P} \mathcal{D}A\; e^{\pi i k S_{X,P}(A)}$$

Thus, we obtain the heuristic result,

$$\bar{Z}_{U(1)}(X,p,k) = Z_{U(1)}(X,p,k). \tag{A.13}$$

On the other hand, we obtain another description of $\bar{Z}_{U(1)}(X,p,k)$ by integrating Φ out. We will briefly review this computation here. Our starting point is the formula for the shifted partition function

$$\bar{Z}_{U(1)}(X,p,k) = \frac{1}{\mathrm{Vol}(\mathcal{S})}\frac{1}{\mathrm{Vol}(\mathcal{G}_P)}\int_{\mathcal{A}_P} DA\,D\Phi\; e^{\pi i k S_{X,P}(A,\Phi)} \tag{A.14}$$

where

$$S_{X,P}(A,\Phi) = S_{X,P}(A) - \int_X [2\kappa\wedge Tr(\hat{\Phi}\wedge F_{\hat{A}}) - \kappa\wedge d\kappa\; Tr(\hat{\Phi}^2)] \tag{A.15}$$

We formally complete the square with respect to $\hat{\Phi}$ as follows:

$$\begin{aligned}
\int_X [\kappa\wedge d\kappa\; Tr(\hat{\Phi}^2) \;&-\; 2\kappa\wedge Tr(\hat{\Phi}\wedge F_{\hat{A}})] \\
&= \int_X \left[Tr(\hat{\Phi}^2) - \frac{2\kappa\wedge Tr(\hat{\Phi}\wedge F_{\hat{A}})}{\kappa\wedge d\kappa}\right]\kappa\wedge d\kappa \\
&= \int_X Tr\left(\hat{\Phi}^2 - \frac{2\kappa\wedge F_{\hat{A}}}{\kappa\wedge d\kappa}\hat{\Phi}\right)\kappa\wedge d\kappa \\
&= \int_X Tr\left(\left[\hat{\Phi} - \frac{\kappa\wedge F_{\hat{A}}}{\kappa\wedge d\kappa}\right]^2 - \left[\frac{\kappa\wedge F_{\hat{A}}}{\kappa\wedge d\kappa}\right]^2\right)\kappa\wedge d\kappa
\end{aligned}$$

We then only need to compute the Gaussian

$$\begin{aligned}
\int D\Phi \quad \exp &\left[\pi i k\int_X \mathrm{Tr}\left(\left[\hat{\Phi} - \frac{\kappa\wedge F_{\hat{A}}}{\kappa\wedge d\kappa}\right]^2\right)\kappa\wedge d\kappa\right] \\
&= \int D\Phi\,\exp\left[\pi i k\int_X Tr(\hat{\Phi}^2)\kappa\wedge d\kappa\right] \\
&= \int D\Phi\,\exp\left[\frac{ik}{4\pi}\int_X \Phi^2\kappa\wedge d\kappa\right] \\
&= \int D\Phi\,\exp\left[-\frac{1}{2}\,(\Phi, A\Phi)\right]
\end{aligned}$$

where we take $A = 2\pi i k \mathbb{I}$ acting on the space of fields Φ and the inner product (Φ, Φ) is defined as in (A.12). We then formally get

$$\int D\Phi \exp\left[-\frac{1}{2}(\Phi, A\Phi)\right] = \sqrt{\frac{(2\pi)^{\Delta\mathcal{G}}}{\det A}} \tag{A.16}$$

$$= \left(\frac{-i}{k}\right)^{\Delta\mathcal{G}/2} \tag{A.17}$$

where the quantity $\Delta\mathcal{G}$ is formally the dimension of the gauge group $\mathcal{G}$. Note that we have abused notation slightly throughout by writing $\frac{1}{\kappa\wedge d\kappa}$. We have done this with the understanding that since $\kappa \wedge d\kappa$ is non-vanishing, then $\kappa \wedge F_{\hat{A}} = \phi\, \kappa \wedge d\kappa$ for some function $\phi \in \Omega^0(X)$, and we identify $\frac{\kappa\wedge F_{\hat{A}}}{\kappa\wedge d\kappa} := \phi$.

Our new description of the partition function is now,

$$\bar{Z}_{U(1)}(X,p,k) = C \int_{\mathcal{A}_P} DA \ \exp\left[\pi i k \left(S_{X,P}(A) - \int_X \frac{Tr[(\kappa\wedge F_{\hat{A}})^2]}{\kappa\wedge d\kappa}\right)\right] \tag{A.18}$$

where $C = \frac{1}{\mathrm{Vol}(\mathcal{S})}\frac{1}{\mathrm{Vol}(\mathcal{G}_P)}\left(\frac{-i}{k}\right)^{\Delta\mathcal{G}/2}$. We may rewrite this partition function after choosing a flat base point A_P in $\mathcal{A}_P$ so that $F_{A_P} = 0$ and identifying $\mathcal{A}_P = A_P + \Omega^1(X)$. We then obtain

$$\bar{Z}_{U(1)}(X,p,k) = C_1 \int_{\mathcal{A}_P} DA \ exp\left[\frac{ik}{4\pi}\left(\int_X A\wedge dA - \int_X \frac{(\kappa\wedge dA)^2}{\kappa\wedge d\kappa}\right)\right] \tag{A.19}$$

where

$$C_1 = \frac{e^{\pi i k S_{X,P}(A_P)}}{\mathrm{Vol}(\mathcal{S})\,\mathrm{Vol}(\mathcal{G}_P)}\left(\frac{-i}{k}\right)^{\Delta\mathcal{G}/2}.$$

We may further simplify (A.19) by reducing $\mathcal{A}_P$ to its quotient under the shift symmetry $\bar{\mathcal{A}}_P := \mathcal{A}_P/\mathcal{S}$, effectively canceling the factor of $\mathrm{Vol}(\mathcal{S})$ out front of the integral. We obtain:

$$\bar{Z}_{U(1)}(X,p,k) = C_2 \int_{\bar{\mathcal{A}}_P} \bar{D}A \ \exp\left[\frac{ik}{4\pi}\left(\int_X A\wedge dA - \int_X \frac{(\kappa\wedge dA)^2}{\kappa\wedge d\kappa}\right)\right] \tag{A.20}$$

where $C_2 = C_1\,\mathrm{Vol}(S)$.

Note that we are justified in excluding the factor $\left(\frac{-i}{k}\right)^{\Delta\mathcal{G}/2}$ from Eq. A.20 since we may redefine the partition function to cancel this factor.

Appendix B. Framing dependence and the gravitational Chern-Simons term

As observed in [**12**, (7.17)], (1.7) can also be rigorously defined by setting

$$Z_{U(1)}(X,p,k) = \frac{e^{\pi i k S_{X,P}(A_P)}}{\mathrm{Vol}\,U(1)} \int_{\mathcal{M}_P} e^{\frac{\pi i}{4}\,\mathrm{sgn}(\star d)}\,\frac{[\det'(d^*d)]^{1/2}}{[\det'(k\star d)]^{1/2}}\nu \tag{B.1}$$

where ν is the metric induced on the moduli space of flat connections on P, $\mathcal{M}_P$. This last expression has rigorous mathematical meaning if the determinants and signatures of the operators are regularized. The signature of the operator $\star d$ on $\Omega^1(X;\mathbb{R})$ is regularized via the eta invariant, so that $\mathrm{sgn}(\star d) = \eta(\star d) + \frac{1}{3}\frac{\mathrm{CS}(A^g)}{2\pi}$, where

$$\eta(\star d) = \lim_{s\to 0}\sum_{\lambda_j\neq 0} \mathrm{sign}\,\lambda_j|\lambda_j|^{-s} \tag{B.2}$$

and λ_j are the eigenvalues of $\star d$, and

$$\text{CS}(A^g) = \frac{1}{4\pi}\int_X \text{Tr}(A^g \wedge dA^g + \frac{2}{3}A^g \wedge A^g \wedge A^g) \tag{B.3}$$

is the gravitational Chern-Simons term, with A^g the Levi-Civita connection on the spin bundle of X. The determinants are regularized as in Remark 7.6 of [**12**].

It is straightforward to see that the the term inside of the integral

$$\frac{1}{\text{Vol}\,U(1)}\frac{[\det{}'(d^*d)]^{1/2}]}{[\det{}'(k\star d)]^{1/2}]} \tag{B.4}$$

may be identified with the Reidemeister torsion of the three-manifold X, T^d_{RS}, [**12**, (7.22)]. We obtain,

$$Z_{U(1)}(X,p,k) = k^{m_X} e^{\pi i k S_{X,P}(A_P)} e^{\pi i\left(\frac{\eta(\star d)}{4}+\frac{1}{12}\frac{\text{CS}(A^g)}{2\pi}\right)} \int_{\mathcal{M}_P} (T^d_{RS})^{1/2} \tag{B.5}$$

where $m_X = \frac{1}{2}(\dim H^1(X;\mathbb{R}) - \dim H^0(X;\mathbb{R}))$. The Atiyah-Patodi-Singer theorem says that the combination

$$\frac{\eta(\star d)}{4} + \frac{1}{12}\frac{\text{CS}(A^g)}{2\pi} \tag{B.6}$$

is a topological invariant depending only on a 2-framing of X. Recall, [**1**], that a two-framing is choice of a homotopy equivalence class π of trivializations of $TX \oplus TX$, twice the tangent bundle of X viewed as a Spin(6)-bundle. The possible 2-framings correspond to $\mathbb{Z}$. The identification with $\mathbb{Z}$ is given by the signature defect defined by

$$\delta(X,\pi) = \text{sign}(M) - \frac{1}{6}p_1(2TX,\pi) \tag{B.7}$$

where M is a 4-manifold with boundary X and $p_1(2TX,\pi)$ is the relative Pontrjagin number associated to the framing π of the bundle $TX \oplus TX$. The canonical 2-framing π^c corresponds to $\delta(X,\pi^c) = 0$. Either we can choose the canonical framing, and work with this throughout, or we can observe that if the framing of X is twisted by s units, then $CS(A^g)$ transforms by

$$CS(A^g) \to CS(A^g) + 2\pi s \tag{B.8}$$

and so the partition function $Z_{U(1)}(X,k)$ is transformed by

$$Z_{U(1)}(X,k) \to Z_{U(1)}(X,k) \cdot \exp\left(\frac{2\pi i s}{24}\right) \tag{B.9}$$

Then $Z_{U(1)}(X,k)$ is a topological invariant of framed, oriented 3-manifolds, with a transformation law under change of framing. This is tantamount to a topological invariant of oriented 3-manifolds without a choice of framing.

References

[1] M. F. Atiyah, *On framings of 3-manifolds*, Topology **29** (1990), no. 1, 1–7.

[2] M. F. Atiyah, V. K. Patodi, and I. M. Singer, *Spectral asymmetry and Riemannian geometry, II*, Proc. Cambridge Philos. Soc. **78** (1975), 405–432.

[3] Chris Beasley and Edward Witten, *Non-abelian localization for Chern-Simons theory*, J. Differential Geom. **70** (2005), 183–323.

[4] Olivier Biquard, Marc Herzlich, and Michel Rumin, *Diabatic limit, eta invariants and Cauchy-Riemann manifolds of dimension 3*, Ann. Scient. Ec. Norm. Sup. **40** (2007), 589–631.

[5] J. M. Bismut and J. Cheeger, *η-invariants and their adiabatic limits*, J. Amer. Math. Soc. **2** (1989), 33–70.

[6] David E. Blair, *Riemannian geometry of contact and symplectic manifolds*, Lecture notes in mathematics, vol. 509, Springer-Verlag, Berlin, 1976.

[7] Charles P. Boyer and Krzysztof Galicki, *Sasakian geometry*, Oxford University Press, 2008.

[8] X. Dai, *Adiabatic limits, nonmultiplicativity of signature, and Leray spectral sequence*, J. Amer. Math. Soc. **4** (1991), 265–321.

[9] V. Guillemin and S. Sternberg, *Geometric asymptotics*, AMS Mathematical Surveys, vol. 14, Providence RI, Amer. Soc., 1977.

[10] G. Guralnik, A. Iorio, R. Jackiw, and S.-Y. Pi, *Dimensionally reduced gravitational Chern-Simons term and its kink*, Ann. Physics **308** (2003), no. 1, 222–236.

[11] Mitsuhiro Itoh, *Odd dimensional tori and contact structure*, Proc. Japan Acad. **72** (1997), 58–59.

[12] Mihaela Manoliu, *Abelian Chern-Simons theory I: A topological quantum field theory*, J. Math. Phys. **39** (1998), no. 1, 170–206.

[13] Joseph Mattes, Micheal Polyak, and Nikolai Reshetikhin, *On invariants of 3-manifolds derived from abelian groups*, Quantum Topology **3** (1993), 324–338.

[14] B. D. K. McLellan, *Gravitational Chern-Simons and the adiabatic limit*, Submitted to the J. Math. Phys. , arXiv:1004.2909.

[15] R. Ponge, *Noncommutative residue for Heisenberg manifolds and applications in CR and contact geometry*, Mem. Amer. Math. Soc. **252** (2007), 399–463.

[16] D. B. Ray and I. M. Singer, *Analytic torsion*, Proc. Symp. Pure math. **23** (1973), 167–182.

[17] M. Rumin, *Formes différentielles sur les variétés de contact*, J. Differential Geometry **39** (1994), no. 2, 281–330.

[18] M. Rumin and N. Seshadri, *Analytic torsions on contact manifolds*, Preprint on ArXiv: 0802.0123v1, 2008.

[19] A. S. Schwarz, *Instantons and fermions in the field of instanton*, Commun. Math. Phys. **64** (1979), 233–268.

[20] ———, *The partition function of degenerate functional*, Commun. Math. Phys. **67** (1979), 1–16.

[21] E. Witten, *Quantum field theory and the Jones polynomial*, Commun. Math. Phys. **121** (1989), no. 3, 351–399.

[22] ———, *Two dimensional gauge theories revisited*, Commun. Math. Phys. **9** (1992), no. 4, 303–368.

Department of Mathematics, University of Toronto, Toronto, Ontario, Canada M5S 2E4

E-mail address: jeffrey@math.toronto.edu

Department of Mathematics, University of Toronto, Toronto, Ontario, Canada M5S 2E4

E-mail address: mclellan@math.toronto.edu

AMS/IP Studies in Advanced Mathematics
Volume 50, 2011

Delta-Groupoids and Ideal Triangulations

Rinat M. Kashaev

ABSTRACT. A Δ-groupoid is an algebraic structure which axiomatizes the combinatorics of a truncated tetrahedron. By considering two simplest examples coming from knot theory, we illustrate how can one associate a Δ-groupoid to an ideal triangulation of a three-manifold. We also describe in detail the rings associated with the Δ-groupoids of these examples.

1. Introduction

This paper is about the algebraic structure called Δ-groupoid introduced in the papers [**5, 6**]. The motivation for studying this structure comes from the combinatorial descriptions of Teichmüller spaces of punctured surfaces [**10, 9, 1**] which were useful for quantization [**2, 3**], and from the fact that it permits one to describe parts of the set of representations of the fundamental group of three-manifolds into some matrix groups over arbitrary rings [**6**].

In this paper, we relate the definition of the Δ-groupoid to the combinatorics of truncated tetrahedra, and, by studying the two examples of knot complements provided by the trefoil and figure eight knots, we illustrate how ideal triangulations lead to presentations of Δ-groupoids. Relations of Δ-groupoids to representations of the fundamental group of three-manifolds into two-by-two matrix groups over arbitrary rings imply that there are at least two pairs of adjoint functors between the categories of (unital) rings and Δ-groupoids. These adjoint functors permit one to associate to any Δ-groupoid a pair of rings to be referred as A' and B' rings. We work out in detail the structure of these rings in our examples. It should be remarked that the isomorphism classes of these rings are not three-manifold invariants as they depend on the choice of triangulation. In a sense, different ideal triangulations correspond to different local charts in the moduli space of representations of the fundamental group of a manifold into a given matrix group. In analogy with two charts related by a transformation on their common part, the rings, associated with two different triangulations of one and the same three-manifold, admit a common localization. The situation here is very similar the one with deformation varieties [**10, 12, 11**], where to cover the moduli space

---

2010 *Mathematics Subject Classification.* Primary 20L05, 57M27; Secondary 16S10.
Key words and phrases. Low-dimensional topology, knot theory.
The work is supported by the Swiss National Science Foundation.

of $PSL_2(\mathbb{C})$-representations, one may need several ideal triangulations. In the case of $PSL_2(\mathbb{R})$-representations of surface groups, this kind of phenomenon has been demonstrated in [**4**].

In general, the A' ring is a quotient ring of the B' ring, so that in principle it is enough to calculate only the B'-ring, but in practice, the A' ring can be simpler to calculate. In the case of the trefoil knot, the two rings are commutative and both are isomorphic to the ring $\mathbb{Z}[t, 3^{-1}]/(\Delta_{3_1}(t))$, where $\Delta_{3_1}(t) = t^2 - t + 1$ is the Alexander polynomial of the trefoil knot. The case of the figure-eight knot is less trivial and reveals interesting features: both rings are non-commutative and not isomorphic to one another. The non-commutativity, however, is rather mild in the sense that the rings are quotients of formalized versions of two-by-two matrix rings over commutative rings. In this example, the B'-ring contains non-trivial torsion, namely, an element ϵ satisfying the equation $5\epsilon = 0$, which vanishes on the level of the A'-ring.

The B'-rings can be considered as non-commutative generalizations of the subring of functions on Penner's decorated Teichmüller spaces of punctured surfaces generated by *ratio coordinates*. These are certain ratios of Penner's λ-coordinates on the decorated Teichmüller space [**9**], which were shown in [**3**] to be useful for quantization purposes. In this light, it would be natural to expect that hyperbolic geometry in three dimensions is related somehow to Δ-groupoids. Nonetheless, it should be pointed out that the rings discussed in this paper do not seem to have any direct relations to hyperbolic geometry of knot complements. As shows the example of the figure-eight knot, the abelianization of the corresponding B'-ring, being isomorphic to the ring (see Theorem 3.6)

$$\mathbb{Z}[a, b]/(a^2 + a - 1, b^2 + b - 1),$$

does not admit ring homomorphisms into the invariant trace field of the figure-eight knot complement, which is known to be isomorphic to the field $\mathbb{Q}(\sqrt{-3})$ [**8**]. The reason for this phenomenon could be in the fact that the ratio coordinates do not parameterize the Teichmüller space itself or its decorated version, but the cartesian product of the projectivized decorated Teichmüller space with the first real cohomology group of the underlying surface, and, due to this cohomological part, the presentation of the B'-ring appears to be too restrictive in three-dimensional context.

The organization of the paper is as follows. In Section 2, by studying certain combinatorial features of a truncated tetrahedron, we naturally arrive at the definition of the Δ-groupoid. After giving a list of examples of Δ-groupoids, we define the notion of a labeled truncated tetrahedron and introduce a graphical notation for it. In Section 3 we consider in detail two of the simplest examples of ideal triangulations (each consisting of only two ideal tetrahedra) of the complements of the trefoil and the figure-eight knots. In the part dedicated to the figure-eight knot, for technical reasons, we introduce the class of formal M_2-rings which contains as a sub-class the rings of two-by-two matrices over commutative rings. The main result of the paper is given by Theorem 3.9.

Acknowledgments. I would like to thank the anonymous referee for stimulating remarks on possible relations of Δ-groupoids to hyperbolic geometry.

2. Truncated tetrahedra and Δ-groupoids

Consider a truncated tetrahedron with its natural cell decomposition drawn in this picture

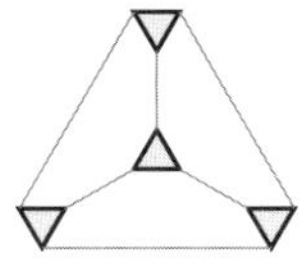

This is a three-dimensional cell complex which consists of twelve 0-cells (vertices), eighteen 1-cells (edges), eight 2-cells (faces), and one 3-cell. There are two types of 2-cells: the triangular faces (those which are shaded in the above picture) and the hexagonal faces which are remnants of the faces of the corresponding non-truncated tetrahedron. Note that there are four 2-cells of each type. There are also two types of edges: twelve *short* edges (those drawn by thick lines in the picture) which bound the triangular faces and six *long* edges (drawn by thin lines) which are remnants of the edges of the non-truncated tetrahedron.

The set of triangular faces constitutes a two-dimensional cell sub-complex whose fundamental edge-path groupoid will be denoted G. Algebraically, this is not a very interesting groupoid: it consists of four connected components (one component for each triangular face) any of which is isomorphic to one and the same coarse groupoid over a three-element set[1], but it becomes more interesting due to the following additional structure.

We remark that the set $H \subset G$ of oriented short edges is naturally a free S_3-set, where S_3 is the permutation group of three elements. Indeed, the elements of H can be grouped into four six-element families associated with hexagonal faces, and a free transitive action of the group S_3 within each family is defined through realization of the three elementary transpositions in S_3 by three involutions $i, j, k\colon H \to H$, where $i(x) = x^{-1}$ while $j(x)$ and $k(x)$ are the unique oriented short edges such that the domain (respectively the codomain) of x belongs to the same long edge as the domain (respectively the codomain) of $j(x)$ (respectively $k(x)$) as is described in this picture:

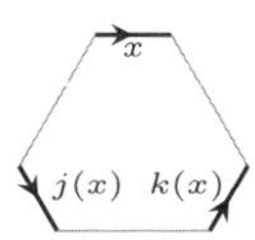

It is easy to check that the defining relations of the group S_3 are satisfied, in particular, the equalities

$$k = iji = jij. \tag{2.1}$$

As the inversion map i and the product operation are related through the identity $i(xy) = i(y)i(x)$, the following question arises: are there other identities in the groupoid G involving the involutions i, j, k and the product operation? To answer this question, let us say that a pair of elements $(x, y) \in H \times H$ is *H-composable* if it is composable in G and $xy \in H$. Not any composable pair is H-composable, for example, for any $x \in H$, the pair (x, x^{-1}) is composable but not H-composable.

[1]A *coarse* groupoid over a set S is the groupoid S^2 where the set of objects is S and the set of morphisms (arrows) is the cartesian square of S with the domain and the codomain maps being the projections to the first and second components, respectively, with the product $(x, y)(y, z) = (x, z)$.

Picking an arbitrary H-composable pair (x, y), we can label all the elements of the set H according to this picture

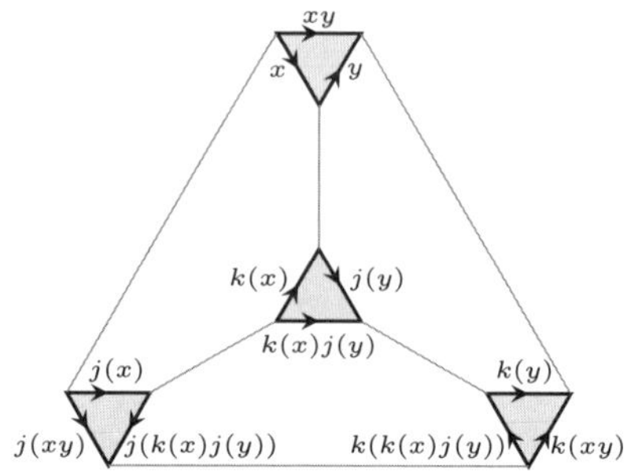

from which, in particular, we observe that the pair $(k(x), j(y))$ is also H-composable and the following identities hold:

$$j(xy) = j(x)j(k(x)j(y)), \quad k(xy) = k(k(x)j(y))k(y).$$

Moreover, taking into account equalities (2.1), the latter two identities are not independent, and in fact one follows from another as shown by the following calculation:

$$\begin{aligned} k(xy) &= iji(xy) = ij(i(y)i(x)) = i(ji(y)j(ki(y)ji(x))) \\ &= ij(ki(y)ji(x))iji(y) = iji(iji(x)iki(y))iji(y) = k(k(x)j(y))k(y). \end{aligned}$$

As any non-trivial product in this groupoid corresponds to a triangular cell, it is clear that there are no other non-trivial relations involving the multiplication.

All of above motivates the following definition [**6**].

DEFINITION 2.1. A Δ-*groupoid* is a groupoid G, a generating subset $H \subset G$, and an involution $j\colon H \to H$, such that

(i) $i(H) = H$, where $i(x) = x^{-1}$;
(ii) the involutions i and j generate an action of the symmetric group S_3 on the set H, i.e. the following equation is satisfied: $iji = jij$;
(iii) if $(x, y) \in H^2$ is a composable pair then $(k(x), j(y))$ is also a composable pair, where $k = iji$;
(iv) if $(x, y) \in H^2$ is an H-composable pair then $(k(x), j(y))$ is also an H-composable pair, and the following identity is satisfied:

$$j(xy) = j(x)j(k(x)j(y)). \tag{2.2}$$

A Δ-*group* is a Δ-groupoid with one object (identity element).

In the category $\mathbf{\Delta Gpd}$ of Δ-groupoids a morphism between two Δ-groupoids is a groupoid morphism $f\colon G \to G'$ such that $f(H) \subset H'$ and $j'f = fj$.

REMARK 2.2. In any Δ-groupoid G there is a canonical involution $A \mapsto A^*$ acting on the set of objects (or the identities) of G. It can be defined as follows. Let $\mathrm{dom}, \mathrm{cod}\colon G \to \mathrm{Ob}G$ be the domain (source) and the codomain (target) maps, respectively. As H is a generating set for G, for any $A \in \mathrm{Ob}G$ there exists $x \in H$ such that $A = \mathrm{dom}(x)$. We define $A^* = \mathrm{dom}(j(x))$. This definition is independent of the choice of x. Indeed, let $y \in H$ be any other element satisfying the same condition. Then, the pair $(i(y), x)$ is composable, and, therefore, so is $(ki(y), j(x))$.

Thus[2],

$$\mathrm{dom}(j(y)) = \mathrm{cod}(ij(y)) = \mathrm{cod}(ki(y)) = \mathrm{dom}(j(x)).$$

2.1. Examples of Δ-groupoids.

EXAMPLE 2.3. Let G be a group. The coarse groupoid G^2 is a Δ-groupoid with $H = G^2$, $j(f,g) = (f^{-1}, f^{-1}g)$.

EXAMPLE 2.4. Let X be a set. The set X^3 can be thought of as a disjoint sum of coarse groupoids X^2 indexed by X: $X^3 \simeq \sqcup_{x\in X}\{x\} \times X^2$. In particular, $\mathrm{Ob}(X^3) = X^2$ with $\mathrm{dom}(a,b,c) = (a,b)$ and $\mathrm{cod}(a,b,c) = (a,c)$. This is a Δ-groupoid with $H = X^3$ and $j(a,b,c) = (b,a,c)$. In the case where X is a three element set this Δ-groupoid is isomorphic to the one associated with the truncated tetrahedron.

EXAMPLE 2.5. Let R be a ring. We define a Δ-group AR as the subgroup of the group R^* of invertible elements of R generated by the subset $H = (1-R^*)\cap R^*$ with $k(x) = 1 - x$. Thus, $j(x) = iki(x) = (1 - x^{-1})^{-1}$.

EXAMPLE 2.6. For a ring R, let $R \rtimes R^*$ be the semidirect product of the additive group R with the multiplicative group R^* with respect to the (left) action of R^* on R by left multiplication. Set theoretically, one has $R \rtimes R^* = R \times R^*$, the group structure being given explicitly by the product $(x,y)(u,v) = (x + yu, yv)$, the unit element $(0,1)$, and the inversion map $(x,y)^{-1} = (-y^{-1}x, y^{-1})$. We define a Δ-group BR as the subgroup of $R \rtimes R^*$ generated by the subset $H = R^* \times R^*$ with $k(x,y) = (y,x)$. Thus, $j(x,y) = kik(x,y) = ki(y,x) = k(-x^{-1}y, x^{-1}) = (x^{-1}, -x^{-1}y)$.

EXAMPLE 2.7. Let $(G, G_\pm, \theta)$ be a symmetrically factorized group of [**7**]. That means that G is a group with two isomorphic subgroups $G_\pm$ conjugated to each other by an involutive element $\theta \in G$, and the restriction of the multiplication map $m\colon G_+ \times G_- \to G_+G_- \subset G$ is a set-theoretical bijection, whose inverse is called the factorization map $G_+G_- \ni g \mapsto (g_+, g_-) \in G_+ \times G_-$. In this case, the subgroup of G_+ generated by the subset $H = G_+ \cap G_-G_+\theta \cap \theta G_+G_-$ is a Δ-group with $j(x) = (\theta x)_+$.

EXAMPLE 2.8. Let A be a group and $B \subset A$ a malnormal subgroup.[3] In this case the (left) action of the group B on the set of non-trivial left cosets $X = \{aB|\ a \in A \setminus B\}$ is free. Let G be the groupoid associated to this free action[4] and let $H \subset G$ be the set of all non-identity morphisms. The triple (G, H, j) is a Δ-groupoid, where $j\colon B(aB, a'B) \mapsto B(a^{-1}B, a^{-1}a'B)$.

REMARK 2.9. The constructions in Examples 2.5 and 2.6 are functorial [**6**], and there exist the respective left adjoint functors $A', B'\colon \mathbf{\Delta Gpd} \to \mathbf{Ring}$ defined as follows. Let $G = (G, H, j)$ be a Δ-groupoid. The ring $A'G$ is the quotient ring of the groupoid ring $\mathbb{Z}[G]$ (generated over $\mathbb{Z}$ by the elements $\{w_x|\ x \in G\}$ with the

---

[2]For compositions in groupoids we use the convention adopted for fundamental groupoids of topological spaces, i.e. $(x,y) \in G^2$ is composable iff $\mathrm{cod}(x) = \mathrm{dom}(y)$, and the product is written xy rather than $y \circ x$.

[3]A subgroup $B \subset A$ is called *mal-normal* if $aBa^{-1} \cap B = \{1\}$ for any $a \in A \setminus B$.

[4]The groupoid associated to a free G-set X has the G-orbits in X as objects and the G-orbits in $X \times X$ (with respect to the diagonal action) as morphisms, the domain and the codomain maps being respectively $G(x,y) \mapsto Gx$ and $G(x,y) \mapsto Gy$, with the product $(G(x,y), G(y,z)) \mapsto G(x,z)$.

defining relations $w_x w_y = w_{xy}$ if x, y are composable) with respect to the additional relations $w_x + w_{k(x)} = 1$ for all $x \in H$. The ring $B'G$ is generated over $\mathbb{Z}$ by the elements $\{u_x, v_x | \ x \in G\}$ with the defining relations $u_x u_y = u_{xy}, v_{xy} = u_x v_y + v_x$ if x, y are composable, and $u_{k(x)} = v_x, v_{k(x)} = u_x$ for all $x \in H$.

REMARK 2.10. There exists a functorial morphism $\alpha \colon B' \to A'$ defined by the relations

$$\alpha_G(u_x) = w_x, \ \alpha_G(v_x) = 1 - w_x.$$

Moreover, α_G induces a ring isomorphism $A'G \simeq B'G/I_G$, where I_G is the two-sided ideal generated by the set $\{u_x + v_x - 1 | \ x \in H\}$.

2.2. Labeled oriented truncated tetrahedra. In what follows, in a Δ-groupoid (G, H, j), we will use the following binary operation on the set of H-composable pairs: $x * y = j(k(x)j(y))$.

Consider a tetrahedron with totally ordered vertices. The vertex order induces natural orientation of long edges of the corresponding truncated tetrahedron, where the orienting arrows point from smaller to bigger vertices. In each hexagonal face we choose the unique oriented short edge according to this picture

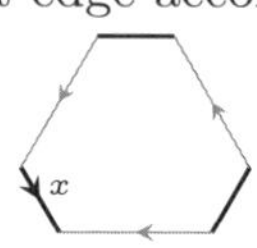

where the short edge x is characterized by the following properties:

(1) it is located near the biggest (truncated) vertex of the face;
(2) its orientation is antiparallel to the orientation of the opposite long edge.

With this choice, our *labeled oriented truncated (l.o.t.)* tetrahedron looks as in this picture

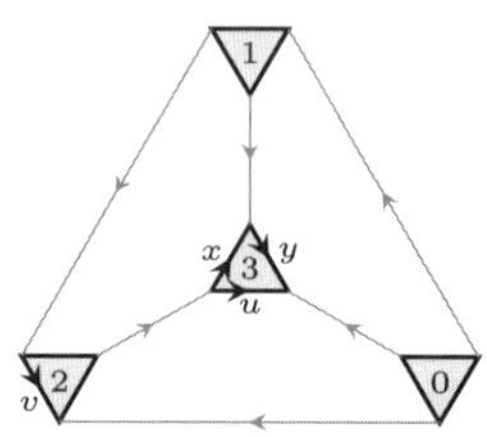

where $u = xy$, $v = x * y$, and the numbers within triangular faces enumerate the vertices (according to their order) of the non-truncated tetrahedron. As each hexagonal face is opposite to a unique vertex of the non-truncated tetrahedron, we have an induced order in the set of hexagonal faces. To express all this information, it will be convenient to use the following graphical notation for such l.o.t. tetrahedra:

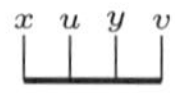

where the vertical labeled segments, ordered from left to right, correspond to the hexagonal faces with their induced order, and the labels correspond to the distinguished oriented short edges.

3. Examples of Δ-groupoids from ideal triangulations

In this section, by considering concrete examples, we show how can one associate Δ-groupoids to ideal triangulations. One can say that an ideal triangulation gives a presentation of a Δ-groupoid in the same sense as a cell complex gives a presentation for its edge-path groupoid. We also calculate the associated rings obtained by applying the adjoint functors A' and B'.

3.1. The trefoil knot. There is an ideal triangulation of the complement of the trefoil knot consisting of two tetrahedra described by the following diagram

(3.1)

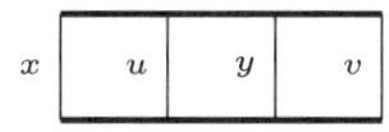

Using our conventions, we obtain a presentation of the associated Δ-groupoid G given by four generators x, y, u, v and four relations:

$$u = xy, \tag{3.2a}$$

$$v = x * y, \tag{3.2b}$$

$$y = vu, \tag{3.2c}$$

$$x = v * u. \tag{3.2d}$$

Excluding generators u and v by using equations (3.2a) and (3.2b), we obtain an equivalent presentation with two generators x, y and two relations:

$$y = (x * y)xy, \tag{3.3a}$$

$$x = (x * y) * (xy). \tag{3.3b}$$

Equation (3.3a) implies that

$$i(x) = x * y = j(k(x)j(y)) \Leftrightarrow ji(x) = k(x)j(y) \Leftrightarrow y = j((ji(x))^2).$$

Combining this with (3.3b) we obtain

$$x = j(k(x * y)j(xy)) = j(ki(x)j(x)i(x)) = ji(x).$$

Thus, all the relations (3.2) of the presentation of our Δ-groupoid reduce to the relations

$$i(x) = j(x) = k(x) = v, \quad y = j(x^2), \quad u = j(x^{-2}).$$

3.1.1. *The A'-ring.*

THEOREM 3.1. *The A'-ring associated to ideal triangulation* (3.1) *is isomorphic to the ring $\mathbb{Z}[t, 3^{-1}]/(\Delta_{3_1}(t))$, where $\Delta_{3_1}(t) = t^2 - t + 1$ is the Alexander polynomial of the trefoil knot.*

PROOF. We have the following presentation

$$A'G = \mathbb{Z}\langle w_x^{\pm 1}, w_y^{\pm 1} |\ w_x^{-1} = 1 - w_x, w_y^{-1} = 1 - w_x^{-2}\rangle.$$

The maps

$$\phi(w_x) = t,\ \phi(w_y) = 3^{-1}(2 - t),$$

specify a well defined ring homomorphism $\phi \colon A'G \to \mathbb{Z}[t, 3^{-1}]/(\Delta_{3_1}(t))$. Indeed, we have

$$\phi(w_x^{-1}) = \phi(w_x)^{-1} = t^{-1} = 1 - t = 1 - \phi(w_x) = \phi(1 - w_x),$$

and

$$\begin{aligned}\phi(w_y^{-1}) &= \phi(w_y)^{-1} = 3(2 - t)^{-1} = 3(1 + t^{-1})^{-1} \\ &= 3((1 + t^{-1})(1 + t))^{-1}(1 + t) = 1 + t = 1 - t^{-2} = 1 - \phi(w_x)^{-2} = \phi(1 - w_x^{-2}).\end{aligned}$$

On the other hand, it is easily verified that the inverse of ϕ has the form:

$$\phi^{-1}(t) = w_x,\ \phi^{-1}(3^{-1}) = w_x w_y^2.$$

□

3.1.2. *The B'-ring.*

THEOREM 3.2. *The B'-ring associated to ideal triangulation* (3.1) *is isomorphic to its A'-ring.*

PROOF. Due to Remark 2.10, it is enough to show that $u_x + v_x = 1$, $x \in H$, in the presentation of the B'-ring. Indeed, we have the presentation

$$B'G = \mathbb{Z}\langle u_x^{\pm 1}, v_x^{\pm 1}, u_y^{\pm 1}, v_y^{\pm 1} | \\ u_x^{-1} = v_x,\ -u_x^{-1}v_x = u_x,\ u_x^2 = -v_y^{-1}u_y,\ (u_x+1)v_x = v_y^{-1}\rangle$$

Denoting $u_x = t$, from the first, fourth and third relations we have

$$v_x = t^{-1},\ v_y = (1+t^{-1})^{-1},\ u_y = -(1+t^{-1})^{-1}t^2,$$

while the second relation implies that

$$t^3 + 1 = (t+1)(t^2 - t + 1) = 0.$$

Taking into account invertibility of v_y, we conclude that $t^2 - t + 1 = 0$. Thus,

$$u_x + v_x = t + t^{-1} = 1, \quad u_y + v_y = (1+t^{-1})^{-1}(t^{-1}+1) = 1.$$

□

3.2. The figure-eight knot. The standard ideal triangulation of the complement of the figure-eight knot [**10**] also consists of two tetrahedra and, in our notation, is described by the following diagram

(3.4)

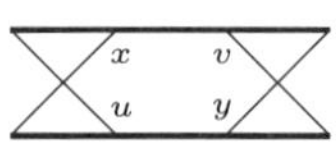

The corresponding presentation of the associated Δ-groupoid is given by four generators x, y, u, v and four relations:

(3.5a) $$u = xy,$$

(3.5b) $$v = x * y,$$

(3.5c) $$v = yx,$$

(3.5d) $$u = y * x.$$

which, by excluding the variables u and v, reduces to a presentation with two generators x and y and two relations:

(3.6a) $$xy = y * x \Leftrightarrow j(xy) = k(y)j(x),$$

(3.6b) $$yx = x * y \Leftrightarrow j(yx) = k(x)j(y).$$

To analyze the associated rings of this Δ-groupoid, we shall find it useful to use a class of rings which contains the rings of two-by-two matrices over commutative rings.

3.2.1. *Formal M_2-rings.* A ring R with center $Z(R)$ is called *formal M_2-ring* if there exists a homomorphism of Abelian groups $L\colon R \to Z(R)$, called *trace function*, such that

$$L(1) = 2, \quad L(x)x - x^2 \in Z(R), \quad \forall x \in R.$$

Thus, a formal M_2-ring is a ring where any element satisfies a quadratic equation over its center. For a given formal M_2-ring the trace function L is not necessarily unique. Indeed, for any Abelian group homomorphism $M\colon R \to Z(R)$ satisfying the conditions $M(x)x \in Z(R)$ for all $x \in R$, and $M(1) = 0$, the function $L + M$ is also a trace function.

In a formal M_2-ring R, the map $Q\colon R \to Z(R)$, defined by the formula $Q(x) = L(x)x - x^2$ is a quadratic form on the $\mathbb{Z}$-module R such that $Q(1) = 1$, and one has the identity

$$xy + yx = -(x, y) + L(x)y + L(y)x, \quad \forall x, y \in R, \tag{3.7}$$

where

$$(x, y) = Q(x + y) - Q(x) - Q(y) \tag{3.8}$$

is (twice) the symmetric $\mathbb{Z}$-bilinear form associated with Q. In particular, specifying $y = 1$ in (3.7), we obtain the equality

$$L(x) = (x, 1), \quad \forall x \in R.$$

A formal M_2-ring R with a trace function L is called *symmetric* if $L \circ L = 2L$ and the following identity is satisfied:

$$(x, y) = L(x)L(y) - L(xy), \quad \forall x, y \in R.$$

In a symmetric formal M_2-ring R one has the additional properties:

$$L \circ Q = 2Q, \quad L(xy) = L(yx), \quad Q(xy) = Q(x)Q(y), \quad \forall x, y \in R.$$

A basic example of a symmetric formal M_2-ring is the ring $M_2(K)$ of two-by-two matrices over a commutative ring K, where the trace function is given by the usual trace: $L(x) = \operatorname{tr}(x)$.

LEMMA 3.3. *Let a ring R be generated over its center $Z(R)$ by two elements a and b. Then, the condition*

$$ab + ba \in Z(R) + Z(R)a + Z(R)b$$

is satisfied if and only if the conditions

$$a^2 \in Z(R) + Z(R)a, \quad b^2 \in Z(R) + Z(R)b$$

are satisfied.

PROOF. For some $x, y \in Z(R)$, consider the elements

$$z = ya + xb - ab - ba, \ p = xa - a^2, \ q = yb - b^2.$$

Denoting $[u, v]_\pm = uv \pm vu$ for any $u, v \in R$, we have

$$\begin{aligned}[a, z]_- = x[a, b]_- - [a, [a, b]_+]_- &= x[a, b]_- - [a, [a, b]_-]_+ \\ &= x[a, b]_- - [a^2, b]_- = [xa - a^2, b]_- = [p, b]_-\end{aligned}$$

and, by a similar calculation, $[b, z]_- = [q, a]_-$. Thus, taking into account the evident equalities $[a, p]_- = [b, q]_- = 0$, we conclude that z is central if and only if p and q are central. □

LEMMA 3.4. *Let x, y, p, q be arbitrary elements in a commutative ring K. Then the ring R, defined by the presentation*

$$R(K, x, y, p, q) = K\langle a, b|\ a^2 = xa - p,\ b^2 = yb - q\rangle, \tag{3.9}$$

is a symmetric formal M_2-ring with the trace function defined by the formulae

$$L(a) = x, \quad L(b) = y. \tag{3.10}$$

PROOF. By Lemma 3.3, the element

$$z = ya + xb - ab - ba$$

is central. Any element of $R(K, x, y, p, q)$ is a linear combination of the elements $1, a, b, ab$ with coefficients in the polynomial ring $K[z]$. Indeed, by using the relation $ba = -ab + ya + xb - z$, any word in letters a and b can be written as a finite sum of the form $\sum_{m,n\in\mathbb{Z}_{\geq 0}} r_{m,n} a^m b^n$ with $r_{m,n} \in K[z]$. Due to the relation $a^2 = xa - p$, any power a^m is a linear combination of 1 and a with coefficients in K. Similarly, due to the relation $b^2 = yb - q$, any power b^n is a linear combination of 1 and b with coefficients in K. Thus, any word in a and b is reduced to a unique linear combination of the words $1, a, b, ab$ with coefficients in $K[z]$. Moreover, by considering a representation into the matrix ring $M_4(K[z])$ of the form

$$a \mapsto \begin{pmatrix} 0 & -p & 0 & 0 \\ 1 & x & 0 & 0 \\ 0 & 0 & 0 & -p \\ 0 & 0 & 1 & x \end{pmatrix}, \quad b \mapsto \begin{pmatrix} 0 & -z & -q & -qx \\ 0 & y & 0 & q \\ 1 & x & y & xy - z \\ 0 & -1 & 0 & 0 \end{pmatrix},$$

it can be verified that the reduced form of any element of $R(K, x, y, p, q)$ is unique.

It can be now straightforwardly verified that R is a symmetric formal M_2-ring with the trace function L defined by the formulae (3.10) and extended linearly over $K[z]$ to the whole ring $R(K, x, y, p, q)$. Indeed, we have

$$L(x) = L(L(a)) = 2L(a) = 2x, \quad L(p) = L(Q(a)) = 2Q(a) = 2p,$$

and similarly for y and q. Since

$$(a, b) = (x + y)(a + b) - (a + b)^2 - p - q = xb + ya - ab - ba = z,$$

we also have $L(z) = 2z$. Thus, for any $u \in \mathbb{Z}[x, y, z, p, q]$, we have $L(u) = 2u$. □

3.2.2. *Ideals in formal M_2-rings.* Let I be a two-sided ideal in a formal M_2-ring R with a trace function $L\colon R \to Z(R)$. Then, evidently, $Q(I) \subset I$ but not necessarily $L(I) \subset I$. However, from the identity

$$xL(y) - (x, y) \in I, \quad x \in R,\ y \in I,$$

it follows that $RL(I) \subset I + Z(R)$. Moreover, in a symmetric formal M_2-ring the relation

$$xL(y) - L(x)L(y) + L(xy) = xL(y) + L(xy - L(x)y) \in I, \quad x \in R,\ y \in I, \tag{3.11}$$

implies that $RL(I) \subset L(I) + I$. Thus, the set $L(I) + I$ is a two-sided ideal in R and its image in the quotient ring R/I is an ideal contained in the center $Z(R/I)$. It is clear that the quotient ring $R/(L(I) + I)$ inherits the structure of a formal M_2-ring with the trace function induced from that of R.

REMARK 3.5. Equation (3.11) implies that if $y \in I$ is such that $L(y) \in I$, then $L(xy) \in I$ for any $x \in R$. This fact can be used for construction of elements of the intersection $I \cap Z(R)$.

3.2.3. *The A'-ring.* The ring $A'G$ has the initial presentation

$$\mathbb{Z}\langle w_x^{\pm 1}, w_y^{\pm 1}, (1-w_x)^{-1}, (1-w_y)^{-1}, (1-w_x w_y)^{-1} | \\ (1-(w_x w_y)^{-1})^{-1} = (1-w_y)(1-w_x^{-1})^{-1}, \\ (1-(w_y w_x)^{-1})^{-1} = (1-w_x)(1-w_y^{-1})^{-1}\rangle,$$

which can be shown to be equivalent to the quotient ring $R(\mathbb{Z}[s], s, s, 1, 1)/I$, where the two-sided ideal I is generated by the elements

$$(s-1)(a+1) - z, \quad (s-1)(b+1) - z.$$

Identification with the initial presentation is given by the formulae:

$$w_x \mapsto a, \quad w_y \mapsto b.$$

The techniques of formal M_2-rings lead to an equivalent presentation of the form

(3.12)
$$\mathbb{Z}[s,z]/((s-1)(z-2), s^2-2z-1, z^2-2z)\langle a, g |\ a^2 - a = z - s,\ g^2 = z-2, \\ ag + ga = 2 - z + g,\ sa = a - s + z + 1,\ sg = g,\ za = s-1,\ zg = 0\rangle,$$

with the identifications

$$w_x \mapsto a, \quad w_y \mapsto a + g.$$

The part $J = \mathbb{Z}(s-1) + \mathbb{Z}z$ in this presentation coincides with the image of the ideal $L(I) + I$. The quotient ring $A'G/J$, having a unique structure of a symmetric formal M_2-ring, is isomorphic to the ring of Hurwitz quaternions.

3.2.4. *The B'-ring.*

THEOREM 3.6. *The B'-ring associated to diagram* (3.4) *admits the following presentation*

$$\mathbb{Z}\left\langle a, b, c^{\pm 1} \,\middle|\, c = a(a+1),\ c^{-1} = b(b+1),\ c = \left(aba^{-1}b^{-1}\right)^2 \right\rangle.$$

PROOF. Relations (3.6) correspond to the presentation

$$B'G = \mathbb{Z}\langle u_x^{\pm 1}, v_x^{\pm 1}, u_y^{\pm 1}, v_y^{\pm 1} | \\ (u_x v_y + v_x)^{-1} u_x u_y = v_y v_x^{-1} u_x,\ (u_x v_y + v_x)^{-1} = (v_y v_x^{-1} + u_y), \\ (u_y v_x + v_y)^{-1} u_y u_x = v_x v_y^{-1} u_y,\ (u_y v_x + v_y)^{-1} = (v_x v_y^{-1} + u_x)\rangle$$

Introducing two new elements

$$a = v_x^{-1} u_x v_y, \quad b = v_y^{-1} u_y v_x,$$

we obtain an equivalent presentation

$$B'G \simeq \mathbb{Z}\langle u_x^{\pm 1}, v_x^{\pm 1}, u_y^{\pm 1}, v_y^{\pm 1}, a^{\pm 1}, b^{\pm 1} |\ u_x = v_x a v_y^{-1},\ u_y = v_y b v_x^{-1}, \\ (a+1)^{-1} a b v_x^{-1} = v_y a v_y^{-1},\ (a+1)^{-1} = v_y(1+b), \\ (b+1)^{-1} b a v_y^{-1} = v_x b v_x^{-1},\ (b+1)^{-1} = v_x(1+a)\rangle. \tag{3.13}$$

The fourth and sixth relations in (3.13) can be used to eliminate the generators v_x and v_y, thus obtaining another presentation

$$(3.14)\quad B'G \simeq \mathbb{Z}\langle u_x^{\pm 1}, v_x^{\pm 1}, u_y^{\pm 1}, v_y^{\pm 1}, a^{\pm 1}, b^{\pm 1} | \ u_x = v_x a v_y^{-1},\ u_y = v_y b v_x^{-1},$$
$$v_y^{-1} = (b+1)(a+1),\ v_x^{-1} = (a+1)(b+1),$$
$$ab(a+1)(b+1) = (b+1)^{-1}a(b+1)(a+1),\ ba(b+1)(a+1) = (a+1)^{-1}b(a+1)(b+1)\rangle.$$

The last two relations in this presentation imply the equality

$$b(b+1)a(a+1) = 1,$$

which means that the invertible element $c = a(a+1)$ is central and $c^{-1} = b(b+1)$. Solving these equations with respect to $a+1$ and $b+1$ and substituting them into either of the last two relations of (3.14), we obtain

$$ba(ab)^{-1} = c^{-1}ab(ba)^{-1} \Leftrightarrow c = (ab(ba)^{-1})^2.$$

□

The ring

$$R = \mathbb{Z}\left\langle a, b, c^{\pm 1} \,\middle|\, c = a(a+1),\ c^{-1} = b(b+1) \right\rangle$$

is isomorphic to the ring $R(\mathbb{Z}[c, c^{-1}], -1, -1, -c, -c^{-1})$ and, due to Lemma 3.4, is a symmetric formal M_2-ring. In fact, it is a four-dimensional algebra over the polynomial ring $\mathbb{Z}[d, c, c^{-1}]$, where the element $d = ab + ba + a + b$ is central, and the elements $1, a, b, ab$ constitute a linear basis. The trace function is given by a $\mathbb{Z}[d, c, c^{-1}]$-linear map defined by the relations

$$L(1) = 2,\ L(a) = L(b) = -1,\ L(ab) = 1 + d.$$

In particular, we have the following identity:

$$(3.15)\quad q + q^{-1} = w \in \mathbb{Z}[d, c, c^{-1}], \quad q = aba^{-1}b^{-1}, \quad w = d^2 + d - c - c^{-1} - 2.$$

Indeed, we have

$$q = ab(a+1)(b+1) = (d+1)ab + (d + c^{-1} + 1)a - cb - c - 1,$$

and, by the evident symmetry $a \leftrightarrow b$, $c \leftrightarrow c^{-1}$, we also have

$$q^{-1} = ba(b+1)(a+1) = (d+1)ba + (d + c + 1)b - c^{-1}a - c^{-1} - 1.$$

Adding these equalities, we obtain

$$q + q^{-1} = (d+1)(ab + ba + a + b) - c - c^{-1} - 2$$

which is equivalent to equality (3.15).

Thus, by Theorem 3.6, the ring $B'G$ is isomorphic to the quotient ring R/I, where I is the two sided ideal generated by the element $\xi = q^2 - c = wq - 1 - c$, Analysis of the structure of the ring $B'G$ gives rise to the following description.

LEMMA 3.7. *The following elements of R are in the ideal I:*

$$(3.16)\quad 5\lambda,\ \lambda^2,\ (a-2)\lambda,\ (b-2)\lambda,\ (w-2)\lambda,\ (d-2)\lambda,\ (c-1)\lambda,\ (q-1)\lambda,$$
$$\lambda + 2(c - c^{-1}),\ w^2 - 2w,\ d^2 + d - 3w,$$

where λ is either of the elements $c^{\pm 1} + 1 - w \in R$.

PROOF. In the ring R we have the following identities

$$(a^{-1}\xi - \xi a^{-1})b^{-1} + (d-a)\xi = (a-d)(c+1-w),$$
$$(b^{-1}\xi b - \xi)a^{-1}b^{-1} + (d-b)\xi = (b-d)(c+1-w).$$

That means that the elements $(a-d)(c+1-w)$ and $(b-d)(c+1-w)$ belong to the ideal I. As this ideal is invariant with respect to the symmetry $a \leftrightarrow b$, $c \leftrightarrow c^{-1}$, the elements $(a-d)(c^{-1}+1-w)$ and $(b-d)(c^{-1}+1-w)$ also belong to I.

Let $\lambda \in \{c^{\pm 1}+1-w\}$. We have the inclusions

$$\{(a-d)\lambda, (b-d)\lambda\} \subset I.$$

Using the definition of the element d in terms of a and b we have

$$(d - 2d(d+1))\lambda \in I. \tag{3.17}$$

On the other hand, from the definition of q, it follows that

$$(q-1)\lambda = ((a-d)b + b(d-a))(ba)^{-1}\lambda \in I,$$

which implies the following sequence of inclusions

$$(c-1)\lambda = -\xi\lambda + (q^2-1)\lambda \in I,$$
$$(d(d+1)-1)\lambda = (d(d+1) - a(a+1))\lambda + (c-1)\lambda \in I,$$
$$(d-2)\lambda = (d-2d(d+1))\lambda + 2(d(d+1)-1)\lambda \in I,$$
$$(w-2)\lambda = ((d(d+1)-6) + (2-c-c^{-1}))\lambda \in I,$$
$$\lambda^2 = ((c^{\pm 1}-1) + (2-w))\lambda \in I,$$

and

$$5\lambda = (6-d(d+1))\lambda + (d(d+1)-1)\lambda \in I.$$

The identity

$$w - 1 - c - 2(c-c^{-1}) = (5+3(q-1))(1+c^{-1}-w) + 3\xi(1-q^{-1}c^{-1})$$

is satisfied due to the relation $q+q^{-1} = w$, and its right hand side is manifestly an element of I. By a similar identity, obtained by the replacements $q \leftrightarrow q^{-1}$, $c \leftrightarrow c^{-1}$, we have $w-1-c^{-1}+2(c-c^{-1}) \in I$. We have two more identities

$$d(d+1) - 3w = (c+1-w+2(c-c^{-1})) + (c^{-1}+1-w-2(c-c^{-1}))$$

and

$$w^2 - 2w = \xi(1-q^{-2}c^{-1}) + (d(d+1)-3w),$$

where the right hand sides are also manifestly in I. □

LEMMA 3.8. *The ideal $L(I)+I$ is generated by the elements $\epsilon = c+1-w$ and $\xi = q^2 - c$.*

PROOF. As elements $1, a, b, ba$ constitute a linear basis in R over its center, and due to the cyclic property of the trace function, the trace of any element in the ideal I is a linear combination of the elements

$$L(\xi),\ L(\xi a),\ L(\xi b),\ L(\xi ba)$$

with coefficients in $Z(R)$. Calculating these four traces, we obtain

$$L(\xi) = -2\epsilon + w^2 - 2w,\ L(\xi a) = \epsilon + 2w - w^2,\ L(\xi b) = \epsilon,\ L(\xi ba) = (1+d)\epsilon.$$

□

Since we shall work with elements of the quotient ring R/I, in order to simplify the notation, for all $x \in R$, we will denote by the same letter x the corresponding class $x + I \in R/I$.

THEOREM 3.9. *Any element $x \in R/I$ can uniquely be written as a linear combination*

$$x = m\epsilon + \sum_{\mu \in \{1,w,d\}} \sum_{\nu \in \{1,a,b,ab\}} n_{\mu,\nu}\, \mu\nu,$$

where $m \in \{0, \pm 1, \pm 2\}$ and $n_{\mu,\nu} \in \mathbb{Z}$. The ring structure of R/I is characterized by the conditions that the elements ϵ, w, d are central, and the multiplication rules:

$$\epsilon^2 = 0,\ 5\epsilon = 0,\ \epsilon\mu\nu = \varepsilon_\mu \varepsilon_\nu\, \epsilon, \quad \varepsilon_x = \begin{cases} 1, & \text{if } x = 1; \\ 2, & \text{if } x \in \{w, d, a, b\}; \\ 4, & \text{if } x = ab; \end{cases}$$

$$a^2 = \epsilon - 1 + w - a,\ b^2 = -\epsilon - 1 + w - b,\ ba = d - a - b - ab;$$

$$w^2\nu = 2w\nu,\ d^2\nu = -d\nu + 3w\nu, \quad \nu \in \{1, a, b, ab\};$$

$$wd\nu = \begin{cases} \epsilon + wa + wb + 2wab, & \text{if } \nu = 1; \\ 2\epsilon + w - wa + 2wb - wab, & \text{if } \nu = a; \\ 2\epsilon + w + 2wa - wb - wab, & \text{if } \nu = b; \\ -\epsilon + 2w - wa - wb, & \text{if } \nu = ab; \end{cases}$$

$$a^2b = 2\epsilon - b + wb - ab,\ ab^2 = -2\epsilon - a + wa - ab;$$

$$aba = 2\epsilon + 1 - w + a + da + b - wb;$$

$$bab = -2\epsilon + 1 - w + a - wa + b + db;$$

$$(ab)^2 = -1 + ab + dab.$$

PROOF. Let $K \subset Z(R/I)$ be the sub-ring generated by the elements w and d. The relations $w^2 = 2w$, $d^2 = -d + 3w$ and $wd = dw$ imply that K is a four-dimensional commutative $\mathbb{Z}$-algebra, with elements 1, w, d, and wd as a linear basis. Any element $x \in R/I$ can be written in the form

$$x = m\epsilon + x_0 + x_1 a + x_2 b + x_3 ab, \quad m \in \{0, \pm 1, \pm 2\}, \quad x_0, x_1, x_2, x_3 \in K, \tag{3.18}$$

but such form is not unique. Indeed, by using Lemma 3.7, we have the identity:

$$w(-d + a + b + 2ab) = w(ab - ba) = w(q - 1)ba = (c + 1 - w)ba = \epsilon ba = 4\epsilon = -\epsilon,$$

which can be solved for the product wd:

$$wd = \epsilon + wa + wb + 2wab.$$

By repeated use of this relation, we can eliminate all the products wd in formula (3.18). The resulting formula for x, called *reduced form*, takes the same form as before, but each coefficient x_i, $i \in \{0, 1, 2, 3\}$, is a $\mathbb{Z}$-linear combination of 1, w and d. To prove that the reduced form is unique it is enough to check that $x = 0$ if and only if its reduced form vanishes. This can be straightforwardly checked by using the representation in the endomorphism ring of the Abelian group

$(\mathbb{Z}/5\mathbb{Z}) \oplus \mathbb{Z}^{12}$ obtained by left multiplications. Namely, with respect to the ordered basis $(\epsilon, 1, w, d, a, wa, da, b, wb, db, ab, wab, dab)$ this representation takes the form

$$a \mapsto \begin{pmatrix} 2 & 0 & 0 & 0 & 1 & 2 & -2 & 0 & 0 & 0 & 2 & -1 & 1 \\ 0 & 0 & 0 & 0 & -1 & 0 & 0 & 0 & 0 & 0 & 0 & 0 & 0 \\ 0 & 0 & 0 & 0 & 1 & 1 & 0 & 0 & 0 & 0 & 0 & 0 & 1 \\ 0 & 0 & 0 & 0 & 0 & 0 & -1 & 0 & 0 & 0 & 0 & 0 & 0 \\ 0 & 1 & 0 & 0 & -1 & 0 & 0 & 0 & 0 & 0 & 0 & 0 & 0 \\ 0 & 0 & 1 & 0 & 0 & -1 & 1 & 0 & 0 & 0 & 0 & 0 & 2 \\ 0 & 0 & 0 & 1 & 0 & 0 & -1 & 0 & 0 & 0 & 0 & 0 & 0 \\ 0 & 0 & 0 & 0 & 0 & 0 & 0 & 0 & 0 & 0 & -1 & 0 & 0 \\ 0 & 0 & 0 & 0 & 0 & 0 & 1 & 0 & 0 & 0 & 1 & 1 & -1 \\ 0 & 0 & 0 & 0 & 0 & 0 & 0 & 0 & 0 & 0 & 0 & 0 & -1 \\ 0 & 0 & 0 & 0 & 0 & 0 & 0 & 1 & 0 & 0 & -1 & 0 & 0 \\ 0 & 0 & 0 & 0 & 0 & 0 & 2 & 0 & 1 & 0 & 0 & -1 & -1 \\ 0 & 0 & 0 & 0 & 0 & 0 & 0 & 0 & 0 & 1 & 0 & 0 & -1 \end{pmatrix},$$

$$b \mapsto \begin{pmatrix} 2 & 0 & 0 & 0 & 0 & 1 & 0 & -1 & -2 & -1 & -2 & -2 & -2 \\ 0 & 0 & 0 & 0 & 0 & 0 & 0 & -1 & 0 & 0 & 1 & 0 & 0 \\ 0 & 0 & 0 & 0 & 0 & 0 & 3 & 1 & 1 & 0 & -1 & 0 & -1 \\ 0 & 0 & 0 & 0 & 1 & 0 & -1 & 0 & 0 & -1 & 0 & 0 & 1 \\ 0 & 0 & 0 & 0 & -1 & 0 & 0 & 0 & 0 & 0 & 1 & 0 & 0 \\ 0 & 0 & 0 & 0 & 0 & 0 & 0 & 0 & 0 & 1 & -1 & 1 & 0 \\ 0 & 0 & 0 & 0 & 0 & 0 & -1 & 0 & 0 & 0 & 0 & 0 & 1 \\ 0 & 1 & 0 & 0 & -1 & 0 & 0 & -1 & 0 & 0 & 1 & 0 & 0 \\ 0 & 0 & 1 & 0 & 0 & 0 & 0 & 0 & -1 & 1 & 0 & 0 & 0 \\ 0 & 0 & 0 & 1 & 0 & 0 & -1 & 0 & 0 & -1 & 1 & 0 & 0 \\ 0 & 0 & 0 & 0 & -1 & 0 & 0 & 0 & 0 & 0 & 0 & 0 & 0 \\ 0 & 0 & 0 & 0 & 0 & 1 & 0 & 0 & 0 & 2 & 0 & -1 & -1 \\ 0 & 0 & 0 & 0 & 0 & 0 & -1 & 0 & 0 & 0 & 0 & 0 & 0 \end{pmatrix},$$

where the numbers on the first lines of these matrices are considered modulo 5. □

3.2.5. *Remarks.*

(1) A $\mathbb{Z}$-linear basis of the center $Z(R/I)$ is given by the elements

$$\epsilon,\ 1,\ w,\ d,\ p = 5wa,\ q = 5wb,\ r = wab - 2wa - 2wb.$$

(2) The quotient ring $R/(L(I) + I)$, which is identified with $(R/I)/\mathbb{Z}\epsilon$, is a symmetric formal M_2-ring with the trace function defined by the formulae

$$L(1) = 2,\ L(a) = L(b) = -1,\ L(ab) = 1 + d,$$

and extended linearly over the polynomial ring $\mathbb{Z}[w, d]$. The elements

$$1,\ w,\ d,\ e = wa + wb + 2wab,\ f = wa,\ g = wab$$

constitute a $\mathbb{Z}$-basis of the center $Z(R/(L(I) + I))$, and their traces are given by the formulae

$$L(\lambda) = \begin{cases} 2\lambda & \text{if } \lambda \in \{1, w, d, e\}; \\ -w & \text{if } \lambda = f; \\ w + e & \text{if } \lambda = g. \end{cases}$$

(3) The kernel of the surjective ring homomorphism $\alpha_G \colon B'G \to A'G$ is the ideal in R/I generated by the elements ϵ and $w - d$.

References

[1] F. Bonahon, *Shearing hyperbolic surfaces, bending pleated surfaces and Thurstons symplectic form,* Ann. Fac. Sci. Toulouse Math **6** (1996), 233–297.

[2] V. V. Fock, L. O. Chekhov, *A quantum Teichmüller space,* Theor. Math. Phys., **120** (1999), no. 3, 1245–1259.

[3] R. M. Kashaev, *Quantization of Teichmüller spaces and the quantum eilogarithm,* Lett. Math. Phys. **43** (1998), no. 2, 105–115.

[4] ———, *Coordinates for the moduli space of flat* $PSL(2,\mathbb{R})$*-connections,* Math. Res. Lett., **12** (2005), no. 1, 23–36.

[5] ———, *On ring-valued invariants of topological pairs.* Preprint arXiv:math/0701543.

[6] ———, *Delta-groupoids in knot theory.* Preprint arXiv:0908.1261, to be published in Geometriae Dedicata.

[7] R. M. Kashaev and N. Yu. Reshetikhin, *Symmetrically Factorizable Groups and Set-Theoretical Solutions of the Pentagon Equation,* Contemporary Mathematics, **433** (2007), 267–279.

[8] C. Maclachlan and A. Reid, *The arithmetic of hyperbolic 3-manifolds.* Graduate Texts in Mathematics, 219. Springer-Verlag, New York, 2003.

[9] R. C. Penner, *The Decorated Teichmiiller Space of Punctured Surfaces,* Commun. Math. Phys. **113** (1987), 299–339.

[10] W. P. Thurston, *The geometry and topology of 3-manifolds.* Lecture notes, Princeton University, 1976–79.

[11] S. Tillmann, *Degenerations of ideal hyperbolic triangulations.* Preprint arXiv:math/0508295.

[12] T. Yoshida, *On ideal points of deformation curves of hyperbolic 3-manifolds with one cusp,* Topology **30** (1991), 155–170.

Université de Genève, Section de mathématiques, 2-4, rue du Lièvre, CP 64, 1211 Genève 4, Suisse

E-mail address: rinat.kashaev@unige.ch

AMS/IP Studies in Advanced Mathematics
Volume 50, 2011

Invariants of Knots and 3–Manifolds Derived from the Equivariant Linking Pairing

Christine Lescop

ABSTRACT. Let M be a closed oriented 3-manifold with first Betti number one. Its equivariant linking pairing may be seen as a two-dimensional cohomology class in an appropriate infinite cyclic covering of the configuration space of ordered pairs of distinct points of M. We show how to define the equivariant cube $\mathcal{Q}(M,\mathbb{K})$ of this Blanchfield pairing with respect to a framed knot $\mathbb{K}$ that generates $H_1(M;\mathbb{Z})/\text{Torsion}$.

We present the invariant $\mathcal{Q}(M,\mathbb{K})$ and some of its properties including a surgery formula.

Via surgery, the invariant $\mathcal{Q}$ is equivalent to an invariant $\hat{\mathcal{Q}}$ of null-homologous knots in rational homology spheres, that is conjecturally equivalent to the two-loop part of the Kontsevich integral.

We generalize the construction of $\hat{\mathcal{Q}}$ to obtain a topological construction for an invariant that is conjecturally equivalent to the whole Kricker rational lift of the Kontsevich integral for null-homologous knots in rational homology spheres.

1. Introduction

1.1. Background. The study of 3–manifold invariants built from integrals over configuration spaces started after the work of Witten on Chern-Simons theory in 1989 [**42**], with work of Axelrod, Singer [**3, 4**], Kontsevich [**20**], Bott, Cattaneo [**9, 10, 12**], Taubes [**39**]. In 1999, in [**23**], G. Kuperberg and D. Thurston announced that some of these invariants, the Kontsevich ones, fit in with the framework of finite type invariants of homology spheres studied by Ohtsuki, Le, J. and H. Murakami, Goussarov, Habiro, Rozansky, Garoufalidis, Polyak, Bar-Natan [**32, 14, 25, 18, 5, 6, 7**] and others. They showed that these invariants together define a universal finite type invariant for homology 3-spheres. I gave specifications on the Kuperberg-Thurston work in [**26**] and generalisations in [**27**].

Similar studies for the knots and links cases had been performed by many other authors including Guadagnini, Martellini, Mintchev [**16**], Bar-Natan [**8**], Kontsevich [**21**], Polyak, Viro [**36**], Bott, Taubes [**11**], Altschüler, Freidel [**2**], D. Thurston [**40**], Poirier [**35**]. See also the Labastida survey [**24**] and the references therein.

2010 *Mathematics Subject Classification.* Primary 57M27, 55R80; Secondary 57N10, 57M25.
Institut Fourier, CNRS, UJF Grenoble.

The above mentioned Kuperberg-Thurston work shows how to write the Casson invariant λ, originally defined by Casson in 1984 as an algebraic number of conjugacy classes of irreducible $SU(2)$-representations [**1, 17, 30**], as

$$\lambda(N) = \frac{1}{6}\int_{(N\setminus\{\infty\})^2\setminus\text{diagonal}} \omega^3$$

for a homology sphere N (a closed oriented 3-manifold with the same integral homology as S^3), a point ∞ in N, and a closed 2-form ω such that for any 2-component link

$$J \sqcup L\colon S^1 \sqcup S^1 \to N \setminus \{\infty\},$$

the linking number of J and L reads

$$lk(J,L) = \int_{J\times L} \omega.$$

In this sense, $6\lambda(N)$ may be viewed as the *cube of the linking form of* N. It can also be expressed as the algebraic triple intersection $\langle F_X, F_Y, F_Z\rangle$ of three codimension 2 cycles F_X, F_Y, F_Z of $(C_2(N), \partial C_2(N))$ (Poincaré dual to the previous ω) for a compactification $C_2(N)$ of $\left((N\setminus\{\infty\})^2 \setminus \text{diagonal}\right)$ that is a 6–manifold with boundary. Here, for any 2-component link (J,L) of $(N\setminus\{\infty\})$ as above, the linking number of J and L is the algebraic intersection of $J\times L$ and F_X, (or F_Y or F_Z) in the compactification $C_2(N)$. A complete definition of λ in these terms is described in the appendix.

1.2. Introduction to the results. In the first part of this article, we shall present a similar construction for an *equivariant cube* $\mathcal{Q}(M,\mathbb{K})$ *of the equivariant linking pairing* for a closed 3–manifold M with $H_1(M;\mathbb{Q}) = \mathbb{Q}$, with respect to a framed knot $\mathbb{K} = (K, K_\parallel)$, that is a knot K equipped with a parallel $K_\parallel$, such that $H_1(M;\mathbb{Z})/\text{Torsion} = \mathbb{Z}[K]$.

Our invariant will live in the field of rational functions $\mathbb{Q}(x,y)$. The simplest example of a pair $(M,\mathbb{K})$ as above is the pair $(S^1\times S^2, S^1\times u)$ where $S^1\times u$ is equipped with a parallel. Note that the choice of the parallel does not affect the diffeomorphism class of the pair $(M,\mathbb{K})$ in this case. We shall have

$$\mathcal{Q}(S^1\times S^2, S^1\times u) = 0.$$

Furthermore, if N is a rational homology sphere, and if $\sharp$ stands for the connected sum,

$$\mathcal{Q}(M\sharp N,\mathbb{K}) = \mathcal{Q}(M,\mathbb{K}) + 6\lambda(N)$$

where λ is the Walker generalization of the Casson invariant normalized like the Casson invariant in [**1, 17, 30**]. If λ_W denotes the Walker invariant normalized as in [**41**], then $\lambda = \frac{\lambda_W}{2}$.

We shall also state a surgery formula in Proposition 1.6 for our invariant, and we shall determine the vector space spanned by the differences $(\mathcal{Q}(M,\mathbb{K}') - \mathcal{Q}(M,\mathbb{K}))$ for other framed knots $\mathbb{K}'$ whose homology classes generate $H_1(M;\mathbb{Z})/\text{Torsion}$, in Proposition 1.8. This determination will allow us to define an induced invariant for closed oriented 3-manifolds with first Betti number one. This latter invariant should be equivalent to a special case (the two-loop case) of invariants combinatorially defined by Ohtsuki in 2008, in [**34**], for 3-manifolds of rank one.

Let $M_{\mathbb{K}}$ be the manifold obtained from M by surgery on $\mathbb{K}$: This manifold is obtained from M by replacing a tubular neighborhood of K by another solid torus

$N(\hat{K})$ whose meridian is the given parallel $K_{\|}$ of K. It is a rational homology sphere and the core $\hat{K}$ of the new torus $N(\hat{K})$ is a null-homologous knot in $M_{\mathbb{K}}$.

Our data $(M, \mathbb{K})$ are equivalent to the data $(M_{\mathbb{K}}, \hat{K})$. Indeed, M is obtained from $M_{\mathbb{K}}$ by 0-surgery on $\hat{K}$. Hence our invariant can be seen as an invariant of null-homologous knots in rational homology spheres. For these (and even for boundary links in rational homology spheres), following conjectures of Rozansky [**37**], Garoufalidis and Kricker defined a rational lift of the Kontsevich integral in [**22, 15**], that generalizes the Rozansky 2–loop invariant of knots in S^3 of [**37**, Section 6, 6.9]. The two-loop part of this Kricker lift for knots is often called the two-loop polynomial. Its history and many of its properties are described in [**33**]. Our invariant shares many features with this two-loop polynomial and is certainly *equivalent* to this invariant, in the sense that if one of the invariants distinguishes two knots with equivalent equivariant linking pairing, then the other one does. It could even be equal to the two-loop polynomial.

In 2005, Julien Marché also proposed a similar "cubic" definition of an invariant equivalent to the two-loop polynomial [**29**].

In terms of Jacobi diagrams or Feynman graphs, the Casson invariant was associated with the graph θ and our equivariant cube is associated with the graph θ with hair or beads.

All the results of the first part of this article are proved in [**28**].

In the second part of this article, we explain how the topological construction of $\mathcal{Q}(M, \mathbb{K})$ generalizes to the construction of an invariant of $(M, \mathbb{K})$ that should be equivalent to the Kricker rational lift of the Kontsevich integral of null-homologous knots in rational homology spheres.

This article is an expansion of the talk I gave at the conference *Chern-Simons Gauge theory: 20 years after, Hausdorff center for Mathematics* in Bonn in August 2009. I thank the organizers Joergen Andersen, Hans Boden, Atle Hahn and Benjamin Himpel of this great conference.

The first part of the article and the appendix are of expository nature and do not contain all the proofs; that first part may be considered as a research announcement for the results of [**28**]. The second part relies on some results of the first part and contains the construction of a more powerful invariant of $(M, \mathbb{K})$ with the proof of its invariance.

I started to work on this project after a talk of Tomotada Ohtsuki for a workshop at the CTQM in Århus in Spring 2008. I thank Joergen Andersen and Bob Penner for organizing this very stimulating meeting, and Tomotada Ohtsuki for discussing this topic with me. Last but not least, I thank the referee for preventing the invariants constructed in the second part from living in a far less interesting space.

1.3. Conventions. All the manifolds considered in this article are oriented. Boundaries are oriented by the *outward normal first* convention. The fiber $N_u(A)$ of the normal bundle $N(A)$ of a submanifold A in a manifold C at $u \in A$ is oriented so that $T_uC = N_u(A) \oplus T_uA$ as oriented vector spaces. For two transverse submanifolds A and B of C, $A \cap B$ is oriented so that $N_u(A \cap B) = N_u(A) \oplus N_u(B)$. When the sum of the dimensions of A and B is the dimension of C, and when $A \cap B$ is finite, the *algebraic intersection* $\langle A, B \rangle$ of A and B in C is the sum of the signs of the points of $A \cap B$, where the sign of an intersection point u of $A \cap B$ is 1 if and only if $T_uC = N_u(A) \oplus N_u(B)$ (that is if and only if $T_uC = T_uA \oplus T_uB$)

as oriented vector spaces. It is (-1) otherwise. The algebraic intersection of n compact transverse submanifolds A_1, A_2 ..., A_n of C whose codimensions sum is the dimension of C is defined similarly. The sign of an intersection point u is 1 if and only if $T_uC = N_u(A_1) \oplus N_u(A_2) \oplus \cdots \oplus N_u(A_n)$ as oriented vector spaces.

1.4. On the equivariant linking pairing. Fix $(M, \mathbb{K})$ as in Subsection 1.2. Let

$$p_M \colon \tilde{M} \to M$$

denote the regular infinite cyclic covering of M, and let θ_M be the generator of its covering group that corresponds to the action of the class of K. The action of θ_M on $H_1(\tilde{M}; \mathbb{Q})$ is denoted as the multiplication by t_M.
The $\mathbb{Q}[t_M^{\pm 1}]$-module $H_1(\tilde{M}; \mathbb{Q})$ reads

$$H_1(\tilde{M}; \mathbb{Q}) = \bigoplus_{i=1}^{k} \frac{\mathbb{Q}[t_M^{\pm 1}]}{\delta_i}$$

for polynomials δ_i of $\mathbb{Q}[t_M^{\pm 1}]$ where δ_i divides δ_{i+1}. Then $\delta = \delta(M) = \delta_k$ is the *annihilator* of $H_1(\tilde{M}; \mathbb{Q})$ and $\Delta = \Delta(M) = \prod_{i=1}^k \delta_i$ is the *Alexander polynomial* of M.

These very classical invariants are normalised so that $\Delta(t_M) = \Delta(t_M^{-1})$, $\Delta(1) = 1$, $\delta(t_M) = \delta(t_M^{-1})$ and $\delta(1) = 1$. (In order to make δ symmetric, we may have to allow it to belong to $\left((t_M^{1/2} + t_M^{-1/2})\mathbb{Q}[t_M^{\pm 1}] \cup \mathbb{Q}[t_M^{\pm 1}]\right)$.) Note that Δ and δ coincide when Δ has no multiple roots.

Let (J, L) be a two-component link of $\tilde{M}$ such that $p_M(J) \cap p_M(L) = \emptyset$. If J bounds a (compact) surface Σ in $\tilde{M}$ transverse to $\theta_M^n(L)$ for all n, define the *equivariant intersection* $\langle \Sigma, L \rangle_e$ as

$$\langle \Sigma, L \rangle_e = \sum_{n \in \mathbb{Z}} t_M^n \langle \Sigma, \theta_M^n(L) \rangle$$

where $\langle \Sigma, \theta_M^n(L) \rangle$ is the usual algebraic intersection. Then the *equivariant linking pairing* of J and L is

$$lk_e(J, L) = \langle \Sigma, L \rangle_e.$$

In general, $\delta(\theta_M)(J)$ bounds a surface $\delta(\theta_M)\Sigma$ and

$$lk_e(J, L) = \frac{\langle \delta(\theta_M)\Sigma, L \rangle_e}{\delta(t_M)}.$$

For any two one-variable polynomials P and Q,

$$lk_e(P(\theta_M)(J), Q(\theta_M)(L)) = P(t_M)Q(t_M^{-1})lk_e(J, L).$$

1.5. The construction of $\mathcal{Q}(M, \mathbb{K})$. Consider the infinite cyclic covering of M^2

$$\widetilde{M^2} = \frac{\tilde{M}^2}{(u, v) \sim (\theta_M(u), \theta_M(v))} \xrightarrow{\quad p \quad} M^2$$

with generating covering transformation θ.

$$\theta(\overline{(u, v)}) = \overline{(\theta_M(u), v)} = \overline{(u, \theta_M^{-1}(v))}.$$

The diagonal of $\left(\tilde{M}^2\right)$ projects to a preferred lift of the diagonal of M^2 in $\widetilde{M^2}$.

$$p^{-1}(\mathrm{diag}(M^2)) = \sqcup_{n \in \mathbb{Z}} \theta^n \left(\overline{\mathrm{diag}(\tilde{M}^2)} = \mathrm{diag}(M^2)\right) = \mathbb{Z} \times \mathrm{diag}(M^2).$$

DEFINITION 1.1. In a smooth n-manifold C, a tubular neighborhood of a compact m-submanifold A locally reads as $\mathbb{R}^{n-m} \times U$ for some open $U \subset A$, where $\mathbb{R}^{n-m} = (]0,\infty[\times S^{n-m-1}) \cup \{0\}$ stands for the fiber of the normal bundle of A, and S^{n-m-1} stands for the fiber of the unit normal bundle of A. In this article, the manifold $C(A)$ obtained by *blowing up* A in C is obtained by replacing A by its unit normal bundle in A. Near U, $\mathbb{R}^{n-m} \times U$ is replaced by $[0,\infty[\times S^{n-m-1} \times U$. The blown-up manifold $C(A)$ is homeomorphic to the complement of an open tubular neighborhood of A in C, but it has a canonical smooth projection onto C and a canonical smooth structure. When C and A are compact, $C(A)$ is a compactification of $C \setminus A$.

The *configuration space* $C_2(M)$ is obtained from M^2 by blowing up the diagonal of M^2 in this sense, and the configuration space $\tilde{C}_2(M)$ is obtained from $\widetilde{M^2}$ by blowing up $p^{-1}(\mathrm{diag}(M^2))$. The transformation θ of $\widetilde{M^2}$ naturally lifts to a transformation of $\tilde{C}_2(M)$ that is still denoted by θ. The quotient of $\tilde{C}_2(M)$ by the action of θ is $C_2(M)$.

Since the normal bundle of the diagonal of M^2 is canonically equivalent to the tangent bundle TM of M via

$$(u,v) \in \frac{TM^2}{\mathrm{diag}(TM^2)} \mapsto (v-u) \in TM,$$

the unit normal bundle of $p^{-1}(\mathrm{diag}(M^2))$ is $\mathbb{Z} \times ST(M)$ where $ST(M)$ is the unit tangent bundle of M so that

$$\partial\tilde{C}_2(M) = \mathbb{Z} \times ST(M).$$

A trivialisation $\tau : TM \to M \times \mathbb{R}^3$ of TM identifies $ST(M)$ with $M \times S^2$.

Like any oriented closed 3-manifold, M bounds an (oriented compact) manifold W^4 with signature 0. Then $TW^4_{|M} = \mathbb{R} \oplus TM$. A trivialisation τ of TM induces a trivialisation of $TW^4 \otimes \mathbb{C}$ on M. The *first Pontrjagin class* $p_1(\tau)$ of such a trivialisation τ of the tangent bundle of M is the obstruction $p_1(W^4;\tau)$ to extend this trivialisation to W^4. It belongs to $H^4(W^4, M; \pi_3(SU(4))) = \mathbb{Z}$. We use the notation and conventions of [**31**], see also [**26**, Section 1.5].

Now, the construction of $\mathcal{Q}(M,\mathbb{K})$ is given by the following theorem.

THEOREM 1.2. *Let $\tau : TM \to M \times \mathbb{R}^3$ be a trivialisation of TM and let $p_1(\tau)$ be its first Pontrjagin class. Assume that τ maps the oriented unit tangent vectors of K to some fixed $W \in S^2$. Then τ induces a parallelisation of K. Let K_X, K_Y, K_Z be three disjoint parallels of K, on the boundary $\partial N(K)$ of a tubular neighborhood of K, that induce the same parallelisation of K as τ.*

Consider the continuous map

$$\begin{array}{llll} \breve{A}(K)\colon & (S^1 = [0,1]/(0\sim 1)) \times [0,1] & \to & C_2(M) \\ & (t, u \in]0,1[) & \mapsto & (K(t), K(t+u)), \end{array}$$

and its lift $A(K)\colon S^1 \times [0,1] \to \tilde{C}_2(M)$ such that the lift of $(K(t), K(t+\varepsilon))$ is in a small neighborhood of the canonical lift of the diagonal, for a small positive ε. Let $A(K)$ also denote the 2-chain $A(K)(S^1 \times [0,1])$.

For $V \in S^2$, let

$$s_\tau(M;V) = \tau^{-1}(M \times V) \subset (ST(M) = \{0\} \times ST(M)) \subset \partial\tilde{C}_2(M).$$

Let

$$I_\Delta(t) = \frac{1+t}{1-t} + \frac{t\Delta'(t)}{\Delta(t)}$$

where $\Delta = \Delta(M)$. *Let* X, Y, Z *be three distinct points in* $S^2 \setminus \{W, -W\}$.

There exist three rational transverse 4–dimensional chains G_X, G_Y *and* G_Z *of* $\tilde{C}_2(M)$ *whose boundaries are*

$$\partial G_X = (\theta - 1)\delta(\theta)\left(s_\tau(M;X) - I_\Delta(\theta)ST(M)_{|K_X}\right),$$

$$\partial G_Y = (\theta - 1)\delta(\theta)\left(s_\tau(M;Y) - I_\Delta(\theta)ST(M)_{|K_Y}\right) \text{ and}$$

$$\partial G_Z = (\theta - 1)\delta(\theta)\left(s_\tau(M;Z) - I_\Delta(\theta)ST(M)_{|K_Z}\right)$$

and such that the following equivariant algebraic intersections in $\tilde{C}_2(M)$ *vanish*

$$\langle G_X, A(K)\rangle_e = \langle G_Y, A(K)\rangle_e = \langle G_Z, A(K)\rangle_e = 0.$$

Define the equivariant algebraic triple intersection in $\tilde{C}_2(M)$

$$\langle G_X, G_Y, G_Z\rangle_e = \sum_{(i,j)\in\mathbb{Z}^2} \langle G_X, \theta^{-i}(G_Y), \theta^{-j}(G_Z)\rangle_{\tilde{C}_2(M)} y^i z^j \in \mathbb{Q}[y^{\pm 1}, z^{\pm 1}].$$

Let $R_\delta = \frac{\mathbb{Q}[x^{\pm 1}, y^{\pm 1}, z^{\pm 1}, \frac{1}{\delta(x)}, \frac{1}{\delta(y)}, \frac{1}{\delta(z)}]}{(xyz=1)}$. *Then*

$$\mathcal{Q}(M,\mathbb{K}) = \frac{\langle G_X, G_Y, G_Z\rangle_e}{(x-1)(y-1)(z-1)\delta(x)\delta(y)\delta(z)} - \frac{p_1(\tau)}{4} \in R_\delta$$

and $\mathcal{Q}(M,\mathbb{K})$ *only depends on the isotopy class of the knot* K *and on its parallelisation. Furthermore,*

$$\mathcal{Q}(M,\mathbb{K})(x,y,z) = \mathcal{Q}(M,\mathbb{K})(y,x,z) = \mathcal{Q}(M,\mathbb{K})(z,y,x) = \mathcal{Q}(M,\mathbb{K})(x^{-1},y^{-1},z^{-1})$$

and $\mathcal{Q}(M,\mathbb{K})$ *does not depend on the orientation of* K.

Of course, the theorem above contains a lot of statements. Let us explain their flavour. Consider the homology of $\tilde{C}_2(M)$ with coefficients in $\mathbb{Q}$ endowed with the structure of $\mathbb{Q}[t,t^{-1}]$-module where the multiplication by t is induced by the action of θ on $\tilde{C}_2(M)$. Let $\mathbb{Q}(t)$ be the field of fractions of $\mathbb{Q}[t,t^{-1}]$ and set

$$H_*(C_2(M);\mathbb{Q}(t)) = H_*(\tilde{C}_2(M);\mathbb{Q}) \otimes_{\mathbb{Q}[t,t^{-1}]} \mathbb{Q}(t).$$

LEMMA 1.3. $H_i(C_2(M);\mathbb{Q}(t)) \cong H_{i-2}(M;\mathbb{Q}) \otimes_{\mathbb{Q}} \mathbb{Q}(t)$ *for any* $i \in \mathbb{Z}$.

$$H_2(C_2(M);\mathbb{Q}(t)) = \mathbb{Q}(t)[ST(M)_{|*}(\cong * \times S^2)]$$

$$H_3(C_2(M);\mathbb{Q}(t)) = \mathbb{Q}(t)[ST(M)_{|K}(\cong K \times S^2)]$$

$$H_4(C_2(M);\mathbb{Q}(t)) = \mathbb{Q}(t)[ST(M)_{|S}(\cong S \times S^2)]$$

where S *is a closed surface of* M *such that* $H_2(M;\mathbb{Z}) = \mathbb{Z}[S]$.

In particular, the statement of the theorem contains the following lemma:

LEMMA 1.4. *The homology class in* $H_3(C_2(M);\mathbb{Q}(t))$ *of a global section of* $ST(M)$ *induced by a trivialisation of* M *is* $I_\Delta[ST(M)_{|K}]$.

Therefore, $I_\Delta[ST(M)_{|K}]$ cannot be removed from our boundaries. Lemma 1.3 can be proved by classical means. Observe that since

$$\langle A(K), ST(M)_{|S}\rangle_e = 1 - t^{-1},$$

the class of $A(K)$ in $H_2(\tilde{C}_2(M), \partial\tilde{C}_2(M))$ detects $[ST(M)_{|S}]$. Thus, the condition $\langle A(K), G_X\rangle_e = 0$ ensures that if G'_X satisfies the same conditions as G_X, $(G'_X - G_X)$ bounds a 5-chain and $\langle G'_X, G_Y, G_Z\rangle_e = \langle G_X, G_Y, G_Z\rangle_e$. Therefore, our algebraic intersection $\langle G_X, G_Y, G_Z\rangle_e$ is well-defined.

The class of $F_X = \frac{G_X}{(t-1)\delta(t)}$ (that is the same as the class of $F_Y = \frac{G_Y}{(t-1)\delta(t)}$ or $F_Z = \frac{G_Z}{(t-1)\delta(t)}$) in $H_4(C_2(M), \partial C_2(M); \mathbb{Q}(t))$ is dual to $[ST(M)_{|*}]$:

$$\langle ST(M)_{|*}, F_X\rangle_e = 1.$$

For a two-component link (J, L) of $\tilde{M}$ such that $p_M(J) \cap p_M(L) = \emptyset$, the class of $J \times L$ in $\tilde{C}_2(M)$ reads $lk_e(J, L)[ST(M)_{|*}]$. By the above equation, this can be rewritten as

$$lk_e(J, L) = \langle J \times L, F_X\rangle_e$$

and the chains F_V represent the equivariant linking number in this sense.

Recall that all the assertions of this section are proved in details in [**28**].

1.6. A few properties of $\mathcal{Q}$. Recall that λ denotes the Casson-Walker invariant normalised like the Casson invariant. I added the following proposition in order to answer a question that George Thompson asked me at the conference *Chern-Simons Gauge theory : 20 years after* in Bonn. I thank him for asking.

PROPOSITION 1.5. *Let $M_{\mathbb{K}}$ denote the rational homology sphere obtained from M by surgery along $\mathbb{K}$. Then*

$$\mathcal{Q}(M, \mathbb{K})(1, 1, 1) = 6\lambda(M_{\mathbb{K}}).$$

For a function f of x, y and z, $\sum_{\mathfrak{S}_3(x,y,z)} f(x, y, z)$ stands for

$$\sum_{\sigma\in\mathfrak{S}_3(x,y,z)} f(\sigma(x), \sigma(y), \sigma(z))$$

where $\mathfrak{S}_3(x, y, z)$ is the set of permutations of $\{x, y, z\}$.

PROPOSITION 1.6. *Let J be a knot of M that bounds a Seifert surface Σ disjoint from K whose H_1 goes to 0 in $H_1(M)/Torsion$.*

Let p/q be a nonzero rational number. Let $(a_i, b_i)_{i=1,\dots,g}$ be a symplectic basis of $H_1(\Sigma)$.

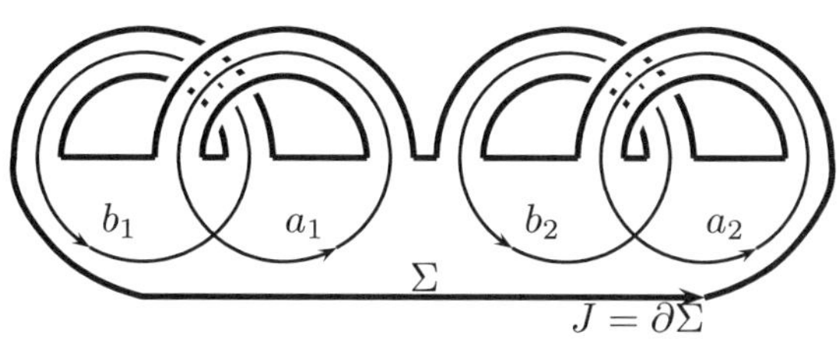

Let

$$\lambda'_e(J) = \frac{1}{12}\sum_{(i,j)\in\{1,\dots,g\}^2}\sum_{\mathfrak{S}_3(x,y,z)} \left(\alpha_{ij}(x, y) + \alpha_{ij}(x^{-1}, y^{-1}) + \beta_{ij}(x, y)\right) \in R_\delta$$

where

$$\alpha_{ij}(x,y) = lk_e(a_i, a_j^+)(x) lk_e(b_i, b_j^+)(y) - lk_e(a_i, b_j^+)(x) lk_e(b_i, a_j^+)(y)$$

and

$$\beta_{ij}(x,y) = \left(lk_e(a_i, b_i^+)(x) - lk_e(b_i^+, a_i)(x)\right)\left(lk_e(a_j, b_j^+)(y) - lk_e(b_j^+, a_j)(y)\right),$$

then

$$\mathcal{Q}(M(J;p/q),\mathbb{K}) - \mathcal{Q}(M,\mathbb{K}) = 6\frac{q}{p}\lambda'_e(J) + 6\lambda(S^3(U;p/q))$$

where $S^3(U;p/q)$ is the lens space $L(p,-q)$ obtained from S^3 by p/q–surgery on the unknot U.

Since $H_1(\Sigma)$ goes to 0 in $H_1(M)/\text{Torsion}$ in the above statement, Σ lifts as homeomorphic copies of Σ and $lk_e(a_i, a_j^+)$ denotes the equivariant linking number of a lift of a_i in $\tilde{M}$ in some lift of Σ and a lift of a_j^+ near the same lift of Σ. The superscript $+$ means that a_j is pushed in the direction of the positive normal to Σ.

In [**28**], we deduce the surgery formula of Proposition 1.6 for surgeries on knots from a surgery formula for Lagrangian-preserving replacements of rational homology handlebodies.

When the above knot J is inside a rational homology ball, $\lambda'_e(J)$ coincides with $\frac{1}{2}\Delta''(J)(1)$, where $\Delta(J)$ is the Alexander polynomial of J, and the right-hand side is nothing but 6 times the variation of the Casson-Walker invariant under a p/q–surgery on J. Since any rational homology sphere can be obtained from S^3 by a sequence of surgeries on null-homologous knots in rational homology spheres with nonzero coefficients, after a possible connected sum with lens spaces, we easily deduce the following proposition from the above surgery formula.

PROPOSITION 1.7. *Let N be a rational homology sphere, then*

$$\mathcal{Q}(M\sharp N,\mathbb{K}) = \mathcal{Q}(M,\mathbb{K}) + 6\lambda(N).$$

Recall $I_\Delta(t) = \frac{1+t}{1-t} + \frac{t\Delta'(t)}{\Delta(t)}$.

PROPOSITION 1.8. *Let $\mathbb{K}'$ be another framed knot of M such that*

$$H_1(M)/\mathit{Torsion} = \mathbb{Z}[K'].$$

Then there exists an antisymmetric polynomial $\mathcal{V}(\mathbb{K},\mathbb{K}')$ in $\mathbb{Q}[t,t^{-1}]$ such that

$$\mathcal{Q}(M,\mathbb{K}') - \mathcal{Q}(M,\mathbb{K}) = \sum_{\mathfrak{S}_3(x,y,z)} \frac{\mathcal{V}(\mathbb{K},\mathbb{K}')(x)}{\delta(x)} I_\Delta(y).$$

Furthermore, for any $k\in\mathbb{Z}$, there exists a framed knot $(\mathbb{K}')$ such that $\mathcal{V}(\mathbb{K},\mathbb{K}') = q(t^k - t^{-k})$ for some nonzero rational number q.

PROPOSITION 1.9. *If $\mathbb{K} = (K, K_\parallel)$ and if $\mathbb{K}' = (K, K'_\parallel)$, where $K'_\parallel$ is another parallel of K such that the difference $(K'_\parallel - K_\parallel)$ is homologous to a positive meridian of K in $\partial N(K)$, then*

$$\mathcal{V}(\mathbb{K},\mathbb{K}')(t) = -\frac{\delta(t)}{2}\frac{t\Delta'(t)}{\Delta(t)}.$$

PROPOSITION 1.10. *If K and K' coincide along an interval, if $K'-K$ bounds a surface B that lifts in $\tilde{M}$ such that $(K'_{\parallel}-K_{\parallel})$ is homologous to a curve of B in the complement of $(\partial B \cup K)$ in a regular neighborhood of B, and if $(a_i,b_i)_{i\in\{1,\dots,g\}}$ is a symplectic basis of $H_1(B;\mathbb{Z})$, then*

$$\frac{\mathcal{V}(\mathbb{K},\mathbb{K}')(t)}{\delta(t)}=\sum_{i=1}^{g}\left(lk_e(a_i,b_i^+)-lk_e(b_i^+,a_i)\right).$$

1.7. The derived 3-manifold invariant.

DEFINITION 1.11. *Definition of an invariant for 3-manifolds of rank one:* Let $Q_k(\delta,\Delta)=\sum_{\mathfrak{S}_3(x,y,z)}\frac{x^k-x^{-k}}{\delta(x)}I_\Delta(y)$ for $k\in(\mathbb{N}\setminus\{0\})$. For a fixed (δ,Δ), define $\overline{\mathcal{Q}}(M)$ as the class of $\mathcal{Q}(M,\mathbb{K})$ in the quotient of R_δ by the vector space generated by the $Q_k(\delta,\Delta)$ for $k\in(\mathbb{N}\setminus\{0\})$.

Thanks to Proposition 1.8, $\overline{\mathcal{Q}}(M)$ is an invariant of M. This invariant is certainly equivalent to a special case (the two-loop case) of invariants combinatorially defined by Ohtsuki in 2008 in [**34**], for 3-manifolds of rank one, when $\delta=\Delta$. The following proposition shows that it often detects the connected sums with rational homology spheres with non-trivial Casson-Walker invariants.

PROPOSITION 1.12. *If Δ has only simple roots and if N is a rational homology sphere such that $\lambda(N)\neq 0$, then $\overline{\mathcal{Q}}(M)\neq\overline{\mathcal{Q}}(M\sharp N)$.*

1.8. An alternative definition for $\mathcal{Q}$ with boundary conditions. Write the sphere S^2 as the quotient of $[0,8]\times S^1$ where $\{0\}\times S^1$ is identified with a single point (the North Pole of S^2) and $\{8\}\times S^1$ is identified with another single point (the South Pole of S^2). When $\alpha\subset[0,8]$, D^2_α denotes the image of $\alpha\times S^1$ via the quotient map q_S. For example, $D^2_{[1,8]}$ is a disk. Embed $D^2_{[1,8]}\times S^1$ as a tubular neighborhood of K, so that $K=\{*_W\}\times S^1$ for some $*_W\in\partial D^2_{[0,5]}$, and $K_\parallel=\{q_X\}\times S^1$ for some $q_X\in\partial D^2_{[0,1]}$, and let $M_{[0,1]}=M\setminus(D^2_{]1,8]}\times S^1)$. More generally, let

$$\begin{array}{llll} r\colon & M & \to & [1,8]\\ & x\in M_{[0,1]} & \mapsto & 1\\ & (q_S(t,z_W),z)\in D^2_{[1,8]}\times S^1 & \mapsto & t.\end{array}$$

When $\alpha\subset[0,8]$, $M_\alpha=r^{-1}(\alpha)$ and $\tilde{M}_\alpha=p_M^{-1}(M_\alpha)$.

Consider a map $f_M\colon M\to S^1$ that coincides with the projection onto S^1 on $D^2_{[1,8]}\times S^1$, and a lift of this map $\tilde{f}_M\colon\tilde{M}\to\mathbb{R}$. Embed $\tilde{M}_{[1,6]}=p_M^{-1}(D^2_{[1,6]}\times S^1)$ in $\mathbb{R}^3$, seen as $\mathbb{C}\times\mathbb{R}$, as $\{z\in\mathbb{C};1\le|z|\le 6\}\times\mathbb{R}$ naturally so that the projection on $\mathbb{R}$ is $\tilde{f}_M$. Here $\mathbb{C}$ is thought of as horizontal and $\mathbb{R}$ is vertical. This embedding induces a trivialisation τ on $TM_{[1,6]}$ that we extend on $TM_{[0,6]}$. This trivialisation respects the product structure with $\mathbb{R}$ on $\tilde{M}_{[1,6]}$, we also extend it on $\tilde{M}_{[1,8]}$ so that it still respects the product structure with $\mathbb{R}$, there.

Construction of a map $\pi\colon p^{-1}\left((M^2_{[0,4]}\setminus M^2_{[0,2[})\setminus\mathrm{diag}(M^2_{[2,4]})\right)\to S^2$

Fix $\varepsilon\in]0,1/2]$. Let

$$\begin{array}{llll} \chi\colon & [-4,4] & \to & [0,1]\\ & t\in[-\varepsilon,4] & \mapsto & 1\\ & t\in[-4,-2\varepsilon] & \mapsto & 0\end{array}$$

be a smooth map. Recall that $\tilde{M}_{[1,4]}$ is embedded in $\mathbb{R}^3$ that is seen as $\mathbb{C}\times\mathbb{R}$. When $(u,v)\in\left(\tilde{M}^2_{[0,4]}\setminus\tilde{M}^2_{[0,2[}\right)\setminus\mathrm{diag}(\tilde{M}^2_{[2,4]})$, set

$$U(u,v)=(1-\chi(r(u)-r(v)))(0,\tilde{f}_M(u))+\chi(r(u)-r(v))u$$

$$V(u,v)=(1-\chi(r(v)-r(u)))(0,\tilde{f}_M(v))+\chi(r(v)-r(u))v$$

so that $(U(u,v),V(u,v))\in(\mathbb{R}^3)^2\setminus\mathrm{diag}$.

Define

$$\begin{array}{rccc}\pi\colon & p^{-1}\left((M^2_{[0,4]}\setminus M^2_{[0,2[})\setminus\mathrm{diag}(M^2_{[0,4]})\right) & \to & S^2\\ & \overline{(u,v)} & \mapsto & \frac{V(u,v)-U(u,v)}{\|V(u,v)-U(u,v)\|}.\end{array}$$

When A is a submanifold of M, $C_2(A)$ denotes the preimage of A^2 under the blowup map from $C_2(M)$ to M^2, and $\tilde{C}_2(A)$ is the preimage of $C_2(A)$ under the covering map of $\tilde{C}_2(M)$. The map π naturally extends to $\tilde{C}_2(M_{[0,4]})\setminus\tilde{C}_2(M_{[0,2[})$.

Set

$$J_\Delta=J_\Delta(t)=\frac{t\Delta'(t)}{\Delta(t)}.$$

Let S^2_H denote the subset of S^2 made of the vectors whose vertical coordinate is in $]0,\frac{1}{50}[$. The following proposition is proved in [**28**, Section 12].

Proposition 1.13. *Let q_X, q_Y and q_Z be three distinct points on $\partial D^2_{[0,1]}$, and let X, Y and Z be three distinct vectors of S^2_H. For $V=X,Y$ or Z, let $K_V=q_V\times S^1$, then there exists a 4-dimensional rational chain C_V of $\tilde{C}_2(M_{[0,2]})$ whose boundary is*

$$\delta(M)\left(\pi^{-1}_{|\partial\tilde{C}_2(M_{[0,2]})\setminus\partial\tilde{C}_2(M_{[0,2[})}(V)\cup s_\tau(M_{[0,2]};V)\cup(-J_\Delta)ST(M)_{|K_V}\right),$$

and that is transverse to $\partial\tilde{C}_2(M_{[0,2]})$, and

$$\delta(M)(x)\delta(M)(y)\delta(M)(z)\mathcal{Q}(M,\mathbb{K})=\langle C_X,C_Y,C_Z\rangle_{e,\tilde{C}_2(M_{[0,2]})}-\frac{p_1(\tau)}{4}.$$

Note that when $M=S^1\times S^2$, $J_\Delta=0$, π extends to $\tilde{C}_2(M_{[0,2]})$ and $C_V=\pi^{-1}_{|\tilde{C}_2(M_{[0,2]})}(V)$ fulfills the conditions. Also note that the proposition implies that

$$\delta(M)(x)\delta(M)(y)\delta(M)(z)\mathcal{Q}(M,\mathbb{K})\in\frac{\mathbb{Q}[x^{\pm1},y^{\pm1},z^{\pm1}]}{(xyz=1)}.$$

2. Construction of more general invariants

The invariant that has been discussed so far corresponds to the graph θ (with hair or beads), where the two vertices of the graph θ correspond to the two points of a configuration in $C_2(M)$, and the three edges are equipped with $\frac{1}{\delta}C_X$, $\frac{1}{\delta}C_Y$, and $\frac{1}{\delta}C_Z$, respectively.

The chains C_V that were used in the definition of $\mathcal{Q}$ can be used to define invariants $\tilde{z}_n(M,\mathbb{K})$ of $(M,\mathbb{K})$ associated to beaded trivalent graphs with $2n$ vertices, and thus to configuration spaces $C_{2n}(M)$ of $2n$ points. We present the construction of these invariants below. Together, they will form a series $(\tilde{z}_n(M,\mathbb{K}))_{n\in\mathbb{N}}$, that should be equivalent to the Kricker rational lift of the Kontsevich integral for null-homologous knots in rational homology spheres, described in [**22, 15**].

2.1. On the target spaces. Here, a *trivalent graph* is a finite trivalent graph. Such a graph will be said to be *oriented* if each of its vertices is equipped with a *vertex orientation*, that is a cyclic order of the three half-edges that meet at this vertex. Such an oriented graph will be represented by the image of one of its planar immersions so that the vertex orientation is induced by the counterclockwise order of the half-edges meeting at this vertex.

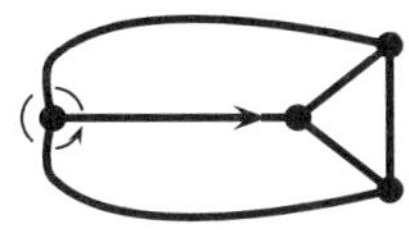

Let $\mathcal{A}_n^h(\delta)$ be the rational vector space generated by oriented trivalent graphs with $2n$ vertices whose edges are oriented and equipped with some rational functions of $\mathbb{Q}[t^{\pm 1}, 1/\delta(t)]$, and quotiented by the following relations:

(1) Reversing the orientation of an edge beaded by $P(t)$ and transforming this $P(t)$ into $P(t^{-1})$ gives the same element in the quotient.

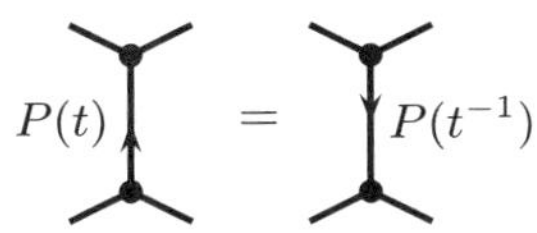

(2) If two graphs only differ by the label of one oriented edge, that is $P(t)$ for one of them and $Q(t)$ for the other one, then the class of their sum is the class of the same graph with label $(P(t) + Q(t))$.

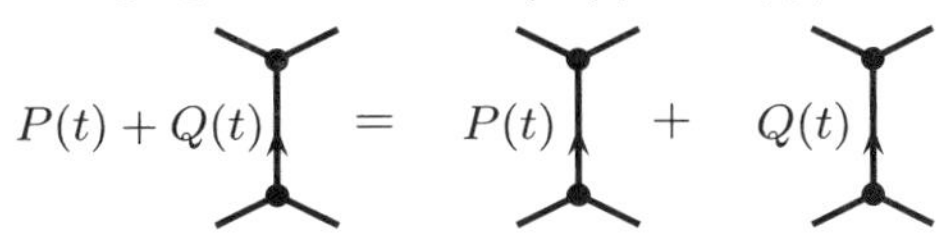

(3) Multiplying by t the three rational functions of three edges adjacent to a vertex, oriented towards that vertex, does not change the element in the quotient.

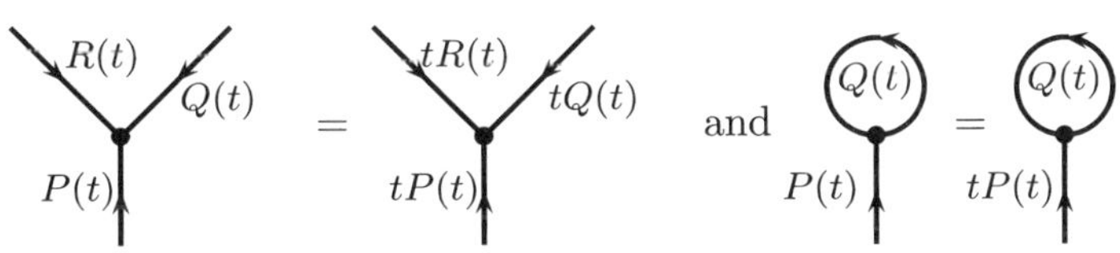

(4) (AS) Changing the orientation of a vertex multiplies the element of the quotient by (-1).

(5) (IHX or Jacobi) The sum of three graphs that coincide outside a disk, where they look as in the picture below, vanishes in the quotient. (The complete edges of the relation are equipped with the polynomial 1 that is not written.)

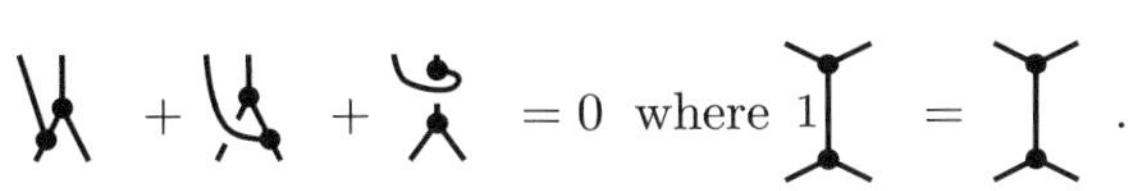

For example, there is a map $\psi\colon \mathcal{A}_1^h(\delta) \to \mathbb{Q}(x,y)$ such that

$\psi($ 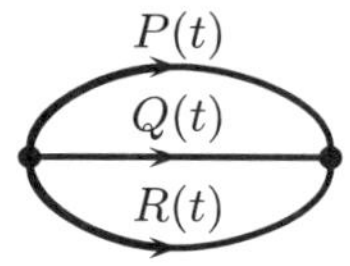

$) = \sum_{\mathfrak{S}_3(x,y,z)} (P(x)Q(y)R(z) + P(x^{-1})Q(y^{-1})R(z^{-1}))$

for elements $P(t)$, $Q(t)$ and $R(t)$ of $\mathbb{Q}[t^{\pm 1}, 1/\delta(t)]$, where $z = (xy)^{-1}$.

A more general version of this space of diagrams was introduced in [**15**, Definition 3.8].

REMARK 2.1. As pointed out to me by the referee, if we replaced $\mathbb{Q}[t^{\pm 1}, 1/\delta(t)]$ by $\mathbb{Q}[t^{\pm 1}, 1/\delta(t), 1/(t-1)]$ in the definition of $\mathcal{A}_n^h(\delta)$, unwanted relations would occur. For example, diagrams with separating edges would become trivial. Indeed, such a diagram with a separating edge beaded by $P/(1-t)$ would be equal to the diagram obtained by changing the beading into $Pt/(1-t)$ so that the difference of these two diagrams would be trivial. Therefore, the dumbbell-shaped diagram 

, that does not vanish in $\mathcal{A}_1^h(\delta)$, would vanish.

2.2. On compactifications of configuration spaces. We describe the main features of the Fulton and MacPherson type compactifications of configuration spaces [**13**] that are used to define configuration space invariants of knots or manifolds. Details can be found in [**26**, Section 3].

Let N be a finite set, and let I be a subset of N with cardinality $\sharp I \geq 2$. Let $\Delta_I(M^N) = \{(m_i)_{i\in N}; m_i = m_j \text{ if } \{i,j\} \subset I\}$. $\Delta_I(M^N)$ is a codimension $3(\sharp I - 1)$-manifold. The fiber of its unit normal bundle at $(m_j)_{j\in N}$ is the space $(T_{m_i}M^I/\Delta_I(T_{m_i}M^I) \setminus \{0\})/\mathbb{R}^{+*}$ where $i \in I$. It is the space $C_I(T_{m_i}M)$ of I-tuples of points of the tangent space $T_{m_i}M$ up to global translation and global homothety with positive ratio. Thus, in the blown-up manifold $M^N(\Delta_I(M^N))$ (see Definition 1.1), a point P that projects to $\Delta_I(M^N)$ under the canonical projection is equipped with the data of the infinitely small configuration $c_I(P) \in C_I(T_{m_i}M)$ of the points m_i indexed by I in $T_{m_i}M$ (up to translation and dilation).

Of course, the manifold $C_2(M)$ reads $M^2(\Delta_{\{1,2\}}(M^2))$ with this notation.

Let J be a subset of I, $J \subsetneq I$, $\sharp J \geq 2$.

$$\Delta_I(M^N) \subset \Delta_J(M^N)$$

The closure of $\Delta_J(M^N) \setminus \Delta_I(M^N)$ in $M^N(\Delta_I(M^N))$ is the submanifold

$$\Delta_J(M^N)(\Delta_I(M^N)),$$

and we can blow up $M^N(\Delta_I(M^N))$ along $\Delta_J(M^N)(\Delta_I(M^N))$ to get a manifold with boundary and with corners, where the relative configuration of the possibly coinciding points of $T_{m_i}M$ indexed by elements of J in an infinitely small configuration $c_I(P)$ of points of $T_{m_i}M$ is known. It is thought of as infinitely smaller.

DEFINITION 2.2. Let Γ be a graph whose vertices are indexed by $\{1, 2, \dots, 2n\}$. When I is a subset of $\{1, 2, \dots, 2n\}$, the graph Γ_I is the subgraph of Γ made of the vertices indexed in I and the edges between two of them. The *configuration space* $C(M,\Gamma)$ is the smooth compact $6n$-manifold with corners obtained by blowing up successively all the $\Delta_I(M^N)$ for which Γ_I is connected and $\sharp I \geq 2$, inductively, starting with maximal not yet treated I with this property. The space $C(M,\Gamma)$ is a compactification of the complement of all the diagonals of M^{2n}, with a canonical projection $p(M,\Gamma)\colon C(M,\Gamma) \to M^{2n}$. The *configuration space* $C_{2n}(M)$ is $C(M,\Gamma_c)$ where Γ_c is the complete graph with $2n$ vertices.

Let P_0 be a point of $C(M,\Gamma)$, it projects to $(m_i)_{i\in\{1,2,\dots,2n\}}$. Consider the maximal connected subgraphs Γ_I of Γ (with $\sharp I \geq 2$) such that $m_j = m_k$ whenever j and k are in I. For each of these subgraphs Γ_I, consider the maximal connected subgraphs Γ_J with $\sharp J \geq 2$ and $J \subsetneq I$ such that the points indexed by J coincide in the infinitely small configuration $c_I(P_0)$. Iterating the process, the "considered" graphs form a family $\mathcal{E}(P_0)$ of connected subgraphs Γ_I of Γ such that the intersection $\Gamma_I \cap \Gamma_J$ of any two graphs of the family is Γ_I, Γ_J or $\emptyset$. The points P such that $\mathcal{E}(P)$ is empty are the points in the interior of $C(M,\Gamma)$. If $\mathcal{E}(P_0) \neq \emptyset$, the points P such that $\mathcal{E}(P) = \mathcal{E}(P_0)$ are in the boundary of $C(M,\Gamma)$ and they form a codimension $\sharp\mathcal{E}(P_0)$ *face* of $C(M,\Gamma)$. In particular, the *codimension* 1 *faces* of $C(M,\Gamma)$ correspond to connected subgraphs Γ_I of Γ.

When Γ has no *loops* (edges whose two ends coincide), for every oriented edge e of Γ from the vertex $v(j)$ of Γ indexed by j to $v(k)$, there is a canonical projection

$$p(\Gamma, e)\colon C(M,\Gamma) \to C_2(M)$$

that maps a point P of $p(M,\Gamma)^{-1}((m_i)_{i\in\{1,2,\dots,2n\}})$ to (m_j, m_k) if $m_j \neq m_k$, and to the infinitely small configuration of $(T_{m_j}M^{\{j,k\}}/\Delta_{\{j,k\}}(T_{m_j}M^{\{j,k\}}) \setminus \{0\})/\mathbb{R}^{+*}$ that can be seen at some scale in $C(M,\Gamma)$, if $m_j = m_k$.

2.3. Definition of the invariants $\tilde{z}_n(M,\mathbb{K})$. Fix $(M,\mathbb{K},\tau)$ as in Subsection 1.8 so that τ maps the tangent vectors of K to $\mathbb{R}W$, where W is the vertical vector of S^2 pointing upward. For any $n \in \mathbb{N}\setminus\{0\}$, the invariant $\tilde{z}_n(M,\mathbb{K},\tau)$ is defined as follows. Consider

- $\varepsilon \in]0, \frac{1}{2n}[$, f_M (involved in the definition of the map π in Subsection 1.8),
- $3n$ disjoint parallels $(K_1,\dots,K_{3n})$ of K on $\partial M_{[0,1]}$, with respect to the parallelisation of $\mathbb{K}$,
- a *regular* $3n$-tuple $(X_1,\dots,X_{3n})$ of $(S^2_H)^{3n}$, where *regular* means in some open dense subset of $(S^2_H)^{3n}$ that will be specified in Subsection 2.4,
- $3n$ integral chains (integral combinations of properly C^∞–embedded 4-simplices) H_i in *general* $3n$*-position* in $\tilde{C}_2(M_{[0,2]})$ whose boundaries are

$$k\delta(t)\left(\pi^{-1}_{|\partial\tilde{C}_2(M_{[0,2]})\setminus\partial\tilde{C}_2(M_{[0,2[})}(X_i) \cup s_\tau(M_{[0,2]};X_i) \cup (-J_\Delta)ST(M)_{|K_i}\right),$$

  respectively, for some positive integer k. The notion of *general* $3n$*-position* will be specified in Definition 2.3.

The chains H_i will be seen as cycles of $(C_2(M_{[0,2]}), \partial C_2(M_{[0,2]}))$ (combinations of simplices) with coefficients in $\mathbb{Q}[H_1(M)/\text{Torsion}]$ by projection as follows. Pick a basepoint $*_C$ in $ST(M_{[0,2]}) \subset \partial C_2(M)$, and consider the equivalence relation on the set of paths of $C_2(M)$ from $*_C$ to some other point in $C_2(M)$ that identifies two paths if they lift to paths with identical ends in $\tilde{C}_2(M)$. For a point of $C_2(M)$ that projects to $(m_1,m_2) \in M^2$, this datum is equivalent to a rational homology class of paths from m_1 to m_2. A *chain of* $C_2(M)$ *with coefficients* in $\mathbb{Q}[H_1(M)/\text{Torsion}]$ is a chain every point P of which is equipped with a class of paths from $*_C$ to P as above, in a continuous way (or every simplex (or contractible cell) of which is equipped with such a class of paths in a way compatible with face identifications).

Let $\mathcal{S}_n$ be the set of connected trivalent graphs Γ with $2n$ vertices numbered from 1 to $2n$ and with $3n$ oriented edges numbered from 1 to $3n$ without loops.

Consider $\Gamma \in \mathcal{S}_n$. Set $C(M_{[0,4]},\Gamma) = p(M,\Gamma)^{-1}(M^{2n}_{[0,4]}) \subset C(M,\Gamma)$. See Definition 2.2. Let $e = e(i)$ be the edge of Γ numbered by i that goes from $v(j)$ to $v(k)$. Consider the map $p(\Gamma,i) = p(\Gamma,e(i))\colon C(M,\Gamma) \to C_2(M)$ that lifts $(m_1,\dots,m_{3n}) \mapsto (m_j,m_k)$, continuously. Let $H_i(\Gamma) \subset C(M_{[0,4]},\Gamma)$ be defined as

$$H_i(\Gamma) = p(\Gamma,i)^{-1}\left(H_i \cup k\delta(t)\pi^{-1}(X_i)\right)$$

where $\pi^{-1}(X_i) = \pi^{-1}_{|\tilde{C}_2(M_{[0,4]})\setminus\tilde{C}_2(M_{[0,2[})}(X_i) \subset \tilde{C}_2(M_{[0,4]})\setminus\tilde{C}_2(M_{[0,2[})$ and H_i are seen as chains with coefficients. Then $H_i(\Gamma)$ is a chain (here, a combination of manifolds with boundaries) whose points (that project to) $(m_1,\dots,m_{3n})$ are continuously equipped with rational homology classes of paths from m_j to m_k. It is cooriented in $C(M,\Gamma)$ (that is oriented like M^{2n}) by the coorientation of H_i in $C_2(M)$.

Then the intersection of the $H_i(\Gamma)$, for $i \in \{1,2,\dots,3n\}$, is a compact subspace $I(\Gamma,\{H_j\})$ of $C(M_{[0,4]},\Gamma)$. Its image $p(\Gamma,i)(I(\Gamma,\{H_j\}))$ in $H_i \cup \pi^{-1}(X_i)$ is a compact subset of $H_i \cup \pi^{-1}(X_i)$.

DEFINITION 2.3. The H_i are said to be in *general $3n$-position* if, for any $\Gamma \in \mathcal{S}_n$,

- $I(\Gamma,\{H_j\})$ is finite,
- $p(\Gamma,i)(I(\Gamma,\{H_j\}))$ is made of points in the interiors of the 4-simplices of H_i or in the interior of $\pi^{-1}(X_i)$, for any $i \in \{1,2,\dots,3n\}$, and,
- all the $H_i(\Gamma)$ intersect transversally at the corresponding points that are in the interior of $C(M_{[0,4]},\Gamma)$.

The fact that such H_i exist will be proved in Subsection 2.5.

Fix $\Gamma \in \mathcal{S}_n$. Under the given assumptions, the cooriented $H_i(\Gamma)$ only intersect transversally at distinct points in the interior of $C(M_{[0,4]},\Gamma)$. We define their equivariant algebraic intersection $I_\Gamma(\{H_i\}) \in \mathcal{A}^h_n(\delta)$ as follows.

Consider an intersection point m, such that $p(M,\Gamma)(m) = (m_1,\dots,m_{2n}) \in M^{2n}$, it is equipped with a sign $\varepsilon(m)$ as usual, and it is also equipped with the following additional data: Associate m_j with $v(j)$. Since m belongs to $H_i(\Gamma)$, the edge $e(i)$ from $v(j)$ to $v(k)$ is equipped with a rational homology class of paths from m_j to m_k. Each edge of Γ is equipped with a rational homology class of paths between its ends in this way. Choose a simply connected graph γ (e. g. a spider) in M that contains a basepoint and all the points m_j of M. A path from m_j to m_k can be composed with paths in the graph γ to become a loop based at the basepoint whose rational homology class is well-determined by the rational homology class of the path and by γ. It reads $n(m,e(i),\gamma)[K]$ for some integer $n(m,e(i),\gamma)$. Let $\Gamma(m,\gamma)$ be the graph obtained from Γ by assigning $t^{-n(m,e(i),\gamma)}/(k\delta(t))$ to the oriented edge $e(i)$ for each edge. Orient the vertices of Γ so that the permutation of the half edges from (first half of first edge, second half of first edge, ..., second half of last edge) to (half-edges of the first vertex ordered in a way compatible with the vertex orientation, ...) is even.

Note that the class of $\Gamma(m,\gamma)$ in $\mathcal{A}^h_n(\delta)$ does not depend on γ. Indeed, changing the path that goes from m_j to the basepoint in γ amounts to add some fix integer to $n(m,e,\gamma)$ for the edges e going to m_j and to remove this integer from $n(m,e,\gamma)$ for the edges starting at m_j. Because of Relations 1 and 3, it does not change the class of $\Gamma(m,\gamma)$ that will be denoted by $[\Gamma(m)]$. Finally, assign $\varepsilon(m)[\Gamma(m)] \in \mathcal{A}^h_n(\delta)$ to m.

Then define $I_\Gamma(\{H_i\}) \in \mathcal{A}^h_n(\delta)$ as the sum over the intersection points of the $\varepsilon(m)[\Gamma(m)]$.

THEOREM 2.4.

$$\tilde{z}_n(M,\mathbb{K},\tau) = \sum_{\Gamma\in\mathcal{S}_n} \frac{I_\Gamma(\{H_i\})}{2^{3n}(3n)!(2n)!} \in \mathcal{A}_n^h(\delta)$$

is an invariant of $(M,\mathbb{K},\tau)$. *Let* ξ_n *be the element of* $\mathcal{A}_n^h(1)$ *defined in* [**26**, Section 1.6]. *It is an element of the subspace of* $\mathcal{A}_n^h(\delta)$ *generated by the diagrams whose edges are beaded by* 1, *and it is zero when* n *is even. Then*

$$\tilde{z}_n(M,\mathbb{K}) = \tilde{z}_n(M,\mathbb{K},\tau) + \frac{p_1(\tau)}{4}\xi_n$$

is an invariant of $(M,\mathbb{K})$.

The collection $(\Delta(M),(\tilde{z}_n(M,\mathbb{K}))_{n\in\mathbb{N}})$ should be equivalent to the Kricker rational lift of the Kontsevich integral.

Note that with the map ψ of the end of Subsection 2.1 we have

$$\mathcal{Q}(M,\mathbb{K}) = \psi(\tilde{z}_1(M,\mathbb{K})).$$

The properties of the chains H_i that are proved in [**28**] will allow us to generalize properties of $\mathcal{Q}(M,\mathbb{K})$ to $(\tilde{z}_n(M,\mathbb{K}))_{n\in\mathbb{N}}$ in a future work.

REMARK 2.5. Though the graphs Γ of $\mathcal{S}_n$ in Theorem 2.4 have no loops, it is shown in [**28**, Section 9.2] that the evaluation of $\tilde{z}_1$ on a dumbbell-shaped clasper, whose looped edges are equipped with the equivariant linking numbers of the corresponding leaves, is the corresponding element  of $\mathcal{A}_h^1(\delta)$. Similar properties are expected for all the $\tilde{z}_n$.

2.4. Regular points of $(S^2_H)^{3n}$. The restriction of p to

$$\left(\tilde{C}_2(M_{[0,4]}) \setminus \tilde{C}_2(M_{[0,2[})\right) \cap \pi^{-1}(\overline{S^2_H})$$

is a diffeomorphism p_H onto its image P_H in $\left(C_2(M_{[0,4]}) \setminus C_2(M_{[0,2[})\right)$. Define $\overline{\pi}$ as $\pi \circ p_H^{-1}$ on P_H.

Consider a graph $\Gamma \in \mathcal{S}_n$, its set of vertices $V(\Gamma)$, a coloring $L\colon V(\Gamma) \to \{M_{[0,2]}, M_{[2,4]}\}$ of its vertices, and a set B of edges of Γ that contains all the edges between two vertices of color $M_{[0,2]}$. Set $A = \{1,\dots,3n\}\setminus B$ and

$$C(\Gamma,L,B) = p(M,\Gamma)^{-1}\left(M_{[0,2]}^{L^{-1}(\{M_{[0,2]}\})} \times M_{[2,4]}^{L^{-1}(\{M_{[2,4]}\})}\right) \cap \cap_{a\in A} p(\Gamma,a)^{-1}(P_H).$$

Consider the product

$$g(\Gamma,L,B) = \prod_{a\in A} \overline{\pi}\circ p(\Gamma,a) \times \text{Identity}(S^2)^B \colon C(\Gamma,L,B)\times (S^2)^B \to (S^2)^A \times (S^2)^B.$$

A *regular value* of this map $g(\Gamma,L,B)$ is a point Y of the target $(S^2)^{3n}$ such that for any point y of $g(\Gamma,L,B)^{-1}(Y)$, for any face $\mathcal{F}$ of $C(\Gamma,L,B)$, the tangent map of $g(\Gamma,L,B)$ at y is surjective, and, if $y\in\mathcal{F}\times(S^2)^B$, the tangent map of $g(\Gamma,L,B)_{|\mathcal{F}\times(S^2)^B}$ at y is surjective.

LEMMA 2.6. *The set of regular values of* $g(\Gamma,L,B)_{|g(\Gamma,L,B)^{-1}((S^2_H)^{3n})}$ *is open and dense in* $(S^2_H)^{3n}$.

PROOF. The density is a direct corollary of the Morse-Sard theorem [**19**, Chapter 3, Section 1]. Since the set where the tangent maps are not all surjective is closed in the compact source of $g(\Gamma, L, B)$, its image is compact, and the set of regular values is open. □

Thus the finite intersection of the sets of regular values of such maps $g(\Gamma, L, B)$ is also dense and open.

DEFINITION 2.7. Let $P_V(X_i)$ be the vertical plane of $\mathbb{R}^3$ that contains X_i. An element $(X_i)_i$ of $(S^2_H)^{3n}$ is *regular* if it is a regular value for all the maps $g(\Gamma, L, B)$, and if, for any pair $\{i,j\}$ of $\{1,2,\ldots,3n\}$, $P_V(X_i) \cap P_V(X_j)$ is the vertical line.

The set of regular $(X_i)_i$ is dense and open in $(S^2_H)^{3n}$.

LEMMA 2.8. *If $(X_1, \ldots, X_{3n})$ is regular, then the $H_i(\Gamma)$ do not intersect over the small diagonal of M^{2n} in $C_{2n}(M)$.*

PROOF. If they did, there would be an intersection point in the small diagonal of $C_{2n}(M)$. Such a point would be a configuration in the tangent space T_mM of M at some point m of M, where T_mM is identified with $\mathbb{R}^3$ via τ. Let D be a 3–ball in $M_{[3,3+\varepsilon]}$. Note that $C(D,\Gamma)$ embeds in $C(\mathbb{R}^3,\Gamma)$ and that $\overline{\pi} \circ p(\Gamma, a)$ reads as the map "direction of $e(a)$", there. This map factors through the global translations and the homotheties with positive ratio. Let $L_{[2,4]}$ be the constant coloring with value $M_{[2,4]}$. Since $(X_1, \ldots, X_{3n})$ is regular, it belongs neither to the –at most $(6n-4)$-dimensional– $p(\Gamma, L_{[2,4]}, \emptyset)(C(D,\Gamma) \cap C(\Gamma, L_{[2,4]}, \emptyset))$, nor to the –at most $(6n-2)$-dimensional– $p(\Gamma, L_{[2,4]}, \{e(i)\})((C(D,\Gamma) \cap C(\Gamma, L_{[2,4]}, \{e(i)\})) \times S^2)$ for some edge e(i), when m belongs to K_i. □

2.5. Transversality. In this subsection, we prove that there exist H_i in general $3n$-position.

It follows from general transversality properties that the $H_i(\Gamma)$ can be perturbed so that they have a finite number of transverse intersection points (see [**19**, Chapter 3]). The subtleties here are

- that we want simultaneous transversality for all the $\Gamma \in \mathcal{S}_n$,
- that we would like to perturb the H_i rather than the $H_i(\Gamma)$[1].

To prove the existence of H_i in general $3n$-position, we proceed as follows.

Choose cycles H_i^0, together with their simplicial decompositions, with the given boundaries, so that the H_i^0 are transverse to $\partial\tilde{C}_2(M_{[0,2]})$.

Assume that

- the 4-dimensional simplices of H_i^0 are C^∞ embeddings indexed by elements j of some finite set $S(i)$, and equipped with local coefficients,
$$\Delta_j^0(i)\colon \Delta^4 \to O_j(i) \subset C_2(M_{[0,2]})$$
of the standard 4-simplex Δ^4 in open regions $O_j(i)$ of $C_2(M_{[0,2]})$ such that there are C^∞ embeddings
$$\Phi_j(i)\colon O_j(i) \to \mathbb{R}^6$$
with open image in some intersection of half-spaces in $\mathbb{R}^6$,

[1]The $H_i(\Gamma)$ are associated with H_i and Γ, and the face identifications, that will be performed later, could not be performed without the fact that $H_i(\Gamma)$ is precisely the preimage of some fixed $H_i \cup k\delta(t)\pi^{-1}(X_i)$.

- there is a set B_i of 3-dimensional faces of the 4-dimensional simplices of H_i^0 that constitute a simplicial decomposition of ∂H_i^0,
- there is a bijection b_i of the set of the 3-dimensional faces of the $\Delta_j^0(i)$ that are not in B_i to $A_i \times \{+,-\}$, for some finite set A_i, such that for any $a \in A_i$, the face labeled by $(a,+)$ has the same image in $C_2(M)$ than the face labeled by $(a,-)$ with opposite boundary orientation and matching local coefficients (this is the *face identification*),
- the simplices $\delta_r^0(i)$ of dimension k ($k \leq 4$) of H_i^0 are codimension 0 submanifolds with boundaries of the preimage of a regular point $N_{i,r}$ under a constraint map $f(\delta_r(i))$ from the intersection of the open sets $O_j(i)$ such that $\delta_r^0(i)$ is a face of $\Delta_j^0(i)$ to S^{6-k}.

Define $\mathcal{M}(H_i^0)$ as the set of deformations $H_i = \{\Delta_j(i) \colon \Delta^4 \to O_j(i)\}_{j \in S(i)}$ of the $H_i^0 = \{\Delta_j^0(i)\}_{j \in S(i)}$, where the set $S(i)$ of simplices, the face identification and the simplices of ∂H_i are fixed.

Equip $\mathcal{M}(H_i^0)$ with the C^∞ topology that is its topology of a subset of

$$\prod_{j \in S(i)} C^\infty(\Delta^4, O_j(i))$$

where the topology of $C^\infty(\Delta^4, O_j(i))$ is defined in [**19**, Chapter 2, Section 1].

Consider the topological space

$$\mathcal{M} = \prod_{i=1}^{3n} \mathcal{M}(H_i^0).$$

Define the subset $\mathcal{M}_*$ of $\mathcal{M}$ made of the $(H_1, H_2, \ldots, H_{3n})$ whose simplices satisfy:
For each $\Gamma \in \mathcal{S}_n$, for any $3n$-tuple $(\delta_{j_i}(i) \subset H_i \cup \pi^{-1}(X_i))_{i \in \{1,\ldots,3n\}}$ of simplices, where $\pi^{-1}(X_i)$ and $\partial\pi^{-1}(X_i)$ are abusively considered as additional simplices $\Delta_0(i)$ and $\partial\Delta_0(i)$ and the constraint map associated to $(\Delta_0(i) \subset (O_0(i) = \overline{\pi}^{-1}(S_H^2)))$ is $\overline{\pi}$,
$*(\Gamma, \delta_{j_i}(i))$: the $p(\Gamma,i)^{-1}(\delta_{j_i}(i))$ are transverse in $C(M_{[0,4]}, \Gamma)$.

PROPOSITION 2.9. *$\mathcal{M}_*$ is open and dense in $\mathcal{M}$. In particular, there exists a $(3n)$-tuple $(H_1, H_2, \ldots, H_{3n})$ in general $3n$-position.*

PROOF. It is enough to prove that the subspace of $\mathcal{M}$ made of the $(H_1, \ldots, H_{3n})$ that satisfy $*(\Gamma, \delta_{j_i}(i))$ is open and dense for each $\Gamma \in \mathcal{S}_n$, for any $3n$-tuple $(\delta_{j_i}(i) \subset H_i)_{i \in \{1,\ldots,3n\}}$ of (generalized) simplices.

It is obviously open and we only need to prove that it is dense.

Consider the regions where the $\delta_{j_i}(i)_{i \in \{1,\ldots,3n\}}$ live. The intersection of their preimages under the $p(\Gamma, i)$ determines an open subspace of

$$A_I = \cap_{i \in I} p(\Gamma,i)^{-1}(C_2(M_{[0,4]}) \setminus C_2(M_{[0,2[})) \cap \cap_{j \in J} p(\Gamma,j)^{-1}(C_2(M_{[0,2]}))$$

for two complementary subsets I and J of $\{1, 2, \ldots, 3n\}$, where the transversality of the $\delta_{j_i}(i)$ reads: $(N_{i,j_i})_{i \in \{1,\ldots,3n\}}$ is a regular value of the constraint map $\prod_{i=1}^{3n} f(\delta_{j_i}(i)) \circ p(\Gamma, i)$. According to the Morse-Sard theorem, the set of regular $(N_i)_{i \in \{1,\ldots,3n\}}$ is dense. It is furthermore open.

Therefore, if none of the $\delta_{j_i}(i)$ touches the boundary, it is easy to slightly move the simplices by changing the preimage of N_{i,j_i} to the preimage of some closed point. (This can be done by a global isotopy of $C_2(M_{[0,2]})$ supported near $\delta_{j_i}(i)$ that moves $\delta_{j_i}(i)$ to the preimage of some close point, when $j_i \neq 0$.)

In general, we have a set of simplices in the boundary, with indices in I_b, a set of simplices that don't douch the boundary with indices in I_e, and the other ones with indices in I_t that have a maximal face $\mathcal{F}_i$ in the boundary. (Here, the special simplex $\pi^{-1}(X_i)$ is considered in the boundary.) Since $(X_1, \dots, X_{3n})$ is regular, the intersection of the $p(\Gamma, i)^{-1}(\partial H_i \cup k\delta(t)\pi^{-1}(X_i))$, for $i \in I_b \cup I_t$, is transverse in the faces of A_I. Therefore, it is a $(6n - 2(\sharp I_b + \sharp I_t) - c)$-submanifold of the codimension c faces of A_I. The intersection of the $\delta_{j_i}(i)$, for $i \in I_e$ will be transverse to this manifold if the restriction of $\prod_{i\in I_e} f(\delta_{j_i}(i)) \circ p(\Gamma, i)$ to this manifold is regular, and this can be achieved as before. Then the transversality holds in a neighborhood of the $\mathcal{F}_i$, and we can make it hold for all the simplices by applying the above argument to $(I_b, I_e \cup I_t)$ instead of $(I_b \cup I_t, I_e)$, the openness allowing us to recorrect the $\delta_{j_i}(i)$ for $i \in I_t$ near the boundaries without losing transversality. □

2.6. Proof of invariance. In this section, we prove that we have defined a topological invariant of $(M, \mathbb{K}, \tau)$. We first prove that our construction yields a well-defined function $\tilde{z}_n$ of $((M, M_{[1,8]}), \mathbb{K}, \tau, f_M, \varepsilon, (K_i)_i)$, by showing that our definition is independent of $(X_1, \dots, X_{3n})$ and $(H_1, \dots, H_{3n})$, i.e. that it is independent of $(H_1, \dots, H_{3n})$ when $(X_1, \dots, X_{3n})$ is not fixed.

Since the invariance is obvious when $(H_1, \dots, H_{3n})$ moves in a neighborhood of $(H_1, \dots, H_{3n})$ in general $3n$-position, it is enough to prove invariance when H_1 is changed to some H_1' such that $(H_1', H_2, \dots, H_{3n})$ is in general $3n$-position.

First pick a C^∞-path $P(X_1, X_1')$ from X_1 to X_1' in S^2_H whose image is some codimension 0 submanifold with boundary of a submanifold defined by $f_P(x) = 0$ for some smooth map f_P that is defined in a regular neighborhood of the image of this path and valued in $[-1, 1]$. Without loss assume that $(0, X_2, X_3, \dots, X_{3n})$ is a regular value for all the compositions of the maps $g(\Gamma, L, B)$ of Section 2.4 by f_P on the first coordinate. (If it is not the case, use the Morse-Sard theorem to find a close regular value and move everything slightly.)

LEMMA 2.10. *There exists a 5-chain C_0 of $\tilde{C}_2(M_{[0,2]})$ in general $3n$-position with H_2, ..., H_{3n} such that*

$$\partial C_0 = H_1' - H_1 + k\delta(t)\left(s_\tau(M_{[0,2]}; P(X_1, X_1')) - \pi^{-1}_{|\partial\tilde{C}_2(M_{[0,2]})\setminus\partial\tilde{C}_2(M_{[0,2[})}(P(X_1, X_1'))\right).$$

PROOF. The wanted boundary ∂C_0 is a 4-cycle and $H_4(C_2(M_{[0,2]}); \mathbb{Q}[t^{\pm 1}]) = 0$. □

Here the notion of general $3n$-position is similar to the previous notion and can be achieved in the same way. Let $C = C_0 \cup k\delta(t)\pi^{-1}(P(X_1, X_1'))$.

The intersection of $p(\Gamma, 1)^{-1}(C)$ and the $H_i(\Gamma)$ for $i \geq 2$ is some C^∞ compact 1-manifold with boundary in $C(M, \Gamma)$. $p(\Gamma, 1)^{-1}(C)$ is locally seen as the zero locus of some real function except near $p(\Gamma, 1)^{-1}(\partial C)$ where some additional independent real function must be positive. Therefore,

$$\partial(p(\Gamma, 1)^{-1}(C) \cap \bigcap_{i=2}^{3n} H_i(\Gamma)) =$$

$$H_1'(\Gamma) \cap \bigcap_{i=2}^{3n} H_i(\Gamma) - H_1(\Gamma) \cap \bigcap_{i=2}^{3n} H_i(\Gamma) + p(\Gamma, 1)^{-1}(C) \cap \partial C(M_{[0,4]}, \Gamma) \cap \bigcap_{i=2}^{3n} H_i(\Gamma).$$

Let us check that the first sign is correct. The two other ones are similar. The normal bundle of the one-manifold reads $N^1_C \oplus \bigoplus_{i=2}^{3n} N^i$ where N^1_C is the normal bundle to $p(\Gamma,1)^{-1}(C)$ in $C(M,\Gamma)$. Since H'_1 is oriented as the boundary of C, C is oriented as the outward normal N_o to C, followed by the orientation of H'_1 along H'_1. Therefore the normal bundle of $p(\Gamma,1)^{-1}(H'_1)$ is oriented by $N^1_C \oplus N_o$.

Use the notation $\langle \frac{1}{\hat{\delta}} H_1(\Gamma), \dots, \frac{1}{\hat{\delta}} H_{3n}(\Gamma)\rangle_{\mathcal{A},C(M,\Gamma)}$, where $\hat{\delta} = k\delta(t)$, to denote $I_\Gamma(\{H_i\})$, that was defined before the statement of Theorem 2.4. With similar notation, set

$$I_\Gamma(C,\{H_i\}) = -\langle \frac{1}{\hat{\delta}} p(\Gamma,1)^{-1}(C) \cap \partial C(M_{[0,4]},\Gamma), \frac{1}{\hat{\delta}} H_2(\Gamma), \dots, \frac{1}{\hat{\delta}} H_{3n}(\Gamma)\rangle_{\mathcal{A},C(M,\Gamma)}.$$

Then $I_\Gamma(C,\{H_i\})$ is the variation of $I_\Gamma(\{H_i\})$ when H_1 is changed to H'_1. The intersection points in $p(\Gamma,1)^{-1}(C) \cap \partial C(M_{[0,4]},\Gamma) \cap H_2(\Gamma) \cap \dots \cap H_{3n}(\Gamma)$ only occur inside codimension 1 faces of $C(M_{[0,4]},\Gamma)$.

LEMMA 2.11. *All these intersection points project to $M^{2n}_{[0,4[}$ under $p(M,\Gamma)$.*

PROOF. Consider the projection $(m_i)_{i=1,2,\dots,2n}$ of such an intersection point. Assume that $r(m_\ell) > 4-\varepsilon$. Construct a subgraph G of Γ such that the edges of G have an orientation independent of the edge orientation in Γ, so that G is maximal with the properties: G contains $v(\ell)$, there is an increasing path (following edge orientations in G) from $v(\ell)$ to every other vertex of G, if an edge $e(a)$ of G goes from $v(i)$ to $v(j)$ then $r(m_j) > r(m_i) - 2\varepsilon$.

Such a graph G can be constructed inductively starting from $v(\ell)$: For each constructed vertex $v(i)$ with less than three incoming edges, add each other edge of Γ adjacent to $v(i)$ whose other end $v(j)$ satisfies $r(m_j) > r(m_i) - 2\varepsilon$ and orient it from $v(i)$ to $v(j)$.

For every $v(i) \in G$, $r(m_i) > 2$.

In G, the valency of a vertex that does not belong to $e(1)$ is 2 or 3 while the valency of a vertex that belongs to $e(1)$ is at least 1. Indeed, if the edge $e(a \neq 1) = \{v(i), v(j)\}$ of Γ is not in G, and if $v(i) \in G$, then $r(m_j) \leq r(m_i) - 2\varepsilon$, and $\overline{\pi}(m_i, m_j) = \pm X_a$ so that $m_i \in P_V(X_a)$. Therefore, since $(X_a)_a$ is regular, we cannot erase more than one edge different from $e(1)$ that contains $v(i)$ from Γ.

Now, look at the constraints we know for the elements $m_j \in M_{[2,4]}$ for which $v(j) \in G$. They read $\overline{\pi}(m_i, m_j) = \pm X_b$ for internal edges $e(b \neq 1) = \{v(i), v(j)\}$ of G, and $m_j \in P_V(X_a)$ for other edges $e(a \neq 1)$ of Γ that contain $v(j)$. The vertical translation induces an action of S^1 on the space $C_{V(G)}(M_{[2,4]})$ of configurations of the vertices of G in $M_{[2,4]}$, and the constraint maps are invariant under this action. Then if no vertex of G belongs to $e(1)$, the product of the above elementary constraints maps, that only depend on the restriction of the configuration to $V(G)$, goes to a space whose dimension is the number of half-edges of Γ that contain an element of G, that is the dimension of $M^{V(G)}_{[2,4]}$, that is greater than the dimension of $C_{V(G)}(M_{[2,4]})/S^1$.

When a vertex of G belongs to $e(1)$, and when the above map is a submersion, it becomes a local diffeomorphism on $C_{V(G)}(M_{[2,4]})/S^1$. Then there are isolated preimages in $C_{V(G)}(M_{[2,4]})/S^1$ that provide isolated S^1-orbits of $C_{V(G)}(M_{[2,4]})$ that must lie in the interior of $C_{V(G)}(M_{[2,4]})$. Then $r(m_\ell) < 4$. □

Thus intersection points in $p(\Gamma,1)^{-1}(C) \cap \partial C(M_{[0,4]};\Gamma) \cap H_2(\Gamma) \cap \dots \cap H_{3n}(\Gamma)$ only occur inside codimension 1 faces of $C(M,\Gamma)$.

Let us study some point P in this intersection. Its codimension one face corresponds to some connected subgraph Γ_I made of the vertices numbered in a set I and the edges between two of these, as in Subsection 2.2. Define the graph $\Gamma(I) = \Gamma/\Gamma_I$ obtained from Γ by identifying Γ_I with a point $v(I)$. The vertices of $\Gamma(I)$ are the vertices of Γ numbered in $\{1, \dots, 2n\} \setminus I$ and the additional vertex $v(I)$, and the edges of $\Gamma(I)$ are the edges of Γ that are not edges of Γ_I. The set $E(\Gamma)$ of edges of Γ is the disjoint union of the sets $E(\Gamma_I)$ and $E(\Gamma(I))$.

Our face $\mathcal{F}(\Gamma, I)$ fibers over the configuration space of the vertices of $\Gamma(I)$ that form the set $V(\Gamma(I))$, and our intersection point projects to $M_{[0,4[}^{V(\Gamma(I))}$. The fiber is made of the infinitesimal configurations of I up to translation and dilation. The dimension of the fiber is $(3\sharp I - 4)$ and the dimension of the base is $3(\sharp V(\Gamma(I))) = 6n - 3\sharp I + 3$.

Since the intersection is transverse, the edges give $(6n-1)$ independent constraints on our intersection point P, where the edge numbered by 1 gives one of these constraints and each of the other ones gives two constraints.

The constraints coming from the edges of $\Gamma(I)$ only concern the base, while the constraints of $E(\Gamma_I)$ only concern the fiber except possibly for one edge $(e(j) \in E(\Gamma_I))$ where the constraint can read: "The point $v(I) \in K_j$."

In this case, this constraint is a constraint for the base, and $j \neq 1$. If we have such an exceptional constraint "$v(I) \in K_j$", set $\chi_k = 1$. Otherwise, set $\chi_k = 0$.

If the edge $e(1)$ belongs to $E(\Gamma_I)$, set $\chi_A = 1$, and if it belongs to $E(\Gamma(I))$, set $\chi_A = 0$.

Then we have $2\sharp E(\Gamma_I) - 2\chi_k - \chi_A$ constraints on the $(3\sharp I - 4)$–dimensional fiber.

The transversality condition tells us that $2\sharp E(\Gamma_I) - 2\chi_k - \chi_A = 3\sharp I - 4$ that is

$$4 - 2\chi_k - \chi_A = 3\sharp I - 2\sharp E(\Gamma_I)$$

where the right-hand side is the valency of the vertex $v(I)$ in $\Gamma(I)$ as a count of half-edges shows. This valency must be 4, 3, 2 or 1.

The transversality also tells us that the map "direction of the edges" from the $(3\sharp I - 4)$–dimensional fiber to the product of the images of the constraint functions must be a local diffeomorphism.

In particular, unless $\sharp I = 2$, there can be no vertex of valency 1 in Γ_I because moving this vertex in the direction of the unique edge incident to such a vertex would not change the image of the above map.

Thus, either Γ_I is an edge, and this case will be treated by the IHX (or Jacobi) identification, or Γ_I is connected, all the vertices of Γ_I have valency 2 or 3 and there are χ vertices of valency 2 where $1 \leq \chi \leq 4$, and this case will be treated by the parallelogram identification.

Let us explain how these classical identifications allow us to prove that the sum of the contributions of all such intersection points to

$$\mathcal{V}(C, \{H_i\}) = \sum_{\Gamma \in \mathcal{S}_n} I_\Gamma(C, \{H_i\})$$

vanishes in our setting. We start with the parallelogram case.

LEMMA 2.12. *Let E be the set of codimension one faces $\mathcal{F}(\Gamma, I)$ of spaces $C(M_{[0,4[}, \Gamma)$ indexed by some (Γ, I) where $\Gamma \in \mathcal{S}_n$, I is a subset of $\{1, 2, \dots, 2n\}$, Γ_I is connected, the vertices of Γ_I have valency 2 or 3 and at least one vertex of*

Γ_I has valency 2. There is an involution $\hat{\rho}$ of E without fixed point such that the sum of the contributions of the intersection points in two faces $\mathcal{F}_1$ and $\hat{\rho}(\mathcal{F}_1)$ to $\mathcal{V}(C, \{H_i\})$ vanishes.

PROOF. Fix Γ and I as above. Let i be the smallest element of I such that the valency of $v(i)$ in Γ_I is 2. Define the labelled graph $\rho(\Gamma, I) \in \mathcal{S}_n$ from Γ by modifying the two edges $e(a)$ and $e(b)$ incident to $v(i)$ in Γ_I as follows: if $e(a)$ reads $\varepsilon(a)\overrightarrow{v(i)v(j)}$ and if $e(b)$ reads $\varepsilon(b)\overrightarrow{v(i)v(k)}$ in Γ, then $e(a)$ reads $\varepsilon(a)\overrightarrow{v(k)v(i)}$ and $e(b)$ reads $\varepsilon(b)\overrightarrow{v(j)v(i)}$ in $\rho(\Gamma, I)$. Define $\hat{\rho}$ so that

$$\hat{\rho}(\mathcal{F}(\Gamma, I)) = \mathcal{F}(\rho(\Gamma, I), I).$$

Consider the orientation-reversing involutive diffeomorphism $\phi(\hat{\rho})$ from $\mathcal{F}(\Gamma, I)$ to $\hat{\rho}(\mathcal{F}(\Gamma, I))$ that maps a point P to the point where the only changed piece of data is the position of $v(i)$ in the infinitesimal configuration of I where $v(i)$ is mapped to its symmetric with respect to the middle of the two (possibly coinciding) other ends of $e(a)$ and $e(b)$. The standard properties of a parallelogram guarantee that for any edge label j, $p(\Gamma, j) = p(\rho(\Gamma, I), j) \circ \phi(\hat{\rho})$ on $\mathcal{F}(\Gamma, I)$.

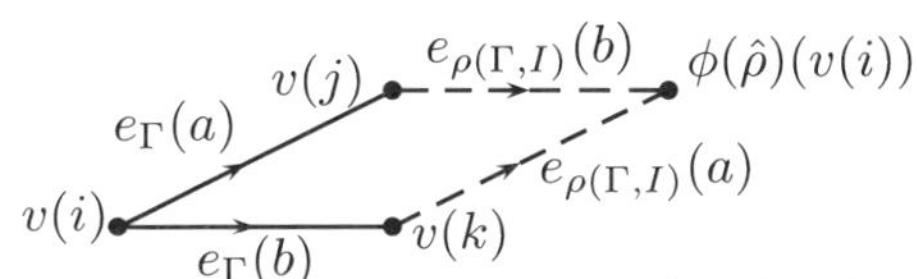

Consider a point P of $\mathcal{F}(\Gamma, I)$ that contributes to $I_\Gamma(C, \{H_i\})$. Recall that the chains C and $H_a \cup k\delta(t)\pi^{-1}(X_a)$ are transverse to $\partial C_2(M_{[0,4[})$. They are made of two kinds of pieces there: $(-J_\Delta)ST(M)_{|K_a}$ and $s_\tau(M; P(X_1, X_1'))$ or $s_\tau(M; X_a)$. Since our assumptions imply that the intersection is transverse, if $\sharp I > 2$, P cannot be in the parts $p(\Gamma, a)^{-1}(ST(M)_{|K_a})$ or $p(\Gamma, b)^{-1}(ST(M)_{|K_b})$ because in this case the constraint coming from the other edge adjacent to P gives the direction of this other edge and stretching this other edge leads to a non trivial kernel of the constraint map. Therefore, if $\sharp I > 2$, the two edges can be assumed to be weighted by $1 = \frac{k\delta(t)}{k\delta(t)}$. Thus, when a point P as above contributes to the variation $I_\Gamma(C, \{H_i\})$, the point $\phi(\hat{\rho})(P)$ contributes with the same beaded graph with the opposite sign to $I_{\rho(\Gamma,I)}(C, \{H_i\})$. (Indeed, the product of the constraint maps only differs by the orientation-reversing diffeomorphism $\phi(\hat{\rho})$ of the faces, and the graphs are obtained from one another by an exchange of two edge labels and a possible simultaneous reverse of two edge orientations, for two edges beaded by 1.) If $\sharp I = 2$, our assumptions imply that Γ_I is made of two edges $e(a)$ and $e(b)$ between the two elements i and j of I like in the figure.

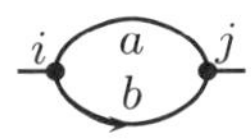

In $\rho(\Gamma, I)$, the orientations of both edges are reversed. Up to exchanging a and b, $P \in p(\Gamma, a)^{-1}(s_\tau(M; X_a)) \cap (-J_\Delta)p(\Gamma, b)^{-1}(ST(M)_{|K_b})$, and the rational functions are ± 1 and $\pm J_\Delta$.

Let us show that we have the same cancellation as before in this case. Let $K_b \times D_b$ denote a trivialised tubular neighborhood of K_b. Then the local constraint map associated to $(-1/(J_\Delta\hat{\delta}))H_b$ near $ST(M)_{|K_b}$ can be viewed as a map $f_b'\colon ST(M) \to D_b$ associated with the natural projection of $K_b \times D_b$ to the disk D_b.

Let P' be a transverse intersection point in

$$p(\Gamma, a)^{-1}(s_\tau(M; X_a)) \cap p(\Gamma, b)^{-1}(ST(M)_{|K_b}).$$

Let η be the local degree of $(\prod_{e\neq b} f_e \circ p(\Gamma, e), f'_b \circ p(\Gamma, b))_{|\mathcal{F}(\Gamma,I)}$ at P', where the f_e are local constraint maps associated to the $H_e \cup k\delta(t)\pi^{-1}(X_e)$. Then P' contributes as

η 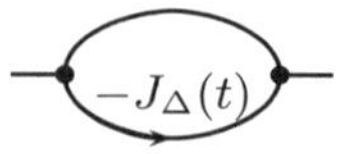 .

As before, the local degree of $(\prod_{e\neq b} f_e \circ p(\rho(\Gamma, I), e), f'_b \circ p(\rho(\Gamma, I), b))_{|\mathcal{F}(\rho(\Gamma,I),I)}$ at $\phi(\hat\rho)(P')$ is $(-\eta)$ since this map is the composition of the former one by $\phi(\hat\rho)^{-1}$.

Let ι denote the orientation-reversing involution of $C_2(M)$ that extends the exchange of points in $(M^2 \setminus \mathrm{diag})$ and let ι_{S^2} denote the antipode of S^2.

Focus on the edges $e(a)$ and $e(b)$. Without loss, consider η as the degree of

$$f_a \times f'_b \colon (ST(D_b) =_\tau S^2 \times D_b) \to S^2 \times D_b$$

where $(ST(M) = \partial C_2(M))$ and $ST(D_b)$ are oriented by the orientation of $e_\Gamma(b)$, and f_a is the projection of $ST(D_b)$ on S^2 if the orientations of $e_\Gamma(a)$ and $e_\Gamma(b)$ coincide, and its composition by ι_{S^2} otherwise.

With these orientations and conventions, $e_{\rho(\Gamma,I)}(b)$ is equipped with

$$-J_\Delta(t)\iota(ST(M)_{|K_b}) = -J_\Delta(t^{-1})ST(M)_{|K_b}$$

while $(-\eta)$ is the degree of $(f_a \circ \iota) \times (f'_b \circ \iota) = (\iota_{S^2} \circ f_a) \times f'_b$. Then $\phi(\hat\rho)(P')$ contributes as $(-\eta)$ 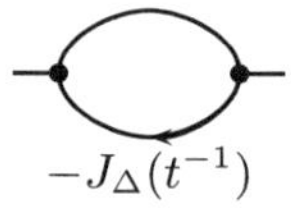 $= (-\eta)$ 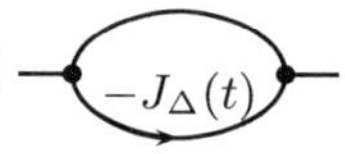 . □

LEMMA 2.13. *The intersection points of the faces $\mathcal{F}(\Gamma, I)$ where Γ_I is an edge do not make the sum $\tilde z_n(M, \mathbb{K}, \tau)$ vary, thanks to the IHX relation.*

PROOF. Let f be the label of the unique edge of Γ_I. Like in the previous proof, notice that an intersection point P in such a face cannot come from the $p(\Gamma, f)^{-1}(ST(M)_{|K_f})$ part. In particular, the edge $e(f) = \Gamma_I$ is beaded by 1. Consider the graph $\Gamma(I) = \Gamma/\Gamma_I$. Its vertex $v(I)$ has 4 incident edges $e(a)$, $e(b)$, $e(c)$, $e(d)$, and, with the natural identifications, in the initial graph Γ, $e(f)$ goes from $v(i)$ to $v(j)$ where $v(i)$ is incident to two edges among $e(a)$, $e(b)$, $e(c)$, $e(d)$ and $v(j)$ is incident to the two other ones. For $g = b$, c or d, let Γ^g be the graph of $\mathcal{S}_n$ such that $\Gamma^g(I = \{i, j\}) = \Gamma(I)$, and $v(i)$ is incident to $e(a)$ and $e(g)$. It is enough to prove that the contributions of the possible intersection points of the three faces indexed by (Γ^g, I) cancel in the sum $\mathcal{V}(C, \{H_i\})$. Our intersection point P of $C(M, \Gamma)$ will give a transverse intersection point in the three faces, with a common sign, and Γ^b, Γ^c and Γ^d are oriented so that their vertex orientations coincide outside the disk of the following picture, and they are given by the picture for the two vertices there. Indeed the roles of the half-edges indexed by b, c and d are cyclically permuted.

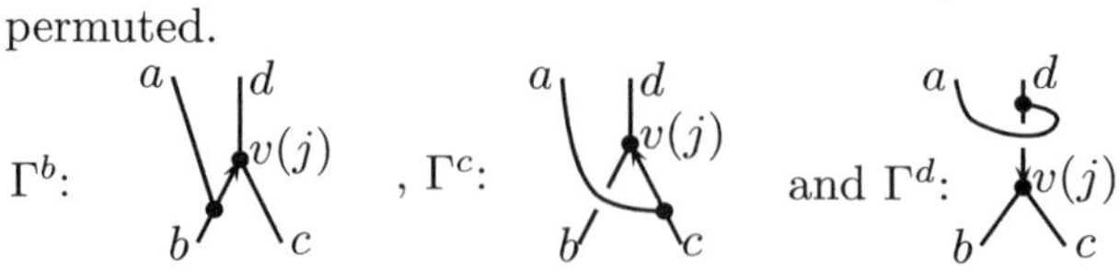

□

Let us now prove that $\tilde z_n((M, M_{[1,8]}), \mathbb{K}, \tau, f_M, \varepsilon, (K_i)_i)$ does not depend on f_M.

PROOF. Let $g_M : M \to S^1$ be a map that coincides with f_M on $M_{[1,8]}$. Then the graphs $G(f_M)$ and $G(g_M)$ of the restrictions of f_M and g_M to $M_{[0,1]}$ cobound a 4-chain C_M in $M_{[0,1]} \times S^1$. $\partial C_M = G(g_M) - G(f_M)$. The intersection of $\pi(f_M)^{-1}(X_i)$ with $M_{[0,1]} \times M_{[2,4]}$ reads $\{(m, (q_S(t, z_W(X_i)), z(t)z(X_i)f_M(m))); m \in M_{[0,1]}, t \in [2,4]\}$. Let $C(C_M, X_i) = \hat{\delta}A(C_M, X_i) + B(C_M, X_i)$ where

$$A(C_M, X_i) = \{(m, (q_S(t, z_W(X_i)), z(t)z(X_i)c)); (m, c) \in C_M, t \in [1.5, 4]\},$$

and

$$B(C_M, X_i) = \{((q_S(t, -z_W(X_i)), (z(t)z(X_i))^{-1}c), m); (m, c) \in C_M, t \in [1.5, 4]\}$$

oriented by (m, c, t). Define H'_i so that $H'_i = (H_i + \partial C(C_M, X_i)) \cap C_2(M_{[0,2]})$. As before, we can assume that, for any $i \in \{1, \dots, 3n\}$, the $H'_j(\Gamma)$ for $j < i$, $p(\Gamma, i)^{-1}(C(C_M, X_i))$ and the $H_j(\Gamma)$ for $j > i$, are transverse. Then the sum over $\Gamma \in \mathcal{S}_n$ of the variations

$$\langle \frac{1}{\hat{\delta}} H'_1, \dots, \frac{1}{\hat{\delta}} H'_{i-1}, -\frac{1}{\hat{\delta}} p(\Gamma, i)^{-1} C(C_M, X_i) \cap \partial C(M_{[0,4]}, \Gamma), \frac{1}{\hat{\delta}} H_{i+1}, \dots, \frac{1}{\hat{\delta}} H_{3n} \rangle_{\mathcal{A}}$$

(where H'_j and H_j stand for $H'_j(\Gamma)$ and $H_j(\Gamma)$) vanishes as above for any $i \in \{1, \dots, 3n\}$, and $\tilde{z}_n$ does not depend on the choice of f_M. □

Now, $\tilde{z}_n(M, \mathbb{K}, \tau, \varepsilon, (K_i)_i)$ is a locally constant function of $(\varepsilon, (K_i)_i)$ that is therefore a topological invariant of $(M, \mathbb{K}, \tau)$.

2.7. The dependence on the trivialisation. We now compute the variation of $\tilde{z}_n(M, \mathbb{K}, \tau)$ under a change of trivialisation. Let τ' be a trivialisation that coincides with τ on the tubular neighborhood $M_{[1,8]}$ of K. We want to compute the variation caused by the replacement of the H_i by chains H'_i of $C_2(M_{[0,2]})$ where $\partial H'_i = k\delta(t)(s_{\tau'}(M; X_i) \cup \pi^{-1}_{|\partial \tilde{C}_2(M_{[0,2]}) \setminus \partial \tilde{C}_2(M_{[0,2[})}(X_i) - J_\Delta(t) ST(M)_{|K_i})$. There exists a cobordism c_i between $s_\tau(M_{[0,1]}; X_i)$ and $s_{\tau'}(M_{[0,1]}; X_i)$ in $ST(M_{[0,1]})$ [**28**, Lemma 5.10], and there is a 5-chain C_i of $C_2(M_{[0,2]})$ such that

$$\partial C_i = H'_i - H_i \cup k\delta(t) c_i.$$

Then

$$\tilde{z}_n(M, \mathbb{K}, \tau') - \tilde{z}_n(M, \mathbb{K}, \tau) = \sum_{\Gamma \in \mathcal{S}_n} \sum_{i=1}^{3n} \mathcal{V}(\Gamma, i)$$

where $\mathcal{V}(\Gamma, i)$ equals

$$\langle \frac{1}{\hat{\delta}} H'_1, \dots, \frac{1}{\hat{\delta}} H'_{i-1}, -\frac{1}{\hat{\delta}} p(\Gamma, i)^{-1}(C_i) \cap \partial C(M_{[0,4]}, \Gamma), \frac{1}{\hat{\delta}} H_{i+1}, \dots, \frac{1}{\hat{\delta}} H_{3n} \rangle_{\mathcal{A}, C(M,\Gamma)}$$

like in the previous subsection. Now the proof of the previous subsection can be applied verbatim to cancel most of the contributions, except for the computation of the valency of $v(I)$ in $\Gamma(I)$, because if the tangent space of a point of M is identified with $\mathbb{R}^3$ via τ, constraints associated with edges of Γ_I coming from chains H'_j or C_i constrain both the fiber and the base. Therefore, the valency of $v(I)$ can be zero and this is the only case that is not covered by the cancellations of the previous subsection. In this case, $\Gamma_I = \Gamma$. Let $\mathcal{F}(\Gamma)$ denote $\mathcal{F}(\Gamma, \{1, \dots, 2n\})$. Then $\mathcal{V}(\Gamma, i)$ reads

$$\pm \langle s_{\tau'}(M; X_1), \dots, s_{\tau'}(M; X_{i-1}), p(\Gamma, i)^{-1}(c_i), s_\tau(M; X_{i+1}), \dots, s_\tau(M; X_{3n}) \rangle_{\mathcal{A}, \mathcal{F}(\Gamma)}.$$

In particular, $(\tilde{z}_n(M,\mathbb{K},\tau')-\tilde{z}_n(M,\mathbb{K},\tau))$, that does not involve any coefficient or non-trivial bead, is the same as the variation in the case of the universal finite type invariant of rational homology spheres, that is $(p_1(\tau)-p_1(\tau'))\frac{\xi_n}{4}$ like in [**26**, Section 1.6]. This finishes the proof of Theorem 2.4.

Appendix: A configuration space definition of the Casson-Walker invariant

For $r\in\mathbb{R}$, let $B(r)$ denote the ball of radius r in $\mathbb{R}^3$ that is equipped with its standard parallelisation τ_s. A rational homology sphere N may be written as $B_N\cup_{B(1)\setminus\mathring{B}(1/2)}B^3$ where B_N is a *rational homology ball*, that is a connected compact (oriented) smooth 3–manifold with boundary S^2 with the same rational homology as a point, B^3 is a 3-ball, B_N contains $(B(1)\setminus\mathring{B}(1/2))$ as a neighborhood of its boundary $\partial B_N=\partial B(1)$, and B^3 contains $(B(1)\setminus\mathring{B}(1/2))$ as a neighborhood of its boundary $\partial B^3=-\partial B(1/2)$. Let $B(N)=B_N(3)$ be obtained from $B(3)$ by replacing the unit ball $B(1)$ of $\mathbb{R}^3$ by B_N. Equip $B(N)$ with a trivialisation τ_N that coincides with τ_s outside B_N.

Let W be a compact connected 4-manifold with signature 0 and with boundary

$$\partial W=B_N(3)\cup_{\{1\}\times\partial B(3)}(-[0,1]\times\partial B(3))\cup_{\{0\}\times\partial B(3)}(-B(3)).$$

Define $p_1(\tau_N)\in(H^4(W,\partial W;\pi_3(SU(4)))=\mathbb{Z})$ as the obstruction to extend the trivialisation of $TW\otimes\mathbb{C}$ induced by τ_s and τ_N on ∂W to W. Again, we use the notation and conventions of [**31**], see also [**26**, Section 1.5].

Let $\mathbb{R}^3(N)$ be obtained from $\mathbb{R}^3$ by replacing its unit ball $B(1)$ by B_N. Let $C_2(\mathbb{R}^3(N))$ be obtained from $\mathbb{R}^3(N)^2$ by blowing up the diagonal as in Definition 1.1. Let $P\colon C_2(\mathbb{R}^3(N))\to\mathbb{R}^3(N)^2$ be the associated canonical projection and let $C_2(B(N))=P^{-1}(B(N)^2)$. Consider a smooth map $\chi\colon\mathbb{R}\to[0,1]$ that maps $]-\infty,-2]$ to 0 and $[-1,\infty[$ to 1. Define

$$\begin{array}{llll}p_{B(3)}\colon & B(3)^2\setminus\text{diagonal} & \to & S^2\\ & (U,V) & \mapsto & \frac{\chi(\|V\|-\|U\|)V-\chi(\|U\|-\|V\|)U}{\|\chi(\|V\|-\|U\|)V-\chi(\|U\|-\|V\|)U\|}\end{array}$$

This map extends to $C_2(B(3))$ to a map still denoted by $p_{B(3)}$, that reads as the projection to S^2 induced by τ_s (see Subsection 1.5) on the unit tangent bundle of $B(3)$. A similar map p_N can be defined on the boundary $\partial C_2(B(N))$: The map p_N is the projection to S^2 induced by τ_N on the unit tangent bundle of $B(N)$, and the map p_N is given by the above formula, where we set $\| U \|=1$ when $U\in B_N$, for the other points of the boundary that are pairs (U,V) of $\big(B(N)^2\setminus\text{diagonal}\big)$ where U or V belongs to $\partial B(3)$ (therefore a possible point of B_N is replaced by $0\in\mathbb{R}^3$ in the formula).

The following theorem, that gives a configuration space definition for the Casson-Walker invariant, is due to Kuperberg and Thurston [**23**] for the case of integral homology spheres (though it is stated in other words). It has been generalised to rational homology spheres in [**27**, Section 6].

THEOREM 2.14. *Let X, Y and Z be three distinct points of S^2. Under the above assumptions, for $V=X$, Y or Z, the submanifold $p_N^{-1}(V)$ of $\partial C_2(B(N))$ bounds a rational chain $F_{N,V}$ in $C_2(B(N))$, and*

$$\lambda(N)=\frac{\langle F_{N,X},F_{N,Y},F_{N,Z}\rangle_{C_2(B(N))}}{6}-\frac{p_1(\tau_N)}{24}.$$

It is easy to see that $\langle F_{N,X}, F_{N,Y}, F_{N,Z}\rangle_{C_2(B(N))}$ is a well-defined invariant of (N, τ_N). Indeed, $C_2(B(N))$ has the same rational homology as $\left((\mathbb{R}^3)^2 \setminus \text{diagonal}\right)$ that is homotopy equivalent to S^2 via the map $(x,y) \mapsto \frac{y-x}{\|y-x\|}$. Therefore, $C_2(B(N))$ has the same rational homology as S^2. In particular, since $H_3(C_2(B(N));\mathbb{Q}) = \{0\}$, the cycle $p_N^{-1}(V)$ bounds a rational chain in $C_2(B(N))$ and, since

$$H_4(C_2(B(N));\mathbb{Q}) = \{0\},$$

$\langle F_{N,X}, F_{N,Y}, F_{N,Z}\rangle_{C_2(B(N))}$ only depends on the non-intersecting boundaries of the $F_{N,V}$.

References

[1] S. Akbulut, J. McCarthy, *Casson's invariant for oriented homology 3-spheres, an exposition,* Mathematical Notes 36, Princeton University Press, Princeton 1990.

[2] D. Altschüler and L. Freidel, *Vassiliev knot invariants and Chern–Simons perturbation theory to all orders,* Comm. Math. Phys. **187** (1997), 261–287.

[3] S. Axelrod and I. M. Singer, *Chern–Simons perturbation theory,* Proceedings of the XXth International Conference on Differential Geometric Methods in Theoretical Physics, Vol. 1, 2 (New York, 1991), World Sci. Publishing, River Edge, NJ (1992), 3–45.

[4] ———, *Chern–Simons perturbation theory. II,* J. Differential Geom. **39** (1994), 173–213.

[5] D. Bar-Natan, S. Garoufalidis, L. Rozansky, and D. P. Thurston, *The Århus integral of rational homology 3-spheres. I. A highly non trivial flat connection on S^3,* Selecta Math. (N.S.) **8** (2002), 315–339.

[6] ———, *The Århus integral of rational homology 3-spheres. II. Invariance and universality,* Selecta Math. (N.S.) **8** (2002), 341–371.

[7] ———, *The Århus integral of rational homology 3-spheres. III. Relation with the Le-Murakami-Ohtsuki invariant,* Selecta Math. (N.S.) **10** (2004), 305–324.

[8] D. Bar-Natan, *Perturbative Chern–Simons Theory,* Journal of Knot Theory and its Ramifications **4** (1995), 503–547.

[9] R. Bott and A. Cattaneo, *Integral invariants of* 3*-manifolds,* J. Differential Geom. **48** (1998), 91–133.

[10] ———, *Integral invariants of* 3*-manifolds. II,* J. Differential Geom. **53** (1999), 1–13.

[11] R. Bott and C. Taubes, *On the self-linking of knots,* Jour. Math. Phys. **35** (1994), 5247–5287.

[12] A. S. Cattaneo, *Configuration space integrals and invariants for* 3*-manifolds and knots,* Low-dimensional topology (Funchal, 1998), Contemp. Math., 233, Amer. Math. Soc., Providence, RI, (1999), 153 165.

[13] W. Fulton and R. MacPherson *A compactification of configuration spaces,* Ann. of Math. **139** (1994), 183–225.

[14] S. Garoufalidis, M. Goussarov, and M. Polyak, *Calculus of clovers and finite type invariants of 3-manifolds,* Geom. Topol. **5** (2001), 75–108.

[15] S. Garoufalidis and A. Kricker, *A rational noncommutative invariant of boundary links,* Geom. Topol. **8** (2004), 115–204.

[16] E. Guadagnini, M. Martellini, and M. Mintchev, *Chern–Simons field theory and link invariants,* Nucl. Phys. **B330** (1990) 575–607.

[17] L. Guillou and A. Marin, *Notes sur l'invariant de Casson des sphères d'homologie de dimension 3,* Enseignement Math. **38** (1992) 233–290.

[18] K. Habiro, *Claspers and finite type invariants of links,* Geom. Topol. **4** (2000) 1–83.

[19] M. W. Hirsch, *Differential Topology,* Graduate Texts in Mathematics, Springer-Verlag **33** (1988).

[20] M. Kontsevich, *Jacobi diagrams and low-dimensional topology,* First European Congress of Mathematics II, Birkhäuser Basel (1994), 97–121.

[21] ———, *Vassiliev's knot invariants,* Adv. Sov. Math **16**(2) (1993) 137–150.

[22] A. Kricker, *The lines of the Kontsevich integral and Rozansky's Rationality Conjecture,* math/GT.0005284.

[23] G. Kuperberg and D. P. Thurston, *Perturbative 3–manifold invariants by cut-and-paste topology,* math.GT/9912167.

[24] J. M. F. Labastida, *Chern-Simons Gauge Theory: Ten Years After* Trends in theoretical physics, II (Buenos Aires, 1998), 1–40, AIP Conf. Proc., 484, Amer. Inst. Phys., Woodbury, NY, 1999. ArXiv:hep-th/9905057.

[25] T. T. Q. Le, J. Murakami, and T. Ohtsuki, *On a universal perturbative invariant of* 3-*manifolds,* Topology **37** (1998), 539–574.

[26] C. Lescop, *On the Kontsevich-Kuperberg-Thurston construction of a configuration-space invariant for rational homology 3-spheres,* math.GT/0411088.

[27] ———, *Splitting formulae for the Kontsevich-Kuperberg-Thurston invariant of rational homology* 3-*spheres,* math.GT/0411431.

[28] ———, *On the cube of the equivariant linking pairing for knots and* 3–*manifolds of rank one,* arXiv:1008.5026.

[29] J. Marché, *An equivariant Casson invariant of knots in homology spheres,* preprint, 2005.

[30] A. Marin, *Un nouvel invariant pour les sphères d'homologie de dimension 3 (d'après Casson),* Semin. Bourbaki, 40ème Année, Vol. 1987/88, Exp. 693, Astérisque 161-162 (1988), 151–164.

[31] J. W. Milnor and J. Stasheff, *Characteristic classes.* Annals of Mathematics Studies **76**, Princeton University Press, 1974.

[32] T. Ohtsuki, *Finite type invariants of integral homology* 3-*spheres.* J. Knot Theory Ramifications **5** (1996), 101–115.

[33] ———, *On the 2-loop polynomial of knots.* Geom. Topol. **11** (2007), 1357–1475.

[34] ———, *Perturbative invariants of 3-manifolds with the first Betti number 1.* Geom. Topol. **14** (2010), 1993–2045.

[35] S. Poirier, *The configuration space integral for links in* R^3, Alg. Geom. Topol. **2** (2002), 1001–1050.

[36] M. Polyak and O. Viro, *On the Casson knot invariant,* Knots in Hellas '98, 3 (Delphi). J. Knot Theory Ramifications **10** (2001), 711–738.

[37] L. Rozansky, *A rationality conjecture about Kontsevich integral of knots and its implications to the structure of the colored Jones polynomial,* In Proceedings of the Pacific Institute for the Mathematical Sciences Workshop (Calgary, AB, 1999), Topology Appl. **127** (2003), 47–76.

[38] ———, *A universal U(1)-RCC invariant of links and rationality conjecture,* arXiv:math.GT/0201139.

[39] C. H. Taubes, *Homology cobordism and the simplest perturbative Chern-Simons* 3-*manifold invariant,* Geometry, topology, and physics, Conf. Proc. Lecture Notes Geom. Topology, IV, Int. Press, Cambridge, MA (1995), 429–538.

[40] D. P. Thurston, *Integral Expressions for the Vassiliev Knot Invariants,* math.QA/9901110.

[41] K. Walker, *An extension of Casson's invariant,* Annals of Math. Studies, Princeton University Press **126** (1992).

[42] E. Witten, *Quantum field theory and the Jones polynomial,* Comm. Math. Phys. **121** (1989), 351–399.

Institut Fourier, BP 74, 38402 Saint-Martin d'Hères cedex, France
E-mail address: `lescop@ujf-grenoble.fr`

AMS/IP Studies in Advanced Mathematics
Volume 50, 2011

Chern–Simons Theory, the $1/N$ Expansion, and String Theory

Marcos Mariño

ABSTRACT. Chern–Simons theory in the $1/N$ expansion has been conjectured to be equivalent to a topological string theory. This conjecture predicts a remarkable relationship between knot invariants and Gromov–Witten theory. We review some basic aspects of this relationship, as well as the tests of this conjecture performed over the last ten years. Particular attention is given to indirect tests based on integrality conjectures, both for the HOMFLY and for the Kauffman invariants of links.

1. Introduction

Twenty years ago, Edward Witten provided a Quantum Field Theory (QFT) formulation of knot and three-manifold invariants [**52**]. The QFT is based on the Chern–Simons action

$$S = \frac{1}{g_s}\int_M \mathrm{Tr}\left(A\wedge \mathrm{d}A + \frac{2}{3}A\wedge A\wedge A\right). \tag{1.1}$$

where A is a connection for a gauge group G, and M is a three-manifold. The partition function of this theory,

$$Z(M) = \int \mathcal{D}A\, \mathrm{e}^{\mathrm{i}S(A)} \tag{1.2}$$

is formally a topological invariant of M. This invariant can be defined rigorously and it is sometimes called the WRT (Witten–Reshetikhin–Turaev) invariant of M. Invariants of knots are obtained in the following way. Let R be a representation of G, and let $\mathcal{K}$ be a framed, oriented knot inside M. The holonomy of A around $\mathcal{K}$ will be denoted by $U_{\mathcal{K}}$. Then, the normalized vacuum expectation value (vev)

$$W_R(\mathcal{K}) = \frac{1}{Z(M)}\int \mathcal{D}A\, \mathrm{e}^{\mathrm{i}S(A)}\, \mathrm{Tr}_R\, U_{\mathcal{K}} \tag{1.3}$$

is an invariant of the framed knot $\mathcal{K}$, and it agrees with the quantum group invariant of $\mathcal{K}$ based on the group G and representation R. This can be extended to framed

2010 *Mathematics Subject Classification.* Primary 57R56; Secondary 57M27, 81T30.
Key words and phrases. Differential geometry, algebraic geometry.
I am supported in part by the Fonds National Suisse. I would like to thank the organizers of the conference *Chern–Simons geuge theory: twenty years after* for a wonderful meeting.

links $\mathcal{L}$ of L components $\mathcal{K}_1, \cdots, \mathcal{K}_L$ and in representations $R_1, \cdots, R_L$,

$$W_{R_1,\cdots,R_L}(\mathcal{L}) = \frac{1}{Z(M)} \int \mathcal{D}A\, \mathrm{e}^{\mathrm{i}S(A)}\, \mathrm{Tr}_{R_1} U_{\mathcal{K}_1} \cdots \mathrm{Tr}_{R_L} U_{\mathcal{K}_L} \tag{1.4}$$

The invariant (1.4) turns out to be the quantum group invariant of the link $\mathcal{L}$, based on the group G, and with "colorings" $R_1, \cdots, R_L$. Our normalization is the natural one from the QFT point of view, namely, the invariant of the unknot with coloring R is the quantum dimension of R:

$$W_R(\bigcirc) = \dim_q R. \tag{1.5}$$

In particular, for $G = U(N)$, the QFT vev (1.4) is the colored HOMFLY invariant of $\mathcal{L}$, and will be denoted by $\mathcal{H}_{R_1,\cdots,R_L}(\mathcal{L})$. It is a rational function of the variables

$$q = \mathrm{e}^{g_s/2}, \qquad \nu = q^N. \tag{1.6}$$

When $G = SO(N)$, (1.4) is the colored Kauffman invariant of $\mathcal{L}$, and will be denoted by $\mathcal{G}_{R_1,\cdots,R_L}(\mathcal{L})$. It is also a rational function of two variables q, ν, where now $q = \nu^{N-1}$. Our conventions are such that

$$\mathcal{G}_{\Box}(\bigcirc) = 1 + \frac{\nu - \nu^{-1}}{q - q^{-1}}. \tag{1.7}$$

Invariants based on $Sp(N)$ are equivalent to the invariants based on $SO(N)$, as we will review below.

One of the most fascinating aspects of this QFT re-interpretation of knot and three-manifold invariants is that one can analyze the vevs (1.4) by using different techniques in QFT, and very often this leads to properties of the invariants which are not obvious at all from their mathematical definition. Let us give a few examples:

(1) The semiclassical analysis of the path integral defining $Z(M)$ leads to an asymptotic expansion of the exact quantum invariant at small g_s. This relates combinatorial and geometric data of M and leads to a conjectural asymptotics of the WRT invariant which has not been proved yet.
(2) The Chern–Simons QFT can be solved exactly by using its relationship with conformal feld theory. This is the point of view put forward by Witten in [**52**] and leads to the standard properties of quantum knot invariants, like for example skein relations.
(3) The perturbative analysis of (1.3) makes contact with the theory of Vassiliev invariants. In particular, different choices of gauge to compute (1.3) lead to different formulations of Vassiliev theory, see [**27**] for a detailed review.

In this talk I will review the results which have been obtained in Chern–Simons theory by using another important QFT technique, namely the $1/N$ expansion [**51**]. This approach can be used when the gauge group is a classical gauge group: $G = U(N)$, $SO(N)$ and $Sp(N)$. Most importantly, when described in this way, Chern–Simons theory is equivalent to a (topological) string theory, and this provides surprising relationships between quantum group invariants of knots and links, and Gromov–Witten invariants. These ideas first took form ten years ago, and I will attempt here to provide a (biased) overview of the subject. In section 2 I explain some QFT aspects of the $1/N$ expansion, and I motivate the string/gauge theory equivalence from a diagrammatic point of view. In section 3, after stating the precise conjectures for the large N string dual description of Chern–Simons theory

FIGURE 1. Thickening an edge.

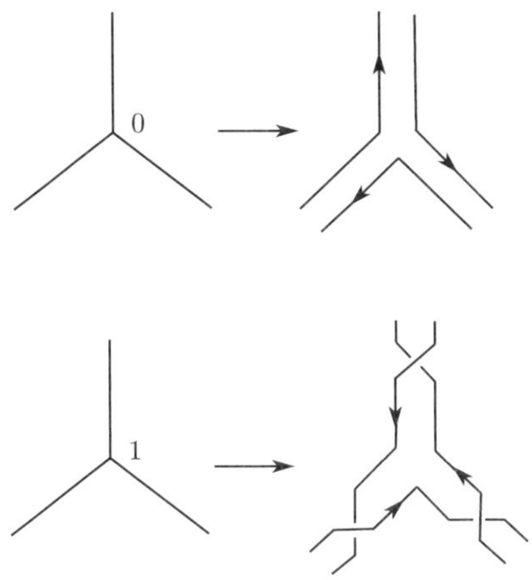

FIGURE 2. Thickening a marked vertex.

in the $U(N)$ case, I review some of its tests involving knots and links. In section 4 I focus on indirect tests based on integrality properties. Finally, in section 5, I explain in some detail the recent generalization of these integrality properties to the SO/Sp case, i.e. to the Kauffman invariant of links. Section 6 contains some concluding remarks.

2. The $1/N$ expansion

The $1/N$ expansion was invented by 't Hooft in [**51**], and it can be used in any gauge theory with a classical gauge group. The general idea is to reorganize the diagrams of standard perturbation theory according to their power of N. The group factor of a Feynman diagram is in general a polynomial in N, so this suggests that we should "split" each diagram in different components, each of them with a well-defined power of N. This leads to the double-line diagrams or fatgraphs of 't Hooft.

Let us now formalize this, in the case of $G = U(N)$, by using a well-known technique to Chern–Simons practicioners, namely the description of the $U(N)$ weight system given in [**10, 4**]. Let us consider the diagrams that appear in the calculation of the LMO invariant. These are trivalent diagrams modulo the IHX relations, and their space is usually denoted by $\mathcal{A}(\emptyset)$. The $U(N)$ weight system is a map

$$W_{U(N)} : \mathcal{A}(\emptyset) \to \mathbb{Z}[N]. \tag{2.1}$$

In order to construct $W_{U(N)}$, one first maps a trivalent diagram D into a formal sum of *fatgraphs* or double-line diagrams, which are in fact Riemann surfaces of genus g with h boundaries. This formal sum is obtained as follows: first, we thicken the edges as shown in Fig. 1. Then, we mark all trivalent vertices of D by 0 or 1. We thicken the marked vertices as shown in Fig. 2. We then obtain

$$D \to \sum_M (-1)^{s_M} \Sigma_{D,M} \tag{2.2}$$

where M is the set of all possible markings, s_M is the sum, over all vertices, of the values of M, and $\Sigma_{D,M}$ is the resulting fatgraph. Each fatgraph leads to a factor $N^{h(\Sigma_{D,M})}$, where $h(\Sigma_{D,M})$ is the number of boundaries of the Riemann surface. In terms of this construction, the weight system is given by

$$W_{U(N)}(D) = \sum_M (-1)^{s_M} N^{h(\Sigma_{D,M})}. \tag{2.3}$$

This procedure just mimicks the structure of the Lie algebra of $U(N)$ [**10, 4**]: edges are associated to Lie algebra generators T^a. These are Hermitian matrices with two indices: one index in the fundamental representation, and another index in the anti-fundamental representation. These two indices lead to the two lines in Fig. 1, with opposite orientation. The vertices are associated to the structure constants of the Lie algebra, and the appearance of two markings in the vertices correspond to the two terms in the commutator of the Lie algebra.

A simple example of the map (2.2) is the theta graph, which leads to two different Riemann surfaces. One of them (the first one shown in Fig. 3) has $g = 0$, $h = 3$, while the second one has $g = 1$, $h = 1$. The trivalent fatgraphs of genus g and with h boundaries can also be regarded, by looking at the dual diagram, as triangulations of a *closed* Riemann surface of genus g. This is in fact the original picture in [**51**].

Let us now consider the expansion of Chern–Simons theory around the trivial flat connection, and let us denote by Z the resulting partition function. By elementary principles of QFT, the *free energy* $F = \log Z$ has a perturbative expansion given as a sum over connected vacuum graphs, which are the trivalent graphs of $\mathcal{A}(\emptyset)$. When this sum is reorganized in terms of fatgraphs, in the way we have explained, it has the structure

$$F = \log\, Z = \sum_{g=0}^{\infty} F_g(t) g_s^{2g-2} \tag{2.4}$$

where

$$t = N g_s \tag{2.5}$$

is the so called *'t Hooft parameter*. In this expression, $F_g(t)$ is in principle a formal power series, and it is obtained by summing the contributions of all fatgraphs of fixed genus g, weigthed by the power t^h, where h is the number of boundaries of the fatgraph. For example, the first graph in Fig. 3 gives a factor $g_s N^3 = g_s^{-2} t^3$, with the power of g_s appropriate for a fatgraph of genus zero. The expansion (2.4) is called the $1/N$ *expansion of the free energy*, since $g_s = t/N$ and the resulting series at fixed t can be also regarded as a series in $1/N$. The leading contribution to this expansion (i.e. for small g_s, or equivalently large N) comes from the fatgraphs of genus zero, also known as *planar graphs*. The quantities $F_g(t)$ are called *genus g free energies* of the manifold M.

The formal power series (2.4) in two variables can be defined rigorously by using the LMO invariant as its starting point, as in [**14**]. Notice that we have restricted ourselves to the contribution of the trivial connection. The $1/N$ expansion for

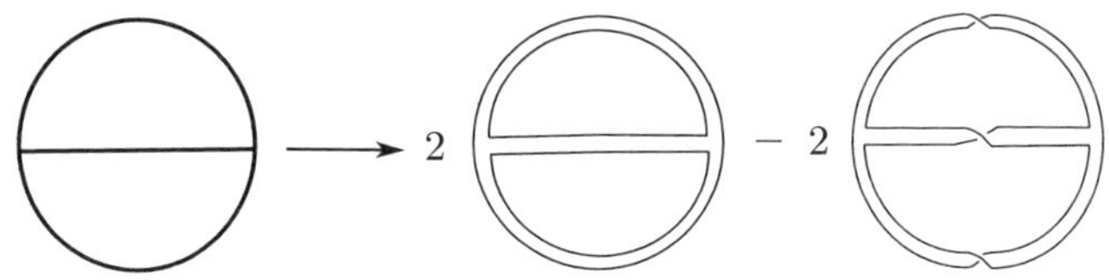

FIGURE 3. The two Riemann surfaces with boundaries associated to a simple trivalent graph.

generic flat connections and for the full WRT invariant is more subtle and we will not discuss it here, see [**1, 39**] for some results in that direction in the case of lens spaces.

So far, the genus g free energies $F_g(t)$ are just formal power series. In the case of $M = \mathbb{S}^3$ they can be computed in closed form for all g [**15**], and the resulting power series can be resummed to an analytic function at $t = 0$. For example, for $g = 0$ one has

$$F_0(t) = \mathrm{Li}_3(\mathrm{e}^{-t}) - \frac{1}{2}t^2 \log t, \tag{2.6}$$

up to a polynomial of degree 3 in t.

In general, perturbative series in QFT have zero radius of convergence. There are two reasons for such a behaviour: the factorial growth of the number of Feynman diagrams contributing at each loop (see [**5**] for a review), and the factorial behavior of some class of diagrams (renormalons). Using the theory of the LMO invariant, it is possible to show that indeed there are no renormalons in Chern–Simons theory, as expected in a superrenormalizable QFT [**13**]. But the perturbative series around the trivial flat connection is still factorially divergent due to the growth in the number of diagrams. It is also expected that the number of fatgraphs of genus g grows only exponentially [**25**], so that in a theory without renormalons quantities like $F_g(t)$ are analytic at the origin, for all g. This expectation from QFT at large N can be actually rigorously proved for Chern–Simons theory, by using the theory of the LMO invariant, and one has the following

THEOREM 2.1. [**14**] *$F_g(t)$ are analytic at $t = 0$ for $g \geq 0$.*

The $1/N$ analysis can be also applied to vevs of Wilson loops. In general, in order to have a well-defined $1/N$ expansion, one has to consider *connected* vevs, and this involves considering complicated combinations of Wilson loops in different representations (see for example [**28, 30**]). The simplest example is the Wilson loop for a knot in the fundamental representation. In this case, the connected vev is equal to the ordinary vev, and one finds the large N structure

$$\mathcal{H}_{\square}(\mathcal{K}) = \sum_{g \geq 0} g_s^{2g-1} \mathcal{H}_g^{\mathcal{K}}(t). \tag{2.7}$$

The quantities $\mathcal{H}_g^{\mathcal{K}}(t)$ also have a diagrammatic representation. The relevant space of diagrams is $\mathcal{A}(\mathbb{S}^1)$, i.e. trivalent diagrams with a single $\mathbb{S}^1$ boundary. The map (2.1) leads to fatgraphs which can be interpreted as triangulations of Riemann surfaces with one boundary, corresponding to the $\mathbb{S}^1$. The formal power series in t $\mathcal{H}_g^{\mathcal{K}}(t)$ is then the contribution of fatgraphs with fixed genus g and one boundary.

FIGURE 4. Connected vev for a two-component link.

The case of links where all components are in the fundamental representation is also relatively simple and important. For a two-component link $L = 2$, one has

$$\mathcal{H}^{(c)}_{\square,\square}(\mathcal{L}) = \mathcal{H}_{\square,\square}(\mathcal{L}) - \mathcal{H}_{\square}(\mathcal{K}_1)\mathcal{H}_{\square}(\mathcal{K}_2). \tag{2.8}$$

This is illustrated in Fig. 4 in the case of the Hopf link. In general, the connected vev of a link in the fundamental representation is obtained by subtracting to the original vev all possible products of vevs of sublinks, with signs [**30**]. The combinatorics is the same one that appears in the cumulant expansion of probabilities. The $1/N$ expansion of these vevs is of the form

$$\mathcal{H}^{(c)}_{\square,\cdots,\square}(\mathcal{L}) = \sum_{g\geq 0} g_s^{2g-2+L}\mathcal{H}_g^{\mathcal{L}}(t). \tag{2.9}$$

This structure is a simple consequence of the diagrammatic expansion, and it has appeared in the work of Przytycki and others on the Vassiliev theory of Brunnian links (see for example [**45**]).

Of course, one can construct more general connected vevs for knots and links, and they involve nontrivial combinations of the standard vevs in representations R (which are in turn given by quantum group invariants). Therefore, the first lesson of the $1/N$ expansion is that the natural invariants of knots and links are not the vevs (1.3), (1.4), but rather their connected versions.

3. The $1/N$ expansion and string theory

3.1. The gauge theory/string theory correspondence. The appearance of Riemann surfaces in the $1/N$ expansion has been regarded as an indication that gauge theories, when reorganized in this way, are described by string theories. We will refer to this idea as the *gauge theory/string theory correspondence or duality.* The correspondence was already suggested in the seminal paper by 't Hooft [**51**], but for a long time it remained a rather speculative idea due to the lack of concrete examples. The situation changed in the early nineties, where some simple string theories were found to be described by matrix models in the $1/N$ expansion [**12**]. But the true turning point was the AdS/CFT correspondence of Maldacena, which postulated a duality between $\mathcal{N} = 4$ super Yang–Mills theory and type IIB string theory on $\mathrm{AdS}_5 \times \mathbb{S}^5$ [**37**]. In the AdS/CFT correspondence, the 't Hooft parameter corresponds to the common radius of the AdS_5 and the five-sphere, i.e. to a geometric parameter of the target space for the string.

A very powerful aspect of the gauge/string theory correspondence is that there is a precise dictionary between gauge theory quantities and string theory quantities, and very often a difficult calculation in gauge theory gets translated into a simple calculation in string theory, and viceversa. For example, the quantities $F_g(t)$, i.e. the free energies of the gauge theories in the $1/N$ expansion on a compact manifold, should have an interpretation as free energies at genus g of a string theory. Another general expectation is that $\mathcal{H}_g^{\mathcal{K}}(t)$, i.e. the genus expansion of a Wilson loop vev in the fundamental representation around a contour $\mathcal{K}$, should correspond to an amplitude in *open* string theory at genus g and with one boundary. The boundary conditions of the open string should be fixed by the geometry of the loop $\mathcal{K}$. Wilson loop vevs involving higher representations should correspond to Riemann surfaces with many boundaries. These ideas have been tested in the AdS/CFT correspondence in great detail. In view of the success of this idea, a natural question is: what is the string theory description of the $1/N$ expansion of Chern–Simons theory?

3.2. String theory and Chern–Simons theory. Building on [**53**], Gopakumar and Vafa proposed in 1998 a conjectual string theory description of Chern–Simons theory on $\mathbb{S}^3$ [**15**]. The relevant string theory is type A topological string theory (see for example [**3**] for more details), and its target is the simplest (non-compact) Calabi–Yau manifold, namely the *resolved conifold*

$$X = \mathcal{O}(-1) \oplus \mathcal{O}(-1) \to \mathbb{P}^1. \tag{3.1}$$

The 't Hooft parameter t is identified with the complexified Kähler parameter of X, i.e. with the complexified area of the $\mathbb{P}^1$ base. The conjecture of Gopakumar and Vafa can be checked by simply computing the free energies of the topological string theory. These are known to be generating functionals of Gromov–Witten invariants $n_{g,d}$ of X,

$$F_g(t) = \sum_{d \geq 1} n_{g,d} \mathrm{e}^{-dt}. \tag{3.2}$$

The invariant $n_{g,d}$ "counts" in an appropriate sense the number of holomorphic curves of genus g and degree d in X. In the case of genus zero, these curves are just multicoverings of $\mathbb{P}^1$, and the Aspinwall–Morrison formula [**2**] gives

$$n_{0,d} = \frac{1}{d^3}. \tag{3.3}$$

We then find,

$$F_0(t) = \sum_{d \geq 1} \frac{\mathrm{e}^{-dt}}{d^3} = \mathrm{Li}_3(\mathrm{e}^{-t}). \tag{3.4}$$

This agrees with the calculation in (2.6). More precisely, the full agreement involves adding a term $t^2 \log t/2$ to (2.6), which indeed comes from the overall measure of the Chern–Simons path integral which we have not taken into account in the calculation. An explicit determination of the F_g for $g > 0$ confirms indeed that the 't Hooft resummation of genus g fatgraphs in Chern–Simons theory on $\mathbb{S}^3$ agrees with the generating function of genus g Gromov–Witten invariants (see for example [**3**] for a detailed exposition).

This is a remarkable agreement, but it only involves a simple quantity, namely the total free energy. This gauge/string correspondence can be considerably enriched by incorporating Wilson loops, as first done by Ooguri and Vafa in [**43**]. Their conjectural correspondence can be stated as follows. Given a link $\mathcal{L}$ in $\mathbb{S}^3$, there exists a Lagrangian submanifold $L_{\mathcal{L}}$ of X providing boundary conditions for open topological strings. The genus expansion of the connected vevs of Wilson loops based on this link are generating functionals for open Gromov–Witten invariants, counting holomorphic maps from Riemann surfaces with boundaries to X. The boundaries map to $L_{\mathcal{L}}$.

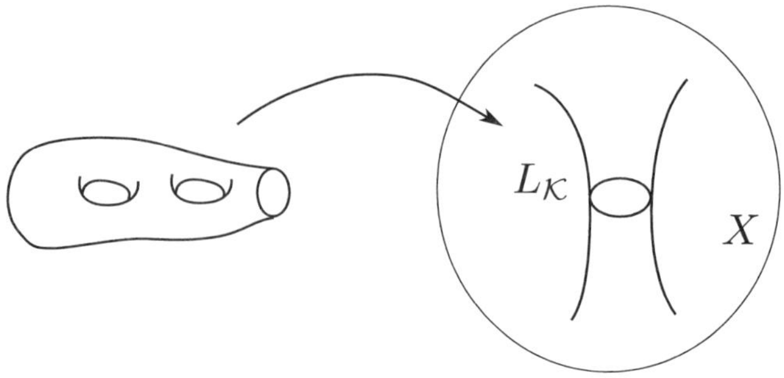

FIGURE 5. In the string theory dual, the HOMFLY polynomial of a knot $\mathcal{K}$ is interpreted in terms of maps of a Riemann surface with one boundary to the resolved conifold X. The boundary is mapped to a Lagrangian submanifold $L_{\mathcal{K}}$.

At this stage, this conjecture is not very precise, but it provides a far-reaching connection between knot invariants and Gromov–Witten invariants. Consider for example a knot $\mathcal{K}$ in $\mathbb{S}^3$, and the Wilson loop vev $\mathcal{H}_\square(\mathcal{K})$. As we explained in (2.7), this vev (which is nothing but the HOMFLY polynomial of $\mathcal{K}$, in the normalization of [**52**]) has a $1/N$ expansion involving functions $\mathcal{H}_g^{\mathcal{K}}(t)$, where t is the 't Hooft parameter. According to [**43**], this function is a generating functional of the form

$$\mathcal{H}_g^{\mathcal{K}}(t) = \sum_Q n_{g,Q}^{\mathcal{K}}\, \mathrm{e}^{Qt} \tag{3.5}$$

where $n_{g,Q}^{\mathcal{K}}$ are now *open* Gromov–Witten invariants. They count holomorphic maps from Riemann surfaces with one boundary which maps to the Lagrangian submanifold $L_{\mathcal{K}}$, see Fig. 5.

Clearly, in order to be more precise we need various ingredients, like for example a detailed construction of the map

$$\mathcal{L} \subset \mathbb{S}^3 \to L_{\mathcal{L}} \subset X. \tag{3.6}$$

We also need a precise definition of open Gromov–Witten invariants. We will now discuss these issues, in the more general context of tests of the correspondence.

3.3. Tests of the correspondence. Since the Gopakumar–Vafa correspondence was proposed, there have been some non-trivial tests involving knots and links. These are the following:

(1) For the (framed) *unknot* $\mathcal{K} = \bigcirc$, there is a candidate Lagrangian submanifold [**43**], and open Gromov–Witten invariants can be defined and calculated in terms of integrals on the Deligne–Mumford moduli space $\overline{M}_{g,n}$ [**22, 31**]. The correspondence leads to a highly nontrivial formula expressing these integrals in terms of quantum dimensions [**40**], which generalizes the celebrated ELSV formula for Hurwitz numbers. This formula is by now well-established [**34, 42**]. It is interesting to point out that the simplest case of this correspondence, at the level of knots, produces already an amazing amount of information! The Hopf link can be also understood in this framework [**35**].

(2) For generic knots and links there are two different proposals for the map (3.6). The construction of [**30, 50**], based on twistor ideas, only applies

to a special class of knots. There is another, more general proposal, which is quite natural [**26**]. It is not clear how these two proposals are related. Unfortunately, the corresponding open Gromov–Witten invariants have not been defined or computed, even at the heuristic level, and very little progress has been made in the last ten years based on direct computations of knot invariant in the Gromov–Witten theory side.

One can also try to consider extensions of the Gopakumar–Vafa conjecture which are amenable to test. There have been three main generalizations:

(1) So far the discussion has involved only Chern–Simons theory with a $U(N)$ gauge group. One can consider other classical gauge groups, like $SO(N)$ (after all, this leads to well-studied link invariants, like the Kauffman polynomial [**23**]) and study their large N dual. We will come back to this later on.
(2) Another possibility is to consider Chern–Simons theory on more general three-manifolds. The lens spaces $L(p, 1)$ are relatively well-understood, and the string/gauge theory correspondence can be extended to these backgrounds [**1**].
(3) One can consider refinements of the standard Chern–Simons invariants, like those coming from Khovanov homology [**24**]. It has been conjectured that this refinement has a string theory counterpart [**18**], and some non-trivial tests of this conjecture have been made for the Hopf link in [**17**].

In this talk I will concentrate on an *indirect* test of the correspondence, namely the integrality properties of link invariants derived from the string theory picture. These properties were conjectured in [**43, 30**] for the invariants associated to the $U(N)$ gauge group, and I will review them here. I will also review these properties in the $SO(N)/Sp(N)$ case, which were partially conjectured in [**7**] and finally spelled out in detail in [**38**].

4. Integrality properties

The basic idea behind the integrality tests of the Gopakumar–Vafa correspondence is the following. The correspondence states that link invariants can be expressed in terms of open Gromov–Witten invariants. If these open invariants enjoy *structural properties* which are not manifest in the link invariants, and if these properties turn out to be true in the knot theory side, we have provided an indirect test of the conjecture.

What these properties could be? There is strong evidence that generating functionals of closed Gromov–Witten invariants, which are rational numbers, can be reexpressed in terms of *integer* invariants. These are called *Gopakumar–Vafa invariants* [**16**], and there has been recent progress in defining them rigorously [**44**]. These integer invariants are interpreted as counting BPS states in compactifications of type II superstring theory. The physics arguments that led to the construction of the Gopakumar–Vafa invariants can be generalized to the open case [**43, 30**], and this leads indeed to conjectural, structural properties of knot and link invariants [**43, 30, 29**].

Let us first present a concrete example. Consider the vev $\mathcal{H}_{\square}(\mathcal{K})$ for an arbitrary knot $\mathcal{K}$ in the fundamental representation of $U(N)$. The standard $1/N$ expansion leads to the representation (2.7), where $\mathcal{H}_g^{\mathcal{K}}(t)$ are generating functions of open Gromov–Witten invariants. The integrality properties obtained with physics arguments in [**16, 43, 30**] say that this vev can be written as

$$\mathcal{H}_{\square}(\mathcal{K}) = \sum_{g\geq 0}\sum_{Q\in\mathbb{Z}} N_{\square;g,Q}(q-q^{-1})^{2g-1}\nu^{Q}, \tag{4.1}$$

where the variables q, ν were defined in (1.6), and $N_{\square;g,Q}$ are BPS *integer* invariants, labeled by two "quantum numbers" g, Q which are related to the genus and degree of embedded Riemann surfaces in the resolved conifold X. In this case, the predicted structure of $\mathcal{H}_{\square}(\mathcal{K})$ is well-known, and it follows from the skein relations for the HOMFLY polynomial. The BPS structural result "explains" in some sense the appearance of this structure, and moreover it gives a heuristic interpretation of the integers appearing in the HOMFLY polynomials in terms of Euler characteristics of appropriate moduli spaces [**30**]. This leads immediately to the possibility of refining these invariants by considering the cohomology of these spaces. In this sense, the BPS description of the HOMFLY polynomial should lead to a natural categorification. According to [**18**], this is nothing but the one proposed by Khovanov.

The prediction (4.1) is only the simplest one in a relatively involved integrality conjecture formulated in [**43, 30, 29**]. Let us state this conjecture in the case of knots. Let $\mathcal{K}$ be a knot, and let $\mathcal{H}_R(\mathcal{K})$ be the $U(N)$ Chern–Simons invariant (1.3) in the representation R. We first define the generating functional

$$Z_{\mathcal{H}}(v) = \sum_{R}\mathcal{H}_R(\mathcal{K})s_R(v), \tag{4.2}$$

understood as a formal power series in Schur polynomials $s_R(v)$. Notice that we are working formally in the limit $N=\infty$, so that all representations R are allowed, and similarly, in constructing the Schur polynomials, we are considering the ring of symmetric polynomials in an infinite number of variables. In (4.2) we sum over all possible colorings, including the empty one $R=\cdot$. We now define the *reformulated HOMFLY invariants* of $\mathcal{K}$, $f_R(q,\nu)$ through the equation

$$\log Z_{\mathcal{H}}(v) = \sum_{k=1}^{\infty}\sum_{R}\frac{1}{k}f_R(q^k,\nu^k)s_R(v^k), \tag{4.3}$$

where the l.h.s. is understood as a formal power series. One can easily prove [**29**] that this equation determines uniquely the reformulated HOMFLY invariants f_R in terms of the colored HOMFLY invariants of $\mathcal{K}$. Explicit formulae for f_R in terms of $\mathcal{H}_R$ for representations with up to three boxes are listed in [**29**].

Let C_μ be the conjugacy class of the symmetric group $S_{\ell(\mu)}$ associated to a partition μ, where $\ell(\mu)$ is the lenght of the partition. We define

$$z_\mu = \frac{\ell(\mu)!}{|C_\mu|}. \tag{4.4}$$

If $\ell(R)=\ell(S)$, we define the matrix

$$M_{RS} = \sum_{\mu}\frac{1}{z_\mu}\chi_R(C_\mu)\chi_S(C_\mu)\frac{\prod_{i=1}^{\ell(\mu)}(q^{\mu_i}-q^{-\mu_i})}{q-q^{-1}} \tag{4.5}$$

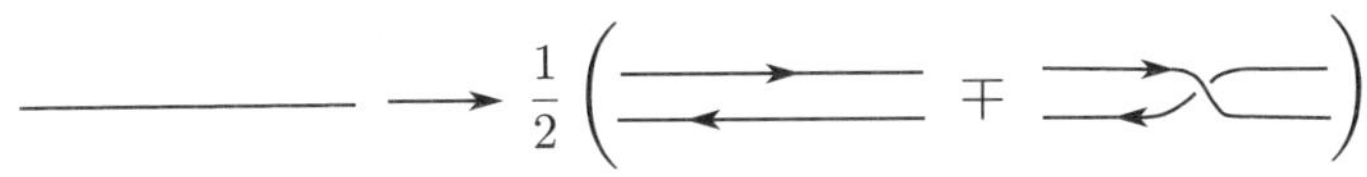

FIGURE 6. Thickening an edge in the weight system of SO/Sp. The $\mp$ sign corresponds to the groups SO, Sp, respectively.

which is zero otherwise. It is easy to show that this matrix is invertible. We finally define

$$\hat{f}_R(q,\nu) = \sum_S M^{-1}_{RS} f_S(q,\nu). \tag{4.6}$$

In principle, $\hat{f}_R(q,\nu)$ are rational functions, and they belong to the ring $\mathbb{Q}[q^{\pm 1}, \nu^{\pm 1}]$ with denominators given by products of $q^r - q^{-r}$. However, we have the following

CONJECTURE 4.1. $\hat{f}_R(q,\nu) \in z^{-1}\mathbb{Z}[z^2, \nu^{\pm 1}]$, where

$$z = q - q^{-1} \tag{4.7}$$

i.e. they have the structure

$$\hat{f}_R(q,\nu) = z^{-1} \sum_{g \geq 0} \sum_{Q \in \mathbb{Z}} N_{R;g,Q} z^{2g} \nu^Q, \tag{4.8}$$

where $N_{R;g,Q}$ are integer numbers and are called the BPS invariants of the knot $\mathcal{K}$. The sum appearing here is finite, i.e. for a given knot and a given coloring R, the $N_{R;g,Q}$ vanish except for finitely many values of g, Q.

This conjecture extracts the BPS invariants from the standard quantum group invariants, and it has been recently proved in [**36**]. It can be extended to links, although the required notation is slightly more cumbersome. One simple consequence of the extension to links is that

$$\mathcal{H}^{(c)}_{\square,\cdots,\square}(\mathcal{L}) \in z^{L-2}\mathbb{Z}[z^2, \nu^{\pm 1}] \tag{4.9}$$

Notice that this implies (2.9), but it is much stronger. The conjecture (4.9) is maybe the simplest consequence of the string theory description of quantum group invariants. It agrees with, and generalizes, previous results by Lickorish–Millett [**33**] and Kanenobu–Miyazawa [**21**] on the structure of the HOMFLY polynomial of links. In particular, it implies the Kanenobu–Miyazawa conjecture on the HOMFLY polynomial of Brunnian links.

5. Generalization to SO/Sp gauge groups

The integrality conjecture of [**43, 30**] (now a theorem) provides a strong indirect test of the string theory description of link invariants. It is an expected property of topological string amplitudes, which on the knot theory side is far from obvious. Moreover, it reproduces in an elegant way simple structural properties of the HOMFLY polynomial. In order to provide further tests of the string theory description, it would be very interesting to extend it to the other classical gauge groups. This is not merely an academic exercise, since the quantum group invariant based on SO/Sp, namely the colored Kauffman invariant of links, is definitely an important object in quantum topology. As we will see, there is such a string

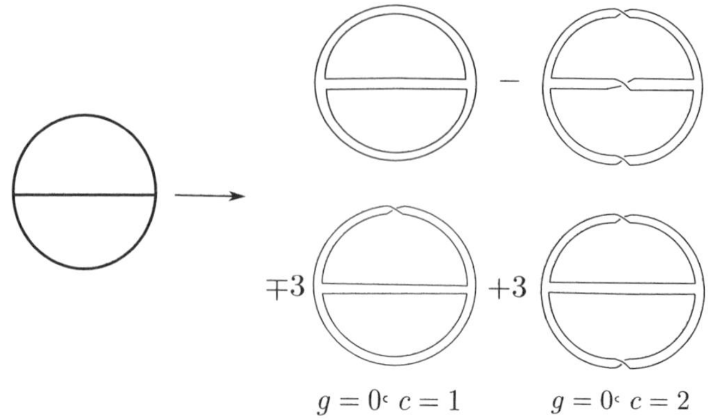

FIGURE 7. The double-line diagrams obtained with the SO/Sp weight system.

theory description, which in particular makes possible to "explain" many mysterious properties of the Kauffman invariant, and in particular its relationships to the HOMFLY invariant.

The best starting point to understand what is the appropriate string theory is to look again at the type of diagrams that appear in the weight system. In the $U(N)$ case, they indicated us the presence of oriented Riemann surfaces. However, in the SO/Sp case, the weight system involves in a crucial way *non-orientable* Riemann surfaces [**9**]. This is due to the fact that the edges are thickened as shown in Fig. 6. Non-orientable surfaces are characterized topologically by their genus g and the number of crosscaps c, which can be one or two. For example, the theta graph considered in Fig. 3, when evaluated with the weight system of SO/Sp, leads to the two orientable Riemann surfaces we had obtained before, together with two non-orientable Riemann surfaces with $c = 1$ and $c = 2$, respectively, as we can see in Fig. 7. The $\mp$ sign in the $c = 1$ diagram corresponds to the two different choices of gauge group SO/Sp. It follows from the rule to thicken the edges that the weight systems of SO and Sp differ just in the sign of the $c = 1$ double-line diagrams. This leads in particular to the $SO(-N) = Sp(N)$ relationship between the SO and the Sp weight systems [**10, 49**], and implies that the Chern–Simons invariants based on $SO(N)$ and $Sp(N)$ are equivalent up to a trivial change of sign in the $c = 1$ contributions. For concreteness we will focus on the $SO(N)$ gauge group.

Notice that the oriented diagrams appear in Fig. 7 with half the coefficient they had in Fig. 3. This is also a general feature of the SO/Sp system, and we can write, schematically,

$$SO(N)/Sp(N) = \frac{1}{2}U(N) + \text{non-orientable}. \tag{5.1}$$

One is then led to conjecture that, for a knot in the fundamental representation of $SO(N)$, we should have

$$\mathcal{G}_\square(\mathcal{K}) = \sum_{g\geq 0} g_s^{2g-1}\mathcal{G}^{\mathcal{K}}_{g,c=0}(t) + \sum_{g\geq 0} g_s^{2g}\mathcal{G}^{\mathcal{K}}_{g,c=1}(t) + \sum_{g\geq 0} g_s^{2g+1}\mathcal{G}^{\mathcal{K}}_{g,c=2}(t), \tag{5.2}$$

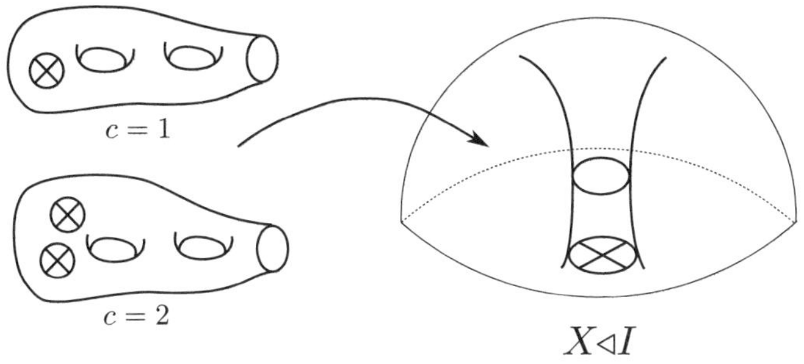

FIGURE 8. In the string theory dual to SO/Sp Chern–Simons theory, knot invariants have a non-orientable contribution given by maps of a Riemann surface with crosscaps to a quotient of the resolved conifold X/I.

The first term gives the contribution of orientable surfaces, and it should be equal to the contribution appearing in (2.7), i.e.

$$\mathcal{G}^{\mathcal{K}}_{g,c=0}(t) = \mathcal{H}^{\mathcal{K}}_g(t). \tag{5.3}$$

The second and third terms in (5.2) are the contributions from non-orientable surfaces with $c = 1$ and $c = 2$ crosscaps, respectively. The counterpart of (5.1) is then

$$\text{Kauffman} = \text{HOMFLY} + \text{non-orientable} \tag{5.4}$$

According to the gauge/string theory correspondence, the contributions with $c \neq 0$ should correspond to maps from non-orientable Riemann surfaces to a quotient of the resolved conifold X by a free involution I. One convenient description of this manifold is as follows. The resolved conifold (3.1) is a toric manifold, defined by the equation

$$|X_1|^2 + |X_2|^2 - |X_3|^2 - |X_4|^2 = t \tag{5.5}$$

and a further quotient by a $U(1)$ action where the coordinates $(X_1, \cdots, X_4)$ have charges $(1, 1, -1, -1)$. In this description, the free involution is

$$\begin{aligned} I : X &\to X, \\ (X_1, X_2, X_3, X_4) &\to (\overline{X}_2, -\overline{X}_1, \overline{X}_4, -\overline{X}_3). \end{aligned} \tag{5.6}$$

Notice that the quotient geometry contains a projective plane $\mathbb{RP}^2 = \mathbb{P}^1/I$. One can now consider holomorphic maps from Riemann surfaces with crosscaps to X/I (as shown in Fig. 8), and count them in an appropriate way. The resulting theory should be a non-orientable version of Gromov–Witten theory. In string theory, such a construction is called an *orientifold*. It is a particular case of a $\mathbb{Z}_2$ orbifold in which the discrete group acts as a orientation reversal on the worldsheet and as an involution I on the target. The above orientifold construction was introduced in [**47**] in order to describe the large N dual of Chern–Simons theory with SO/Sp gauge groups. It was further studied in [**11, 6, 7**].

Like before, direct tests of this string/gauge theory duality are difficult, but one could try to make an indirect test based on integrality properties. It is easy to generalize the construction of BPS invariants to the non-orientable case. The simplest situation occurs again for the SO/Sp vevs in the fundamental representation, and we have

$$\mathcal{G}_\square(\mathcal{K}) = \sum_Q \sum_{g\geq 0} \left(N_{\square;g,Q} z^{2g-1} + N^{c=1}_{\square;g,Q} z^{2g} + N^{c=2}_{\square;g,Q} z^{2g+1} \right) \nu^Q, \tag{5.7}$$

where $N^{c=1,2}_{\square;g,Q}$ are BPS invariants labelled, in addition to the quantum number g, by a quantum number c which corresponds to the number of crosscaps. The BPS invariant $N_{\square;g,Q}$ agrees with the one appearing in the HOMFLY invariant (4.1). A simple consequence of the structural result (5.7) is that the polynomial in $\nu^{\pm 1}$ which multiplies z^{-1} is common to the HOMFLY invariant and the Kauffman invariant of a knot (in the fundamental representation). This is a well-known property of the Kauffman invariant [**32**].

We can now clarify two subtle issues concerning the comparison between the Kauffman and the HOMFLY invariants. First of all, in the comparison of the weight systems, we saw that the $U(N)$ system is multiplied by $1/2$. Where did this factor go in the comparison between the invariants, in (5.2)? A second issue is the following: the Kauffman invariant is an invariant of *non-oriented* knots and links, while the HOMFLY invariant is an invariant of *oriented* knots and links. How are we going to compare both objects? It turns our that these two issues are related to each other. In fact, the right version of (5.4) is rather

$$\text{Kauffman} = \frac{1}{2} \sum_{\text{orientations}} \text{HOMFLY} + \text{non-orientable} \tag{5.8}$$

i.e. there is a factor of $1/2$ inherited from (5.1), but in comparing both invariants we have to sum over all possible orientations of the link in the HOMFLY side. This produces in fact an invariant of non-oriented links, and the second issue is also resolved. In the case of a knot, there are only two possible orientations, and the HOMFLY invariants are equal, therefore the r.h.s. of (5.8) gives back (5.4). However, in the case of links the sum over possible orientations involves different invariants, as we will make more explicit in a moment.

Some ingredients in the integrality conjecture for SO/Sp link invariants were uncovered in [**6, 7**]. The final picture was found in [**38**]. One of the key ingredients of this conjecture, inspired by the results of Morton and Ryder [**41**], is that the HOMFLY contribution to the Kauffman invariant involves in general *composite representations* of $U(N)$. In mathematical terms, this implies that we need the full HOMFLY skein of the annulus, which has been much developed in the work of Morton and collaborators (see for example [**19**]). Composite representations are labelled by two Young diagrams (R, S), and they can be expressed in terms of tensor products $V \otimes \overline{W}$, where $\overline{W}$ are conjugate representations. The precise formula is

$$(R, S) = \sum_{U,V,W} (-1)^{\ell(U)} N^R_{UV} N^S_{U^T W} \, (V \otimes \overline{W}), \tag{5.9}$$

and N^{ST}_R Littlewood–Richardson coefficients. The HOMFLY invariant of a knot $\mathcal{L}$ whose components are colored with the composite representations (R_1, S_1), $\cdots$, (R_L, S_L) will be denoted by

$$\mathcal{H}_{(R_1,S_1),\cdots,(R_L,S_L)}(\mathcal{L}). \tag{5.10}$$

One has the property

$$\mathcal{H}_{(R_1,S_1),\cdots,(S_j,R_j),\cdots,(R_L,S_L)}(\mathcal{L}) = \mathcal{H}_{(R_1,S_1),\cdots,(R_j,S_j),\cdots,(R_L,S_L)}(\overline{\mathcal{L}}_j), \tag{5.11}$$

where $\overline{\mathcal{L}}_j$ is the link obtained from the link $\mathcal{L}$ by reversing the orientation of the j-th component. In particular,

$$\mathcal{H}_{(R,S)}(\mathcal{K}) = \mathcal{H}_{(S,R)}(\overline{\mathcal{K}}), \tag{5.12}$$

where $\mathcal{K}$ is an oriented knot and $\overline{\mathcal{K}}$ is the knot with opposite orientation. Therefore, summing over all possible composite representations of a link involves in a particular summing over all its possible orientations.

We are now ready to state the integrality conjecture for the Kauffman invariants proposed in [**38**]. Like before, we will restrict ourselves to the case of knots, and we will make some indications in the case of links. We introduce the two generating functionals

$$\begin{aligned} Z_{H\overline{H}}(v) &= \sum_{R,R_1,R_2} N_{R_1 R_2}^{R} \mathcal{H}_{(R_1,R_2)}(\mathcal{K}) s_R(v), \\ Z_{\mathcal{G}}(v) &= \sum_{R} \mathcal{G}_R(\mathcal{K}) s_R(v). \end{aligned} \tag{5.13}$$

Notice that, due to the property (5.12), both are invariants of non-oriented knots. We define g_R through

$$\log Z_{\mathcal{G}}(v) - \frac{1}{2}\log Z_{H\overline{H}}(v) = \sum_{k\,\text{odd}} \sum_{R} \frac{1}{k} g_R(q^k, \nu^k) s_R(v^k). \tag{5.14}$$

Here the sum over k is over all positive *odd* integers. The factor $1/2$ in the l.h.s. is closely related to the factor $1/2$ in (5.1). We also define, as in (4.6),

$$\hat{g}_R(q,\nu) = \sum_{S} M_{RS}^{-1} g_S. \tag{5.15}$$

The analogue of the conjecture 4.1 is

CONJECTURE 5.1. $\hat{g}_R(q,\nu) \in \mathbb{Z}[z,\nu^{\pm 1}]$, i.e. they have the structure

$$\hat{g}_R(q,\nu) = \sum_{g\geq 0} \sum_{Q\in\mathbb{Z}} \left(N_{R;g,Q}^{c=1} z^{2g} + N_{R;g,Q}^{c=2} z^{2g+1} \right) \nu^Q, \tag{5.16}$$

where $N_{R;g,Q}^{c=1,2}$ are integers. They are the BPS invariants associated to the non-orientable case.

When $R = \square$, we recover the result (5.7). For representations with two boxes, we have that

$$\begin{aligned} g_{\square\square} &= \mathcal{G}_{\square\square} - \frac{1}{2}\mathcal{G}_{\square}^2 - \mathcal{H}_{\square\square} + \mathcal{H}_{\square}^2 - \frac{1}{2}\mathcal{H}_{(\square,\square)}, \\ g_{\substack{\square\\\square}} &= \mathcal{G}_{\substack{\square\\\square}} - \frac{1}{2}\mathcal{G}_{\square}^2 - \mathcal{H}_{\substack{\square\\\square}} + \mathcal{H}_{\square}^2 - \frac{1}{2}\mathcal{H}_{(\square,\square)}. \end{aligned} \tag{5.17}$$

The integrality conjecture 5.1 implies in particular that

$$\mathcal{G}_{\square}^2(\mathcal{K}) = \mathcal{H}_{(\square,\square)}(\mathcal{K}), \qquad \text{mod } 2. \tag{5.18}$$

This is an intriguing theorem due to Rudolph [**46**], recently generalized by Morton and Ryder to higher representations [**41**], which finds a "natural" explanation in the context of BPS integrality properties in string theory. The generalization of [**41**] also seems to follow from the conjecture.

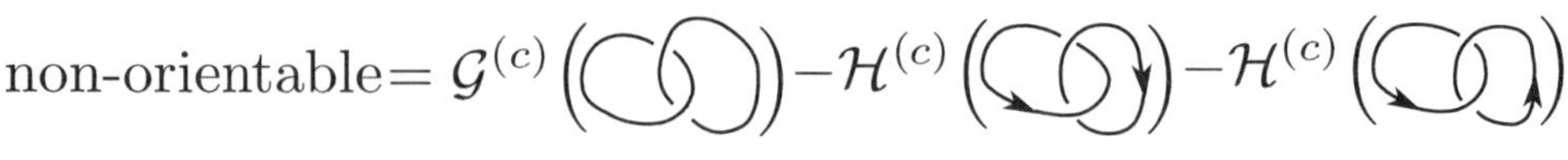

FIGURE 9. In order to obtain the non-orientable contribution to the connected Kauffman invariant of a two-component, non-oriented link, one has to subtract the sum of the HOMFLY invariants for all possible orientations of the link.

The above integrality conjecture for SO/Sp invariants can be extended to links. The simplest case of this conjecture says that
(5.19)

$$\mathcal{G}^{(c)}_{\square,\cdots,\square}(\mathcal{L}) = \sum_{\alpha=1}^{2^{L-1}} \mathcal{H}^{(c)}_{\square,\cdots,\square}(\mathcal{L}_\alpha) + g_{\square,\cdots,\square}(\mathcal{L}), \qquad g_{\square,\cdots,\square}(\mathcal{L}) \in z^{L-1}\mathbb{Z}[z,\nu^{\pm 1}].$$

In this equation, the sum over α is over all possible non-equivalent orientations of the link $\mathcal{L}$, denoted by $\mathcal{L}_\alpha$ (two orientations are equivalent if they are related by an overall reversal of orientation of all the components in the link). If $\mathcal{L}$ has L components, there are 2^{L-1} non-equivalent orientations. The quantity $g_{\square,\cdots,\square}(\mathcal{L})$ gives the non-orientable contribution to the invariant of the link, and as shown in (5.19), it involves the sum over all possible orientations of $\mathcal{L}$. The case of a two-component link is illustrated in Fig. 9. The conjecture (5.19) implies some results of Kanenobu for the Kauffman invariant of links [**20**], and on top of that it leads to new relationships between the Kauffman polynomial of non-oriented links and the sum of the HOMFLY polynomials of all its possible orientations, see [**38**] for a detailed explanation and for examples. Further tests of the conjecture 5.1 were performed in [**48, 8**].

6. Conclusions

The string theory description of Chern–Simons theory is extremely interesting, both from a physical and from a mathematical point of view. From a physical point of view, it provides one of the most detailed examples of a large N string duality, a subject of paramount importance in modern theoretical physics. From a mathematical point of view, it provides a bridge between two major areas of modern mathematics.

Unfortunately, progress in this area has been difficult and slow. Many important aspects of the correspondence still lack a sound mathematical basis, and there are no direct tests for nontrivial knots. Indirect tests based on integrality properties are so far our best indication that the conjectural string theory/Gromov–Witten description of Chern–Simons theory holds for general knots and links. One important ingredient in order to make further progress is to provide precise definitions of all the objects and invariants involved in the Gromov–Witten side.

Our limited knowledge of this large N duality has already given us very interesting results: new structural properties of quantum group invariants of knots and links, new results for intersection theory on the Deligne–Mumford moduli space, new perspectives on categorification... It is likely that further progress in this area will unveil even more treasures.

References

[1] M. Aganagic, A. Klemm, M. Mariño, and C. Vafa, *Matrix model as a mirror of Chern–Simons theory*, J. High Energy Phys. (2004), no. 2, 010, 46 pp.

[2] P. S. Aspinwall and D. R. Morrison, *Topological field theory and rational curves*, Comm. Math. Phys. **151** (1993), 245–262.

[3] D. Auckly and S. Koshkin, *Introduction to the Gopakumar-Vafa large N duality*, The interaction of finite-type and Gromov-Witten invariants (BIRS 2003), 195–456, Geom. Topol. Monogr., 8, Geom. Topol. Publ., Coventry, 2006.

[4] D. Bar-Natan, *On the Vassiliev knot invariants*, Topology **34** (1995), 423–72.

[5] C. M. Bender, *Perturbation theory in large order*, Adv. in Math. **30** (1978), 250–267.

[6] V. Bouchard, B. Florea , and M. Mariño, *Counting higher genus curves with crosscaps in Calabi-Yau orientifolds*, J. High Energy Phys. (2004), no. 12, 035, 35 pp.

[7] ———, *Topological open string amplitudes on orientifolds*, J. High Energy Phys. (2005), no. 2, 002, 35 pp.

[8] P. Chandrima, B. Pravina, and P. Ramadevi, *Composite Invariants and Unoriented Topological String Amplitudes*, arXiv:1003.5282 [hep-th].

[9] G. M. Cicuta, *Topological expansion for* SO_N *and* Sp_{2N} *gauge theories*, Lett. Nuovo Cimento **35** (1982), 87–92.

[10] P. Cvitanovic, *Group theory. Birdtracks, Lie's, and exceptional groups*, Princeton University Press, Princeton, 2008.

[11] D.-E. Diaconescu, B. Florea, and A. Misra, *Orientifolds, unoriented instantons and localization*, J. High Energy Phys. (2003), no. 7, 041, 30 pp.

[12] P. Di Francesco, P. Ginsparg, and J. Zinn-Justin, *2D gravity and random matrices*, Phys. Rep. **254** (1995), 1–133.

[13] S. Garoufalidis and T.T.Q. Le, *Gevrey series in quantum topology*, J. Reine Angew. Math., **618** (2008), 169–195.

[14] S. Garoufalidis, T.T.Q. Le, and M. Mariño, *Analyticity of the free energy of a closed 3-manifold*, SIGMA **4** (2008), 080, 20 pages.

[15] R. Gopakumar and C. Vafa, *On the gauge theory/geometry correspondence*, Adv. Theor. Math. Phys. **3** (1999) 1415–1443.

[16] ———, *M-theory and topological strings, 2*, arXiv:hep-th/9812127.

[17] S. Gukov, A. Iqbal, C. Kozcaz, and C. Vafa, *Link homologies and the refined topological vertex*, arXiv:0705.1368 [hep-th].

[18] S. Gukov, A. Schwarz, and C. Vafa, *Khovanov-Rozansky homology and topological strings*, Lett. Math. Phys. **74** (2005), 53–74.

[19] R.J. Hadji and H. R. Morton, *A basis for the full Homfly skein of the annulus*, Math. Proc. Cambridge Philos. Soc. **141** (2006), 81–100.

[20] T. Kanenobu, *The first four terms of Kauffman's link polynomial*, Kyungpook Math. J. **46** (2006), 509–525.

[21] T. Kanenobu and Y. Miyazawa, *The second and third terms of the HOMFLY polynomial of a link*, Kobe J. Math. **16** (1999), 147–159.

[22] S. Katz and C.-C. M. Liu, *Enumerative geometry of stable maps with Lagrangian boundary conditions and multiple covers of the disc*, Adv. Theor. Math. Phys. **5** (2001), 1–49.

[23] L. H. Kauffman, *An invariant of regular isotopy*, Trans. Amer. Math. Soc. **318** (1990), 417–471.

[24] M. Khovanov, *A categorification of the Jones polynomial*, Duke Math. J. **101** (2000), 359–426.

[25] J. Koplik, A. Neveu, and S. Nussinov, *Some aspects of the planar perturbation series*, Nucl. Phys. B **123** (1977), 109–131.

[26] S. Koshkin, *Conormal bundles to knots and the Gopakumar-Vafa conjecture*, Adv. Theor. Math. Phys. **11** (2007), 591–634.

[27] J. M. F. Labastida, *Chern-Simons gauge theory: Ten years after*, arXiv:hep-th/9905057.

[28] J. M. F. Labastida and M. Mariño, *Polynomial invariants for torus knots and topological strings*, Commun. Math. Phys. **217** (2001) 423–449.

[29] ———, *A new point of view in the theory of knot and link invariants*, J. Knot Theory Ramifications **11** (2002), 173–197.

[30] J. M. F. Labastida, M. Mariño, and C. Vafa, *Knots, links and branes at large N*, J. High Energy Phys. (2000), no. 11, Paper 7, 42 pp.

[31] J. Li and Y. S. Song, *Open string instantons and relative stable morphisms*, Adv. Theor. Math. Phys. **5** (2001), 67–91.

[32] W. B. R. Lickorish, *An introduction to knot theory*, Springer-Verlag, 1997.

[33] W. B. R. Lickorish and K. Millett, *A polynomial invariant of oriented links*, Topology **26** (1987), 107–141.

[34] C.-C. M. Liu, K. Liu, and J. Zhou, *A proof of a conjecture of Mariño-Vafa on Hodge integrals*, *J. Differential Geom.* **65** (2003), 289–340.

[35] ______, *A formula of two-partition Hodge integrals*, J. Amer. Math. Soc. **20** (2007), 149–184.

[36] K. Liu and P. Peng, *Proof of the Labastida-Mariño-Ooguri-Vafa Conjecture*, arXiv:0704.1526 [math.QA].

[37] J. Maldacena, *The large N limit of superconformal field theories and supergravity*, Adv. Theor. Math. Phys. **2** (1998), 231–252.

[38] M. Mariño, *String theory and the Kauffman polynomial*, arXiv:0904.1088 [hep-th].

[39] M. Mariño, S. Pasquetti, and P. Putrov, *Large N duality beyond the genus expansion*, arXiv:0911.4692 [hep-th].

[40] M. Mariño and C. Vafa, *Framed knots at large N*, Orbifolds in mathematics and physics (Madison, WI, 2001), 185–204, Contemp. Math., **310**, Amer. Math. Soc., Providence, RI, 2002.

[41] H. R. Morton and N. D. A. Ryder, *Relations between Kauffman and Homfly satellite invariants*, arXiv:math.GT/0902.1339.

[42] A. Okounkov and R. Pandharipande, *Hodge integrals and invariants of the unknot*, Geom. Topol. **8** (2004), 675–699.

[43] H. Ooguri and C. Vafa, *Knot invariants and topological strings*, Nuclear Phys. B **577** (2000), 419–438.

[44] R. Pandharipande and R. Thomas, *Curve counting via stable pairs in the derived category*, Invent. Math. **178** (2009), 407–447.

[45] J. H. Przytycki and K. Taniyama, *The Kanenobu-Miyazawa conjecture and the Vassiliev-Gusarov skein modules based on mixed crossings*, Proc. Amer. Math. Soc. **129** (2001), 2799–2802.

[46] L. Rudolph, *A congruence between link polynomials*, Math. Proc. Cambridge Philos. Soc. **107** (1990), 319–327.

[47] S. Sinha and C. Vafa, *SO and Sp Chern-Simons at large N*, arXiv:hep-th/0012136.

[48] S. Stevan, *Chern-Simons Invariants of Torus Knots and Links*, arXiv:1003.2861 [hep-th].

[49] T. Takata, *On the $SO(N)$ and $Sp(N)$ free energy of a rational homology 3-sphere*, preprint.

[50] C. H. Taubes, *Lagrangians for the Gopakumar-Vafa conjecture*, Adv. Theor. Math. Phys. **5** (2001), 139–163.

[51] G. 't Hooft, *A planar diagram theory for strong interactions*, Nucl. Phys. B **72** (1974), 461–473.

[52] E. Witten, *Quantum field theory and the Jones polynomial*, Comm. Math. Phys. **121** (1989), 351–399.

[53] ______, *Chern-Simons gauge theory as a string theory*, The Floer memorial volume, 637–678, Progr. Math., **133**, Birkhaüser, Basel, 1995.

Département de Physique Théorique et Section de Mathématiques, Université de Genève, CH 1211 Switzerland

E-mail address: marcos.marino@unige.ch

AMS/IP Studies in Advanced Mathematics
Volume 50, 2011

Global Lorentzian Geometry from Lightlike Geodesics: What Does an Observer in (2+1)-Gravity See?

Catherine Meusburger

ABSTRACT. We show how an observer could measure the non-local holonomy variables that parametrise the flat Lorentzian 3d manifolds arising as spacetimes in (2+1)-gravity. We consider an observer who emits lightrays that return to him at a later time and performs several realistic measurements associated with such returning lightrays: the eigentime elapsed between the emission of the lightrays and their return, the directions into which the light is emitted and from which it returns and the frequency shift between the emitted and returning lightray. We show how the holonomy variables and hence the full geometry of these manifolds can be reconstructed from these measurements in finite eigentime.

1. Introduction

Gravity in (2+1) dimensions is of interest due to its rich mathematical structure and as a toy model for quantum gravity in higher dimensions [**1**]. Important progress in the understanding of the classical theory and its quantisation followed the discovery that the theory can be formulated as a Chern-Simons gauge theory [**2, 3**]. This made it possible to apply gauge theoretical methods to the theory and related its quantisation to quantum groups, knot and link invariants and topological quantum field theory [**4, 5**].

However, it is not straightforward to interpret the resulting quantum theories as quantum geometry or quantised general relativity. Although the manifolds arising as spacetimes in (2+1)-dimensional gravity have a rich geometrical features involving Teichmüller theory and hyperbolic geometry [**6, 7**], it has been difficult to interpret the associated quantum theories in geometrical terms. This hindered the physical interpretation of the theory and made it difficult to extract interesting physics from (2+1)-gravity. In particular, it remained unclear how the fundamental variables that parametrise the classical solutions and play a central role in the

2010 *Mathematics Subject Classification.* Primary 83C80, 53C50; Secondary 57M60, 30F60.
Key words and phrases. General relativity in three dimensions, Lorentzian geometry, Teichmüller space.
C. Meusburger is supported by the Emmy Noether fellowship ME 3425/1-1 of the German research Foundation (DFG).

quantisation of the theory are related to concrete observations by an observer in the spacetime.

In this paper, we address this issue for classical (2+1)-gravity with vanishing cosmological constant and without matter. The relevant spacetimes are flat, maximally globally hyperbolic three-dimensional Lorentzian manifolds with a Cauchy surface of genus $g \geq 2$. We consider an observer who investigates the geometry of these spacetimes by emitting lightrays that return to him at a later time. The observer can perform several measurements associated with such returning lightrays: He can determine the return time, the directions into which the lightrays are emitted or from which they return and the relative frequencies of the lightray at its emission and return. We give explicit expressions for these measurements in terms of the holonomy variables which play a central role in the quantisation of the theory. Moreover, we demonstrate how these measurements allow the observer to determine the full geometry of the spacetime in finite eigentime.

The paper is structured as follows: Section 2 summarises the relevant properties of flat Lorentzian 3d manifolds, their classification and their parametrisation by holonomy variables. In Section 3 we investigate the measurements associated with returning lightrays. We show how the relevant concepts such as observers, lightrays and returning lightrays are realised in the universal cover. We then derive explicit expressions for these measurements in terms of the holonomy variables and discuss their geometrical interpretation. In Section 4, we show how these measurements can be used to reconstruct the holonomy variables and thus the full geometry of the spacetime in finite eigentime. Section 5 contains our concluding remarks.

2. Vacuum spacetimes in (2+1)-gravity

2.1. Notations and conventions. We denote by $\mathbb{M}^3$ three-dimensional Minkowski space with the Minkowski metric $\eta = \mathrm{diag}(-1,1,1)$. Throughout the paper, we write $\boldsymbol{x} \cdot \boldsymbol{y}$ for $\eta(\boldsymbol{x},\boldsymbol{y})$ and $\boldsymbol{x}^2$ for $\eta(\boldsymbol{x},\boldsymbol{x})$. Vectors $\boldsymbol{x} \in \mathbb{R}^3$ are called timelike, lightlike and spacelike if $\boldsymbol{x}^2 < 0$, $\boldsymbol{x}^2 = 0$ and $\boldsymbol{x}^2 > 0$, respectively. The group of orientation and time orientation preserving isometries of $\mathbb{M}^3$ is the three-dimensional Poincaré group $P_3 = ISO^+(2,1) = SO^+(2,1) \ltimes \mathbb{R}^3$, which is the semidirect product of the three-dimensional proper, orthochronous Lorentz group $SO^+(2,1) \cong PSL(2,\mathbb{R}) \cong PSU(1,1)$ with the abelian translation group $\mathbb{R}^3$. We parametrise elements of P_3 as $(v,\boldsymbol{a})$ with $v \in SO^+(2,1)$ and $\boldsymbol{a} \in \mathbb{R}^3$. In terms of this parametrisation, the group multiplication law then takes the form

$$(v_1,\boldsymbol{a}_1) \cdot (v_2,\boldsymbol{a}_2) = (v_1 v_2, \boldsymbol{a}_1 + v_1 \boldsymbol{a}_2), \tag{2.1}$$

where $v\boldsymbol{a}$ denotes the action of an element $v \in SO^+(2,1)$ on a vector $\boldsymbol{a} \in \mathbb{R}^3$. Elements of $SO^+(2,1)$ are called parabolic, elliptic, hyperbolic, respectively, if the stabilise the orthogonal complement $\boldsymbol{y}^\perp$ of a timelike, lightlike or spacelike vector $\boldsymbol{y} \in \mathbb{R}^3$.

2.2. Classification and properties of (2+1)-spacetimes. As the Ricci tensor of a three-dimensional manifold determines its sectional curvature, gravity in three dimensions has no local gravitational degrees of freedom. Any vacuum solution of the three-dimensional Einstein equations without cosmological constant is a flat Lorentzian 3d manifold which is locally isometric to Minkowski space $\mathbb{M}^3$. In contrast to the four-dimensional case, this allows one to give an explicit classification of the diffeomorphism equivalence classes of solutions of Einstein's equations.

A complete classification has been achieved for *maximally globally hyperbolic* (MGH) flat Lorentzian 3d manifolds with complete Cauchy surfaces [**6, 7**]. The assumption of global hyperbolicity selects spacetimes with acceptable causality behaviour (no closed timelike curves, the intersection of the future of a point with the past of another is always compact). The condition is equivalent to imposing that the manifold contains a Cauchy-surface, a spacelike two-surface S, which every inextensible, causal curve intersects exactly once. The completeness condition excludes Cauchy surfaces with singularities, while the maximality condition is a technical condition imposed to avoid overcounting of spacetimes.

Throughout the paper, we restrict attention to the simplest case, namely to MGH flat 3d Lorentzian manifolds M with a *compact* Cauchy surface S of genus $g \geq 2$[1]. The properties of these manifolds have been determined in [**6**], for a detailed and accessible review see also [**9**]. It is shown there that they have topology $M \approx \mathbb{R}^+ \times S$, that they are future complete, but not past-complete and that they have an initial "big bang" singularity. Moreover, they are equipped with a canonical cosmological time function $T : M \to \mathbb{R}^+$ [**10**] that tends to zero along every past-directed inextensible causal curve and is given by

$$T(p) = \sup\{L(c) \mid c : [a,b] \to M, c(a) = p,\ c \text{ past-directed and causal}\}. \tag{2.2}$$

The surfaces M_T of constant cosmological time T are Cauchy surfaces and foliate M

$$M = \bigcup_{T \in \mathbb{R}^+} M_T. \tag{2.3}$$

The classification of flat MGH (2+1)-spacetimes makes use of their description as quotients of their universal covers. It is shown in [**6, 11**], that their universal covers $\tilde{M}$ can be identified with regular domains in Minkowski space $\mathbb{M}^3$. These are open, future complete regions $\tilde{M} \subset \mathbb{M}^3$ that are domains of dependence, i. e. given as the future of a set $\tilde{M}_0 \subset \mathbb{M}^3$. The cosmological time function on M lifts to a cosmological time function $T : \tilde{M} \to \mathbb{R}^+$, which gives the geodesic distance of points $p \in \tilde{M}$ from M_0 and whose level surfaces foliate the domain

$$\tilde{M} = \bigcup_{T \in \mathbb{R}^+} \tilde{M}_T. \tag{2.4}$$

The fundamental group $\pi_1(M) \cong \pi_1(S)$ acts on $\tilde{M}$ via deck transformations, which are given by a group homomorphism $h : \pi_1(M) \to P_3$, in the following referred to as holonomies. This group action is free and properly discontinuous and preserves each surface $\tilde{M}_T$ of constant cosmological time. Moreover, the Lorentzian component of the holonomies $h_L : \pi_1(M) \to SO^+(2,1)$ defines a faithful and discrete representation of $\pi_1(M)$. For the case of a Cauchy surface of genus $g \geq 2$, this implies that the image of h_L is a cocompact Fuchsian group $\Gamma \cong \pi_1(M) \cong \pi_1(S)$ of genus $g \geq 2$. This is a discrete subgroup of the three-dimensional Lorentz group $SO^+(2,1) \cong PSL(2,\mathbb{R})$ with $2g$ generators and a defining relation

$$\Gamma = \langle v_{a_1}, v_{b_1}, ..., v_{a_g}, v_{b_g} \mid [v_{b_g}, v_{a_g}] \cdots [v_{a_1}, v_{b_1}] = 1 \rangle \subset PSL(2,\mathbb{R}). \tag{2.5}$$

Note that all elements of such a cocompact Fuchsian group are hyperbolic, i. e. they stabilise planes in $\mathbb{M}^3$ with *spacelike* normal vectors.

---

[1]The case of Cauchy surfaces of genus one (torus universe) can be treated along similar lines, see [**1, 8**], but the geometrical properties of these spacetimes differ from the higher genus cases.

It has been shown by Mess [**6**] that the holonomies characterise flat MGH Lorentzian 3d manifolds with compact, complete Cauchy surfaces uniquely. Given a group homomorphism $h : \pi_1(M) \to P_3$ whose Lorentzian component $h_L : \pi_1(M) \to SO^+(2,1)$ defines a cocompact Fuchsian group of genus g, there exists a unique domain $\tilde{M} \subset \mathbb{M}^3$ on which $\pi_1(M) \cong \pi_1(S)$ acts freely and properly discontinuously in such a way that each constant cosmological time surface is preserved.

Two flat MGH spacetimes with Cauchy surfaces of genus $g \geq 2$ are isometric if and only if the associated holonomy maps are related by conjugation with a constant element of the Poincaré group P_3. This implies that the physical (or reduced) phase space of the theory, the set of diffeomorphism equivalence classes of solutions of the three-dimensional Einstein equations, can be identified with

$$\mathcal{P} = \mathrm{Hom}_0(\pi_1(M), P_3)/P_3, \tag{2.6}$$

where the index 0 indicates that the Lorentzian component of the group homomorphism must define a Fuchsian group of the appropriate genus. Note that $\mathcal{P}$ is a connected component of the moduli space of flat P_3-connections on S and coincides with the cotangent bundle $T^*\tau(S)$ of Teichmüller space on S. The holonomy variables $h(\lambda)$, $\lambda \in \pi_1(M)$, thus play a central role in the description of the solutions of the theory and in its quantisation.

2.3. Geometry of (2+1)-spacetimes. It is shown in [**6, 9**] that the description of these manifolds as quotients of their universal covers gives rise to a concrete and explicit description of their geometry. One distinguishes two cases, conformally static spacetimes, for which the translational components of the holonomies are trivial, and evolving ones for which this is not the case.

2.3.1. *Conformally static spacetimes.* For conformally static spacetimes M^s, the universal cover $\tilde{M}^s$ is the interior of the future lightcone of a point $\boldsymbol{p} \in \mathbb{M}^3$

$$\tilde{M}^s = \{\boldsymbol{y} \in \mathbb{M}^3 \mid (\boldsymbol{y} - \boldsymbol{p})^2 < 0, y^0 - p^0 > 0\}. \tag{2.7}$$

The cosmological time function gives the Lorentzian distance from the tip of the lightcone

$$T(\boldsymbol{y}) = \sqrt{|(\boldsymbol{y} - \boldsymbol{p})^2|} \qquad \forall \boldsymbol{y} \in \tilde{M}^s, \tag{2.8}$$

and the foliation of $\tilde{M}^s$ by constant cosmological time surfaces is the standard foliation of the lightcone by hyperboloids. Each constant cosmological time surface $\tilde{M}^s_T$ is thus a copy of two-dimensional hyperbolic space $\mathbb{H}^2$, rescaled with the cosmological time T

$$\tilde{M}^s_T = \{\boldsymbol{y} \in \mathbb{M}^3 \mid (\boldsymbol{y} - \boldsymbol{p})^2 = -T^2,\ y^0 - p^0 > 0\} \cong T \cdot \mathbb{H}^2. \tag{2.9}$$

As the surfaces $\tilde{M}^s_T$ must be preserved by the action of $\pi_1(M^s)$ via the holonomies $h : \pi_1(M^s) \to P_3$, the translational component of the holonomies is trivial. It is given by conjugation with a global translation to the tip of the lightcone

$$h(\lambda) = (1, \boldsymbol{p}) \cdot (v_\lambda, 0) \cdot (1, -\boldsymbol{p}) \qquad \forall \lambda \in \pi_1(M^s), \tag{2.10}$$

while the Lorentzian components v_λ, $\lambda \in \pi_1(M^s)$, define a cocompact Fuchsian group Γ of genus g. The action of $\pi_1(M^s)$ on the constant cosmological time surfaces $\tilde{M}^s_T$ agrees with the canonical action of Γ on $\mathbb{H}^2$. Each constant cosmological time surface M^s_T in the quotient spacetime is thus a copy of the same Riemann surface

$\Sigma_\Gamma = \mathbb{H}^2/\Gamma$, rescaled with the cosmological time and equipped with a metric of constant curvature $-1/T^2$

$$(2.11) \qquad M^s = \bigcup_{T\in\mathbb{R}^+} M^s_T = \bigcup_{T\in\mathbb{R}^+} T\cdot\Sigma_\Gamma \qquad\qquad g_s = -dT^2 + T^2 g_{\Sigma_\Gamma}.$$

As the geometry of the constant cosmological time surfaces M_T does not evolve with T and the metric takes the form (2.11), these spacetimes are called conformally static.

2.3.2. *Evolving spacetimes via grafting.* It has been shown by Mess [**6**] that any evolving flat MGH 3d Lorentzian manifold with a Cauchy surface of genus $g \geq 2$ can be obtained from a conformally static one via grafting along a measured geodesic lamination on the associated Riemann surface Σ_Γ. We sketch the grafting construction for measured geodesic laminations which are weighted multicurves, i. e. sets of non-intersecting, simple geodesics on Σ_Γ, each associated with a positive number, the weight. General measured geodesic laminations on a Riemann surface Σ_Γ are obtained as limits of suitable sequences of weighted multicurves [**6, 9**].

Schematically, grafting on a Riemann surface Σ_Γ amounts to cutting Σ_Γ along all geodesics in the weighted multicurve and gluing in strips whose width is given by the weight, as shown in Figure 1. While Σ_Γ is equipped with a Riemannian metric of constant curvature -1, the metric on the grafted strips is a Riemannian metric of vanishing curvature.

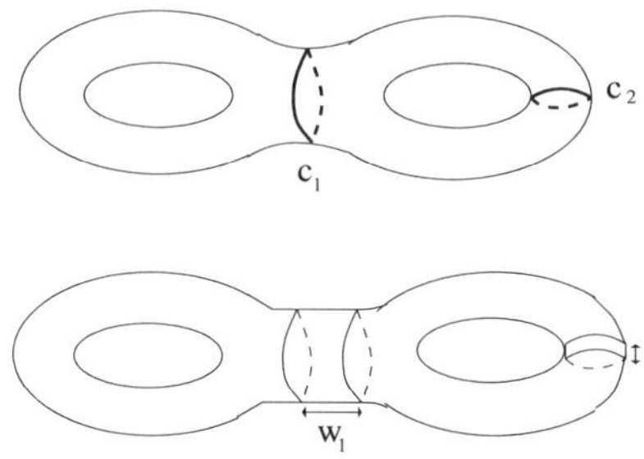

FIGURE 1. Grafting along two geodesics on a genus two surface.

Grafted (2+1)-spacetimes are obtained by applying the grafting construction simultaneously to all constant cosmological time surfaces $M^s_T = T\cdot\Sigma_\Gamma$ such that the weight of the grafting geodesics is the same for all surfaces M^s_T. The construction is performed in the universal cover. For this, one lifts all grafting geodesics on Σ_Γ to a weighted multicurve on each constant cosmological time surface $\tilde{M}^s_T \cong T\cdot\mathbb{H}^2$ by taking one lift of each geodesic on Σ_Γ and acting on it with the holonomies. This yields an infinite set of non-intersecting weighted geodesics on each hyperboloid $\tilde{M}^s_T$. These geodesics are given as the intersection of the lightcone $\tilde{M}^s$ with planes with spacelike normal vectors as shown in Figure 2 a).

To construct the grafted domain, one selects a basepoint in $\tilde{M}^s$ outside all geodesics in the lifted multicurves. One then cuts the lightcone along all planes defined by these geodesics and shifts the pieces that do not contain the basepoint away from the basepoint, in the direction of the planes' normal vectors and by a distance given by the weights as shown in Figure 2 b). Finally, one joins the translated pieces by straight lines as indicated in Figure 2 c).

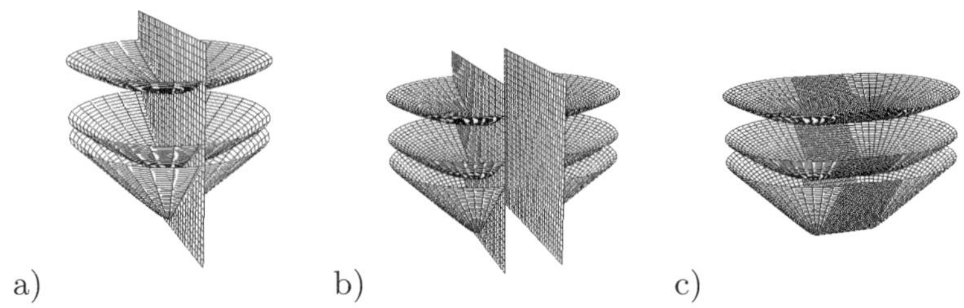

FIGURE 2. The grafting construction in the lightcone for a single geodesic.

This yields a deformed domain, which is no longer the future of a point but of a graph $\tilde{M}_0$. The cosmological time function of the deformed domain gives the geodesic distance of points in $\tilde{M}$ from $\tilde{M}_0$. The surfaces $\tilde{M}_T$ of constant cosmological time are the images of the hyperboloids $\tilde{M}^s_T$ under the grafting construction. They are deformed hyperboloids with strips glued in along the geodesics in the multicurve.

The fundamental group $\pi_1(M)$ acts on the deformed domain $\tilde{M}$ in such a way that two points on a constant cosmological time surface are identified if and only if the corresponding points on the hyperboloid are identified for the associated conformally static spacetime. The holonomies $h : \pi_1(M) \to P_3$ thus acquire a non-trivial translational component which takes into account the translations in the grafting construction.

The spacetime M is obtained by taking the quotient of the deformed domain $\tilde{M}$ by this action of the fundamental group. One finds that its constant cosmological time surfaces M_T undergo a non-trivial evolution with the cosmological time T, which is indicated in Figure 3. While the hyperbolic part of the constant cosmological time surfaces, i. e. the part outside of the grafting strips, is rescaled with the cosmological time T, the width of the grafted strips remains constant. The effect of grafting is thus dominant near the initial singularity for $T \to 0$ and vanishes in the limit $T \to \infty$.

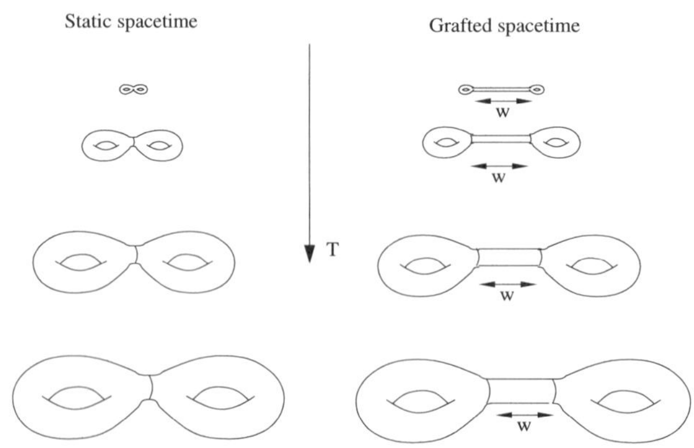

FIGURE 3. Evolution of static and grafted spacetimes with the cosmological time. The hyperbolic part of each constant cosmological time surface is rescaled with the cosmological time T, the width of the grafted strips remains constant.

3. Measurements associated with lightrays

3.1. Spacetime geometry via returning lightrays. While the description of flat 3d MGH Lorentzian spacetimes as quotients of their universal covers allows one to classify these spacetimes and to give an explicit description of their geometry, it does not provide a clear physical interpretation. To extract interesting physics from the theory, one needs to relate the variables which characterise these manifolds to concrete geometrical quantities that could be measured by an observer in the spacetime. These measurements should allow the observer to distinguish different spacetimes and to determine their geometry in finite eigentime.

This is difficult, since measurements are required to be *local*, i. e. take place in a small neighbourhood of a point in the spacetime, while the manifolds under consideration are locally isometric to Minkowski space $\mathbb{M}^3$ and can be distinguished only through their *global* geometrical properties. Moreover, as the spacetimes contain no matter and their curvature vanishes everywhere, it is a priori not clear what physically meaningful measurements could be performed at all.

The idea that allows one to address this problem is to consider an observer who probes the geometry of the spacetime by emitting lightrays. As we will see in the following, lightrays emitted in certain directions return to the observer at a later time. The observer can then perform several measurements associated with such returning lightrays. He can record their return time, the amount of eigentime elapsed between the emission of the lightray and its return. He can determine the directions into which light is emitted and from which it returns, and he can compare the frequencies of the emitted and returning lightrays.

3.2. Physics with returning lightrays. To derive explicit expressions for the measurements associated with returning lightrays in a flat MGH 3d Lorentzian manifold M, it is advantageous to work in its universal cover, the associated domain $\tilde{M} \subset \mathbb{M}^3$. This requires a clear definition of the relevant concepts such as observers, lightrays and returning lightrays in terms of the universal cover. We start with the notion of an observer, restricting attention to observers in free fall.

DEFINITION 3.1 (Observer). An *observer* (in free fall) in M is characterised uniquely by a timelike, future-directed geodesic $g : [a, \infty) \to M$, his *worldline*. Equivalently, an observer can be defined as a $\pi_1(M)$-equivalence class of timelike, future-directed geodesics $\tilde{g} : [a, \infty) \to \tilde{M}$ in the universal cover with the equivalence relation $\tilde{g}_1 \sim_h \tilde{g}_2$ if there exists an element $\lambda \in \pi_1(M)$ such that $\tilde{g}_1(t) = h(\lambda)\tilde{g}_2(t)$ for all $t \in [a, \infty)$. The worldline $g : [a, \infty) \to M$ is *parametrised according to eigentime* if $\dot{g}(t)^2 = -1\ \forall t \in [a, \infty)$ or, equivalently, $\dot{\tilde{g}}(t)^2 = -1\ \forall t \in [a, \infty)$ for all lifts $\tilde{g}$ of g.

If an observer's worldline $g : t \mapsto g(t)$ is parametrised according to eigentime, the parameter t gives the time as perceived by the observer, i. e. the time that would be shown by a clock carried with him. Note that the eigentime is unique up to a constant time shift $t \mapsto t + t_0$ with $t_0 \in \mathbb{R}$.

Analogously, we can define a lightray as a $\pi_1(M)$-equivalence class of geodesics in the universal cover. This yields the following definition.

DEFINITION 3.2 (Lightray). A *lightray* in M is a lightlike, future-directed geodesic $c : [p, \infty) \to M$ or, equivalently, a $\pi_1(M)$-equivalence class of lightlike, future-directed geodesics in $\tilde{M}$ with equivalence relation $\sim_h$ of Def. 3.1. A lightray emitted (received) by an observer with worldline $g : [a, \infty) \to M$ at eigentime t is a lightlike, future-directed geodesic $c : [p, q] \to M$ with $c(p) = g(t)$ $(c(q) = g(t))$ or, equivalently, the $\pi_1(M)$-equivalence class of lightlike, future-directed geodesics $\tilde{c} : [p, q] \to \tilde{M}$ for which there exists a lift $\tilde{g} : [a, \infty) \to \tilde{M}$ of g such that $\tilde{c}(p) = \tilde{g}(t)$ $(\tilde{c}(q) = \tilde{g}(t))$.

Note that the picture obtained by lifting timelike or lightlike curves to the universal cover differs from the one habitually encountered in Riemannian geometry, where a *closed* curve on a Riemannian manifold lifts to a single *open* curve in its universal cover. As the spacetimes under consideration are globally hyperbolic, they do not exhibit any closed time- or lightlike curves. Instead, open time- or lightlike curves in M lift to a $\pi_1(M)$-equivalence class of time- or lightlike curves in $\tilde{M}$. This accounts for the possibility of *returning lightrays*, lightrays emitted by an observer that return to him at a later time.

DEFINITION 3.3 (Returning lightray). A *returning lightray* with respect to an observer with worldline $g : [a, \infty) \to M$ is a lightlike, future-directed geodesic $c : [p, q] \to M$ that intersects g in $c(p)$ and $c(q)$. Equivalently, a returning lightray is given as a $\pi_1(M)$-equivalence class of lightlike, future-directed geodesics $\tilde{c} : [p, q] \to \tilde{M}$ such that there exists an element $\lambda \in \pi_1(M)$ and a lift $\tilde{g} : [a, \infty) \to \tilde{M}$ of g with $\tilde{c}(p) \in \tilde{g}$, $\tilde{c}(q) \in h(\lambda)\tilde{g}$.

The description of returning lightrays in the universal cover $\tilde{M}$ is pictured in Figure 4. As a returning lightray relates a lift of the observer's worldline to one of its images under the action of $\pi_1(M)$, it defines a unique element $\lambda \in \pi_1(M)$. However, it is a priori not guaranteed that for each observer and each element of $\pi_1(M)$ there exists an associated returning lightray. This is a consequence of the geometry of Minkowski space and the fact that the domains $\tilde{M}$ are future-complete.

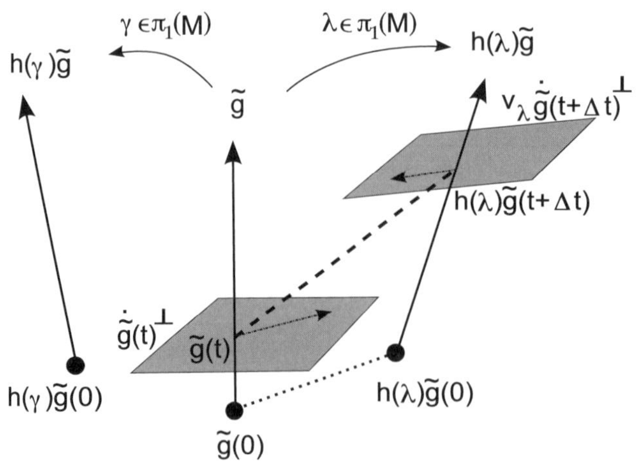

FIGURE 4. Lifts of the observer's worldline to $\tilde{M}$ with returning lightray (dashed line), orthogonal complements $\dot{\tilde{g}}^{\perp}(t)$, $(v_\lambda\dot{\tilde{g}})(t+\Delta t)^{\perp}$ (grey planes) and projection of the returning lightray to $\dot{\tilde{g}}^{\perp}(t)$, $(v_\lambda\dot{\tilde{g}})(t+\Delta t)^{\perp}$ (dashed arrows).

LEMMA 3.4. *Let $g : [a, \infty) \to M$ be the worldline of an observer in free fall. Then for all $t \in [a, \infty)$ the returning lightrays $c : [p, q] \to M$ with $c(p) = g(t)$ are in one-to-one correspondence with elements of the fundamental group $\pi_1(M)$.*

PROOF. We consider a lift $\tilde{g} : [a, \infty) \to \tilde{M}$ of the observer's worldline. Returning lightrays $c : [p, q] \to M$ with $c(p) = g(t)$ are in one-to-one correspondence with lightlike geodesics $\tilde{c}_\lambda : [p, q] \to \tilde{M}$ such that $\tilde{c}_\lambda(p) = \tilde{g}(t)$ and $\tilde{c}_\lambda(q) \in h(\lambda)\tilde{g}$ for a certain $\lambda \in \pi_1(M)$. As $\tilde{g}$ and $h(\lambda)\tilde{g}$ are timelike, future oriented geodesics in Minkowski space, for all choices of $\lambda \in \pi_1(M)$, $t \in [a, \infty)$, there exists a a lightlike geodesic $\tilde{c}_\lambda : [p, q] \to \mathbb{M}^3$ such that $\tilde{c}_\lambda(p) = \tilde{g}(t)$ and $\tilde{c}_\lambda(q) \in h(\lambda)\tilde{g}$. As $\tilde{g}(t) \in \tilde{M} \subset \mathbb{M}^3$ and $\tilde{M}$ is future complete, we have $\tilde{c}_\lambda(t) \in \tilde{M}$ for all $t \in [p, q]$, and $\tilde{c}_\lambda$ defines a returning lightray. □

After formulating the relevant concepts in terms of the universal cover $\tilde{M} \subset \mathbb{M}^3$, we are ready to consider the measurements associated with returning lightrays. We start with the return time, the interval of eigentime elapsed between the emission of a returning lightrays and its return as measured by the observer.

DEFINITION 3.5 (Return time). Let $g : [a, \infty) \to M$ be the worldline of an observer in free fall, parametrised according to eigentime and $\tilde{g} : [a, \infty) \to \tilde{M}$ a lift of g. Then by Lemma 3.4 for each $t_e \in [a, \infty)$ and each $\lambda \in \pi_1(M)$ there exists a unique $t_r \in (t_e, \infty)$ and a unique lightlike geodesic $\tilde{c}_\lambda : [0, 1] \to \tilde{M}$ with $\tilde{c}_\lambda(0) = \tilde{g}(t_e)$ and $\tilde{c}_\lambda(1) = h(\lambda)\tilde{g}(t_r)$. The return time is given by $\Delta t = t_r - t_e$ and obtained as the unique positive solution of the quadratic equation

$$(h(\lambda)\tilde{g}(t_e + \Delta t) - \tilde{g}(t_e))^2 = 0. \tag{3.1}$$

To obtain the directions in which the light needs to be emitted in order to return to the observer and from which it returns, we recall that the directions an observer perceives as "spatial" are given by the orthogonal complement $\dot{g}(t)^\perp$, where $g : [a, \infty) \to M$ is the observer's wordline. By considering the associated quantities in the universal cover, we obtain the following definition.

DEFINITION 3.6 (Directions of emission and return). Let $g : [a, \infty) \to M$ the worldline of an observer, parametrised according to eigentime, and $\tilde{g} : [a, \infty) \to \tilde{M}$ a lift of g. Let $\tilde{c} : [p, q] \to \tilde{M}$ be a future-directed, lightlike geodesic with $\tilde{g}(t) = \tilde{c}(p)$ ($\tilde{g}(t) = \tilde{c}(q)$). Then the direction into which the lightray associated with $\tilde{c}$ is emitted (from which the lightray lightray associated with $\tilde{c}$ returns) as perceived by the observer is given by the spacelike unit vector

$$\hat{\boldsymbol{p}}_e = \Pi_{\dot{\tilde{g}}(t)^\perp}(\dot{\tilde{c}}(p))/|\Pi_{\dot{\tilde{g}}(t)^\perp}(\dot{\tilde{c}}(p))| \qquad \left(\hat{\boldsymbol{p}}_r = \Pi_{\dot{\tilde{g}}(t)^\perp}(\dot{\tilde{c}}(q))/|\Pi_{\dot{\tilde{g}}(t)^\perp}(\dot{\tilde{c}}(q))|\right). \tag{3.2}$$

Finally, the observer can determine the relative frequencies of a returning lightray at its emission and return. In the universal cover, this problem is analogous to the relativistic Doppler effect. The only difference is that here the two timelike geodesics correspond to a lift of the observer's wordline and its image instead of two different observers for the relativistic Doppler effect. This yields the following definition.

DEFINITION 3.7 (Frequency shift). Let $g : [a, \infty) \to M$ be the worldline of an observer parametrised according to eigentime and $\tilde{g} : [a, \infty) \to \tilde{M}$ a lift of g. Let $\tilde{c} : [p, q] \to \tilde{M}$ a lightlike geodesic associated with a returning lightray with $\tilde{c}(p) = \tilde{g}(t_e)$, $\tilde{c}(q) = h(\lambda)\tilde{g}(t_r)$ for an element $\lambda \in \pi_1(M)$. Then the quotient of frequencies of the lightray at its emission and return as measured by the observer is given by

$$\frac{f_r}{f_e} = \frac{h(\lambda)\dot{\tilde{g}} \cdot (h(\lambda)\tilde{g}(t_r) - \tilde{g}(t_e))}{\dot{\tilde{g}} \cdot (h(\lambda)\tilde{g}(t_r) - \tilde{g}(t_e))}. \tag{3.3}$$

3.3. Explicit results. To obtain explicit results for the return time, the directions of emission and return and the frequency shift, we make use of the fact that the universal cover $\tilde{M}$ is a future-complete region in Minkowski space $\mathbb{M}^3$. The geodesics characterising the observer and the lightrays therefore take a particularly simple form. In the following, we parametrise timelike, future-directed geodesics in $\tilde{M}$ in terms of an element $\boldsymbol{x} \in \mathbb{H}^2$, the *velocity vector* and a vector $\boldsymbol{x}_0 \in \tilde{M}$, its *initial position* at $t = 0$

$$\tilde{g}(t) = t\boldsymbol{x} + \boldsymbol{x}_0 \qquad \boldsymbol{x}^2 = -1, x^0 > 0, \boldsymbol{x}_0 \in \tilde{M}. \tag{3.4}$$

Note that the parametrisation is unique up to a time shift

$$t \mapsto t - t_0 \qquad \boldsymbol{x}_0 \mapsto \boldsymbol{x}_0 + t_0\boldsymbol{x}. \tag{3.5}$$

Similarly, each lightlike, future-directed geodesic $\tilde{c} : [0, \infty) \to \tilde{M}$ is given by a lightlike vector $\boldsymbol{y}$ and a initial position vector $\boldsymbol{y}_0 \in \tilde{M}$

$$\tilde{c}(s) = s\boldsymbol{y} + \boldsymbol{y}_0 \qquad \boldsymbol{y}^2 = 0, y^0 > 0, \boldsymbol{y}_0 \in \tilde{M}. \tag{3.6}$$

To obtain explicit expressions for the measurements performed by the observer, it is advantageous to introduce additional parameters, which are given as functions of the velocity vector $\boldsymbol{x}$, the initial position $\boldsymbol{x}_0$ and the holonomies $h(\lambda) = (v_\lambda, \boldsymbol{a}_\lambda)$. For $\lambda \in \pi_1(M) \setminus \{1\}$ and $\tilde{g}, \boldsymbol{x}, \boldsymbol{x}_0$ as in (3.4), we define

$$\cosh \rho_\lambda = -\boldsymbol{x} \cdot v_\lambda \boldsymbol{x} \tag{3.7}$$

$$h(\lambda)\tilde{g}(0) - \tilde{g}(0) = \sigma_\lambda(v_\lambda\boldsymbol{x} - \boldsymbol{x}) + \tau_\lambda v_\lambda \boldsymbol{x} + \nu_\lambda \boldsymbol{x} \wedge v_\lambda \boldsymbol{x}. \tag{3.8}$$

The parameter ρ_λ, which depends only on the velocity vector $\boldsymbol{x}$ and the Lorentzian component of the holonomy has a direct interpretation as the geodesic distance $\rho_\lambda = d_{\mathbb{H}^2}(\boldsymbol{x}, v_\lambda\boldsymbol{x})$ of $\boldsymbol{x}$ and $v_\lambda\boldsymbol{x}$ in hyperbolic space $\mathbb{H}^2$. It coincides with the length of the associated geodesic on the Riemann surface $\mathbb{H}^2/\Gamma$. The parameters $\sigma_\lambda, \tau_\lambda, \nu_\lambda$ characterise the relative initial position of the geodesic $\tilde{g}$ and its image $h(\lambda)\tilde{g}$. They depend on the velocity vector $\boldsymbol{x}$, the initial position $\boldsymbol{x}_0$ as well as both components of the holonomies. They are invariant under Poincaré transformations $(v, \boldsymbol{a})$ acting simultaneously on the geodesic $\tilde{g}$ and on all holonomies by conjugation

$$\boldsymbol{x} \mapsto v\boldsymbol{x},\ \boldsymbol{x}_0 \mapsto v\boldsymbol{x}_0 + \boldsymbol{a} \qquad h(\lambda) \mapsto (v, \boldsymbol{a}) \cdot h(\lambda) \cdot (v, \boldsymbol{a})^{-1}. \tag{3.9}$$

Using the Definitions 3.5, 3.6, 3.7 one can derive explicit expressions for the measurements associated with returning lightrays in terms of the parameters $\rho_\lambda, \sigma_\lambda, \tau_\lambda, \nu_\lambda$. These expressions are derived in [**12**] and summarised in the following theorem.

THEOREM 3.8 ([**12**]). *Let $\tilde{g} : [a, \infty) \to \tilde{M}$ be a lift of the worldline of an observer in free fall, parametrised as in* (3.4)*. Consider a returning lightray associated with an element $\lambda \in \pi_1(M)$ that is emitted by the observer at eigentime t and returns at $t + \Delta t$. Then the eigentime Δt elapsed between the emission and return of the lightray is given by*

$$\Delta t(t, \boldsymbol{x}, \boldsymbol{x}_0, h(\lambda)) = (t + \sigma_\lambda)(\cosh \rho_\lambda - 1) - \tau_\lambda + \sinh \rho_\lambda \sqrt{(t + \sigma_\lambda)^2 + \nu_\lambda^2}, \tag{3.10}$$

where $\rho_\lambda, \sigma_\lambda, \tau_\lambda, \nu_\lambda$ are functions of $\boldsymbol{x}, \boldsymbol{x}_0$ and $h(\lambda)$ defined by (3.7), (3.8). The direction into which the lightray is emitted is given by the spacelike unit vector

$$\hat{\boldsymbol{p}}^e_\lambda = \cos\phi_e \, \frac{v_\lambda \boldsymbol{x} + (\boldsymbol{x}\cdot v_\lambda \boldsymbol{x})\boldsymbol{x}}{|v_\lambda \boldsymbol{x} + (\boldsymbol{x}\cdot v_\lambda \boldsymbol{x})\boldsymbol{x}|} + \sin\phi_e \, \frac{\boldsymbol{x}\wedge v_\lambda \boldsymbol{x}}{|\boldsymbol{x}\wedge v_\lambda \boldsymbol{x}|} \tag{3.11}$$

$$\tan\phi_e(t, \boldsymbol{x}, \boldsymbol{x}_0, h(\lambda)) = \frac{\nu_\lambda}{\sinh\rho_\lambda \sqrt{(t+\sigma_\lambda)^2 + \nu_\lambda^2}} \tag{3.12}$$

and the direction from which it returns by

$$\hat{\boldsymbol{p}}^r_\lambda(t) = \cos\phi_r \, \frac{v_\lambda^{-1} \boldsymbol{x} + (\boldsymbol{x}\cdot v_\lambda^{-1} \boldsymbol{x})\boldsymbol{x}}{|v_\lambda^{-1} \boldsymbol{x} + (\boldsymbol{x}\cdot v_\lambda^{-1} \boldsymbol{x})\boldsymbol{x}|} + \sin\phi_r \, \frac{\boldsymbol{x}\wedge v_\lambda^{-1} \boldsymbol{x}}{|\boldsymbol{x}\wedge v_\lambda^{-1} \boldsymbol{x}|} \tag{3.13}$$

$$\tan\phi_r(t, \boldsymbol{x}, \boldsymbol{x}_0, h(\lambda)) = \frac{\nu_\lambda}{(t+\sigma_\lambda)}. \tag{3.14}$$

The relative frequencies of the lightray at its emission and return as measured by the observer are given by

$$f_r/f_e(t, \boldsymbol{x}, \boldsymbol{x}_0, h(\lambda)) = \frac{\sqrt{(t+\sigma_\lambda)^2 + \nu_\lambda^2}}{\cosh\rho_\lambda \sqrt{(t+\sigma_\lambda)^2 + \nu_\lambda^2} + \sinh\rho_\lambda (t+\sigma_\lambda)} < 1. \tag{3.15}$$

As indicated by the notation, the return time, directions and frequency shift are given as functions of the emission time t, the two vectors $\boldsymbol{x}, \boldsymbol{x}_0$ characterising the observer's worldline and the holonomies $h(\lambda)$, $\lambda \in \pi_1(M)$. The fact that they depend only on the sum $t+\sigma_\lambda$, but not on t and σ_λ individually, reflects the invariance under a time shift (3.5). Moreover, as they are given in terms of the parameters ρ_λ, σ_λ, τ_λ, ν_λ formulas (3.10),(3.12), (3.14) and (3.15) are invariant under Poincaré transformations (3.9) acting simultaneously on the observer's geodesic and on the holonomies. In particular, this implies that they are invariant under change of the choice of the lift combined with an inner automorphism of $\pi_1(M)$

$$\tilde{g} \mapsto h(\eta)\tilde{g} \qquad \lambda \mapsto \eta\cdot\lambda\cdot\eta^{-1} \quad \forall \lambda \in \pi_1(M). \tag{3.16}$$

3.4. Interpretation. To understand how the expressions (3.10) to (3.15) reflect the geometry of the underlying spacetime, we consider a conformally static spacetime and an observer whose worldline starts at $t = 0$ at the initial singularity. In this case, the universal cover can be identified with the lightcone based at the origin, and the translational components of the holonomies as well as the initial position vector $\boldsymbol{x}_0$ in (3.4) can be set to zero. This implies that the parameters $\sigma_\lambda, \tau_\lambda, \nu_\lambda$ in (3.8) vanish for all $\lambda \in \pi_1(M)$ and the expressions for the return time, directions and the frequency shift take the form

$$\Delta t(t, \boldsymbol{x}, \boldsymbol{v}_\lambda) = t(e^{\rho_\lambda} - 1) \qquad \phi^{e,r}(t, \boldsymbol{x}, \boldsymbol{v}_\lambda) = 0 \qquad f_r/f_e(t, \boldsymbol{x}, \boldsymbol{v}_\lambda) = e^{-\rho_\lambda}. \tag{3.17}$$

The return time is thus a linear function of eigentime at which the lightrays was emitted, with a coefficient related to the length ρ_λ of the associated geodesic on a constant cosmological time surface. The directions of emission and return as well as the frequency shift are independent of the emission time. Note that the frequency shift is a red shift $f_r/f_e < 1$, as expected for an expanding spacetime. It depends only on the length ρ_λ of the associated geodesic on a constant cosmological time surface.

For a general spacetimes and general observers, these values of the return time, directions and frequency shift are approached in the limit $t \to \infty$

$$\lim_{t\to\infty} \Delta t/t = e^{\rho_\lambda} - 1 \qquad \lim_{t\to\infty} \phi^{e,r} = 0 \qquad \lim_{t\to\infty} f_r/f_e = e^{-\rho_\lambda} \tag{3.18}$$

This reflects the fact that for all observers, the cosmological time tends to infinity $T(\tilde{g}(t)) \to \infty$ as $t \to \infty$. In this limit, the effects of grafting become negligible and the spacetime approaches the associated conformally static spacetime.

Generally, for a given element $\lambda \in \pi_1(M)$, the return time (3.10) is a linear function of the emission time and the directions of emission (3.11) and return (3.13) as well as the frequency shift (3.15) are constant if and only if the parameter ν_λ defined by (3.8) vanishes.

To investigate the geometrical interpretation of this condition, one considers an evolving spacetime obtained by grafting along a single geodesic on the associated Riemann surface $\Sigma_\Gamma = \mathbb{H}^2/\Gamma$ and an observer whose wordline starts at $t = 0$ at the initial singularity. It is shown in [**12**] that the parameter ν_λ vanishes in this situation if and only if the geodesic associated with $\lambda \in \pi_1(M)$ either does not cross the grafting geodesic or crosses it orthogonally as shown in Figure 5 a). In this case, the geodesics associated with λ on each constant cosmological time surface are not deflected at the grafted strip and their length increases by a constant. The frequency shift of the associated lightray and its directions of emission and return therefore do not depend on the emission time, while the return time depends on it linearly.

In contrast, if the geodesics on the constant cosmological time surfaces associated with $\lambda \in \pi_1(M)$ cross the grafting geodesic non-orthogonally as shown in Figure 5 b), they are deflected at the grafting strip. As the width of the strip remains constant, while the rest of the surface is rescaled with the cosmological time, this deflection depends on the cosmological time T and vanishes for $T \to \infty$. Their length increase through grafting thus depends non-linearly on the cosmological time and hence the emission time. Consequently, the directions of emission and return of the lightray and its frequency shift depend on the emission time, and the return time becomes a non-linear function of the emission time.

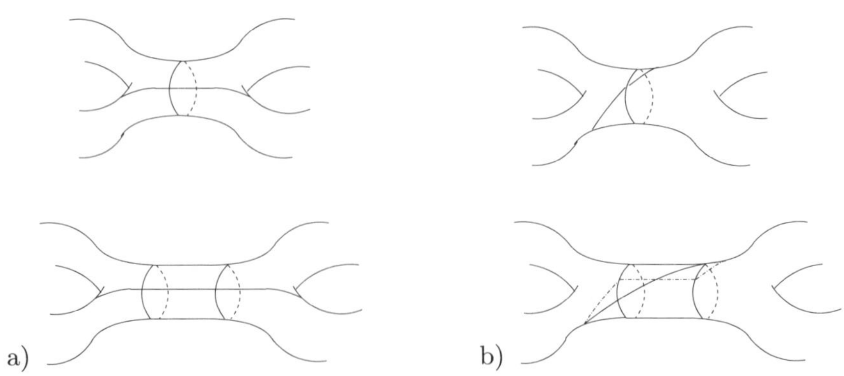

FIGURE 5. Deflection of geodesics through grafting. Case a) corresponds to the situation where $\nu_\lambda = 0$, while $\nu_\lambda \neq 0$ in b).

4. Reconstructing spacetime geometry from measurements

As demonstrated in the last section, the measurements associated with returning lightrays reflect the geometry of the underlying spacetime and allow the observer to determine some of its geometrical properties. We will now show that they allow the observer to reconstruct the *full* geometry of the spacetime in *finite* eigentime. For this we recall that a spacetime is determined uniquely by the holonomies $h : \pi_1(M) \to P_3$ modulo simultaneous conjugation with P_3. Reconstructing the geometry of the spacetime is thus equivalent to determining the holonomies for a set of generators $\{\lambda_1, .., \lambda_n\}$ of $\pi_1(M)$ up to simultaneous conjugation with P_3.

4.1. Conformally static spacetimes. We start by considering the case of conformally static spacetimes and observers whose worldlines start at the initial singularity M_0. In this case, the translational components of the holonomies and the parameters σ_λ, τ_λ, ν_λ in (3.8) can be set to zero for all $\lambda \in \pi_1(M)$. Reconstructing the holonomies for a set of generators of $\pi_1(M)$ therefore amounts to determining a set of generators of the Fuchsian group Γ defined by the Lorentzian components of the holonomies. A way to achieve this is to reconstruct the Dirichlet region of Γ.

This is the set of points in $\mathbb{H}^2$ whose geodesic distance from a given point $\boldsymbol{x} \in \mathbb{H}^2$ is less or equal than the geodesic distance from all its images [**13**]

$$D_\Gamma(\boldsymbol{x}) = \{\boldsymbol{y} \in \mathbb{H}^2 \mid d_{\mathbb{H}^2}(\boldsymbol{y}, \boldsymbol{x}) \leq d_{\mathbb{H}^2}(\boldsymbol{y}, v\boldsymbol{x})\ \forall v \in \Gamma\}. \tag{4.1}$$

It is obtained by constructing the perpendicular bisectors of the geodesic segments $[\boldsymbol{x}, v\boldsymbol{x}]$ for $v \in \Gamma$, and intersecting the associated half-planes as shown in Figure 6 a). The Dirichlet region of a cocompact Fuchsian group Γ of genus $g \geq 2$ is a geodesic arc $2k$-gon with $k \geq 2g$, whose sides are identified pairwise by a set of generators of Γ. Reconstructing a set of generators of a Fuchsian group Γ is thus equivalent to reconstructing a Dirichlet region of Γ together with the information about the identification of its sides[2].

To see how the observer can determine the Dirichlet region of the Fuchsian group Γ and the identification of its sides in finite eigentime, we consider the following procedure:

(1) The observer emits light in all directions at a given eigentime t. The returning lightrays associated with elements $\lambda \in \pi_1(M)$ return to the observer one by one at different eigentimes $t_\lambda = t + \Delta t_\lambda$. For each returning lightray, the observer measures the eigentime Δt_λ elapsed since the emission and the direction from which the light returns. The former allows him to determine the geodesic distance $\rho_\lambda = d_{\mathbb{H}^2}(\boldsymbol{x}, v_\lambda \boldsymbol{x})$ between his velocity vector $\boldsymbol{x} \in \mathbb{H}^2$ and its image $v_\lambda \boldsymbol{x} \in \mathbb{H}^2$ via (3.17). The latter corresponds to the tangent vector at $v_\lambda \boldsymbol{x}$ to the geodesic segment $[\boldsymbol{x}, v_\lambda \boldsymbol{x}]$ in $\mathbb{H}^2$. Hence, given the observer's velocity vector $\boldsymbol{x}$, the measurement of the return time and direction of return allow him to determine the image $v_\lambda \boldsymbol{x}$.

[2]In the generic case, the sides of the Dirichlet region $D_\Gamma(\boldsymbol{x})$ have different lengths, which allows one to determine directly, which of them are identified by the action of Γ. However, this is not the case if the Dirichlet region has a high degree of symmetry. There exist Fuchsian groups which are not isomorphic, have the same Dirichlet region and differ only in the way in which the sides of the Dirichlet region are identified. In this case, additional information about the identification of sides is needed in order to determine a set of generators of a Fuchsian group Γ from its Dirichlet region. I thank R. C. Penner for pointing this out to me.

(2) For each returning lightray, the observer can thus construct the perpendicular bisector of the geodesic segment $[\boldsymbol{x}, v_\lambda \boldsymbol{x}]$ in $\mathbb{H}^2$ as shown in Figure 6 a). After a finite number of returning lightrays, the perpendicular bisectors of the associated geodesic segments $[\boldsymbol{x}, v_\lambda \boldsymbol{x}]$ close to form a geodesic arc polygon $P \subset \mathbb{H}^2$ as shown in Figure 6 a). If $r = \max\{d_{\mathbb{H}^2}(\boldsymbol{x}, \boldsymbol{p}) \mid \boldsymbol{p}$ corner of $P\}$, the perpendicular bisectors of images $v_\lambda \boldsymbol{x}$ with $d_{\mathbb{H}^2}(\boldsymbol{x}, v_\lambda \boldsymbol{x}) > 2r$ cannot intersect P and therefore do not affect the Dirichlet region. This implies that lightrays returning after a time $\Delta t = t(e^{2r} - 1)$ are irrelevant, and the observer can reconstruct the Dirichlet region $D_\Gamma(\boldsymbol{x})$ in finite eigentime $\Delta t = t(e^{2r} - 1)$.

(3) After determining the Dirichlet region, the observer sends out a finite number of additional lightrays into the directions associated with the geodesic arcs that form the boundary of the Dirichlet region and records from which directions the associated lightrays return. This allows him to conclude which sides of the Dirichlet region are identified by Γ. Together with the Dirichlet region, this information allows him to reconstruct a set of generators of Γ[3].

This procedure allows the observer to determine a set of generators of the Fuchsian group Γ and hence to reconstruct the full geometry of the spacetime in finite eigentime. Moreover, it is shown in **[12]** that the observer does not need to know his velocity vector $\boldsymbol{x}$ to do so, as a change of the vector $\boldsymbol{x} \mapsto v\boldsymbol{x}$ with $v \in SO^+(2,1)$ amounts to simultaneous conjugation of all elements of Γ with v.

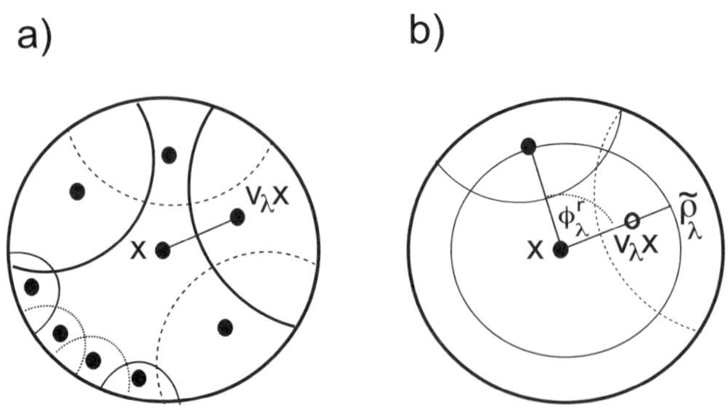

FIGURE 6. a) Dirichlet region for a Fuchsian group of genus $g = 2$. b) Translated images for evolving spacetimes with deviation angle ϕ^r_λ, modified distance $\tilde{\rho}_\lambda$.

4.2. Evolving spacetimes. We now consider the case of a general observer in an evolving spacetime. In this case, the observer can proceed as if the spacetime was conformally static and measure the return time and direction for each returning lightray as outlined in the last subsection. The return direction then depends on the emission time and is modified with respect to the return direction of the associated conformally static spacetime by the angle

$$\phi^r_\lambda(t) = \arctan\left(\nu_\lambda/(t + \sigma_\lambda)\right) = \nu_\lambda/t + O(1/t^2). \tag{4.2}$$

[3]I thank R. C. Penner and several other participants of the workshop "Chern-Simons Gauge Theory: 20 years after" for discussions of this issue.

Moreover, the observer can use expression (3.17) for the return time in a conformally static spacetime to assign to each returning lightray a modified distance $\tilde{\rho}_\lambda(t)$ defined by

$$\Delta t(t, \boldsymbol{x}, \boldsymbol{x}_0, h(\lambda)) = t(e^{\tilde{\rho}_\lambda(t)} - 1). \tag{4.3}$$

In terms of the hyperbolic distance variable $\rho_\lambda = d_{\mathbb{H}^2}(\boldsymbol{x}, v_\lambda \boldsymbol{x})$ of the associated static spacetime this modified distance is given by

$$\begin{aligned}\tilde{\rho}_\lambda(t) =& \ln\left(e^{\rho_\lambda} - \tau_\lambda/t + \sinh\rho_\lambda\left(\sqrt{(1+\sigma_\lambda/t)^2 + \nu_\lambda^2/t^2} - 1\right)\right)\\ =&\rho_\lambda + (\sigma_\lambda(e^{\rho_\lambda} - 1) - \tau_\lambda)/t + O(1/t^2).\end{aligned} \tag{4.4}$$

The observer can now act as if the spacetime was conformally static and construct a geodesic arc polygon as outlined in the last subsection, using the variables $\tilde{\rho}_\lambda(t)$ and the return directions obtained from his measurements. The images of the velocity vector $\boldsymbol{x} \in \mathbb{H}^2$ obtained this way and, consequently, the resulting perpendicular bisectors will be translated with respect to the conformally static case as indicated in Figure 6 b). The observer thus constructs a deformed geodesic arc polygon $\tilde{P}(t)$ which approaches the Dirichlet region of the associated conformally static spacetime in the limit $t \to \infty$.

To determine the holonomies along a set of generators of the fundamental group, the observer can now repeat the measuring procedure outlined in the last subsection several times and obtains a sequence of deformed polygons $\tilde{P}(t_1), \tilde{P}(t_2), ..., \tilde{P}(t_n)$ in $\mathbb{H}^2$. By observing the change of the polygons with the emission time, the observer can extrapolate to the limit $t \to \infty$ to recover the Dirichlet region of the associated conformally static spacetime and, after a finite number of additional measurements, the identification of its sides. Using formulas (4.2) to (4.4), he can then determine the associated parameters $\sigma_\lambda, \tau_\lambda, \nu_\lambda$. Via (3.8) he then obtains the holonomies $h(\lambda)$ for a set of generators of the fundamental group $\pi_1(M)$ and hence the full geometry of the spacetime in finite eigentime.

5. Concluding remarks

We showed how the description of flat MGH 3d Lorentzian manifolds in terms of their universal cover can be used to obtain interesting physics in (2+1)-dimensional gravity. By considering an observer who probes the geometry of the spacetime by emitting returning lightrays, we defined several measurements that could be made by such an observer: the eigentime elapsed between the emission of the lightray and its return, the directions into which light is emitted and from which it returns as well as the frequency shift between the emitted and returning lightray.

We gave explicit expressions for these measurements in terms of the holonomy variables which parametrise the flat Lorentzian 3d manifolds arising in (2+1)-gravity and play a central role in the quantisation of the theory. Moreover, we demonstrated how an observer can use these measurements to determine the holonomy variables and thus reconstruct the full geometry of the spacetime in finite eigentime.

The results serve a concrete and non-trivial example in which concrete physics questions and conceptual issues of (quantum) gravity can be investigated [**12**]. It would also be interesting to generalise them to more realistic scenarios relevant to cosmology such as observations of external light sources or background radiation emitted near the initial singularity. Finally, one of the main motivations of this work is its application to a quantum theory of (2+1)-gravity. This would offer the prospect of investigating realistic physical measurements in a fully and rigourously quantised theory of gravity.

Acknowledgments. I thank the organisers of the workshop "Chern-Simons Gauge Theory: 20 years after" in which this work was presented, and I am grateful to its participants for helpful comments and suggestions. In particular, I thank R. C. Penner for helpful remarks regarding the measurement of the Dirichlet region. This work was supported by the Emmy Noether fellowship ME 3425/1-1 of the German Research Foundation (DFG).

References

[1] S. Carlip, *Quantum gravity in 2+1 dimensions*, Cambridge University Press, Cambridge, 1998.

[2] A. Achucarro and P. Townsend, *A Chern–Simons action for three-dimensional anti-de Sitter supergravity, theories*, Phys. Lett. B **180** (1986), 85–100.

[3] E. Witten, *2+1 dimensional gravity as an exactly soluble system*, Nucl. Phys. B **311** (1988), 46–78.

[4] ———, *Quantum field theory and the Jones polynomial*, Comm. Math. Phys. **121** 3 (1989), 351–399.

[5] N. Reshetikhin and V. Turaev, *Invariants of 3-manifolds via link-polynomials and quantum groups*, Invent. Math.**103** (1991), 547–597.

[6] G. Mess, *Lorentz spacetimes of constant curvature, preprint IHES/M/90/28* (1990), Geometriae Dedicata **126:1** (2007), 3–45.

[7] L. Andersson, T. Barbot, R Benedetti, F. Bonsante, W. M. Goldman, F. Labourie, K. P. Scannell, and J.-M. Schlenker, *Notes on a paper of Mess*, Geometriae Dedicata **126:1** (2007), 47–70.

[8] R. Franzosi and E. Guadagnini, Topology and classical geometry in (2 + 1) gravity, Class. Quant. Grav. **13** (1996), 433–460.

[9] R. Benedetti and F. Bonsante, *Canonical Wick rotations in 3-dimensional gravity*, Memoirs AMS **926**, vol. 198, Amer. Math. Soc. (2009).

[10] R. Benedetti and E. Guadagnini, *Cosmological time in (2+1)-gravity*, Nucl. Phys. B **613** (2001), 330–352.

[11] T. Barbot, *Globally hyperbolic flat spacetimes*, Journ. Geom. Phys. **53** (2005), 123–165.

[12] C. Meusburger, *Cosmological measurements, time and observables in (2+1)-dimensional gravity*, Class. Quant. Grav. **26** (2009), 055006.

[13] S. Katok, *Fuchsian Groups*, The University of Chicago Press, Chicago, 1992.

Department Mathematik, Universität Hamburg, Bundesstrasse 55, D-20146 Hamburg, Germany

E-mail address: catherine.meusburger@uni-hamburg.de

AMS/IP Studies in Advanced Mathematics
Volume 50, 2011

Spin Foam State Sums and Chern-Simons Theory

Aleksandar Miković and Joao Faria Martins

ABSTRACT. We review the spin foam state-sum invariants of 3-manifolds, and explain their relationship to manifold invariants coming from the Chern-Simons theory. We also explain the relationship between the known invariants of spin networks by using the Chain-Mail formalism of J. Roberts. This formalism can be understood as a quantum-group regularization of the corresponding BF theory path integrals.

1. Spin foams

Spin foams are colored two-complexes and they naturally appeared in the context of Loop Quantum Gravity, see [**1**]. They can be viewed as world-sheets of spin networks propagating in time, where the spin networks are colored graphs, originally invented by Penrose [**13**], representing the states of quantum geometry on a three-dimensional space. One can then imagine a spin foam interpolating between two fixed spin networks. By associating a complex number to this spin foam, such that this number depends on the spin foam labels, and then summing over the labels, one can obtain an invariant of the spin networks, provided that the sum over the labels is independent of the triangulation of the four manifold whose dual two-complex is the given spin foam. The sum over the labels is called the spin foam state sum, and the rules of how to find a triangulation independent spin foam state sum can be inferred from the path integrals for BF theories, see [**1**].

An n-dimensional BF theory can be defined by the action

$$S = \int_M Tr\,(B \wedge F)\,, \tag{1.1}$$

where M is an n-dimensional compact manifold, $F = dA + A \wedge A$ is the curvature 2-form for a connection A on the principal bundle $P_G(M)$ for a compact Lie group G, B is a $\mathbf{g}$-valued $(n-2)$-form, where $\mathbf{g}$ is the Lie algebra of G. The path integral

---

2010 *Mathematics Subject Classification.* Primary 57M27, 57R56; Secondary 81T25.

Key words and phrases. State-sum invariants, BF theory, path integrals.

The first author was supported in part by the FCT grant PTDC/MAT/69635/2006 and by Mathematical Physics Group of the University of Lisbon.

The second author was supported by CMA, through Financiamento Base 2009 ISFL-1-297 from FCT/MCTES/PT .

for this BF theory can be written as a functional integral

$$Z(M) = \int DA\,DB \exp\left(i\int_M Tr\,(B\wedge F)\right), \tag{1.2}$$

where $DA\,DB$ denotes a measure on the space of fields A and B. $Z(M)$ can be formally written as a functional integral over the flat connections on M

$$Z(M) = \int DA\,\delta(F), \tag{1.3}$$

where $\delta(F)$ denotes the product $\prod_{x\in M}\delta(F(x))$ and $\delta(x)$ is the Dirac delta-function, see [**1**].

All these expressions are formal and have to be defined, and the standard way is through a triangulation $\Delta(M)$ of M. The guiding idea is that $Z(M)$ should be a manifold invariant, since the corresponding BF theory is a topological theory, so that a discretization of the path integral (1.3) should not depend on a chosen triangulation of M. Let us replace $\Delta(M)$ with its dual simplical complex $\Delta^*(M)$ and consider a discretization of (1.3) given by

$$Z(M) = \int \prod_l dg_l \prod_f \delta(g_f), \tag{1.4}$$

where l are the dual edges and f are the dual faces, g_l is the holonomy along an edge l and $g_f = \prod_{l\in\partial f} g_l$ is the holonomy along the boundary loop ∂f of f.

The group integrations in (1.4) can be performed by using the Peter-Weyl theorem

$$\delta(g) = \sum_\Lambda \dim\Lambda\, Tr\left(D^{(\Lambda)}(g)\right), \tag{1.5}$$

and

$$\int_G dg\, D^{(\Lambda_1)}(g)\otimes\cdots\otimes D^{(\Lambda_m)}(g) = \sum_\iota C^{(\iota)}\otimes\overline{C^{(\iota)}}, \tag{1.6}$$

where $D^{(\Lambda)}(g)$ is the matrix of the group element g in the irreducible representation Λ and $C^{(\iota)}$ is the tensor for the intertwiner ι from the tensor product of representations $\Lambda_1,\cdots,\Lambda_m$. Then

$$Z(M) = \sum_{\Lambda,\iota}\prod_f \dim\Lambda_f \prod_v A_v(\Lambda,\iota), \tag{1.7}$$

where the sum is over the colorings $\Lambda = (\Lambda_1,\cdots,\Lambda_N)$ of the faces of the dual simplical complex by the irreps of G. The sum in (1.7) also includes a sum over the intertwiners $(\iota_1,\cdots,\iota_L)$, which are associated to the dual edges of $\Delta(M)$ and A_v is given by the classical evaluation of the n-simplex spin network, see [**1**].

In the case when $G = SU(2)$, the irreps are called spins, and hence the name "spin foam" is used for the colored dual two-complex. For a three-dimensional manifold, the dual faces of a triangulation correspond to the edges, and the dual vertices correspond to the tetrahedrons, so that A_v is the $6j$ symbol.

The spin-foam state sum (1.7) can be shown to be independent of the triangulation, due to the invariance under the n-dimensional Pachner moves, see [**14, 12**]. However, the sum (1.7) is infinite and therefore divergent in general case, so that one still does not have a well-defined invariant. Fortunately, the formal expression (1.7) can be regularized by replacing the irreps of G by the irreps of the quantum group $U_q(\mathbf{g})$, where q is a root of unity. This can be done because there is a one-to-one correspondence between the irreps of a compact Lie group and the irreps of the corresponding quantum group when q is not a root of unity. When q becomes a root of unity, this correspondence still holds for the irreps of non-zero quantum dimension. Consequently, all the spin network evaluations, which are the simplex weights in (1.7), can be replaced by the corresponding quantum spin network evaluations. The invariance under the Pachner moves is maintained and (1.7) becomes a finite sum, because when q is a root of unity there are only finitely many irreps Λ of non-zero quantum dimension, see [**4**] for the case $n = 3$.

Therefore, it can be conjectured that a finite sum

$$Z(M) = \kappa^N \sum_{\Lambda,\iota} \prod_f \dim_q \Lambda_f \prod_l A_l^{(q)}(\Lambda,\iota) \prod_v A_v^{(q)}(\Lambda,\iota) \tag{1.8}$$

is a manifold invariant for $n \geq 2$, where the label q denotes the quantum group evaluations of the relevant spin networks. The factor κ is related to q, N is the number of 0-simplices in the triangulation and a dual-edge amplitude A_l appears[1]. In the three-manifold case and $G = SU(2)$, (1.8) coincides with the Turaev-Viro invariant [**18**]. In the 4-manifold case, (1.8) coincides with the Crane-Yetter invariant [**5**].

2. Chern-Simons and BF theory path integrals

A BF theory is a theory of flat connections on a manifold, and in three dimensions one also has the Chern-Simons (CS) theory, which is also a topological theory of flat connections. Given that the spin-foam state sum invariants are derived from the BF theory, the question is what is the relation to the manifold invariants derived from the CS theory? Witten showed that the CS invariants can be derived from the CS path integral [**20**], and the path-integral approach can be used to find the relation between the spin foam and CS invariants.

A BF theory with a cubic interaction is given by the action

$$S_\lambda(A,B) = \int_M Tr\,(B \wedge F + \lambda B \wedge B \wedge B)\,, \tag{2.1}$$

where λ is a parameter. In the $SU(2)$ group case, this action corresponds to the three-dimensional (3d) General Relativity (GR) theory with a cosmological constant λ. It is easy to show that

$$S_\lambda(A,B) = S_{CS}(A_+) - S_{CS}(A_-)\,, \tag{2.2}$$

where $A_\pm = A \pm \frac{B}{\sqrt{\lambda}}$, $S_{CS}(A)$ is the Chern-Simons action for a connection A

$$S_{CS}(A) = \frac{k}{2\pi}\int_M Tr\left(A \wedge dA + \frac{2}{3} A \wedge A \wedge A\right)\,, \tag{2.3}$$

and $k = 2\pi/\sqrt{\lambda}$. Witten has showed that the CS action is well-defined if k is an integer [**20**], and this integer is related to the order of the root of unity taken for q in the quantum group $U_q(\mathbf{g})$. In the $SU(2)$ case, $q = \exp(\frac{\pi i}{k+2})$.

By using (2.2) one can formally write

$$\int DA\, DB e^{iS_\lambda(A,B)} = \int DA_+ DA_- e^{iS_{CS}(A_+) - iS_{CS}(A_-)} = |Z_{CS}(M)|^2\,, \tag{2.4}$$

[1] One can set $A_l = 1$ by changing the normalization of the spin network evaluations.

where $Z_{CS}(M)$ is the Chern-Simons path integral. Witten, Turaev and Reshetikhin have shown that this path integral can be rigorously defined [**20, 15**], and the corresponding manifold invariant is proportional to the quantum group evaluation of a surgery link of M. We denote this invariant as $Z_{WRT}(M)$.

On the other hand, the Turaev-Walker theorem [**17**] implies

$$Z_{TV}(M) = |Z_{WRT}(M)|^2 , \tag{2.5}$$

where $Z_{TV}(M)$ is the spin foam state-sum invariant (1.7) for the 3-manifold M. From (2.5) and (2.4) it follows that Z_{TV} is the path integral for the three-dimensional euclidean General Relativity theory with a cosmological constant.

Note that the parameter λ can take only the restricted values $4\pi^2/k^2$, $k \in \mathbf{N}$ in the case of the path integral $Z(\lambda, M)$ for 3d GR. A natural question is how to define this path integral for the case when $\lambda \neq 4\pi^2/k^2$. This can be done by using the spin foam perturbation theory [**8**], and the result can be again expressed in terms of the $Z_{WRT}(M)$ invariant. Namely

$$Z(\lambda, M) = Z_{TV}(k, M)\big|_{k=\frac{2\pi}{\sqrt{\lambda}}} ,$$

where the right-hand side of the equation represents the function $Z_{TV}(k, M) = |Z_{WRT}(k, M)|^2$ evaluated at a non-integer point $k = 2\pi/\sqrt{\lambda}$.

3. Spin network invariants

Witten showed in [**20**] that the Jones polynomial $J_K(q)$ of a knot K embedded in a three-sphere S^3 can be obtained by analysing the following path-integral invariant of K

$$\langle K \rangle = \int DA e^{iS_{CS}(A)} W_K(A) , \tag{3.1}$$

where $W_K(A)$ is the trace of the holonomy of a connection A around K. The invariant $\langle K \rangle$ depends on the representation of the $SU(2)$ group used to define the holonomy. When K is an unknot, then

$$\langle K \rangle = \dim_q j = (-1)^{2j} \frac{q^{2j+1} - q^{-2j-1}}{q - q^{-1}} ,$$

where j is the spin of the $SU(2)$ representation and the Jones polynomial corresponds to $j = \frac{1}{2}$.

Beside colored loops, one can consider more general objects, like spin networks Γ, which are oriented graphs whose edges are colored by the spins. In analogy to (3.1), one can consider a spin-network invariant

$$\langle \Gamma \rangle = \int DA \, e^{iS_{CS}(A)} W_\Gamma(A) , \tag{3.2}$$

where $W_\Gamma(A)$ is given by the product of the edge holonomies contracted by the intertwiner tensors $C^{(\iota)}$

$$W_\Gamma(A) = \prod_{l \in E(\gamma)} \mathrm{Hol}_l(A) \prod_{v \in V(\gamma)} C^{(\iota_v)} ,$$

where $E(\gamma)$ and $V(\gamma)$ are the sets of the edges and the vertices of the graph γ corresponding to the spin network Γ.

One can be even more general, and consider that Γ is embedded in an arbitrary closed 3-manifold M. In any case, it can be shown that the path-integral (3.2) can be rigorously defined as the Witten-Reshetikhin-Turaev invariant $Z_{WRT}(M,\Gamma)$, which is proportional to the quantum group evaluation of the link $L_M \cup \gamma$ embedded in S^3, where L_M is any surgery link of M [**20, 15**].

Given that (3.2) can be rigorously defined in CS theory, a natural question is how to define the analogous invariant in the BF theory case. One way is to start from the path-integral expression

$$\langle \Gamma \rangle_{BF} = \int DADB\, e^{i\int_M Tr\,(B\wedge F)} W_\Gamma(A)\,,$$

and to use the spin-foam discretization to define it. By repeating the procedure from section 1 for Γ embedded in the handle-body H obtained from thickening the dual one-complex of the triangulation of M, such that each edge of Γ goes along some one-handle of H and each vertex of Γ is placed in some zero-handle of H, then a Turaev-Viro type state sum is obtained [**9**]. In the $SU(2)$ case this sum takes the form

$$Z_{TV}(\Gamma, M) = \sum_{J,\iota}\prod_f \dim_q j_f \prod_{v\notin V_\gamma}\{6j\}_v^{(q)} \prod_{v\in V_\gamma} A_v^{(q)}(j, j_\Gamma, \iota, \iota_\Gamma)\,, \tag{3.3}$$

where V_γ are the vertices of the dual 1-complex whose thickenings contain the vertices of the graph γ of Γ and $A_v^{(q)}(j, j_\Gamma, \iota, \iota_\Gamma)$ is the quantum group evaluation of a tetrahedron spin network with an insertion corresponding to the spin network vertex residing in the 3-handle of v, where $\{j_\Gamma, \iota_\Gamma\}$ are the spins and intertwiners of Γ, see [**9**].

By using the Chain-Mail formalism of J. Roberts [**16**], it can be shown that $Z_{TV}(\Gamma, M)$ can be represented as the quantum group evaluation of a link embedded in S^3 [**7**], and hence a relationship to the invariant $Z_{WRT}(\Gamma, M)$ can be obtained. The reason is that a Chain-Mail link CHL_M for M can be constructed by using the handle-body H, which was a thickening of the dual one-complex $\Delta_1^*(M)$ of a triangulation of M. Let N_p be the number of p-handles of H, then the link CHL_M consists of N_2 loops running along the one-handles of H, corresponding to the boundaries of the faces of $\Delta^*(M)$, and N_1 loops running around the middle cross-sections of the one-handles of H.

It can be shown for $G = SU(2)$ and a trivalent spin network[2] that

$$Z_{TV}(\Gamma, M) = \eta^{-N_0-N_2}\langle CHL_M \cup \gamma, \Omega^{N_1+N_2}, j_\Gamma\rangle\,, \tag{3.4}$$

where

$$\eta^2 = \sum_{j=0}^{k/2} (\dim_q j)^2 = \frac{k+2}{\sin^2\left(\frac{\pi}{k+2}\right)}$$

and Ω is the coloring of each loop of CHL_M [**7**]. The Ω-element is a linear combination of all possible colorings of a loop, and it is given by the sum

$$\Omega = \sum_{j=0}^{k/2} (\dim_q j)\, C(j)\,,$$

where $C(j)$ denotes the coloring of a loop by spin j.

[2]In the $SU(2)$ group case, a spin network vertex of valence $m > 3$ requires an intertwiner. However, such a vertex can be represented as a tri-valent spin network with $m-2$ vertices.

The proof of (3.4) is formally similar to the proof that the Lie group state sum (1.7) follows from the group integral (1.4), since both proofs are based on the same graphical calculus, which is associated to the equation (1.6), see [**1, 9**].

Given that the pair (CHL_M, Γ) is a surgery link for the spin-network Γ embedded in the connected sum $M\#\overline{M}$, where $\overline{M}$ is the manifold M with the opposite orientation and Γ is embedded in the obvious way in the M part of $M\#\overline{M}$, it follows from (3.4) that

$$Z_{TV}(\Gamma, M) = Z_{WRT}(\Gamma, M)\overline{Z_{WRT}(M)}\,, \tag{3.5}$$

see [**7**].

The formula (3.5) has several interesting consequences:

- It gives a state-sum representation of $Z_{WRT}(\Gamma, M)$, since
$$Z_{WRT}(\Gamma, M) = \left(\overline{Z_{WRT}(M)}\right)^{-1} Z_{TV}(\Gamma, M)\,.$$
- It allows for a path-integral interpretation, which is
$$Z_{TV}(\Gamma, M) = \int DADB\, e^{iS_\lambda(A,B)} W_\Gamma(A_+)\,,$$
where $\lambda = 4\pi^2/k^2$.
- It gives a relation to Turaev's shadow-world invariants [**17**]. Namely, Turaev has defined a state-sum invariant of a spin network Γ embedded in M, which is based on the $Z_{TV}(M)$ sum in the following way. Let M' be a 3-manifold obtained by excising a tubular neighborhood of γ from M, where γ is the graph of Γ. Triangulate M' and construct the corresponding $Z_{TV}(M')$ such that the $6j$ symbols for the tetrahedrons which have a triangle on the triangulation of the boundary surface Σ of M' are excluded. Instead of the boundary tetrahedrons weights, one includes in the state sum the weight associated with a graph $\gamma' \cup \delta$ where γ' is a natural projection of γ onto Σ and δ is the dual one-simplex of the triangulation of Σ induced by the triangulation of M'.

  The graph $\gamma' \cup \delta$ divides the surface Σ into disjoint discs, which can be colored by spins s. One also has the spins associated to the edges of γ', which are the spins of the spin network Γ, and the spins associated to the edges of δ, which are the spins of a coloring J of $\Delta(M')$. There is a natural way to associate the weights to such a coloring of Σ in terms of the quantum $6j$ symbols, see [**17, 7**]. By making the product of the surface weights and summing over the surface spins s one obtains the weight $w_\Sigma(\Gamma, j)$ so that
$$\tilde{Z}_{TV}(\Gamma, M) = \sum_J \prod_{j\in J} \dim_q j \prod_{v\in M'} \{6j\}_v\, w_\Sigma(\Gamma, j)\,.$$

  Turaev has shown that $\tilde{Z}_{TV}(\Gamma, M)$ is related to Z_{WRT} by the same formula as Z_{TV} to Z_{WRT}, which means that $\tilde{Z}_{TV}$ and Z_{TV} coincide. One can also show this directly, by using the Chain-Mail techniques, and one obtains that $Z_{TV} = \tilde{Z}_{TV}$, see [**7**].
- In the case $M = \Sigma \times I$, where $I = [0,1]$, $\tilde{Z}_{TV}$ coincides with the spin foam state-sum invariants constructed in [**10**] for spin networks embedded in Σ. In particular, when $\Sigma = S^2$ then $\tilde{Z}_{TV}(\Gamma, S^2 \times I)$ is proportional to

the quantum group evaluation of the spin network Γ, which means that the evaluation of a spin network can be expressed as a sum of products of $6j$ symbols. This can be used for obtaining the large-spin asymptotics of the quantum group evaluation of a spin network, since the large-spin asymptotics of the quantum $6j$ symbols is known [**11, 19**].

4. Conclusions

The spin foam state sum invariants of 3-manifolds and embedded graphs can be constructed by using the essentially finite category of representations of the quantum group $U_q(\mathbf{g})$ for q a root of unity. In the $SU(2)$ group case these state sums can be identified with the path integrals for the 3d Euclidean GR with a non-zero cosmological constant λ. An interesting question is how to define these state sums when $\lambda = 0$, since then the finite category of the quantum group representations is replaced by the infinite category of the Lie group representations. It turns out that it is possible to define the path integral in this case, by regularizing the group integral (1.4) [**3**]. A finite invariant can be obtained only in special cases, when the second twisted cohomology group vanishes. This is in agreement with the results of Witten [**21**], who showed that the $\lambda = 0$ path integral can be represented as an integral over the space of flat connections of the Reidemeister torsion, and the integral exists if the B field does not have certain zero-modes.

Note that we have discussed the spin-foam and related CS theory invariants for the case of a compact Lie group G. The case when G is non-compact is an outstanding problem, and several strategies have been suggested so far:

- Repeat the steps of section 1 for the category of unitary irreps of G. Because the unitary irreps of non-compact Lie groups are infinite-dimensional, just defining the classical spin network evaluations is a non-trivial task. Then one has to regularize the resulting formal sums by finding a finite category of quantum group representations. However, it is still unclear what is an appropriate category of quantum group irreps, see [**6, 2**].
- Try to regularize the group integral (1.4), by using the approach of [**3**].
- Try to find an analytic continuation of the compact group results, see [**6, 22**].

References

[1] J. Baez, *An introduction to spin foam models of quantum gravity and BF theory*, Lect. Notes Phys. **543** (2000), 25–94.

[2] J. W. Barrett and L. Crane, *A lorentzian signature model for quantum general relativity*, Class. Quant. Grav. **17** (2000), 3101–3118.

[3] J. W. Barrett, I. Naish-Guzman, *The Ponzano-Regge model*, Class. Quant. Grav. **26** (2009), 155014.

[4] J. S. Carter, D. E. Flath, and M. Saito, *The classical and quantum 6j-symbols*, Mathematical Notes 43, Princeton University Press, Princeton, NJ, 1995.

[5] L. Crane and D. N. Yetter, *A categorical construction of 4D topological Quantum Field Theories*, in Quantum Topology, L. H. Kauffman and R. A. Baadhio eds., World Scientific, Singapore (1993), pp. 120–130.

[6] J. Faria Martins, *On the analytic properties of the z-coloured Jones polynomial*, J. Knot Theory Ramifications **14** (2005), 435–466.

[7] J. Faria Martins and A. Miković, *Invariants of spin networks embedded in three-manifolds*, Commun. Math. Phys. **279** (2008), 381–399.

[8] ———, *Spin foam perturbation theory for three-dimensional Quantum Gravity*, Commun. Math. Phys. 288 (2009), 745–772.

[9] A. Miković, *Quantum gravity vacuum and invariants of embedded spin networks*, Class. Quant. Grav. **20** (2003) 3483–3492.

[10] ———, *String theory and quantum spin networks*, preprint arXiv:hep-th/0307141 (2003).

[11] S. Mizoguchi and T. Tada, *Three-dimensional gravity from the Tuarev-Viro invariant*, Phys. Rev. Lett. **68** (1992), 1795–1798.

[12] H. Ooguri, *Topological lattice models in four dimensions*, Mod. Phys. Lett. A7 (1992), 2799–2810.

[13] R. Penrose, *Angular momentum: an approach to combinatorial space-time*, Quantum Theory and Beyond, T. Bastin, ed., Cambridge University Press, Cambridge, 1971.

[14] G. Ponzano and T. Regge, *Semiclassical limit of Racah coefficients*, in Spectroscopic and Group Theoretical Methods in Physics, F. Block, ed., North Holland, Amsterdam, 1968,

[15] N. Reshetikhin and V. G. Turaev, *Invariants of 3-manifolds via link polynomials and quantum groups*, Invent. Math. **103** (1991), 547–597.

[16] J. Roberts, *Skein theory and Turaev-Viro invariants*, Topology **34** (1995), no. 4, 771–787.

[17] V. G. Turaev, *Quantum invariants of knots and 3-manifolds*, de Gruyter Studies in Mathematics, 18. Walter de Gruyter & Co., Berlin, 1994.

[18] V. Turaev and O. Viro, *State sum invariants of 3-manifolds and quantum 6j-symbols*, Topology **31** (1992), 865–902.

[19] Y. Taylor and C. Woodward, *6j symbols for $U_q(sl_2)$ and non-Euclidean tetrahedra*, Selecta Math. **11** (2005), 539–571.

[20] E. Witten, *Quantum field theory and the Jones polynomial*, Commun. Math. Phys. **121** (1989), 351–399.

[21] ———, *Topology-changing amplitudes in (2+1)-dimensional gravity*, Nucl. Phys. B **323** (1989), 113–140.

[22] ———, *Analytic continuation of Chern-Simons theory*, preprint arXiv:1001.2933 (2010).

Departamento de Matemática, Faculdade de Engenharias e Ciências Naturais, Universidade Lusófona, 1749-024 Lisboa, Portugal

E-mail address: amikovic@ulusofona.pt

Departmento de Matemática, Faculdade de Ciências e Tecnologia, Universidade Nova de Lisboa, 2829-516 Caparica, Portugal

E-mail address: jn.martins@fct.unl.pt

AMS/IP Studies in Advanced Mathematics
Volume 50, 2011

Representations of the Ptolemy Groupoid, Johnson Homomorphisms, and Finite Type Invariants

Robert C. Penner

ABSTRACT. This paper revisits our joint work with J. E. Andersen, A. J. Bene, and J.-B. Meilhan that relates 2d geometry with 3d quantum topology emphasizing the 2d, specifically, our representations of the Ptolemy groupoid acting on various quantum invariants. Aspects of the 3d theory are surveyed as well admittedly along a path of least resistance in an attempt to keep this paper reasonably self-contained.

1. Introduction and background

Following our joint work [**1**] with Jørgen Ellegaard Andersen, Alex James Bene, and Jean-Baptiste Meilhan, we shall present new connections between 2d geometry and 3d quantum topology, specifically, between the decorated Teichmüller theory of surfaces with boundary [**25, 27**] and finite type invariants of 3-manifolds or links within them [**24**]. In effect, this paper is an extraction and elaboration of certain aspects of [**1**], where more specifically, we have suppressed technical aspects of the Kontsevich integral and the Le-Murakami-Ohsuki invariant [**18**] in order to give here a presentation more heavily emphasizing the 2d theory and absent the 3d proofs. Nevertheless for completeness, this paper does include the basic definitions and should be reasonably self-contained for the non-expert and 2d or 3d expert alike. At any rate, we are grateful to our colleagues and co-authors just mentioned to indulge us this quick revisitation of our joint work if not the outright pilfering of certain figures and statements of results from [**1**]. Special thanks to J.-B. in particular for his comments on an earlier draft of this paper.

Emblematic of what we have in mind, there is a basic combinatorial *Whitehead move* $G \xrightarrow{W} G'$ on a graph G producing a graph G' embedded in a surface as illustrated in Figure 1, which also portrays the corresponding dual move on the dual polygons, where one diagonal of a quadrilateral is replaced by another; let us think of the dual quadrilateral before and after a Whitehead move as the respective bottom and top faces of a 3d tetrahedron as is also depicted in the figure, or in

2010 *Mathematics Subject Classification.* Primary 57M27, 32G15, 57N05 20F99; Secondary.
Key words and phrases. Finite type invariants, fatgraphs, Ptolemy groupoid, Johnson homomorphisms, mapping class group.
Thanks to the organizers for the opportunity to present this work at the marvelous conference commemorated in this volume.

other words, regard a Whitehead move as a triangulated cobordism of triangulated surfaces. (The orientations and labels in the figure will be useful later.) In fact, we are able to read off 3d quantum invariants one Whitehead move at a time on a triangulated 3-manifold in this sense, but we have achieved this in a manner that is natural or readily tractable only in very special cases including all of the higher Johnson homomorphisms as we shall discuss later

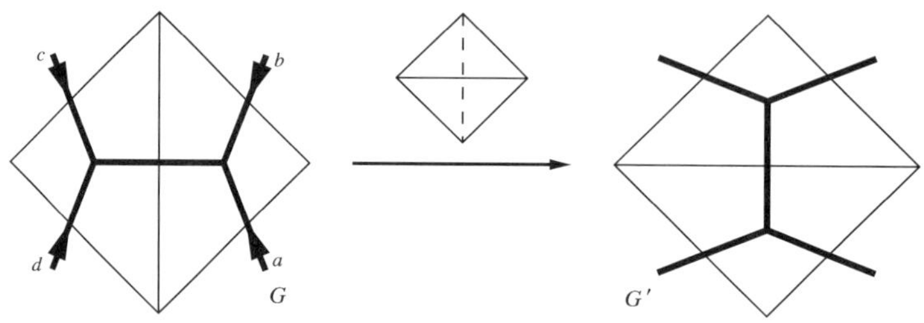

FIGURE 1. Whitehead move $G \xrightarrow{W} G'$ as tetrahedron.

More precisely, a *fatgraph* G is a finite 1d CW complex together with cyclic orderings on the half-edges incident on each vertex, cf. [**26**]; let us adopt the usual terminology for graphs such as vertices, edges, and so on for G. The fatgraph G determines a topological *skinny surface* $F(G)$ with boundary as follows: to each vertex u of G of valence $k \geq 2$ is associated an oriented polygon P_u of $2k$ sides containing in its interior a single vertex of valence k each of whose incident edges is also incident on a univalent vertex contained in alternating sides of P_u, which are identified with the half-edges of G incident on u so that the induced counter-clockwise cyclic ordering on the boundary of P_u agrees with the fattening of G about u; for a vertex u of valence $k = 1$, the corresponding polygon P_u contains u in its boundary. See Figure 2 for examples. The surface $F(G)$ is the quotient of the disjoint union $\sqcup_u P_u$ as u ranges over the vertices of G, where the frontier edges, which are oriented with the polygons on their left, are identified by an orientation-reversing homeomorphism whenever the corresponding half-edges lie in a common edge of G as is also illustrated in the figure. In particular, $F(G)$ contains G as a strong deformation retract, and we call G a *spine* of $F(G)$.

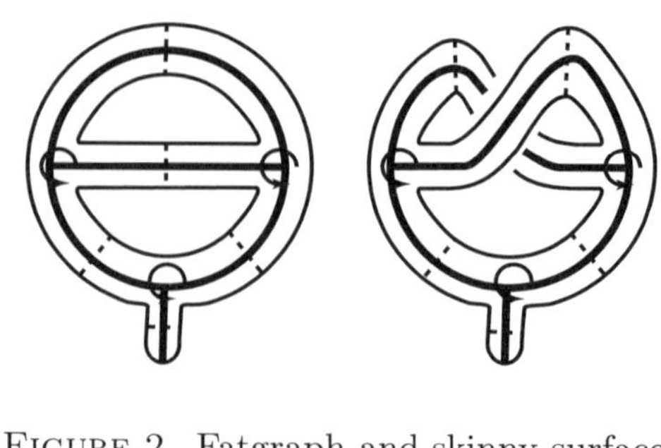

FIGURE 2. Fatgraph and skinny surface.

In several different contexts, appropriate complexes of fatgraphs describe ideal cell decompositions of certain decorated versions of Teichmüller spaces. To be precise in this regard in our current setting, we shall rely upon smooth oriented surfaces F with exactly one boundary component ∂F, $s \geq 0$ distinguished points and of some genus $g \geq 0$ referred to as a *once-bordered* surface. Provided $2g+s > 1$, we may uniformize hyperbolically with a complete finite-area Riemannian metric of constant Gauss curvature minus one so that the boundary is geodesic and the distinguished points are removed and arise as punctures.

The Teichmüller space $T(F)$ as usual is the space of all such metrics on F up to push-forward by diffeomorphisms isotopic to the identity, where the isotopy is required to pointwise fix ∂F. (An easy way to understand $T(F)$ is to take Fenchel-Nielsen coordinates as usual for a pants decomposition of F, where ∂F is among the pants curves; in particular $T(F)$ is homeomorphic to an open ball of real dimension 6g+2s-2.) A *decoration* on F is the further specification of s horocycles in F, one about each of the punctures, as conveniently parametrized by their s tuple of hyperbolic lengths. The *decorated Teichmüller space* is the total space $\tilde{T}(F)$ of the trivial bundle $\tilde{T}(F) = T(F) \times \mathbb{R}^s_{>0} \to T(F)$, where the fiber over a point is identified with the space of all decorations on F, so that $\tilde{T}(F)$ is homeomorphic to a ball of dimension $6g + 3s - 2$.

Just as a point of later notation, we shall sometimes think of this surface F of genus g with s punctures and one boundary component as a surface $\Sigma_{g,r}$ of genus g with $r = s + 1$ ideal boundary components, one of which is distinguished.

The *mapping class group* $MC(F)$ of orientation-preserving homeomorphisms of F which pointwise fix ∂F acts on Teichmüller space by push-forward, and this action extends in the natural way to an action on $\tilde{T}(F)$. The quotient *moduli space* $M(F) = T(F)/MC(F)$ is a manifold since $MC(F)$ is torsion free in this bordered case, which is refreshingly simple compared to the familiar cases of closed or punctured surfaces. The *Torelli group* $I(F) < MC(F)$ is the subgroup acting identically on the integral homology of F.

The fatgraph G corresponding to a generic decorated once-bordered surface has only trivalent vertices except for a single univalent one, and the unique edge of G incident upon it is called the *tail.* Real numbers associated to the non-tail edges of G, which are $6g + 3r - 5$ in number, can be taken as coordinates on $\tilde{T}(F)$, both local and global, cf. [**25, 27**].

It is convenient to make two choices once and for all: choose a basepoint for the fundamental group of F to lie in ∂F and choose a distinct point of ∂F where the univalent end of the tail of a fatgraph spine in F is required to lie.

THEOREM 1.1. [**25, 29**] *There is a $MC(F)$-invariant ideal cell decomposition $\mathcal{G}(F)$ of the decorated Teichmüller space $\tilde{T}(F)$ whose cells are indexed by isotopy classes of fatgraph spines with tail in F, where the face relation is generated by collapse of non-tail edges with distinct endpoints.*

In contrast to regarding a Whitehead move as a triangulated cobordism as in the current work, the more usual interpretation is as an oriented edge of the dual complex $\widehat{\mathcal{G}}(F)$ of $\mathcal{G}(F)$. Namely, define two groupoids:

$$\mathcal{P}t(F) = \textit{Ptolemy groupoid},$$
$$MD(F) = \textit{mapping class groupoid}$$

with objects

$$\mathbf{Ob}(\mathcal{P}t(F)) = \text{isotopy classes of trivalent fatgraph spines with tail in } F,$$
$$\mathbf{Ob}(MD(F)) = MC(F) - \text{orbits of such}$$

and morphisms

$$\mathbf{Mor}(\mathcal{P}t(F)) = \text{ordered pairs of elements of } \mathbf{Ob}(\mathcal{P}t(F)),$$
$$\mathbf{Mor}(MD(F)) = MC(F) - \text{orbits of such.}$$

Insofar as $\mathcal{P}t(F)$ is naturally identified with the one-skeleton of $\widehat{\mathcal{G}}(F)$, the Ptolemy groupoid is a model for the fundamental path groupoid of decorated Teichmüller space, which is simply connected and supports the free mapping class action. The mapping class groupoid is therefore likewise a model for the fundamental path groupoid of the moduli space $M(F) = T(F)/MC(F)$.

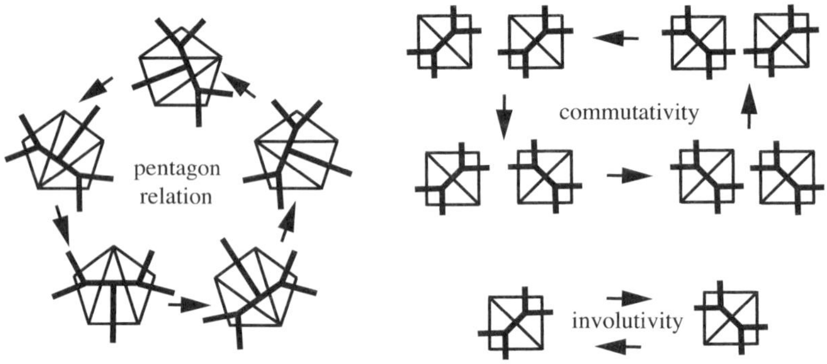

FIGURE 3. Relations in the Ptolemy groupoid.

COROLLARY 1.2. *$\mathcal{P}t(F)$ is generated by Whitehead moves, and a complete set of relations is illustrated in Figure 3. $MD(F)$ is furthemore finitely presented by the $MC(F)$-orbits of such.*

The key utility of this corollary is that the groupoid $MD = MD(F)$ admits this very simple finite presentation, and the vaunted mapping class group $MC = MC(F)$ occurs as a finite-index subgroup(oid) of MD as the self-morphisms of any object; any two of these copies of MC in MD are conjugate since $\tilde{T}(F)$ is connected. For the proof of the corollary, connectivity of $\tilde{T}(F)$, Theorem 1.1 and general position of paths with respect to the codimension one faces of $\mathcal{G}(F)$ immediately prove that Whitehead moves generate $\mathcal{P}t = \mathcal{P}t(F)$ or MD. A similar further analysis of the possible codimension two phenomena in $\mathcal{G}(F)$ likewise provides the relations.

This general point of view has been useful for understanding representations $X \to K$ of subgroups $X < MC$ in some other group K. Upon choosing a basepoint, i.e., an isotopy class of uni-trivalent fatgraph spine with tail for F, there is an induced inclusion $MC < MD$ and hence a commutative diagram

$$\begin{array}{ccccc} K & \leftarrow & - & - & \mathcal{P}t \\ \uparrow & & & & \downarrow \\ X & < & MC & < & MD \end{array}$$

We seek some explicit expression for the top arrow, which can conveniently be described one Whitehead move at a time, in order to better understand or illuminate the representation $X \to K$ or at any rate, explicitly express it purely combinatorially.

For any group K, we define a K-marking on a fatgraph G to be an assignment of $k(e) \in K$ to each oriented edge e of G with reverse $\bar{e}$ so that $k(\bar{e}) = [k(e)]^{-1}$ and so that for any vertex of G with incident oriented edges $e_1, \ldots e_n$, in this linear order compatible with the cyclic order determined by the fattening, we must have the product $k(e_1) \cdots k(e_n) = 1 \in K$. In particular, a K-marking on a fatgraph determines a homomorphism $\pi_1(F(G)) \to K$ in the natural way, and this gives one useful method of describing targets of representations, for example:

THEOREM 1.3. [**22**] *In the notation of Figure 1, where the labels next to the oriented edges are taken as the natural $H = H_1(F;\mathbb{Z})$-marking, the cochain j which assigns to that Whitehead move the quantity*

$$j(W) = a \wedge b \wedge c \in \Lambda^3 H$$

is a $MC(F)$-equivariant cocycle in $Z^1(\hat{\mathcal{G}}(F), \Lambda^3 H)$. Thus, j descends to a cocycle on $\hat{\mathcal{G}}(F)/I(F)$, and the associated homomorphism in the first cohomology group $H^1(\hat{\mathcal{G}}(F)/I(F), \Lambda^3 H) \approx$ Hom$(I(F), \Lambda^3 H)$ is 6 times the first Johnson homomorphism τ_1. More explicitly, given a sequence of Whitehead moves $G \xrightarrow{W_1} G_1 \xrightarrow{W_2} \ldots \xrightarrow{W_k} G_k = \varphi(G)$ representing $\varphi \in I(F)$, we have $\tau_1(\varphi) = \frac{1}{6} \sum_{i=1}^{k} j(W_i)$.

Similarly combinatorial but more intricate expressions have been derived for the higher Johnson homomorphisms [**9**], for the Morita homomorphisms [**20**], for the symplectic representation $MC \to Sp(H)$ and the Nielsen representation $MC \to$ Aut $\pi_1(F)$ [**2**]. This latter extension $\mathcal{P}t \to$ Aut $\pi_1(F)$ depends upon a greedy algorithm for building a maximal tree in a fatgraph G with tail, which we shall recall later, that is the starting point for a series of algorithms. Indeed, further rather intricate algorithms in [**2**] produce a map $\mathcal{P}t \to I(F)$, which extends the identity $I(F) < MC(F) < \mathcal{P}t(F) \to I(F)$ as well as a map $\mathcal{P}t(F) \to MC(\Lambda) = \{\varphi \in MC(F)$ preserving the Lagrangian $\Lambda \subset H\}$ likewise extending the identity which we shall also require for certain applications here. Alex Bene has furthermore defined an explicit complexity and discovered a complexity-reducing algorithm [**8**] that yields factorization of arbitrary mapping classes in terms of Whitehead moves. (Perhaps this extension to the Ptolemy groupoid business has become a bit of an obsession for me, and perhaps I have brought a few friends along as well for better or worse as fellow Ptolemists.)

These considerations furthermore extend to the current paper where we shall answer the following:

Basic Question In what sense does the LMO-invariant lift to the Ptolemy groupoid?

More precisely (and using terminology that will be explained in the next section), the LMO-invariant is a universal rational finite type invariant of integral or rational homology 3-spheres taking values in a suitable algebra of Jacobi diagrams. The Torelli group acts on Heegaard decompositions of integral or rational homology 3-spheres by composing with the gluing maps, which we may further compose with the LMO-invariant and ask: In what sense is there a mapping from the Ptolemy groupoid to the automorphism algebra of Jacobi diagrams that extends the LMO invariant? The answer we shall find in Theorem 6.3 is that there is a natural weakened version of the TQFT formalism for this extension. We shall furthermore discover other natural representations of the Ptolemy groupoid as well in Section 6, which closely follows [**1**].

Of course, these finite type invariants are the coefficients of such quantum invariants as arise in Chern-Simons theory explaining the possible relevance of these considerations to the main topic of this volume.

2. The LMO and AMR invariants

The mechanism for relating fatgraphs to finite type invariants of 3-manifolds involves applying the methods of Le-Murakami-Ohtsuki (LMO) [**18**] to a reduced case of the invariant of Andersen-Mattes-Reshetikhin (AMR) [**3**] of tangles in the product of a surface with an interval. This is in contrast to the usual LMO invariant, which applies the LMO method to the Kontsevich integral [**24**] invariant of tangles in the 3-ball or 3-sphere to produce quantum and finite type invariants of 3-manifolds. We next discuss the spaces which provide the targets for all the invariants, cf. [**5**].

Fix an oriented 1-manifold X as well as a linear ordering on its component circles and segments. We shall consider fatgraphs whose vertices are either uni- or tri-valent, so-called uni-trivalent fatgraphs, whose univalent vertices are disjointly embedded in X. Let $\mathcal{A}(X)$ denote the rational vector space generated by the isotopy classes of all such maps of fatgraphs, or *Jacobi diagrams with core* X, where we furthermore take the quotient modulo the AS, IHX, and STU relations depicted in Figure 4. The *degree* of a Jacobi diagram is half its total number of vertices. Ordered disjoint union of linearly ordered sets induces a tensor product on Jacobi diagrams.

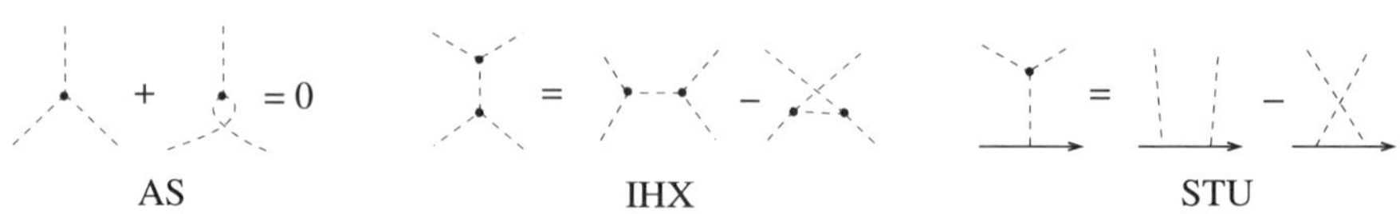

FIGURE 4. AS, IHX,and STY relations on Jacobi diagrams.

More generally for any linearly ordered set S, it is useful to consider the rational vector space $\mathcal{A}(X, S)$ on all *S-labeled Jacobi diagrams with core* X, where each univalent vertex is either embedded in X as before or else labeled by some element of S. In fact, $\mathcal{A}(X)$ and $\mathcal{A}(X, S)$ also denote the respective degree completions, i.e., formal infinite sums of Jacobi diagrams which are finite in each degree.

Consider the 3-manifold $M = \Sigma \times I$, for $\Sigma = \Sigma_{g,r}$, and let X be a 1-manifold as before possibly with boundary. A *tangle with core* X *in* M is a proper embedding of X in M such that all boundary points of X lie on specified segments $I_i \subset \Sigma \times \{i\}$, for $i = 0, 1$. We shall identify such an embedding with the (isotopy class relative to the boundary) of its image. A *framed tangle* is a tangle together with a non-vanishing normal vector field on it. A *q-tangle* is a framed tangle plus a "bracketing", i.e., a consistent collection of parentheses, on each of the sets of boundary points in the segments I_0 and I_1. There is the following diagram of sets and invariants:

$$
\begin{array}{ccccc}
\left\{\begin{array}{c}\text{q-tangles with core } X \\ \text{in } \Sigma_{0,1}\times I\end{array}\right\} & \xrightarrow[\text{integral } Z]{\text{Kontsevich}} & & & \mathcal{A}(X) \\
& & & \nearrow & \downarrow \\
& & \begin{array}{c}\text{reduced AMR} \\ \text{invariant } V\end{array} & & \begin{array}{c}\text{LMO} \\ \text{construction}\end{array} \\
& \nearrow & & & \downarrow \\
\left\{\begin{array}{c}\text{q-tangles with core } X \\ \text{in } \Sigma_{g,r}\times I\end{array}\right\} & & & & \mathcal{A}(X - \left\{\begin{array}{c}\text{component} \\ \text{circles of } X\end{array}\right\})
\end{array}
$$

The AMR invariant [**3**] actually takes values in the set of all homotopy classes of chord diagrams in $\Sigma = \Sigma_{g,r}$ modulo the 4T relations (which is isomorphic as a Poisson ring to the associated graded to the vector space generated by all isotopy classes of links in $\Sigma \times I$ with its standard grading [**4**]) and gives a universal Vassiliev invariant of links in $\Sigma \times I$. Furthermore, appropriate markings on these chord diagrams by elements of suitable Lie groups give a model for the C^∞ functions on their moduli spaces of flat connections on Σ. However, we forget all this beautiful homotopy and other data here and just take values in the quotient algebra $\mathcal{A}(X)$ of Jacobi diagrams.

The LMO construction depends upon a map ι_n which is a key tool for their 3-manifold invariants and for ours as well. It "replaces circles by sums of trees" in the rough sense that a core circle component can be erased by suitably summing over all trees spanning the endpoints of a Jacobi diagram in that component. The more precise description is a necessary evil/beautiful complication of deriving 3-manifold invariants from the link invariants of Kontsevich or AMR, and we next explicate this rather complicated map ι_n for completeness.

Given some linearly ordered set S, for any pair (m, n) of positive integers, any 1-manifold X without circle components and any linearly ordered $S = \{s_1, ..., s_m\}$, first define the auxiliary map

$$j_n : \mathcal{A}(X, S) \to \mathcal{A}(X)$$

$$j_n(D) = \begin{cases} O_n(< D >) & \text{if } D \text{ has exactly } 2n \text{ vertices labeled with each color,} \\ 0 & \text{otherwise,} \end{cases}$$

where $< D >$ is the sum of all possible Jacobi diagrams obtained by pairwise identifying univalent vertices of D having the same color, and O_n serially removes all isolated loops, one at a time and each with a compensatory factor $(-2n)$.

In standard notation, let $\bigcirc^m$ and $\uparrow^m$, respectively, denote m disjoint circles and intervals with orientations and linear ordering on components as before. Given $x \in \mathcal{A}(X \sqcup \bigcirc^m)$, where X has no circle components and those of $\bigcirc^m$ are labeled by $S = \{1, \ldots, m\}$ in the natural way, choose an element $y \in \mathcal{A}(X \sqcup \uparrow^m)$ such that x arises from y by identifying to a distinct point the two endpoints of each core segment, and consider $\chi^{-1}(y) \in \mathcal{A}(X, S)$, where $\chi : \mathcal{A}(\emptyset, S) \to \mathcal{A}(\uparrow^m, \emptyset)$ is the Poincaré–Birkhoff–Witt isomorphism (PBW) [**5**], which maps a diagram to the average of all possible combinatorially distinct ways of attaching its s_i-colored univalent vertices to the ith interval, for $i = 1, \ldots, m$.

The assignment $\iota_n(x) := \big(j_n(\chi^{-1}(y))\big)_{\leq n}$ yields a well-defined map

$$\iota_n : \mathcal{A}(X \sqcup \circledcirc^m) \to \mathcal{A}_{\leq n}(X),$$

where the subscript $\leq n$ denotes truncation to degree at most n. Note that the definition of ι_n given here is small reformulation of Le's simplified version [**15**] of the original in LMO.

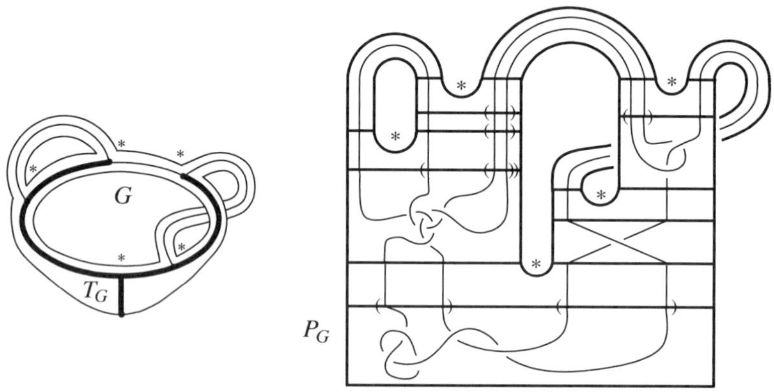

FIGURE 5. A fatgraph G with tail in the twice-punctured torus $\Sigma_{1,2}$ with its maximal tree T_G determined by the greedy algorithm and a knot in admissible position with respect to the polygonal decomposition P_G.

A key point relating the AMR invariant and its reduction V which we study here with 2d geometry is that the definition of V depends upon a polygonal decomposition of the surface into hexagons and quadrilaterals plus further data, and the required structure –and much more–is determined by a uni-trivalent fatgraph spine G with tail in $\Sigma = \Sigma_{g,r}$. As a point of convenient terminology, we shall refer to the product with I of one of these quadrilaterals in Σ as a *box* in $\Sigma \times I$, and to the box containing the tail as the *preferred box.* According to the AMR formalism, each hexagon in the decomposition must come equipped with a "forbidden" boundary edge, to be indicated with an asterisk. Knots or links in $\Sigma \times I$ are put into a suitable admissible position, requiring in particular that the restriction to each box is a q-tangle and that the projection of the restriction of the knot or link to any hexagon consists of arcs connecting distinct boundary edges none of which is parallel to this forbidden edge. The reduced AMR invariant then just performs the usual Kontsevich construction [**24**] (where as a technical point, the associator must be taken to be even) in each box, and the constituent Jacobi diagrams are glued together using the arcs in the hexagons.

An illustrative example is given in Figure 5, indicating first of all that the hexagons arise from the vertices and the quadrilaterals from the edges of the fatgraph G in the natural way, and we shall let P_G denote this polygonal decomposition of Σ determined by G. Consider traversing the boundary component of $F(G)$ containing the tail beginning from the tail with $F(G)$ on the left. Further boundary components, if any, are likewise serially traversed in a well-defined order according to some convention so that the oriented edges arise in a linear ordering starting from the tail one boundary component at a time compatibly with the natural cyclic orderings. The first traversal near a vertex naturally determines the opposite sector between edges of G and hence the required forbidden edges of hexagons required for the AMR invariant as illustrated in the figure.

First occurrence of underlying unoriented edge furthermore induces a linear ordering on the unoriented edges and hence a corresponding greedy algorithm for producing a maximal tree T_G in G: traverse the boundary starting from the tail (oriented so as to point at the rest of the fatgraph) and add traversed edges to a putative maximal tree whenever possible as depicted with bold lines in the figure for example. Moreover, first traversal determines not only a linear ordering on the set of edges of G complementary to the maximal tree but also an orientation on each such edge. In particular, a fatgraph with tail determines a canonical ordered basis X_G for the fundamental group of $F(G)$ manifest as a linearly ordered set of oriented edges of G complementary to the maximal tree in G determined by the greedy algorithm.

The upshot is that a trivalent fatgraph spine G with tail in a bordered surface Σ specifies a polygonal decomposition P_G of Σ into hexagons with forbidden edges and rectangles, thus giving the data required for the definition of an AMR invariant to be denoted $V_G(L)$. The fatgraph with tail furthermore determines an ordered collection X_G of generators of $\pi_1(\Sigma)$. In the spirit of LMO, we shall find in the next section that a suitably normalized version of $\iota_n(V_G(L))$, which depends upon further structure also determined by the fatgraph, produces 3-manifold invariants.

3. The invariant

Fix $\Sigma = \Sigma_{g,r}$, let $h = 2g + r - 1$ denote the rank of the integral first homology group H of Σ, and consider some framed link L in the interior of the cylinder $\Sigma \times I$. Surgery on L produces a 3-manifold $M = (\Sigma \times I)_L$, whose boundary is canonically identified with the boundary of $\partial(\Sigma \times I) = (\partial\Sigma \times I) \cup (\Sigma \times \{0, 1\})$; we may think of $\partial\Sigma \times [0, \frac{1}{2}]$ and $\partial\Sigma \times [\frac{1}{2}, 1]$ as collars of the boundary in the two copies $\Sigma \times \{0\}$ and $\Sigma \times \{1\}$ of Σ. This explains why 3-manifolds of this type are called *cobordisms over* Σ, and the natural monoid structure of gluing them top-to-bottom and then reparametrizing the interval explains why $\Sigma \times I$ itself is denoted 1_Σ.

In addition to the data discussed before, the fatgraph G in Σ furthermore determines a link $K_G \subset 1_\Sigma$ and a tangle $I_G \subset 1_\Sigma$ as follows. A *linking pair* is a 2-component link arising as the core longitude and meridian of an embedded solid torus, and the link K_G is a collection of pairwise disjoint linking pairs. As before, G determines a linearly ordered collection X_G of oriented edges of G complementary to the canonical maximal tree, namely, each such edge determines a unique cycle in G, oriented by first traversal, and we take a linking pair in K_G whose longitude lies in this free homotopy class. The ith longitude is framed along $\Sigma \times \{1\}$ and pushed off in the I direction in 1_Σ to height $1 - i\varepsilon$ for some tiny ε, for $i = 1, \ldots, h$, and we take a tiny 0-framed meridian linking it. The framed link K_G consists of one such linking pair for each of the h many edges of G complementary to the maximal tree, and the components of K_G are oriented and arise in pairs which occur in a specified linear order. The family I_G of arcs is comprised of one arc for each edge in X_G, called a "latch", embedded in the boundary of 1_Σ as illustrated in Figure 6, and again the h many latches come with canonical orientations and framings in a specified linear ordering.

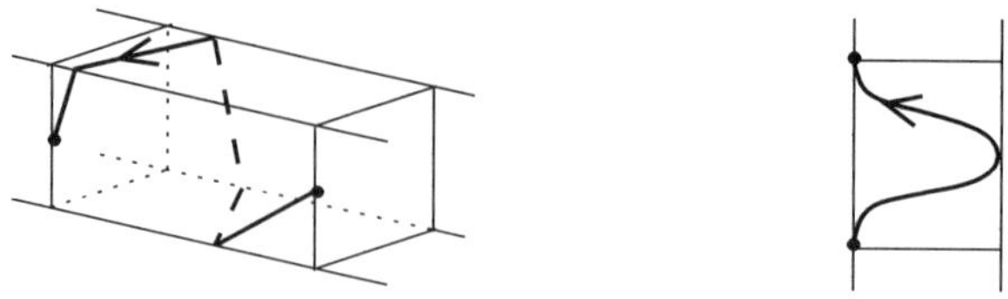

FIGURE 6. Latches in the boxes corresponding to generators.

We may perform handle slides in the Kirby calculus[1] over the framed links in K_G to alter other links $L \subset 1_\Sigma$. Since the longitudes in K_G contain (free homotopy classes of) generators of the fundamental group, we can move L with relative freedom except that it "gets caught" on the meridians, which can however be isotoped into some standard location, for example, into the preferred box. In effect, classical LMO constructions and arguments in $\Sigma_{0,1} \times I$ can then be performed in the preferred box. Moreover, the latches serve as the core of the space $\mathcal{A}(\uparrow^h)$ in which our invariant takes its values.

Given an oriented framed link $L \subset 1_\Sigma$, we may choose a representative L disjoint from $K_G \cup I_G$ and define

$$\hat{\nabla}_n^G(L) := \frac{\iota_n(\check{V}_G(L \cup K_G \cup I_G))}{\iota_n(\check{V}_G(U_+))^{\sigma_+^{L\cup K_G}} \iota_n(\check{V}_G(U_-))^{\sigma_-^{L\cup K_G}}} \in \mathcal{A}_{\leq n}(\uparrow^h), \tag{3.1}$$

where $U_\pm$ denotes the ± 1-framed unknot in 1_Σ, $\check{V}_G(\gamma)$ arises from $V_G(\gamma)$ for any framed tangle γ by taking connected sum with $\nu = Z(\bigcirc)$ on each *closed* component, $\sigma_\pm^{L\cup K_G}$ is the number of $\substack{\text{positive}\\ \text{negative}}$ eigenvalues of the linking matrix[2] of $L \cup K_G$, here using that $\{1, \ldots, h\}$ is in canonical bijection with X_G.

In fact, the need for this kind of elaborate normalization of the basic reduced AMR invariant is an adaptation of a standard part of LMO theory–daunting as it may seem–which is formulated to guarantee that $\hat{\nabla}_n^G(L)$ is independent of the standard Kirby moves. Indeed, the LMO-invariant Z^{LMO} is formulated in precisely this same way replacing V_G by the Kontsevich integral Z; what is novel here beyond using V_G instead of Z is our inclusion of linking pairs and latches.

THEOREM 3.1. *For any uni-trivalent fatgraph G with tail in $\Sigma_{g,r}$, $\hat{\nabla}_n^G(L)$ is invariant under the three Kirby moves on L as well as under the change of orientation on L. Thus, $\hat{\nabla}_n^G(L)$ depends only upon G and the 3-manifold $M = (\Sigma \times I)_L$.*

Just as our ∇_G is a variant of Z^{LMO} with linking pairs and latches, so too the proof is a variant of the standard one after moving L to the preferred box. A further standard aspect in our setting of the LMO yoga is to pass to the infinite series

$$\hat{\nabla}_G(M) = 1 + (\hat{\nabla}_1^G(M))_1 + (\hat{\nabla}_2^G(M))_2 + \cdots \in \mathcal{A}(\uparrow^h),$$

---

[1]In the current context of 3-manifolds with boundary, there is a "third" move of adding/removing a linking pair required for the Kirby calculus, cf. [**28**]

[2]The linking number of two disjoint oriented knots $K_1, K_2 \subset 1_\Sigma$ in generic position is defined as follows: project K_1, K_2 to $\Sigma \approx \Sigma \times \{0\}$ and sum over all crossings of the projections a sign ± 1 associated to each crossing, where the sign is positive if and only if the projections of the tangent vectors to the over- and under-crossing in this order agree with the given orientation on Σ.

where the outer subscript denotes the projection onto a particular degree. In particular for $g = 0, r = 1$, our $\hat{\nabla}_G(M)$ is exactly the LMO invariant of M.

It is convenient to re-index using the PBW map and project out the "struts" of Jacobi diagrams, i.e., the components which are isomorphic to intervals, as indicated in the following diagram

$$\begin{array}{ccc} \left\{ \begin{array}{c}\text{cobordisms}\\ \text{over } \Sigma\end{array} \right\} & \overset{\hat{\nabla}_G}{\rightarrow} & \mathcal{A}(\uparrow^h) \\ \nabla_G \downarrow & & \downarrow PBW \\ \mathcal{A}(\emptyset, \{1, \ldots, h\})/\text{struts} & \overset{\text{projection}}{\leftarrow} & \mathcal{A}(\emptyset, \{1, \ldots, h\}) \end{array},$$

in order to define our basic invariant

$$\nabla_G : \left\{ \begin{array}{c}\text{cobordisms}\\ \text{over } \Sigma\end{array} \right\} \rightarrow \mathcal{A}_h := \mathcal{A}(\emptyset, \{1, \ldots, h\})/\text{struts}.$$

COROLLARY 3.2. *For any uni-trivalent fatgraph spine G of $\Sigma = \Sigma_{g,r}$ with tail, ∇_G is an invariant of cobordisms over Σ taking values in $\mathcal{A}_h$.*

4. Claspers and finite type invariants of homology cylinders

We briefly review the Goussarov-Habiro theory [**14, 11, 12**] of finite type invariants for compact oriented 3-manifolds, which essentially generalizes Ohtsuki's theory [**23**] for integral homology spheres.

A *clasper* C in a 3-manifold M is an embedding in M of the skinny surface of a (possibly disconnected) uni-trivalent fatgraph having a framed copy of S^1 attached to each univalent vertex. The copies of S^1 are called the *leaves* of C, and the trivalent vertices are called the *nodes* of C. We tacitly demand that each connected component of a clasper contains at least one node. The number of connected components of C is denoted $|C|$, and its *degree* is the total number of nodes. A connected clasper of degree 1 is often called a *Y-graph.*

A clasper C of degree k in M determines a framed link $L(C)$ in M, and *surgery along* C means surgery along $L(C)$. To construct $L(C)$ from C, first split edges as illustrated on the left-hand side of Figure 7 to arrange that C becomes a disjoint union of k many Y-graphs. Next in a regular neighborhood, replace each Y-graph by a 6-component framed link as shown on the right-hand side of Figure 7 to finally produce $L(C)$, which has $6k$ components if C has degree k.

FIGURE 7. The edge splitting rule and the surgery link associated to a Y-graph.

A *homology cylinder* over a compact surface Σ is a 3-manifold $M = (1_\Sigma)_C$ that arises from surgery on some clasper C in 1_Σ, cf. [**19**]. Likewise, a *homology handlebody of genus* g is a homology cylinder over $\Sigma_{0,g+1}$, i.e., it arises from clasper surgery in a handlebody $\Sigma_{0,g+1} \times I$ whose boundary has genus g, cf. [**13**].

The set of homology cylinders over Σ up to orientation-preserving diffeomorphism is denoted $\mathcal{HC}(\Sigma)$. There is a natural *stacking product* on $\mathcal{HC}(\Sigma)$ that arises by identifying the top of one homology cylinder with the bottom of another and reparametrizing the interval, i.e., by stacking one clasper on top of another. This induces a monoid structure on $\mathcal{HC}(\Sigma)$ with 1_Σ as unit element.

Let $\mathcal{H}_\Sigma$ be the $\mathbb{Q}$-vector space freely generated by elements of $\mathcal{HC}(\Sigma)$ with its descending *Goussarov-Habiro filtration* given by

$$\mathcal{H}_\Sigma \supset \mathcal{F}_1(\Sigma) \supset \mathcal{F}_2(\Sigma) \supset \ldots, \tag{4.1}$$

where for $k \geq 1$, $\mathcal{F}_k(\Sigma)$ denotes the subspace generated by elements

$$[M; C] := \sum_{C' \subseteq C} (-1)^{|C'|} M_{C'},$$

with $M \in \mathcal{HC}(\Sigma)$, C a degree $\geq k$ clasper in M, and the sum running over all subsets C' of the set of connected components of C.

A *finite type invariant of degree* $\leq k$ is a map $f : \mathcal{HC}(\Sigma) \to V$, where V is a $\mathbb{Q}$-vector space, whose natural extension to $\mathcal{H}_\Sigma$ vanishes on $\mathcal{F}_{k+1}(\Sigma)$. Let

$$\overline{\mathcal{H}}_\Sigma := \text{(degree completion of } \mathcal{H}_\Sigma)/(\cap_k \mathcal{F}_k(\Sigma))$$

denote the space of all formal infinite sums of elements of $\mathcal{H}_\Sigma$, which are finite in each degree as before, modulo the intersection $\cap_k \mathcal{F}_k(\Sigma)$ of all terms in the Goussarov-Habiro filtration, and note that it is a fundamental open question whether this intersection is trivial.

Our main result in [**1**] relating homology cylinders to Jacobi diagrams is:

THEOREM 4.1. *For any uni-trivalent fatgraph spine G of $\Sigma = \Sigma_{g,r}$ with tail, ∇_G induces a filtered isomorphism*

$$\overline{\mathcal{H}}_\Sigma \to \mathcal{A}_h,$$

i.e., ∇_G is a universal finite type invariant for homology cylinders.

5. Stacking, shelling, and pairing

There is the natural "stacking" product of gluing homology cylinders top-to-bottom which induces

$$\cdot : \overline{\mathcal{H}}_{\Sigma_{g,1}} \times \overline{\mathcal{H}}_{\Sigma_{g,1}} \to \overline{\mathcal{H}}_{\Sigma_{g,1}},$$

and likewise the "pairing" of gluing two homology handlebodies boundary-to-boundary to produce a homology ball which induces

$$\cup_\iota : \overline{\mathcal{H}}_{\Sigma_{0,g+1}} \times \overline{\mathcal{H}}_{\Sigma_{0,g+1}} \to \overline{\mathcal{H}}_{\Sigma_{0,1}}$$

and the "shelling" product of gluing a homology cylinder to a homology handlebody bottom-to-boundary to produce another homology handlebody which induces

$$* : \overline{\mathcal{H}}_{\Sigma_{g,1}} \times \overline{\mathcal{H}}_{\Sigma_{0,g+1}} \to \overline{\mathcal{H}}_{\Sigma_{0,g+1}}.$$

These natural operations on homology cylinders and homology handlebodies can be computed purely diagrammatically in terms of a basic contraction operation $\odot$ on Jacobi diagrams, to be discussed presently, and specific diagrammatic constants T_g, S_g, R_g given as q-tangles. Namely, given Jacobi diagrams $D, D' \in \mathcal{A}_g$ and $E, E' \in \mathcal{A}_{2g}$, define operations

$$\begin{array}{lll} <,>: \mathcal{A}_g \times \mathcal{A}_g \to \mathcal{A}(\emptyset), & & < D, D' > := (D \otimes D') \odot T_g, \\ \bullet : \mathcal{A}_{2g} \times \mathcal{A}_{2g} \to \mathcal{A}_{2g}, & \text{by} & E \bullet E' := (E \otimes E') \odot S_g \ , \\ \star : \mathcal{A}_{2g} \times \mathcal{A}_g \to \mathcal{A}_g & & E \star D := (E \otimes D) \odot R_g, \end{array}$$

where the diagrammatic constants are given in Figure 8 in terms of the basic q-tangles $T_{m,n}$, which are also depicted.

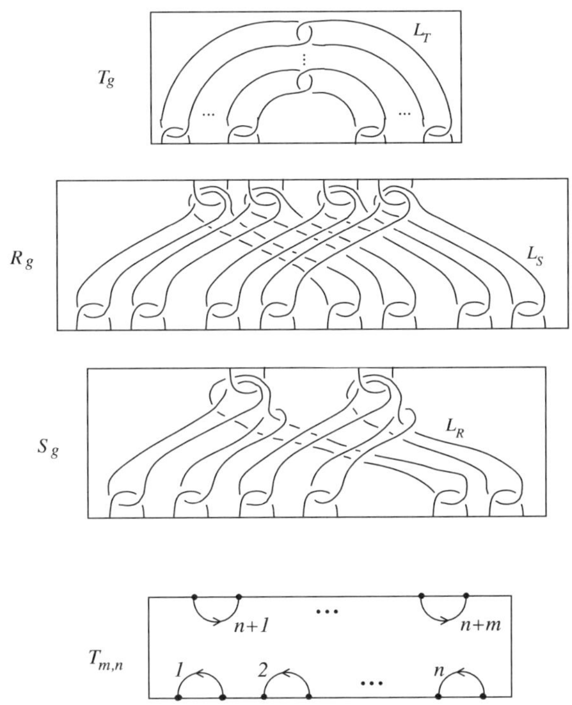

FIGURE 8. The q-tangles $T_g = T_{0,2g} \cup L_T$, $S_g = T_{2g,4g} \cup L_S$ and $R_g = T_{g,3g} \cup L_R$ in terms of the basic q-tangle $T_{m,n}$, where the bracketing on both sets of boundary points in $T_{m,n}$ is of the form $(\bullet\bullet)\Big((\bullet\bullet)((\bullet\bullet)\cdots((\bullet\bullet)(\bullet\bullet))\cdots)\Big)$.

There is a "rigid" version ∇^r_G of our invariant, whose direct definition in the spirit of ∇_G itself is somewhat involved, cf. the original source, so instead, we formulate it here in terms of the standard extension Z^{LMO} of the LMO invariant to q-tangles in homology balls. Namely, suppose that $M = (1_\Sigma)_L$ is a cobordism over Σ for some framed link L in 1_Σ. Consider the cube $C = [0,1]^3$ arising as the preferred box. We may take representatives of the longitudes in K_G that meet this cube in a single unknotted arc and representative meridians in K_G as small linking circles in C. Sliding L along the longitudes in the Kirby calculus, we may produce another link $L_0 \subset C \subset 1_\Sigma$ with $M = (1_\Sigma)_L = (1_\Sigma)_{L_0}$. Our modified invariant can be defined as

$$\hat{\nabla}^r_G(M) := Z^{LMO}(B_M, \gamma_M) \in \mathcal{A}(\uparrow^m),$$

where (B_M, γ_M) denotes the result of surgery on $(C, K_G \cap C)$ along the link L_0 with the same bracketing as for the the tangle $T_{m,n}$ in Figure 5; B_M is indeed a homology ball if M is a homology cylinder, and the corresponding $\nabla^r_G(L) \in \mathcal{A}(\uparrow^h)$ is defined using PBW and projecting out struts as before. In this formulation, ∇^r_G shares many properties with the invariant of so-called Lagrangian cobordisms studied by Cheptea-Habiro-Massuyeau [**10**], which depends upon choices similar to certain of those determined by a fatgraph that we have discussed here.

THEOREM 5.1. *Let H and H' be two genus g homology handlebodies, and let M, M' be two homology cylinders over $\Sigma_{g,1}$. Then*

$$Z^{LMO}(H \cup_\iota H') =< \nabla^r_{\overline{C}_g}(H), \nabla^r_{\overline{C}_g}(H') >, \tag{5.1}$$

$$\nabla^r_{C_g}(M \cdot M') = \nabla^r_{C_g}(M) \bullet \nabla^r_{C_g}(M'), \tag{5.2}$$

$$\nabla^r_{\overline{C}_g}(M * H) = \nabla^r_{C_g}(M) \star \nabla^r_{\overline{C}_g}(H), \tag{5.3}$$

where the "standard" uni-trivalent fatgraph spines with tail $\bar{C}_g$ in $\Sigma_{0,g+1}$ and C_g in $\Sigma_{g,1}$ are illustrated in Figure 9.

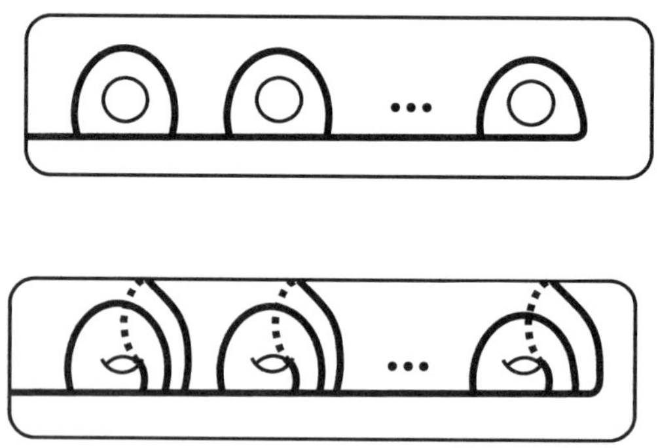

FIGURE 9. Preferred marked fatgraphs with tail $\overline{C}_g \hookrightarrow \Sigma_{0,g+1}$ and $C_g \hookrightarrow \Sigma_{g,1}$.

To complete our discussion of this result, the rest of this section is dedicated to the somewhat technical definition of the binary contraction operation $\odot$ on Jacobi diagrams. For convenience, adopt the standard notation $\mathcal{B}(S) = \mathcal{A}(\emptyset, S)$ for any linearly ordered set S. Let $D \in \mathcal{B}(S)$ and $D' \in \mathcal{B}(S')$ be diagrams, for some finite sets S and S', and let $R \subseteq S \cap S'$. Define the *contraction product* $D \circ_R D' \in \mathcal{B}((S \cup S') - R)$, as follows. If $R = \emptyset$ or if for some $x \in R$ the number of x-colored vertices of D and D' is not the same, then set $D \circ_R D' = 0$, and otherwise, $D \circ_R D'$ is defined to be the sum of all possible ways of pairwise gluing the univalent vertices of D and D' labeled by the same element of R. By linear extension, this defines a contraction map

$$\circ_R : \mathcal{B}(S) \times \mathcal{B}(S') \to \mathcal{B}((S \cup S') - R),$$

which we shall call the *contraction over R*.

Actually, we require two variants of the standard PBW construction as follows: χ_S means apply χ to all segment components of the core and compose with the natural bijection $i \mapsto s_i$ of $\{1, \ldots, k\}$ with $S = \{s_1, \ldots, s_k\}$, and $\chi_{T_{0,n},U}$ means apply χ only to the n segments of the core coming from $T_{0,n}$ and compose as before with the natural bijection.

Let $-_s \in \mathcal{A}(\uparrow, \{s\})$ be the Jacobi diagram consisting of a single strut with one vertex on $\uparrow$ and one vertex colored by s. Set

$$\lambda(s,u,v) := \chi^{-1}_{\{v\}}(\exp(-_s) \cdot \exp(-_u)) \in \mathcal{B}(\{s,u,v\}),$$

where the exponential is with respect to the stacking product of Jacobi diagrams. As explained in [**6**, Proposition 5.4] and [**10**, Remark 4.8], $\lambda(s,u,v)$ can also be defined in terms of the Baker-Campbell-Hausdorff series. If S, U and V respectively denote the sets $\{s_1, ..., s_n\}$, $\{u_1, ..., u_n\}$ and $\{v_1, ..., v_n\}$, then define

$$\Lambda^n(S,U,V) := \sqcup_{i=1}^n \lambda(s_i, u_i, v_i) \in \mathcal{B}(S \cup U \cup V).$$

Consider the q-tangle $\gamma' = T_{m,n} \cup L'$ in the standard cube $C = [0,1]^3$, where L' is some framed link disjoint from $T_{m,n}$. For any element $D \in \mathcal{A}(\uparrow^n)$, define

$$D \overset{k}{\odot} \gamma' = \chi_S^{-1} D \circ_S \frac{\iota_k j_k^V \left(\Lambda^n(S,U,V) \circ_U \chi^{-1}_{T_{0,n},U} \hat{Z}(C,\gamma')\right)}{\iota_k(\check{Z}(U_+))^{\sigma_+(\gamma')} \iota_k(\check{Z}(U_-))^{\sigma_-(\gamma')}} \in \mathcal{A}(\uparrow^m), \tag{5.4}$$

and set

$$D \odot \gamma' := 1 + \sum_{k \geq 1} (D \overset{k}{\odot} \gamma')_k \in \mathcal{A}(\uparrow^m), \tag{5.5}$$

where: $\hat{Z}(C,\gamma')$ is obtained from $\check{Z}(C,\gamma')$ by taking connected sum with ν on each component of $T_{0,n} \subset \gamma'$, and $\check{Z}(C,\gamma)$ is obtained from the Kontsevich integral $Z(C,\gamma)$ by taking connected sum of circle components with ν as before; $\sigma_\pm(\gamma') := \sigma_\pm^{L' \cup T_{0,n}}$ denotes the number of positive and negative eigenvalues of the linking matrix of the tangle $(T_{0,n} \cup L') \subset \gamma'$; and the map $j_k^V : \mathcal{B}(S \cup V) \to \mathcal{B}(S)$ is defined by applying the LMO map j_k to the V-colored vertices leaving the S-colored vertices unchanged.

6. Representations of the Ptolemy groupoid

In this section, we shall extend to the Ptolemy groupoid various actions of the mapping class groups and to this end begin with a discussion of the latter. As a point of terminology, by a "representation" of the Ptolemy groupoid $\mathcal{P}t$ in set X supporting a binary operation, we mean simply a map $\mathbf{Mor}(\mathcal{P}t) \to X$ that preserves compositions.

6.1. Classical group actions. We shall consider Heegaard decompositions of homology 3-spheres $M = H' \cup_\iota H$ into two genus g homology handlebodies H, H', where $\iota : \partial H \to \partial H'$ is an orientation-reversing homeomorphism. A mapping class $f \in MC(\Sigma_{g,r})$ extends naturally to $f \in MC(\Sigma_{g,0})$ by capping off each boundary component with a disk and extending by the identity. We may therefore let f act upon the Heegaard decomposition

$$f \mapsto M_f = H' \cup_{\iota \circ f} H, \tag{6.1}$$

and if f is in the Torelli group, then M_f is again a homology 3-sphere.

In fact, the mapping cylinder of $\varphi \in MC(\Sigma)$, denoted $C(\varphi) = (1_\Sigma, \varphi, Id)$, is a special case of cobordism over Σ, and restricting to $\varphi \in I(\Sigma)$, we obtain a homomorphism of monoids

$$\begin{aligned} I(\Sigma) &\to \mathcal{HC}(\Sigma), \\ \varphi &\mapsto C(\varphi). \end{aligned} \tag{6.2}$$

The action (6.1) is given equivalently by $f \mapsto M_f = H' \cup_\iota (C(f) * H)$ in terms of the operations in the previous section. More generally, (6.2) induces an action of $I(\Sigma)$ on the vector space generated by homology cylinders over Σ by stacking, i.e.,

$$M \mapsto M \cdot C(\varphi)$$

for $M \in \mathcal{HC}(\Sigma)$ and $\varphi \in I(\Sigma)$.

Similarly, there is the adjoint action

$$M \mapsto C(\varphi) \cdot M \cdot C(\varphi^{-1})$$

of $\varphi \in MC(\Sigma)$ on homology cylinders $M = (1_\Sigma)_L$ over Σ where

$$C(\varphi) \cdot (1_\Sigma)_L \cdot C(\varphi^{-1}) = C(\varphi) \cdot C(\varphi^{-1}) \cdot (1_\Sigma)_{\varphi^{-1}(L)} = (1_\Sigma)_{\varphi^{-1}(L)} \in \mathcal{HC}(\Sigma). \tag{6.3}$$

Analogously, there is a shelling action

$$H \mapsto C(\varphi) * H$$

of the Lagrangian-preserving subgroup $MC(\Lambda)$ of the mapping class group on the set of genus g homology handlebodies with Lagrangian Λ, where Λ is given by the homology kernel of the inclusion of its boundary into the handlebody.

The preferred fatgraph spines with tail C_g and $\overline{C}_g$ in Figure 9 induce isomorphisms

$$\nabla_{C_g} : \overline{\mathcal{H}}_{\Sigma_{g,1}} \xrightarrow{\cong} \mathcal{A}_{2g} \quad \text{and} \quad \nabla_{\overline{C}_g} : V(\Sigma_g, \Lambda^{st}) = \overline{\mathcal{H}}_{\Sigma_{0,g+1}} \xrightarrow{\cong} \mathcal{A}_g,$$

where the standard Lagrangian Λ^{st} arises by regarding $\Sigma_{g,1} \approx \Sigma_{0,g+1} \times I$ and identifying the two illustrations in Figure 9. We thus obtain representations

$$\xi \colon MC(\Sigma_{g,1}) \to \mathrm{Aut}(\mathcal{A}_{2g}) \quad \text{and} \quad \zeta : MC(\Lambda^{st}) \to \mathrm{Aut}(\mathcal{A}_g)$$

respectively induced by conjugation and shelling.

6.2. The adjoint representation. Any marked fatgraph spine G with tail in Σ provides a graded isomorphism

$$\nabla_G \colon \overline{\mathcal{H}}_\Sigma \xrightarrow{\cong} \mathcal{A}_h$$

by universality of our basic invariant, and this immediately gives an extension to $\mathcal{P}t(\Sigma)$ of the adjoint action of $MC(\Sigma)$.

THEOREM 6.1. *The map*

$$(G \xrightarrow{W} G') \mapsto \nabla_{G'} \circ \nabla_G^{-1}.$$

defines a representation

$$\hat{\xi} \colon \mathcal{P}t(\Sigma) \to \mathrm{Aut}(\mathcal{A}_h)$$

of the Ptolemy groupoid acting on $\mathcal{A}_h$. *This representation extends the representation* $\xi \colon MC(\Sigma_{g,1}) \to \mathrm{Aut}(\mathcal{A}_{2g})$ *induced by the adjoint action in the sense that for any sequence* $C_g \xrightarrow{W_1} \cdots \xrightarrow{W_k} \varphi(C_g)$ *of Whitehead moves representing* $\varphi \in MC(\Sigma_{g,1})$, *we have the identity* $\hat{\xi}(W_1) \circ \cdots \circ \hat{\xi}(W_k) = \xi(\varphi)$.

PROOF. Since the relations in the Ptolemy groupoid begin and end at one and the same isotopy classes of fatgraph spines, this indeed gives a representation. For any link $L \subset 1_\Sigma$, we have $\nabla_n^G(L) = \nabla_n^{\varphi(G)}(\varphi(L))$ by construction, so that

$$\nabla_{\varphi(G)}((1_\Sigma)_L) = \nabla_G((1_\Sigma)_{\varphi^{-1}(L)}).$$

Thus by (6.3) for any $M \in \overline{\mathcal{H}}_\Sigma$, we have

$$\xi(\varphi)(\nabla_G(M)) = \nabla_G(C(\varphi) \cdot M \cdot C(\varphi^{-1})) = \nabla_{\varphi(G)}(M),$$

and setting $G = C_g$, the result follows. □

There is the following topological interpretation of the automorphism associated to a morphism from G to G' in $\mathcal{P}t(\Sigma)$: pull back an element of $\mathcal{A}_h$ using ∇_G^{-1} to an element of $\overline{\mathcal{H}}_\Sigma$, represented by a formal series of framed links in 1_Σ in admissible position with respect to the polygonal decomposition P_G; evolve G by a sequence of Whitehead moves to a new marked fatgraph G', and isotope the links in L accordingly to put them in admissible position with respect to the new polygonal decomposition $P_{G'}$; finally, evaluate $\nabla_{G'}$ on the resulting series of links to land back in $\mathcal{A}_h$.

6.3. The shelling representation. We may similarly extend the shelling action ζ to the Ptolemy groupoid based on an algorithm in [**2**] that gives a representation

$$\mathcal{P}t(\Sigma_{g,1}) \to MC(\Lambda^{st})$$

of the Ptolemy groupoid extending the identity on $MC(\Lambda^{st})$.

PROPOSITION 6.2. *Let Σ_g be a closed genus g surface. Fix a disc in Σ_g and let $\Sigma_{g,1}$ be its complement. Then we have an explicit algorithmically defined representation*

$$\hat{\zeta} : \mathcal{P}t(\Sigma_{g,1}) \to \mathrm{Aut}(\mathcal{A}_g)$$

which extends the shelling action $\zeta : MC(\Lambda^{st}) \to \mathrm{Aut}(\mathcal{A}_g)$.

Owing to its dependence on the elaborate algorithms in [**2**], this action on $\mathcal{A}_g$ is much more complicated than the action on $\mathcal{A}_h$ described in the previous section.

6.4. Extension of the LMO invariant. We next give a kind of Ptolemy groupoid action on finite type invariants of homology spheres which extends the usual action of the Torelli group via Heegaard decomposition. More precisely, we give a Ptolemy groupoid action on finite type invariants of homology cylinders over Σ which extends the stacking action of $I(\Sigma)$ on $\overline{\mathcal{H}}_\Sigma$, and which in the case of $\Sigma = \Sigma_{g,1}$ induces a map from $\mathcal{P}t(\Sigma_{g,1})$ to $\mathcal{A}(\emptyset)$ extending the analogous map of the Torelli group $I(\Sigma_{g,1})$.

By [**2**], any choice of a fatgraph spine G with tail in $\Sigma = \Sigma_{g,1}$ determines a representation

$$\hat{id}_G \colon \mathcal{P}t(\Sigma_{g,1}) \to I(\Sigma_{g,1})$$

of the Ptolemy groupoid which extends the identity homomorphism of $I(\Sigma_{g,1})$. Let $\hat{id} = \hat{id}_{C_g}$ be the representation provided by the standard fatgraph C_g in $\Sigma_{g,1}$ illustrated in Figure 9. Define a representation

$$\rho : \mathcal{P}t(\Sigma_{g,1}) \to \mathcal{A}_{2g},$$

where the target is imbued with the stacking product $\bullet$, by setting

$$\rho(W) := \nabla^r_{C_g}\left(C(\hat{id}(W))\right).$$

THEOREM 6.3. *The representation $\rho : \mathcal{P}t(\Sigma_{g,1}) \to \mathcal{A}_{2g}$ of the Ptolemy groupoid of $\Sigma_{g,1}$ provides an extension of the LMO invariant of integral homology spheres to the Ptolemy groupoid in the following sense: Let $f \in I(\Sigma_{g,1})$ and let*

$$G \xrightarrow{W_1} G_1 \xrightarrow{W_2} \ldots \xrightarrow{W_k} G_k = f(G)$$

be a sequence of Whitehead moves representing f. Let $M = H \cup_\iota H'$ be a genus g Heegaard splitting of an integral homology sphere M. Then the LMO invariant of the integral homology 3-sphere $M_f = H' \cup_{\iota \circ f} H$ is given by

$$Z^{LMO}(M_f) = \langle v, (\rho(W_1) \bullet \rho(W_2) \bullet \cdots \bullet \rho(W_k)) \star v' \rangle,$$

where $v = \nabla^r_{\overline{C}_g}(H) \in \mathcal{A}_g$ and $v' = \nabla^r_{\overline{C}_g}(H') \in \mathcal{A}_g$.

PROOF. Since $\hat{id}$ extends the identity homomorphism of $I(\Sigma_{g,1})$, we therefore have $\hat{id}(W_1) \circ \hat{id}(W_2) \circ \ldots \circ \hat{id}(W_k) = f$, hence $M_{W_1} \cdot M_{W_2} \cdot \ldots \cdot M_{W_k} = C(f)$, where M_W denotes $C(\hat{id}(W))$. Since $M_f = H \cup_\iota (C(f) * H')$, the formula follows from Theorem 5.1. □

We should emphasize that explicit purely diagrammatic expression for the LMO invariant is thus provided in conjunction with Theorem 5.1. In the special case of a Heegaard decomposition of the 3-sphere, $\nabla^r_{\overline{C}_g}(H_g) = Z^{LMO}(C, T_{g,0})$ by definition, where $T_{g,0}$ is the q-tangle in Figure 8 and $v = v'$. By definition, the Kontsevich integral $Z(T_{g,0}) \in \mathcal{A}(\uparrow^g)$ of this tangle is given by including a $\sqrt{\nu}$ on each copy of $\uparrow$, and it follows that the constant is given by

$$v = \sqcup_{i=1}^g (\chi^{-1}_{\{i\}} \sqrt{\nu}) \in \mathcal{A}_g,$$

with an explicit formula for ν given in [**7**].

6.5. Algebraic representations and the first Johnson homomorphism. As a general point, algebraic operations on the target $\mathcal{A}_h$ of our invariant induce corresponding operations on specializations of our invariant. For example, insofar as elements of $\mathcal{A}(\uparrow^h)$ that are non-zero in degree zero have multiplicative inverses (which can be computed recursively) we may use this multiplicative structure to immediately define another representation

$$\mathcal{P}t(\Sigma_{g,r}) \to \mathcal{A}(\uparrow^h)$$
$$\left(G \xrightarrow{W} G'\right) \mapsto \nabla_{G'}(1_{\Sigma_{g,r}})/\nabla_G(1_{\Sigma_{g,r}}).$$

Indeed, the same comment applies to the additive structure, for example, and either the multiplicative or additive versions would suffice for the computation of the first Johnson homomorphism we shall give here in this low degree.

In fact, to make contact with cocycle j in Theorem 1.3, we shall next generalize our basic invariant by allowing various systems of latches as opposed to the fixed set of latches I_G in the definition of ∇_G. To this end, define a general latch for G to be an embedded interval in the boundary of $\Sigma \times I$ with endpoints lying in $(\partial\Sigma) \times \{\frac{1}{2}\}$ which can be isotoped relative to its boundary so that its projection is in admissible position with respect to the polygonal decomposition P_G. A collection of h disjoint latches in the boundary of 1_Σ whose homotopy class relative to the boundary induces a free basis for $H_1(1_\Sigma, \partial 1_\Sigma; \mathbb{Q})$ is a *general system of latches* for G. In fact, substituting for I_G in (3.1) any general system of latches yields an invariant of cobordisms and indeed of homology cylinders over Σ.

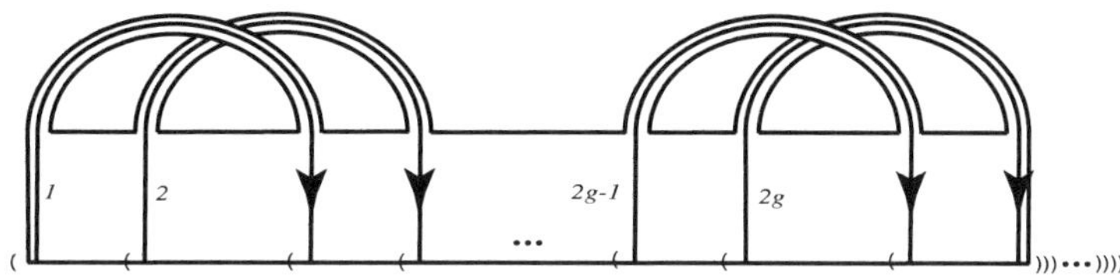

FIGURE 10. The general system I_g of latches.

Restricting for the rest of this section to the case of a surface $\Sigma = \Sigma_{g,1}$, let I_g denote the $2g$-component q-tangle in $\Sigma \times I$ depicted in Figure 10. By isotoping I_g so that it is contained in $(\Sigma \times \{1\}) \cup (\partial\Sigma \times I)$, we may consider I_g as a general system of latches for any choice of fatgraph spine G with tail in Σ. In effect, rather than have the latches intrinsically determined by the fatgraph as before, now the latches depend upon the extrinsic data of how the fatgraph is embedded in the surface. We may isotope so that the endpoints of I_g lie in the boundary of the preferred box.

Let G be a fatgraph with tail in Σ and let L be a framed link in $\Sigma \times I$ which is disjoint from both K_G and I_g, and define

$$\nabla_n^{G,I_g}(L) := \frac{\iota_n(\check{V}_G(L \cup K_G \cup I_g))}{\iota_n(\check{V}_G(U_+))^{\sigma_+^{L\cup K_G}} \, \iota_n(\check{V}_G(U_-))^{\sigma_-^{L\cup K_G}}} \in \mathcal{A}_{\leq n}(\uparrow^{2g}), \tag{6.4}$$

where the notation is as before in the definition of the basic invariant. This is likewise an invariant of the surgered manifold $M = (\Sigma_{g,1} \times I)_L$, and we set

$$\nabla_G^{I_g}(M) := 1 + \left(\nabla_1^{G,I_g}(L)\right)_1 + \ldots + \left(\nabla_n^{G,I_g}(L)\right)_n + \ldots \in \mathcal{A}(\uparrow^{2g})$$

as always in the Tao of LMO and finally define the modified multiplicative representation

$$\mathcal{J} : \mathcal{P}t(\Sigma) \to \mathcal{A}(\uparrow^{2g})$$

$$(G \xrightarrow{W} G') \mapsto \nabla_{G'}^{I_g}(1_\Sigma) / \nabla_G^{I_g}(1_\Sigma)$$

Insofar as $H_1(\Sigma; \mathbb{Z})$ and $H_1(\Sigma, \partial\Sigma; \mathbb{Z})$ are isomorphic via Poincaré duality, we may define a map $\mathfrak{h} : \{1, ..., 2g\} \to H = H_1(\Sigma; \mathbb{Z})$ sending i to the element of H dual to the class of the ith component of I_g in $H_1(\Sigma, \partial\Sigma; \mathbb{Z})$. More explicitly letting $\{A_i, B_i\}_{i=1}^{2g}$ denote the standard symplectic basis of Σ, we have

$$\mathfrak{h}(2k) = A_k, \quad \mathfrak{h}(2k-1) = B_k \quad \text{for } k = 1, ..., g.$$

There are several spaces of Jacobi diagrams relevant to our calculations: First, there is the space $\mathcal{B}(2g) = \mathcal{A}(\emptyset, \{1, \ldots, 2g\})$ considered before, where we shall say that a component diagram is Y-shaped if it has exactly one trivalent vertex. Next, there is the subspace $\mathcal{B}^{\mathsf{Y}}(2g) \subset \mathcal{B}(2g)$ generated by all Jacobi diagrams each of whose components is Y-shaped, which in turn contains the subspace $\mathcal{B}_1^{\mathsf{Y}}(2g)$ generated by connected diagrams. There is the obvious isomorphism $\mathcal{B}_1^{\mathsf{Y}}(2g) \cong \Lambda^3 H_{\mathbb{Q}}$, where $H_{\mathbb{Q}} = H_1(\Sigma; \mathbb{Q})$, defined by sending a Y-shaped diagram colored by i, j, k in this cyclic order from the fattening to $\mathfrak{h}(i) \wedge \mathfrak{h}(j) \wedge \mathfrak{h}(k) \in \Lambda^3 H_{\mathbb{Q}}$.

We project out the target of our representation $\mathcal{J}$ by composing with the series of maps given by

$$(6.5)\qquad Y : \mathcal{A}(\uparrow^{2g}) \to \mathcal{B}(2g) \to \mathcal{B}^{\mathsf{Y}}(2g) \to \mathcal{B}_1^{\mathsf{Y}}(2g) \cong \Lambda^3 H_{\mathbb{Q}},$$

where the first map is the inverse of the PBW isomorphism and the second and third maps are the natural projections. This finally provides our desired representation of the Ptolemy groupoid

$$\mathcal{J}^{\mathsf{Y}} : \mathcal{P}t(\Sigma) \to \Lambda^3 H_{\mathbb{Q}}.$$

THEOREM 6.4. *The representation $\mathcal{J}^{\mathsf{Y}}$ extends the first Johnson homomorphism τ_1 to the Ptolemy groupoid. More precisely, given a sequence $G \xrightarrow{W_1} G_1 \xrightarrow{W_2} \dots \xrightarrow{W_k} G_k = \varphi(G)$ of Whitehead moves representing $\varphi \in I(\Sigma)$, we have $\tau_1(\varphi) = -4\sum_{i=1}^k \mathcal{J}^{\mathsf{Y}}(W_i)$.*

The computation of the invariant $\mathcal{J}^{\mathsf{Y}}$ is considerably simplified by the following observation.

LEMMA 6.5. *For any marked bordered fatgraph G in Σ, we have*

$$Y\left(\nabla_G^{I_g}(1_\Sigma)\right) = Y\left(\check{V}_G(I_g)\right) \in \Lambda^3 H_{\mathbb{Q}},$$

where Y is the sequence of maps in (6.5).

In other words, the Y-shaped part of $\nabla_G^{I_g}(1_\Sigma)$ comes purely from the tangle I_g, and the system of linking pairs K_G can simply be ignored in the computation.

Assuming this lemma from [**1**], we may proceed with the proof of Theorem 6.4 and calculate the representation $\mathcal{J}^{\mathsf{Y}}$ on a Whitehead move W. To this end, for any marked fatgraph G in Σ, we can arrange that the q-tangle I_g is in admissible position and intersects each box except the preferred one in a trivial q-tangle. For each oriented edge of G, we may equip each strand of the trivial q-tangle in the corresponding box with a sign, according to whether its orientation agrees (plus sign) or disagrees (minus sign) with the specified one. For each oriented edge of G, assign an element of H to quadrilateral as follows: label all the strands of I_g using $\mathfrak{h}$ intersecting the box by elements of the symplectic basis $\{A_i, B_i\}_{i=1}^{2g}$ and take the signed sum of these labels in H to produce an H-marking as before.

As illustrated in the upper part of Figure 11, the boxes are labeled $A, B, C \in H$ as just explained and each of the three strands depicted there represents a collection of parallel strands of $I_g \cup K_G$. The bracketing $(C, (B, A))$ in the bottom-left box in each case is imposed by the forbidden edges of the hexagons. After the Whitehead move, we have one of the three situations represented in the lower part of Figure 11 depending on the order of first traversal by the boundary starting from the tail of the sectors near the edge on which the move has been performed. In each case, the bracketing of the three strands in the bottom left box is changed to $((C, B), A)$. Also, in the last two cases, we get an extra cap or cup due to the evolution of the forbidden edges, and these are the only changes; in particular, there are no crossing changes.

Standard (non-trivial) calculations show that the caps and cups cannot contribute in this degree, so the only contribution comes from the change in bracketing, i.e., from the associator. Even associators always begin as 1 plus $\frac{1}{24}$ times the connected Y-shaped diagram plus higher-degree terms, and it follows from Lemma 6.5

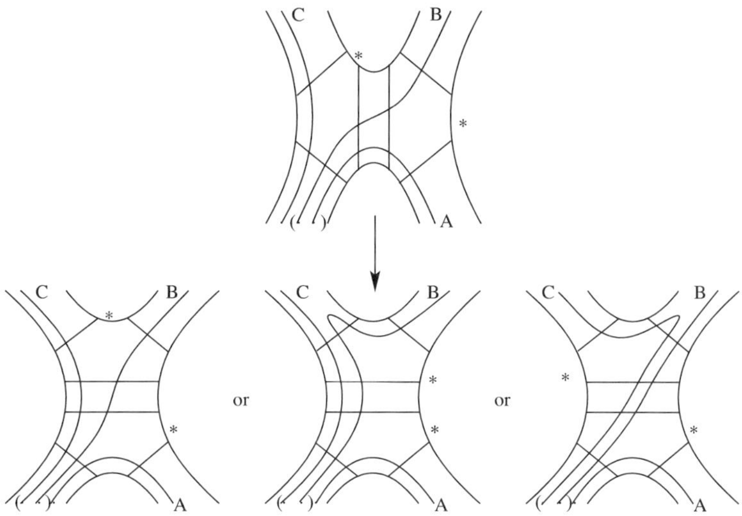

FIGURE 11. The three possible evolutions of the forbidden edges under a Whitehead move.

that

$$\mathcal{J}^{\mathsf{Y}}(W) = \frac{1}{24}\, A \wedge B \wedge C \in \Lambda^3 H_{\mathbb{Q}}.$$

This formula coincides with $-\frac{1}{4}$ times the cocycle j in Theorem 1.3. □

7. Concluding remarks

From the outset of this discussion, we have sought a quantum invariant interpretation of the Whitehead move and have succeeded in this with Theorem 6.3 at the considerable expense of interpreting the Whitehead move as an element of the Torelli group relying on cumbersome algorithms from [**2**]. Several of our other representations of the Ptolemy groupoid do not suffer from such dependence upon those algorithms (other than the greedy algorithm used to define our basic invariant), and we are hopeful that these in particular may be still more explicitly calculable a degree at a time; for example, the next degree term for the multiplicative algebraic representation "should" make contact with the Casson invariant.

We have already in [**1**] discussed various open problems and questions associated with this work, for example, the faithfulness of the induced representations of the mapping class group. To that list, we would just add the following. AMR invariants can likewise be computed based upon the more general, non-generic and not necessarily uni-trivalent fatgraphs that arise for the decorated Teichmüller space of once-bordered surfaces. What is the meaning of these higher codimension cells as finite type? Is there a corresponding continuous mapping from decorated Teichmüller space into the collection of all finite type or quantum invariants interpolating what we do here on just the one-skeleton, and if so, why on earth should it be so?

References

[1] J. E. Andersen, A. J. Bene, J.-B. Meilhan, and R. C. Penner, *Finite type invariants and fatgraphs*, math.GT/0907.2827

[2] J. E. Andersen, A. J. Bene, and R. C. Penner, *Groupoid extensions of mapping class representations for bordered surfaces*, preprint (2007), arXiv:0710.2651, to appear Top. Applns.

[3] J. E. Andersen, J. Mattes, and N. Reshetikhin, *Quantization of the algebra of chord diagrams*, Math. Proc. Camb. Phil. Soc. **124** (1998), 451–467.

[4] ———, *The Poisson structure on the moduli space of flat connections and chord diagrams*, Topology **35** (1996), 1069–1083.

[5] D. Bar-Natan, *On the Vassiliev knot invariants*, Topology **34** (1995), no. 2, 423–472.

[6] D. Bar-Natan, S. Garoufalidis, L. Rozansky, and D.P. Thurston, *The Århus integral of rational homology 3-spheres II. Invariance and universality*, Selecta Math. (N.S.) **8** (2002), no. 3, 341–371.

[7] D. Bar-Natan, T. T. Q. Le, and D. P. Thurston, *Two applications of elementary knot theory to Lie algebras and Vassiliev invariants*, Geom. Topol. **7** (2003), 1–31.

[8] A. J. Bene, *Mapping class factorization via fatgraph Nielsen reduction*, math.GT.0904.4067.

[9] A. J. Bene, N. Kawazumi, and R. C. Penner, *Canonical extensions of the Johnson homomorphisms to the Torelli groupoid*, Advances Math. **221** (2009), 627–659.

[10] D. Cheptea, K. Habiro, and G. Massuyeau, *A functorial LMO invariant for Lagrangian cobordisms*, Geom. Topol. **12**, No. 2 (2008), 1091–1170.

[11] S. Garoufalidis, M. Goussarov, and M. Polyak, *Calculus of clovers and finite type invariants of 3-manifolds*, Geom. Topol. **5** (2001), 75–108.

[12] M. Goussarov, *Finite type invariants and n-equivalence of 3-manifolds*, Compt. Rend. Acad. Sc. Paris, **329** Série I (1999), 517–522.

[13] N. Habegger, *Milnor, Johnson and tree-level perturbative invariants*, preprint (2000).

[14] K. Habiro, *Claspers and finite type invariants of links*, Geom. Topol. **4** (2000), 1–83.

[15] T. T. Q. Le, *On denominators of the Kontsevich integral and the universal perturbative invariant of* 3-*manifolds*, Invent. Math. **135** (1999), 689–722.

[16] ———, *An invariant of integral homology* 3-*spheres which is universal for all finite type invariants*, in "Solitons, geometry an topology: on the crossroad", (V. Buchstaber and S. Novikov, eds.) AMS Translations Series 2 **179** (1997), 75–100.

[17] T. T. Q. Le and J. Murakami, *Parallel version of the universal Vassiliev-Kontsevich invariant*, J. Pure Appl. Algebra **121** (1997), no. 3, 271–291.

[18] T. T. Q. Le, J. Murakami, and T. Ohtsuki, *On a universal perturbative quantum invariant of 3-manifolds*, Topology **37** (1998), 539–574.

[19] J. Levine, *Homology cylinders: an enlargement of the mapping class group*, Alg. Geom. Topol. **1** (2001), 243–270.

[20] G. Massuyeau, private communication.

[21] S. Morita, *Casson's invariant for homology 3-spheres and characteristic classes of surface bundles I*, Topology **28** (1989), 305–323.

[22] S. Morita and R. C. Penner, *Torelli groups, extended Johnson homomorphisms, and new cycles on the moduli space of curves*, Math. Proc. Camb. Philos. Soc. **144**, No. 3 (2008), 651–671.

[23] T. Ohtsuki, *Finite type invariants of integral homolgy spheres*, J. Knot Theory Ram. **5** (1996), 101-115.

[24] ———, *Quantum invariants. A study of knots, 3-manifolds, and their sets*, Series on Knots and Everything, 29, World Scientific, 2002.

[25] R. C. Penner, *The decorated Teichmüller space of punctured surfaces*, Comm. Math. Phys. **113** (1987), 299–339.

[26] ———, *Perturbative series and the moduli space of Riemann surfaces*, J. Diff. Geom. **27** (1988), 35–53.

[27] ———, *Decorated Teichmüller theory of bordered surfaces*, Comm. Anal. Geom. **12** (2004), 793–820.

[28] J. Roberts, *Kirby calculus in manifolds with boundary*, Turkish J. Math. **21** (1997), no. 1, 111–117.

[29] K. Strebel, Quadratic Differentials, *Ergebnisse der Math. und ihrer Grenzgebiete*, Springer-Verlag, Berlin (1984).

Department of Mathematics, Aarhus University, DK 8000 C, Aarhus Denmark
E-mail address: rpenner@imf.au.dk

AMS/IP Studies in Advanced Mathematics
Volume 50, 2011

Yang-Mills in Two Dimensions and Chern-Simons in Three

Ambar N. Sengupta

ABSTRACT. Mathematically rigorous approaches to functional integrals of two dimensional Yang-Mills gauge theory and three-dimensional Chern-Simons theory are described.

1. Introduction

Feynman's path integral method connects time evolution under a quantum Hamiltonian to integration over a space of classical paths with a weighting obtained from the classical theory. The power, mystery and beauty of this method endures in countless instances, and especially so in the context of geometric and topological field theories. The most celebrated expression of a topological quantity in terms of a functional integral is given by Witten's formula [**38**] expressing topological invariants of links in three-dimensional space in terms of functional integrals associated to Chern-Simons field theory. Formulating such integrals and establishing the relationship with topological or geometric quantities rigorously is a challenge for mathematicians. We refer to the books of Mazzucchi [**27**] and Johnson and Lapidus [**20**] for wide-ranging accounts of rigorous methods for studying Feynman path integrals. In this article we shall focus on two field theories, Yang-Mills in two dimensions and Chern-Simons in three dimensions, and review progress in rigorously formulating functional integrals for these theories.

2. Functional integrals: Real and imagined

The typical functional integral arising in quantum field theory has the form

$$Z^{-1}\int_{\mathcal{A}} f(A)e^{\beta S(A)}\,DA$$

where $S(\cdot)$ is an action functional, β a physical constant (real or complex), f is some function of the field A of interest, DA signifies 'Lebesgue integration' on an infinite-dimensional space $\mathcal{A}$ of field configurations, and Z a 'normalizing constant'. The reason for putting quotes on 'Lebesgue integration' and 'normalizing constant' is that these objects are more imagined than mathematically realizable. In this

2010 *Mathematics Subject Classification.* Primary 81T13, 14E20; Secondary 46E25, 20C20.
Key words and phrases. Yang-Mills, Chern-Simons, Functional Integrals, White Noise Analysis.
Research supported in part by NSF Grant DMS-0601141.

section we will examine the trouble with these notions and how the trouble can be fixed in the special case of Gaussian measures.

2.1. Infinite product Lebesgue measure. 'Lebesgue measure' in infinite dimensions is a natural, yet seriously troublesome notion. To be specific, suppose we wish to work with a measure l_∞ on the countably infinite product space $\mathbb{R}^{\mathbb{N}}$, where $\mathbb{N} = \{1, 2, ..\}$, which assigns measure $\prod_{j=1}^{\infty}(b_j - a_j)$ to the infinite box $\prod_{j\in\mathbb{N}}[a_j, b_j]$, where $a_j < b_j$ are real numbers for each $j \in \{1, 2, ...\}$. Then the measure of the box $[0, a]^{\mathbb{N}}$ is 0 if $a \in [0, 1)$, and ∞ if $a > 1$. It would be very difficult to do any analysis with this kind of background measure. If we were to focus simply on $[0, 1]^{\mathbb{N}}$ the situation would not be as bad, and indeed the product measure on $[0, 1]^{\mathbb{N}}$ is a useful probability measure. Nonetheless, experience suggests that a measure formally given by a density such as $e^{\beta S(A)}$ times a background measure is not even absolutely continuous with respect to the background measure, when both are rigorously meaningful. All of this is simply a reminder that while working formally with an infinite dimensional Lebesgue measure can be very useful, it is a challenge to make the formal work mathematically rigorous.

2.2. Gaussian measure in infinite dimensions. The first truly useful, and rigorously meaningful, measure in ifninite dimensions is Gaussian measure. Non-linear quantum field theories generally involve more difficult functional integrals, some of which have been put on rigorous mathematical foundations, as explained in Jonathan Weitsman's lecture. Fortunately, in two interesting field theory models, two dimensional Yang-Mills and three-dimensional Chern-Simons, functional integrals can be worked out meaningfully by Gaussian measure techniques. In the case of Chern-Simons, this is still a developing story, in that the conjectured relationships between the mathematically defined functional integrals and topological invariants still remain to be established completely.

Let's now take a look at Gaussian measure in infinite dimensions. To start with, in a formal sense, this is a measure on an infinite dimensional, separable, real Hilbert space $\mathbb{H}$ given formally by the expression

$$d\mu(x) = \frac{1}{Z}e^{-\|x\|^2/2}Dx,$$

where Dx is 'Lebesgue measure' on $\mathbb{H}$ and Z a 'normalizing constant.'

2.3. A fast construction of Gaussian measure. Choosing an orthonormal basis $u_1, u_2, ...$ in $\mathbb{H}$, and writing x_j for the j-th coordinate $\langle x, u_j\rangle$ of x, the measure μ, viewed on coordinate space $\mathbb{R}^{\mathbb{N}}$, is the product measure

$$\prod_{j\in\mathbb{N}}(2\pi)^{-1/2}e^{-x_j^2/2}dx_j.$$

This is a meaningful measure; for entertainment, let's take a run through one construction. A sequence $(s_n) \in \{0, 1\}^{\mathbb{N}}$ of $0 - 1$ coin tosses encodes a real number $\sum_{n=1}^{\infty} s_n 2^{-n} \in [0, 1]$, and the mapping $\{0, 1\}^{\mathbb{N}} \to [0, 1]$, handled with some care, transfers Lebesgue measure on the unit interval $[0, 1]$ to the coin-tossing product measure $\{0, 1\}^{\mathbb{N}}$ which associates weight 2^{-k} to any subset obtained by specifying the results of k particular tosses, i.e., a subset of the form $\{s \in \{0, 1\}^{\mathbb{N}} : s_{i_1} = j_1, ..., s_{i_k} = j_{i_k}\}$ for any $j_1, ..., j_k \in \{0, 1\}$ and distinct $i_1, ..., i_k \in \mathbb{N}$. In this way, the measure space $\{0, 1\}^{\mathbb{N}}$ is essentially a copy of $[0, 1]$ with Lebesgue measure. The infinite-product $[0, 1]^{\mathbb{N}}$ is viewable as a copy of $\{0, 1\}^{\mathbb{N}\times\mathbb{N}} \simeq \{0, 1\}^{\mathbb{N}}$, by identifying

$\mathbb{N} \times \mathbb{N}$ with $\mathbb{N}$. Thus, $[0,1]^{\mathbb{N}} \simeq \{0,1\}^{\mathbb{N}}$, which, again, is viewable as a copy of $[0,1]$ with Lebesgue measure. The measure on $[0,1]^{\mathbb{N}}$, obtained in this way from $\{0,1\}^{\mathbb{N}\times\mathbb{N}}$, is product measure giving weight $\prod_{j\in\mathbb{N}}(b_j - a_j)$ to the box $\prod_{j\in\mathbb{N}}[a_j, b_j]$ if $0 \leq a_j \leq b_j \leq 1$ for all $j \in \mathbb{N}$. Next, we transfer this product measure on $[0,1]^{\mathbb{N}}$ to $\mathbb{R}^{\mathbb{N}}$, using the Gaussian distribution function $\Phi : \mathbb{R} \to (0,1) : x \mapsto \int_{-\infty}^{x}(2\pi)^{-1/2}e^{-t^2/2}\,dt$ on each coordinate, to obtain the product Gaussian measure μ' on $\mathbb{R}^{\mathbb{N}}$. Each coordinate projection $X_j : \mathbb{R}^{\mathbb{N}} \to \mathbb{R} : (x_n)_{n\in\mathbb{N}} \mapsto x_j$ becomes a Gaussian random variable:

$$\mu'[(x_n) : x_j \in [a,b]] = (2\pi)^{-1/2}\int_a^b e^{-t^2/2}\,dt,$$

with mean 0 and variance 1, i.e., $\int X_j\,d\mu'$ is 0 and $\int X_j^2\,d\mu'$ is 1. Finally, we transfer the measure to the Hilbert space $\mathbb{H}$, using the map

$$\mathbb{H} \to \mathbb{R}^{\mathbb{N}} : x \mapsto (\langle x, u_j\rangle)_{j\in\mathbb{N}}$$

We face a problem here: the μ' measure of the image of $\mathbb{H}$ is 0. One has to resort to some subterfuge at this point: introduce some sequence of positive reals $c_n \in (0,\infty)$ with $\sum_n c_n^2 < \infty$. Then $\int \sum_n c_n^2 X_n^2\,d\mu' = \sum_n c_n^2 < \infty$, and so it is sensible to restrict μ' to the subset of $\mathbb{R}^{\mathbb{N}}$ on which $\sum_n c_n^2 X_n^2$ is finite. To this end, consider the new Hilbert space $\mathbb{H}_1$ obtained by completing $\mathbb{H}$ with respect to the norm arising from the inner product which keeps the vectors u_n orthogonal to each other but with new lengths $\|u_n\|_1 = c_n$. An element of $\mathbb{H}_1$ is of the form $\sum_{n=1}^{\infty} x_n u_n$ with $\sum_{n=1}^{\infty} c_n^2 x_n^2 < \infty$. So, in the end, we do have Gaussian measure on an infinite dimensional Hilbert space, except it isn't the original one $\mathbb{H}$ but a slightly larger one $\mathbb{H}_1 \supset \mathbb{H}$.

2.4. Frameworks for measures in infinite dimensions. We have gone through a quick and dirty construction of Gaussian measure in infinite dimensions. There are two very convenient settings for doing analysis in infinite dimensions with a Gaussian background measure: (i) the Abstract Wiener Space formalism of Gross [**14**], and (ii) the setting of nuclear spaces. The construction of non-Gaussian measures in the setting of nuclear spaces is heavily dependent on a result of Minlos which guarantees the existence of a probability measure on such spaces with specified Fourier transform.

3. The Yang-Mills functional integral for two dimensions

The functional integral for quantum Yang-Mills on the plane $\mathbb{R}^2$ is realized mathematically by means of Gaussian measure. Consider gauge theory over $\mathbb{R}^2$ with gauge group a compact matrix group $G \subset U(N)$, whose Lie algebra $L(G)$ is equipped with the Ad-invariant inner product specified by $\langle a, b\rangle = -\mathrm{Tr}(ab)$. Formally, the functional measure is given by

$$\frac{1}{Z}e^{-\frac{1}{2g^2}\|F^A\|_{L^2}^2}DA$$

where F^A is the curvature of a generic connection $A = A_x dx + A_y dy$, the integration element DA is a formal Lebesgue measure on the infinite dimensional space $\mathcal{A}$ of connections, Z is a formal normalizing constant, and g a physical constant. (See section 5 for a compendium of relevant geometric notions.) More precisely, this measure should be considered on the space of connections modulo gauge transformations. For convenience, it is best to work with $\mathcal{G}_o$, the infinite dimensional group of all gauge transformations which are the identity over the fixed basepoint $o = (0,0) \in \mathbb{R}^2$, and with the quotient $\mathcal{A}/\mathcal{G}_o$. For notational convenience, we take the gauge group G to be a compact matrix group.

For the base manifold $\mathbb{R}^2$, the quotient $\mathcal{A}/\mathcal{G}_0$ can be identified with the infinite dimensional *linear space* $\mathcal{A}_0$ of all connections $A = A_x dx + A_y dy$ for which $A_y = 0$, A_x is 0 along the x-axis, and A vanishes at o; the point is that every connection is gauge transformable to such a special connection in $\mathcal{A}_0$. Arguing formally, the 'Lebesgue measure' DA goes over to the formal Lebesgue measure DA on $\mathcal{A}_0$. The best feature of the setting $\mathcal{A}_0$ is that on $\mathcal{A}_0$ the curvature F^A depends linearly on A:

$$F^A = dA + A \wedge A = -\underbrace{\partial_y A_x}_{f^A} \, dx \wedge dy.$$

The function A_x can be recovered from f^A here by integration, and so f^A itself can be used as a coordinate on $\mathcal{A}_0$. Consequently, the functional measure takes the form

$$\frac{1}{Z} e^{-\frac{1}{2g^2}\|f^A\|_{L^2}} \, df^A.$$

This is, formally, a Gaussian measure, plain and simple. Thus, the functional measure for Yang-Mills on the plane is rigorously meaningful as Gaussian measure for the Hilbert space of $L(G)$-valued L^2-functions f on $\mathbb{R}^2$. From the probabilistic point of view, this is a white-noise with values in the Lie algebra $L(G)$. Planar Yang-Mills theory was developed rigorously in [**15**] and by Driver [**10**], to which we refer for much more as well as bibliography on this topic.

Next, still staying in two dimensions, it is possible to construct the functional measure for the Yang-Mills field over compact surfaces. One approach [**29, 30, 31**] is to take the Gaussian measure described above and *condition* it to satisfy necessary topological constraints. Fine [**11, 12**] developed the Yang-Mills functional integral for compact surfaces. Lévy [**24, 25**] developed the full probabilistic theory for the measure. Once the measure is constructed, the Yang-Mills action itself appears somehow far removed, and one may well ask what precisely is the relationship between the measure and the Yang-Mills action; this is answered by Lévy and Norris [**26**] who show that the large deviation principle for the Yang-Mills measure for compact surfaces is the Yang-Mills action. This parallels a more classical result for Brownian motion, where the large deviation principle is the action/energy of the path. For more insights into the two dimensional Yang-Mills measure we refer to Witten [**36, 37**] and Singer [**35**].

In yet another direction, it is fruitful to inquire into the large-N limit of planar $U(N)$ Yang-Mills functional integrals, specifically Wilson loop expectations. The remarkable fact is that large-N limits, with suitable scaling of the coupling constant g^2, do exist and correspond to a theory which can be described in terms of free probabilty theory. For more on the underlying ideas see Singer [**35**] and the papers [**32, 34**] for more bibliography on this topic.

Let us note very briefly that quantum Yang-Mills theory in three dimensions remains an outstanding challenge. The works by Karabali, Nair et al. [**21, 22**] and Rajeev [**28**] provide ideas for establishing some important properties, such as the existence of a mass gap, in this theory.

4. The Chern-Simons functional integral

We shall work first with Chern-Simons theory over $\mathbb{R}^3$, with gauge group a compact matrix group G whose Lie algebra is denoted $L(G)$. The formal Chern-Simons functional integral has the form

$$\frac{1}{Z}\int_{\mathcal{A}} f(A)e^{i\mathrm{CS}(A)}\,DA$$

where f is a function of interest on the linear space $\mathcal{A}$ of all $L(G)$-valued 1-forms A on $\mathbb{R}^3$, and $\mathrm{CS}(\cdot)$ is the Chern-Simons action given by

$$\mathrm{CS}(A) = \frac{\kappa}{4\pi}\int_{\mathbb{R}^3} \mathrm{Tr}(A\wedge dA + \frac{2}{3}A\wedge A\wedge A), \tag{4.1}$$

involving a parameter κ. Working on the flat space $\mathbb{R}^3$ allows us to ignore topological issues at this point and focus on the analytic problem of constructing the integral; indeed, the factor of $\kappa/(4\pi)$ could be dropped for this purpose, but we will hold on to it. We will describe a non-perturbative method to the functional integral (perturbative methods, starting with Axelrod and Singer's work [**5, 6**], have been enormously successful).

The trick used for two dimensional Yang-Mills is also effective for the Chern-Simons action: we choose a gauge in which one component of $A = a_0dx_1 + a_1dx_1 + a_2dx_2$ vanishes, say $a_2 = 0$. This makes the triple wedge term $A\wedge A\wedge A$ disappear, and we end up with a quadratic expression

$$\mathrm{CS}(A) = \frac{\kappa}{4\pi}\int_{\mathbb{R}^3} \mathrm{Tr}(A\wedge dA) \qquad \text{for } A = a_0dx_0 + a_1dx_1. \tag{4.2}$$

Then the functional integral has the form

$$\frac{1}{Z}\int_{\mathcal{A}_0} \phi(A)e^{i\frac{\kappa}{4\pi}\int_{\mathbb{R}^3}\mathrm{Tr}(A\wedge dA)}\,DA \tag{4.3}$$

where $\mathcal{A}_0$ consists of all A for which $a_2 = 0$. (Note that the functions ϕ on $\mathcal{A}$ of ultimate interest are gauge invariant.) As in the two dimensional case, the integration element remains DA after gauge-fixing. Having a quadratic in the exponent in the integrand (4.3) makes us happy, but the factor i throws a spanner in the works: we no longer have a measure here (not even a complex measure).

All is not lost, however. The map

$$\phi \mapsto \langle\phi\rangle_{\mathrm{CS}} = \frac{1}{Z}\int_{\mathcal{A}_0} \phi(A)e^{i\frac{\kappa}{4\pi}\int_{\mathbb{R}^3}\mathrm{Tr}(A\wedge dA)}\,DA, \tag{4.4}$$

whatever it is rigorously, would be a linear functional on a space of functions ϕ over $\mathcal{A}_o$. As it turns out, it is possible to give rigorous meaning to this functional as a *distribution* on a space of test functions ϕ in infinite dimensions.

Now for $A = a_0dx_0 + a_1dx_1 \in \mathcal{A}_0$, decaying fast enough at infinity, we have, on integrating by parts,

$$\mathrm{CS}(a_0dx_0 + a_1dx_1) = -\frac{\kappa}{2\pi}\int_{\mathbb{R}^3} \mathrm{Tr}(a_0f_1)\,dx_0dx_1dx_2 \tag{4.5}$$

where

$$f_1 = \partial_2 a_1. \tag{4.6}$$

So now the original functional integral is reformulated as an integral of the form

$$\langle\phi\rangle_{CS} = \frac{1}{Z}\int e^{i\frac{\kappa}{2\pi}\langle a_0,f_1\rangle}\phi(a_0,f_1)\,Da_0Df_1 \tag{4.7}$$

where

$$\langle a,f\rangle = -\int_{\mathbb{R}^3}\mathrm{Tr}(af)\,d\mathrm{vol}.$$

and Z always denotes the relevant formal normalizing constant.

Taking ϕ to be of the special form

$$\phi_j(a_0,f_1) = e^{ia_0(j_0)+if_1(j_1)} \tag{4.8}$$

where j_0 and j_1 are, say, rapidly decreasing $L(G)$-valued smooth functions on $\mathbb{R}^3$, we find, from a formal calculation,

$$\langle\phi_j\rangle_{CS} = e^{-i\frac{2\pi}{\kappa}\langle j_0,j_1\rangle}. \tag{4.9}$$

This finally brings us to a point where a rigorous framework can be installed.

Let

$$\mathcal{E} = \{(a,f) : \text{all smooth rapidly decreasing } a,f:\mathbb{R}^3\to L(G)\} \tag{4.10}$$

with the Schwartz topology and let $\mathcal{E}'$ be the topological dual.

For $z\in\mathcal{E}_{\mathbb{C}}$ (the complexification of $\mathcal{E}$), let

$$(\cdot,z):\mathcal{E}'\to\mathbb{C}:\xi\mapsto\xi(z) \tag{4.11}$$

where we have taken $\xi(z)$ to mean $\xi(x)+i\xi(y)$ if $z=x+iy$ with $x,y\in\mathcal{E}$.

Consider

$$\mathcal{P} = \text{linear span of the functions } e^{(\cdot,z)} \text{ as } z \text{ runs over } \mathcal{E}_{\mathbb{C}}, \tag{4.12}$$

which is an algebra under pointwise operations,

If $\psi\in\mathcal{P}$ is of the form $\psi_z=e^{(\cdot,z)}$ where $z\in\mathcal{E}_{\mathbb{C}}$ is (z_0,z_1), we set

$$\langle e^{(\cdot,z)}\rangle_{CS} = e^{i\frac{2\pi}{\kappa}(z_0,z_1)_0} \tag{4.13}$$

where

$$(z_0,z_1)_0 = \int_{\mathbb{R}^3}\mathrm{Tr}(z_0z_1)\,d\mathrm{vol} \tag{4.14}$$

Then by linearity, and independence of exponentials, $\langle\psi\rangle_{CS}$ is meaningful for all $\psi\in\mathcal{P}$.

So now we have a well-defined linear functional

$$\mathcal{P}\to\mathbb{C}:\psi\mapsto\langle\psi\rangle_{CS} \tag{4.15}$$

It would be a very minimal theory if all we had was the value of $\langle\psi\rangle_{CS}$ for $\psi\in\mathcal{P}$. A continuity property of $\langle\cdot\rangle_{CS}$ needs to be established to extend its scope to a larger class of functions of interest. And to have any notion of continuity we need a topology on a suitable space of functions on the space $\mathcal{E}'$.

Topologies on spaces of 'test functions' on the infinite dimensional space $\mathcal{E}'$ are used in white noise analysis [**23**]. We state here a rough and rapid summary. Let $\mathcal{E}_0$ be the real Hilbert space of $L(G)\oplus L(G)$-valued square-integrable functions on $\mathbb{R}^3$. Sobolev norms are obtained using a suitable differential operator T for which T^{-1} is a Hilbert-Schmidt operator on $\mathcal{E}_0$. Let $\mathcal{E}_p$ be the range $\mathrm{Im}(T^{-p})$, and $\langle u,v\rangle_p=\langle T^pu,T^pv\rangle$. Identifying $\mathcal{E}_0$ with its dual $\mathcal{E}_0^*$, there is a chain of inclusions:

$$\mathcal{E}=\cap_p\mathcal{E}_p\subset\cdots\subset\mathcal{E}_2\subset\mathcal{E}_1\subset\mathcal{E}_0\simeq\mathcal{E}_0^*\subset\mathcal{E}_{-1}\subset\mathcal{E}_{-2}\cdots\subset\mathcal{E}^*\stackrel{\text{def}}{=}\cup_p\mathcal{E}_p \tag{4.16}$$

where $\mathcal{E}_{-p} = \mathcal{E}_p^*$, for $p \in \{0, 1, 2, ...\}$. The operator T is chosen so that the Schwartz space $\mathcal{E}$ is the intersection of the completions $\mathcal{E}_p$. White noise analysis provides inner-products $\langle\langle\cdot,\cdot\rangle\rangle_p$ on $\mathcal{P}$ obtained from the Schwartz/Sobolev inner products $\langle\cdot,\cdot\rangle_p$ on $\mathcal{E}$. These inner-products are obtained quite naturally through identification of $L^2(\mathcal{E}', \mu)$, where μ is standard Gaussian measure for $\mathcal{E}_0$ realized on $\mathcal{E}'$, with the symmetric Fock space over $\mathcal{E}_0$. Leaving details aside, the inner-product $\langle\langle\cdot,\cdot\rangle\rangle_p$ on $\mathcal{P}$ works out to

$$\langle\langle e^{(\cdot,z_1)}, e^{(\cdot,z_2)}\rangle\rangle_p = e^{\langle z_1,z_2\rangle_p + \frac{1}{2}(z_1,z_1)_0 + \frac{1}{2}(\overline{z}_2,\overline{z}_2)_0} \tag{4.17}$$

where $(\cdot,\cdot)_0$ is the complex bilinear extension of the L^2-inner-product on $\mathcal{E} = \mathcal{S}(\mathbb{R}^3, L(G)) \oplus \mathcal{S}(\mathbb{R}^3, L(G))$. The completion $[\mathcal{E}]_p$ of $\mathcal{P}$ under the norm $\|\cdot\|_p$ is a space of functions on $\mathcal{E}'$, and the intersection $[\mathcal{E}] = \cap_{p\geq 0}[\mathcal{E}]_p$ provides a good test function space over $\mathcal{E}'$. There is then also the dual space $[\mathcal{E}]^*$ of all continuous linear functionals on $[\mathcal{E}]$; these are *distributions* over the infinite dimensional space $\mathcal{E}'$.

In [**3**] it is shown that there is a continuous linear functional

$$[\mathcal{E}] \to \mathbb{C} : \psi \mapsto \langle\psi\rangle_{\mathrm{CS}} \tag{4.18}$$

which agrees with $\langle\psi\rangle_{CS}$, as defined in (4.13), for $\psi \in \mathcal{P}$. Thus, (4.18) provides a rigorous realization for the Chern-Simons functional integral with axial gauge fixing.

In the case of abelian gauge groups the functional integral was developed rigorously and studied in detail by Albeverio and Schäfer [**1, 2**].

What is disappointing about (4.18) is that it appears hardly likely that holonomy (Wilson loop) variables are in the test function space $[\mathcal{E}]$. A smearing process was developed by Hahn [**4, 17**] to define the functional integral on Wilson loop variables. The smearing of a loop into a thickened tube requires making certain directional choices of the smearing; this encodes a *framing* of the loop.

Everything we have done depends very crucially on the possibility of a gauge choice which made the Chern-Simons action become a quadratic expression in the connection. For manifolds such as $\Sigma \times S^1$, where Σ is a compact surface, no such gauge choice is possible. However, instead of setting one directional component of the connection form A to 0, we can choose this component to fall in the Lie algebra of a maximal torus in the gauge group. This leads to a considerable simplification in the functional integral. This method of torus-gauge fixing has been developed by Hahn [**18, 19**] and appears promising for a rigorous formulation of the Chern-Simons functional integral for compact surfaces. The method of torus-gauge fixing goes back to the work of Blau and Thompson [**7**].

The work of Freed [**13**] describes the geometry of the Chern-Simons functional as well as the functional integral in relation to other topological and geometric field theories, including two dimensional Yang-Mills theory. For this see also [**33**]. The book of Guadagnini [**16**] gives an overview of many aspects of the Chern-Simons functional integral and its relationship with topology.

5. A summary of geometric notions

For ease of consultation, we summarize some standard notions and notation from differential geometry here (for more see Bleecker [**8**]).

Let M be a smooth manifold of dimension $n \geq 1$, and G a compact Lie group with Lie algebra $L(G)$. A principal G-bundle over M is a smooth manifold P, on which there is a smooth right action $P \times G \to P : (p, g) \mapsto R_g p = pg$ of G, along with a smooth surjection $\pi : P \to M$ with the local triviality property: every $m \in M$ has a neighborhood U such that there is a diffeomorphism $\phi : U \times G \to \pi^{-1}(U)$ satisfying $\pi\phi(u, g) = u$ and $\phi(u, g)h = \phi(u, gh)$ for all $u \in U$ and $g, h \in G$. The subset $\pi^{-1}(m)$ is the *fiber* over the point m. A *gauge transformation* is a diffeomorphism $f : P \to P$ which maps each fiber into itself, i.e. $\pi \circ f = \pi$, and is equivariant under the action of G on P. For $v \in T_pP$ and $g \in G$ we denote by vg the vector $(R_g)_* v \in T_{pg}P$. The subspace $V_p = \ker d\pi_p \subset T_pP$ is called the vertical subspace at p, and a vector in V_p is called *vertical.* The map

$$L(G) \to T_pP : H \mapsto pH = \frac{d}{dt}\Big|_{t=0} p \exp(tH)$$

is a linear isomorphism of $L(G)$ onto V_p.

A *connection* on the principal bundle $\pi : P \to M$ is an $L(G)$-valued 1-form ω on P for which: (i) $\omega(vg) = \mathrm{Ad}(g^{-1})\omega(v)$ for all $g \in G$ and $v \in TP$, and (ii) $\omega(pH) = H$ for all $p \in P$ and $H \in L(G)$.

A k-form η on P with values in a vector space W on which there is a left-action ρ of G is said to be *ρ-equivariant* if

$$\eta(v_1 g, ..., v_k g) = \rho(g^{-1})\eta(v_1, ..., v_k)$$

for all $v_1, ..., v_k \in T_pP$, all $p \in P$, and all $g \in G$ (when $k = 0$ read this as $\eta(pg) = \rho(g^{-1})\eta(p)$ for all $p \in P$ and $g \in G$). For example, a connection form is Ad-equivariant. Many forms of interest are equivariant and also vanish on vertical vectors. The *curvature* of a connection ω is the LG-valued 2-form Ω^ω given by

$$\Omega^\omega(v, w) = d\omega(v, w) + [\omega(v), \omega(w)] \tag{5.1}$$

for all $v, w \in T_pP$ and $p \in P$. This form is Ad-equivariant and vanishes on (v, w) if v or w is vertical.

The subspace

$$H_p^\omega = \ker \omega_p \tag{5.2}$$

is called the *horizontal* subspace for the connection ω at p, and a vector in H_p^ω is called *ω-horizontal.*

A piecewise differentiable path in P is said to be *horizontal* if its tangent vector, wherever defined, is ω-horizontal. A horizontal path $\tilde{c} : [a, b] \to P$ is said to be a horizontal *lift* of a C^1 path $c : [a, b] \to M$ if $c = \pi \circ \tilde{c}$; a unique C^1 horizontal lift exists for each choice of initial point $\tilde{c}(a)$ on the fiber over $c(a)$. If c is a loop, and $\tilde{c}_u$ the horizontal lift with initial point u, then the unique $h \in G$ for which the end point of $\tilde{c}$ is uh is called the *holonomy* $h_u(c; \omega)$ of ω around c with initial point u. Horizontal paths remain horizontal under the right action of G, and this implies that $h_{ug}(c; \omega)$ is $g^{-1}h_u(c; \omega)g$. Now suppose M has a metric, and suppose also that the Lie algebra $L(G)$ has a metric $\langle \cdot, \cdot \rangle_{L(G)}$ which is invariant under the adjoint action of G on $L(G)$. Then the function $\langle \Omega^\omega, \Omega^\omega \rangle$ is the function on M whose value at any point $m \in M$ is $\sum_{1 \leq j < k \leq n} \langle \Omega^\omega(e_j, e_k), \Omega^\omega(e_j, e_k) \rangle_{LG}$ where $e_1, ..., e_n \in T_pP$, with $p \in \pi^{-1}(m)$, project to an orthonormal basis of T_mM; this value is determined by ω, the point m, and the metrics on T_mM and $L(G)$, and is independent of the choice of p in the fiber $\pi^{-1}(m)$ and the basis $\{e_i\}$. The *Yang-Mills action* of a connection ω is given by

$$S_{\mathrm{YM}}(\omega) = \frac{1}{2g^2}\int_M \langle \Omega^\omega, \Omega^\omega\rangle \, d\mathrm{vol} \tag{5.3}$$

where the integration is with respect to the volume measure induced by the metric on M, and the parameter g is a physical coupling constant. A *section* of the bundle $\pi : P \to M$ is a smooth map $s : M \to P$ such that $\pi(s(m)) = m$ for all $m \in M$. Such an s exists if $M = \mathbb{R}^2$, and then it is easier to work with $A = s^*\omega$, which an LG-valued 1-form on $\mathbb{R}^2$. The curvature pulls down to the LG-valued 2-form $F^A = s^*\Omega^\omega$ whose relation to A is conveniently expressed, with G a group of matrices, by $F^A = dA + A \wedge A$

Acknowledgment. My thanks first to the referee for several useful suggestions. Research support from U.S. National Science Foundation Grant DMS-0601141 is gratefully acknowledged.

References

[1] S. Albeverio and J. Schäfer, *Abelian Chern-Simons theory and linking numbers via oscillatory integrals*, J. Math. Phys. **36**(5) (1994), 2135–2169.

[2] ———, *A mathematical model of Abelian Chern-Simons theory*, Proc. Locarno Conf. Stochastic Processes – Physics and Geometry II, S. Albeverio, U. Cattaneo, D. Merlini, eds., World Scientific, Singapore, 1995, pp. 86–95.

[3] S. Albeverio and A. N. Sengupta, *A mathematical construction of the non-abelian Chern-Simons functional integral*, Comm. Math. Phys. **186** (1997), 563–579.

[4] S. Albeverio, A. Hahn, and A. N. Sengupta, *Rigorous Feynman path integrals, with applications to quantum theory, gauge fields, and topological invariants*, Stochastic Analysis and Mathematical Physics, R. Rebolledo, J. Rezende, and J.-C. Zambrini, eds., World Scientific, 2004, pp. 1–60.

[5] S. Axelrod and I. M. Singer, *Chern-Simons perturbation theory*, Differential geometric methods in mathematical physics. Proceedings of the 20th international conference, June 3-7, 1991, Catto, Sultan et al., eds., New York City, NY, World Scientific, Singapore and Teaneck, NJ 1992, pp. 3–45.

[6] ———, *Chern-Simons perturbation theory II*. J. Differential Geom. **39** (1994), 173–213.

[7] M. Blau and G. Thompson, *On diagonalization in* Map(M, G). Commun. Math. Phys. **171** (1995), 639–660.

[8] D. Bleecker, *Gauge theory and variational principles*, Addison-Wesley Pub. Co. 1981.

[9] S.-S. Chern and J. Simons, *Characteristic forms and geometric invariants*, Annals of Math. **99** (1974), 48–69.

[10] B. K. Driver, YM_2 : *Continuum Expectations, Lattice Convergence, and Lassos*, Commun. Math. Phys. **123** (1989), 575–616.

[11] D. Fine, *Quantum Yang-Mills on the Two-Sphere*, Commun. Math. Phys. **134** (1990), 273–292.

[12] ———, *Quantum Yang-Mills on a Riemann Surfaces*, Commun. Math. Phys. **140** (1991), 321–338.

[13] D. Freed, *Classical Chern-Simons theory, Part 1* , Advances in Math. **113** (1995), 237–303.

[14] L. Gross, *Abstract Wiener spaces*. Proc. Fifth Berkeley Sympos. Math. Statist. and Probability (Berkeley, Calif., 1965/66), Vol. II: Contributions to Probability Theory, Part 1, University California Press, Berkeley, CA, 1967, pp. 31–42.

[15] L. Gross, C. King, and A. Sengupta, *Two Dimensional Yang-Mills Theory via Stochastic Differential Equations*, Ann. Phys. **194** (1989), 65–112.

[16] E. Guadagnini, *The link invariants of the Chern-Simons field theory: New developments in topological quantum field theory*, De Gruyter Expositions in Mathematics, 1993.

[17] A. Hahn, *The Wilson loop observables of Chern-Simons theory on* $\mathbb{R}^3$ *in axial gauge*, Commun. Math. Phys. **248**(3) (2004), 467–499.

[18] ———, *Chern-Simons models on $S^2 \times S^1$, torus gauge fixing, and link invariants I*, J. Geom. Phys. **53**(3) (2005), 275–314.

[19] ———, *An analytic approach to Turaev's shadow invariant*, J. Knot Th. Ram. **17**(11) (2008), 1327–1385.

[20] G. W. Johnson and M. L. Lapidus, *The Feynman integral and Feynman's operational calculus*, Oxford Mathematical Monographs, 2002.

[21] D. Karabali, C. Kim, and V. P. Nair, *Planar Yang-Mills Theory: Hamiltonian, Regulators, and Mass Gap*, Nucl. Phys. **B524** (1998), 661–694.

[22] ———, *On the vacuum wave function and string tension of Yang-Mills theories in (2+1) dimensions*, Phys. Lett. B **434** (1998), 103.

[23] H.-H. Kuo, *White noise distribution theory*, CRC Press, 1996.

[24] T. Lévy, *The Yang-Mills measure for compact surfaces*, Memoirs Amer. Math. Soc. **166**(790) (2003).

[25] ———, *Discrete and continuous Yang-Mills measure for non-trivial bundles over compact surfaces*, Probability Theory and Related Fields **136**(2) (2006), 171–202.

[26] T. Lévy and J. Norris, *Large deviations for the Yang-Mills measure on a compact surface*, Commun. Math. Phys. **261**(2) (2006), 405-450 .

[27] S. Mazzucchi, *Mathematical Feynman integrals and their applications*, World Scientific, 2009.

[28] S. G. Rajeev, *Three dimensional simplicial Yang-Mills theory: An approach to the mass gap*, (2008). `http://arxiv.org/abs/hep-th/0401202`

[29] A. N. Sengupta, *Quantum gauge theory on compact surfaces*, Ann. Phys. **221** (1993), 17-52.

[30] ———, *Gauge theory on compact surfaces*, Memoirs of the Amer. Math. Soc. **126** No. 600 (1997).

[31] ———, *Yang-Mills on surfaces with boundary :Quantum theory and symplectic limit*, Commun. Math. Phys. **183** (1997), 661–706.

[32] ———, *Traces in two-dimensional QCD: The large-N limit*, Traces in Gemeotry, Number Theory and Quantum Fields, S. Albeverio, M. Marcolli, S. Paycha, eds., Vieweg, 2008, pp. 193–212.

[33] ———, *Gauge theory in two dimensions: Topological, geometric and probabilistic aspects*, Stochastic Analysis in Mathematical Physics, Gerard Ben Arous, Ana Bela Cruzeiro, Yves Le Jan, and Jean-Claude Zambrini, eds., World Scientific, 2008, pp. 109–129.

[34] ———, *The large-N Yang-Mills field on the plane and free noise*, Geometric Methods in Physics - Proceedings of the XVII Workshop on Geometric Methods in Physics, P. Kielanowski, A. Odzijewicz, M. Schlichenmaier, and T. Voronov, eds., American Institute of Physics, 2008, pp. 121–132.

[35] I. M. Singer, *On the master field in two dimensions*, Functional Analysis on the Eve of the 21st Century, Volume I, S. Gindikin et al., eds., Birkhäuser, 1995.

[36] E. Witten, *On quantum gauge theories in two dimensions*, Commun. Math. Phys. **141** (1991), 153–209.

[37] ———, *Two dimensional quantum gauge theory revisited*, J. Geom. Phys. **9** (1992), 303–368.

[38] ———, *Quantum field theory and the Jones polynomial*, Commun. Math. Phys. (1989), 121–351.

Department of Mathematics, Louisiana State University, Baton Rouge, Louisiana 70803 USA

E-mail address: `sengupta@math.lsu.edu`

AMS/IP Studies in Advanced Mathematics
Volume 50, 2011

Intersection Pairings on Spaces of Connections and Chern-Simons Theory on Seifert Manifolds

George Thompson

ABSTRACT. Let M be a $U(1)$ bundle over a smooth Riemann surface. I show that for Chern-Simons theory on M, with structure group G, the path integral is an integral over the space of G-connections on the Riemann surface involving characteristic classes as well as a certain 4-dimensional class that comes from a universal bundle. When M is the product of a Riemann surface with a circle the 4-dimensional class does not enter and the path integral takes the form of a Riemann-Roch formula albeit in infinite dimensions. The discussion is generalised to include Wilson lines along the fibre direction in M.

1. Introduction

Let M be a compact closed 3-manifold and $G \to P \to M$ a trivial principal G bundle over M. We denote by $\mathcal{A}_3$ the space of connections on P and $\mathcal{G}_3$ the gauge group. Let $Z_{k,G}[M]$ denote the Chern-Simons path integral at level k and for group G (taken as above),

$$Z_{k,G}[M] = \frac{1}{\mathrm{Vol}(\mathcal{G}_3)} \int_{\mathcal{A}_3} \exp\left(I(\mathscr{A})\right)$$

where the Chern-Simons action, for $\mathscr{A} \in \mathcal{A}_3$, is

$$I(\mathscr{A}) = i\frac{k}{4\pi} \int_M \mathrm{Tr}\left(\mathscr{A} \wedge d\mathscr{A} + \frac{2}{3}\mathscr{A} \wedge \mathscr{A} \wedge \mathscr{A}\right)$$

and Tr is normalized so that under large gauge transformations (recall that $\pi_0(\mathcal{G}_3) = \mathbb{Z}$) $I(\mathscr{A}^{\mathscr{G}}) = I(\mathscr{A}) + 2\pi i n$, $\mathscr{G} \in \mathcal{G}_3$ and $n \in \mathbb{Z}$ so that the exponential is invariant.

One can also consider the inclusion of 'Wilson lines' in the path integral. A Wilson line is a combination of a knot K and an irreducible representation R of the group G, and is defined to be the holonomy

$$W_R(K) = \mathrm{Tr}_R\, P\exp \oint_K \mathscr{A}$$

The relevant path integral is

$$Z_{k,G}[M, (K_i, R_i)] = \frac{1}{\mathrm{Vol}(\mathcal{G}_3)} \int_{\mathcal{A}_3} \exp\left(I(\mathscr{A})\right) \prod_{i=1} W_{R_i}(K_i)$$

2010 *Mathematics Subject Classification.* Primary 57R56, 53D99, 19L10.

Let Σ be a smooth genus g curve and ω a unit volume Kähler form on Σ. Let $\mathfrak{M}$ be the moduli space of flat G connections on Σ. The moduli space has a natural symplectic, infact Kähler, structure and we let $\mathfrak{L} \to \mathfrak{M}$ be the fundamental line bundle whose first Chern class agrees with the natural symplectic form on $\mathfrak{M}$ (for $G = SU(n)$ this will be the determinant line bundle). E. Witten [**15**] has shown that the quantum Hilbert space of states of Chern-Simons theory on Σ is $\mathrm{H}^0(\mathfrak{M}, \mathfrak{L}^k)$. Quite generally the Chern-Simons invariant, $Z_{k,G}[M]$, for a 3-manifold M, by Heegard splitting, will be the inner product of two vectors in the Hilbert space. However, the dimension of the Hilbert space is the Chern-Simons invariant of the 3-manifold $\Sigma \times S^1$.

If one includes Wilson lines, then the Hilbert space is $\mathrm{H}^0(\mathfrak{M}, \mathfrak{L}^k)$ where now $\mathfrak{M}$ is the moduli space of parabolic bundles on Σ, which is still naturally Kähler, and $\mathfrak{L}^k$ is the prequantum line bundle over $\mathfrak{M}$ associated to Chern-Simons theory at level k in this situation.

The Hirzebruch-Riemann-Roch theorem tells us that

$$\sum_{q=0}(-1)^q \dim \mathrm{H}^q(\mathfrak{M}, \mathfrak{L}^k) = \int_{\mathfrak{M}} \mathrm{Todd}(\mathfrak{M}) \wedge \mathrm{Ch}(\mathfrak{L}^k)$$

If the canonical bundle of $\mathfrak{M}$ is negative, as will be the case in the examples we discuss, then the higher cohomology groups are trivial by Kodaira vanishing and we have

$$\dim \mathrm{H}^0(\mathfrak{M}, \mathfrak{L}^k) = \int_{\mathfrak{M}} \mathrm{Todd}(\mathfrak{M}) \wedge \mathrm{Ch}(\mathfrak{L}^k)$$

E. Verlinde [**14**] provides us with a concrete formula for the dimension of $\mathrm{H}^0(\mathfrak{M}, \mathfrak{L}^k)$. However, as we have seen, Chern-Simons theory provides us with another formulation of E. Verlinde's dimension count, namely

$$Z_{k,G}[\Sigma \times S^1, (K_i, R_i)] = \int_{\mathfrak{M}} \mathrm{Todd}(\mathfrak{M}) \wedge \mathrm{Ch}(\mathfrak{L}^k)$$

The moduli spaces, which thus far have been generically denoted by $\mathfrak{M}$, are singular at the points where there are reducible connections. Nevertheless, suitably interpreted, the index theorem still yields a topological expression for the invariants $Z_{k,G}[\Sigma \times S^1, (K_i, R_i)]$. This raises a

Question: Are there other 3-manifolds whose Chern-Simons invariants, or parts of them, can be expressed as intersection pairings on an appropriate moduli space $\mathfrak{M}$?

The question has been partially answered in the affirmative by Beasley and Witten [**4**]. From now on set $M \equiv M_{(g,p)}$ to denote the Seifert manifold that is presented as a degree $-p$, $U(1)$ bundle over a Riemann surface Σ of genus g. Using non-Abelian localization for the Seifert 3-manifolds M Beasley and Witten are able to show that (equation (5.176) in their paper with $n = -p$ due to a different choice of orientation and noting that there is a slightly different normalization of Θ),

PROPOSITION 1.1. (Beasley-Witten) Let M be as above. The portion of the Chern-Simons invariant which is localized on the smooth part, $\mathfrak{M}$, of the moduli space of Yang-Mills connections is

$$Z_{k,G}[M]|_{\mathfrak{M}} = \frac{1}{|\Gamma|}\exp\left(i\frac{\pi}{2}\eta_0\right)\int_{\mathfrak{M}} \mathrm{Todd}(\mathfrak{M})\wedge \mathrm{Ch}(\mathfrak{L}^k)\wedge \exp\left(-i\frac{p}{2\pi}(k+c_{\mathbf{g}})\Theta(\mathfrak{M})\right)$$

In this formula, $\Theta(\mathfrak{M})$ is a certain degree 4 cohomolgy class on $\mathfrak{M}$, $c_{\mathbf{g}}$ and Γ are the dual Coxeter number and centre of G respectively while η_0 is the framing of M. Notice that this formula is quite analogous to the Hirzebruch-Riemann-Roch formula and reduces to it when $p = 0$.

Let $\mathcal{A}$ be the space of connections on the trivial G bundle on Σ and $\mathcal{G}$ the associated group of gauge transformations. I will show that

PROPOSITION 1.2. The Chern-Simons path integral on M, $Z_{k,G}[M]$, is equal to a path integral on the space of connections over Σ, namely

$$Z_{k,G}[M] = \exp\left(i\frac{\pi}{2}\eta_0\right)\frac{1}{\mathrm{Vol}(\mathcal{G})}\int_{\mathcal{A}} \mathrm{Todd}(\mathcal{A})\wedge \mathrm{Ch}(\mathfrak{L}^k)\wedge \exp\left(-i\frac{p}{2\pi}(k+c_{\mathbf{g}})\Theta(\mathcal{A})\right)$$

in this formula all the classes have the same form as those in Proposition 1.1.

As there is a striking resemblance between the formulae presented in Propositions 1.1 and 1.2 it is worthwhile, at this point, to make some remarks.

REMARK 1.3. The formula in Proposition 1.2 is an exact expression for $Z_{k,G}[M]$ unlike that in Proposition 1.1 which is a part of the answer.

REMARK 1.4. It is quite straightforward to evaluate the path integral in Proposition 1.2. This has been done for general Σ in [**8**] by Abelianization and for $\Sigma = S^2$ in [**4**] by non-Abelian localization. The Reshetikhin-Turaev-Witten invariants for Seifert 3-manifolds may be obtained by surgery prescriptions as in [**10**] and [**11**] and agree with the path integral results. However, I leave the path integral 'un-integrated' as it brings the geometry to the fore.

REMARK 1.5. I am not at all implying that Proposition 1.1 follows easily from Proposition 1.2, though one would reasonably expect that it does. Presumably an application of non-Abelian localization to the path integral appearing in Proposition 1.2 is what is required.

REMARK 1.6. Proposition 1.2 has already been established, somewhat indirectly in [**1**] and more directly in [**8**], though the language is somewhat different and the classes were not identified in either of these works and so the geometric significance of the right hand side of Proposition 1.2 was not appreciated.

I will prove a rather more general result involving knots in the fibre direction which are located at points $x_i \in \Sigma$ on the base of the fibration and which run along the fibre. The point of view adopted here is that to the parabolic points x_i one 'attaches' a co-adjoint orbit M_{R_i} defined by the representation R_i.

PROPOSITION 1.7. The Chern-Simons path integral on M, with knots in the fibre direction at the points $x_i \in \Sigma$ and associated representations R_i, $Z_{k,G}[M,(x_i,R_i)]$, is given by

$$\begin{aligned} Z_{k,G}[M,(x_i,R_i)] &= \exp\left(i\frac{\pi}{2}\eta_0\right)\frac{1}{\mathrm{Vol}(\mathcal{G})}\int_{\mathcal{A}\times\prod_i M_{R_i}} \widehat{A}(\mathcal{A}\times\prod_i M_{R_i})\wedge \\ &\quad .\exp\left((k+c_{\mathbf{g}})\Omega(\mathcal{A})+\sum_i \omega(M_{R_i}) - i\frac{p}{2\pi}(k+c_{\mathbf{g}})\Theta(\mathcal{A})\right) \end{aligned}$$

where $\widehat{A}$ is the A hat genus.

Acknowledgments. This paper is an outcome of a line of research on abelianization in low dimensions that Matthias Blau and I have been pursuing over many years. It is a pleasure to thank him for this very enjoyable collaboration. It is also a pleasure to thank M.S. Narasimhan and T. Ramadas for helping me to formulate the results presented here. Finally a thank you to the organizers of the Bonn workshop "Chern-Simons Gauge Theory: 20 Years After" and to Chris Beasely for discussions on the relationship between his work and mine.

2. Background and strategy

While it is clear from the description of the Hilbert space of Chern-Simons theory that the moduli spaces are built in, what is not at all obvious is that Chern-Simons should, in general, have knowledge of the cohomology ring of the moduli space. This paragraph is intended to motivate such a connection. Recall that Witten [**16**] had established that the topological field theory analogue of Donaldson theory on a curve can be mapped to Yang-Mills theory on the curve. Since the topological field theory is designed to probe the cohomology ring of the moduli space we learn that Yang-Mills theory will do just that. The action of Yang-Mills theory is

$$S(F_A,\psi,\phi)=\frac{1}{4\pi^2}\int_\Sigma \mathrm{Tr}\left(i\phi F_A+\frac{1}{2}\psi\wedge\psi\right)+\frac{\epsilon}{8\pi^2}\int_\Sigma \omega\ \mathrm{Tr}\,\phi^2 \tag{2.1}$$

where A is a connection on P, $\phi\in\Gamma(\Sigma,\mathrm{ad}\ P)$ and ψ is interpreted as a one form on the space $\mathcal{A}/\mathcal{G}$ (in terms of the universal bundle construction described in the next section the elements of the action are all cohomology classes on $\mathcal{A}/\mathcal{G}$).

In [**4**] Beasley and Witten recall that every 3-manifold has a contact structure κ. Here we use the $U(1)$ bundle structure and the associated nowhere vanishing vector field ξ with $\iota_\xi\kappa=1$. (This has the advantage that it corresponds to the obvious structure on $\Sigma\times S^1$.) One may, therefore, decompose connections as

$$\mathscr{A}=A+\kappa\frac{1}{2\pi}\phi,\quad \iota_\xi A=0$$

The Chern-Simons action, on choice of the contact structure, is

$$I(A,\phi)=i\frac{k}{4\pi^2}\int_M\left(\pi\kappa\,\mathrm{Tr}\,A\iota_\xi dA+\kappa\,\mathrm{Tr}\,\phi F_A+\kappa d\kappa\frac{1}{4\pi}\mathrm{Tr}\,\phi^2\right) \tag{2.2}$$

and this has some resemblance to (2.1). The first term in (2.2) is the 'time' derivative (in the direction of the vector field dual to κ). There are some differences and perhaps the most glaring is that the term proportional to $\psi\wedge\psi$ is absent. However, a variation on the theme is now available. The structure group of a general 3-manifold is $SO(3)$, however, on fixing the bundle structure the compatible structure group reduces to $SO(2)$ (the structure group of Σ). There is a minimal, so called $N=1$, supersymmetric version of Chern-Simons theory which is not obviously topological. However, when the structure group is reduced to $SO(2)$ one may 'twist' the $N=1$ supersymmetric Chern-Simons theory to obtain a manifestly topological theory (up to choice of contact structure). The twisted version has action (2.2) augmented with

$$\frac{k}{8\pi^2}\int_M\kappa\ \mathrm{Tr}\,\psi\wedge\psi,\quad \iota_\xi\psi=0$$

The $N = 1$ action becomes

$$I(A, \phi, \psi) = \frac{k}{4\pi^2} \int_M \left(i\pi\kappa \operatorname{Tr} A \wedge \iota_\xi dA + \kappa \operatorname{Tr} \left(i\phi F_A + \frac{1}{2}\psi \wedge \psi \right) + \frac{i}{4\pi}\kappa d\kappa \operatorname{Tr} \phi^2 \right)$$

so that now the resemblance of the two theories is rather remarkable. The supersymmetry transformations are closely linked to those in [**4**], equation (3.48),

$$QA = i\psi, \quad Q\psi = -d_A\phi - 2\pi\iota_\xi dA \tag{2.3}$$

The fact that the fields ψ are free means that partition function and knot invariants of the $N = 1$ and the usual Chern-Simons theory agree.

One possible novelty that there is a new set of observables in the theory namely those that are 3-forms on the moduli space

$$\int_M J_C \wedge \operatorname{Tr} \phi . \psi$$

with J_C a de Rham current with delta function support on the 1-dimensional cycle $C \subset M$ is, unfortunately, spoilt by the second part of the transformation of ψ in (2.3).

In any case it would appear that Chern-Simons theory is some supersymmetric quantum mechanics on the moduli spaces of interest. Since supersymmetric quantum mechanics is related to index theory one may reasonably expect a positive response to the question.

The strategy we will follow here is to

(1) Identify the objects that appear in the path integral in terms of cohomology classes on the space of connections modulo the gauge group. This is done in Section 3.
(2) Next we relate the classes that one obtains from the universal bundle construction to classes associated with the tangent bundle of $\mathcal{A}/\mathcal{G}$. These are denoted by Todd($\mathcal{A}$) and $\widehat{A}(\mathcal{A})$ in Section 4 for reasons that will become apparent as we go along.
(3) One can now express Witten's localisation of the Yang-Mills path integral on Σ (Proposition 5.1 below) in terms of classes that come from the universal bundle on $\mathcal{A}/\mathcal{G}$ on the one hand and the same classes restricted to $\mathfrak{M}$ on the other.
(4) Having established the relationship with the cohomology ring in $\mathcal{A}/\mathcal{G}$ our final task is to show that the integral over $\mathcal{A}_3/\mathcal{G}_3$ reduces to that of certain classes on $\mathcal{A}/\mathcal{G}$ on integrating out sections which are not $U(1)$ invariant.

3. A universal bundle

Let P be a principal G bundle over some smooth manifold X, $\mathcal{A}$ the space of connections on P and $\mathcal{G}$ the group of gauge transformations (bundle automorphisms). We have the following universal bundle construction due to Atiyah and Singer [**2**]. There is an action of $\mathcal{G}$ on P so that we may form the space (this is not smooth unless one makes further assumptions and as it stands it is a stack)

$$\mathcal{Q} = \mathcal{A} \times_{\mathcal{G}} P$$

Now G operates on $\mathcal{Q}$ and infact $\mathcal{Q}$ is itself the total space of a principle bundle

$$\mathcal{Q} \to \mathcal{Q}/G = \mathcal{A}/\mathcal{G} \times X$$

There is a natural connection on $\mathcal{Q}$ and from it we can define a curvature 2-form and then Chern classes, via Chern-Weil theory, for an associated rank n vector bundle $\mathcal{E} = \mathcal{Q} \times_G \mathbb{C}^n \to \mathcal{A}/\mathcal{G} \times X$. Finally we restrict to some (hopefully smooth) moduli space $\mathfrak{M} \subset \mathcal{A}/\mathcal{G}$, for which the above construction makes sense.

REMARK 3.1. A detailed description of this bundle and its relationship to topological field theories can be found in [**6**] chapter 5, especially sections 5.1 and 5.3.

REMARK 3.2. The path integral we need to deal with is the one over all of $\mathcal{A}/\mathcal{G}$ and not some smooth finite dimensional subspace $\mathfrak{M}$. Consequently, we will be integrating over the stack.

From now on we take X to be Σ. Decompose the curvature 2-form on $\mathcal{E}$ into its Kunneth components as

$$1 \otimes F_A + \Psi + \Phi \otimes 1 \in \mathrm{H}^2(\mathcal{A} \times \Sigma, \mathrm{Lie}\, G)$$

(If $\mathfrak{M}$ is simply connected we have that $\mathrm{Tr}\,\Psi$, restricted to $\mathfrak{M}$, is cohomologously trivial in which case $c_1(\mathcal{E}) = 1 \otimes c_1(E) + \frac{i}{2\pi}\,\mathrm{Tr}\,\Phi \otimes 1$ where $\mathrm{Tr}\,\Phi \in \mathrm{H}^2(\mathfrak{M})$.) If we fix on $G = SU(n)$ then $c_1(\mathcal{E})$ vanishes and the second Chern class decomposes as

$$\begin{aligned} c_2(\mathcal{E}) &= \frac{1}{4\pi^2}\,\mathrm{Tr}\left(\Phi \otimes F_A + \frac{1}{2}\Psi \wedge \Psi\right) + \frac{1}{4\pi^2}\,\mathrm{Tr}\,\Phi \wedge \Psi + \frac{1}{8\pi^2}\,\mathrm{Tr}\,\Phi^2 \otimes 1 \\ &= \Omega(\mathcal{E}) + \gamma(\mathcal{E}) + \Theta(\mathcal{E}) \qquad (3.1) \end{aligned}$$

We have a differential Q on the space $\mathcal{A} \times \Sigma$ which satisfies

$$QA = \Psi, \quad Q\Psi = d_A\Phi, \quad Q\Phi = 0, \quad Q^2 = \mathcal{L}_\Phi$$

so that Ψ is a (basis) one form on $\mathcal{A}$ and

$$(Q - d_A)(1 \otimes F_A + \Psi + \Phi \otimes 1) = 0.$$

Hence associated to Q we have equivariant cohomology on $\mathcal{A}/\mathcal{G} \times \Sigma$ and the Chern classes $c_n(\mathcal{E})$ are Q closed.

Note the second Chern class appears as the action of Yang-Mills theory (2.1) if we make the identifications $\Phi = i\phi$ and $\Psi = \psi$, and we consider these as forms on $\mathcal{A}/\mathcal{G} \times \Sigma$

$$S(F_A, \psi, \phi) = \pi_*\left(\Omega(\mathcal{E}) - \epsilon\Theta(\mathcal{E}) \otimes \omega\right) \simeq \Omega(\mathcal{A}) - \epsilon\Theta(\mathcal{A}) \qquad (3.2)$$

where $\pi : \mathcal{A}/\mathcal{G} \times \Sigma \to \mathcal{A}/\mathcal{G}$ is projection onto the first factor.

REMARK 3.3. The identification (3.2) shows us that the Yang-Mills action is a Q closed form (of mixed degree). It is also quite clearly not Q exact.

In general for a vector bundle V $c_2(\mathrm{End}\, V) = 2rc_2(V) - (r-1)c_1(V)^2$ and consequently $c_2(\mathrm{End}\,\mathcal{E}) = 2rc_2(\mathcal{E})$. The main interest here will be on the trace free part $\mathrm{End}_0\,\mathcal{E}$ (which in any case has the same second Chern class as $\mathrm{End}\,\mathcal{E}$). Note that the classes on $\mathcal{E}$ are in the 'fundamental' representation while they are taken to be in the 'adjoint' representation on $\mathrm{End}_0\,\mathcal{E}$.

In the rank 2 case, Newstead [**12**], writes the second Chern Class of his universal bundle as

$$c_2(\mathrm{End}\, U) = 2\alpha \otimes \omega + 4\gamma - \beta \otimes 1 \qquad (3.3)$$

Thus to make contact with that work restrict to $\mathfrak{M} \subset \mathcal{A}/\mathcal{G}$ and set,

$$\alpha = \frac{1}{4\pi^2}\int_\Sigma \operatorname{Tr}\Psi\wedge\Psi, \quad \beta = -\frac{1}{2\pi^2}\operatorname{Tr}\Phi^2, \quad \gamma = \frac{1}{4\pi^2}\operatorname{Tr}\Phi\wedge\Psi$$

4. The Todd and the $\widehat{A}$ genera

In order to express the Todd genus in terms of the classes arising from the universal bundle we follow the approach of Newstead [**12**] for determining the Pontrjagin class in the rank 2 case. Then we use an observation of Thaddeus [**13**] to give the $\widehat{A}$ class in terms of the Pontrjagin roots and from there Todd.

The tangent bundle, $\mathrm{T}_{\mathfrak{M}}$, of $\mathfrak{M}$ is given by

$$\mathrm{T}_{\mathfrak{M}} \simeq R^1\pi_*\mathrm{End}_0\,\mathcal{E}$$

where R^i denotes the i-th direct image sheaf under the map $\pi : \mathfrak{M}\times\Sigma \to \mathfrak{M}$ onto the first factor. The Grothendieck-Riemann-Roch theorem states that

$$\mathrm{Ch}(\mathrm{T}_{\mathfrak{M}}) - \mathrm{Ch}(\pi_*\mathrm{End}_0\,\mathcal{E}) = -\pi_*\left(\mathrm{Ch}(\mathrm{End}_0\,\mathcal{E})(1-(g-1)\omega)\right)$$

For the spaces that we are interested in the direct image sheaf $R^0\pi_*\mathrm{End}_0\,\mathcal{E}$ is trivial so

$$\mathrm{Ch}(\mathrm{T}_{\mathfrak{M}}) = -\pi_*\left(\mathrm{Ch}(\mathrm{End}_0\,\mathcal{E})(1-(g-1)\omega\right) \tag{4.1}$$

Denote the complexification of the Lie algebra of G by $\mathbf{g}_{\mathbb{C}}$, the complexification of the Cartan subalgebra by $\mathfrak{t}_{\mathbb{C}}$ and the space of roots by $\mathbf{k}$ then

$$\mathbf{g}_{\mathbb{C}} = \mathfrak{t}_{\mathbb{C}} \oplus \mathbf{k}$$

We also set $\mathbf{k}_+$ to be the space of positive roots.

LEMMA 4.1. The Pontrjagin class of the tangent bundle $P(T_{\mathfrak{M}})$ is given by

$$P(T_{\mathfrak{M}}) = \det_{\mathbf{k}}(1+\operatorname{ad}\Phi/2\pi)^{2g-2} = \prod_{\mathbf{k}_+}\left(1-\left(\frac{\alpha(\Phi)}{2\pi}\right)^2\right)^{2g-2}$$

Proof: Quite generally the only classes that contribute to $\mathrm{Ch}_{2m}(\mathrm{T}_{\mathfrak{M}})$ from (4.1) are powers of Φ, so we have $\mathrm{Ch}_{2m}(\mathrm{T}_{\mathfrak{M}}) = (g-1)\operatorname{Tr}_{\mathrm{Ad}}(\exp(i\Phi/2\pi))_{2m}$, and consequently $\mathrm{Ch}(\mathrm{T}_{\mathfrak{M}}\oplus\mathrm{T}^*_{\mathfrak{M}}) = 2(g-1)\operatorname{Tr}_{\mathrm{Ad}}(\exp(i\Phi/2\pi))$. This is the same as having the direct sum of $2g-2$ copies of a vector bundle with curvature 2-form $\operatorname{ad}\Phi$. The Chern class for one copy is $\det_{\mathbf{g}}(1+i\operatorname{ad}\Phi/2\pi) = \det_{\mathbf{k}}(1+i\operatorname{ad}\Phi/2\pi)$, the direct sum formula for Chern classes gives us that $c(\mathrm{T}_{\mathfrak{M}}\oplus\mathrm{T}^*_{\mathfrak{M}}) = \prod_{\alpha\in\mathbf{k}_+}(1+(\alpha(\Phi)/2\pi)^2)^{2g-2}$. On the otherhand we have that the Pontrjagin classes are related to Chern classes by $p_m(E) = (-1)^m c_{2m}(E\oplus E^*)$ and this completes the proof. □

LEMMA 4.2. The Todd class of the tangent bundle of $\mathfrak{M}$ is

$$\mathrm{Todd}(\mathfrak{M}) = \exp\frac{1}{2}c_1(T_{\mathfrak{M}})\,.\left(\frac{\det_{\mathbf{k}}\sin(\operatorname{ad}\Phi/4\pi)}{\det_{\mathbf{k}}(\operatorname{ad}\Phi/4\pi)}\right)^{1-g}$$

Proof: Thaddeus (p147 in [**13**]) notes that on writing, $P = \prod_{i=1}(1+y_i)$ where the y_i are the Pontrjagin roots, then $\widehat{A} = \prod_{i=1}(\sqrt{y_i}/2/\sinh\sqrt{y_i}/2)$. From Lemma 4.1 the roots are $-(\alpha(\Phi)/2\pi)^2$ and they come with a multiplicity of $2g-2$ so that

$$\widehat{A}(\mathfrak{M}) = \prod_{\alpha\in\mathbf{k}_+}\left(\frac{\sinh(i\alpha(\Phi)/4\pi)}{i\alpha(\Phi)/4\pi}\right)^{2-2g}$$

and the standard relation between Todd and $\widehat{A}$ completes the proof. □

We can also make use of (4.1) to determine the first Chern class of the moduli space. In this case the term proportional to $(g-1)$ cannot contribute since, as we have seen, it contributes to even classes. We have

$$\mathrm{Ch}(\mathrm{End}_0\mathcal{E}) = -c_2(\mathrm{End}_0\mathcal{E}) + \cdots = -2rc_2(\mathcal{E}) + \ldots \tag{4.2}$$

where the ellipses represent higher degree forms.

THEOREM 4.3. (J-M. Drezet and M.S. Narasimhan [**9**]) The first Chern class of the tangent bundle of $\mathfrak{M}(r,d)$ the moduli space of holomorphic vector bundles of rank r and determinant of degree d is

$$c_1(T\mathfrak{M}(r,d)) = 2\,\mathrm{g.c.d}(r,d)\,\Omega(\mathfrak{M}(r,d))$$

where $\Omega(\mathfrak{M}(r,d))$ is the class of the basic line bundle.

In our case $d=0$ so that $c_1(T\mathfrak{M}) = 2r\Omega(\mathfrak{M})$ and we set, on comparing with (4.2), $\Omega(\mathfrak{M}) = \pi_* c_2(\mathcal{E})$ that is

$$\Omega(\mathfrak{M}) = \frac{1}{4\pi^2}\int_\Sigma \left(i\,\mathrm{Tr}\,\phi F_A + \frac{1}{2}\,\mathrm{Tr}\,\Psi\wedge\Psi \right) \tag{4.3}$$

which is the form one would expect on $F_A = 0$.

5. Intersection pairings on moduli spaces

By comparing to the second Chern class of the universal bundle we see that, quite generally, one should think of $S(F_A,\psi,\phi)$ as a form on the space of connections $\mathcal{A}$. With abuse of notation we denote those classes on $\mathcal{A}$ by the same symbols as those on $\mathfrak{M}$,

$$\frac{1}{\mathrm{Vol}(\mathcal{G})}\int_{\mathcal{A}\otimes\Omega^0(\Sigma,\mathbf{g})} \exp\left(S(F_A,\psi,\phi)\right) \equiv \int_{\mathcal{A}/\mathcal{G}} \exp\left(\Omega(\mathcal{A}) - \epsilon\Theta(\mathcal{A})\right) \tag{5.1}$$

Expressing the path integral in this way hides certain things, like the fact that there is a Gaussian integration over the degree four class or, as the space $\mathcal{A}/\mathcal{G}$ is infinite dimensional, one cannot expand the exponential out to pick the top form term. Indeed a detailed analysis of the path integral shows that some care needs to be exercised in the interpretation of the right hand side (and on the left). Nevertheless expressing the path integral in this way is also very suggestive.

E. Witten [**16**] shows that this path integral essentially devolves to one on the moduli space,

PROPOSITION 5.1. (E. Witten [**16**]) The path integral localises onto the moduli space of flat connections,

$$\int_{\mathcal{A}/\mathcal{G}} \mathrm{e}^{\left(\Omega(\mathcal{A}) - \epsilon\Theta(\mathcal{A})\right)} = \int_{\mathfrak{M}} \mathrm{e}^{\left(\Omega(\mathfrak{M}) - \epsilon\Theta(\mathfrak{M})\right)} + \text{ terms non} - \text{analytic in } \epsilon$$

and the non-analytic terms vanish as $\epsilon \to 0$ (provided $\mathfrak{M}$ is not singular). The non-analytic terms arise from higher fixed points of the action, that is, from non-flat solutions to $d_A * F_A = 0$.

The intersection pairings on the moduli space of flat G connections on Σ as presented by Witten [**16**] agree with those derived by Thaddeus [**13**] in the rank 2 case and predicts those for higher rank.

6. Supersymmetric quantum mechanics and passing from Chern-Simons to Yang-Mills

In [**8**] one begins with the Chern-Simons theory, then integrates out modes in the bundle direction, to be left with a theory on the Riemann surface. This Abelian theory is then solved to finally provide one with the invariant for the Seifert manifold. However, it is also pointed out that the same result could be obtained by considering instead Yang-Mills theory on Σ with the inclusion of some observables one of which (equation (6.8) there) written in the current notation is

$$j_{\mathbf{g}}(\phi)^{(1-g)} = \widehat{A}(\mathcal{A})$$

This observation was critical for the present study.

In this section the approach of [**8**] is followed 'half-way' to the point where we have non-Abelian Yang-Mills theory on Σ. In this way we are able to obtain the classes on $\mathcal{A}/\mathcal{G}$ that one must integrate. I will not repeat the entire calculation but, rather, explain the essential ingredients especially those which go beyond [**8**]. I should point out that, when needed, I will consider the section ϕ to be momentarily constant on Σ and this will simplify the calculation of the determinants that we will come across presently. After the determinants are calculated ϕ will be allowed to be non-constant once more. The justification for this simplification really comes from knowing that had we abelianized then ϕ would be forced to be constant on Σ and so consequently we lose nothing in making this asumption.

6.1. Fourier modes and a gauge choice. In order to begin the calculation we note that, as there is a non degenerate S^1 action on M, we may decompose all the sections in terms of characters of that action (a Fourier series). We may therfore write

$$\begin{aligned}
\mathcal{A}_3 &= \mathcal{A} \oplus \Omega^0(\Sigma, \mathrm{ad}\, P) \oplus_{n\neq 0} \Omega^1(\Sigma, L^{\otimes -np} \otimes \mathrm{ad}\, P) \oplus_{n\neq 0} \Omega^0(\Sigma, L^{\otimes -np} \otimes \mathrm{ad}\, P) \\
\varphi &= \sum_{n=-\infty}^{\infty} \varphi_n, \quad \iota_\xi d\varphi_n = -2\pi i n \varphi_n, \quad \varphi_n \in \Omega^*(\Sigma, L^{\otimes -np} \otimes \mathrm{ad}\, P)
\end{aligned}$$

where L is the linc bundle that defines M.

Now note that there is enough gauge symmetry to impose the condition that the section ϕ is constant in the fibre direction, that is $\iota_\xi . d\phi = 0$, and we do this. Alternatively put, we make the identification,

$$\mathcal{A}_3/\mathcal{G}_3 \simeq \left(\mathcal{A} \oplus \Omega^0(\Sigma, \mathrm{ad}\, P) \oplus_{n\neq 0} \Omega^1(\Sigma, L^{\otimes -np} \otimes \mathrm{ad}\, P)\right)/\mathcal{G}$$

There is a caveat here as the 'natural' measures do not coincide since the non-constant components of the section ϕ, in $\Omega^0(\Sigma, L^{\otimes -np} \otimes \mathrm{ad}\, P)$ are tangent vectors to the orbit of $\mathcal{G}_3$. To correct for this mismatch one introduces the Faddeev-Popov ghost determinant, $\Delta_{FP}(\phi)$, which is essentially the ratio of the volume of the orbit to that of the group.

In any case the choice of gauge simplifies the path integral immensely.

6.2. Integrating over non-trivial characters. Our aim is to integrate out all those Fourier modes of fields such that $n \neq 0$. As, by the gauge condition, ϕ has no such modes and the integral over ψ is Gaussian, we concentrate on the integral of the A_n for $n \neq 0$. Note that (with A_0 denoted by A again)

$$I(A, \phi, \psi) = kS(F_A, \psi, \phi) + \Delta I \tag{6.1}$$

where $S(F_A, \psi, \phi)$ is the Yang-Mills action with $\epsilon = ip/2\pi$ and

$$\Delta I = \frac{k}{4\pi^2} \int_\Sigma \sum_{n \neq 0} \mathrm{Tr}\,(A_{-n} \wedge (2\pi n + \mathrm{ad}\,\phi) A_{-n} + \psi_n \wedge \psi_{-n}) \tag{6.2}$$

Let

$$\exp i\Gamma(A,\phi) = \int \prod_{n\neq 0} dA_n\, d\psi_n\ \Delta_{\mathrm{FP}}(\phi)\ \exp i\Delta I$$

where the Faddeev-Popov determinant Δ_{FP}, as I mentioned above, takes into account the gauge condition on ϕ.

DEFINITION 6.1. We will say that two gauge invariant functions on $\mathcal{A}$ are equivalent if their integrals over $\mathcal{A}/\mathcal{G}$ agree and we will denote that equivalence by $\simeq$.

PROPOSITION 6.2. The supersymmetric quantum mechanics path integral gives, for ϕ valued in the Cartan subalgebra,

$$\exp i\Gamma(A,\phi) \simeq \exp\left(i\frac{\pi}{2}\eta_0\right) \widehat{A}(i\phi) \wedge \exp\left(i\frac{c_{\mathbf{g}}}{4\pi^2}\int_\Sigma \mathrm{Tr}\,(\phi.F_A + \frac{p}{4\pi}\phi^2\omega)\right)$$

where η_0 is the framing correction.

Proof: As one can see the action is such that the part of the connection $A^{\mathbf{k}}$, only enters in a Gaussian fashion and so may easily be integrated out. The integration gives rise to a determinant which requires regularization (a definition). This calculation has been performed in [**8**] but includes the constant mode (see (B.23) there) and the ϕ there should be rescaled to $\phi/2\pi$ to agree with the definition here. So on putting back the constant mode contribution in that work and changing the normalization of ϕ we obtain the (square root of the) determinant as

$$\prod_{\alpha\in\mathbf{k}} \left(\frac{\prod_n (2\pi n + i\alpha(\phi)/2\pi)}{i\alpha(\phi)/2\pi}\right)^{1-g} = \prod_{\alpha\in\mathbf{k}_+} \left(\frac{\sin^2 i\alpha(\phi)/4\pi}{(i\alpha(\phi)/4\pi)^2}\right)^{1-g} \equiv j_{\mathbf{g}}(\phi)^{1-g}$$

together with the phase (B.31)-(B.34) in [**8**]. The integral over those $A^{\mathfrak{t}}$ which are non-constant leads to a simple overall factor in front of the path integral. The integral over the non-constant parts of the symplectic volume give rise to a normalization which compensates that of the connections. The calculation presumes ϕ constant (not just along the fibre) as in the path integral over $\mathcal{A}$ it may be taken to be so. □

There are still 2 issues that we need to deal with:

(1) Extend Proposition 6.2 to general sections $\phi \in \Gamma(\Sigma, \mathrm{ad}\,P)$.
(2) Make sure that supersymmetry is preserved.

REMARK 6.3. Both issues are resolved by recalling a basic tennet of renormalizable field theory: Upon regularising a theory it may be neccessary to add to the Lagrangian local counterterms in order to restore symmetries broken by the choice of regularization.

We begin with the first issue. We know that $\Gamma(\phi)$ is formally gauge invariant,

$$\Gamma(g^{-1}\,\phi\, g) = \Gamma(\phi), \quad g \in \mathcal{G}$$

However, the curvature 2-form F_A that appears in the formula in Proposition 6.2 lies in the Cartan direction, $F_A^{\mathfrak{t}} = dA^{\mathfrak{t}}$, and as it stands the result is only gauge invariant under gauge transformations in the maximal torus, $g \in \mathrm{Map}(\Sigma, T)$. The source for this is that the regularization adopted in [**8**] was only designed to preserve the Torus invariance.

It is straightforward to check that, in general, the absolute value of the determinant is a function of ϕ^2 and indeed agrees with the function $\widehat{A}(\sqrt{\phi^2})$ so that this is invariant under $\mathcal{G}$. Our difficulty, therefore, rests with the phase and we perform a finite renormalization to put in the complete non-Abelian curvature 2-form which re-instates $\mathcal{G}$ invariance.

This is still not quite the end of the story. The one loop correction is not supersymmetric. Or put another way we have not maintained the original supersymmetry (2.3) at the level of the zero modes, which is now,

$$QA = i\psi, \quad Q\psi = -d_A\phi, \quad Q\phi = 0, \tag{6.3}$$

We can add another finite renormalization

$$\exp \frac{c_{\mathbf{g}}}{8\pi^2} \int_\Sigma \mathrm{Tr}\, \psi \wedge \psi$$

to correct this.

PROPOSITION 6.4. The gauge invariant and supersymmetric evaluation of the path integral along the fibres of M is

$$\exp i\Gamma(A, \phi, \psi) \simeq \exp\left(i\frac{\pi}{2}\eta_0\right) \widehat{A}(i\phi) \wedge \exp\left(\frac{c_{\mathbf{g}}}{4\pi^2} \int_\Sigma \mathrm{Tr}\,(i\phi.F_A + \frac{1}{2}\psi \wedge \psi + i\frac{p}{4\pi}\phi^2\omega)\right)$$

REMARK 6.5. With the identifications that $\psi \simeq \Psi$ and $\phi \simeq -i\Phi$ and as $c_{\mathbf{g}} = r$ we have

$$\exp i\Gamma(A, \phi, \psi) \simeq \exp\left(i\frac{\pi}{2}\eta_0\right) \mathrm{Todd}(\mathcal{A}) \wedge \exp\left(-i\frac{p}{2\pi} c_{\mathbf{g}} \Theta(\mathcal{A})\right)$$

The path integral now becomes one over objects defined on the Riemann surface directly and we have established Proposition 1.2. □

7. Wilson lines and parabolic points

How is the picture that we have obtained in the previous sections affected by the inclusion of Wilson lines? Since our manifold is a S^1 fibration there is a special class of knots which are located at point $x \in \Sigma$ on the base of the fibration and which run along the fibre. To such a fibre knot we associate

$$W_R(x) = \mathrm{Tr}_R\, P \exp\left(\int_{S^1} \kappa\phi/2\pi\right) = \mathrm{Tr}_R \exp\left(\phi(x)/2\pi\right)$$

the second equality following from the condition that $\iota_\xi d\phi = 0$. As the only addition to path integral involves functions without dependence on the fibre, the calculation of the previous section goes through unchanged.

A geometric way in which to add such traces is through the introduction of co-adjoint orbits. Let $\lambda \in \mathbf{g}^*$ ($\mathbf{g}^*$ is the dual of $\mathbf{g}$, however, we identify the two so that an invariant inner product $< f, \phi > \equiv \mathrm{Tr}\, f\phi$, $f \in \mathbf{g}^*$, $\phi \in \mathbf{g}$) then the orbit through λ is

$$M_\lambda = \left\{g^{-1}\lambda g;\ \forall g \in G\right\}$$

while the stabilizer of λ is

$$G(\lambda) = \left\{g \in G:\ g^{-1}\lambda g = \lambda\right\}$$

If λ is regular ($\det_{\mathbf{k}}(\mathrm{ad}\,\lambda) \neq 0$) then $G(\lambda) = T$ and we consider this case for now so that $M_\lambda = G/G(\lambda) = G/T$.

The homogeneous space G/T comes equipped with a natural G invariant symplectic 2-form (the Kirillov-Konstant form) Ω_λ given by

$$\Omega_\lambda(X,Y) =< \lambda, [X,Y] >= \mathrm{Tr}\,(\lambda\,[X,Y]) \quad X, Y \in \mathbf{g}$$

Kirillov tells us that for $\lambda = \Lambda + \rho$ regular, Λ an element of the weight lattice and ρ the Weyl vector then

$$\mathrm{Tr}_\lambda\,(\exp\phi/2\pi) = j_{\mathbf{g}}^{-1/2}(\phi/2\pi) \int_{M_\lambda} \exp\left(i\frac{1}{2\pi} < \lambda, \phi > +\Omega_\lambda\right)$$

Now we see that geometrically we should product in the co-adjoint orbits so consider the space $\mathcal{A}_3 \times \prod_i M_{R_i}$, and we have

$$Z_{k,G}[M,(x_i,R_i)] = \frac{1}{\mathrm{Vol}(\mathcal{G}_3)} \int_{\mathcal{A}_3\times\prod_i M_{R_i}} \exp\left(I(\mathscr{A})\right) \prod_{i=1} j_{\mathbf{g}}^{-1/2}(\phi(x_i)/2\pi)\, \exp\omega(M_{R_i})$$

where, in analogy with $\Omega(\mathcal{A})$,

$$\omega(M_{R_i}) = \frac{i}{2\pi}\,\mathrm{Tr}\,\lambda_i\phi(x_i) + \Omega_{R_i}$$

We have the following:

LEMMA 7.1. (Lemma 8.5 [**5**]) The equivariant $\widehat{A}$-genus, $\widehat{A}_{\mathbf{g}}(X, G/T)$, of the Riemannian manifold G/T and $j_{\mathbf{g}}^{-1/2}(X)$ represent the same class in equivariant deRham cohomology.

Consequently Proposition 1.7 is proved. □

REMARK 7.2. C. Beasley [**3**] has computed, in the spirit of [**4**], the localization formula for $Z_{k,G}[M,(x_i,R_i)]|_{\mathfrak{M}}$. This formula agrees with that in Proposition 1.7 when restricted to $\mathfrak{M}$.

References

[1] M. Aganacic, H. Ooguri, N. Saulina, and C. Vafa, *Black holes, q-deformed 2d Yang-Mills and non-perturbative string theory*, **arXiv:hep-th/0411280**.

[2] M. Atiyah and I. Singer, *Dirac operators coupled to vector potentials*, Proc. Nadl. Acad. Sci. USA **81** (1984), 2597–2600.

[3] C. Beasley, presentation at the *Chern-Simons Gauge Theory: 20 years after*, meeting at the Hausdorff Centre, Bonn August (2009) and *Localization For Wilson Loops In Cher-Simons Theory*, **arXiv:0911.2687**.

[4] C. Beasley and E. Witten, *Non-abelian localization for Chern-Simons theory*, J. Differential Geometry **70** (2005), 183–323.

[5] N. Berline, E. Getzler, and M. Vergne, *Heat kernels and Dirac operators*, Grundlehren der Math. Wissen. **298** Springer, Berlin, 1996.

[6] D. Birmingham, M. Blau, M.Rakowski, and G. Thompson, *Topological field theory*, Physics Reports **209** (1991), 129–340.

[7] M. Blau and G. Thompson, *Lecture notes on 2d Gauge theories: Topological aspects and path integral techniques*, ICTP series in theoretical physics-Volume **10** (1993), 175–244. hep-th/9412056.

[8] ———, *Chern-Simons theory on S^1 bundles: Abelianization and q-deformed Yang-Mills theory*, JHEP 0605 (2006) 003, hep-th/0601068.

[9] J-M. Drezet and M. S. Narasimhan, *Groupe de Picard des variétés de modules fibrés semi-stables sur le courbes algébraiques*, Invent. Math **97** (1989), 43–59.

[10] R. Lawrence and L. Rozansky, *Witten-Reshitekhin-Turaev invariants of Seifert manifolds*, Commun. Math. Phys. **205** (1999), 287–314.

[11] M. Mariño, *Chern-Simons theory matrix integrals and perturbative three-manifold invariants*, Commun. Math. Phys. **253** (2004), 25–49.

[12] P. Newstead, *Characteristic classes of stable bundles of rank 2 over an algebraic curve*, Trans. AMS. **169** (1972), 337–345.

[13] M. Thaddeus, *Conformal field theory and the moduli space of stable bundles*, J. Diff. Geom. **35** (1992), 131–149.

[14] E. Verlinde, *Fusion rules and modular transformations in 2d conformal field theory*, Nuclear Phys. **B200** (1988), 360–376.

[15] E. Witten, *Quantum field theory and the Jones polynomial*, Comm. Math. Phys. **121** (1989), 351–399.

[16] ———, *Two dimensional Gauge theories revisited*, J. Geom. Phys. **9** (1992), 303–368.

ICTP, Strada Costiera 11, 34100 Trieste, Italy
E-mail address: thompson@ictp.it

AMS/IP Studies in Advanced Mathematics
Volume 50, 2011

Fermionization and Convergent Perturbation Expansions in Chern-Simons Gauge Theory

Jonathan Weitsman

ABSTRACT. We show that Chern-Simons gauge theory with appropriate cutoffs is equivalent, term by term in perturbation theory, to a Fermionic theory with a nonlocal interaction term. When an additional cutoff is placed on the Fermi fields, this Fermionic theory gives rise to a convergent perturbation expansion.

1. Introduction

Chern-Simons gauge theory was studied by Witten [**9**] as a geometric context for the Jones polynomial using formal path integrals as follows. Let M be a compact three-manifold, and let G be a compact simple Lie group. Choose an invariant inner product on $\mathfrak{g} = Lie(G)$. The space $\Omega^1(M, \mathfrak{g})$ of $\mathfrak{g}$-valued one-forms on M can be identified with the space $\mathcal{A}(M)$ of connections on the trivialized principal G-bundle on $P = M \times G \to M$. In these terms, the Chern-Simons invariant of a connection $A \in \mathcal{A}(M)$ is given by

$$CS(A) = \frac{1}{4\pi}\text{tr} \int_M A \wedge dA + \frac{2}{3}A^3.$$

Given $\lambda \in \mathbb{Z}$, the partition function of the Chern-Simons quantum field theory is given schematically by

$$\int_{\mathcal{A}(M)} \mathcal{D}A e^{-i\lambda CS(A)}, \tag{1.1}$$

where integration on $\mathcal{A}(M)$ is a formal—and mysterious—operation.

The integrand in (1.1) is invariant under the group $\mathcal{G} = \text{Aut }(P) = \text{Map }(M, G)$ of automorphisms of the bundle P. The gauge fixed action was studied by Axelrod and Singer [**1**]. Suppose there exists a flat connection $A_0 \in \mathcal{A}(M)$ such that $H^*(\Omega^*(M, \mathfrak{g}), d_{A_0})$ vanishes.[1] Choose a Riemannian metric on M. Choose also an orthonormal basis e_α for $\mathfrak{g}$, and denote by $f_{\alpha\beta\gamma}$ the corresponding structure constants. The gauge-fixed action is a function of a connection $A \in \ker d^*_{A_0}$ and of two

1991 *Mathematics Subject Classification.* Primary 57R56, 81T13, 81T08.
Supported in part by NSF grant DMS 04/05670.

[1] Here d_{A_0} denotes the de Rham operator in the twisted de Rham complex corresponding to the bundle $M \times \mathfrak{g} = ad(P)$ and the connection A_0.

Fermi fields $c \in \Omega^0(M, \mathfrak{g})$ and $C \in \ker\,(d^*_{A_0}) \subset \Omega^2(M, \mathfrak{g})$. It is given by [**1**]

$$\begin{aligned} S(A,c,C) &= \frac{1}{2\pi}\int_M \sum_\alpha \frac{1}{2}(A_\alpha \wedge (d_{A_0}A)_\alpha) - C_\alpha \wedge (d_{A_0}c)_\alpha \\ &\quad + \frac{1}{6}\sum_{\alpha,\beta,\gamma} f_{\alpha\beta\gamma}(A_\alpha \wedge A_\beta \wedge A_\gamma - 6C_\alpha \wedge A_\beta \wedge c_\gamma), \end{aligned} \tag{1.2}$$

and the gauge fixed partition function is given by

$$Z_\lambda(M) = \int \mathcal{D}A\mathcal{D}c\mathcal{D}Ce^{-i\lambda S(A,c,C)}. \tag{1.3}$$

Now the formal path integral appearing in (1.3) is not in any sense well-defined. However, it does give rise to a perturbation series by a variant of the usual Feynman procedure. Axelrod and Singer show that each of the terms in this series is finite– in other words that the usual divergences appearing in perturbative quantum field theory do not appear in this case. They also show that appropriate combinations of the terms in the perturbation series give rise to topological invariants of the three-manifold M. The methods of [**1**] do not address convergence of the perturbation series, and hence their results do not give a mathematical definition of the path integral. Indeed the general expectation in Bosonic quantum field theories is that the perturbation series has radius of convergence equal to zero.

However, in [**7**] we showed that a cut-off version of Yang-Mills theory in four dimensions is equivalent, term-by-term in perturbation theory, to a Fermionic theory with nonlocal interactions. This Fermionic theory, when given a further cutoff, gives rise to a *convergent* perturbation series. The purpose of the present paper is to show that the methods of [**7**] apply also to Chern-Simons theory. That is, a cut-off version of the action (1.2) is equivalent, term-by-term in perturbation theory, to a theory where the connection A is replaced by a bilinear in Fermion fields (there is obviously no need to Fermionize c and C since they are already Fermions); and a further momentum cutoff placed on the Fermion fields yields a convergent perturbation series. Since our estimates are not uniform in the cutoff, this does not imply convergence of the perturbation series when the cutoff is removed. Nevertheless, our results may hint at possible unexpected analytic structure in the perturbation series.

1.1. The results of Axelrod and Singer. We first describe in some more detail the results of Axelrod and Singer [**1**]; we refer the reader to [**1**] for more information.

Let $\Delta_{A_0} = d^*_{A_0}d_{A_0} + d_{A_0}d^*_{A_0}$ be the Laplacian on $\Omega^*(M, \mathfrak{g})$, and let $L : \Omega^*(M,\mathfrak{g}) \to \Omega^*(M,\mathfrak{g})$ be the operator defined by

$$L = d^*_{A_0}(\Delta_{A_0})^{-1}.$$

Denote the component of $L \circ *$ (where $*$ denotes the Hodge star operator) acting on p–forms by L_p.

If we choose an orthonormal framing of the tangent bundle TM, we may view the kernels of L_0 and L_1 as a smooth functions on $M \times M - \Delta$ with values in $(\mathbb{R}^3 \otimes \mathfrak{g}) \otimes (\mathbb{R}^3 \otimes \mathfrak{g})$; here $\Delta \subset M \times M$ denotes the diagonal. Denote these functions by $L_0(x,y)$ and $L_1(x,y)$ for $x, y \in M$.

Let $\chi \in C^\infty(\mathbb{R})$ satisfy

- $1 \geq \chi \geq 0$.

- $\chi' \geq 0$.
- $\chi(x) = 0$ if $x \leq 1$.
- $\chi(x) = 1$ if $x \geq 2$.

For $\epsilon > 0$ let $\chi_\epsilon(x) := \chi(\epsilon x)$, and for all $x, y \in M \times M - \Delta$, and $i = 0, 1$, let

$$L_i^\epsilon(x,y) := L_i(x,y)\chi_\epsilon(d(x,y)),$$

where $d(x,y)$ denotes the distance between x and y given by the Riemannian metric on M. Then the functions L_i^ϵ extend to smooth functions on $M \times M$, which we continue to denote by L_i^ϵ.

The cut-off perturbation series of the action (1.2) is given by the formal power series

$$Z_{sc}(A_0;\lambda)\sum_{n=0}^{\infty}\frac{1}{\lambda^n}\Xi_n(\epsilon) \tag{1.4}$$

where $Z_{sc}(A_0;\lambda)$ is the semi-classical approximation to the partition function, involving Chern-Simons and torsion invariants of A_0 (see [**9**]), and where

$$\Xi_n(\epsilon) := \left.\frac{R_0{}^{3n}}{(3n)!}\frac{R_I{}^{2n}}{(2n)!}\right|_{A=0,c=0,C=0}, \tag{1.5}$$

and R_0, R_I are defined as follows. In terms of formal even variables $A_\alpha^i(x)$ and formal odd variables $c_\alpha(x), C_\alpha^{i,j}(x)$, $x \in M$, $i,j = 1,2,3$, $i < j$, and $\alpha = 1,\dots,\dim\mathfrak{g}$, the polynomial $R_I(A,c,C)$ is given by

$$R_I(A,c,C) := \frac{-i}{2\pi}\sum_{i,j,k,\alpha,\beta,\gamma}\int_M dx\ \epsilon_{ijk}f_{\alpha\beta\gamma}\Big(\frac{1}{6}A_\alpha^i(x)A_\beta^j(x)A_\gamma^k(x) - A_\alpha^i(x)c_\beta(x)C_\gamma^{j,k}(x)\Big),$$

and the formal differential operator R_0 is given by

$$\begin{aligned} R_0 \quad := \quad -2\pi i \sum_{i,j,\alpha,\beta}\int_{M\times M} dxdy\ \Big((L_1^\epsilon(x,y))_{i,j;\alpha,\beta}\frac{\delta}{\delta A_\alpha^i(x)}\frac{\delta}{\delta A_\beta^j(y)} \\ -2(L_0^\epsilon(x,y))_{i,j;\alpha,\beta}\frac{\delta}{\delta c_\alpha(x)}\frac{\delta}{\delta C_\beta^{i,j}(y)}\Big); \end{aligned}$$

here we have used the notation $(L_k^\epsilon(x,y))_{i,j;\alpha,\beta}, k = 0,1$ for the matrix elements of $L_k^\epsilon(x,y)$ in the basis given by the framing of the tangent bundle and the chosen basis e_α of $\mathfrak{g}$.

Then the key result of Axelrod and Singer is the following

THEOREM 1.1 (Axelrod and Singer [**1**]). *The limit*

$$\Xi_n = \lim_{\epsilon\to 0}\Xi_n(\epsilon)$$

is finite for every n.

Axelrod and Singer then show that the quantities Ξ_n are topological invariants of M. These "finite-type" invariants have been the focus of intensive research since the publication of [**1**].

1.2. Fermionization. We now Fermionize Chern-Simons gauge theory, by a method similar to the one used in [**7**] in the case of Yang-Mills theory. Morally, we replace the connection A^i_α with a bilinear in Fermi fields. Let $H_i(x)$, $i = 1,2,3$, and $\Psi_\alpha(x)$, $\alpha = 1,\ldots,\dim\mathfrak{g}$, be complex Fermi fields. The Fermi action is given by

(1.6)
$$S^\epsilon_F(H_i,\Psi_\alpha,c_\alpha,C^{i,j}_\alpha,\bar{H}_i,\bar{\Psi}_\alpha,\bar{c}_\alpha,\bar{C}^{i,j}_\alpha)$$
$$= S_{F,0}(H_i,\Psi_\alpha,c_\alpha,C^{i,j}_\alpha,\bar{H}_i,\bar{\Psi}_\alpha,\bar{c}_\alpha,\bar{C}^{i,j}_\alpha) + S^\epsilon_{F,I}(H_i,\Psi_\alpha,c_\alpha,C^{i,j}_\alpha,\bar{H}_i,\bar{\Psi}_\alpha,\bar{c}_\alpha,\bar{C}^{i,j}_\alpha)$$

where

(1.7)
$$S_{F,0}(H_i,\Psi_\alpha,c_\alpha,C^{i,j}_\alpha,\bar{H}_i,\bar{\Psi}_\alpha,\bar{c}_\alpha,\bar{C}^{i,j}_\alpha)$$
$$= \int_M dx\Big(\sum_i |H_i(x)|^2 + \sum_\alpha |\Psi_\alpha(x)|^2 + \sum_\alpha |c_\alpha(x)|^2 + \sum_{\alpha,i,j} |C^{i,j}_\alpha(x)|^2\Big),$$

and

$$S^\epsilon_{F,I}(H_i,\Psi_\alpha,c_\alpha,C^{i,j}_\alpha,\bar{H}_i,\bar{\Psi}_\alpha,\bar{c}_\alpha,\bar{C}^{i,j}_\alpha) =$$
$$\frac{i}{2\pi\sqrt{\lambda}}\sum_{i,j,k,\alpha,\beta,\gamma}\int_M dx\ \epsilon_{ijk}f_{\alpha\beta\gamma}\Big(\frac{1}{(3!)^2}\bar{H}_i(x)\bar{\Psi}_\alpha(x)\bar{H}_j(x)\bar{\Psi}_\beta(x)\bar{H}_k(x)\bar{\Psi}_\gamma(x)$$
$$-\bar{H}_i(x)\bar{\Psi}_\alpha(x)\bar{c}_\beta(x)\bar{C}^{j,k}_\gamma(x)\Big)$$
$$-2\pi i\sum_{i,j,\alpha,\beta}\int_{M\times M} dxdy\quad\Big((L^\epsilon_1(x,y))_{i,j;\alpha,\beta}H_i(x)\Psi_\alpha(x)H_j(y)\Psi_\beta(y)$$
$$-2(L^\epsilon_0(x,y))_{i,j;\alpha,\beta}c_\alpha(x)C^{i,j}_\beta(y)\Big).$$

To make further progress, we impose a cutoff on the Fermi fields, as in [**7**]. It is convenient to do this by convolutions with approximate delta functions and step functions, as follows.

Let $\zeta \in C^\infty(\mathbb{R})$ be an even function satisfying

- $\zeta \geq 0$.
- $\int_0^\infty x^2\zeta(x)dx = \frac{1}{4\pi}$.
- $\zeta'(x) \leq 0$ for $x > 0$.
- $\zeta(x) = 0$ for $x \geq 1$.

Given $h > 0$, define $\delta_h : M \times M \to \mathbb{R}$ by

$$\delta_h(x,y) := (2/h)^3\zeta(2d(x,y)/h).$$

Similarly let $Z \in C^\infty(\mathbb{R})$ satisfy

- $Z \geq 0$.
- $Z(x) = 1$ if $x \in [-1,1]$
- $Z'(x) \leq 0$ if $x > 0$.
- $Z(x) = 0$ for $|x| \geq 2$.

For all $h > 0$, define $D_h : M \times M \to \mathbb{R}$ by

$$D_h(x,y) := Z(d(x,y)/2h).$$

We now define the cut-off Fermi fields by

$$\Psi_\alpha^h(x) = \int_M \Psi_\alpha(y)\delta_h(y,x)dy,$$

$$H_i^h(x) = \int_M H_i(y)\delta_h(y,x)dy,$$

$$c_\alpha^h(x) = \int_M c_\alpha(y)\delta_h(y,x)dy,$$

$$C_\alpha^{i,j;h}(x) = \int_M C_\alpha^{i,j}(y)\delta_h(y,x)dy,$$

$$\bar{\Psi}_\alpha^h(x) = \int_M \bar{\Psi}_\alpha(y)\delta_h(y,x)dy,$$

$$\bar{H}_i^h(x) = \int_M \bar{H}_i(y)D_h(y,x)dy,$$

$$\bar{c}_\alpha^h(x) = \int_M \bar{c}_\alpha(y)\delta_h(y,x)dy,$$

$$\bar{C}_\alpha^{i,j;h}(x) = \int_M \bar{C}_\alpha^{i,j}(y)\delta_h(y,x)dy,$$

and define the cut-off Fermi action by

(1.8)

$$S_F^{\epsilon,h}(H_i,\Psi_\alpha,c_\alpha,C_\alpha^{i,j},\bar{H}_i,\bar{\Psi}_\alpha,\bar{c}_\alpha,\bar{C}_\alpha^{i,j})$$
$$= S_{F,0}(H_i,\Psi_\alpha,c_\alpha,C_\alpha^{i,j},\bar{H}_i,\bar{\Psi}_\alpha,\bar{c}_\alpha,\bar{C}_\alpha^{i,j}) + S_{F,I}^\epsilon(H_i^h,\Psi_\alpha^h,c_\alpha^h,C_\alpha^{i,j;h},\bar{H}_i^h,\bar{\Psi}_\alpha^h,\bar{c}_\alpha^h,\bar{C}_\alpha^{i,j;h}).$$

Then the analog of the Fermionization theorem of [**7**] is the following

THEOREM 1.2. *Each term of the perturbation series of the Fermi action $S_F^{\epsilon,h}$ coincides in the limit $h \to 0$ with the corresponding term $\Xi_n(\epsilon)$ of the perturbation series (1.4) of the cut-off, gauge-fixed Chern-Simons gauge theory.*

REMARK 1.9. As in [**7**], it is possible to write down a Fermi theory which gives rise to a perturbation series identical with that of (1.4). However this theory does not arise from a Lagrangian; the free correlation functions for the Fermi fields H_i in this theory are given by

$$< \bar{H}_i(x)H_j(y) >= \delta_{ij} \text{ if } x = y;$$

$$< \bar{H}_i(x)H_j(y) >= 0 \text{ if } x \neq y,$$

as might be expected from the limiting behavior of the correlations of the cut-off fields $\bar{H}_i^h, H_i^h$.

1.3. Convergent perturbation theory. As in other examples of purely Fermionic theories [**3, 4, 7**], we expect the cut-off Fermionic action (1.8) to give rise to a convergent perturbation series. The analog of Theorem 2 of [**7**] is the following.

THEOREM 1.3. *The perturbation series corresponding to the action $S_F^{\epsilon,h}$ converges for all $\lambda \neq 0$.*

REMARK 1.10. In fact it is not necessary to place an additional cut-off on the fields c and C; the action

$$S_{F,0}(H_i, \Psi_\alpha, c_\alpha, C_\alpha^{i,j}, \bar{H}_i, \bar{\Psi}_\alpha, \bar{c}_\alpha, \bar{C}_\alpha^{i,j}) + S_{F,I}^\epsilon(H_i^h, \Psi_\alpha^h, c_\alpha, C_\alpha^{i,j}, \bar{H}_i^h, \bar{\Psi}_\alpha^h, \bar{c}_\alpha, \bar{C}_\alpha^{i,j})$$

also gives rise to a convergent perturbation series.

The convergence estimates used to prove Theorem 1.3 are not uniform in ϵ and h,[2] as is indeed the case for the Gross-Neveu model [**3, 4**] and for Yang-Mills theory [**7**]. In those asymptotically free theories, the coupling constant is adjusted to approach zero as the ultraviolet cutoff is removed, which in our case would correspond to the limit where ϵ and h approach zero. It is this fact which makes non-uniform estimates useful for the construction of the theory in [**3, 4**]. The situation is different for Chern-Simons gauge theory, which is finite to all orders in perturbation theory. Still, Theorem 1.3 may be a hint that that the perturbation series of Chern-Simons gauge theory has unexpected analytical properties. In particular:

CONJECTURE 1.11. *The perturbation series*

$$\sum_{n=0}^{\infty} \frac{1}{\lambda^n} \Xi_n(\epsilon) \tag{1.12}$$

is convergent for all ϵ and all $\lambda \neq 0$.

Evidently a proof of Conjecture 1.11 would depend on finer determinant estimates than the ones used by [**3, 4, 7**] and applied to our case in the proof of Theorem 1.3.

2. Proof of Theorem 1.2

Recall that the terms $\Xi_n(\epsilon)$ of the perturbation series of Chern-Simons gauge theory with cutoff are given by

$$\Xi_n(\epsilon) := \frac{{R_0}^{3n}}{(3n)!} \frac{(R_I)^{2n}}{(2n)!} |_{A=0,c=0,C=0}, \tag{2.1}$$

where R_0 and R_I are given in terms of formal even variables $A_\alpha^i(x)$ and formal odd variables $c_\alpha(x), C_\alpha^{i,j}(x)$ by

---

[2] A slight variation of our methods allows one to take the limit $\epsilon \to 0$ as long as h remains finite; however, the theory remains a theory with cut-offs.

$$R_I^\epsilon(A,c,C) := \frac{-i}{2\pi}\sum_{i,j,k,\alpha,\beta,\gamma}\int_M dx\,\epsilon_{ijk}f_{\alpha\beta\gamma}\Big(\frac{1}{6}A^i_\alpha(x)A^j_\beta(x)A^k_\gamma(x) - A^i_\alpha(x)c_\beta(x)C^{j,k}_\gamma(x)\Big)$$

and

$$R_0 \;:=\; -2\pi i\sum_{i,j,\alpha,\beta}\int_{M\times M} dxdy\,\Big((L_1^\epsilon(x,y))_{i,j,\alpha,\beta}\frac{\delta}{\delta A^i_\alpha(x)}\frac{\delta}{\delta A^j_\beta(y)}$$
$$-2(L_0^\epsilon(x,y))_{i,j,\alpha,\beta}\frac{\delta}{\delta c_\alpha(x)}\frac{\delta}{\delta C^{i,j}_\beta(y)}\Big).$$

Similarly the perturbation series of the Fermionic action $S_F^{\epsilon,h}$ is given by

$$\sum_{n=0}^{\infty}\frac{1}{\lambda^n}\Theta_n(\epsilon,h),$$

where

$$\Theta_n(\epsilon,h) := \int \mathcal{D}H\mathcal{D}\bar{H}\mathcal{D}\Psi\mathcal{D}\bar{\Psi}\mathcal{D}c\mathcal{D}\bar{c}\mathcal{D}C\mathcal{D}\bar{C}e^{S_{F,0}}\frac{(T_0^\epsilon)^{3n}}{(3n)!}\frac{(T_I^h)^{2n}}{(2n)!}, \tag{2.2}$$

and where the polynomials T_I^h and T_0^ϵ are given by

$$T_I^h(H,\Psi,c,C) =$$
$$\frac{i}{2\pi}\sum_{i,j,k,\alpha,\beta,\gamma}\int_M dx\;\epsilon_{ijk}f_{\alpha\beta\gamma}\Big(\frac{1}{(3!)^2}\bar{H}^h_i(x)\bar{\Psi}^h_\alpha(x)\bar{H}^h_j(x)\bar{\Psi}^h_\beta(x)\bar{H}^h_k(x)\bar{\Psi}^h_\gamma(x)$$
$$-\bar{H}^h_i(x)\bar{\Psi}^h_\alpha(x)\bar{c}^h_\beta(x)\bar{C}^{j,k;h}_\gamma(x)\Big)$$

and

$$T_0^\epsilon \;:=\; -2\pi i\sum_{i,j,\alpha,\beta}\int_{M\times M} dxdy\,\Big((L_1^\epsilon(x,y))_{i,j;\alpha,\beta}H^h_i(x)\Psi^h_\alpha(x)H^h_j(y)\Psi^h_\beta(y)$$
$$-2(L_0^\epsilon(x,y))_{i,j;\alpha,\beta}c^h_\alpha(x)C^{i,j;h}_\beta(y)\Big).$$

Thus

(2.3)

$$
\begin{aligned}
\Theta_n(\epsilon,h) = {} & \frac{1}{(2n)!(3n)!}\int \mathcal{D}H\mathcal{D}\bar{H}\mathcal{D}\Psi\mathcal{D}\bar{\Psi}\mathcal{D}c\mathcal{D}\bar{c}\mathcal{D}C\mathcal{D}\bar{C}e^{S_{F,0}} \\
& (-2\pi i)^{3n}\sum_{k=1}^{6n}\sum_{i_k,j_k=1,2,3}\sum_{\alpha_k,\beta_k=1}^{\dim\mathfrak{g}}\int_{M^{6n}}dx_1\ldots dx_{6n}\int_{M^{6n}}dz_1\ldots dz_{6n} \\
& \Big(\prod_{l=1}^{3n}\Big[(L_1^\epsilon(x_{2l},x_{2l-1}))_{i_{2l},i_{2l-1};\alpha_{2l},\alpha_{2l-1}}H^h_{i_{2l}}(x_{2l})\Psi^h_{\alpha_{2l}}(x_{2l})H^h_{i_{2l-1}}(x_{2l-1})\Psi^h_{\alpha_{2l-1}}(x_{2l-1}) \\
& \qquad -2(L_0^\epsilon(x_{2l},x_{2l-1}))_{i_{2l},i_{2l-1};\alpha_{2l},\alpha_{2l-1}}c^h_{\alpha_{2l}}(x_{2l})C^{i_{2l},i_{2l-1};h}_{\alpha_{2l-1}}(x_{2l-1})\Big]\Big) \\
& \Big(\frac{i}{2\pi}\Big)^{2n}\Big(\prod_{m=1}^{2n}\epsilon_{j_{3m}j_{3m-1}j_{3m-2}}f_{\beta_{3m}\beta_{3m-1}\beta_{3m-2}}\delta(z_{3m},z_{3m-1})\delta(z_{3m},z_{3m-2}) \\
& \Big[\frac{1}{(3!)^2}\bar{H}^h_{j_{3m}}(z_{3m})\bar{\Psi}^h_{\beta_{3m}}(z_{3m})\bar{H}^h_{j_{3m-1}}(z_{3m-1})\bar{\Psi}^h_{\beta_{3m-1}}(z_{3m-1})\bar{H}^h_{j_{3m-2}}(z_{3m-2})\bar{\Psi}^h_{\beta_{3m-2}}(z_{3m-2}) \\
& \qquad -\bar{H}^h_{j_{3m}}(z_{3m})\bar{\Psi}^h_{\beta_{3m}}(z_{3m})\bar{c}^h_{\beta_{3m-1}}(z_{3m-1})\bar{C}^{j_{3m-1},j_{3m-2};h}_{\beta_{3m-2}}(z_{3m-2})\Big]\Big).
\end{aligned}
$$

Expanding the products, we have

$$
\begin{aligned}
\Theta_n(\epsilon,h) = {} & \frac{(-2\pi i)^{3n}}{(2n)!(3n)!}\Big(\frac{i}{2\pi}\Big)^{2n}\int \mathcal{D}H\mathcal{D}\bar{H}\mathcal{D}\Psi\mathcal{D}\bar{\Psi}\mathcal{D}c\mathcal{D}\bar{c}\mathcal{D}C\mathcal{D}\bar{C}e^{S_{F,0}} \\
& \sum_{k=1}^{6n}\sum_{i_k,j_k=1,2,3}\sum_{\alpha_k,\beta_k=1}^{\dim\mathfrak{g}}\int_{M^{6n}}dx_1\ldots dx_{6n}\int_{M^{6n}}dz_1\ldots dz_{6n} \\
& \sum_{q=1}^{2n}\binom{2n}{q}\binom{3n}{q+n}\Big(\prod_{l=1}^{q+n}\Big[(L_1^\epsilon(x_{2l},x_{2l-1}))_{i_{2l},i_{2l-1};\alpha_{2l},\alpha_{2l-1}} \\
& \qquad H^h_{i_{2l}}(x_{2l})\Psi^h_{\alpha_{2l}}(x_{2l})H^h_{i_{2l-1}}(x_{2l-1})\Psi^h_{\alpha_{2l-1}}(x_{2l-1})\Big]\Big) \\
& \prod_{l=q+n+1}^{3n}\Big[-2(L_0^\epsilon(x_{2l},x_{2l-1}))_{i_{2l},i_{2l-1};\alpha_{2l},\alpha_{2l-1}}c^h_{\alpha_{2l}}(x_{2l})C^{i_{2l},i_{2l-1};h}_{\alpha_{2l-1}}(x_{2l-1})\Big] \\
& \Big(\prod_{m=1}^{2n}\epsilon_{j_{3m}j_{3m-1}j_{3m-2}}f_{\beta_{3m}\beta_{3m-1}\beta_{3m-2}}\delta(z_{3m},z_{3m-1})\delta(z_{3m},z_{3m-2})\Big) \\
& \Big(\prod_{m=1}^{q}\Big[\frac{1}{(3!)^2}\bar{H}^h_{j_{3m}}(z_{3m})\bar{\Psi}^h_{\beta_{3m}}(z_{3m})\bar{H}^h_{j_{3m-1}}(z_{3m-1})\bar{\Psi}^h_{\beta_{3m-1}}(z_{3m-1}) \\
& \qquad \bar{H}^h_{j_{3m-2}}(z_{3m-2})\bar{\Psi}^h_{\beta_{3m-2}}(z_{3m-2})\Big]\Big) \\
& \Big(\prod_{m=q+1}^{2n}\Big[-\bar{H}^h_{j_{3m}}(z_{3m})\bar{\Psi}^h_{\beta_{3m}}(z_{3m})\bar{c}^h_{\beta_{3m-1}}(z_{3m-1})\bar{C}^{j_{3m-1},j_{3m-2};h}_{\beta_{3m-2}}(z_{3m-2})\Big]\Big).
\end{aligned}
$$

Standard Feynman diagram techniques allow us to write $\Theta_n(\epsilon,h)$ in as a sum of terms corresponding to trivalent graphs. When the ghost lines in such a graph are cut, we obtain a pair of graphs, one of which corresponds to the Ψ and $\bar{\Psi}$ fields and one of which to the H and $\bar{H}$ fields. We use this fact to write

$$\Theta_n(\epsilon,h)=\Theta_n^1(\epsilon,h)+\Theta_n^2(\epsilon,h)$$

where $\Theta_n^1(\epsilon,h)$ is the sum of those terms in the diagrammatic expansion of $\Theta_n(\epsilon,h)$ corresponding to pairs of *identical* Feynman diagrams—that is, Feynman diagrams where the combinatorics of the pairings of the Ψ and $\bar{\Psi}$ fields are the same as those of the H and $\bar{H}$ fields; we will prove that in the limit $h\to 0$ the terms appearing in $\Theta_n^1(\epsilon,h)$ approach the corresponding terms in the diagrammatic expansion of $\Xi_n(\epsilon)$. The sum of the remaining terms in the diagrammatic expansion of $\Theta_n(\epsilon,h)$, which we denote by $\Theta_n^2(\epsilon,h)$, consists of terms corresponding to pairs of Feynman diagrams where at least one H propagator is not matched by a corresponding Ψ propagator. We will see that in the limit $h\to 0$, such an "unmatched" propagator will give rise to a factor of order $O(h^3)$, so that $\lim_{h\to 0}\Theta_n^2(\epsilon,h)=0$. More precisely, we write

$$\begin{aligned}
&\Theta_n^1(\epsilon,h)\\
&=\frac{(-2\pi i)^{3n}}{(2n)!(3n)!}\Big(\frac{i}{2\pi}\Big)^{2n}\sum_{k=1}^{6n}\sum_{i_k,j_k=1,2,3}\sum_{\alpha_k,\beta_k=1}^{\dim\mathfrak{g}}\int_{M^{6n}}dx_1\dots dx_{6n}\int_{M^{6n}}dz_1\dots dz_{6n}\\
&\quad\sum_{q=1}^{2n}\binom{2n}{q}\binom{3n}{q+n}\sum_{\sigma\in S_{2q+2n}}\prod_{l=1}^{q+n}(L_1^\epsilon(x_{2l},x_{2l-1}))_{i_{2l},i_{2l-1};\alpha_{2l},\alpha_{2l-1}}\\
&\quad\prod_{l=q+n+1}^{3n}(-2L_0^\epsilon(x_{2l},x_{2l-1}))_{i_{2l},i_{2l-1};\alpha_{2l},\alpha_{2l-1}}\\
&\quad\Big(\prod_{m=1}^{2n}\epsilon_{j_{3m}j_{3m-1}j_{3m-2}}f_{\beta_{3m}\beta_{3m-1}\beta_{3m-2}}\delta(z_{3m},z_{3m-1})\delta(z_{3m},z_{3m-2})\Big)\\
&\quad\Big(\prod_{m=1}^{q}\Big[\frac{1}{(3!)^3}\int\mathcal{D}H\mathcal{D}\bar{H}\mathcal{D}\Psi\mathcal{D}\bar{\Psi}e^{S_{F,0}}\\
&\qquad H^h_{i_{\sigma(3m)}}(x_{\sigma(3m)})\Psi^h_{\alpha_{\sigma(3m)}}(x_{\sigma(3m)})H^h_{i_{\sigma(3m-1)}}(x_{\sigma(3m-1)})\Psi^h_{\alpha_{\sigma(3m-1)}}(x_{\sigma(3m-1)})\\
&\qquad H^h_{i_{\sigma(3m-2)}}(x_{\sigma(3m-2)})\Psi^h_{\alpha_{\sigma(3m-2)}}(x_{\sigma(3m-2)})\bar{H}^h_{j_{3m}}(z_{3m})\bar{\Psi}^h_{\beta_{3m}}(z_{3m})\\
&\qquad \bar{H}^h_{j_{3m-1}}(z_{3m-1})\bar{\Psi}^h_{\beta_{3m-1}}(z_{3m-1})\bar{H}^h_{j_{3m-2}}(z_{3m-2})\bar{\Psi}^h_{\beta_{3m-2}}(z_{3m-2})\Big]\Big)\\
&\quad\Big(\prod_{m=q+1}^{2n}\Big[\int\mathcal{D}H\mathcal{D}\bar{H}\mathcal{D}\Psi\mathcal{D}\bar{\Psi}e^{S_{F,0}}\\
&\qquad H^h_{i_{\sigma(2q+m)}}(x_{\sigma(2q+m)})\Psi^h_{\alpha_{\sigma(2q+m)}}(x_{\sigma(2q+m)})\bar{H}^h_{j_{3m}}(z_{3m})\bar{\Psi}^h_{\beta_{3m}}(z_{3m})\Big]\Big)\\
&\quad\int\mathcal{D}c\mathcal{D}\bar{c}\mathcal{D}C\mathcal{D}\bar{C}e^{S_{F,0}}\Big(\prod_{l=q+n+1}^{3n}c^h_{\alpha_{2l}}(x_{2l})C^{i_{2l},i_{2l-1};h}_{\alpha_{2l-1}}(x_{2l-1})\Big)\\
&\qquad\Big(\prod_{m=q+1}^{2n}\Big[-\bar{c}^h_{\beta_{3m-1}}(z_{3m-1})\bar{C}^{j_{3m-1},j_{3m-2};h}_{\beta_{3m-2}}(z_{3m-2})\Big]\Big).
\end{aligned}$$

To obtain an explicit expression for $\Theta_n^2(\epsilon,h)$, we proceed as in [**7**]. Given $\sigma,\tau\in S_{2q+2n}$, we say $\sigma\sim\tau$ if for every $m=1,\dots,3q$, there exists $k\in 1,\dots,q$ such that

$$\sigma(m)\tau(m) \in \{3k, 3k+1, 3k+2\} \tag{2.4}$$

and if for every $m = 3q+1, \dots, 2q+2n$, we have

$$\sigma(m) = \tau(m). \tag{2.5}$$

If $\sigma \not\approx \tau$, there exists a smallest integer $m = m(\sigma, \tau)$ such that either (2.4) or (2.5) is false. Then, as in [**7**],

(2.6)

$$\begin{aligned}
&\Theta_n^2(\epsilon, h) = \\
&\quad \frac{(-2\pi i)^{3n}}{(2n)!(3n)!}\Big(\frac{i}{2\pi}\Big)^{2n} \sum_{k=1}^{6n} \sum_{i_k, j_k = 1,2,3} \sum_{\alpha_k, \beta_k = 1}^{\dim \mathfrak{g}} \int_{M^{6n}} dx_1 \dots dx_{6n} \int_{M^{6n}} dz_1 \dots dz_{6n} \\
&\quad \sum_{q=1}^{2n} \binom{2n}{q} \binom{3n}{q+n} \Big(\frac{1}{3!}\Big)^{2q} \sum_{\sigma, \tau \in S_{2q+2n}} \operatorname{sgn}(\sigma, \tau) \\
&\quad \prod_{l=1}^{q+n} (L_1^{\epsilon}(x_{2l}, x_{2l-1}))_{i_{2l}, i_{2l-1}; \alpha_{2l}, \alpha_{2l-1}} \prod_{l=q+n+1}^{3n} (-2L_0^{\epsilon}(x_{2l}, x_{2l-1}))_{i_{2l}, i_{2l-1}; \alpha_{2l}, \alpha_{2l-1}} \\
&\quad \Big(\prod_{m=1}^{2n} \epsilon_{j_{3m} j_{3m-1} j_{3m-2}} f_{\beta_{3m}\beta_{3m-1}\beta_{3m-2}} \delta(z_{3m}, z_{3m-1}) \delta(z_{3m}, z_{3m-2})\Big) \\
&\quad \prod_{l=1}^{2q+2n} \delta_{\alpha_{\sigma(l)}, \beta_l} \delta_{i_{\tau(l)}, j_l} \\
&\quad \Big(\prod_{\substack{l=1 \\ l \neq m(\sigma,\tau), \sigma^{-1}\circ\tau(m(\sigma,\tau))}}^{2q+2n} \tilde{\delta}_h(x_{\sigma(l)}, z_{f_{n,q}(l)}) \tilde{D}_h(x_{\tau(l)}, z_{f_{n,q}(l)}) \Big) \\
&\quad \tilde{\delta}_h(x_{\sigma(m(\sigma,\tau))}, z_{f_{n,q}(m(\sigma,\tau))}) \tilde{D}_h(x_{\tau\circ\sigma^{-1}\circ\tau(m(\sigma,\tau))}, z_{f_{n,q}(\sigma^{-1}\circ\tau(m(\sigma,\tau)))}) \\
&\quad \tilde{D}_h(x_{\tau(m(\sigma,\tau))}, z_{f_{n,q}(m(\sigma,\tau))}) \tilde{\delta}_h(x_{\tau(m(\sigma,\tau))}, z_{f_{n,q}(\sigma^{-1}\circ\tau(m(\sigma,\tau)))}) \\
&\quad \int \mathcal{D}c \mathcal{D}\bar{c} \mathcal{D}C \mathcal{D}\bar{C} e^{S_{F,0}} \Big(\prod_{l=q+n+1}^{3n} c_{\alpha_{2l}}^h(x_{2l}) C_{\alpha_{2l-1}}^{i_{2l}, i_{2l-1}; h}(x_{2l-1}) \Big) \\
&\qquad \Big(\prod_{m=q+1}^{2n} \Big[-\bar{c}_{\beta_{3m-1}}^h(z_{3m-1}) \bar{C}_{\beta_{3m-2}}^{j_{3m-1}, j_{3m-2}; h}(z_{3m-2}) \Big] \Big),
\end{aligned}$$

where, as in [**7**], $\operatorname{sgn}(\sigma, \tau)$ is a sign we will not compute explicitly, where $f_{n,q} : \{1, \dots, 2q+2n\} \to \mathbb{Z}$ is defined by

$$f_{n,q}(l) = \begin{cases} l & \text{if } l \leq 3q \\ 3q + 3(l - 3q) & \text{if } l > 3q, \end{cases} \tag{2.7}$$

and where we have written

$$\tilde{\delta}_h(x, y) := \int_M \delta_h(x, z) \delta_h(z, y) dz$$

and

$$\tilde{D}_h(x,y) := \int_M \delta_h(x,z) D_h(z,y) dz.$$

LEMMA 2.8. *The functions* $\tilde{\delta}_h$ *and* $\tilde{D}_h$ *are positive. Furthermore, for* h *sufficiently small,*

- $\lim_{h\to 0} \tilde{\delta}_h = \delta$ *(as elements of* $\mathcal{D}'(M)$*).*
- $\tilde{\delta}_h \tilde{D}_h = \tilde{\delta}_h$.
- $||\tilde{D}_h(x,\cdot)||_\infty = 1$ *for any* $x \in M$.
- $||\tilde{\delta}_h(x,\cdot)||_1 \leq C$ *for any* $x \in M$*, where* C *is a constant independent of* h *and* x.
- $||\tilde{\delta}_h \star \tilde{D}_h(x,\cdot)||_1 = O(h^3)$ *for any* $x \in M$*,*

where $\tilde{\delta}_h \star \tilde{D}_h$ *is the convolution*

$$\tilde{\delta}_h \star \tilde{D}_h(x,y) := \int_M \tilde{\delta}_h(x,z) \tilde{D}_h(z,y) dz.$$

PROPOSITION 2.9. *We have*

$$\lim_{h\to 0} \Theta_n^2(\epsilon, h) = 0.$$

Proof. By (2.6), and using the fact that M is compact and the L_i^ϵ's are bounded,

(2.10)

$$|\Theta_n^2(\epsilon,h)| \leq C \sup_q \left(\sup_{z\in M} ||\tilde{\delta}_h(\cdot,z)||_1\right)^{6n-1} \left(\sup_{z\in M} ||\tilde{D}_h(\cdot,z)||_\infty\right)^{2q+2n-1} \sup_{z\in M} ||\tilde{\delta}_h \star \tilde{D}_h(\cdot,z)||_1.$$

By Lemma 2.8,

$$|\Theta_n^2(\epsilon,h)| = O(h^3).$$

It remains to show

PROPOSITION 2.11.

$$\lim_{h\to 0} \Theta_n^1(\epsilon,h) = \Xi_n(\epsilon).$$

Proof. We note that

$$\lim_{h\to 0} \int \mathcal{D}H\mathcal{D}\bar{H}\mathcal{D}\Psi\mathcal{D}\bar{\Psi} e^{S_{F,0}} H_i^h(x)\Psi_\alpha^h(x)\bar{H}_j^h(y)\bar{\Psi}_\beta^h(y) = -\frac{\delta}{\delta A_\alpha^i(x)} A_\beta^j(y).$$

Hence

$$
\begin{aligned}
&\lim_{h\to 0}\Theta_n^1(\epsilon,h) = \\
&\quad \frac{(-2\pi i)^{3n}}{(2n)!(3n)!}\Big(\frac{-i}{2\pi}\Big)^{2n}\sum_{k=1}^{6n}\sum_{i_k,j_k=1,2,3}\sum_{\alpha_k,\beta_k=1}^{\dim\,\mathfrak{g}}\int_{M^{6n}}dx_1\dots dx_{6n}\int_{M^{6n}}dz_1\dots dz_{6n} \\
&\quad \sum_{q=1}^{2n}\binom{2n}{q}\binom{3n}{q+n}\sum_{\sigma\in S_{2q+2n}}\prod_{l=1}^{q+n}(L_1^\epsilon(x_{2l},x_{2l-1}))_{i_{2l},i_{2l-1};\alpha_{2l},\alpha_{2l-1}} \\
&\qquad\qquad \prod_{l=q+n+1}^{3n}(-2L_0^\epsilon(x_{2l},x_{2l-1}))_{i_{2l},i_{2l-1};\alpha_{2l},\alpha_{2l-1}} \\
&\quad \Big(\prod_{m=1}^{2n}\epsilon_{j_{3m}j_{3m-1}j_{3m-2}}f_{\beta_{3m}\beta_{3m-1}\beta_{3m-2}}\delta(z_{3m},z_{3m-1})\delta(z_{3m},z_{3m-2})\Big) \\
&\quad \Big(\prod_{m=1}^{q}\Big[\frac{1}{(3!)^2}\frac{\delta}{\delta A^{i_{\sigma(3m)}}_{\alpha_{\sigma(3m)}}(x_{\sigma(3m)})}\frac{\delta}{\delta A^{i_{\sigma(3m-1)}}_{\alpha_{\sigma(3m-1)}}(x_{\sigma(3m-1)})}\frac{\delta}{\delta A^{i_{\sigma(3m-2)}}_{\alpha_{\sigma(3m-2)}}(x_{\sigma(3m-2)})} \\
&\qquad\qquad A^{j_{3m}}_{\beta_{3m}}(z_{3m})A^{j_{3m-1}}_{\beta_{3m-1}}(z_{3m-1})A^{j_{3m-2}}_{\beta_{3m-2}}(z_{3m-2})\Big]\Big) \\
&\quad \Big(\prod_{m=q+1}^{2n}\Big[\frac{\delta}{\delta A^{i_{\sigma(2q+m)}}_{\alpha_{\sigma(2q+m)}}(x_{\sigma(2q+m)})}A^{j_{2q+m}}_{\beta_{2q+m}}(z_{3m})\Big]\Big) \\
&\quad \int \mathcal{D}c\mathcal{D}\bar{c}\mathcal{D}C\mathcal{D}\bar{C}e^{S_{F,0}}\Big(\prod_{l=q+n+1}^{3n}c_{\alpha_{2l}}(x_{2l})C^{i_{2l},i_{2l-1}}_{\alpha_{2l-1}}(x_{2l-1})\Big) \\
&\qquad\qquad \Big(\prod_{m=q+1}^{2n}\Big[-\bar{c}_{\beta_{3m-1}}(z_{3m-1})\bar{C}^{j_{3m-1},j_{3m-2}}_{\beta_{3m-2}}(z_{3m-2})\Big]\Big) \\
&= \Xi_n(\epsilon).
\end{aligned}
$$

3. Proof of Theorem 1.3

Recall the explicit expression for the terms of the perturbation series of the action $S_F^{\epsilon,h}$. This perturbation series is given by

$$\sum \frac{1}{\lambda^n}\Theta_n(\epsilon,h)$$

where (see (2.2) and (2.3))

$$\begin{aligned}
&\Theta_n(\epsilon, h) = \\
&\quad \frac{1}{(2n)!(3n)!}\int \mathcal{D}\Psi\mathcal{D}H\mathcal{D}c\mathcal{D}C\mathcal{D}\bar{\Psi}\mathcal{D}\bar{H}\mathcal{D}\bar{c}\mathcal{D}\bar{C}\exp(S_{F,0}) \\
&\quad \Big(\frac{i}{2\pi\sqrt{\lambda}}\sum_{i,j,k,\alpha,\beta,\gamma}\int_M dx\ \epsilon_{ijk}f_{\alpha\beta\gamma}\Big[\frac{1}{(3!)^2}\bar{H}_i^h(x)\bar{\Psi}_\alpha^h(x)\bar{H}_j^h(x)\bar{\Psi}_\beta^h(x)\bar{H}_k^h(x)\bar{\Psi}_\gamma^h(x) \\
&\qquad\qquad -\bar{C}_\alpha^{i,j;h}(x)\bar{H}_k^h(x)\bar{\Psi}_\beta^h(x)\bar{c}_\gamma^h(x)\Big]\Big)^{2n} \\
&\quad \Big(-2\pi i\sum_{i,j,\alpha,\beta}\int_{M\times M} dxdy\ \Big[(L_1^\epsilon(x,y))_{i,j;\alpha,\beta}H_i^h(x)\Psi_\alpha^h(x)H_j^h(y)\Psi_\beta^h(y) \\
&\qquad\qquad -2(L_0^\epsilon(x,y))_{i,j;\alpha,\beta}c_\alpha^h(x)C_\beta^{i,j;h}(y)\Big]\Big)^{3n}.
\end{aligned}$$

Theorem 1.2 follows from the following estimate.

PROPOSITION 3.1. *There exists a constant $C = C(\epsilon, h) > 0$ such that*

$$|\Theta_n(\epsilon, h)| \leq \frac{C^n}{(2n)!(3n)!}.$$

To prove Proposition 3.1, we note that $\Theta_n(\epsilon, h)$ is a sum of $O(C^n)$ terms, each of which is (up to a constant of order C^n) of the form

(3.2)

$$\begin{aligned}
&\frac{1}{(2n)!(3n)!}\int \mathcal{D}\Psi\mathcal{D}H\mathcal{D}c\mathcal{D}C\mathcal{D}\bar{\Psi}\mathcal{D}\bar{H}\mathcal{D}\bar{c}\mathcal{D}\bar{C}\exp(S_{F,0}) \\
&\int_{M^{6n}} dx_1dy_1\dots dx_{3n}dy_{3n}\int_{M^{2n}} dz_1\dots dz_{2n} \\
&\prod_{l=1}^{\frac{3q+p}{2}}(L_1^\epsilon(x_l,y_l))_{i_l,j_l;\alpha_l,\beta_l}H_{i_l}^h(x_l)\Psi_{\alpha_l}^h(x_l)H_{j_l}^h(y_l)\Psi_{\beta_l}^h(y_l) \\
&\prod_{l=\frac{3q+p}{2}+1}^{3n}(-2L_0^\epsilon(x_l,y_l))_{i_l,j_l;\alpha_l,\beta_l}c_{\alpha_l}^h(x_l)C_{\beta_l}^{i_l,j_l;h}(y_l) \\
&\prod_{m=1}^{q}\bar{H}_{p_m}^h(z_m)\bar{\Psi}_{\theta_m}^h(z_m)\bar{H}_{q_m}^h(z_m)\bar{\Psi}_{\iota_m}^h(z_m)\bar{H}_{r_m}^h(z_m)\bar{\Psi}_{\sigma_m}^h(z_m) \\
&\prod_{m=q+1}^{2n}\bar{H}_{p_m}^h(z_m)\bar{\Psi}_{\theta_m}^h(z_m)\bar{c}_{\iota_m}^h(z_m)\bar{C}_{\sigma_m}^{q_m,r_m;h}(z_m),
\end{aligned}$$

where $p + q = 2n$.

The Berezin integral appearing in (3.2) is

$$
\begin{aligned}
B \; := \; & \int \mathcal{D}\Psi\mathcal{D}H\mathcal{D}c\mathcal{D}C\mathcal{D}\bar{\Psi}\mathcal{D}\bar{H}\mathcal{D}\bar{c}\mathcal{D}\bar{C}\exp(S_{F,0}) \\
& \qquad \int_{M^{6n}} dx_1 dy_1 \ldots dx_{3n} dy_{3n} \int_{M^{2n}} dz_1 \ldots dz_{2n} \\
& \prod_{l=1}^{\frac{3q+p}{2}} H^h_{i_l}(x_l)\Psi^h_{\alpha_l}(x_l)H^h_{j_l}(y_l)\Psi^h_{\beta_l}(y_l) \prod_{l=\frac{3q+p}{2}+1}^{3n} c^h_{\alpha_l}(x_l)C^{i_l,j_l;h}_{\beta_l}(y_l) \\
& \prod_{m=1}^{q} \bar{H}^h_{p_m}(z_m)\bar{\Psi}^h_{\theta_m}(z_m)\bar{H}^h_{q_m}(z_m)\bar{\Psi}^h_{\iota_m}(z_m)\bar{H}^h_{r_m}(z_m)\bar{\Psi}^h_{\sigma_m}(z_m) \\
& \prod_{m=q+1}^{2n} \bar{H}^h_{p_m}(z_m)\bar{\Psi}^h_{\theta_m}(z_m)\bar{c}^h_{\iota_m}(z_m)\bar{C}^{q_m,r_m;h}_{\sigma_m}(z_m).
\end{aligned}
$$

This Berezin integral is the inner product of two elements of

$$\bigwedge^{6q+4p} \left(L_2(M)\otimes\left(\mathbb{R}^3\oplus\mathfrak{g}\oplus\mathfrak{g}\oplus\left(\mathfrak{g}\otimes\mathbb{R}^3\otimes\mathbb{R}^3\right)\right)\right),$$

and is bounded by

$$|B| \leq \left(\sup_{x\in M}||\delta_h(x,\cdot)||_{L_2(M)}\right)^{9q+7p}\left(\sup_{x\in M}||D_h(x,\cdot)||_{L_2(M)}\right)^{3q+p}.$$

Since the kernels L^ϵ_i are smooth, Proposition 3.1 follows.

4. Remarks

4.1. Correlation functions. As in [**7**], the generating function for correlation functions of the gauge fields, which is obtained by adding a term of the form $\int_M \sum_{i,\alpha} J^i_\alpha(x)A^i_\alpha(x)dx$ to the Chern-Simons Lagrangian, can be Fermionized by adding the term $\int_M \sum_{i,\alpha} J^i_\alpha(x)\bar{H}_i(x)\bar{\Psi}_\alpha(x)dx$ to the Fermionized action.

4.2. Yang-Mills and QCD in three dimensions. Our techniques apply just as well to a Lagrangian obtained by adding a Yang-Mills term

$$S(A) = \frac{1}{\lambda^2}\int_M |F(A)|^2$$

to the Chern-Simons Lagrangian. As in [**7**], such a Lagrangian is equivalent to the Lagrangian

$$S(A,F) = \frac{1}{\lambda^2}(||F||^2 + 2i\langle F, dA\rangle + 2i\langle F,[A,A]\rangle) + ikCS(A)$$

where $F \in \Omega^2(M,\mathfrak{g})$ is a conjugate field. The theory then has a cubic interaction term and can be Fermionized by the same method we have used for pure Chern-Simons gauge theory. These ideas work also for pure Yang-Mills theory (with no Chern-Simons term) in three dimensions. The addition of Fermionic matter fields can likewise be accommodated by the same techniques.

Similar techniques should also apply to two-dimensional gauge theories.

4.3. String field theory. I believe that our techniques should also give a Fermionization of Witten's string field theory. Recall that the string field theory Lagrangian is given by

$$S_{\text{sft}}(A) = \int (A * QA + \frac{2}{3} A * A * A),$$

where $A = A(\varphi, b, c)$ is the string field, which is a function of a bosonic field φ and two ghosts b and c, and the operator Q and the operations $\int$ and $*$ are defined in [**10**]. Imposing a gauge condition reduces the quadratic part of S_{sft} (up to a constant) to a positive-definite form. One can then write, as in this paper and in [**7**]

$$A = H(\pi_+\varphi, \pi_+ b, \pi_+ c)\Psi(\pi_-\varphi, \pi_- b, \pi_- c)$$

where H and Ψ are fermionic fields, π_+ is the operator on Fock space induced by the projection $\pi_+ : L_2([0,1]) \to L_2([0, \frac{1}{2}])$, and π_- is the operator induced on Fock space by the projection $\pi_- : L_2([0,1]) \to L_2([\frac{1}{2}, 1])$.[3] There are various technical problems associated with the ghost current anomaly, but I hope that with a proper cutoff (such as that of [**8**]) this theory can also be shown to yield a convergent perturbation series.

References

[1] S. Axelrod and I. M. Singer. J. Differential Geom. 39 (1994), no. 1, 173–213. – Proceedings of the XXth International Conference on Differential Geometric Methods in Theoretical Physics, Vol. 1, 2 (New York, 1991), pp. 3–45, World Sci. Publ., River Edge, NJ, 1992.

[2] D. Bar Natan, Topology, **34** (1995), 423–472 – math.GT/0408182.

[3] J. Feldman, J. Magnen, V. Rivasseau, and R. Seneor, Commun. Math. Phys. **103** (1986), 67–105.

[4] K. Gawedzki and A. Kupiainen, Commun. Math. Phys. **102** (1985), 1–30.

[5] J. Glimm and A. Jaffe, *Quantum Physics*, Springer-Verlag, 1987.

[6] M. Salmhofer, *Renormalization*, Springer-Verlag, 1999.

[7] J. Weitsman, Fermionization, *Convergent perturbation theory, and correlations in the Yang-Mills quantum field theory in four dimensions.* Preprint arxiv:0902.0096.

[8] ______, arXiv:0807.2069; Pure and Applied Mathematics Quarterly (Special Issue: In Honor of Michael Atiyah and Isadore Singer) **6** (2010), 623–642.

[9] E. Witten, *Commun. Math. Phys.* **121** (1988), 359.

[10] ______, *Nucl. Phys.* **B268** (1986), 253.

Department of Mathematics, Northeastern University, Boston, MA 02115

E-mail address: j.weitsman@neu.edu

[3]The addition of a gauge group to the string field by factors attached to the string edges can be accommodated (in the case of $SU(n)$ or $SO(n)$) by taking the fermionic string fields H and Ψ to have values in the fundamental representation of the gauge group.

AMS/IP Studies in Advanced Mathematics
Volume 50, 2011

Analytic Continuation of Chern-Simons Theory

Edward Witten

ABSTRACT. The title of this article refers to analytic continuation of three-dimensional Chern-Simons gauge theory away from integer values of the usual coupling parameter k, to explore questions such as the volume conjecture, or analytic continuation of three-dimensional quantum gravity (to the extent that it can be described by gauge theory) from Lorentzian to Euclidean signature. Such analytic continuation can be carried out by generalizing the usual integration cycle of the Feynman path integral. Morse theory or Picard-Lefschetz theory gives a natural framework for describing the appropriate integration cycles. An important part of the analysis involves flow equations that turn out to have a surprising four-dimensional symmetry. After developing a general framework, we describe some specific examples (involving the trefoil and figure-eight knots in S^3). We also find that the space of possible integration cycles for Chern-Simons theory can be interpreted as the "physical Hilbert space" of a twisted version of $\mathcal{N} = 4$ super Yang-Mills theory in four dimensions.

1. Introduction

This paper is devoted to the analytic continuation of three-dimensional Chern-Simons gauge theory as a function of its coupling parameters. One motivation for this study concerns three-dimensional quantum gravity, which plausibly has at least some sort of relationship to three-dimensional Chern-Simons theory with gauge group $SL(2, \mathbb{C})$. In that context, Euclidean quantum gravity is obtained by analytic continuation to imaginary values of the coupling parameter s that is introduced in section 2.[1] However, for imaginary s, the Chern-Simons path integral, like the Euclidean quantum gravity path integral in any dimension, does not appear to be convergent. Our analysis in this paper will explain how to make sense of this sort of path integral, at least in the Chern-Simons case.

Another motivation is to understand aspects of the Jones polynomial of knots[2] and its generalizations that have not yet been explained from the point of view

2010 *Mathematics Subject Classification.* Primary 70S15; Secondary 81T13.

[1]And the singularity at $s = \pm i\ell$ that we will discuss in section 2.4 is at least a cousin of the "chiral point" of three-dimensional gravity, which has been much studied recently [**1**].

[2]The usual mathematical terminology is that a knot is an embedded circle in a three-manifold (usually a three-sphere) while a union of disjoint embedded circles is called a link. Except in section 4.2.7, all statements about knots in this paper have rather immediate analogs for links, but for brevity we will refer to knots.

of three-dimensional gauge theory. The Jones polynomial of a knot was originally defined as a Laurent polynomial in a complex variable q – a rational function of q with poles only at $q = 0$ and $q = \infty$. Chern-Simons gauge theory with gauge group $SU(2)$ (or another compact Lie group) reproduces the Jones polynomial, and its generalizations, at values of q of the form $\exp(2\pi i/(k+h))$, where k is a positive integer, and h (which is 2 for $SU(2)$) is the dual Coxeter number of the gauge group. The Jones polynomial and related invariants are determined by their values for these special values of q, but it would be desirable to have a gauge theory explanation of the existence of a natural analytic continuation. It has generally been expected that the explanation has something to do with the relation between Chern-Simons theories with gauge groups $SU(2)$ or $SL(2,\mathbb{C})$, but a really clear picture has not been available. A better understanding appears to require analytic continuation of Chern-Simons theory beyond the values of the parameters at which the path integral can be most simply understood.

Finally, in the last decade or more, striking new results have been obtained [**2**–**21**] about the analytic continuation of invariants related to the Jones polynomial. (A useful review is [**17**]; partial physics-based explanations with some relevance to the present paper are [**6**, **18**, **19**]. While it is impractical to give complete references on Chern-Simons theory, some contributions comparing the asymptotic or in some cases exact behavior of the Chern-Simons partition function to geometry, but not focusing on analytic continuation, are [**22**–**34**].) The colored Jones polynomial $J_n(q)$ of a knot is interpreted in Chern-Simons gauge theory in terms of the expectation value of a Wilson loop operator associated to the n-dimensional representation of $SU(2)$. Let $q^n = \exp(2\pi i\gamma)$, and consider the behavior of $J_n(q)$ in a scaling limit with n tending to infinity for fixed γ. The behavior of this limit for certain values of γ can be naturally understood in $SU(2)$ Chern-Simons theory, but many of the more recent results certainly appear to be more directly related to the theory with gauge group $SL(2,\mathbb{C})$. For example, for γ near 1, and at least for some knots, the large n behavior appears to be dominated by an $SL(2,\mathbb{C})$-valued flat connection that does not take values in $SU(2)$. Although there has not been a completely clear picture of how and why the scaling limit of $J_n(q)$ should be related to $SL(2,\mathbb{C})$ Chern-Simons theory, the idea that there is such a relationship has been a fruitful source of new ideas and generalizations [**6**]. Again, it has been clear that a fuller picture will require a better understanding of the analytic continuation of Chern-Simons theory.

In this paper, we will definitely not obtain complete answers to all of the above questions, and among other things we do not have a general answer concerning the volume conjecture. (It is true, as will become clear, if a certain integer-valued coefficient is always nonzero, but we do not know why this would always be so.) But we do hope to make some things clearer. We show that critical points of the analytically continued Chern-Simons functional should be used to determine – by steepest descent – suitable integration cycles for the quantum path integral. For a given value of the coupling parameters, not all critical points should be included; doing so would lead to immediate contradiction with known facts about the knot polynomials. Instead, one should start with a real integration cycle for real coupling and then, as one varies the coupling into the complex plane, one varies the integration cycle so that the integral remains convergent. In the process, one encounters subtle Stokes phenomena that govern the appearance and disappearance of critical points.

The Stokes phenomena can be described, in principle, by solving steepest descent equations that in the context of Chern-Simons theory amount to four-dimensional elliptic differential equations that generalize the instanton equation. This gives a suitable framework for analytically continuing the $SL(2,\mathbb{C})$ Chern-Simons theory as a function of its coupling parameter, and for better understanding its relation to the $SU(2)$ theory.

Section 2 of the paper is devoted to a more complete overview of some of these matters.

In section 3, we describe the analytic continuation of oscillatory integrals in finite dimensions and their analysis by steepest descent. This is an application of the framework of [**35**–**40**], which is reviewed here for convenience. Various aspects of this framework have been applied to quantum field theory from different points of view [**41**–**45**]. Floer theory (for reviews see for instance [**46, 47**]) is a prototype for the application of Morse theory to middle-dimensional cohomology of function spaces, which appears in our approach to Chern-Simons gauge theory.

In section 4, we apply what we have learned to the analytic continuation of Chern-Simons gauge theory on a generic three-manifold. A striking result emerges: the gradient flow equations relevant to analytic continuation of three-dimensional Chern-Simons theory actually have four-dimensional symmetry.

In section 5, we consider the special case of knot invariants in S^3, where stronger conclusions are possible because, in the absence of the knot, there are no non-trivial flat connections. The main examples considered in section 5 are the trefoil knot and the figure-eight knot. In each case, we consider the semi-classical limit of the colored Jones polynomial, keeping fixed the conjugacy class of the monodromy of a flat connection around the knot. (In other words, we consider the limit mentioned above of $n \to \infty$ with fixed γ.) The technique that we use to describe the Stokes phenomena relies on analyzing the singularities of the moduli space of flat connections – that is, the special values of the monodromy at which flat connections that are generically distinct become coincident. The semiclassical behavior near such a special value is interesting in itself; it can be modeled by a finite-dimensional integral associated to the relevant singularity. We analyze several examples of such singularities and compare the results to what has been found in the mathematical literature in studies of the semiclassical limit of the colored Jones polynomial. For the trefoil knot, though many interesting questions remain, the Stokes phenomena that we analyze suffice to determine the integration cycle of the analytically continued Chern-Simons path integral for all values of γ. For the figure-eight knot, we primarily determine what happens close to the real γ axis. In both cases, our analysis of the Stokes phenomena suffices to explain the validity of the volume conjecture.

Finally, in section 6, we make a few remarks on a four or five-dimensional interpretation of what we have learned, with the possible goal of a new understanding of Khovanov homology.

2. Overview

We begin by considering Chern-Simons gauge theory with the gauge group G a complex Lie group such as $SL(2,\mathbb{C})$. Let $\mathcal{A}$ be a connection on a G-bundle E over a three-manifold M. Such a connection has a complex-valued Chern-Simons invariant

$$W(\mathcal{A}) = \frac{1}{4\pi}\int_M \mathrm{Tr}\,\left(\mathcal{A}\wedge \mathrm{d}\mathcal{A} + \frac{2}{3}\mathcal{A}\wedge\mathcal{A}\wedge\mathcal{A}\right), \tag{2.1}$$

which we have normalized to be gauge-invariant modulo 2π. (For $G = SL(N,\mathbb{C})$, the symbol Tr denotes the trace in the N-dimensional representation. In general, it denotes a suitably normalized quadratic form on the Lie algebra $\mathfrak{g}$ of G.) Just as in the case of a compact group, $W(\mathcal{A})$ is gauge-invariant modulo 2π. The indeterminacy in $W(\mathcal{A})$ is real, and is the same as it would be if G were replaced by a maximal compact subgroup H, because G is contractible onto H.[3]

Thus the imaginary part of $W(\mathcal{A})$ is actually a well-defined real number, while its real part takes values in $\mathbb{R}/2\pi\mathbb{Z}$. To obtain a quantum field theory, we wish to start with a classical action I that is a linear combination of $\mathrm{Re}\,W$ and $\mathrm{Im}\,W$. The quantum theory is based upon integrating the expression $\exp(iI)$, and for this function to be well-defined, I must be defined mod 2π. Because of the indeterminacy in $\mathrm{Re}\,W$, its coefficient must be an integer, while (if the only desired condition is to ensure that $\exp(iI)$ is a well-defined complex-valued function) the coefficient of $\mathrm{Im}\,W$ may be an arbitrary complex number. The action therefore has the general form

$$I = -s\,\mathrm{Im}\,W + \ell\,\mathrm{Re}\,W,\ \ s\in\mathbb{C},\ \ell\in\mathbb{Z}. \tag{2.2}$$

Alternatively, we can write

$$\begin{aligned} I = \frac{tW}{2} + \frac{\widetilde{t}\,\overline{W}}{2} \ &= \ \frac{t}{8\pi}\int_M \mathrm{Tr}\,\left(\mathcal{A}\wedge \mathrm{d}\mathcal{A} + \frac{2}{3}\mathcal{A}\wedge\mathcal{A}\wedge\mathcal{A}\right) \\ &\quad + \frac{\widetilde{t}}{8\pi}\int_M \mathrm{Tr}\,\left(\overline{\mathcal{A}}\wedge \mathrm{d}\overline{\mathcal{A}} + \frac{2}{3}\overline{\mathcal{A}}\wedge\overline{\mathcal{A}}\wedge\overline{\mathcal{A}}\right), \end{aligned} \tag{2.3}$$

with

$$t = \ell + is,\ \widetilde{t} = \ell - is. \tag{2.4}$$

There is more to life, however, than making sure that $\exp(iI)$ is a well-defined complex-valued function. We also want to integrate this function over the space $\mathcal{Y}$ of gauge fields modulo gauge transformations. This makes sense straightforwardly only if s is real, for then I is real and the integrand $\exp(iI)$ of the path integral is bounded. The path integral

$$\int_{\mathcal{Y}} D\mathcal{A}\,D\overline{\mathcal{A}}\,\exp(iI) \tag{2.5}$$

is then an infinite-dimensional oscillatory integral. In this respect, it is like any Lorentz signature Feynman integral (though here we are doing topological field theory rather than quantum field theory on a manifold of Lorentz signature). If s is not real, the path integral is not convergent; our task is to make sense of it anyway.

---

[3]The indeterminacy in $W(\mathcal{A})$ comes from its change in going around a loop in the space of connections modulo gauge transformations, and this change only depends on the homotopy class of the loop. Since G is contractible to H, any family of $\mathfrak{g}$-valued connections on M can be deformed to a family of $\mathfrak{h}$-valued connections, where $\mathfrak{g}$, $\mathfrak{h}$ are the Lie algebras of G and H. So the indeterminacy in the Chern-Simons function for gauge group G reduces to that for gauge group H.

2.1. Finite-dimensional analogs. A much simpler prototype of this problem is a finite-dimensional oscillatory integral such as the one that defines the Airy function:

$$Z_\lambda = \int_{-\infty}^{\infty} \mathrm{d}x \, \exp\left(i\lambda(x^3/3 - x)\right). \tag{2.6}$$

For real λ, this integral converges, though not absolutely so, because of rapid oscillations at infinity; when λ becomes complex, it must be defined by a process of analytic continuation.

The Airy integral will serve as a useful practice case in section 3. The following slightly more elaborate example is also useful. Introduce a complex variable z and a complex-valued polynomial

$$g(z) = \sum_{j=0}^{n} a_j z^j. \tag{2.7}$$

Now consider the integral

$$Z_g = \int |\mathrm{d}^2 z| \exp\left(g(z) - \overline{g(z)}\right). \tag{2.8}$$

This again is a convergent oscillatory integral and a closer analog of complex Chern-Simons theory, with $-ig(z)$ and $i\overline{g(z)}$ corresponding to the terms tW and $\widetilde{t}\,\overline{W}$ in the action (2.3).

We stress that no contour integral is intended in (2.8) – a contour integral could scarcely be intended here as the integrand is not holomorphic! Rather, if $z = u+iv$, with real u, v, we integrate separately over u and v, the integration measure being $|\mathrm{d}^2 z| = 2\,\mathrm{d}u\,\mathrm{d}v$.

A finite-dimensional analog of analytically continuing Chern-Simons theory with complex gauge group as a function of s is to analytically continue the Airy function Z_λ to complex λ. An even closer analog is to analytically continue the integral Z_g to let $\overline{g}$ be independent of g. By this, we mean the following. Define a new polynomial

$$\widetilde{g}(z) = \sum_{j=0}^{n} \widetilde{a}_j z^j, \tag{2.9}$$

and generalize the integral Z_g to

$$Z_{g,\widetilde{g}} = \int |\mathrm{d}^2 z| \exp\left(g(z) - \widetilde{g}(\overline{z})\right). \tag{2.10}$$

Then $Z_{g,\widetilde{g}}$ coincides with the original Z_g if $\widetilde{g} = \overline{g}$ (by which we mean that $\widetilde{a}_j = \overline{a}_j$ for all j).

The natural way to analytically continue the Airy function in λ is to let x become a complex variable and to deform the integration contour away from the real axis so that the integral remains convergent as λ varies. Similarly, to extend the definition of $Z_{g,\widetilde{g}}$ to $\widetilde{g} \neq \overline{g}$, we analytically continue, treating z and $\overline{z}$ – or equivalently u and v, defined by $z = u + iv$, $\overline{z} = u - iv$ – as independent complex variables. Denoting the independent complex variables as z and $\widetilde{z}$ (rather than z and $\overline{z}$), the analytically continued integral is

$$Z_{g,\widetilde{g}} = \int_{\mathcal{C}} \mathrm{d}z \, \mathrm{d}\widetilde{z} \exp\left(g(z) - \widetilde{g}(\widetilde{z})\right). \tag{2.11}$$

The integral is over a two-dimensional real integration cycle $\mathcal{C}$, which is the real slice $\widetilde{z} = \overline{z}$ if $\widetilde{g} = \overline{g}$, and in general must be deformed as g and $\widetilde{g}$ vary so that the integral remains convergent.

This then is the framework that we will have to imitate to understand the analytic continuation of Chern-Simons theory with a complex gauge group. Just as we promoted $\overline{z}$ to a complex variable $\widetilde{z}$ that is independent of z, we will have to promote $\overline{\mathcal{A}}$ to a new G-valued connection $\widetilde{\mathcal{A}}$ that is independent of $\mathcal{A}$. Thus we consider the classical theory with independent G-valued connections $\mathcal{A}$ and $\widetilde{\mathcal{A}}$ (we recall that G is a complex Lie group such as $SL(2,\mathbb{C})$) and action

$$
\begin{aligned}
I(\mathcal{A},\widetilde{\mathcal{A}}) &= \frac{t}{8\pi}\int_M \mathrm{Tr}\left(\mathcal{A}\wedge \mathrm{d}\mathcal{A}+\frac{2}{3}\mathcal{A}\wedge\mathcal{A}\wedge\mathcal{A}\right) \\
&\quad +\frac{\widetilde{t}}{8\pi}\int_M \mathrm{Tr}\left(\widetilde{\mathcal{A}}\wedge \mathrm{d}\widetilde{\mathcal{A}}+\frac{2}{3}\widetilde{\mathcal{A}}\wedge\widetilde{\mathcal{A}}\wedge\widetilde{\mathcal{A}}\right).
\end{aligned}
\tag{2.12}
$$

Then we have to find an appropriate integration cycle $\mathcal{C}$ in the path integral

$$
\int_{\mathcal{C}} D\mathcal{A}\, D\widetilde{\mathcal{A}}\, \exp\left(iI(\mathcal{A},\widetilde{\mathcal{A}})\right). \tag{2.13}
$$

The cycle $\mathcal{C}$ must be equivalent to $\widetilde{\mathcal{A}} = \overline{\mathcal{A}}$ if s is real, and in general must be obtained by deforming that one as s varies so that the integral remains convergent. (We also have to learn how to deal with gauge invariance when $\mathcal{A}$ and $\widetilde{\mathcal{A}}$ are treated as independent.)

2.2. Role of Morse theory. Finding a suitable integration cycle in the infinite-dimensional path integral of Chern-Simons gauge theory may seem hopeless at first sight. What saves the day is that suitable integration contours can be found by Morse theory and steepest descent (or in a different language, via Picard-Lefschetz theory). The facts we need for understanding oscillatory integrals have been described in [**35**–**40**] (and the aspiration of applying this approach to Feynman integrals is stated in [**37**], p. 321). Section 3 of the present paper is devoted to a hopefully self-contained exposition of this approach, with an emphasis on the aspects that we need in Chern-Simons theory. The basic idea of the theory is to associate potential integration cycles to components of the critical point set of the extended action $I(\mathcal{A},\widetilde{\mathcal{A}})$.

What makes it practical to use this formalism in Chern-Simons theory is that the critical points are accessible. As long as t and $\widetilde{t}$ are both non-zero, the critical point equation derived from the analytically continued action (2.12) is

$$
\mathcal{F} = \widetilde{\mathcal{F}} = 0, \tag{2.14}
$$

where $\mathcal{F} = \mathrm{d}\mathcal{A}+\mathcal{A}\wedge\mathcal{A}$ and $\widetilde{\mathcal{F}} = \mathrm{d}\widetilde{\mathcal{A}}+\widetilde{\mathcal{A}}\wedge\widetilde{\mathcal{A}}$ are the curvatures of the two connections. Hence a critical point is just a pair of flat connections, corresponding to a pair of homomorphisms from the fundamental group of M to the complex Lie group G. (We postpone to section 4 an important detail about the value of the Chern-Simons invariant at a critical point.) Hence in Chern-Simons theory on a given three-manifold M, the critical points can be effectively described; this is typically not true for oscillatory Feynman integrals.

For simplicity, we assume until section 3.3.2 that the critical points are isolated points. Let Σ be the set of critical points and denote the critical points as p_σ, $\sigma\in\Sigma$. Then to each p_σ, we attach an integration cycle $\mathcal{J}_\sigma$ that is obtained by downward or gradient flow from σ, taking the real part of the action as a Morse function. (Details are explained in section 3.) The $\mathcal{J}_\sigma$, which are known as Lefschetz thimbles, are constructed so that the integral over $\mathcal{J}_\sigma$ always converges, regardless of the choice of σ or of the parameters in the action. Any suitable integration cycle $\mathcal{C}$ for the sort of integral that we are interested in is equivalent to a linear combination of the $\mathcal{J}_\sigma$ with integer coefficients:

$$\mathcal{C} = \sum_\sigma \mathfrak{n}_\sigma \mathcal{J}_\sigma, \ \mathfrak{n}_\sigma \in \mathbb{Z}. \tag{2.15}$$

The coefficients $\mathfrak{n}_\sigma$ can themselves be computed via Morse theory. To do this, one attaches to each critical point p_σ another cycle $\mathcal{K}_\sigma$ by upwards gradient flow starting from p_σ. The expansion coefficients in eqn. (2.15) can be computed as the intersection pairing of the integration cycle $\mathcal{C}$ with the $\mathcal{K}_\sigma$:

$$\mathfrak{n}_\sigma = \langle \mathcal{C}, \mathcal{K}_\sigma \rangle. \tag{2.16}$$

One might naively think that all critical points would contribute to the integral with $\mathfrak{n}_\sigma = 1$ (or with $\mathfrak{n}_\sigma = \pm 1$, since in general the $\mathcal{J}_\sigma$ do not have natural orientations), but this cannot be so. For example, in Chern-Simons theory with real s, there are complex critical points at which the action I has a negative imaginary part; their inclusion would lead to a pathologically wrong exponential growth of the partition function for large real s. In fact, for real s, with $\mathcal{C}$ taken to be the real cycle $\mathcal{C}_\mathbb{R}$ defined by $\widetilde{\mathcal{A}} = \overline{\mathcal{A}}$, real critical points (whose contributions to the path integral are oscillatory) all have $\mathfrak{n}_\sigma = 1$, but critical points that would make exponentially large contributions to the path integral have $\mathfrak{n}_\sigma = 0$. These statements will be deduced from (2.16). However, there is no simple formula for $\mathfrak{n}_\sigma$ for critical points whose contributions to the real s path integral are exponentially small.

2.3. Stokes phenomena. Once one expresses the integration cycle $\mathcal{C}$ in terms of critical point cycles for real values of the Chern-Simons coupling constant s (or of the appropriate parameters in the other examples) via a formula $\mathcal{C} = \sum_\sigma \mathfrak{n}_\sigma \mathcal{J}_\sigma$, the next step is to make an analytic continuation in s. There is no problem in defining the critical points p_σ and the corresponding Lefschetz thimbles $\mathcal{J}_\sigma$ for any generic value of s. In fact, in Chern-Simons theory, the critical point equation (2.14) is independent of s (except at $s = \pm i\ell$, where t or $\widetilde{t}$ vanishes). The $\mathcal{J}_\sigma$ do depend on s (their definition by steepest descent makes use of the action, which depends on s), but their definition ensures that the path integral over $\mathcal{J}_\sigma$ is always convergent, even when s has an imaginary part.

One can find the right coefficients in the expansion $\mathcal{C} = \sum_\sigma \mathfrak{n}_\sigma \mathcal{J}_\sigma$ by starting at real s, where one knows what to do, and then continuously evolving in s. Here, however, one runs into the main subtlety of the theory: Stokes phenomena. As one varies s, one crosses Stokes curves at which both the Lefschetz thimbles $\mathcal{J}_\sigma$ and the coefficients $\mathfrak{n}_\sigma$ jump, in such a way that the homology cycle $\sum_\sigma \mathfrak{n}_\sigma \mathcal{J}_\sigma$ is locally constant. Such jumping can occur when there are paths of steepest descent that connect two critical points. The flow by steepest descent has a conserved quantity (the imaginary part of the holomorphic function whose real part is taken as the Morse function), and a flow between two critical points is possible only when this quantity takes equal values at the two critical points. This is reviewed in section 3.

After taking the Stokes phenomena into account, one can obtain a recipe for how the $\mathfrak{n}_\sigma$ evolve as s varies in the complex plane. This then gives, finally, a way to define the analytic continuation in s of the Chern-Simons path integral for the case that the gauge group is a complex Lie group.

2.4. Singularities of the analytically continued theory. At this stage, it is very illuminating to ask what sort of singularities the analytically continued theory has as a function of s. Generically, the analytic continuation of an oscillatory integral grows exponentially as the appropriate complex parameter goes to infinity in certain directions in the complex plane. The appropriate parameters are s for complex Chern-Simons theory, λ for the Airy function, and the coefficients of the polynomials g and $\widetilde{g}$ in our other example.

Because of this exponential growth, the analytically continued Chern-Simons path integral has an essential singularity at $s = \infty$. Generically, there is in addition an ordinary monodromy around $s = \infty$; it results from the fact that, by virtue of the Stokes phenomena, the integration contour $\mathcal{C}$ does not return to itself under continuation around $s = \infty$.

Apart from an essential singularity at infinity, will analytically continued Chern-Simons theory have other singularities as a function of s? The answer to this question is that one should expect singularities at $s = \pm i\ell$, where the action (2.3) becomes holomorphic or antiholomorphic. An oversimplified prototype of this singularity is the Gaussian integral

$$\int |dz \wedge d\overline{z}| \exp(az^2 + b\overline{z}^2) = \frac{\pi}{\sqrt{-ab}}, \tag{2.17}$$

with singularities at $a = 0$ or $b = 0$. (The integral is defined by analytic continuation starting from an oscillatory Gaussian integral at $b = -\overline{a}$.)

This example is oversimplified because the analytically continued exponent, which is $az^2 + b\widetilde{z}^2$, has only a unique critical point, the one at $z = \widetilde{z} = 0$. In Chern-Simons theory, there are in general many critical points p_σ. In general there are distinct critical points that have the same action at $s = \pm i\ell$, and as a result many Stokes curves pass through the points $s = \pm i\ell$. Hence the singularities at $s = \pm i\ell$ will be much more complicated than in the Gaussian example. These singularities will not be analyzed in this paper, but it will not be surprising if the analytic continuation of Chern-Simons theory in general should be defined on the universal cover of the complex s-plane, with the points at $s = \pm i\ell$ (and ∞) omitted. Thus the partition function would be really a function not of s but of a certain dilogarithm of s.

2.5. Analytic continuation for compact gauge groups. Now we will discuss the analytic continuation of Chern-Simons theory with a compact gauge group. We consider it first in a naive way and then in relation to the theory with a complex gauge group.

In Chern-Simons gauge theory on a three-manifold M with a compact gauge group H, the gauge field is a connection A on an H-bundle $E \to M$. In terms of the Chern-Simons function $W(A)$, which we normalize as usual to be well-defined modulo $2\pi\mathbb{Z}$, the action is $I = kW$, where k is taken to be an integer so that the integrand $\exp(iI) = \exp(ikW)$ of the path integral is well-defined. Then one defines the path integral as the integral over the space[4] $\mathcal{U}$ of connections mod gauge

---

[4] A variant in which one takes $\mathcal{U}$ to be the quotient of the space of connections by the group of based gauge transformations is also useful. (A based gauge transformation is one that equals 1 at a prescribed point; the advantage of dividing only by this group is that it acts freely, so

transformations:

$$Z_H(k) = \int_{\mathcal{U}} DA \, \exp(ikW(A)). \tag{2.18}$$

In the presence of a knot K labeled by a representation R, one includes a holonomy factor and considers the integral:

$$Z_H(K;k) = \int_{\mathcal{U}} DA \, \exp(ikW(A)) \, \mathrm{Tr}_R P \exp\left(-\oint_K A\right). \tag{2.19}$$

As written, the integral does not make sense unless k is an integer. Indeed, as $W(A)$ is defined only modulo 2π, the expression $\exp(ikW(A))$ makes sense as a function on $\mathcal{U}$ if and only if $k \in \mathbb{Z}$. Can we evade this limitation? A naive way to try to do so is to replace $\mathcal{U}$ by its universal cover[5] $\widehat{\mathcal{U}}$. $\widehat{\mathcal{U}}$ is the quotient of the space of connections by the group of topologically trivial gauge transformations. The expression $\exp(ikW(A))$ is invariant under such gauge transformations, so it makes sense as a function on $\widehat{\mathcal{U}}$. However, its integral vanishes for $k \notin \mathbb{Z}$, since the integrand transforms with a non-trivial phase under the deck transformations associated to the cover $\widehat{\mathcal{U}} \to \mathcal{U}$.

To cure this problem, we complexify the gauge field, replacing A by a complex-valued connection $\mathcal{A}$. $\mathcal{A}$ is now a connection on an $H_{\mathbb{C}}$-bundle $E_{\mathbb{C}} \to M$. We replace $\mathcal{U}$ and $\widehat{\mathcal{U}}$ by their complexifications $\mathcal{U}_{\mathbb{C}}$ and $\widehat{\mathcal{U}}_{\mathbb{C}}$, which respectively parametrize complex-valued connections modulo complex-valued gauge transformations or modulo topologically trivial complex-valued gauge transformations. Finally, in $\widehat{\mathcal{U}}_{\mathbb{C}}$, we pick an integration cycle $\mathcal{C}$ and define the integral

$$Z_{H,\mathcal{C}}(k) = \int_{\mathcal{C}} DA \, \exp(ikW(\mathcal{A})). \tag{2.20}$$

(When a knot K is present, we denote the corresponding integral as $Z_{H,\mathcal{C}}(K;k)$.) The analytic continuation of the path integral measure DA on $\widehat{\mathcal{U}}$ is a holomorphic volume form on $\widehat{\mathcal{U}}_{\mathbb{C}}$ that we continue to denote simply as DA. In section 2.7.1, it will be expressed in terms of the holomorphic volume form $D\mathcal{A}$ used in the complex path integral (2.13).

We will pick $\mathcal{C}$ so that the integral (2.20) converges even when k becomes a complex number. It will be constructed from Lefschetz thimbles in the usual way,

$$\mathcal{C} = \sum_{\sigma} \mathfrak{n}_\sigma \mathcal{J}_\sigma. \tag{2.21}$$

Here the p_σ correspond to flat connections $\mathcal{A}_\sigma$ with structure group $H_{\mathbb{C}}$, each with a chosen lift to $\widehat{\mathcal{U}}_{\mathbb{C}}$. (The sum in (2.21) will generally be a finite sum, but see the

dividing by it produces no singularities.) In this variant, one must divide the path integral by a constant factor, the volume of the residual group of gauge transformations. Another important variant arises in working out the contribution to the path integral from a given gauge orbit of flat connections on M; here one may take $\mathcal{U}$ to be a transversal to the given orbit in the space of all connections on E. When we speak loosely of $\mathcal{U}$ as the space of connections mod gauge transformations, we refer to any of these variants that may be most convenient in a given context.

[5]To be more precise, we can take $\widehat{\mathcal{U}}$ to be the smallest cover of $\mathcal{U}$ on which the Chern-Simons function $W(A)$ makes sense as a real-valued function. If H is connected and simply-connected, this is the same as the universal cover.

end of section 5.2.4 for an exception.) The analytically continued path integral will therefore be

$$Z_H(k) = \sum_\sigma \mathfrak{n}_\sigma Z_{H,\sigma}(k) \tag{2.22}$$

with

$$Z_{H,\sigma}(k) = \int_{\mathcal{J}_\sigma} D\mathcal{A}\, \exp(ikW(\mathcal{A})). \tag{2.23}$$

A finite sum (2.21) is not invariant under the deck transformations of the cover $\widehat{\mathcal{U}}_{\mathbb{C}} \to \widehat{\mathcal{U}}$, so it is not possible to use those deck transformations to show that $Z_{H,\mathcal{C}}(K,k) = 0$ for $k \notin \mathbb{Z}$. (Invariance under deck transformations would require including every lift of a given $\mathcal{A}_\sigma$.) As a cycle in the appropriate homology, $\mathcal{C}$ will be locally constant in k (Stokes jumping of cycles $\mathcal{J}_\sigma$ is compensated by jumping of coefficients $\mathfrak{n}_\sigma$), ensuring that $Z_{H,\mathcal{C}}(K,k)$ will be holomorphic in k.

2.5.1. *Downward flow again.* To get an analytic continuation of the original Chern-Simons integral (2.18), $\mathcal{C}$ should obey one more condition: it should be chosen so that $Z_{H,\mathcal{C}}(k) = Z_H(k)$ for integer k.

In a finite dimensional problem of this type, a suitable $\mathcal{C}$ can be found as follows. (For a prototype, see the review in section 3.5 of the analytic continuation of the Bessel function.) First, lift the real cycle $\mathcal{U} \subset \mathcal{U}_{\mathbb{C}}$ to a non-closed chain $\mathcal{C}_0 \subset \widehat{\mathcal{U}}_{\mathbb{C}}$ as follows. Pick a real number α and define a point in $\mathcal{C}_0$ to be a real connection A whose Chern-Simons function takes values in the closed interval $[\alpha, \alpha+2\pi]$. (We are using the fact that on $\widehat{\mathcal{U}}_{\mathbb{C}}$, the Chern-Simons function is an ordinary complex-valued function.) We have to use a closed interval here or $\mathcal{C}_0$ would not be closed. Though closed, $\mathcal{C}_0$ is not a cycle because it has a non-zero boundary $\partial\mathcal{C}_0$ – consisting of real connections with Chern-Simons function α or $\alpha + 2\pi$. Of course, $\partial\mathcal{C}_0$ is a union of two disjoint pieces, which we will loosely call components (we do not need to know if they are connected), corresponding to those two values of the Chern-Simons function.

We want to complete $\mathcal{C}_0$ to a candidate integration cycle $\mathcal{C} \subset \widehat{\mathcal{U}}_{\mathbb{C}}$. For this, we add to $\mathcal{C}_0$ an additional piece $\mathcal{C}_1$ defined as follows. We define $\mathcal{C}_1$ by downward Morse flow from the boundary $\partial\mathcal{C}_0$, with respect to the usual Morse function $h = \mathrm{Re}\,(ikW(\mathcal{A}))$. Assume first that downward flow from $\partial\mathcal{C}_0$ leads to no critical points. Then $\mathcal{C}_1$ is closed and, if we orient correctly its two components, then $\mathcal{C} = \mathcal{C}_0 \cup \mathcal{C}_1$ has no boundary and is the integration cycle we want. (In the prototype of fig. 7 in section 3.5, the two components of $\mathcal{C}_1$ correspond to the two half-lines that start at $w = 0$ and $w = 2\pi i$.)

$\mathcal{C}_1$ fails to be closed only if a downward flow from $\partial\mathcal{C}_0$ approaches a critical point for $t \to +\infty$. (This critical point is then in the closure of $\mathcal{C}_1$, but not in $\mathcal{C}_1$ itself.) This possibility is severely limited because the downward flow equations have a conserved quantity, which for Chern-Simons theory is $\mathrm{Im}\,(ikW(\mathcal{A}))$, while in addition, $\mathrm{Re}(ikW(\mathcal{A}))$ decreases along a flow. So $\mathcal{C}_1$ may fail to be closed only if there is a critical point at which $W(\mathcal{A})$ has positive imaginary part and real part α or $\alpha + 2\pi$. For a generic choice of α, there is no such critical point and the above definition works and gives a suitable cycle $\mathcal{C}$.

Now consider the integral

$$\int_{\mathcal{C}} D\mathcal{A} \exp(ikW(\mathcal{A})) = \int_{\mathcal{C}_0} D\mathcal{A}\, \exp(ikW(\mathcal{A})) + \int_{\mathcal{C}_1} D\mathcal{A}\, \exp(ikW(\mathcal{A})). \tag{2.24}$$

Here $\mathcal{C}_1$ consists of two components (downward flow from real connections with $W(\mathcal{A})$ equal to α or $\alpha + 2\pi$) that differ from each other by a deck transformation that shifts $W(A)$ by 2π. For $k \in \mathbb{Z}$, the integrand $\exp(ikW(\mathcal{A}))$ is invariant under this deck transformation, but on the other hand, the two components of $\mathcal{C}_1$ are oppositely oriented (that is, the deck transformation maps the orientation of one to minus the orientation of the other). So for $k \in \mathbb{Z}$, we have $\int_{\mathcal{C}_1} DA \exp(ikW(\mathcal{A})) = 0$ and $\int_{\mathcal{C}} DA \exp(ikW(\mathcal{A})) = \int_{\mathcal{C}_0} DA \exp(ikW(\mathcal{A}))$. But for $k \in \mathbb{Z}$, the function $\exp(ikW(\mathcal{A}))$ is a pullback from the original space $\mathcal{U}_{\mathbb{C}}$ of complex-valued connections modulo gauge transformations, and the cycle $\mathcal{C}_0$ pushes down in $\mathcal{U}_{\mathbb{C}}$ to the original real integration cycle $\mathcal{U}$. In short, $\int_{\mathcal{C}} DA \exp(ikW(\mathcal{A}))$ is equivalent to the original Chern-Simons path integral when k is an integer. We have accomplished our goal of expressing the Chern-Simons path integral $Z_H(k)$ with a compact gauge group in a form that makes sense when k is not an integer.

$\mathcal{C}$ is an unfamiliar sort of integration cycle for a path integral, but actually, it is straightforward in concept to use (2.16) to re-express $\mathcal{C}$ in terms of Lefschetz thimbles. We just count the upward flows from a given critical point $\mathcal{A}_\sigma$ to $\mathcal{C}$. If k is real, the flow equations conserve $\mathrm{Im}\,(iW(\mathcal{A}))$, so, given the condition that we have imposed on α, $\mathfrak{n}_\sigma$ is nonzero only for critical points for which $\mathrm{Im}\,(iW(\mathcal{A}))$ takes values in the open interval $(\alpha, \alpha + 2\pi)$.

2.5.2. *Indeterminacy of $\mathcal{C}$.* The only problem with the analytic continuation that we have obtained for the Chern-Simons path integral is that it is not unique.

The choice of $\mathcal{C}$ and therefore the analytic continuation of the Chern-Simons path integral $Z_H(k)$ will in general jump in crossing the exceptional values of α at which $\mathcal{C}_1$ fails to be closed. This is related to a very basic fact. The contribution of any given critical point p_σ to the path integral can be multiplied by $\exp(2\pi imk)$, for any integer m, without changing its value for $k \in \mathbb{Z}$. This multiplicative factor has a simple interpretation: it results from shifting the Chern-Simons invariant of the flat connection $\mathcal{A}_\sigma$ by $2\pi m$. So, without more input, we cannot expect analytic continuation in k to be unique.

In a situation (such as that of section 5) in which the Chern-Simons path integral depends on another parameter γ in addition to k, it may not be possible to make our almost canonical choice of $\mathcal{C}$ uniformly in γ. The reason is that it may not be possible to choose α so that it varies continuously with γ and never coincides with the value of $\mathrm{Im}\,(iW(\mathcal{A}))$ for any flat connection $\mathcal{A}$. If we choose $\mathcal{C}$ as above for some range of γ, then upon varying γ and letting $\mathcal{C}$ evolve continuously, we will arrive at a more general $\mathcal{C}$ and hence a more general form of the analytic continuation.

A rather general $\mathcal{C}$ would entail multiplying the contribution to the path integral from each representation of the fundamental group by a finite sum

$$\sum_{m \in \mathbb{Z}} a_m \exp(2\pi imk), \tag{2.25}$$

where the integers a_m are almost all zero and $\sum_m a_m = 1$. In general, the a_m may be different for different representations of the fundamental group. A function such as this will naturally arise in an example in section 5.2.3.

2.5.3. *Knots.* There is an important case in which there is a completely natural choice of the analytic continuation. This is the case or knots or links in S^3. As is known from any of their standard characterizations, the usual Wilson loop invariants for knots or links can be analytically continued to a Laurent polynomial in $q = \exp(2\pi i/(k+h))$ and with this analytic continuation, they have no essential singularity at $k = \infty$. Since the function $\exp(2\pi ik)$ does have an essential singularity at $k = \infty$, the cycle $\mathcal{C}$ used in analytically continuing the knot invariants is unique if one wants the analytically continued function to be free of an essential singularity at infinity.

Perhaps we should spell out a detail concerning the last paragraph. The Jones polynomial and its generalizations are usually normalized to equal 1 for the unknot. From the point of view of quantum field theory, this definition amounts to defining the Jones polynomial of a given knot K by analytic continuation of a ratio $Z_H(K;k)/Z_H(K_0;k)$, where $Z_H(K;k)$ was defined in (2.19) and K_0 is the unknot. The ratio is certainly 1 if $K = K_0$. It is such ratios that analytically continue to Laurent polynomials in q. The numerator and denominator, when taken separately, have an analytic continuations in k with no essential singularity at $k = \infty$, but they are not Laurent polynomials in q. For example, for $H = SU(2)$, the denominator is $Z_H(K_0;k) = \sqrt{2/(k+2)}\sin(\pi/(k+2))$, which is not a Laurent polynomial in q, but has no essential singularity at $k = \infty$. The theory developed in the present paper applies more naturally to a path integral than to a ratio of path integrals, and therefore we will simply study the functions $Z_H(K;k)$ rather than ratios of such functions.

2.6. Comparing compact and complex Lie groups. If we do not require that the generalized integral $Z_{H,\mathcal{C}}(k)$ should agree with the original Chern-Simons integral $Z_H(k)$ when k is an integer, we have much more freedom in the choice of the cycle $\mathcal{C}$. We can take $\mathcal{C}$ to be any integer linear combination of Lefschetz thimbles

$$\mathcal{C} = \sum_{\sigma\in\Sigma} \mathfrak{n}_\sigma \mathcal{J}_\sigma, \tag{2.26}$$

with coefficients $\mathfrak{n}_\sigma$ that jump in crossing Stokes lines to ensure that the homology cycle $\mathcal{C}$ varies smoothly.

The functions $Z_{H,\mathcal{C}}(k)$ of eqn. (2.20), as $\mathcal{C}$ runs over the possible choices, form a vector space $\mathcal{V}$. (This space is actually infinite-dimensional when one allows for the choices of lifts to $\mathbb{C}$ of Chern-Simons invariants of critical points.) What are these functions good for, apart from analytic continuation of Chern-Simons theory with gauge group H? They play a role somewhat analogous to conformal blocks in two-dimensional conformal field theory. Let us compare Chern-Simons theory of the compact group H to Chern-Simons theory of the complex Lie group $G = H_\mathbb{C}$. For reasons that will become clear, we consider first the case of a three-manifold M without knots.

As we have explained, the analytic continuation of the complex Chern-Simons theory as a function of its coupling parameters is made using the same machinery as for a compact Lie group. However, the complexification of a complex Lie group G is isomorphic to the product $G\times G$, and accordingly, the analytic continuation of Chern-Simons theory of gauge group G is described by a pair of complex connections $\mathcal{A}, \widetilde{\mathcal{A}}$. This assertion was part of the derivation of eqn. (2.12), where we also showed that the analytically continued action is the sum of a Chern-Simons action for $\mathcal{A}$ and a Chern-Simons action for $\widetilde{\mathcal{A}}$. We can thus build an integration cycle for the analytically continued G-valued theory by taking the product of two integration cycles for the analytically continued H-valued theory. Concretely, we can consider an integration cycle for the analytically continued G-valued theory of the form $\mathcal{J}_\sigma \times \mathcal{J}_\tau$, where $\mathcal{J}_\sigma$ and $\mathcal{J}_\tau$ are Lefschetz thimbles for the analytically continued H-valued theory. This gives a basis for the space of appropriate cycles, and hence the partition function $Z_G(s,\ell)$ of the G-valued theory can be expanded as a bilinear form in the functions $Z_{H,\sigma}(k)$ of eqn. (2.23):

$$Z_G(s,\ell) = \sum_{\sigma,\tau} \mathfrak{n}_{\sigma,\tau} Z_{H,\sigma}(k_1) Z_{H,\tau}(k_2). \tag{2.27}$$

The coefficients $\mathfrak{n}_{\sigma,\tau}$ jump in the usual way in crossing Stokes curves.

From the formulas (2.3) and (2.4) for the analytically continued action, one might expect the coupling parameters to be $k_1 = t/2 = (\ell + is)/2$, $k_2 = \widetilde{t}/2 = (\ell - is)/2$. Actually there is a shift by the dual Coxeter number h of H and one has

$$k_1 = t/2 - h, \;\; k_2 = \widetilde{t}/2 - h. \tag{2.28}$$

This shift enters in comparing the integration measures $D\mathcal{A}$ and DA used in defining Z_G and Z_H, or, differently put, it reflects the fact that in Chern-Simons theory with a compact gauge group H, a one-loop correction shifts the effective value of k in many formulas to $k+h$, while as shown in [**48**], there is no such shift for a complex Lie group. For more on this, see section 2.7.1.

2.6.1. *Knots again.* It is instructive to consider what happens to this formalism if we specialize to knots in S^3, again in the theory with gauge group $G = H_{\mathbb{C}}$. In the presence of a knot K, we must include in the integrand of the path integral an extra factor that is the holonomy of the connection $\mathcal{A}$ around K, in some representation R. What happens depends very much on what sort of representation we pick. For example, one might pick an irreducible unitary representation of G; these of course are infinite-dimensional. In this case, the simplification that will be explained momentarily does not arise.

We will consider here a finite-dimensional and irreducible but non-unitary representation that is related to an irreducible representation of the compact group H. The action of H on an irreducible representation R can be extended to an action of $G = H_{\mathbb{C}}$ in two ways; in one case the matrix elements of the representations are holomorphic functions on G and in the other case they are antiholomorphic. We will consider the case of a holomorphic representation. Given a knot K and a G-valued connection $\mathcal{A}$, holomorphy means after analytic continuation that the trace of the holonomy around K is a function only of $\mathcal{A}$ and not of $\widetilde{\mathcal{A}}$; as usual we denote this trace as $\mathrm{Tr}_R P \exp\left(-\oint_K \mathcal{A}\right)$. Now consider the path integral:

$$Z_G(K;t,\widetilde{t}) = \int_{\mathcal{C}} D\mathcal{A}\, D\widetilde{\mathcal{A}} \exp\left(itW(\mathcal{A})/2 + i\widetilde{t}W(\widetilde{\mathcal{A}})/2\right) \mathrm{Tr}_R P \exp\left(-\oint_K \mathcal{A}\right). \tag{2.29}$$

We want to express this integral in terms of Lefschetz thimbles. So consider the equations for a critical point. (We will here proceed somewhat informally; the details of how to include the holonomy factor in the Morse theory framework are explained in section 4.2.) Since the holonomy factor only depends on $\mathcal{A}$, the equation for $\widetilde{\mathcal{A}}$ is unaffected by the presence of the knot. This equation says simply that the curvature of $\widetilde{\mathcal{A}}$ vanishes, and as S^3 is simply-connected, this means that up to a gauge transformation we can take $\widetilde{\mathcal{A}} = 0$.

Consequently, there is essentially only one possible integration cycle for $\widetilde{\mathcal{A}}$; it is the Lefschetz thimble $\widetilde{\mathcal{Q}}$ associated to the critical point at $\widetilde{\mathcal{A}} = 0$. Hence, the integration cycle $\mathcal{C}$ is simply the product of $\widetilde{\mathcal{Q}}$ with some integration cycle $\mathcal{Q}$ for $\mathcal{A}$. Since the integrand of the path integral is also a product, it follows that the path integral is the product of an integral over $\mathcal{A}$ and an integral over $\widetilde{\mathcal{A}}$ (rather than a sum of such products, as in the more typical case (2.27)). Moreover, the integral over $\widetilde{\mathcal{A}}$ does not depend on the knot K. So the general form of $Z_G(K;t,\widetilde{t})$ is

$$Z_G(K;t,\widetilde{t}) = \widetilde{B}(\widetilde{t})B(K;t), \tag{2.30}$$

where only the second factor depends on the knot.

What will be $B(K,t)$? Rather like the knot invariant $Z_H(K;k)$ for the compact group H, $B(K,t)$ is defined by an integral of the form

$$\int_{\mathcal{Q}} D\mathcal{A}\, \exp(i(t/2)W(\mathcal{A}))\, \mathrm{Tr}_R P \exp\left(-\oint_K \mathcal{A}\right) \tag{2.31}$$

for some integration cycle $\mathcal{Q} \subset \widehat{\mathcal{U}}_{\mathbb{C}}$. What integration cycle shall we use?

In perturbation theory, $Z_G(K;t,\widetilde{t})$ is equal to $Z_H(K;k)$ (with $k = t/2 - h$, and apart from a factor $\widetilde{B}$ that depends only on $\widetilde{t}$ and not on K). This relation was presented as exercise 6.32 in [**49**]. As a perturbative statement, it simply reflects the fact that we started with a holomorphic representation whose holonomy depends on $\mathcal{A}$ and not $\widetilde{\mathcal{A}}$. When we compute the expectation value of $\mathrm{Tr}_R\, P \exp\left(-\oint_K \mathcal{A}\right)$ in perturbation theory, it does not matter if $\widetilde{\mathcal{A}}$ is absent (the case of a compact gauge group H) or present but decoupled from $\mathcal{A}$ (the case of the complex gauge group G, after analytic continuation to decouple $\overline{\mathcal{A}}$ from $\mathcal{A}$).

This relation between the G and H theories will hold as an exact statement, not just an asymptotic one, if we simply take $\mathcal{Q}$, the integration cycle used in constructing the theory with complex gauge group G, to be the same integration cycle that was used in analytically continuing the theory with compact gauge group H. This is a satisfying result, but it is perhaps best understood as a definition. The original path integral (2.29) is not convergent, even for the usual real values of the coupling parameters s and ℓ, if regarded as an integral over the usual real cycle $\widetilde{\mathcal{A}} = \overline{\mathcal{A}}$. The reason for this is that the holonomy factor $\mathrm{Tr}_R P \exp\left(-\oint_K \mathcal{A}\right)$ is exponentially growing in some directions. So the choice of $\mathcal{Q}$ that reduces the knot invariants for G to those of H can be regarded as a definition of the knot invariants for a holomorphic representation of G, rather than something we deduce from a more primitive framework. By contrast, in the case of a unitary representation, the holonomy can be expressed as an oscillatory integral by adding more integration variables; this is explained in section 4.2. After doing so, the integral over the usual real cycle makes sense.

2.7. Other real forms. Let H' be a noncompact real form of the complex Lie group $G = H_{\mathbb{C}}$. (For example, if $H = SU(2)$, H' might be $SL(2,\mathbb{R})$.) We sometimes describe this more briefly by calling H' a noncompact real form of H. Chern-Simons theory with gauge group H' is formally defined by the familiar sort of integral

$$Z_{H'}(k') = \int_{\mathcal{U}'} DA'\, \exp(ik'W(A')), \tag{2.32}$$

where now A' is a connection on an H' bundle $E' \to M$ and $\mathcal{U}'$ is the space of such connections, up to gauge transformation.

Since the complexification of H' is simply $G = H_{\mathbb{C}}$, the complexification of $\mathcal{U}'$ is the familiar space $\mathcal{U}_{\mathbb{C}}$ of complex-valued connections. Consequently, $\mathcal{U}'$ can be viewed as an integration cycle in $\mathcal{U}_{\mathbb{C}}$ on which the usual Chern-Simons integral is an (oscillatory) convergent integral. In this respect, it is no different from $\mathcal{U}$, the space of connections modulo gauge transformations for the compact Lie group H. Either $\mathcal{U}$ or $\mathcal{U}'$ can be regarded as a possible integration cycle in $\mathcal{U}_{\mathbb{C}}$ for the Chern-Simons path integral. ($\mathcal{U}$ and $\mathcal{U}'$ are both fixed points of real involutions acting on $\mathcal{U}_{\mathbb{C}}$, as explained more fully in section 2.7.1.) As such, each can be expressed as a linear combination of the usual Lefschetz thimbles $\mathcal{J}_\sigma$.

Hence, the framework for analytic continuation of $Z_{H'}(k')$ is the same as the framework for analytic continuation of $Z_H(k)$. We seek to expand it as a linear combination with integer coefficients of the functions $Z_{H,\sigma}(k)$ of eqn. (2.23):

$$Z_{H'}(k') = \sum_\sigma \mathfrak{n}'_\sigma Z_{H,\sigma}(k). \tag{2.33}$$

(The coefficients $\mathfrak{n}'_\sigma$ will have to jump in the usual way in crossing Stokes lines.) However, there is a subtlety in the relation between the parameters in $Z_{H'}(k')$ and $Z_H(k)$. This relation is not $k' = k$ but

$$k' + h' = k + h, \tag{2.34}$$

where h' is a constant[6] that has been computed in different ways [**48, 50**]. Like (2.28), which can actually be viewed as a special case, (2.34) arises from a comparison between the complexification of the integration measures used in defining Z_H and $Z_{H'}$. It is further discussed in section 2.7.1.

Just as in the discussion of analytic continuation of $Z_H(k)$, we wish to pick a linear combination of critical point cycles so that the formula (2.33) will agree with $Z_{H'}(k')$ for $k' \in \mathbb{Z}$. The general analysis of this is similar to what it is for the case of compact gauge group, but again there is some simplification for the case of a knot in S^3. If we label our knot by a finite-dimensional irreducible representation of H' (as opposed to a unitary representation of H', which would be infinite-dimensional), then it is natural to pick the same integration cycle that we used in analytically continuing $Z_H(K, k)$. With the identification of parameters $k' + h' = k + h$, this leads to $Z_{H'}(K, k') = Z_H(K, k)$. That relationship is already known in perturbation theory, which is insensitive to the difference between H and H' (except for a one-loop shift in the parameters [**48**]). That fact motivates our choice of cycle.

Our discussion in section 2.6.1 can be viewed as a special case of this, since $G = H_{\mathbb{C}}$ is a noncompact real form of the compact Lie group $H \times H$. The properties of Chern-Simons theory with compact gauge group that we have used do not depend on whether the gauge group is simple, so they hold for $H \times H$.

2.7.1. *Comparison of holomorphic volume forms.* We will now briefly provide a context for eqn. (2.34).

[6] Decompose the Lie algebra of H' as $\mathfrak{h}' = \mathbf{k} \oplus \mathfrak{p}$, where $\mathbf{k}$ is the Lie algebra of a maximal compact subgroup of H', and $\mathfrak{p}$ is its orthocomplement. After defining the dual Coxeter number h as the trace in $\mathfrak{h}'$ of the square of a suitable element of $\mathfrak{h}'$, write $h = h_+ + h_-$, where h_+ and h_- come from traces in $\mathbf{k}$ and $\mathfrak{p}$, respectively. Then $h' = h_+ - h_-$.

The complex Lie group $G = H_{\mathbb{C}}$ admits two real involutions that are important here: one leaves fixed H and one leaves fixed H'. Similarly, writing $\mathfrak{h}_{\mathbb{C}}$, $\mathfrak{h}$, and $\mathfrak{h}'$ for the Lie algebras of $H_{\mathbb{C}}$, H, and H', the space $\mathcal{U}_{\mathbb{C}}$ of $\mathfrak{h}_{\mathbb{C}}$-valued connections modulo gauge transformation admits two relevant real involutions: one leaves fixed the space $\mathcal{U}$ of $\mathfrak{h}$-valued connections modulo gauge transformations, and one leaves fixed the space $\mathcal{U}'$ of $\mathfrak{h}'$-valued connections modulo gauge transformations.

Accordingly, the complexification of either $\mathcal{U}$ or $\mathcal{U}'$ gives the same space $\mathcal{U}_{\mathbb{C}}$. Now let DA and DA' be the path integral measures for the theories with gauge groups H and H', respectively. Either one of these can be analytically continued to a holomorphic volume form on $\mathcal{U}_{\mathbb{C}}$; we denote these holomorphic forms simply as DA and DA'. Our discussion of analytic continuation has concerned the integration of these volume forms over various real cycles in $\mathcal{U}_{\mathbb{C}}$ (or more precisely the integration of their pullbacks to the universal cover $\widehat{\mathcal{U}}_{\mathbb{C}}$ over cycles in $\widehat{\mathcal{U}}_{\mathbb{C}}$).

DA and DA' are holomorphic volume forms on the same space, so their ratio is an invertible holomorphic function of the complexified connection $\mathcal{A}$. The relationship is

$$DA = DA' \exp(i(h - h')W(\mathcal{A})). \tag{2.35}$$

This relationship is computed in [**48**] by comparing η-invariants for the theories with gauge group H and H'. A different language is used there and in most of the literature. In this language, a one-loop correction (coming from the η-invariant) shifts the effective Chern-Simons coupling of the H theory by h and that of the H' theory by h'. The ratio of the two statements is (2.35).

The relation (2.34) is an immediate consequence of (2.35). Similarly (2.28) reflects the fact that the ratio between the holomorphic volume forms on $\mathcal{U}_{\mathbb{C}}$ that we have called DA and $D\mathcal{A}$ is

$$DA = D\mathcal{A} \exp(ihW(\mathcal{A})). \tag{2.36}$$

Alternatively, (2.28) is the special case of (2.34) for $G = H_{\mathbb{C}}$ regarded as a noncompact real form of the compact Lie group $H \times H$.

A finite-dimensional analog of the problem of comparing such holomorphic volume forms would involve a complex manifold X with two different real structures. For example, X could be the complex x-plane, with two real structures determined by the involutions $x \to \overline{x}$ and $x \to -\overline{x}$. If we set $x = u + iv$, the fixed set of the first involution is the u axis $\mathcal{U}$, with real volume form $\mathrm{d}u$, and the fixed point of the second is the v axis $\mathcal{U}'$, with real volume form $\mathrm{d}v$. On the other hand, on X we have the natural holomorphic one-form $\mathrm{d}x = \mathrm{d}u + i\mathrm{d}v$. The analytic continuation of $\mathrm{d}u$ from $\mathcal{U}$ to X is $\mathrm{d}x$, and the analytic continuation of $\mathrm{d}v$ is $-i\mathrm{d}x$. So the relation between the two analytically continued volume forms is $\mathrm{d}v = -i\mathrm{d}u$. In the context of the infinite-dimensional Chern-Simons path integral, such a comparison between volume forms leads to an η invariant, as explained in [**48, 51**], and this η invariant can be expressed in terms of the Chern-Simons function.

3. Finite-dimensional oscillatory integrals, Morse theory, and steepest descent

The goal of the present section is to describe in more detail the finite-dimensional case of the theory that was sketched in section 2.

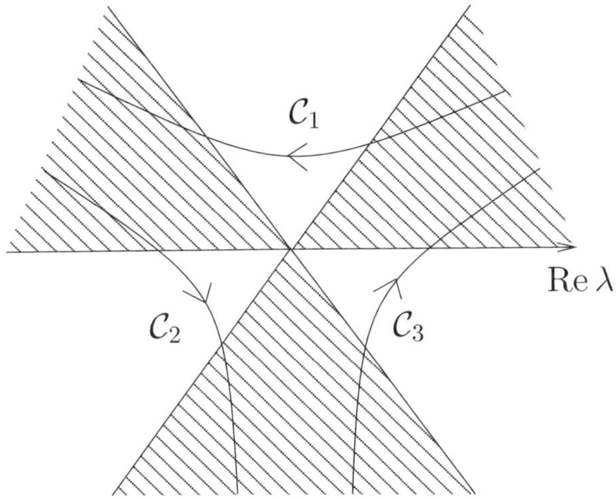

FIGURE 1. The complex λ plane, with the real axis from left to right. The hatched regions are the three "good" regions in which the Airy integrand vanishes at infinity. The contour $\mathcal{C}_i$, $i = 1, 2, 3$, connects the ith good region to the $i + 1^{th}$.

3.1. The Airy function. We begin by examining the Airy integral, initially defined for real λ:

$$Z_\lambda = \int_{-\infty}^{\infty} \mathrm{d}x \exp\left(i\lambda(x^3/3 - x)\right). \tag{3.1}$$

We want to analytically continue the integral to complex λ. This will require deforming the integration contour away from the real x axis, so we will regard the Airy exponent

$$\mathcal{I} = i\lambda(x^3/3 - x) \tag{3.2}$$

as a function of complex variables x and λ. For real λ, the integration is over the real axis. As λ varies, we will deform the integration cycle $\mathcal{C}$ so as to vary smoothly with λ and in such a way that the integral always converges.

In the case of the Airy function, it is not hard to see by hand how to do this. We divide the x-plane into "good" regions in which the imaginary part of λx^3 is positive and "bad" regions in which it is negative. For positive λ, this is shown in fig. 1. The Airy integrand is exponentially small in the good regions. The original oscillatory integral over the real x axis is an integral over the boundary of two good regions. We can deform the contour slightly by giving x a positive imaginary part so that it will connect two good regions. Now as we vary λ (keeping away from $\lambda = 0$), the good regions rotate continuously to the left or the right, and we simply rotate the integration contour $\mathcal{C}$ as λ varies so that $\mathcal{C}$ always connects two good regions. This gives an analytic continuation of Z_λ as a function of λ, but it also makes it clear that there is a monodromy around $\lambda = \infty$ (or 0). In fact, as λ changes in phase by 2π, the three good regions are cyclically permuted. So Z_λ has a monodromy of order 3.

Because the contour $\mathcal{C}$ does undergo a non-trivial monodromy as λ varies, it is useful to consider all possible contours in this problem. Denote the three good regions as $\mathcal{V}_i$, $i = 1, 2, 3$, as in the figure. For each i, let $\mathcal{C}_i$ be a contour connecting region $\mathcal{V}_i$ to region $\mathcal{V}_{i+1}$. The integrals $Z_{i,\lambda} = \int_{\mathcal{C}_i} \mathrm{d}x \exp(i\lambda(x^3/3 - x))$ converge for all i and λ, and any reasonable contour along which one might integrate the Airy function $\exp(i\lambda(x^3/3 - x))$ (or the product of this function and one with slower growth at infinity) is equivalent to a linear combination of the $\mathcal{C}_i$. However, the integrals $Z_{i,\lambda}$ are not independent. They obey

$$Z_{1,\lambda} + Z_{2,\lambda} + Z_{3,\lambda} = 0. \tag{3.3}$$

This reflects the fact that the sum of contours $\mathcal{C}_1 + \mathcal{C}_2 + \mathcal{C}_3$ can be deformed to zero.

3.1.1. *Morse theory.* The Airy function is simple enough that we do not really need a more sophisticated theory. Nevertheless, we will use this simple example as a practice case for describing a more sophisticated approach that is useful in generalizations. To formalize the sense in which $\mathcal{C}_1 + \mathcal{C}_2 + \mathcal{C}_3 = 0$, it is useful to regard the $\mathcal{C}_i$ as relative homology cycles. Let X denote the complex x-plane, and for real U, let X_U be the part of X with $\mathrm{Re}(i\lambda(x^3/3 - x)) \leq U$. Any reasonable integration cycle $\mathcal{C}$ for the Airy integral should represent an element of the relative homology $H_1(X, X_{-T}; \mathbb{Z})$ for very large T. In other words, $\mathcal{C}$ is a closed cycle, and though it may not be compact, its infinite ends should lie in X_{-T}. Differently put, $\mathrm{Re}(i\lambda(x^3/3 - x))$ must tend to $-\infty$ at infinity in $\mathcal{C}$.

The $\mathcal{C}_i$ have this property, and represent elements of the relative homology. Indeed, they generate this relative homology, subject to the relation

$$\mathcal{C}_1 + \mathcal{C}_2 + \mathcal{C}_3 = 0. \tag{3.4}$$

The fact that the $\mathcal{C}_i$ generates the relative homology is fairly obvious geometrically in this simple example. X is contractible, but X_{-T} is the union of three disjoint pieces. A one-cycle that is non-trivial in $H_1(X, X_{-T}; \mathbb{Z})$ must connect distinct pieces of X_{-T}. The relation (3.4) is also clear geometrically.

The integration cycle that we started with for real λ – the real x-axis, which we will call $\mathcal{C}_\mathbb{R}$ – is not quite an element of the relative homology, because at large $|x|$ it lies on the boundary of, rather than in the interior of, the good regions of fig. (1). However, $\mathcal{C}_\mathbb{R}$ becomes an element of the relative homology once we displace it slightly by giving x a small imaginary part ϵ with the same sign as λ. This multiplies the Airy integral by a convergence factor $\exp(-\lambda\epsilon x^2)$. The displaced cycle therefore represents an element of the relative homology. And indeed since the displaced cycle connects two of the good regions at infinity, it is equivalent to one of the $\mathcal{C}_i$.

For generalizations, we need a more powerful way to analyze the relative homology. We consider the real part of the Airy exponent

$$h = \mathrm{Re}(\mathcal{I}) \tag{3.5}$$

as a Morse function. A Morse function is simply a real-valued function whose critical points are nondegenerate. The critical points are the points at which all first derivatives of h vanish, and nondegeneracy of a critical point means that the matrix of second derivatives is invertible at that point. The number of negative eigenvalues of this matrix is called the Morse index of the critical point.

In general, because of the Cauchy-Riemann equations, the critical points of the real part of a holomorphic function $\mathcal{I}$ are the same as the critical points of $\mathcal{I}$. So in this case, there are two critical points which we call $p_\pm$, located at $x = \pm 1$. The values of $\mathcal{I}$ and h at $p_\pm$ are

$$\mathcal{I}_\pm = \mp \frac{2i\lambda}{3} \tag{3.6}$$

and

$$h_\pm = \pm \frac{2\,\mathrm{Im}\lambda}{3}. \tag{3.7}$$

The critical points are nondegenerate, since $\partial^2\mathcal{I}/\partial x^2 \neq 0$ at $x = \pm 1$. In general, in a complex manifold of complex dimension n, if h is a Morse function that is the real part of a holomorphic function $\mathcal{I}$, its isolated and nondegenerate critical points are all of index n. (The general form of $\mathcal{I}$ near an isolated nondegenerate critical point is $\mathcal{I} = c + \sum_{i=1}^n z_i^2$, with a constant c and local complex coordinates z_i; if $z_i = u_i + iv_i$ then $h = \mathrm{Re}\, c + \sum_{i=1}^n (u_i^2 - v_i^2)$ and the matrix of its second derivatives is diagonal with n negative eigenvalues.) So in our case, the critical points have index 1.

In general, any Morse function h on a manifold X can be used to compute upper bounds on the ranks of the real homology (or cohomology) groups of X: the rank of the q-dimensional homology of X is at most the number of critical points of h of Morse index q. If the differences between the indices of distinct critical points of h are never equal[7] to ± 1, then h is a perfect Morse function and the ranks of the homology groups are equal to the upper bounds just stated. In fact, in this case even the integral homology can be described via the critical points. Our Morse function is perfect, since both critical points have index 1.

If X is a compact manifold, h is automatically bounded above and below, and the critical points of a perfect Morse function determine the ordinary homology of X. In our case, X is not compact, and we are dealing with a Morse function h that is unbounded above and below. In such a case, the critical points of h determine relative homology groups, namely the homology groups $H_k(X, X_{-T})$, where T is a large constant. In our example, h is a Morse function with two critical points both of index 1, so $H_1(X, X_{-T})$ is of rank 2, and the other relative homology groups vanish. (Without using Morse theory, we would observe that $H_0(X, X_{-T}) = 0$ since any zero-cycle is a sum of points, which can be deformed into X_{-T}. And $H_2(X, X_{-T}) = 0$ since X has non-compact ends even after collapsing X_{-T} to a point.)

For a generic perfect Morse function h, Morse theory also gives a recipe for constructing a relative one-cycle $\mathcal{J}_\sigma$ attached to each critical point, and representing an element of the appropriate relative homology group. Moreover, these cycles generate the relative homology with integer coefficients.

In general, on any manifold X with real coordinates u^i, pick a Riemannian metric g_{ij} and consider the downward flow or gradient flow equations

$$\frac{\mathrm{d}u^i}{\mathrm{d}t} = -g^{ij}\frac{\partial h}{\partial u^j}. \tag{3.8}$$

[7]This condition ensures that in computing the homology of X, there are no cancellations between critical points of adjacent index.

These are called downward flow equations, because except for a trivial solution that sits at a critical point for all t, h is always strictly decreasing along a flow:

$$\frac{\mathrm{d}h}{\mathrm{d}t} = \sum_i \frac{\partial h}{\partial u^i}\frac{\mathrm{d}u^i}{\mathrm{d}t} = -\sum_i \left(\frac{\partial h}{\partial u^i}\right)^2. \tag{3.9}$$

An important property of the flow equation is that if $u^i(t)$ equals a critical point at some t, then the flow equation implies that $u^i(t)$ is constant for all t. So a nonconstant flow can only reach a critical point at $t = \pm\infty$.

Let p be a nondegenerate critical point of h, and consider the downward flow equations on the half-line $(-\infty, 0]$ with the boundary condition that $u^i(t)$ approaches p for $t \to -\infty$. If p has index k, the moduli space $\mathcal{J}_p$ of such solutions is a k-dimensional manifold, since there are k independent directions of downward flow from the critical point p. We think of $\mathcal{J}_p$ as a submanifold of X, by mapping a downward flow line $u^i(t)$ to the corresponding point $u^i(0) \in X$. (This gives an embedding of $\mathcal{J}_p$ in X since, as the downward flow equation is first order in time, a flow is uniquely determined by its value at $t = 0$.) An equivalent definition of $\mathcal{J}_p$ is that it is the submanifold of X consisting of points that can be reached at any t by a flow that starts at p at $t = -\infty$. Given a flow line $u^i(t)$ that reaches a point $x \in X$ at $t = t^*$, the flow line $u^i(t - t^*)$ arrives at x at $t = 0$. A flow line defined on the full line $(-\infty, \infty)$ will be called a complete flow line.

Once we pick an orientation of $\mathcal{J}_p$, $\mathcal{J}_p$ will define a cycle[8] in the relative homology $H_1(X, X_{-T})$ if it is closed, meaning that any sequence of points in $\mathcal{J}_p$ has a subsequence that either converges or tends to $h = -\infty$. This fails precisely if there is a complete flow line l that starts at p at $t = -\infty$ and ends at another critical point q at $t = +\infty$. In that case, $\mathcal{J}_p$ is not closed, since l is contained in $\mathcal{J}_p$, but a sequence of points in l can converge to q, which is not contained in $\mathcal{J}_p$.

We will determine shortly a necessary condition for a flow line to connect two distinct critical points p and q. When this condition is not satisfied, $\mathcal{J}_p$ defines an element of the relative homology $H_k(X, X_{-T})$. $\mathcal{J}_p$ is known as a Lefschetz thimble.

In our problem, the Morse function is the real part of a holomorphic function $\mathcal{I}$. In such a case, if we pick a Kahler metric $\mathrm{d}s^2 = |\mathrm{d}x|^2$, the flow equation can be written

$$\frac{\mathrm{d}x}{\mathrm{d}t} = -\frac{\partial\overline{\mathcal{I}}}{\partial\overline{x}}, \quad \frac{\mathrm{d}\overline{x}}{\mathrm{d}t} = -\frac{\partial\mathcal{I}}{\partial x}. \tag{3.10}$$

Using the chain rule and (3.10), we find

$$\frac{\mathrm{d}\,\mathrm{Im}\,\mathcal{I}}{\mathrm{d}t} = \frac{1}{2i}\frac{\mathrm{d}(\mathcal{I} - \overline{\mathcal{I}})}{\mathrm{d}t} = \frac{1}{2i}\left(\frac{\partial\mathcal{I}}{\partial x}\frac{\mathrm{d}x}{\mathrm{d}t} - \frac{\partial\overline{\mathcal{I}}}{\partial\overline{x}}\frac{\mathrm{d}\overline{x}}{\mathrm{d}t}\right) = 0. \tag{3.11}$$

So $\mathrm{Im}\,\mathcal{I}$ is conserved along a flow.

A more conceptual explanation of the existence of this conserved quantity is as follows. Associated to the Kahler metric $|\mathrm{d}x|^2$ is the Kahler form $\omega = -i\mathrm{d}x \wedge \mathrm{d}\overline{x}$. Viewing ω as a symplectic form on the x-plane, the gradient flow with respect to

---

[8]Our definition of $\mathcal{J}_p$ is slightly different from the definition usually used in Morse theory on a compact manifold M. In such a case, unless p is a local minimum of h, there are always complete flow lines starting at p and ending at some "lower" critical point, and one takes the closure of $\mathcal{J}_p$ to define a Morse theory cycle. In the situation we consider, as is explained shortly, there generically are no complete flow lines between critical points, and it is most convenient to define $\mathcal{J}_p$ as we have done.

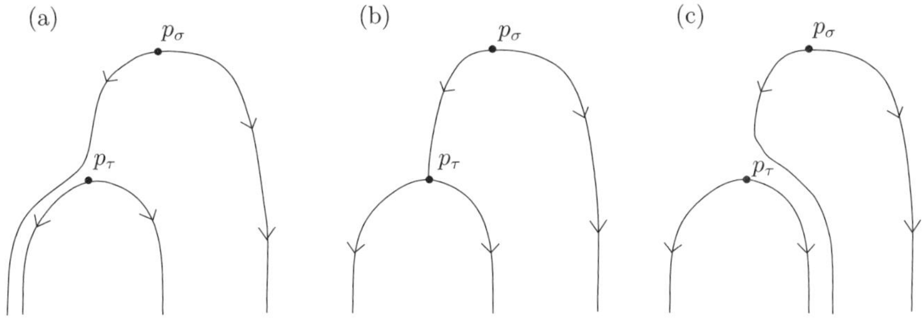

FIGURE 2. This figure illustrates the behavior of flow lines when one crosses a Stokes ray. Sketched are the downward flows from two critical points p_σ and p_τ. The lines are sketched as flowing downward, in the direction of smaller h (though it is not literally true that h corresponds to the height in the figure, as this function has no critical points). The behavior at a Stokes ray is depicted in (b); there is a downward flow from p_σ to p_τ. The flows "before" and "after" crossing the Stokes ray are depicted in (a) and (c). The downward flowing cycle $\mathcal{J}_\tau$ from the lower critical point is unaffected by the Stokes ray. As for the cycle $\mathcal{J}_\sigma$ defined by flow from the upper critical point, it is ill-defined in (b) and jumps by $\mathcal{J}_\sigma \to \mathcal{J}_\sigma + \mathcal{J}_\tau$ between (a) and (c).

the Morse function $\mathrm{Re}\,\mathcal{I}$ is the same as the Hamiltonian flow with Hamiltonian $\mathrm{Im}\,\mathcal{I}$. Naturally, $\mathrm{Im}\,\mathcal{I}$ is conserved in this flow.

From (3.6), we see that in the Airy case, the values of $\mathrm{Im}\,\mathcal{I}$ at the critical points are $\mathrm{Im}\,\mathcal{I}_\pm = \mp(2/3)\mathrm{Re}\,\lambda$. These are equal if and only if λ is purely imaginary. Hence a flow can connect the two critical points at $x = \pm 1$ only if λ is imaginary. Conversely, if λ is imaginary, there is indeed a flow that connects the two critical points (the flow line is simply the part of the real axis with $-1 < x < 1$). Since we always assume that $\lambda \neq 0$, the locus of imaginary λ is not connected but consists of two open rays, differing by the sign of $\mathrm{Im}\,\lambda$. We refer to these as Stokes rays.

3.1.2. *Crossing a Stokes ray.* Away from the Stokes rays, a downward flow that starts at one critical point cannot end at the other; instead it always flows to $h = -\infty$. So away from the Stokes rays, the downward flow procedure attaches to each critical point p_+ or p_- a relative homology one-cycle $\mathcal{J}_\pm$. Let us see what happens to the Lefschetz thimbles $\mathcal{J}_\pm$ in crossing a Stokes ray.

From (3.6), $\mathcal{I}_+$ and $\mathcal{I}_-$ are always unequal (since we always assume that $\lambda \neq 0$). On a Stokes ray $\mathrm{Im}\,\mathcal{I}_+ = \mathrm{Im}\,\mathcal{I}_-$, so necessarily $\mathrm{Re}\,\mathcal{I}_+ \neq \mathrm{Re}\,\mathcal{I}_-$, that is $h_+ \neq h_-$. In fact, $h_+ > h_-$ if $\mathrm{Im}\,\lambda > 0$, and otherwise $h_- > h_+$. As we approach a Stokes ray, let us write q and q' for the critical points with the larger and smaller values of h, and $\mathcal{J}$ and $\mathcal{J}'$ for the corresponding one-cycles. Even when we reach and cross the Stokes ray, nothing happens to $\mathcal{J}'$, because downward trajectories that start at q' can only flow to $h = -\infty$. But a trajectory that starts at q may flow down to q'. (As we noted above, in the case of the Airy function, there is such a trajectory, namely the segment of the real axis connecting q and q'.) When there is such a flow line, the definition of the cycle $\mathcal{J}$ attached to q fails. Since $\mathcal{J}$ is potentially ill-defined when λ is on a Stokes ray, it might jump in crossing a Stokes ray, and that is actually what happens. How this happens is sketched in Fig. 2. The jump is $\mathcal{J} \to \mathcal{J} \pm \mathcal{J}'$, where the sign depends on the orientations of the two cycles and the direction in which λ crosses the Stokes ray. The change in the pair $\mathcal{J}$, $\mathcal{J}'$ can thus be written

$$\begin{pmatrix}\mathcal{J}\\ \mathcal{J}'\end{pmatrix} \to \begin{pmatrix}1 & \pm 1\\ 0 & 1\end{pmatrix}\begin{pmatrix}\mathcal{J}\\ \mathcal{J}'\end{pmatrix}. \tag{3.12}$$

In section 3.1.4, we explain in another way why only $\mathcal{J}$ and not $\mathcal{J}'$ changes in crossing the Stokes ray.

Let us compute the monodromy of the cycles $\mathcal{J}_+$ and $\mathcal{J}_-$ around $\lambda = \infty$ (or 0). In crossing the Stokes ray with $\operatorname{Im}\lambda > 0$, the monodromy is strictly upper triangular:

$$U_+ = \begin{pmatrix}1 & 1\\ 0 & 1\end{pmatrix}. \tag{3.13}$$

On the Stokes ray with $\operatorname{Im}\lambda < 0$, the monodromy is strictly lower triangular:

$$U_- = \begin{pmatrix}1 & 0\\ -1 & 1\end{pmatrix}. \tag{3.14}$$

(If the cycles $\mathcal{J}_\pm$ are oriented so that the upper right matrix element in (3.13) is $+1$, then the lower left matrix element in (3.14) is -1.) The monodromy around $\lambda = \infty$ is not simply the product U_-U_+, as there is also a minus sign involving the orientation of the cycles $\mathcal{J}_\pm$. Indeed, the function $\mathcal{I}$ can be approximated near $x = \pm 1$ by $\mp 2i\lambda/3 + i\lambda(x \mp 1)^2$, and an approximation to the downward flowing cycles $\mathcal{J}_\pm$ is given by $x = \pm 1 + t(-i\lambda)^{-1/2}$, with real t. To orient the cycles, we need to pick a sign of $(-i\lambda)^{-1/2}$, and under monodromy around $\lambda = \infty$, this sign is reversed. So the monodromy of the pair $\begin{pmatrix}\mathcal{J}_+\\ \mathcal{J}_-\end{pmatrix}$ around $\lambda = \infty$ is

$$M = -U_-U_+ = \begin{pmatrix}-1 & -1\\ 1 & 0\end{pmatrix}. \tag{3.15}$$

In particular, the monodromy obeys $M^3 = 1$. The fact that the monodromy is of order 3 is clear without Morse theory: under monodromy around $\lambda = \infty$, the three good regions in Fig. 1 undergo a cyclic permutation of order 3. What we have gained from Morse theory is a powerful framework for generalizations.

In this analysis, the dependence of the orientation of the $\mathcal{J}_\pm$ on a sign of $\sqrt{-i\lambda}$ is inelegant, and the resulting jumping in the sign of the $\mathfrak{n}_\sigma$ is unthematic. This can be avoided by multiplying the original definition of the Airy integral by $\sqrt{-i\lambda}$. The product of this function times the orientation of $\mathcal{J}_\pm$ has no jumping in sign. With this factor included, the jumping of $\mathfrak{n}_\sigma$ occurs only in crossing Stokes lines, and M has the opposite sign to what is given in (3.15) and obeys $M^3 = -1$.

3.1.3. *Asymptotic behavior of integrals over Lefschetz thimbles.* The cycles $\mathcal{J}_\pm$ are defined in such a way that, for any λ, the integrals

$$Z_{\pm,\lambda} = \int_{\mathcal{J}_\pm} \mathrm{d}x \exp(\mathcal{I}) \tag{3.16}$$

always converge. Indeed, we have arranged so that $\operatorname{Re}\mathcal{I} \to -\infty$ at infinity along $\mathcal{J}_\pm$.

Not only are the integrals $Z_{\pm,\lambda}$ always convergent, but it is straightforward to determine their asymptotic behavior as $\lambda \to \infty$ in a fixed direction in the complex plane. The maximum of $h = \operatorname{Re}\mathcal{I}$ on the cycle $\mathcal{J}_\pm$ is precisely at the critical point

$x = \pm 1$. Moreover, the phase $\operatorname{Im}\mathcal{I}$ is stationary at $x = \pm 1$, since those are critical points of $\mathcal{I}$ (and in fact the phase $\operatorname{Im}\mathcal{I}$ is entirely constant along the cycles $\mathcal{J}_\pm$). So the asymptotic behavior of $Z_{\pm,\lambda}$ is dominated by a contribution from the critical point, giving

$$Z_{\pm,\lambda} \sim \exp(\mathcal{I}_\pm)(-i\lambda)^{-1/2}\sum_{t=0}^{\infty} b_t\lambda^{-t}, \tag{3.17}$$

where $\mathcal{I}_\pm = \mp 2i\lambda/3$ and the coefficients b_t in the asymptotic series can be computed by expanding around the critical point. If we include a factor of $(-i\lambda)^{1/2}$ in the definition of the Airy integral, as suggested at the end of section 3.1.1, the factor of $(-i\lambda)^{-1/2}$, which comes from a Gaussian integral, would be canceled.

3.1.4. *Sum of asymptotic series.* An integral $Z_{\mathcal{C},\lambda}$ over a general cycle $\mathcal{C} = \mathfrak{n}_+\mathcal{J}_+ + \mathfrak{n}_-\mathcal{J}_-$ can be expressed asymptotically as a sum of two asymptotic expansions:

$$Z_{\mathcal{C},\lambda} \sim \mathfrak{n}_+ \exp(\mathcal{I}_+)(-i\lambda)^{-1/2}\sum_{t=0}^{\infty} b_t\lambda^{-t} + \mathfrak{n}_- \exp(\mathcal{I}_-)(-i\lambda)^{-1/2}\sum_{t=0}^{\infty} c_t\lambda^{-t}. \tag{3.18}$$

However, generically one of these series suffices, since in a generic direction in the complex λ plane, h_+ and h_- are unequal and one of the two series is exponentially larger than the other. If $h_+ = h_-$ (which happens for real λ), the two series compete and must both be included, and if λ varies so that $h_+ - h_-$ changes sign, the two series exchange dominance.

On a Stokes ray, h_+ and h_- are definitely unequal, so one series in (3.18) is exponentially larger than the other. Consider for definiteness a Stokes ray with $h_+ > h_-$. The integral in (3.18) must vary holomorphically with λ, so the coefficient $\mathfrak{n}_+$ of the dominant exponential series must be continuous. But, since the two series are only asymptotic series that leave exponentially small corrections undetermined, the expansion (3.18) in a region with $h_+ > h_-$ is not powerful enough to ensure that $\mathfrak{n}_-$ is continuous.

Let us compare this expectation to (3.12), which in the present notation says that the jumping of $\mathcal{J}_+$ and $\mathcal{J}_-$ is

$$\begin{pmatrix}\mathcal{J}_+\\ \mathcal{J}_-\end{pmatrix} \to \begin{pmatrix}1 & \pm 1\\ 0 & 1\end{pmatrix}\begin{pmatrix}\mathcal{J}_+\\ \mathcal{J}_-\end{pmatrix}. \tag{3.19}$$

In order for $\mathfrak{n}_+\mathcal{J}_+ + \mathfrak{n}_-\mathcal{J}_-$ to be continuous, the $\mathfrak{n}_\pm$ transform by

$$\begin{pmatrix}\mathfrak{n}_+ & \mathfrak{n}_-\end{pmatrix} \to \begin{pmatrix}\mathfrak{n}_+ & \mathfrak{n}_-\end{pmatrix}\begin{pmatrix}1 & \mp 1\\ 0 & 1\end{pmatrix}. \tag{3.20}$$

In particular, as expected, $\mathfrak{n}_+$ is constant and only $\mathfrak{n}_-$ jumps.

An equivalent way to say some of the same things is to observe that since the integral over $\mathcal{J}_-$ is exponentially smaller than that over $\mathcal{J}_+$, $\mathcal{J}_-$ cannot jump by a multiple of $\mathcal{J}_+$ in crossing the Stokes ray, but $\mathcal{J}_+$ can jump by a multiple of $\mathcal{J}_-$.

3.1.5. *Expressing the real cycle in terms of Lefschetz thimbles.* The function Z_λ whose analytic continuation we actually want to understand is defined for real λ by integration over the real cycle $\mathcal{C}_\mathbb{R}$. So to use the above results, we have to express $\mathcal{C}_\mathbb{R}$ as a linear combination of the $\mathcal{J}_\pm$. The coefficients will be integers, since the $\mathcal{J}_\pm$ are a basis of the relative homology with integer coefficients. For the Airy integral, we can find the right coefficients by inspection, but we want to describe a more general method.

On a compact Riemann surface X, given a basis $\mathcal{J}_\pm$ of $H_1(X)$, to express another class $\mathcal{C} \in H_1(X)$ in the form $\mathcal{C} = \mathfrak{n}_+\mathcal{J}_+ + \mathfrak{n}_-\mathcal{J}_-$, we would find the coefficients $\mathfrak{n}_+$ and $\mathfrak{n}_-$ via intersection pairings. In the present case, we cannot do precisely this, because there is no intersection pairing on the relative homology $H_1(X, X_{-T})$. (The intersection of two one-cycles on the two-manifold X should be a zero-cycle or sum of points; but a point vanishes in the relative homology $H_0(X, X_{-T})$, since it can be deformed into X_{-T}.)

There is a good substitute, however. Indeed, Morse theory tells us what are the duals of the downward flowing cycles from the critical points: they are the upward flowing cycles. The upward flowing cycle attached to a critical point p is defined by reversing the sign in the flow equation (3.8). That is, one considers solutions of the upward flowing equation

$$\frac{\mathrm{d}u^i}{\mathrm{d}t} = +g^{ij}\frac{\partial h}{\partial u^j} \tag{3.21}$$

on the half-line $(-\infty, 0]$, again requiring that $u^i(t)$ approaches p for $t \to -\infty$. The space of possible values of $u^i(t)$ at $t = 0$ now defines the upward flowing cycle $\mathcal{K}_p$ attached to p. It is an $n-k$-dimensional cycle if p has index k (so that the Morse function $-h$ has index $n-k$ at p).

The upward flowing cycles take values in a different relative homology. For a real number T, define X^T as the part of the complex x-plane with $h \geq T$. The upward flowing cycle from a critical point p of the type we are considering has its infinite ends contained in X^T for large positive T (as long as we keep away from Stokes rays). So it defines an element of the relative homology $H_k(X, X^T)$.

For r and s of complementary dimensions, there is a natural intersection pairing between $H_r(X, X_{-T})$ and $H_s(X, X^T)$. This pairing is easily evaluated for the case of interest to us – a perfect Morse function with no flows between distinct critical points. The absence of such flows means that if p_σ and p_τ are distinct critical points, the corresponding cycles $\mathcal{J}_\sigma$ and $\mathcal{K}_\tau$ do not intersect – such an intersection would be a downward flow from p_σ to p_τ. On the other hand, for any one critical point p_σ, the downward cycle $\mathcal{J}_\sigma$ and the upward cycle K_σ have precisely one point of intersection – the trivial constant flow, starting and ending at p_σ. With suitable relative orientations of downwards and upward cycles, the trivial flow contributes $+1$ to the diagonal intersection pairing $\langle \mathcal{J}_\sigma, \mathcal{K}_\sigma \rangle$. So the intersection pairings are

$$\langle \mathcal{J}_\sigma, \mathcal{K}_\tau \rangle = \delta_{\sigma\tau}. \tag{3.22}$$

Given any element $\mathcal{C}$ of the downward flowing relative homology, we now can give a formula to determine the coefficients in the expansion $\mathcal{C} = \sum_\sigma \mathfrak{n}_\sigma \mathcal{J}_\sigma$. By intersecting with $\mathcal{K}_\tau$, we find simply

$$\mathfrak{n}_\tau = \langle \mathcal{C}, \mathcal{K}_\tau \rangle \tag{3.23}$$

for all τ.

3.1.6. *The asymptotic behavior of the Airy function.* We will now give some examples of this procedure. Return to the original Airy function

$$Z_\lambda = \int_{\mathcal{C}_\mathbb{R}} \mathrm{d}x \exp\left(i\lambda(x^3/3 - x)\right) \tag{3.24}$$

with $\mathcal{C}_\mathbb{R}$ being the real axis. The critical points $p_\pm$ at $x = \pm 1$ both lie on $\mathcal{C}_\mathbb{R}$. A nontrivial upward flow starting at a critical point will end at a point with a strictly positive value of $h = \mathrm{Re}\,\mathcal{I}$; such a point will not lie on $\mathcal{C}_\mathbb{R}$, since h vanishes identically on $\mathcal{C}_{\mathbb{R}}$. Hence the intersection of $\mathcal{C}_{\mathbb{R}}$ with either upward flowing cycle $\mathcal{K}_\pm$ consists precisely of the critical point $p_\pm$. Modulo a judicious choice of orientations, this tells us the coefficients when we express the real cycle $\mathcal{C}_{\mathbb{R}}$ in terms of Lefschetz thimbles associated to critical points:

$$\mathcal{C}_{\mathbb{R}} = \mathcal{J}_+ + \mathcal{J}_-. \tag{3.25}$$

This formula expresses the integration cycle in terms of Lefschetz thimbles when λ is real. But when λ is not real, we must integrate over a more general cycle $\mathcal{C}$. As λ varies in the complex plane, we vary the cycle $\mathcal{C}$ smoothly so that it always connects one good region at infinity to another. On the other hand, the cycles $\mathcal{J}_+$ and $\mathcal{J}_-$ have jumps described in eqns. (3.13), (3.14) and a jump in sign involving the orientations. Allowing for these jumps, we get, after analytic continuation from real λ to a wedge-shaped region of the complex λ plane that does not contain Stokes rays, a formula

$$\mathcal{C} = \mathfrak{n}_+\mathcal{J}_+ + \mathfrak{n}_-\mathcal{J}_-. \tag{3.26}$$

(Instead of keeping track of Stokes rays and orientations, one can in this particular problem let $\mathcal{C}$ evolve continuously so that the Airy integral remains convergent, and then directly use (3.23) to determine the coefficients in any wedge-shaped region.) Hence in each wedge-shaped region, the Airy integral can be expressed as

$$Z_\lambda = \mathfrak{n}_+ Z_{+,\lambda} + \mathfrak{n}_- Z_{-,\lambda}. \tag{3.27}$$

Together with (3.17), this determines the asymptotic behavior of Z_λ in each wedge-shaped region. In particular, Z_λ grows exponentially at infinity if and only if the expansion (3.26) has a nonzero coefficient for a critical point at which $h > 0$.

Let us consider a special case in which there is *not* exponential growth at infinity. This is the Airy function with a reversed sign for the linear term in the exponent,

$$\widetilde{Z}_\lambda = \int_{\mathcal{C}_{\mathbb{R}}} \mathrm{d}x \exp\left(i\lambda(x^3/3 + x)\right), \tag{3.28}$$

with λ real and with $\mathcal{C}_{\mathbb{R}}$ again equal to the real axis. This integral is actually a special case of what one gets by analytic continuation of the original Airy integral, since[9] (3.24) is mapped to (3.28) by $\lambda \to i\lambda$, $x \to ix$. However, we will simply treat this slightly modified example directly according to the general recipe.

The critical points are now located at $x = \pm i$. We denote these points as $\widetilde{p}_\pm$. The value of $\widetilde{\mathcal{I}} = i\lambda(x^3/3 + x)$ at $\widetilde{p}_\pm$ is now $\mp 2\lambda/3$, so for either sign of λ, the function $\widetilde{h} = \mathrm{Re}\,\widetilde{\mathcal{I}}$ is positive at one critical point and negative at the other critical point. An upward flow from a critical point with $\widetilde{h} > 0$ will never meet the real axis $\mathcal{C}_{\mathbb{R}}$, since $\widetilde{h} = 0$ on $\mathcal{C}_{\mathbb{R}}$. However, from the critical point with $\widetilde{h} < 0$, there is an upward flow to the real axis.[10] Write q^+ and q^- for the critical points at which $\widetilde{h}$

[9] To be more exact, to map the original Airy function to (3.28) by analytic continuation, one must increase the argument of λ by $2\pi(n + 1/4)$, where the integer n is chosen so that the real cycle of the original Airy function maps back to the real cycle in (3.28).

[10] For example, if λ is real, the part of the imaginary axis between $x = i$ and $x = -i$ is a flow line that intersects the real axis. In one direction or the other, this is an upward flow from the critical point with negative $\widetilde{h}$. If we give λ a small imaginary part to get away from the Stokes ray, the upward flow from that same critical point still intersects the real axis (it then continues to $\widetilde{h} = +\infty$ rather than to the second critical point).

is positive or negative, and $\widetilde{\mathcal{J}}_\pm$ for the corresponding Lefschetz thimbles. Then the expression for $\mathcal{C}_\mathbb{R}$ in terms of Lefschetz thimbles is simply

$$\mathcal{C}_\mathbb{R} = \widetilde{\mathcal{J}}_-. \tag{3.29}$$

The asymptotic behavior of $\widetilde{Z}_\lambda$ for large real λ is hence

$$\widetilde{Z}_\lambda \sim \frac{\exp\left(\widetilde{\mathcal{I}}(q^-)\right)}{\lambda^{1/2}} \sum_{t=0}^\infty a_t \lambda^{-t}. \tag{3.30}$$

In particular, $\widetilde{Z}_\lambda$ vanishes exponentially for large real λ, reflecting the fact that there are no critical points on the real x-axis.

Actually, in this example, the Stokes rays are the positive and negative real λ axis. For real λ, there is a flow line between the two critical points, as noted in footnote 10. Hence the cycles $\widetilde{\mathcal{J}}_\pm$ jump in crossing the real λ axis. However, the jumping only affects the cycle $\widetilde{\mathcal{J}}_+$ attached to the critical point with $\widetilde{h} > 0$. This cycle changes by $\widetilde{\mathcal{J}}_+ \to \widetilde{\mathcal{J}}_+ \pm \widetilde{\mathcal{J}}_-$. The cycle $\widetilde{\mathcal{J}}_-$ is unchanged in crossing the Stokes ray, and hence there is no problem with the formula (3.29).

3.2. Generalization to dimension n. In our analysis of the Airy function, we introduced much more machinery than was really needed. The advantage of having done so is that the generalization to an oscillatory integral of higher dimension is now straightforward.

We consider a real-valued polynomial function $f(x^1, \dots, x^n)$ of n real variables, and we assume that f is sufficiently generic that (even after analytic continuation to complex values of the x^i) it has only finitely many critical points, which are all nondegenerate. We want to study the analytic continuation of the oscillatory integral[11]

$$Q_\lambda = (-i\lambda)^{n/2} \int_{\mathcal{C}_\mathbb{R}} \mathrm{d}^n x \, \exp\left(i\lambda f(x^1, \dots, x^n)\right). \tag{3.31}$$

The integration is over the cycle $\mathcal{C}_\mathbb{R}$ defined by taking $x^1, \dots, x^n$ to be real. We wish to analytically continue the integral to (nonzero) complex λ, and to determine the asymptotic behavior of the analytically continued integral. Depending on f, it may be difficult to see how to vary the integration cycle $\mathcal{C}$ as λ is varied so that the integral remains convergent. So it is difficult in general to imitate the direct approach to the Airy function by rotating $\mathcal{C}$ as λ varies. But the approach via Morse theory can be straightforwardly generalized.

As in the one-variable case, we regard the function $\mathcal{I} = i\lambda f(x^1, \dots, x^n)$ as a function of complex variables, and for fixed complex λ, we let Σ be the set of its critical points. If p_σ, $\sigma \in \Sigma$, is a critical point, we write $\mathcal{I}_\sigma$ for $\mathcal{I}(p_\sigma)$. Now take $h = \mathrm{Re}\,\mathcal{I}$ as a Morse function, and write h_σ for $h(p_\sigma)$. The critical points are all of the same index n, as explained in section 3.1.1. So h is a perfect Morse function, and can be used to describe the appropriate relative homology group in which the possible integration cycles take values. This group is $H_n(X, X_{-T}; \mathbb{Z})$, where $X = \mathbb{C}^n$, and X_{-T} is defined by the condition $h \leq -T$, for large T.

---

[11]Including the elementary prefactor $(-i\lambda)^{n/2}$ avoids sign changes in the orientations of the Lefschetz thimbles, as explained at the end of section 3.1.1. This lets us concentrate on the more interesting Stokes phenomena.

Introducing on $\mathbb{C}^n$ the Kahler metric $\mathrm{d}s^2 = \sum_i |\mathrm{d}x^i|^2$, the downward flow equation (3.8) can be written as in (3.10):

$$\frac{\mathrm{d}x^i}{\mathrm{d}t} = -\frac{\partial \overline{\mathcal{I}}}{\partial \overline{x}^i}. \tag{3.32}$$

For generic λ, there are no flows between critical points. This follows from the fact that $\mathrm{Im}\,\mathcal{I}$ is conserved along a critical point, as shown in eqn. (3.11). (That computation generalizes immediately to n dimensions, as long as the metric on X is hermitian. Just as in the one-dimensional case, if X is Kahler, and we use the Kahler form of X as a symplectic form, then the gradient flow with respect to the Morse function $\mathrm{Re}\,\mathcal{I}$ is the same as the Hamiltonian flow with Hamiltonian $\mathrm{Im}\,\mathcal{I}$; naturally, $\mathrm{Im}\,I$ is conserved in this flow.) If p_σ and p_τ are critical points with distinct values of $\mathcal{I}$, then for generic λ they have distinct values of $\mathrm{Im}\,\mathcal{I}$ and there is no flow between them. But if $\mathcal{I}_\sigma = \mathcal{I}_\tau$, there is never a flow between p_σ and p_τ, since $h = \mathrm{Re}\,\mathcal{I}$ is always strictly decreasing along a non-trivial flow, as explained in eqn. (3.9).

We define a Stokes ray to be a ray along which $\mathrm{Im}\,\mathcal{I}_\sigma = \mathrm{Im}\,\mathcal{I}_\tau$ for some $\sigma, \tau \in \Sigma$. There are only finitely many Stokes rays (we have assumed that Σ is a finite set, but even without this assumption, for any polynomial f, Σ has only finitely many components, and $\mathcal{I}$, which is constant along each component, has only finitely many critical values).

Just as in the one-variable case, we attach to each critical point p_σ a cycle $\mathcal{J}_\sigma$, known as a Lefschetz thimble, by taking the union of all downward flows that start at p_σ, or equivalently the values at $t = 0$ of downward flows $u^i(t)$ that start at p_σ at $t = -\infty$. As long as λ is not on a Stokes ray, $\mathcal{J}_\sigma$ is closed and has its ends at $h = -\infty$. Hence, $\mathcal{J}_\sigma$ defines an element of the relative homology $H_n(X, X_{-T}; \mathbb{Z})$. Moreover, according to Morse theory, the $\mathcal{J}_\sigma$ give a basis of this relative homology.

In crossing a Stokes ray along which $\mathrm{Im}\,\mathcal{I}_\sigma = \mathrm{Im}\,\mathcal{I}_\tau$, the cycles $\mathcal{J}_\sigma$ and $\mathcal{J}_\tau$ may jump. Such jumping involves nontrivial flows between p_σ and p_τ. A downward flow from p_σ to p_τ is only possible if $h_\sigma > h_\tau$. The general form of the jumping across a Stokes ray with $\mathrm{Im}\,\mathcal{I}_\sigma = \mathrm{Im}\,\mathcal{I}_\tau$, $h_\sigma > h_\tau$ is

$$\begin{aligned} \mathcal{J}_\sigma &\to \mathcal{J}_\sigma + \mathfrak{m}_{\sigma\tau}\mathcal{J}_\tau \\ \mathcal{J}_\tau &\to \mathcal{J}_\tau. \end{aligned} \tag{3.33}$$

Here $\mathfrak{m}_{\sigma\tau}$ receives a contribution of ± 1 from each downward flow line from p_σ to p_τ. (The sign of the contribution is the sign of a certain determinant, described at the end of section 4.1.2.)

Whenever the Lefschetz thimble $\mathcal{J}_\sigma$ is defined, the integral

$$Q_{\sigma,\lambda} = (-i\lambda)^{n/2} \int_{\mathcal{J}_\sigma} \mathrm{d}^n x \, \exp(\mathcal{I}) \tag{3.34}$$

is always convergent. Moreover, the phase of the function $\exp(\mathcal{I})$ is constant along the integration cycle, and its modulus has a unique maximum at the critical point p_σ. Hence, for large λ, this integral has an asymptotic expansion

$$Q_{\sigma,\lambda} \sim \exp(\mathcal{I}_\sigma) \sum_{t=0}^{\infty} a_t \lambda^{-t}. \tag{3.35}$$

To find the asymptotic behavior of the analytically continued integral, we will use the expansion

$$\mathcal{C} = \sum_\sigma \mathfrak{n}_\sigma \mathcal{J}_\sigma. \tag{3.36}$$

For real λ, $\mathcal{C}$ is simply the original real integration cycle $\mathcal{C}_\mathbb{R}$. Unlike the case of the Airy function, we do not necessarily have a convenient way to directly determine the dependence of $\mathcal{C}$ on λ. However, once the coefficients $\mathfrak{n}_\sigma$ are known for real λ, we can determine their dependence on λ by using the "wall-crossing" formulas (3.33).

To determine the coefficients for real λ, we proceed as in the one-variable case. For each critical point p_σ, we define a cycle $\mathcal{K}_\sigma$ that consists of upward-going flow lines that start at p_σ. Away from Stokes rays, $\mathcal{K}_\sigma$ is an element of $H_n(X, X^T)$ for very large T, where $X^T \subset X$ is defined by the condition $h \geq T$. Just as for $n = 1$, the groups $H_n(X, X_{-T})$ and $H_n(X, X^T)$ are naturally dual, and moreover the pairings of the $\mathcal{J}_\sigma$ and $\mathcal{K}_\tau$ are

$$\langle \mathcal{J}_\sigma, \mathcal{K}_\tau \rangle = \delta_{\sigma\tau}. \tag{3.37}$$

As a result, the coefficients in (3.36) are

$$\mathfrak{n}_\sigma = \langle \mathcal{C}, \mathcal{K}_\sigma \rangle. \tag{3.38}$$

Thus, $\mathfrak{n}_\sigma$ can be computed by counting (with signs) upward flows from p_σ to $\mathcal{C}$.

Let us analyze this formula for real λ and the original integration cycle $\mathcal{C}_\mathbb{R}$. We divide the set Σ of critical points into three subsets: $\Sigma_\mathbb{R}$ consists of critical points with all x^i real, Σ_- consists of critical points with x^i not all real and $h < 0$, and $\Sigma_{\geq 0}$ consists of critical points with x^i not real and $h \geq 0$. Note that $h = 0$ along $\mathcal{C}_\mathbb{R}$, and in particular $h_\sigma = 0$ for $\sigma \in \Sigma_\mathbb{R}$. If $\sigma \in \Sigma_\mathbb{R}$, then the unique upward flow from p_σ to $\mathcal{C}_\mathbb{R}$ is the trivial one, as h would increase along a nonconstant flow. So in this case $\mathfrak{n}_\sigma = 1$. If $\sigma \in \Sigma_{\geq 0}$, there are no upward flow lines from p_σ to $\mathcal{C}_\mathbb{R}$, as this would contradict the fact that h increases along such lines. Finally, if $\sigma \in \Sigma_-$, there may or may not be ascending flow lines from p_σ to $\mathcal{C}_\mathbb{R}$.

The general structure of the expression for $\mathcal{C}_\mathbb{R}$, with real λ, is therefore

$$\mathcal{C}_\mathbb{R} = \sum_{\sigma \in \Sigma_\mathbb{R}} \mathcal{J}_\sigma + \sum_{\sigma \in \Sigma_-} \mathfrak{n}_\sigma \mathcal{J}_\sigma, \tag{3.39}$$

with unknown integer coefficients $\mathfrak{n}_\sigma$. Correspondingly, for real λ, our original integral Q_λ is

$$Q_\lambda = \sum_{\sigma \in \Sigma_\mathbb{R}} Q_{\sigma,\lambda} + \sum_{\sigma \in \Sigma_-} \mathfrak{n}_\sigma Q_{\sigma,\lambda}. \tag{3.40}$$

Since each term on the right hand side has the asymptotic expansion (3.35), this formula makes manifest the behavior for large real λ. For complex λ, there is a similar formula, but of course one must use the appropriate coefficients.

The significance of the formula (3.40) is as follows. In studying the function Q_λ for large real λ, the obvious critical points that one would expect to contribute are the ones that lie on the real cycle $\mathcal{C}_\mathbb{R}$. They indeed all contribute with a coefficient of 1, as one would expect. Critical points in $\Sigma_{\geq 0}$ would make contributions to the large λ behavior that typically would grow exponentially and dominate the real critical points (if $h_\sigma > 0$) and that in the exceptional case $h_\sigma = 0$ would compete for large real λ with the real critical points (which also have $h = 0$). However, such

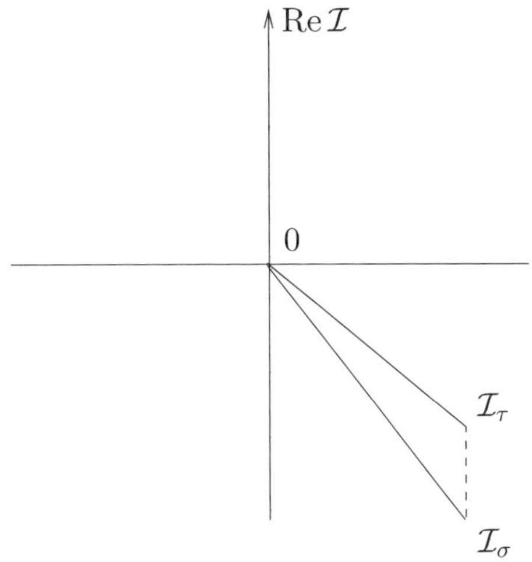

FIGURE 3. The complex $\mathcal{I}$ plane is sketched here with $h = \mathrm{Re}\,\mathcal{I}$ running vertically. The exponents $\mathcal{I}_\tau$ and $\mathcal{I}_\sigma$ of two critical points p_τ and p_σ have arguments between π and $3\pi/2$. If $\mathrm{Arg}\,\mathcal{I}_\tau > \mathrm{Arg}\,\mathcal{I}_\sigma$, then at a Stokes ray with $\mathrm{Im}\,\mathcal{I}_\tau = \mathrm{Im}\,\mathcal{I}_\sigma$, one has $h_\tau > h_\sigma$. This ensures that the coefficient $\mathfrak{n}_\tau$ is unaffected in crossing the Stokes line, though $\mathfrak{n}_\sigma$ may jump. The analytically continued integral will grow exponentially when λ is varied so that $\mathrm{Arg}\,\mathcal{I}_\tau$ exceeds $3\pi/2$.

critical points do not contribute to the integral. Finally, there may be contributions from complex critical points with h strictly negative. Their contributions are exponentially small for large real λ. But there is no easy recipe to determine which such critical points do contribute exponentially small corrections to Q_λ, and with what weights. Rather, for $\sigma \in \Sigma_-$, one simply has to determine the coefficient $\mathfrak{n}_\sigma$ by counting (with signs) the upward flow lines from p_σ to $\mathcal{C}_\mathbb{R}$.

3.2.1. *Essential singularities.* Now we wish to address the following question: Under what conditions does the analytically continued integral Q_λ have an essential singularity at $\lambda = \infty$?

There is one special case in which there is no such singularity. Suppose that, near infinity in the complex λ plane, the critical points p_σ for which the coefficients $\mathfrak{n}_\sigma$ are nonzero all have $\mathcal{I}_\sigma = 0$. Then Q_λ is bounded for $\lambda \to \infty$, in any direction. In particular, it has no essential singularity at infinity.

The hypothesis that the $\mathcal{I}_\sigma = 0$ whenever $\mathfrak{n}_\sigma \neq 0$ is very strong, and there are not many obvious examples, other than an oscillatory Gaussian integral with $f(x) = x^2$. However, something similar is possible in Chern-Simons theory.

Now we want to explain a converse. For simplicity, assume that the values of $\mathcal{I}_\sigma$ at critical points are distinct. (The purpose of this is to avoid possible cancellations.) Let p_σ be a critical point with $\mathcal{I}_\sigma \neq 0$, and suppose that in some wedge-shaped region of the complex λ plane, $\mathfrak{n}_\sigma \neq 0$. We claim that in this case, the integral Q_λ has an essential singularity at $\lambda = \infty$. This result is immediate if $h_\sigma > 0$, for then the critical point p_σ makes an exponentially large contribution to Q_λ in the wedge-shaped region of the λ plane in which we have started.

Suppose, though, that in the initially considered region of the λ plane, all critical points with $\mathfrak{n}_\sigma \neq 0$ have $h_\sigma \leq 0$. This means that for such critical points, $\mathcal{I}_\sigma = r_\sigma \exp(i\phi_\sigma)$, with r_σ positive and $\pi/2 \leq \phi_\sigma \leq 3\pi/2$. Either the largest of the ϕ_σ is at least π or the smallest is no greater than π; suppose for definiteness we are in the former situation. (Otherwise, one makes the same argument, rotating λ in a counterclockwise rather than clockwise direction.) Let τ be such that, in the initially chosen wedge in the complex λ plane, ϕ_τ is maximal among critical points with nonzero $\mathfrak{n}_\tau$. (If there is more than one critical point with maximal ϕ, pick the one of smallest $|\mathcal{I}|$.) Increasing the argument of λ increases all ϕ_σ's by the same amount. Increase the argument of λ until ϕ_τ exceeds $3\pi/2$, so that h_τ becomes positive. In the process, one may cross Stokes rays, but the fact that ϕ_τ is maximal and is at least π ensures that $\mathfrak{n}_\tau$ does not jump in crossing such rays. (See fig. 3; if $\phi_\tau \geq \phi_\sigma$ and $\operatorname{Im} \mathcal{I}_\tau = \operatorname{Im} \mathcal{I}_\sigma$, then $h_\tau \geq h_\sigma$ so $\mathfrak{n}_\tau$ does not jump.) Once h_τ becomes positive, Q_λ grows exponentially with λ, ensuring that it has an essential singularity at $\lambda = \infty$.

3.2.2. *A special situation.* Now we will consider a special situation that was described in section 2.1. We assume that $n = 2w$, and that the polynomial $f(x)$ is actually the imaginary part of a polynomial g in w complex variables $z^1, \ldots, z^w$; we suppose that g is generic enough to have only finitely many critical points, which are all nondegenerate. Then we consider the integral

$$Q_\lambda = \int \mathrm{d}^w z\, \mathrm{d}^w \overline{z}\, \exp(\lambda(g(z) - \overline{g(z)}), \tag{3.41}$$

initially for real λ.

Actually, we can generalize the problem slightly without any difficulty; the generalization is relevant to Chern-Simons theory. As λ is initially real, we can absorb it in the definition of g. Then instead of simply analytically continuing with respect to λ, we can analytically continue with respect to all of the coefficients in $\overline{g}$. In other words, the following integral

$$Q_{g,\widetilde{g}} = \int_{\mathcal{C}_\mathbb{R}} \mathrm{d}^w z\, \mathrm{d}^w \widetilde{z}\, \exp(g(z) - \widetilde{g}(\widetilde{z})), \tag{3.42}$$

is a convergent oscillatory integral if the polynomial $\widetilde{g}$ is the complex conjugate of the polynomial g (meaning that $\widetilde{g}$ is obtained from g by complex conjugating all coefficients) and the integration cycle $\mathcal{C}_\mathbb{R}$ is defined by $\widetilde{z}^i = \overline{z}^i$. We would like to analytically continue with respect to the coefficients in g and $\widetilde{g}$ to get a holomorphic function of those coefficients. (A similar generalization could be made in section 3.2; we could absorb the real number λ in the definition of the polynomial f, and then analytically continue with respect to all of the coefficients in f, not just λ.)

In our previous analysis, we had to avoid $\lambda = 0$, and similarly in the present context, we must impose a condition on how g and $\widetilde{g}$ are allowed to vary. The condition we want is that as g and $\widetilde{g}$ vary, the critical points of the polynomials g and $\widetilde{g}$ should remain nondegenerate and their number should remain fixed. For example, we do not allow leading coefficients in g or $\widetilde{g}$ to vanish in a way that would cause the number of critical points to become infinite or would cause some critical points to go to infinity. This restriction is assumed in what follows.

The basic idea behind the analytic continuation with respect to g and $\widetilde{g}$ is the same as we have already described. For $\widetilde{g} \neq \overline{g}$, we will have to relax the condition

$\widetilde{z} = \overline{z}$ and integrate over a more general cycle $\mathcal{C}$:

$$Q_{\mathcal{J};g,\widetilde{g}} = \int_{\mathcal{C}} \mathrm{d}^w z \, \mathrm{d}^w \widetilde{z} \ \exp(g(z) - \widetilde{g}(\widetilde{z})). \tag{3.43}$$

Of course, the integration cycle $\mathcal{C}$ must be chosen so that the integral converges. By now we know how to describe cycles of the appropriate form. We set $\mathcal{I} = g(z) - \widetilde{g}(\widetilde{z})$ and $h = \mathrm{Re}\,\mathcal{I}$. We treat h as a Morse function. To each critical point p_σ of h, we attach a Lefschetz thimble $\mathcal{J}_\sigma$. These cycles form a basis for the appropriate relative homology group. In the original case $\widetilde{g} = \overline{g}$, we express the original real integration cycle $\mathcal{C}_{\mathbb{R}}$ as a linear combination of the $\mathcal{J}_\sigma$'s, by the usual formula $\mathcal{C}_{\mathbb{R}} = \sum_\sigma \mathfrak{n}_\sigma \mathcal{J}_\sigma$. The integrals $Q_{\mathcal{J}_\sigma;g,\widetilde{g}}$ are convergent for all σ, g and $\widetilde{g}$, and the desired analytic continuation $Q_{g,\widetilde{g}}$ of our original integral is given by

$$Q_{g,\widetilde{g}} = \sum_\sigma \mathfrak{n}_\sigma Q_{\mathcal{J}_\sigma;g,\widetilde{g}}. \tag{3.44}$$

The only subtlety is the usual one. As one varies g and $\widetilde{g}$ away from the locus $\widetilde{g} = \overline{g}$, one may cross surfaces on which the cycles $\mathcal{J}_\sigma$ and therefore the coefficients $\mathfrak{n}_\sigma$ jump. This happens when there are flows between different critical points, as a result of which the downward flow from the critical point with greater h fails to define a relative homology cycle. Since $\mathrm{Im}\,\mathcal{I}$ is a conserved quantity in the flow equation, the jumping can only occur when two distinct critical points have the same value of $\mathrm{Im}\,\mathcal{I}$. The jumping takes the form $\mathcal{J}_\sigma \to \mathcal{J}_\sigma + \mathfrak{m}_{\sigma\tau}\mathcal{J}_\tau$, where $h_\sigma > h_\tau$ and $\mathrm{Im}\,\mathcal{I}_\sigma = \mathrm{Im}\,\mathcal{I}_\tau$. We will refer to surfaces on which distinct critical points have the same value of $\mathrm{Im}\,\mathcal{I}$ as Stokes surfaces. We also use the term Stokes curve for a Stokes surface of real dimension 1 (thus, a Stokes ray is a special case of a Stokes curve).

What is rather special about this problem is that $\mathcal{I} = g(z) - \widetilde{g}(\widetilde{z})$ is the sum of a function on one space (which we will call $\mathbb{C}^w$, parametrized by the z^i) and a function on another space (which we will call $\widetilde{\mathbb{C}}^w$, a second copy of $\mathbb{C}^w$ parametrized by the $\widetilde{z}^i$). $\mathcal{I}$ is defined on the product space $X = \mathbb{C}^{2w} = \mathbb{C}^w \times \widetilde{\mathbb{C}}^w$. The fact that $\mathcal{I}$ is the sum of functions on the two factors has obvious implications for the critical points.

Let q_σ, $\sigma \in \Sigma_g$ be the critical points of g, as a function on $\mathbb{C}^w$, and let r_τ, $\tau \in \Sigma_{\widetilde{g}}$, be the critical points of $\widetilde{g}$, as a function on $\widetilde{\mathbb{C}}^w$. Then the critical points of $\mathcal{I}$ are the points $p_{\sigma,\tau} = q_\sigma \times r_\tau \in \mathbb{C}^w \times \widetilde{\mathbb{C}}^w = X$. Similarly, the Lefschetz thimble $\mathcal{J}_{\sigma,\tau}$ associated to a critical point $p_{\sigma,\tau}$ in X is the product of cycles $\mathcal{J}^g_\sigma \in \mathbb{C}^w$ and $\mathcal{J}^{\widetilde{g}}_\tau \in \widetilde{\mathbb{C}}^w$. Here $\mathcal{J}^g_\sigma$ is the Lefschetz thimble in $\mathbb{C}^w$ associated to the critical point q_σ of the Morse function $\mathrm{Re}\,g$, and similarly $\mathcal{J}^{\widetilde{g}}_\tau$ is the Lefschetz thimble in $\widetilde{\mathbb{C}}^w$ associated to the critical point r_τ of the Morse function $\mathrm{Re}\,(-\widetilde{g})$.

We can therefore rewrite (3.44) as follows:

$$Q_{g,\widetilde{g}} = \sum_{\sigma\in\Sigma_g} \sum_{\tau\in\Sigma_{\widetilde{g}}} \mathfrak{n}_{\sigma,\tau} \int_{\mathcal{J}^g_\sigma} \mathrm{d}^w z \ \exp(g(z)) \int_{\mathcal{J}^{\widetilde{g}}_\tau} \mathrm{d}^w \widetilde{z} \ \exp(-\widetilde{g}(\widetilde{z})). \tag{3.45}$$

Thus $Q_{g,\widetilde{g}}$ is everywhere a finite sum of functions of g times functions of $\widetilde{g}$, obtained from integrals over the z's and the $\widetilde{z}$'s, respectively. The coefficients in the sum, however, can jump as one crosses Stokes surfaces.

As was explained in section 2, the problem that we have just described is a prototype for analytic continuation of Chern-Simons gauge theory with complex gauge group.

3.3. Symmetries. Since gauge symmetry is an essential ingredient in Chern-Simons theory, including symmetries in our finite-dimensional models will improve their analogy with Chern-Simons. Before taking this step, we consider a few preliminaries.

3.3.1. *Oscillatory integrals on more general manifolds.* One generalization is to consider, instead of an oscillatory integral on $\mathbb{R}^n$, an oscillatory integral on some other space Y. Y must have a suitably nice complexification X, but we will not attempt to be precise about this. Instead we just describe a typical example. We take Y to be a real affine variety such as the one defined by the equation

$$x^2 + y^2 = z^2 + 1. \tag{3.46}$$

Then we introduce a sufficiently generic polynomial $f(x, y, z)$, and consider the oscillatory integral $Z_\lambda = \int_Y \mathrm{d}\theta \, \exp(i\lambda f)$, where $\mathrm{d}\theta = \mathrm{d}x \, \mathrm{d}y/z$ is the natural volume form on Y.

Analytic continuation in λ can now be carried out by an obvious generalization of what we have said for $\mathbb{R}^n$. We embed Y in the complex manifold X defined by the same equation (3.46) but now with x, y, and z regarded as complex variables. The top degree form $\mathrm{d}\theta$ on Y extends to a holomorphic form of top degree on X (given by the same formula $\mathrm{d}x \, \mathrm{d}y/z$), which can be integrated over any middle-dimensional cycle in X. Introducing the Morse function $h = \mathrm{Re}\,(i\lambda f)$, where f is regarded now as a holomorphic function on X, we associate to each critical point a relative homology cycle obtained by downward flow. The analytic continuation is made as usual by expressing the original integration cycle Y in terms of these cycles, and then taking Stokes phenomena into account.

3.3.2. *Manifolds of critical points.* Another generalization is to consider a polynomial f on $\mathbb{R}^n$ (or on some more general space, as just described) that has not isolated critical points but a manifold $\mathcal{N}$ of critical points. The analytic continuation of f to a holomorphic function on $\mathbb{C}^n$ likewise has a manifold $\mathcal{M}$ of critical points; $\mathcal{M}$ is a complex manifold, of course, and its intersection with $\mathbb{R}^n$ coincides with $\mathcal{N}$. (We will consider only the case that $\mathcal{M}$ is smooth, and we assume that the matrix of second derivatives of f in the normal direction to $\mathcal{M}$ is invertible.) We want to use Morse theory on $\mathbb{C}^n$ to study the analytic continuation of the usual integral $Z_\lambda = \int \mathrm{d}^n x \, \exp(i\lambda f)$.

Suppose that $\mathcal{M}$ is a union of components $\mathcal{M}_\alpha$ of complex dimension t_α. $\mathcal{I} = i\lambda f$ has a constant value $\mathcal{I}_\alpha$ on $\mathcal{M}_\alpha$, since $\mathcal{M}_\alpha$ is a critical point set. As usual, we consider $h = \mathrm{Re}\mathcal{I}$ as a Morse function. Then $\mathcal{M}_\alpha$ has Morse index $n - t_\alpha$. (This is shown by essentially the same argument used in section 3.1.1 for the case $t_\alpha = 0$; one can choose local coordinates in the normal direction to $\mathcal{M}_\alpha$ so that $\mathcal{I} = c + \sum_{i=1}^{n-t_\alpha} z_i^2$.) The explanation in section 3.1.1 for why h is a perfect Morse function assumed that $t_\alpha = 0$ for all components, so that all critical points have the same Morse index. Here is an alternative argument that does not assume $t_\alpha = 0$. In general, a Morse function is perfect if there are no flows between distinct critical points. (This is only possible for Morse functions that are unbounded below, like the ones we consider here, since otherwise any downward flow from one critical point ends at another critical point.) As we have found in eqn. (3.11), $\mathrm{Im}\,\mathcal{I}$ is a conserved quantity for the flow equation. Moreover, there never are flows between different critical points with the same value of $\mathcal{I}$. For generic λ, no two distinct components of $\mathcal{M}_\alpha$ with different values of $\mathcal{I}_\alpha$ have the same value of $\operatorname{Im}\mathcal{I}_\alpha$, so generically there are no flows between distinct critical points and h is a perfect Morse function.

There is a general recipe, given a perfect Morse function on any manifold X, for describing the relative homology of X in terms of downwards flows from critical points. Let $\mathcal{S}$ be a component of the critical point set that has Morse index s. Let $\mathcal{W} \subset \mathcal{S}$ be a cycle of dimension r. For any point $w \in \mathcal{W}$, the points in X that lie on downwards flows from w form an s-dimensional subspace of X. So the points in X that can be reached by downward flow from somewhere on $\mathcal{W}$ are a family of dimension $r + s$. This family is a cycle $\mathcal{J}_\mathcal{W}$ of dimension $r + s$ in the relative homology. These cycles, which we view as generalized Lefschetz thimbles, generate the relative homology of X (subject only to the relations that hold in the homology of the critical point set).

Applying this to our problem, if $\mathcal{M}_\alpha$ is a component of the critical point set that has complex dimension t_α and Morse index $n - t_\alpha$, then to get an integration cycle on X – which should be a cycle of dimension n – we need to start with a cycle on $\mathcal{M}_\alpha$ of real dimension t_α, that is, a middle-dimensional cycle.

So the analytic continuation of the integral Z_λ can be described using relative homology cycles attached to the components $\mathcal{M}_\alpha$ of the critical point set. But in general to a component $\mathcal{M}_\alpha$ we attach not just one cycle but a number equal to the rank of the middle-dimensional homology of $\mathcal{M}_\alpha$.

This construction may have various applications to Chern-Simons theory, since in general the moduli space of representations of the fundamental group of a three-manifold has components of positive dimension. However, the only application we will make in this paper concerns reducible critical points, that is, flat connections that preserve some gauge symmetry. Moreover, this application concerns a situation in which every relevant component of $\mathcal{M}_\alpha$ has middle-dimensional cohomology of rank 1, and hence contributes only one cycle to the middle-dimensional cohomology. In a sense, therefore, our application involves a situation that is as simple as the case of only isolated critical points.

3.3.3. *Group actions.* Why might critical points not be isolated? One simple answer to this question will be important in our study of gauge theory. In the presence of a group action, the critical points will form orbits of the group, and this may force them to be non-isolated.

We consider a manifold Y with a chosen action of a compact connected Lie group H. We also assume that Y has a good complexification X, which possesses a real involution that leaves Y fixed. We endow Y with an H-invariant function f whose critical points consist of finitely many H orbits $\mathcal{O}_\sigma$, $\sigma \in \Sigma$.

As usual, we want to study the analytic continuation of the integral $Z_\lambda = \int_Y \mathrm{d}\theta \, \exp(i\lambda f)$, where $\mathrm{d}\theta$ is the restriction to Y of a holomorphic top form on X. Assuming that f can be analytically continued to a holomorphic function on Y (which we also denote as f), we analytically continue the integral Z_λ by embedding Y in X and using the Morse function $h = \operatorname{Re}(i\lambda f)$ on X. We assume that the H action on Y analytically continues to an action on X of a complex Lie group G that is the complexification of H; in this case, the critical points of h are orbits of G. We assume that there are only finitely many such critical orbits, near each of which h is nondegenerate in the normal directions. For clarity, we will write $\mathcal{O}^G$ or $\mathcal{O}^H$ for, respectively, a critical orbit of G acting on X or a critical orbit of H acting on Y.

To orient ourselves, let us first consider the case that H acts freely on Y, and G acts freely on X. Then we can write Z_f as an integral over $Y' = Y/H$:

$$Z_f = \int_{Y'} \mathrm{d}\theta' \exp(i\lambda f), \tag{3.47}$$

where the volume form $\mathrm{d}\theta'$ is obtained by integrating $\mathrm{d}\theta$ over the fibers of $Y \to Y'$. From a "downstairs" point of view, to analytically continue Z_f, we must embed Y' in a suitable complex manifold X'. Roughly speaking, the appropriate choice is $X' = X/G$, which – as G is a complexification of H and X is a complexification of Y – is a complexification of $Y' = Y/H$. (Unstable orbits – the notion is discussed below – should be discarded in defining the quotient X/G.)

Since f has only finitely many critical orbits as a function on X, it follows that as a function on X', f has only finitely many critical points. We are thus in the familiar situation of isolated critical points only, and the relative homology of X' therefore has a basis consisting of a Lefschetz thimble $\mathcal{J}'_\sigma$ for every critical point p'_σ. We can analytically continue the integral Z_f in the usual way by expressing the real cycle $\mathcal{C}_\mathbb{R} = Y' \subset X'$ as a linear combination of the $\mathcal{J}'_\sigma$, and keeping track of Stokes phenomena.

How would we get an equivalent result by computing upstairs on Y and X rather than downstairs on Y' and X'? Every critical point p'_σ on X' corresponds to a critical orbit $\mathcal{O}^X_\sigma$ on X. Since we have assumed that G acts freely on X, the orbits are all copies of G. Topologically, the complex Lie group G is isomorphic to the cotangent bundle T^*H of its maximal compact subgroup H. The middle-dimensional homology of T^*H is of rank 1, generated by the zero section of the cotangent bundle. Hence, from an upstairs point of view, the middle-dimensional relative homology of X has a basis with one generator for every critical orbit. These generators correspond in an obvious way to the generators of the relative homology of X'. The associated Lefschetz thimbles also have an obvious correspondence, and the Stokes surfaces are the same whether one computes upstairs or downstairs. So we can perform more or less equivalent computations upstairs or downstairs.

In the downstairs description, the critical points are isolated, so we do not really need the theory sketched in section 3.3.2 as long as H acts freely. Now let us consider a more general case in which H does *not* act freely on Y. This being so, the quotient $Y' = Y/H$ is singular, and the general theory would not readily apply to it. So instead we will calculate on Y. We still assume that the critical point set of f on Y is a finite union of H orbits $\mathcal{O}^H_\sigma$, and that the critical point set of f on X is an analogous (but perhaps larger[12]) finite union of G orbits $\mathcal{O}^G_\tau$.

Any critical orbit $\mathcal{O}^H_\sigma$ in Y has a complexification that is one of the $\mathcal{O}^G_\sigma$. The complexification of an H orbit $\mathcal{O}^H \subset Y$ is, from a topological point of view, simply $T^*\mathcal{O}^H$. Its middle-dimensional homology is of rank 1, generated by the zero section of the cotangent bundle. Each critical orbit in Y therefore contributes one generator of the middle-dimensional relative homology of X, just as in the case of a free action.

[12]This is possible even in the absence of a group action. For example, the function $f(x) = x^3/3 + x$ has complex critical points that do not lie on the real axis.

What about critical orbits in X that do not intersect Y? We will show that not all critical orbits need to be taken into account, but only certain "semistable" ones. The semistable orbits are all topologically of the form $T^*\mathcal{O}^H$ for some H orbit $\mathcal{O}^H$, and again they have middle-dimensional homology of rank 1. (Every critical orbit that intersects Y is semistable.) So in short, the analytic continuation of Z_f is made using a middle-dimensional relative homology of X that has one generator for every semistable critical G orbit.

Now we will explain how semistability comes in. First of all, to construct the downward flowing cycles associated to a critical point, we need to pick a metric on X. We generally cannot pick this metric to be G-invariant, but it is commonly possible to pick an H-invariant Kahler metric, such that the Kahler form ω is odd under the real involution of X that leaves Y fixed. (For example, if $Y = \mathbb{R}^n$ with H-invariant metric $\sum_i (\mathrm{d}x^i)^2$, then we take on $X = \mathbb{C}^n$ the H-invariant Kahler metric $\sum_i |\mathrm{d}x^i|^2$. In this case, $\omega = -i\sum_k \mathrm{d}x^k \wedge \mathrm{d}\overline{x^k}$, and is odd under $x^k \leftrightarrow \overline{x^k}$.) We assume the existence of such a metric. H acts on X preserving the symplectic form ω, and it is possible to define a moment map $\vec{\mu}$ for the action of H on X. Here $\vec{\mu}$ maps a vector field V that generates a one-parameter subgroup of H to a function μ_V that obeys

$$\mathrm{d}\mu_V = \iota_V\omega, \tag{3.48}$$

where ι_V is the operation of contraction with V. In indices

$$\frac{\partial \mu_V}{\partial x^i} = V^{\bar{j}}\omega_{\bar{j}i}, \quad \frac{\partial \mu}{\partial \overline{x^i}} = V^j\omega_{j\bar{i}}. \tag{3.49}$$

(The map from V to μ_V is a linear map from the Lie algebra of H to the space of real-valued smooth functions on X.) We fix possible additive constants in the μ_V by requiring that these functions vanish identically on Y. This also ensures that the μ_V are odd under complex conjugation. This follows from (3.48), since the vector fields V are even under complex conjugation and ω is odd.

The downward flow equations of Morse theory have the beautiful property that $\vec{\mu}$ is conserved, in other words μ_V is conserved for all V. This follows from a straightforward computation. We have

$$\frac{\mathrm{d}\mu_V}{\mathrm{d}t} = \frac{\partial \mu_V}{\partial x^i}\frac{\mathrm{d}x^i}{\mathrm{d}t} + \frac{\partial \mu_V}{\partial \overline{x^i}}\frac{\mathrm{d}\overline{x^i}}{\mathrm{d}t} = -V^{\bar{j}}\omega_{\bar{j}i}g^{i\bar{k}}\frac{\partial h}{\partial \overline{x^k}} - V^j\omega_{j\bar{i}}g^{\bar{i}k}\frac{\partial h}{\partial x^k}. \tag{3.50}$$

Since h is the real part of a holomorphic function $\mathcal{I}$, we have $\partial h/\partial x^i = i\partial \mathrm{Im}\,\mathcal{I}/\partial x^i$, $\partial h/\partial \overline{x^i} = -i\partial\,\mathrm{Im}\,\mathcal{I}/\partial \overline{x^i}$. On the other hand, the Kahler form ω and Kahler metric g obey $\omega_{\bar{j}i}g^{i\bar{k}} = i\delta^{\bar{k}}_{\bar{j}}$, $\omega_{j\bar{i}}g^{\bar{i}k} = -i\delta^k_j$. So

$$\frac{\mathrm{d}\mu_V}{\mathrm{d}t} = -V^{\bar{j}}\frac{\partial\,\mathrm{Im}\,\mathcal{I}}{\partial \overline{x^j}} - V^j\frac{\partial\,\mathrm{Im}\,I}{\partial x^j} = -\iota_V\mathrm{d}\,\mathrm{Im}\,\mathcal{I}. \tag{3.51}$$

The right hand side vanishes, since $\mathcal{I}$ is H-invariant.

A less computational explanation of this result can be found by observing that the gradient flow with respect to the Morse function $\mathrm{Re}\,\mathcal{I}$ coincides with the Hamiltonian flow with Hamiltonian $\mathrm{Im}\,\mathcal{I}$. (Here we regard X as a symplectic manifold with symplectic form ω.) As usual, the moment map for H is a conserved quantity for the flow generated by an H-invariant Hamiltonian.

We will call a critical G-orbit semistable if it contains at least one point with $\vec{\mu} = 0$ and unstable otherwise. (If a semistable orbit is free or has only a finite stabilizer, we call it stable.) Since $\vec{\mu}$ is conserved along flows, and vanishes identically on Y, any critical G-orbit that is connected to Y by a flow is semistable. So when we express the real integration cycle $\mathcal{C}_{\mathbb{R}} = Y$ in terms of generalized Lefschetz thimbles associated to critical orbits, only semistable critical orbits enter. Similarly, as we explain momentarily, in crossing a Stokes surface, a cycle in the relative homology of X associated to a semistable critical orbit is always re-expressed in terms of other cycles associated to semistable critical orbits. So the analytic continuation of the integral Z_f can always be described using the semistable critical orbits only.

It remains to explain why Stokes phenomena connect semistable critical orbits only to each other. Suppose that $\mathcal{O}^G$ is a semistable critical G-orbit. It contains a distinguished H orbit, namely the orbit $\mathcal{W} = \mathcal{O}^H$ with $\vec{\mu} = 0$. $\mathcal{O}^G$ is the complexification of $\mathcal{O}^H$ and is topologically the cotangent bundle $T^*\mathcal{O}^H$. The middle-dimensional homology of $\mathcal{O}^H$ is generated by $\mathcal{W}$. Since $\vec{\mu}$ is conserved along flows, any critical orbit to which $\mathcal{W}$ can be connected by a flow has a point with $\vec{\mu} = 0$ and therefore is semistable.

In contrast to the middle-dimensional cohomology of a semistable G orbit, which is always of rank 1, an unstable G orbit might have vanishing or more complicated middle-dimensional cohomology. For $G = SL(2, \mathbb{C})$, an example of an unstable orbit is $\mathbb{CP}^1$, whose middle-dimensional cohomology vanishes. (Since $\mathbb{CP}^1$ is a homogeneous space for $H = SU(2)$, and the condition $\vec{\mu} = 0$ is H-invariant, $\vec{\mu}$ would have to vanish everywhere on $\mathbb{CP}^1$ if it vanishes anywhere; but if $\vec{\mu}$ is identically zero, then (3.48) implies that $V = 0$ for all generators of H, and the H action is trivial.) More generally, for any G, the flag manifold, which in general can have a middle-dimensional cohomology of high rank, is an example of an unstable orbit.

3.3.4. *More on group actions.* Continuing the discussion of the last subsection, let $\mathcal{J}_\sigma$ be the Lefschetz thimble associated to a semistable critical orbit $\mathcal{O}^G_\sigma$. $\mathcal{J}_\sigma$ consists of the points that can be reached by downward flow from $\mathcal{O}^H_\sigma$, the points in $\mathcal{O}^G_\sigma$ with $\vec{\mu} = 0$.

Suppose that we are given an integration cycle $\mathcal{C}$ and we wish to express it in terms of the $\mathcal{J}_\sigma$. $\mathcal{C}$ is a middle-dimensional, H-invariant element of the relative homology, and in practice one is usually interested in the case that $\mathcal{C}$ lies in $\vec{\mu}^{-1}(0)$.

To find the coefficients in an expansion

$$\mathcal{C} = \sum_\sigma \mathfrak{n}_\sigma \mathcal{J}_\sigma, \tag{3.52}$$

one wants to intersect $\mathcal{C}$ with a dual set of cycles $\mathcal{K}_\sigma$. The dual cycles should have a natural pairing with the $\mathcal{J}_\sigma$, such that

$$\langle \mathcal{J}_\sigma, \mathcal{K}_\tau \rangle = \delta_{\sigma\tau}. \tag{3.53}$$

As we have defined $\mathcal{J}_\sigma$ by downward flow from $\mathcal{O}^H_\sigma \subset \mathcal{O}^G_\sigma$, one's first thought is to define $\mathcal{K}_\sigma$ by upward flow from $\mathcal{O}^H_\sigma$. This does not work well, since in this case the intersection of $\mathcal{J}_\sigma$ with $\mathcal{K}_\sigma$, rather than a single point, would be the orbit $\mathcal{O}^H_\sigma$. (The algebraic intersection number would equal the Euler characteristic of this orbit; for instance, it would vanish for the case of a free orbit.) What one must do instead is to define $\mathcal{K}_\sigma$ by upward flow starting from a suitable dual to $\mathcal{O}^H_\sigma \subset \mathcal{O}^G_\sigma$. The appropriate dual is a fiber of the cotangent bundle. In other words, identifying $\mathcal{O}^G_\sigma$ as $T^*\mathcal{O}^H_\sigma$, we pick an arbitrary point $q_\sigma \in \mathcal{O}^H_\sigma$ and write $\widetilde{\mathcal{O}}^H_\sigma$ for the fiber of $T^*\mathcal{O}^H_\sigma$ at q_σ. Then $\mathcal{K}_\sigma$, defined by upward flow from $\widetilde{\mathcal{O}}^H_\sigma$, meets $\mathcal{J}_\sigma$ precisely at q_σ. ($\mathcal{K}_\sigma$ lies in a dual version of the relative homology, spanned by cycles that are allowed to go to infinity along G orbits as well as in the region where the Morse function becomes large.) Moreover, away from a Stokes curve, $\mathcal{K}_\sigma$ meets $\mathcal{J}_\tau$ not at all if $\sigma \neq \tau$. So $\langle \mathcal{J}_\sigma, \mathcal{K}_\tau \rangle = \delta_{\sigma\tau}$, and the coefficients in (3.52) are $\mathfrak{n}_\sigma = \langle \mathcal{C}, \mathcal{K}_\sigma \rangle$.

Similar questions arise when one considers the jumping phenomena $\mathcal{J}_\sigma \to \mathcal{J}_\sigma + \sum_\tau \mathfrak{m}_{\sigma\tau} \mathcal{J}_\tau$ in crossing a Stokes curve. The coefficients $\mathfrak{m}_{\sigma\tau}$ is computed by counting downward flow lines from $\mathcal{O}^H_\sigma$ to $\widetilde{\mathcal{O}}^H_\tau$. (Each such flow line is weighted by the sign of a certain determinant, as described in section 4.1.2.) The asymmetry between the definitions of downwards and upwards flowing cycles has an interesting consequence, which we will describe for the special case that there are two types of critical orbits – trivial orbits and free orbits. Suppose that the critical point set of an H-invariant Morse function $h = \mathrm{Re}\,\mathcal{I}$ contains a critical point p (located at $\vec{\mu} = 0$) and a free stable G orbit $\mathcal{O}^G$. In general (when allowed by conservation of $\mathrm{Im}\, I$), there may be flows from p to $\widetilde{\mathcal{O}}^H$, and the coefficient $\mathfrak{m}_{p\mathcal{O}}$ that appears in a jumping formula $\mathcal{J}_p \to \mathcal{J}_p + \mathfrak{m}_{p\mathcal{O}} \mathcal{J}_\mathcal{O}$ may be nonzero. However, the coefficient $\mathfrak{m}_{\mathcal{O}p}$ that counts flows from $\mathcal{O}^H$ to p always vanishes. A quick way to show this is to observe that by replacing $\mathcal{O}^H$ by an everywhere nonzero section of the cotangent bundle $\mathcal{O}^G = T^*\mathcal{O}^H \to \mathcal{O}^H$, such as the section $\vec{\mu}^{-1}(c)$, where c is a nonzero constant, one can eliminate all flows from $\mathcal{O}^H$ to p. So the algebraic invariant counting such flows must vanish.

In a more general case in which the critical point set consists of several isolated points p_σ and several free orbits $\mathcal{O}_\alpha$, the coefficients $\mathfrak{m}_{\mathcal{O}_\alpha, p_\sigma}$ will vanish by the argument just indicated. (All other coefficients may be nonzero in general.) This means that the subspace of the relative homology generated by the cycles $\mathcal{J}_{\mathcal{O}_\alpha}$ is invariant under all jumping phenomena. The subspace generated by the cycles $\mathcal{J}_{p_\sigma}$ is not invariant under jumping. This perhaps surprising fact will be demonstrated in an explicit and useful example in section 3.4.2.

3.4. Singularities. In studying an oscillatory integral $\int \mathrm{d}^n x \, \exp(i\lambda f)$, for a single and fairly generic polynomial f, it is reasonable to expect that the critical points of f are isolated and irreducible, as we have assumed so far. But as soon as f depends on one or more parameters, one should expect to meet singularities, where two or more critical points meet.

Singularities play an important role because Stokes surfaces always pass through the locus (in the parameter space of the polynomial f) at which two or more critical points meet. If we vary a parameter in f so that two critical points p_σ and p_τ coincide, then $f(p_\sigma) = f(p_\tau)$, which certainly ensures that $\mathrm{Im}\,(i\lambda f(p_\sigma)) = \mathrm{Im}\,(i\lambda f(p_\tau))$. So a Stokes surface always passes through the singular locus with $p_\sigma = p_\tau$.

Moreover, the flow equations simplify at a singularity. To understand what happens to the Lefschetz thimbles in crossing a Stokes surface, in general we must solve the flow equation in n complex dimensions. However, near a singularity, most of the n complex variables are unimportant. If a singularity can be modeled by the behavior of a complex polynomial in k variables – where in practice k will often be 1 or 2 – then near the singularity, the flow equations reduce effectively to equations for those k variables; the other modes are "massive" and decay quickly in solving the flow equations.

The relation between singularities and Stokes phenomena is therefore very useful in practice. Indeed, the subject reviewed in this section was developed partly for analyzing singularities and their relation to oscillatory integrals [**35**].

In section 5, where we study the analytic continuation of the colored Jones polynomial, we will vary one parameter (the ratio of the highest weight of a representation to the Chern-Simons level) and we will encounter three rather generic types of singularity. These will be described here.

3.4.1. *More on the Airy function.* The only singularity that arises generically in varying a single complex parameter in the absence of any symmetries is a meeting of two critical points. Even if this phenomenon happens in complex dimension n, only one complex variable is essential in describing it. We simply consider a cubic polynomial that depends on a parameter ϵ:

$$f(x) = \frac{x^3}{3} - \epsilon x. \tag{3.54}$$

There are two critical points at $x = \pm\sqrt{\epsilon}$; they meet at $\epsilon = 0$.

The corresponding oscillatory integral $I(\lambda, \epsilon) = \int \mathrm{d}x \exp(i\lambda f)$ is essentially the Airy function. We can use the scaling $x \to x\lambda^{-1/3}$ to express this integral in terms of a function of one variable:

$$I(\lambda, \epsilon) = \lambda^{-1/3} I(1, \lambda^{2/3}\epsilon) = \lambda^{-1/3} \int \mathrm{d}x \, \exp(i(x^3 - \lambda^{2/3}\epsilon x)). \tag{3.55}$$

Setting $\epsilon = 0$, a special case is $I(\lambda, 0) \sim \lambda^{-1/3}$. Comparing this to a Gaussian integral $\widetilde{I}(\lambda) = \int \mathrm{d}x \exp(i\lambda x^2) \sim \lambda^{-1/2}$, the ratio is

$$\frac{I(\lambda, 0)}{\widetilde{I}(\lambda)} \sim \lambda^{1/6}, \ \lambda \to \infty. \tag{3.56}$$

These facts will be useful in section 5.3.2.

For our remaining remarks, we use the above scaling to reduce to the case $\lambda = 1$. The good regions at infinity in the complex x-plane where $\exp(if(x))$ decays are then independent of ϵ and were sketched in fig. 1. The Lefschetz thimble associated to a critical point always connects two of the good regions.

Since the values of $\mathcal{I} = if$ at the critical points are $\mp 2\epsilon^{3/2}/3$, the condition for a Stokes curve is that $\epsilon^{3/2}$ should be imaginary; thus $\operatorname{Arg}\epsilon$ should equal π or $\pm\pi/3$. In particular, several Stokes curves meet at $\epsilon = 0$. It is clear why this is so: at $\epsilon = 0$, the two critical points meet, and therefore have the same value of $\mathcal{I}$ and in particular the same value of $\operatorname{Im}\mathcal{I}$.

Our application will involve the case that ϵ is real. If ϵ is positive, we write $p_\pm$ for the critical points at $x = \pm\sqrt{\epsilon}$, and $\mathcal{J}_\pm$ for the corresponding Lefschetz thimbles. If ϵ is negative, we write $\widetilde{p}_\pm$ for the critical points at which the action is positive or negative, respectively. Thus the points $\widetilde{p}_\pm$ are located at $x = \mp i\sqrt{-\epsilon}$, and the action takes the values $\mathcal{I}_\pm = \pm(2/3)(-\epsilon)^{3/2}$ at those points. We write $\widetilde{\mathcal{J}}_-$ for the Lefschetz thimble associated to $\widetilde{p}_-$. The Lefschetz thimble $\widetilde{\mathcal{J}}_+$ associated to $\widetilde{p}_+$ is ill-defined if $\epsilon < 0$, because the ray $\epsilon < 0$ is a Stokes ray. We define $\widetilde{\mathcal{J}}_+$ by taking the limit as ϵ approaches the negative axis from above.

It is not difficult to describe the cycles $\mathcal{J}_\pm$ and $\widetilde{\mathcal{J}}_\pm$ qualitatively as curves in the complex x-plane (fig. 4). From the figure, we see that the relation between these two bases for the relative homology is

$$\begin{aligned} \mathcal{J}_+ &= -\widetilde{\mathcal{J}}_+ + \widetilde{\mathcal{J}}_- \\ \mathcal{J}_- &= \widetilde{\mathcal{J}}_+ \end{aligned} \tag{3.57}$$

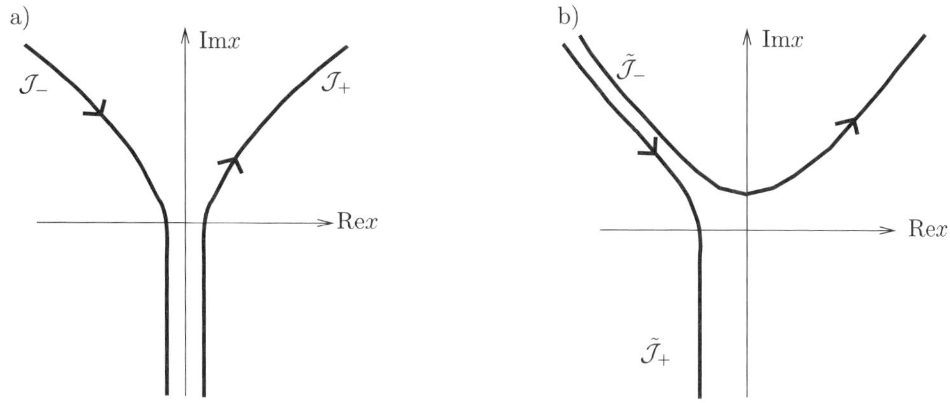

FIGURE 4. Qualitative behavior of Lefschetz thimbles for the Airy function for $\epsilon > 0$ (a) and $\epsilon < 0$ (b). In case (b), ϵ has been given a small imaginary part, since the negative ϵ axis is a Stokes curve.

or equivalently

$$\begin{aligned} \widetilde{\mathcal{J}}_+ &= \mathcal{J}_- \\ \widetilde{\mathcal{J}}_- &= \mathcal{J}_+ + \mathcal{J}_- . \end{aligned} \tag{3.58}$$

3.4.2. *An example with $SO(2)$ symmetry.* Now we will consider a singularity that can arise by varying one parameter in the presence of $SO(2)$ symmetry. A minimal case is a singularity that can be modeled by two variables x, y, with $SO(2)$ acting on the xy plane by rotations. The basic invariant is $x^2 + y^2$. The origin $x = y = 0$ is a critical point of any $SO(2)$-invariant polynomial. There may also be critical orbits characterized by nonzero values of $x^2 + y^2$. A meeting of two critical orbits both with $x^2 + y^2 \neq 0$ can be modeled again by the Airy integral – the $SO(2)$ action just factors out. The new case is a collision between a critical orbit with $x^2 + y^2 \neq 0$ and the critical point at the origin.

Like a collision of two critical points in the absence of any symmetry, this can happen naturally as a result of adjusting one parameter. We simply consider the polynomial $f(x, y) = (x^2 + y^2)^2/2 - \epsilon(x^2 + y^2)$. There is a critical point at $x = y = 0$ and a critical orbit $x^2 + y^2 = \epsilon$. We denote these as p and $\mathcal{O}$ respectively. $\mathcal{O}$ contains real points if $\epsilon > 0$ but not if $\epsilon < 0$. As a real critical orbit, $\mathcal{O}$ thus disappears in passing through $\epsilon = 0$. Setting $\mathcal{I} = i\lambda f$ with $\lambda > 0$, the values of $\mathcal{I}$ on the critical orbits are $\mathcal{I}_p = 0$ and $\mathcal{I}_{\mathcal{O}} = -i\lambda\epsilon^2/2$. So the Stokes curves are given by $\operatorname{Re}\epsilon^2 = 0$, which corresponds to the pair of lines $\epsilon = r\exp(\pm\pi i/2)$, with real r (fig. 5). As expected, they meet at the singularity at $\epsilon = 0$.

A simple scaling reduces $I(\lambda, \epsilon) = \int \mathrm{d}x\,\mathrm{d}y\,\exp\Big(i\lambda\big((x^2+y^2)^2 - \epsilon(x^2+y^2)\big)\Big)$ to a function of one variable. Setting $x, y \to \lambda^{-1/4}x, \lambda^{-1/4}y$, we get

$$I(\lambda, \epsilon) = \lambda^{-1/2} I(1, \lambda^{1/2}\epsilon) = \lambda^{-1/2} \int \mathrm{d}x\,\mathrm{d}y\,\exp\Big(i\big((x^2+y^2)^2 - \lambda^{1/2}\epsilon(x^2+y^2)\big)\Big). \tag{3.59}$$

Setting $\epsilon = 0$, a special case is $I(\lambda, 0) \sim \lambda^{-1/2}$. Let us compare this to the large λ behavior for $\epsilon \neq 0$. Assuming that (as in our application in eqn. (5.24)), the

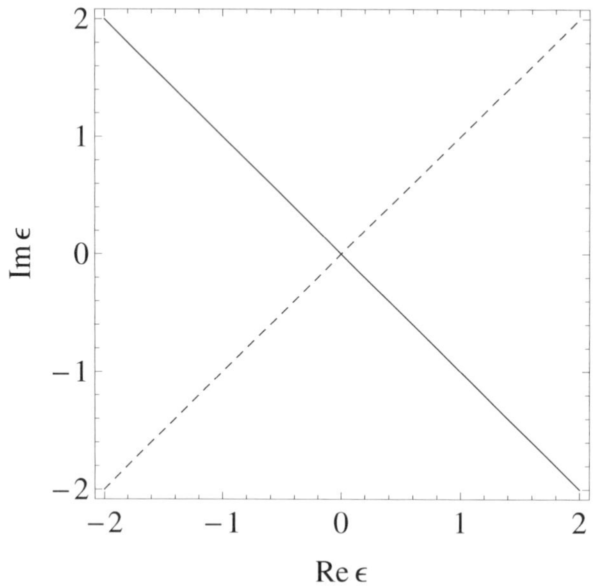

FIGURE 5. The two lines $\epsilon = r\exp(\pm\pi i/2)$, with real r, are Stokes lines in the sense that flows between the critical point p and the critical orbit $\mathcal{O}$ are possible on these lines. Jumping, however, occurs only in crossing the line $\epsilon = r\exp(-\pi i/2)$, shown here as a solid line. There is no jumping of Lefschetz thimbles in crossing the dotted line.

dominant contribution comes from the isolated, nondegenerate critical point at $x = y = 0$, a simple scaling shows that $I(\lambda,\epsilon) \sim 1/\lambda$ for $\epsilon \neq 0$, so the analog of (3.56) is

$$\frac{I(\lambda,0)}{I(\lambda,\epsilon)|_{\epsilon\neq 0}} \sim \lambda^{1/2},\ \lambda \to \infty. \tag{3.60}$$

By introducing polar coordinates in the xy plane, $I(\lambda,\epsilon)$ can be expressed in terms of the error function.

The Lefschetz thimbles associated to critical orbits can be explicitly described in this example, because the symmetry can be used to reduce everything to complex dimension 1. Picking the Kahler metric $\mathrm{d}s^2 = |\mathrm{d}x|^2 + |\mathrm{d}y|^2$, the moment map for the action of $SO(2)$ becomes $\mu = -i(x\overline{y} - y\overline{x})$. Setting $\mu = 0$, we find that by an $SO(2)$ rotation we can set $y = 0$. The Lefschetz thimbles that lie at $\mu = 0$ are therefore surfaces of revolution that can be constructed by acting with the group $SO(2)$ on a real curve in the complex x plane. For the rest of our analysis, we scale x to set $\lambda = 1$.

The flow equation has a conserved quantity $\mathrm{Im}\,(if)$, so to find the right curve in the x-plane, we do not need to solve flow equations; it suffices to set $\mathrm{Im}\,(if)$ to a constant. Equivalently, we must set $\mathrm{Re}\,(x^2-\epsilon)^2$ to a constant c. Setting $x^2 = \epsilon + u + iv$, the condition becomes $u^2 - v^2 = c$. For the Lefschetz thimble $\mathcal{J}_{\mathcal{O}}$ that flows down from the critical orbit $x^2 + y^2 = \epsilon$, we set $c = 0$, and choose the branch $u = v$ to ensure that the real part of $if(x,y)$ is negative. The Lefschetz thimble $\mathcal{J}_p$ that flows down from the critical point $x = y = 0$ passes through $\epsilon + u + iv = 0$, so the constant $u^2 - v^2 = \mathrm{Re}((u+iv)^2)$ is equal to $\mathrm{Re}(\epsilon^2)$. The

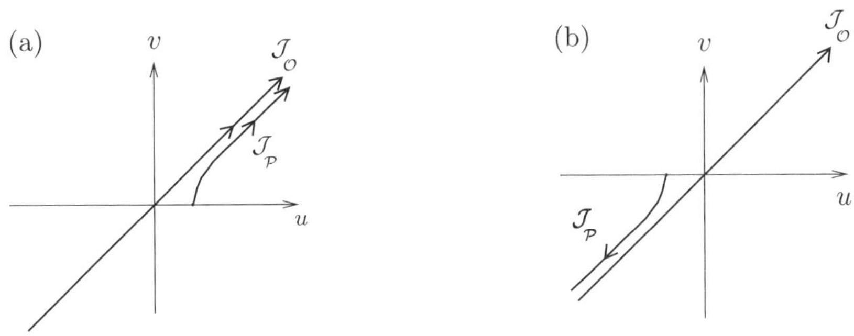

FIGURE 6. For real ϵ, the cycle $\mathcal{J}_p$ is derived from a curve in the uv plane that starts at $(u, v) = (-\epsilon, 0)$ and consists of half of one branch of the hyperboloid $u^2 - v^2 = \epsilon^2$. We pick the half that is asymptotic to the curve $u = v$. For $\epsilon < 0$, the relevant half is asymptotic to $u = v = \infty$, as sketched in (a). For $\epsilon > 0$, it is asymptotic to $u = v = -\infty$, as sketched in (b). Accordingly, the cycle $\mathcal{J}_p$ jumps when ϵ changes sign. The jumping is by $\mathcal{J}_p \to \mathcal{J}_p + \mathcal{J}_\mathcal{O}$, where $\mathcal{J}_\mathcal{O}$ is associated to the line $u = v$ in the uv plane, irrespective of the sign of ϵ. In other words, $\mathcal{J}_p$ in (a) or $\mathcal{J}_p + \mathcal{J}_\mathcal{O}$ in (b) is a path from $(u, v) = (-\epsilon, 0)$ to $u = v = \infty$.

hyperboloid $u^2 - v^2 = \mathrm{Re}\,(\epsilon^2)$ has two branches. We pick the branch that passes through $(u, v) = (-\epsilon, 0)$ (in other words, $x = 0$), and then $\mathcal{J}_p$ is derived from the half of this branch that is asymptotic to $u = v$. As shown in fig. 6, this branch is asymptotic to $u, v >> 0$ or $u, v << 0$ depending on ϵ. In particular, if ϵ is real, $\mathcal{J}_p$ has different asymptotic behavior depending on the sign of ϵ.

From the figure, we can read off that between $\epsilon < 0$ and $\epsilon > 0$, the Lefschetz thimbles change by

$$\begin{aligned} &\mathcal{J}_p \to \mathcal{J}_p + \mathcal{J}_\mathcal{O} \\ &\mathcal{J}_\mathcal{O} \to \mathcal{J}_\mathcal{O}. \end{aligned} \tag{3.61}$$

The meaning of the second statement is that the curve $\mathcal{J}_\mathcal{O}$, which is the diagonal $u = v$, is unchanged in going from fig. 6(a) to fig. 6(b). But $\mathcal{J}_p$ in fig. 6(a) is a path from $(u, v) = (-\epsilon, 0)$ to $u = v = +\infty$, while in fig. 6, it runs from $(-\epsilon, 0)$ to $u = v = -\infty$. In fig. 6(b), to get a path from $(-\epsilon, 0)$ to $(+\infty, +\infty)$, we need to take the sum $\mathcal{J}_p + \mathcal{J}_\mathcal{O}$. So $\mathcal{J}_p$ jumps in passing through $\epsilon = 0$, though $\mathcal{J}_\mathcal{O}$ does not. Invariance of $\mathcal{J}_\mathcal{O}$ has a more general explanation that was described in section 3.3.4.

To recover the Lefschetz thimbles $\mathcal{J}_p$ and $\mathcal{J}_\mathcal{O}$, we have to "rotate" (via $SO(2)$) the curves that are sketched in fig. 6. In the case of $\mathcal{J}_p$, the relevant curve is topologically a half-line whose endpoint is $SO(2)$-invariant. Rotating it by $SO(2)$, we find that as a manifold with $SO(2)$ action, $\mathcal{J}_p$ is equivalent to $\mathbb{R}^2$; $SO(2)$ acts on $\mathcal{J}_p$ with a fixed point at the origin that corresponds to the endpoint of the half-line. On the other hand, in the case of $\mathcal{J}_\mathcal{O}$, the relevant curve is a copy of $\mathbb{R}$, with its ends at infinity. Rotating it by $SO(2)$, we find that as a manifold with $SO(2)$ action, $\mathcal{J}_\mathcal{O}$ is equivalent to $\mathbb{R} \times S^1$ with a free action of $SO(2)$ on the second factor.

It is convenient to express (3.61) in a dual language. Given a general cycle $\mathcal{C} = \mathfrak{n}_p \mathcal{J}_p + \mathfrak{n}_{\mathcal{O}} \mathcal{J}_{\mathcal{O}}$, to compensate for the jumping (3.61) and ensure that $\mathcal{C}$ varies smoothly, the coefficients jump by

$$\begin{aligned} \mathfrak{n}_p &\to \mathfrak{n}_p \\ \mathfrak{n}_{\mathcal{O}} &\to -\mathfrak{n}_p + \mathfrak{n}_{\mathcal{O}}. \end{aligned} \tag{3.62}$$

Finally, let us return to the Stokes lines of fig. 5. We write l_+ and l_- for the two Stokes lines given by $\epsilon = r\exp(\pm i\pi/2)$ with real r. Scaling to $\lambda = 1$, the Morse function $\mathrm{Re}\,(if)$ vanishes at p and equals $\pm r^2/2$ on $l_\pm$. So on l_+, there can be a downward flow from $\mathcal{O}$ to p, and on l_- there can be a downward flow from p to $\mathcal{O}$. Flows of each type actually exist, as the reader can verify by slightly extending the analysis of fig. 6. ($\mathcal{J}_{\mathcal{O}}$ always corresponds to the line $u = v$, and $\mathcal{J}_p$ to part of a hyperboloid through $u + iv = -\epsilon$.) In general, the jumping in eqn. (3.61) occurs in crossing l_-, but there is no such jumping in crossing l_+. The algebraic invariant that "counts" the downward flows from $\mathcal{O}$ to p along l_+ vanishes (even though such flows exist), for a reason explained in section 3.3.4.

3.4.3. *One more case.* The last singularity that we will encounter is more surprising. We will describe it first in a language with $SO(1,1)$ rather than $SO(2)$ symmetry. We consider $SO(1,1)$ acting on real variables u, v by $u \to ru$, $v \to r^{-1}v$, with r real. The basic invariant is uv. We consider an integral $\int \mathrm{d}u\, \mathrm{d}v \exp(i\lambda f(uv))$ where f is a function of uv. The obvious case is that f is a quadratic function. However, and surprisingly, in the world of knots in S^3, there can be a natural reason for f to be an odd function of uv.

Under these conditions, the basic example of a function exhibiting critical behavior is

$$f(u,v) = (uv)^3/3 - \epsilon(uv). \tag{3.63}$$

Apart from the critical point p at $u = v = 0$, there are two critical orbits $\mathcal{O}_\pm$ at $uv = \pm\epsilon^{1/2}$. For real ϵ, these critical orbits are real or imaginary depending on the sign of ϵ. (As real orbits, these orbits are hyperboloids with two topological components.)

A closely related problem is the case of $SO(2)$ action on the xy plane with $f(x,y) = (x^2+y^2)^3/3 - \epsilon(x^2+y^2)$. The relation becomes apparent upon complexification; one problem can be mapped into the other by taking $u = x + iy$, $v = x - iy$. The problem of $SO(1,1)$ acting on the real uv plane and the problem of $SO(2)$ acting on the real xy plane correspond to two real slices of the same situation with complex variables. One real slice is defined by $u = \overline{u}$, $v = \overline{v}$ and the other by $u = \overline{v}$. To study analytic continuation, as usual we must complexify the variables u and v, and then the $SO(1,1)$ and $SO(2)$ problems become equivalent. It is convenient to work with x and y. (We introduced u and v because the real slice with $SO(1,1)$ symmetry will appear naturally in our application.)

First, we will determine the λ dependence of $I(\lambda,\epsilon) = \int \mathrm{d}x\ \mathrm{d}y \exp(i\lambda f(x,y))$ using scaling symmetry. An obvious scaling of x and y shows that

$$I(\lambda,\epsilon) = \lambda^{-1/3} I(1, \lambda^{2/3}\epsilon) = \lambda^{-1/3} \int \mathrm{d}x\, \mathrm{d}y \exp(i((x^2+y^2)^3/3 - \lambda^{2/3}\epsilon(x^2+y^2)). \tag{3.64}$$

A special case is that $I(\lambda, 0) \sim \lambda^{-1/3}$. On the other hand, for $\epsilon \neq 0$, and choosing $\epsilon < 0$ so that the only real critical point is the one at $x = y = 0$, a similar scaling gives $I(\lambda, \epsilon) \sim \lambda^{-1}$. So

$$\frac{I(\lambda,0)}{I(\lambda,\epsilon)|_{\epsilon<0}} \sim \lambda^{2/3}, \ \lambda \to \infty. \tag{3.65}$$

By introducing polar coordinates in the xy plane, one can express $I(\lambda, \epsilon)$ as a multiple of the Airy function. The main difference is that there is an extra possible integration cycle for $I(\lambda, \epsilon)$, as we now explain.

To study the analytic continuation systematically, we consider the integration cycles associated to critical orbits. Let $\mathcal{J}_0$, $\mathcal{J}_+$, and $\mathcal{J}_-$ be the Lefschetz thimbles associated, respectively, to the critical point at $x = y = 0$ and to the critical orbits $x^2 + y^2 = \pm\epsilon^{1/2}$. The same reasoning as in section 3.4.2 can be used to describe them by applying an $SO(2)$ rotation to suitable curves in the complex x^2-plane. In fact, apart from a substitution $x \to x^2$, the polynomial $f(x, 0) = (x^2)^3/3 - \epsilon x^2$ is familiar from our study of the Airy function. Accordingly, the Lefschetz thimbles $\mathcal{J}_\pm$ obtained by flowing from the critical orbits $\mathcal{O}_\pm$, can be derived from the curves that were depicted in fig. 4, but now in the x^2 plane. On the other hand, the third Lefschetz thimble $\mathcal{J}_0$ is associated to a similar curve connecting the point $x^2 = 0$ to infinity. By an analysis similar to that of section 3.4.2, one can determine the jumping that occurs between $\epsilon > 0$ and $\epsilon < 0$. The change in $\mathcal{J}_\pm$ is precisely the same as it is for the Airy function; what is new is that $\mathcal{J}_0$ picks up a linear combination of $\mathcal{J}_\pm$. We leave the details to the reader.

3.5. Analytic continuation in k: A prototype. We have discussed at great length the analytic continuation of the integral of an exponential $\exp(i\lambda f)$ from real to complex values of λ. For a more precise analog of Chern-Simons theory, we should consider analytic continuation of an *integer* parameter (such as the Chern-Simons level k) to complex values.

For an example, we consider the integral representation of the Bessel function:

$$I(\kappa, \lambda) = \frac{1}{2\pi i}\oint \frac{\mathrm{d}z}{z} z^\kappa \exp(\lambda(z - z^{-1})), \tag{3.66}$$

with an integer κ that in our analysis of section 5 will correspond roughly[13] to the Chern-Simons level k.

We take the integration cycle to be the circle $|z| = 1$. Alternatively, we can write

$$I(\kappa, \lambda) = \frac{1}{2\pi i}\oint \frac{\mathrm{d}z}{z} \exp(\mathcal{I}), \tag{3.67}$$

with

$$\mathcal{I} = \lambda(z - z^{-1}) + \kappa \ln z. \tag{3.68}$$

For real λ and integral κ, this integral is an oscillatory integral of the general type that we have been considering, except that the integration cycle is compact, so that analytic continuation in λ poses no problem. However, we would like to analytically continue the integral to complex values of κ. This is a rough analog of analytically continuing Chern-Simons gauge theory with respect to the level k.

Let us recall the steps that were described somewhat abstractly in section 2.5.1. The holomorphic form that we are trying to integrate is naturally defined in the

[13] In that analysis, λ corresponds roughly to the index n of the colored Jones polynomial $J_n(q)$, so the analogy would be closer if also λ were constrained to be an integer.

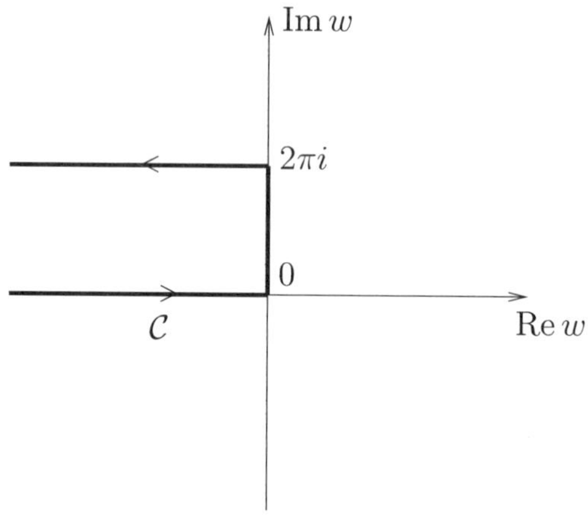

FIGURE 7. The contour $\mathcal{C}$ in the complex t-plane that is used in analytic continuation of the Bessel function.

punctured z-plane, with the point $z = 0$ omitted. To analytically continue in κ, we want to lift the integral from the punctured z-plane to its universal cover, which is the complex w-plane where $z = e^w$. The integration cycle $|z| = 1$ can be lifted to the line segment in the w-plane from αi to $(2\pi + \alpha)i$, for an arbitrary α. Picking $\alpha = 0$, the integral becomes

$$I(\kappa, \lambda) = \frac{1}{2\pi i} \int_0^{2\pi i} dw \, \exp(2\lambda \sinh w + \kappa w). \tag{3.69}$$

The integration is now over an open contour whose ends are the two points $w = 0$ and $w = 2\pi i$. If we stop here, Bessel's equation (whose proof requires integration by parts), will fail when $\kappa \notin \mathbb{Z}$. An analogous procedure in quantum field theory would lead to a failure of the Ward identities. Instead, following a classical procedure, and assuming that $\mathrm{Re}\,\lambda > 0$, we add to the contour two more half lines given (fig.7) by the negative real axis and its translate by $2\pi i$, suitably oriented to give a closed although noncompact integration contour $\mathcal{C}$, which represents a cycle in the appropriate relative homology. The integral over this cycle

$$I(\kappa, \lambda) = \frac{1}{2\pi i} \int_{\mathcal{C}} dw \, \exp(2\lambda \sinh w + \kappa w) \tag{3.70}$$

converges (if $\mathrm{Re}\,\lambda > 0$), and in this integral, κ can now be treated as a complex variable. This gives an analytic continuation of the Bessel function to complex values of κ, since if κ is an integer, the two half lines that have been added to the integration contour cancel each other and the extended integral reduces to the original one. To analytically continue beyond the region $\mathrm{Re}\,\lambda > 0$, one must shift $\mathcal{C}$ vertically in the w-plane as the argument of λ varies. But we will mainly study the dependence on κ with $\mathrm{Re}\,\lambda > 0$.

If λ and κ are real and obey suitable inequalities (λ and $2\lambda + \kappa$ should be positive), the two half-lines that we added to complete the integration contour are precisely as described in section 2.5.1: they are the downward flow lines from the endpoints $w = 0$ and $w = 2\pi i$ of the naive integration contour in the complex w

plane. If we vary λ and κ slightly, the flow lines move but without changing the relative homology class of the integration cycle $\mathcal{C}$.

We have described a simple procedure for analytic continuation away from integer values of κ, but there is no avoiding a basic fact discussed in section 2.5.2: analytic continuation away from integer values is not unique. For an illustrative special case of this, note that we could have begun by lifting the circle $|z| = 1$ to a line segment in the w-plane connecting the points $w = \pm\pi i$. This would be analogous to a different choice of α in section 2.5.1. Provided λ and κ are real and obey suitable inequalities (λ and $2\lambda+\kappa$ should be negative), the downward flow lines from $w = \pm i$ are again half-lines parallel to the real axis, but they extend in the opposite direction to $\mathrm{Re}\, w >> 0$. The two analytic continuations are qualitatively different. The original Bessel integral (3.66) obeys $I(-\kappa, \lambda) = (-1)^\kappa I(\kappa, \lambda)$, as follows from the change of variables $z \leftrightarrow -1/z$. This symmetry is broken by the choice of whether to extend the contour to $\mathrm{Re}\, w << 0$, as in fig. 7, or to $\mathrm{Re}\, w >> 0$. With either choice, it turns out that when κ is not an integer, the large κ behavior depends very much on the sign of $\mathrm{Re}\,\kappa$.

In the rest of our analysis, we will stick with the analytic continuation corresponding to the contour $\mathcal{C}$ of fig. 7. We want to study the behavior of the integral (3.70) as κ and λ become large with a fixed value of $\rho = \kappa/2\lambda$. First we will learn what we can by inspection, taking κ real and λ positive, and then we will reconsider the matter using Morse theory.

The critical point equation is $z^2 + 2\rho z + 1 = 0$ or $\cosh w + \rho = 0$ and the critical points are at

$$z = -\rho \pm i\sqrt{1-\rho^2}. \tag{3.71}$$

or

$$w = \pm\mathrm{arccosh}(-\rho) + 2\pi i n = \ln(-\rho \pm \sqrt{\rho^2 - 1}) + 2\pi i n, \; n \in \mathbb{Z}. \tag{3.72}$$

First let us suppose that κ is an integer, so that $\mathcal{I}_{\kappa,\lambda}$ can be defined by the original integral on the contour $|z| = 1$, and we can use (3.71) to identify the critical points. If $|\rho| \le 1$, the critical points are at $|z| = 1$, on the integration contour. They make oscillatory contributions to the integral, and this gives the behavior of the integral for large κ and fixed ρ in this region.

Now, still with integer κ, suppose that $|\rho| > 1$. The critical points move off the integration contour $|z| = 1$, and the integral therefore vanishes exponentially.

Next, suppose that κ is not an integer. We must consider the effects of the two half-lines that have been added to the integration contour. If $|\rho| \le 1$, the addition of these half-lines does not lead to a qualitative change; the integral is still dominated by oscillatory contributions from critical points at $|z| = 1$. We must be more careful if $|\rho| > 1$, and the two cases $\rho > 1$ and $\rho < -1$ turn out to be completely different.[14] If $\rho > 1$, there is no critical point on the integration contour and $I(\kappa, \lambda)$ vanishes exponentially whether κ is an integer or not. (The most precise analysis is by a saddle point analysis that we describe shortly.) For $\rho < -1$ (which means negative κ, since we have taken $\lambda > 0$), the behavior is more delicate. There are two critical points on our chosen integration cycle $\mathcal{C}$, namely at $w = \ln(-\rho - \sqrt{\rho^2 - 1})$ and $w = \ln(-\rho - \sqrt{\rho^2 - 1}) + 2\pi i$. The two critical points

[14]This difference results entirely from the choice of how to analytically continue the function $I(\kappa, \lambda)$ to nonintegral values of κ, since for $\kappa \in \mathbb{Z}$, the Bessel function has a symmetry $I(-\kappa, \lambda) = (-1)^\kappa I(\kappa, \lambda)$, as already noted.

have the same positive value of $h = \mathrm{Re}\,\mathcal{I}$, and they both make contributions that are exponentially large for $\kappa \to -\infty$. (Indeed, $h \sim -\kappa f(\rho)$ for $-\kappa$ large, where the function $f(\rho) = (\rho^2 - 1)^{1/2}/\rho + \ln(-\rho + (\rho^2 - 1)^{1/2})$ is positive for $\rho < -1$.) On the other hand, $\mathrm{Im}\,\mathcal{I}$ vanishes for one critical point and is $\kappa\,\mathrm{Im}\,w = 2\pi\kappa$ at the other. Allowing for the opposite orientations of the two half-lines that are part of the contour $\mathcal{C}$, the two real critical points contribute

$$\exp(-\kappa f(\rho))\,(1 - \exp(2\pi i\kappa)) \tag{3.73}$$

times the usual sort of asymptotic series. For integer κ, these contributions cancel, but for generic $\kappa << 0$, their sum grows exponentially with $|\kappa|$.

3.5.1. *Description by Lefschetz thimbles.* Finally, we will seek a more precise description by expressing the cycle $\mathcal{C}$ as a sum of Lefschetz thimbles $\mathcal{J}_\sigma$ associated to critical points p_σ. Let us formulate the most parsimonious hypothesis that is consistent with the observations above:

(1) For $\rho > 1$, it is reasonable to guess that $\mathcal{C}$ is equivalent to the Lefschetz thimble associated to the critical point at $w = \ln(-\rho + \sqrt{\rho^2 - 1})$ (w has negative real part and we pick the branch of the logarithm such that $\mathrm{Im}\,w = \pi$). This is the unique critical point enclosed by $\mathcal{C}$ if $\rho > 1$, and as h is negative at this critical point, the associated Lefschetz thimble makes an exponentially small contribution to the integral, as expected.

(2) For $1 > \rho > -1$, the two critical points at $|z| = 1$ must certainly be included. There is no obvious need for more.

(3) For $\rho < -1$, the situation is more tricky. The two real critical points that account for the exponentially growing asymptotics in (3.73) must certainly contribute when $\mathcal{C}$ is expressed in terms of Lefschetz thimbles. But as they cancel when κ is an integer, while $I(\kappa, \lambda)$ is not zero in that case, the expression for $\mathcal{C}$ in terms of Lefschetz thimbles must receive a contribution from at least one more critical point. This critical point must have $h < 0$, since $I(\kappa, \lambda)$ is exponentially small for integer κ and $\rho < -1$.

In summary, in regions (1), (2), or (3), $\mathcal{C}$ must be a sum of at least 1, 2, or 3 critical point contours, respectively. This most parsimonious interpretation is actually correct, as one can see from fig.8, where the relevant Lefschetz thimbles are sketched.

3.5.2. *Singularities at $\rho = \pm 1$.* The most interesting values of ρ are at $\rho = \pm 1$, where critical points meet. Let us first examine the behavior at these points on the z-plane, and then on the w-plane. The analysis on the z-plane is simpler, but it only tells us what happens when κ is an integer. To analyze the behavior for non-integral κ, we need to work on the w-plane.

There are precisely two critical points on the z-plane, located at $z = -\rho \pm i\sqrt{1 - \rho^2}$. Precisely for $\rho = \pm 1$, the two critical points meet. This is reminiscent of the most basic example of critical behavior, the cubic polynomial $f(x) = x^3/3 - \epsilon x$ related to the Airy function, whose two critical points meet at $\epsilon = 0$. The critical points of f are real for $\epsilon > 0$ and form a complex conjugate pair for $\epsilon < 0$. The consequences of the singularity at $\epsilon = 0$ were explored in section 3.4.1.

Our problem on the z-plane is exactly analogous. We endow the punctured z-plane with the antiholomorphic involution $z \to 1/\overline{z}$, so that the real points (the fixed points of the involution) are the circle $|z| = 1$. The two critical points at $z = -\rho \pm i\sqrt{1 - \rho^2}$ are real for ρ real and $|\rho| < 1$ but form a complex conjugate

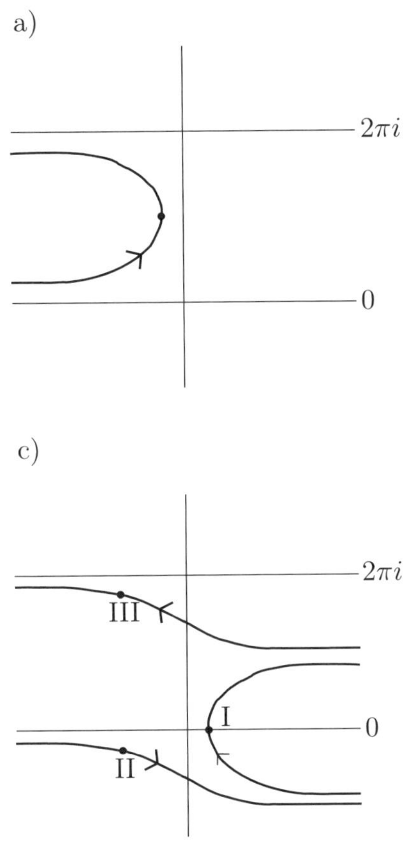

FIGURE 8. Qualitative behavior of Lefschetz thimbles in analytic continuation of the Bessel function, for (a) $\rho > 1$, (b) $1 > \rho > -1$, and (c) $-1 > \rho$. In each case, the critical points are marked by black dots. In (a), the integration cycle of fig. 7 is equivalent to a single Lefschetz thimble associated to a complex critical point that makes an exponentially decaying contribution to the integral. In (b), it is equivalent to the sum of two Lefschetz thimbles that make oscillatory contributions. And in (c), it is equivalent to the sum of a Lefschetz thimble associated to a critical point (labeled as I) that makes an exponentially decaying contribution and two critical points (labeled II and III) whose contributions are exponentially growing, but cancel if κ is an integer. In (c), ρ has been given a small imaginary part to avoid a Stokes line.

pair for ρ real and $|\rho| > 1$. The singularities at $\rho = \pm 1$ are just analogous to the Airy function singularity at $\epsilon = 0$.

As long as κ is an integer and we work on the z-plane, the behavior of the Bessel integral that we found in section 3.5.1 precisely mirrors the behavior of the Airy integral

$$I(\epsilon) = \int_{-\infty}^{\infty} \mathrm{d}x \, \exp\left(i(x^3/3 - \epsilon x)\right). \tag{3.74}$$

Region (2) of the Bessel integral corresponds to the case $\epsilon > 0$ for the Airy integral: the integration cycle is equivalent to the sum of two Lefschetz thimbles associated to two real critical points. Regions (1) and (3) of the Bessel integral correspond to the case $\epsilon < 0$ for the Airy integral: the integral is equivalent to an integral over a single Lefschetz thimble that makes an exponentially small contribution. In section 3.5.1, we found that three critical points are needed in region (3) to describe the integral on the w-plane, but two of them (the ones that make exponentially growing contributions) cancel upon projection to the z-plane.

The analysis on the w-plane is more complicated because each critical point on the z-plane lifts to infinitely many critical points on the w-plane (differing by $w \to w + 2\pi i n$, $n \in \mathbb{Z}$). Nevertheless, it is very illuminating to repeat the analysis of the Bessel function emphasizing its interpretation in terms of Airy function singularities:

($1'$) We start for $\rho > 1$ with a single critical point at $\widetilde{w}_- = \ln\left(-\rho + \sqrt{\rho^2 - 1}\right)$, picking the branch of the logarithm so that $\widetilde{w}_- = \pi i$ at $\rho = 1$. (Choosing a different branch would multiply the analytically continued Bessel integral by an integer power of $\exp(2\pi i \kappa)$.) The name $\widetilde{w}_-$ is motivated by the fact that in section 3.4.1, a complex critical point that makes an exponentially small contribution to the Airy function for $\epsilon < 0$ was called $\widetilde{p}_-$. As in section 3.4.1, we write $\widetilde{\mathcal{J}}_-$ for the Lefschetz thimble corresponding to this critical point. For $\rho > 1$, as we have learned in section 3.5.1, the integration cycle $\mathcal{C}$ that we have used to analytically continue the Bessel function is equivalent to $\widetilde{\mathcal{J}}_-$.

($2'$) The $\rho > 1$ critical point at $\widetilde{w}_- = \ln(-\rho + \sqrt{\rho^2 - 1})$, when continued to $\rho = 1$, meets two real critical points that for $\rho < 1$ are located at $w_\pm = \ln(-\rho \pm i\sqrt{1 - \rho^2})$. Here branches of the logarithm are chosen so that at $\rho = 1$, we have $w_+ = w_- = \widetilde{w}_- = \pi i$. As in the Airy case, we denote as $\mathcal{J}_+$ and $\mathcal{J}_-$ the Lefschetz thimbles that correspond to the real critical points $w_\pm$. We can use the Airy relation (3.58) to express the region (1) integration cycle $\widetilde{\mathcal{J}}_-$ in terms of Lefschetz thimbles $\mathcal{J}_\pm$ appropriate to region (2). The relation is $\widetilde{\mathcal{J}}_- = \mathcal{J}_+ + \mathcal{J}_-$. So the right integration cycle in region (2) is $\mathcal{J}_+ + \mathcal{J}_-$.

($3'$) So far working on the w-plane instead of the z-plane has not given anything essentially new. Now we study how the critical points $w_\pm$ evolve as we vary ρ from 1 to -1. We have chosen branches of the logarithm so that $w_\pm = \ln(-\rho \pm i\sqrt{1 - \rho^2})$ are both equal to πi at $\rho = 1$. However, when continued (along the real ρ axis) to $\rho = -1$, they are not equal. Rather,

$$w_+(\rho = -1) = 0, \; w_-(\rho = -1) = 2\pi i. \tag{3.75}$$

Hence, we cannot analyze what is happening at $\rho = -1$ in terms of a single Airy singularity. We must consider two different Airy singularities, located at $w = 0$ and at $w = 2\pi i$. This will make the analysis a little longer, though the basic idea is just to use what we know about the Airy integral.

Near $w = 0$, let us write $\mathcal{J}'_\pm$ for the Lefschetz thimbles associated to real critical points at $\rho > -1$, and $\widetilde{\mathcal{J}}'_\pm$ for the Lefschetz thimbles associated to complex critical points at $\rho < -1$. We write $\mathcal{J}''_\pm$ and $\widetilde{\mathcal{J}}''_\pm$ for Lefschetz thimbles associated to the analogous real and complex critical points near $w = 2\pi i$. (The reason for the notation $\mathcal{J}'$ and $\mathcal{J}''$ is to distinguish these cycles from the ones we have already defined near $\rho = 1$.)

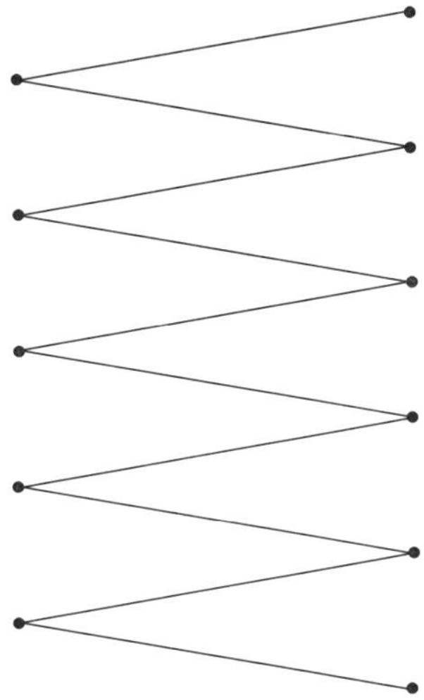

FIGURE 9. Qualitative behavior of the conserved quantity $\mathrm{Im}\, I/\kappa$ for critical points in the region $-1 \leq \rho \leq 1$. ρ is plotted horizontally and $\mathrm{Im}\, \mathcal{I}/\kappa$ vertically. A black dot represents a pair of real critical points that become coincident at $\rho = 1$ or $\rho = -1$. At $\rho = 1$ (right), there are a pair of real critical points for every value of $\mathrm{Im}\, \mathcal{I}/\kappa$ of the form $(2w+1)\pi$, $w \in \mathbb{Z}$. Reducing ρ, $\mathrm{Im}\, \mathcal{I}/\kappa$ increases for one of these critical points and reduces for the other. At $\rho = -1$, the critical points recombine in pairs, but now with $\mathrm{Im}\, \mathcal{I}/\kappa$ of the form $2\pi w$, $w \in \mathbb{Z}$. In all cases, these critical points are found by solving the equation $\cosh w = -\rho$.

We have already found that for ρ just less than 1, the integration cycle is $\mathcal{J}_+ + \mathcal{J}_-$. Continuing along the real axis to ρ just greater than -1 with the aid of eqn. (3.75), we see that $\mathcal{J}_+ + \mathcal{J}_-$ evolves to $\mathcal{J}'_+ + \mathcal{J}''_-$. To continue past $\rho = -1$, we use the Airy relations (3.57), which give $\mathcal{J}'_+ = -\widetilde{\mathcal{J}}'_+ + \widetilde{\mathcal{J}}'_-$, and $\mathcal{J}''_- = \widetilde{\mathcal{J}}''_+$. So $\mathcal{J}'_+ + \mathcal{J}''_- - -\widetilde{\mathcal{J}}'_+ + \widetilde{\mathcal{J}}''_+ + \widetilde{\mathcal{J}}'_-$, and this is the integration cycle for $\rho < -1$. For integer κ, $-\widetilde{\mathcal{J}}'_+$ and $\widetilde{\mathcal{J}}''_+$ cancel and the integration cycle is equivalent to $\widetilde{\mathcal{J}}'_-$. For general κ, the contribution from $-\widetilde{\mathcal{J}}'_+ + \widetilde{\mathcal{J}}''_+$ dominates and is described qualitatively in eqn. (3.73).

With a view to our application in section 5.3, the key point to remember is that the two real critical points on the z-plane, if lifted to the w-plane so that they coincide at $\rho = 1$, do not coincide at $\rho = -1$. The flow of critical points is sketched in fig. 9.

4. Application to Chern-Simons theory

Can the framework of section 3 be applied in the infinite-dimensional context of Chern-Simons gauge theory? The critical points p_σ certainly make sense. The Euler-Lagrange equations simply say that the gauge field should be flat (perhaps when restricted to the complement of a knot). Moreover, by standard quantum field theory methods, one certainly knows how to expand the path integral around any given critical point, and hence in effect the Lefschetz thimbles $\mathcal{J}_\sigma$ are known at least perturbatively.

The aspects of this framework that are not standard in quantum field theory are the use of the flow equation to define the critical point cycles nonperturbatively, to determine the coefficients when the real integration cycle is expressed in terms of critical point cycles, making analytic continuation possible, and to describe the Stokes jumping phenomena that are essential for the consistency of the framework, once one aims for nonperturbative results. All of these questions involve the flow equations, so the question of whether the Morse theory machinery can be applied to Chern-Simons theory is largely the question of whether the flow equation for Chern-Simons makes sense and has properties similar to what it has in finite dimensions.

This is the topic of section 4.1. The flow equations in ordinary Morse theory are ordinary differential equations that describe flow in an auxiliary "time" direction. Chern-Simons theory is three-dimensional even before introducing an auxiliary direction, so the flow equations become partial differential equations in four dimensions. Prototypes for such equations are the Yang-Mills instanton equations, which are interpreted in Floer theory as flow equations for a real connection A with the Chern-Simons function $W(A)$ as a Morse function. In that example, the flow equation has a four-dimensional symmetry, and this is a hint of what we will find in section 4.1.

Our problem is analogous to the case just mentioned but is more elaborate because we want the flow for a complex connection $\mathcal{A}$ with a Morse function derived from the complex Chern-Simons function $W(\mathcal{A})$. Also our flow equations depend on a coupling parameter. It turns out that the flow equations in our problem are essentially equations that have appeared before, in the application of supersymmetric gauge theory to the geometric Langlands program [**53**]. The reason for this is not yet fully clear, but at any rate the elliptic nature of the flow equations for Chern-Simons theory means that they have properties similar to the Morse theory flow equations with finitely many variables.

4.1. The flow equation for Chern-Simons. Let M be a three-manifold. Let H be a compact Lie group and set $G = H_{\mathbb{C}}$. Let $\mathcal{T}$ be the space of connections A on a fixed H-bundle $E \to M$, and let $\mathcal{T}_{\mathbb{C}}$ be the space of complex-valued connections $\mathcal{A}$ on the complexification of E. We can write $\mathcal{A} = A + i\phi$ where A and ϕ are the real and imaginary parts of $\mathcal{A}$; A is an $\mathfrak{h}$-valued connection, and ϕ is a one-form on M with values in the adjoint bundle $\mathrm{ad}(E)$. The curvature of A will be denoted as $F = \mathrm{d}A + A \wedge A$, and that of $\mathcal{A}$ as $\mathcal{F} = \mathrm{d}\mathcal{A} + \mathcal{A} \wedge \mathcal{A}$.

$\mathcal{T}_{\mathbb{C}}$ is a complex manifold; to define a flow equation, we need to endow it with a Kahler metric. We pick a Riemannian metric g on M, which determines a Hodge star operator[15] $\star_M$. We define the metric on $\mathcal{T}$ by

$$\mathrm{d}s^2 = -\int_M \mathrm{Tr}\, \delta\overline{\mathcal{A}} \wedge \star_M \delta\mathcal{A}. \tag{4.1}$$

This is a Kahler metric, with Kahler form

$$\omega = \int_M \mathrm{Tr}\, \delta\phi \wedge \star_M \delta A. \tag{4.2}$$

[15] If $\mathrm{d}x^1, \mathrm{d}x^2, \mathrm{d}x^3$ is an oriented orthonormal frame, we pick the sign of $\star_M$ so that $\star_M(\mathrm{d}x^1) = \mathrm{d}x^2 \wedge \mathrm{d}x^3$, etc. Similarly, later the star operator of $I \times M$ is defined so that $\star(\mathrm{d}t \wedge \mathrm{d}x^1) = \mathrm{d}x^2 \wedge \mathrm{d}x^3$, etc.

The metric ds^2 is invariant under H-valued local gauge transformations, though not under G-valued ones. The moment map for the H-valued local gauge transformations is

$$\mu = \mathrm{d}_A \star_M \phi, \tag{4.3}$$

where $\mathrm{d}_A = \mathrm{d} + [A, \cdot]$ is the covariant derivative with respect to A.

Critical "points" of the Chern-Simons function $W(\mathcal{A})$, understood as a function on the space of connections, actually form orbits under the group of complex gauge transformations. For reasons described in section 3.3.3, we will only consider critical orbits that are semistable, meaning that they admit a locus with $\mu = 0$. The condition for a flat connection $\mathcal{A}$, that is a connection obeying $0 = \mathcal{F} = \mathrm{d}\mathcal{A} + \mathcal{A} \wedge \mathcal{A}$, to be gauge-equivalent to a connection for which $\mu = 0$ is [**52**] that the holonomy of the flat connection $\mathcal{A}$ is not strictly triangular. In other words, if the holonomies of $\mathcal{A}$ around one-cycles in M can be simultaneously put in upper triangular form

$$\begin{pmatrix} \alpha & \beta \\ 0 & \gamma \end{pmatrix}, \tag{4.4}$$

then the upper right elements β all vanish. Strictly triangular flat connections are unstable and we omit them. (A block diagonal flat connection with $\beta = 0$ is considered semistable, and one that cannot be put in a triangular form at all is stable.)

The flow equation for a Morse function that is the real part of a holomorphic function $\mathcal{I}$ is

$$\frac{\mathrm{d}\mathcal{A}}{\mathrm{d}t} = - \star_M \frac{\delta \overline{\mathcal{I}}}{\delta \overline{\mathcal{A}}}. \tag{4.5}$$

We are interested in the case that $\mathcal{I} = \lambda W(\mathcal{A})$ where W is the Chern-Simons functional and λ is a complex number. By rescaling t, we can reduce to the case that $\lambda = 2\pi \exp(-i\alpha)$ with real α. The flow equation then becomes

$$\frac{\mathrm{d}\mathcal{A}}{\mathrm{d}t} = - \exp(-i\alpha) \star_M \overline{\mathcal{F}}, \tag{4.6}$$

where $\overline{\mathcal{F}}$ is the complex conjugate of the curvature $\mathcal{F} = \mathrm{d}\mathcal{A} + \mathcal{A} \wedge \mathcal{A}$. More explicitly, this is

$$\begin{aligned} \frac{\mathrm{d}A}{\mathrm{d}t} &= - \star_M \big(\cos\alpha(F - \phi \wedge \phi) - \sin\alpha \, d_A\phi\big) \\ \frac{\mathrm{d}\phi}{\mathrm{d}t} &= \star_M \big(\sin\alpha(F - \phi \wedge \phi) + \cos\alpha \, \mathrm{d}_A\phi\big). \end{aligned} \tag{4.7}$$

The time t parametrizes a real one-manifold I (which is the real line or a portion of it). We pull back E to a bundle over $I \times M$ and view A as a connection on this pullback. As such, A has vanishing component in the I direction. We view this as a gauge condition. Relaxing this gauge condition, we can write $\mathrm{d}A_i/\mathrm{d}t$ (i is an index tangent to M) in gauge-invariant language as a component F_{ti} of the curvature $F = \mathrm{d}A + A \wedge A$. Similarly, we view ϕ as an $\mathrm{ad}(E)$-valued one-form on $I \times M$; as such, it is not generic, since its component ϕ_t in the I direction vanishes. We endow $I \times M$ with a Riemannian metric that is the sum of the metric $\mathrm{d}t^2$ on I with the metric g on M, and we write simply $\star$ for the Hodge star operator on $I \times M$ (normalized as in footnote 15). After taking linear combinations of the equations in (4.7), one finds that these equations can be put in a form with four-dimensional symmetry

$$(4.8)\qquad \begin{aligned}(F-\phi\wedge\phi)^+ &= uD\phi^+\\ (F-\phi\wedge\phi)^- &= -u^{-1}D\phi^-,\end{aligned}$$

with

$$(4.9)\qquad u=\frac{1-\cos\alpha}{\sin\alpha},\ u^{-1}=\frac{1+\cos\alpha}{\sin\alpha}.$$

(For a two-form v, we define the projections $v^\pm=(v\pm\star v)/2$.) To be more precise, the four-dimensional symmetry is broken in our context by the condition $\phi_t=0$. But if we relax this condition, the equations have full four-dimensional symmetry.

The moment map $\mu=\mathrm{d}_A\star_M\phi$ is as usual a conserved quantity for the flow equations. As explained in section 3.3.3, we are primarily interested in solutions of the flow equations with $\mu=0$. Bearing in mind that we anyway have the condition $\phi_t=0$, $\mathrm{d}_A\star_M\phi$ can be written in four-dimensional language as $\mathrm{d}_A\star\phi$. So we want to supplement the equations (4.8) with an additional condition

$$(4.10)\qquad \mathrm{d}_A\star\phi=0.$$

Henceforth, the phrase "flow equations" will refer to this combined system of equations.

The flow equations have been first encountered [**53**] in the gauge theory approach to geometric Langlands.[16] These are elliptic equations, modulo the gauge symmetry, and this means that their general properties are much like those of flow equations for finitely many variables in ordinary Morse theory. The relevant properties are stated in section (4.1.2).

The condition $\phi_t=0$ can be reinterpreted in terms of a vanishing theorem similar to those of section 3.3 of [**53**]. This argument proceeds as follows. Set $\mathcal{V}^+(u)=(F-\phi\wedge\phi)^+-uD\phi^+$, $\mathcal{V}^-(u)=(F-\phi\wedge\phi)^-+u^{-1}D\phi^-$, $\mathcal{V}_0=\mathrm{d}_A\star\phi$, so the flow equations are $\mathcal{V}^+(u)=\mathcal{V}^-(u)=\mathcal{V}_0=0$. Now consider the following sum of squares of these equations:

$$(4.11)\qquad \begin{aligned}&-\int_{I\times M}\mathrm{Tr}\left(\frac{u^{-1}}{u+u^{-1}}\mathcal{V}^+(u)\wedge\mathcal{V}^+(u)+\frac{u}{u+u^{-1}}\mathcal{V}^-(u)\wedge\mathcal{V}^-(u)+\mathcal{V}_0\wedge\star\mathcal{V}_0\right)\\ &=-\int_{I\times M}\mathrm{d}^4x\sqrt{g}\,\mathrm{Tr}\left(\frac{1}{2}F_{\mu\nu}F^{\mu\nu}+D_\mu\phi_\nu D^\mu\phi^\nu+R_{\mu\nu}\phi^\mu\phi^\nu+\frac{1}{2}[\phi_\mu,\phi_\nu]^2\right)\\ &+\frac{u-u^{-1}}{u+u^{-1}}\int_{I\times M}\mathrm{Tr}\,F\wedge F.\end{aligned}$$

(Indices μ,ν are tangent to $I\times M$, and $R_{\mu\nu}$ is the Ricci tensor of $I\times M$, which is a pullback from M.) Clearly, the left hand side of (4.11) is stationary when the flow equations are obeyed. Therefore the right hand side is also stationary. The part of the right hand side that depends on ϕ_t is simply

$$(4.12)\qquad Z(\phi_t)=-\int_{I\times M}d^4x\sqrt{g}\,\mathrm{Tr}\left(\sum_\mu D_\mu\phi_t D^\mu\phi_t+\sum_{\mu\neq t}[\phi_t,\phi_\mu]^2\right).$$

[16] The parameter u is called t in [**53**], but here we prefer to avoid confusion with the time coordinate.

The flow equations therefore imply that ϕ_t obeys the Euler-Lagrange equations derived from (4.12), or

$$-\sum_{\mu} D_\mu D^\mu \phi_t + \sum_{\mu \neq t} [\phi_\mu, [\phi_\mu, \phi_t]] = 0. \tag{4.13}$$

(Of course, this can also be proved by differentiating the equations (4.8) and (4.10).) Provided that one can integrate by parts, which is true if ϕ_t obeys either Dirichlet or Neumann boundary conditions at each end of I, (4.13) implies that $Z(\phi_t) = 0$. Given the definition of $Z(\phi_t)$ as a sum of squares, this immediately implies that ϕ_t is covariantly constant and commutes with all ϕ_μ. Thus, ϕ_t generates a symmetry of the solution.

We will always choose boundary conditions such that ϕ_t vanishes at (at least) one end of I, and integration by parts is possible at the other end. The above argument shows that in a solution of the flow equations obeying such boundary conditions, ϕ_t is identically zero. There is no need to spoil the four-dimensional symmetry, or the elliptic nature of the equations, by postulating in advance that ϕ_t is identically zero. This is a consequence of the equations plus boundary conditions.

4.1.1. *The index.* For applications, we need to know the expected value of the dimension of the moduli space $\mathcal{M}$ of solutions of the above equations, with appropriate boundary conditions and modulo gauge transformations. This equals the index of the linear elliptic operator L that arises from those equations, after linearization and gauge fixing. We will show that on a four-manifold of the form $I \times M$, with certain types of boundary condition at the ends of I, the index vanishes. This will lead in section 4.1.2 to behavior similar to that of Morse theory with finitely many variables.

The index is independent of u, so it suffices to set $u = 1$. The relevant linear system can be described as follows. We expand around a classical solution (A_0, ϕ_0) by setting $(A, \phi) = (A_0 + \delta A, \phi_0 + \delta\phi)$. A convenient gauge-fixing condition is $\mathrm{d}_{A_0} \star \delta A = 0$. For a four-manifold X, let $\Omega^q_X(\mathrm{ad}(E))$ be the space of q-forms on X with values in $\mathrm{ad}(E)$.

After acting on $\delta\phi$ with the $\star$ operator, we can think of δA and $\delta\phi$ as elements of $\Omega^1_X(\mathrm{ad}(E))$ and $\Omega^3_X(\mathrm{ad}(E))$, respectively. We can view the gauge-fixing condition as asserting the vanishing of the element $\star \mathrm{d}_{A_0} \star \delta A$ of $\Omega^0_X(\mathrm{ad}(E))$, while the equations $\mathcal{V}^+ = \mathcal{V}^- = 0$ assert the vanishing of the element $\mathcal{V}^+ + \mathcal{V}^-$ of $\Omega^2_X(\mathrm{ad}(E))$, and the equation $\mathcal{V}_0 = 0$ asserts the vanishing of an element of $\Omega^4_X(\mathrm{ad}(E))$.

All told, our linear operator L maps $\Omega^1_X(\mathrm{ad}(E)) \oplus \Omega^3_X(\mathrm{ad}(E))$ to $\Omega^0_X(\mathrm{ad}(E)) \oplus \Omega^2_X(\mathrm{ad}(E)) \oplus \Omega^4_X(\mathrm{ad}(E))$. A familiar operator that does this is the operator $L_0 = \mathrm{d}_A + \mathrm{d}_A^*$ mapping $\mathrm{ad}(E)$-valued differential forms of odd degree to those of even degree. L is actually equal to L_0 modulo zeroth order terms that do not affect the index. As we explain momentarily, the index of the operator L_0, or of L, is unaffected by the twisting by the gauge bundle E. If X is compact, this index is simply the product $-\chi(X)\dim(H)$, where $\chi(X)$ is the Euler characteristic of X and $\dim(H)$ is the dimension of the Lie group H. For $X = I \times M$, the topological Euler characteristic of X vanishes, and boundary contributions to the index are also zero, so the index of L vanishes.

The value of the index for compact X and its vanishing for $X = I \times M$ can be understood as follows. Let S_+ and S_- be the two spin bundles of X; each is of rank 2. It is convenient to express the operator L_0 as a Dirac operator mapping sections of $S_+ \otimes V$ to sections of $S_- \otimes V$ for some vector bundle or difference of vector bundles V; the index of L_0 can then be computed from the index theorem for the Dirac operator. To put $L_0 = \mathrm{d}_A + \mathrm{d}_A^*$ in the desired form, we must take V to be a formal difference of vector bundles $V = (S_+ \ominus S_-) \otimes \mathrm{ad}(E)$. Taking the formal difference $S_+ \ominus S_-$ where S_+ and S_- have the same rank causes all bulk and boundary contributions to the index that depend on E to cancel. (As X is a four-manifold and E has vanishing first Chern class, the only bulk E-dependent contribution is a multiple of the second Chern class of E; it cancels when we take the difference $S_+ \ominus S_-$. Boundary contributions are further discussed in the next paragraph.) Hence $\mathrm{ad}(E)$ can be replaced by a trivial bundle of the same rank, namely $\dim(H)$, and L_0 then reduces to the sum of that number of copies of the standard operator $\mathrm{d}+\mathrm{d}^*$ mapping ordinary differential forms of odd degree to those of even degree. The index of this operator on a four-manifold without boundary is $-\chi(X)$.

Since the vanishing of boundary corrections to the index is important, let us spell it out in more detail. If I has a boundary at a finite distance – for example the right boundary of the half-line $(-\infty, 0]$ – then we will pick local boundary conditions. With local boundary conditions, there is no E-dependent boundary contribution to the index, and as the metric of $I \times M$ is a product, there is also no boundary contribution involving the extrinsic curvature of the boundary. Alternatively, if I has an "end" at infinity – for example, the left end of the half-line – then we want normalizable solutions of the equation $L\Psi = 0$. Asking for normalizable solutions is equivalent to cutting off the half-line at a finite distance and imposing global Atiyah-Patodi-Singer boundary conditions. In general, in such a situation there is a boundary contribution to the index of a Dirac-like operator, but it cancels when we take the formal difference $(S_+ \ominus S_-) \otimes \mathrm{ad}(E)$ because S_+ and S_- are equivalent when restricted to M.

The operator L is real, so it has a real determinant line bundle. This determinant line bundle is topologically trivial[17] and after fixing a trivialization, L has a natural real-valued determinant function $\det L$.

4.1.2. *Applications of the flow equation.* The flow equation has two primary applications. The first is to determine the coefficients when a given cycle $\mathcal{C}$ in the relative homology is expanded as a sum of Lefschetz thimbles $\mathcal{J}_\sigma$ in the usual way,

$$\mathcal{C} = \sum_\sigma \mathfrak{n}_\sigma \mathcal{J}_\sigma. \tag{4.14}$$

We are only interested in cycles $\mathcal{C}$ that lie at $\mu = 0$.

The coefficient $\mathfrak{n}_\sigma$ is obtained by counting, with suitable signs, the solutions of the upward flow equation on the half-line $\mathbb{R}^- = (-\infty, 0]$. As usual, the upward flow equations are obtained from the downward ones by reversing the sign of the time. The boundary conditions are that the flow starts at a flat connection $\mathcal{A}_\sigma$ (corresponding to the cycle $\mathcal{J}_\sigma$) at $t = -\infty$ and ends on $\mathcal{C}$ at $t = 0$. Equivalently, we can consider downward flow equations on the half-line $\mathbb{R}^+ = [0, \infty)$. In this case, we consider flows that start on $\mathcal{C}$ at $t = 0$ and end at $\mathcal{A}_\sigma$ at $t = \infty$.

We write $\mathcal{A}_M = A_M + i\phi_M$ for the restriction of $\mathcal{A} = A + i\phi$ to M, at a fixed t. The boundary condition that we want at $t = 0$ is that $\mathcal{A}_M$ should lie in $\mathcal{C}$.

[17]In Donaldson theory, it is shown that the determinant line bundle of the Dirac operator twisted by $S_+ \otimes \mathrm{ad}(E)$ is trivial under a mild topological restriction that is satisfied in our applications. In our case, because we are twisting by the difference $(S_+ \ominus S_-) \otimes \mathrm{ad}(E)$, the topological condition in question is not needed.

Since we assume that $\mathcal{C}$ lies at $\mu = 0$, the condition that $\mathcal{A}_M$ lies in $\mathcal{C}$ implies in particular that $\mu = \mathrm{d}_A \star_M \phi_M = 0$ at $t = 0$. Subtracting this from the bulk equation $\mathrm{d}_A \star \phi = 0$, which is one of the flow equations, we find that the covariant normal derivative of ϕ_t vanishes at $t = 0$: $D_t \phi_t = 0$. This is enough to justify integration by parts in showing that $Z(\phi_t)$, defined in (4.12), vanishes. So it will lead to global vanishing of ϕ_t after we also pick a suitable boundary condition at $t = -\infty$.

The most basic example is the case that $\mathcal{C}$ is the real integration cycle of Chern-Simons theory with compact gauge group H. In this case, the condition for $\mathcal{A}_M$ to lie in $\mathcal{C}$ is simply $\phi_M = 0$. This certainly implies that $\mathrm{d}_A \star \phi_M = 0$.

At $t = -\infty$ (or $t = +\infty$ if we formulate the flow on $\mathbb{R}^+$), we ask that $\mathcal{A}_M$ approaches a specified flat connection $\mathcal{A}_\sigma$ and that ϕ_t vanishes. Given this vanishing, and the fact that we have chosen a boundary condition that allows integration by parts at $t = 0$, the vanishing theorem implies that ϕ_t will vanish everywhere.

Let $\mathcal{M}$ be the moduli space of solutions of the flow equations with boundary conditions just described. Since the index in the linearized problem vanishes, the expected dimension of $\mathcal{M}$ is zero and for a generic metric on M, $\mathcal{M}$ consists of a finite set of points. We count these points with a weight that is 1 or -1 depending on the sign[18] of $\det L$. Summing the contributions of all critical points, we get the integer $\mathfrak{n}_\sigma$ that appears in the expansion $\mathcal{C} = \sum_\sigma \mathfrak{n}_\sigma \mathcal{J}_\sigma$.

Let us compare the definition of $\mathfrak{n}_\sigma$ to the definition of the most basic invariant in Donaldson theory of smooth four-manifolds. This invariant is defined in a situation in which the index of the operator L_- that arises in linearizing the Yang-Mills instanton equation on X is zero. Under this circumstance, for a generic metric on X, the instanton equation has finitely many solutions, and Donaldson's invariant is defined by summing over all such solutions with a weight given by the sign of $\det L_-$. Obviously, $\mathfrak{n}_\sigma$ is a close cousin of this, with the usual instanton equations replaced by the flow equations for a complex-valued connection.

So far, the elliptic nature of the flow equations has enabled us to express a given cycle $\mathcal{C}$ in terms of the Lefschetz thimbles $\mathcal{J}_\sigma$, according to eqn. (4.14). The other important application of the flow equation is to describe Stokes phenomena. Here we must consider complete flow lines between two critical points corresponding to semistable flat connections $\mathcal{A}_\sigma$ and $\mathcal{A}_\tau$ on M. We must define an invariant $\mathfrak{m}_{\sigma\tau}$ that controls what happens in crossing a Stokes ray at which there are flows from $\mathcal{A}_\sigma$ to $\mathcal{A}_\tau$.

We consider the flow equations on $X = \mathbb{R} \times M$, with the requirement that $\mathcal{A}_M \to \mathcal{A}_\sigma$ for $t \to -\infty$ and $\mathcal{A}_M \to \mathcal{A}_\tau$ for $t \to +\infty$. Further we require that $\phi_t \to 0$ at both ends. The index of L still vanishes, but now there is a new ingredient: the flow equations are invariant under the group $\mathbf{T} \cong \mathbb{R}$ of translations in the $\mathbb{R}$ direction, but a solution with $\sigma \neq \tau$ cannot be translation-invariant. So any solution with $\sigma \neq \tau$ will have to come in a one-parameter family, and hence it is impossible to have an isolated solution. This clashes with the fact that the index is 0, and means that generically there are no flows from $\mathcal{A}_\sigma$ to $\mathcal{A}_\tau$.

We know an independent reason for this: the flow equation has a conserved quantity, namely the imaginary part of the holomorphic function $\mathcal{I} = \lambda W(\mathcal{A})$.

[18]The overall sign of the function $\det L$ depends on a choice of orientation of the relevant determinant line bundle. This amounts to a choice of relative orientations of the cycles $\mathcal{C}$ and $\mathcal{J}_\sigma$ that we are comparing. Such choices of sign and orientation are discussed in, for example, [**46**] in the context of finite-dimensional Morse theory and Floer cohomology.

For generic λ, conservation of $\mathrm{Im}\,\mathcal{I}$ makes a flow from $\mathcal{A}_\sigma$ to $\mathcal{A}_\tau$ impossible. The condition for a Stokes ray is precisely that a flow from $\mathcal{A}_\sigma$ to $\mathcal{A}_\tau$ is consistent with the conservation law. Taking $\lambda = 2\pi \exp(-i\alpha)$, a Stokes ray is characterized by some condition $\alpha = \alpha_0$. Instead of regarding α as a constant in our flow equations (4.8) and (4.10), regard it as a variable that takes values in a small interval near $\alpha = \alpha_0$. This adds one variable to the problem without adding any new equations, and the index of the linearized problem is now 1. Thus the moduli space of solutions should generically be one-dimensional – corresponding to families of solutions, obtained from each other by the action of $\mathbf{T}$, that appear at $\alpha = \alpha_0$. To get back to a problem with zero index, we can identify two field configurations that differ by the action of $\mathbf{T}$. This removes one variable, and the number of solutions is now generically finite. Weighting the solutions by the sign of the determinant of the operator $\widetilde{L}$ that appears in the linearization around a solution, we arrive at the desired invariant $\mathfrak{m}_{\sigma\tau}$.

Concretely, adding one variable and removing another and replacing L by $\widetilde{L}$ has the following effect. Suppose we set $\alpha = \alpha_0$ and find a solution of the flow equations. The linearization L around a solution of the flow equations will have a zero mode ψ, generated by the action of $\mathbf{T}$ on the given solution. Hence $\det L = 0$. Now vary α away from α_0. The eigenvector ψ of L can be deformed with α so as to remain an eigenvector, but its eigenvalue $\lambda(\alpha)$ generically vanishes only at $\alpha = \alpha_0$. Replacing L by $\widetilde{L}$ means that in evaluating the determinant at $\alpha = \alpha_0$, the factor that comes from the eigenvector ψ is not $\lambda(\alpha_0) = 0$ but $(\mathrm{d}\lambda/\mathrm{d}\alpha)|_{\alpha=\alpha_0}$, which is generically not zero. So $\det \widetilde{L}$ is generically nonzero and its sign is the contribution of a given $\mathbf{T}$-orbit of flow lines to $\mathfrak{m}_{\sigma\tau}$.

We have defined an invariant $\mathfrak{n}_\sigma$ associated to the flow equations on a half-line, and another invariant $\mathfrak{m}_{\sigma\tau}$ associated to the same equations on the full real line. The relation between these invariants is that in crossing a Stokes ray at which there is a net number $\mathfrak{m}_{\sigma\tau}$ of downward flow lines from $\mathcal{A}_\sigma$ to $\mathcal{A}_\tau$, the invariant $\mathfrak{n}_\tau$ jumps by $\mathfrak{n}_\tau \to \mathfrak{n}_\tau - \mathfrak{n}_\sigma\mathfrak{m}_{\sigma\tau}$. As explained in the discussion of eqn. (3.20), where $\mathfrak{m}_{+-}$ is taken to be ± 1, this jumping of the coefficients $\mathfrak{n}_\tau$ compensates for the jumping in the cycles $\mathcal{J}_\sigma$ and ensures that the sum $\sum_\sigma \mathfrak{n}_\sigma \mathcal{J}_\sigma$ is invariant. In infinite dimensions, it seems most concise to describe the jumping phenomenon in terms of the behavior of the coefficients.

Just as in Morse theory with finitely many variables, the jumping happens because at $\alpha = \alpha_0$ the set of points that can be reached by downward flow from $\mathcal{A}_\sigma$ ceases to be closed – its closure contains $\mathcal{A}_\tau$. A downward flow from $\mathcal{C}$ to $\mathcal{A}_\tau$ can split into a pair of flows, first from $\mathcal{C}$ to $\mathcal{A}_\sigma$ and then from $\mathcal{A}_\sigma$ to $\mathcal{A}_\tau$. This leads to the "wall-crossing" behavior $\mathfrak{n}_\tau \to \mathfrak{n}_\tau - \mathfrak{n}_\sigma\mathfrak{m}_{\sigma\tau}$. See for example [**41**] in the context of flows with finitely many variables. Ellipticity means that such statements about Morse theory carry over to the infinite-dimensional case. This is an important fact in the context of Floer theory (of three-manifolds and of loop spaces of symplectic manifolds). See for example [**46, 47**] for reviews. Hopefully the usual proofs in the context of Floer theory can be adapted to the present situation.

4.2. Knots and Morse theory. Beginning in section 2.5.3, we described the analytic continuation of Chern-Simons knot invariants in a rather informal way. To make this more precise, we need to incorporate in the language of oscillatory integrals the holonomy functions that are included in the path integral to define knot invariants. How to do this has been briefly explained in [**51**], section 3.3, and in [**54**], pp. 1218–1219, and in much more detail in [**34**]. We consider first the case of a compact gauge group H with connection A.

For a knot K labeled by a representation R, the basic factor is

$$\mathrm{Tr}_R\, P \exp\left(-\oint_K A\right). \tag{4.15}$$

According to the Borel-Weil-Bott theorem, letting T denote a maximal torus in H, the representation R can be constructed as the space of holomorphic sections of a certain holomorphic line bundle $\mathcal{L}_R$ over the flag manifold H/T. This means that there is a quantum mechanical problem with H symmetry for which the space of physical states is the representation R. Writing a path integral representation of this system and coupling it to the gauge field A will give the representation that we need of the holonomy function.

The line bundle $\mathcal{L}_R \to H/T$ admits an essentially unique H-invariant connection that we will call a_R. We can describe as follows a classical problem with R symmetry whose quantization leads, in view of the Borel-Weil-Bott theorem, to the representation R. Let S be a one-manifold and let $\rho : S \to H/T$ be a map. On the space of such maps, we define an action $I_R(\rho) = \int_S \rho^*(a_R)$. For S a circle, this is defined mod 2π, so that $\exp(iI_R(\rho))$ is well-defined. (If instead S has boundaries, $\exp(iI_R(\rho))$ is not a complex number but a transition amplitude between quantum wavefunctions; we will not need the details here.) Since $I_R(\rho)$ is invariant under reparametrizations of S, this system is a topological field theory in the one-dimensional sense; the Hamiltonian vanishes.

The basic path integral for this system is the integral over maps $\rho : S \to H/T$:

$$\int D\rho\, \exp(iI_R(\rho)). \tag{4.16}$$

For S a circle, the value of the path integral can be determined by observing that the path integral determines a trace in the physical Hilbert space. The physical Hilbert space is simply the representation R, and as the Hamiltonian vanishes, the trace is simply the trace of the identity operator. So the value of the path integral is d_R, the dimension of the representation R.

We want to generalize this to couple to a gauge field A and to compute the trace not of the identity operator but of the holonomy operator $\mathrm{Tr}_R P \exp\left(-\oint_S A\right)$. To do this, we simply couple the gauge field A to the ρ field in a gauge-invariant fashion. Quantum mechanically, we would couple A to the matrices that generate the action of H in the representation R. The classical limit of these matrices is the moment function μ_R associated to the action of H on H/T. μ_R is a map from H/T to the dual of the Lie algebra $\mathfrak{h}$. So for $h \in \mathfrak{h}$, we have a real valued function $\mu_R(h)$ on H/T. It obeys $\mathrm{d}\mu_R(h) = \iota_h \mathrm{d}a_R$ (where ι_h is contraction with the vector field on H/T that corresponds to h). As A is $\mathfrak{h}$-valued, there is a natural pairing (A, μ_R). The gauge-invariant extension of the action $I_R(\rho)$ is $I_R(\rho, A) = I_R(\rho) - \int_S (A, \mu_R)$. (A more intrinsic description of the geometry involved in defining this action can be found in [**54**]; for more detail see also [**34**].) Thus we arrive at a path integral representation of the holonomy:

$$\mathrm{Tr}_R P \exp\left(-\oint_S A\right) = \int D\rho\, \exp\left(iI_R(\rho, A)\right). \tag{4.17}$$

Now we can rewrite the basic formula (2.19) for the Chern-Simons invariant of a knot in a convenient form:

$$Z_H(K;k) = \int_{\mathcal{U}} DA \int D\rho \, \exp(ikW(A) + iI_R(\rho, A)). \tag{4.18}$$

What we have gained is that now, although we are integrating over a larger number of variables, the integrand is simpler: it is just the exponential of a local functional of the fields. In other words, we have reduced the problem to an oscillatory integral (with more variables), and we can apply the Morse theory machinery rather as if no knot were present.

As usual, the main step in analytic continuation is to extend the action to a holomorphic function of complex variables. We analytically continue the connection A to a complex-valued connection $\mathcal{A}$ and we replace the flag manifold H/T with $H_{\mathbb{C}}/T_{\mathbb{C}}$, where $T_{\mathbb{C}}$ is a complex maximal torus in $H_{\mathbb{C}}$. The analytically continued integral thus takes the form

$$Z_H(K;k) = \int_{\mathcal{C}} D\mathcal{A} \int D\rho \, \exp(ikW(\mathcal{A}) + iI_R(\rho, \mathcal{A})), \tag{4.19}$$

where $\mathcal{C}$ is an integration cycle in the space of fields $\mathcal{A}, \rho$.

Any integration cycle $\mathcal{C}$, including the original real integration cycle of Chern-Simons theory in the presence of a knot, can be expressed in terms of Lefschetz thimbles in the usual way. Such cycles are convenient for several reasons. One reason, discussed in section 3.1.3, is that it is straightforward to determine the asymptotic value of the integral over such a cycle, in the semiclassical region in which the action is large. What is this region? In the absence of a knot, the action is proportional to k, so the semiclassical region is the region of large k. In the presence of a knot labeled by a representation R with highest weight λ_R, the action $I_R(\rho, A)$ is proportional to λ_R, since it is linear in a_R and μ_R which are both proportional to λ_R. So in the presence of a knot, the natural semiclassical limit is obtained by taking k and λ_R to infinity with a fixed ratio. The Morse theory machinery will enable us to understand the asymptotic behavior of the analytically continued Chern-Simons path integral in this limit.

4.2.1. *Classical equations in the presence of a knot.* In the presence of a knot K, the classical equation for A receives an extra contribution and no longer asserts simply that F is flat. Rather, the equation receives an extra contribution that is supported on K:

$$\frac{kF}{2\pi} = \delta_K \mu_R. \tag{4.20}$$

(Here δ_K is a delta function that is Poincaré dual to K, and we use the linear form Tr to identify the Lie algebra $\mathfrak{h}$ with its dual; this is also done in eqn. (4.22) below.) The fact that the curvature is a delta function supported on K means that the connection is flat away from K and has a monodromy around K. Exponentiating the delta function in the curvature, this monodromy is

$$U = \exp(2\pi\mu_R/k). \tag{4.21}$$

To understand this result, recall that H/T is a homogeneous space, and that the conjugacy class of μ_R (as an element of $\mathfrak{h}^*$) is constant and depends only on R. So eqn. (4.21) really defines a conjugacy class in H; the monodromy around the knot must lie in this conjugacy class. We note that the conjugacy class in question depends only on the ratio μ_R/k.

Let us make this more explicit for $H = SU(2)$. Take R to be the n-dimensional irreducible representation of $SU(2)$. Its highest weight is

$$\frac{n-1}{2}\begin{pmatrix} i & 0 \\ 0 & -i \end{pmatrix}. \tag{4.22}$$

(To keep the structure constants real, we represent a real Lie algebra such as $\mathfrak{su}(2)$ by anti-hermitian matrices.) Up to conjugacy, this is μ_R for the representation R. So the monodromy around a knot labeled by this representation is

$$\exp\left(\frac{i\pi(n-1)}{k}\begin{pmatrix} 1 & 0 \\ 0 & -1 \end{pmatrix}\right). \tag{4.23}$$

We will now make a remark on quantum corrections to this classical formula. In physics in general, quantum corrections cannot be entirely understood by shifting the values of classical parameters, but many quantum results in Chern-Simons theory can be understood more readily from a semiclassical point of view if one shifts $n-1$ to n and k to $k+2$ in (4.23). (The shift in k was discussed in section 2.7.1, and the shift in n has a similar interpretation. For the analogous shifts for noncompact Lie groups, see [**48**].) For our purposes, it is not necessary to take these shifts into account explicitly. Our goal is to determine the correct integration cycle for the path integral in the limit of large k and n with fixed ratio n/k. The answer to this question is not affected by shifts $n \to n+1$ or $k \to k+2$. Those shifts enter when, after having found the right integration cycle for the path integral, one carries out the integral including one-loop effects. This will generate the shifts in n and k in the usual way.

4.2.2. *More on the classical equations.* There are some important subtleties concerning the gauge fields obeying (4.20), especially after we analytically continue to a complex gauge field $\mathcal{A}$. In that case, the equation becomes

$$\frac{k\mathcal{F}}{2\pi} = \delta_K \mu_R, \tag{4.24}$$

where $\mathcal{F} = \mathrm{d}\mathcal{A} + \mathcal{A}\wedge\mathcal{A}$. We want to supplement this equation with a condition for vanishing of the moment map,

$$\mathrm{d}_A \star_M \phi = 0, \tag{4.25}$$

where $\mathcal{A} = A + i\phi$. To describe what sort of gauge fields obey this equation, it suffices to work in a two-plane normal to K. We take polar coordinates r, θ on this two-plane, with K understood to lie at $r = 0$. To obey (4.24) and (4.25), we can take

$$\begin{aligned} A &= \alpha\,\mathrm{d}\theta + \dots \\ \phi &= -\gamma\,\mathrm{d}\theta + \dots, \end{aligned} \tag{4.26}$$

where α and γ are commuting elements of the real Lie algebra $\mathfrak{h}$ with $\alpha - i\gamma = \mu_R/2k$, and the ellipses refer to terms less singular than $1/r$. (The angular form $\mathrm{d}\theta$ has a $1/r$ singularity at $r = 0$.) This type of singularity in gauge theory was studied in detail[19] in [**55**], starting in section 2.1. The most natural way to proceed

[19]In that context, there was another term in ϕ of the form $\beta\,\mathrm{d}r/r$; β can be removed by a gauge transformation (which is singular at $r = 0$), and tentatively it seems that in the present problem we can just as well take $\beta = 0$.

is to fix a gauge in which μ_R lies in the Lie algebra of a specified complex maximal torus $T_{\mathbb{C}}$. This can always be done globally along K. Then we require $\mathcal{A}$ to have a singularity of the sort described in (4.26), and we only allow gauge transformations that preserve the gauge condition – in other words, gauge transformations that are $T_{\mathbb{C}}$-valued when restricted to K.

One important point, discussed in [**55**], is that by a gauge transformation that is singular at $r = 0$, one can shift α by a coroot of H. This means that the formalism is invariant under shifting μ_R by k times a coroot. For $H = SU(2)$, this corresponds to the symmetry

$$n \to n + 2k, \tag{4.27}$$

which actually undergoes a quantum correction to $n \to n + 2(k+2)$.

There is an important subtlety, which we will describe first for $H = SU(2)$. Suppose that we vary μ_R and k so that μ_R/k goes to zero, or more generally so that it approaches a cocharacter (which we then remove by a singular gauge transformation, as mentioned in the last paragraph). This means that α and γ go to zero, and a naive look at (4.26) might make one believe that in the limit, A and ϕ become nonsingular. The correct conclusion, however, because of the ellipses in (4.26), which represent terms less singular than $1/r$, is only that in such a case, A and ϕ are less singular than $1/r$. The equations $\mathcal{F} = \mathrm{d}_A\phi = 0$ have a solution with a singularity at $r = 0$ that is (slightly) milder than $1/r$, and this less singular solution may be the result if one varies a family of solutions with the $1/r$ singularity (4.26) so that the coefficient of $1/r$ approaches zero. The singular solution in question, which was essentially first studied by Nahm [**56**], takes the form

$$\begin{aligned} A &= \frac{t_1\,\mathrm{d}\theta}{\ln r} \\ \phi &= \frac{t_2\,\mathrm{d}r}{r\ln r} - \frac{t_3\,\mathrm{d}\theta}{\ln r}, \end{aligned} \tag{4.28}$$

where $[t_1, t_2] = t_3$, and cyclic permutations thereof. This condition means that the t_i define a homomorphism from $\mathfrak{su}(2)$ to $\mathfrak{h}$. For $H = SU(2)$, there is up to conjugation only one nontrivial homomorphism and we pick this one. (For other groups, all possible homomorphisms can arise, in general.) For this solution, the monodromy of $\mathcal{A}$ around the singularity at $r = 0$ is not trivial but is in the conjugacy class that contains the non-diagonalizable, unipotent element

$$V = \begin{pmatrix} 1 & 1 \\ 0 & 1 \end{pmatrix}. \tag{4.29}$$

This is demonstrated by directly showing that, with A, ϕ given in (4.28), the holonomy of $\mathcal{A} = A + i\phi$ along any path that loops around $r = 0$ is in the stated conjugacy class. For example, if the path is the circle $r = r_0$ for some r_0, the monodromy is $\exp(2\pi(t_1 - it_3)/\ln r_0)$, which is in the stated conjugacy class. (For more detail, see section 3.8 of [**55**].)

What is happening here is that in the complex Lie group $G = H_{\mathbb{C}}$, as $\mu_R/k \to 0$, the conjugacy class that contains the element $U = \exp(2\pi\mu_R/k)$ does not approach the conjugacy class of the identity; rather it approaches the conjugacy class of the element V. This may become clear if one considers the family of $SL(2,\mathbb{C})$ elements

$$\begin{pmatrix} \lambda & 1 \\ 0 & \lambda^{-1} \end{pmatrix}, \tag{4.30}$$

which are conjugate to the diagonal matrix $\mathrm{diag}(\lambda,\lambda^{-1})$ for generic λ, but to V if $\lambda = 1$. So if we consider a family of connections $\mathcal{A}$ representing Chern-Simons critical points, parametrized by μ_R/k, with monodromy $U = \exp(2\pi\mu_R/k)$, then as $\mu_R/k \to 0$, the monodromy of $\mathcal{A}$ around the knot might approach 1, but it also might approach a unipotent element conjugate to V. The latter behavior is generic, since the conjugacy class of V has dimension two, while that of the identity consists of a single point.

Modulo a singular gauge transformation that shifts α to zero,[20] the behavior is exactly the same if μ_R/k approaches a value for which $U \to -1$. Given a family of flat connections $\mathcal{A}$ of monodromy U, consider a limit in which $U \to -1$. The monodromy of $\mathcal{A}$ may approach -1, but instead it may approach

$$-V = -\begin{pmatrix} 1 & 1 \\ 0 & 1 \end{pmatrix}. \tag{4.31}$$

We will encounter both cases in section 5.

We have described this exceptional behavior for $SL(2,\mathbb{C})$, but it has a close analog for any complex Lie group G. As long as $U = \exp(2\pi\mu_R/k)$ is regular (that is, at long as the subgroup of G that it commutes with is a copy of $T_{\mathbb{C}}$), Chern-Simons critical points in the presence of a knot correspond to flat connections whose monodromy around the knot is conjugate to U. But in general, the monodromy may be in any conjugacy class whose closure contains U. For more on this, see section 3.8 of [**55**].

The flow equations in the presence of a knot

Now let us briefly describe the flow equations in the presence of a knot. As usual, to formulate the flow equations, we introduce an extra "time" coordinate. So to study Chern-Simons theory on a three-manifold M in the presence of a knot K, we consider flow equations on the four-manifold $X = \mathbb{R}\times M$, with the knot now filling out the two-manifold $\mathbb{R}\times K$. The flow equations are the same as before, but now we must consider solutions with a singularity along $\mathbb{R}\times K$. The singularity is given by (4.28) modulo terms less singular than $1/r$. In particular, for special values of μ_R/k, we must allow a singularity that (perhaps after a singular gauge transformation to set α to zero) takes the form (4.28). This seemingly exceptional case is important in applications, because the flat $SL(2,\mathbb{C})$ connection associated to a hyperbolic metric on the knot complement has this sort of monodromy.

In effect, the flow equations have a codimension two singularity in four dimensions that can be modeled by the same surface operators studied in [**55**].

4.2.3. *Analog for noncompact Lie groups.* For the case of a compact Lie group H, we have described how to incorporate a knot labeled by a representation R in the framework of oscillatory integrals and Morse theory.

What happens if the gauge group is not compact, for example if it is $G = H_{\mathbb{C}}$? If the representation R is unitary and arises by quantization of a coadjoint orbit $\mathcal{O}$ of G, then there is a close analog of (4.17), with H/T replaced by $\mathcal{O}$. The resulting knot invariants, however, have not been seriously studied.

[20]Such a gauge transformation is single-valued in $SO(3)$, but not in $SU(2)$. It is not quite a symmetry of $SU(2)$ Chern-Simons theory, but it is useful for analyzing the classical equations and describing critical points. For more on this, see section 4.2.7.

Let us consider instead the case that R is a finite-dimensional holomorphic representation of G, obtained by analytic continuation of an irreducible representation of H, as described in section 2.6.1. It is straightforward to bring such a representation into the Morse theory framework. We represent the holonomy around the knot in such a representation by the same factor as in (4.19), namely $\int D\rho \exp(iI_R(\rho, \mathcal{A}))$. So the path integral definition (2.29) of knot invariants associated to such a representation can be rewritten

(4.32)
$$Z_G(K;t,\widetilde{t}) = \int_C D\mathcal{A}\, D\widetilde{\mathcal{A}}\, D\rho \exp\left(itW(\mathcal{A})/2 + i\widetilde{t}\,W(\widetilde{\mathcal{A}})/2\right) \exp(iI_R(\rho, \mathcal{A})).$$

Morse theory can now be used in the usual way to describe possible integration cycles for this integral. After analytic continuation, $\mathcal{A}$ and $\widetilde{\mathcal{A}}$ are independent. Since $I_R(\rho, \mathcal{A})$ is independent of $\widetilde{\mathcal{A}}$, the presence of the knot does not affect the Euler-Lagrange equation for $\widetilde{A}$. The equation for a critical point thus tells us simply that $\widetilde{\mathcal{A}}$ is flat, while $\mathcal{A}$ is a flat connection on the complement of K whose monodromy around K is in the conjugacy class of the element $U = \exp(2\pi\mu_R/k)$ described in eqn. (4.21). (When U is not regular, this statement must be refined in a way described in section 4.2.2.) Of course, now we are supposed to take the conjugacy class of U in G rather than in H.

One point deserves note. As $I_R(\rho, \mathcal{A})$ is not real, the exponential $\exp(iI_R(\rho, \mathcal{A}))$ grows exponentially in some regions. As a result, even if t and $\widetilde{t}$ are complex conjugates, the path integral (4.32) does not converge if evaluated on the usual real cycle $\widetilde{\mathcal{A}} = \overline{\mathcal{A}}$. Perhaps it is possible to slightly displace the real cycle to make the integral converge, or perhaps the integral can only be defined using Lefschetz thimbles. As explained in section 2.6.1, in the case of knots in S^3, a suitable integration cycle can essentially be borrowed from the case of a compact gauge group.

A similar analysis applies if the gauge group is a noncompact real form H' of $G = H_{\mathbb{C}}$. If R is a unitary representation, one can imitate (4.17) using the appropriate coadjoint orbit. If R is a finite-dimensional representation of H' (obtained by analytic continuation from H to $G = H_{\mathbb{C}}$ and then restriction to H'), one can represent the knot by the now familiar factor $\int D\rho \exp(iI_R(\rho, A'))$, where now A' is $\mathfrak{h}'$-valued. Prior to analytic continuation, ρ is a map to H/T, which admits an action of G and hence of H'. After analytic continuation, the fields and the integrand of the path integral are the same as they are in the H-valued theory, but the integration cycle is different. This last statement holds exactly as it does in the absence of the knot.

4.2.4. *Classification of critical points.* To study the Chern-Simons path integral via the Morse theory machinery, a first step is to classify the critical points and evaluate the classical action at a critical point. As long as one considers only physical values of the coupling parameters, critical points correspond to flat connections modulo gauge transformation (perhaps with monodromy around a knot), and the Chern-Simons invariant $W(A)$ of such a flat connection is understood as an element of $\mathbb{R}/2\pi\mathbb{Z}$ (or $\mathbb{C}/2\pi\mathbb{Z}$ in the case of a complex Lie group).

Upon analytic continuation away from physical values of the coupling, one must be more careful. Let us consider first the case that the gauge group is a compact Lie group H and one wishes to continue away from integer values of k. In this case, the integrand $\exp(ikW(A))$ is invariant only under the subgroup of gauge transformations that leave $W(A)$ fixed, rather than shifting it by a multiple of 2π. If H is connected and simply-connected, this subgroup is the connected component of the group of gauge transformations. At any rate, $\exp(ikW(A))$ should be understood as a function not on the space $\mathcal{U}$ of gauge fields modulo gauge transformations, but on a cover $\widehat{\mathcal{U}}$ on which the Chern-Simons invariant is well-defined. For connected and simply-connected H, $\widehat{\mathcal{U}}$ is the universal cover of $\mathcal{U}$.

So we need to consider critical points of the Chern-Simons function on $\widehat{\mathcal{U}}$. Such a critical point is simply a flat connection together with a lift of its Chern-Simons invariant to $\mathbb{R}$ (or $\mathbb{C}$, if we consider complex critical points).

As an example, let us consider Chern-Simons theory on S^3 in the presence of a knot K. Typically there are only finitely many critical orbits in the usual sense. In other words, up to gauge transformation, there are typically only finitely many flat connections on the knot complement $S^3\backslash K$ with prescribed monodromy about K. However, when we analytically continue away from integer values of k, we need a real-valued Chern-Simons function, meaning that a critical point must be understood as a pair consisting of a flat connection and a lift of its Chern-Simons function to $\mathbb{R}$. This combined data determines a critical point of the function $W(A)$ on $\widehat{\mathcal{U}}$. There always are infinitely many of these critical points, as the Chern-Simons invariant of any given flat connection can be lifted to $\mathbb{R}$ in infinitely many ways. In section 5, we will study in detail the analytic continuation of the Chern-Simons path integral for some specific knots (the trefoil and the figure-eight knot) and we will see the important role played by distinct critical points that differ only in the value of their Chern-Simons invariant. This should come as no surprise, since in eqn. (3.73) we have already seen similar behavior in the analytic continuation of the Bessel function away from integer values of κ.

The same ideas hold if the gauge group is a non-compact real form H' of $G = H_{\mathbb{C}}$. However, the case that the gauge group is the complex Lie group G involves a new wrinkle. The Chern-Simons theory depends naturally on an integer-valued parameter ℓ and a real parameter s, as shown in eqn. (2.2). As explained in section 2, to analytically continue to complex s, we view $\widetilde{\mathcal{A}}$, which begins life as the complex conjugate of the gauge field $\mathcal{A}$, as an independent $\mathfrak{g}$-valued connection in its own right. The action is a linear combination of the two Chern-Simons actions $W(\mathcal{A})$ and $W(\widetilde{\mathcal{A}})$,

$$I = \frac{t\,W(\mathcal{A})}{2} + \frac{\widetilde{t}\,W(\widetilde{\mathcal{A}})}{2}, \quad t = \ell + is, \ \widetilde{t} = \ell - is. \tag{4.33}$$

If we impose invariance under all gauge transformations, a critical point is a pair of flat connections $\mathcal{A}$, $\widetilde{\mathcal{A}}$, up to gauge equivalence. In that case, $W(\mathcal{A})$ and $W(\widetilde{\mathcal{A}})$ are each naturally defined as elements of $\mathbb{C}/2\pi\mathbb{Z}$. But then (for complex s) the indeterminacy of I is not just an integer multiple of 2π and the integrand $\exp(iI)$ of the path integral is not defined.

One might think that to make sense of the path integral, we would have to lift both $W(\mathcal{A})$ and $W(\widetilde{\mathcal{A}})$ from $\mathbb{C}/2\pi\mathbb{Z}$ to $\mathbb{C}$. We indeed have to do this if the aim is to analytically continue to generic values of both ℓ and s. However, there is a fairly natural problem, described in section 2, in which ℓ remains an integer and only s is continued to complex values. In this case, $(t+\widetilde{t})/2$ remains an integer, so adding 2π to both $W(\mathcal{A})$ and $W(\widetilde{\mathcal{A}})$ causes no change in $\exp(iI)$. The most natural way to proceed is to regard $\exp(iI)$ as a function on the smallest cover of $\mathcal{U}_{\mathbb{C}} \times \mathcal{U}_{\mathbb{C}}$

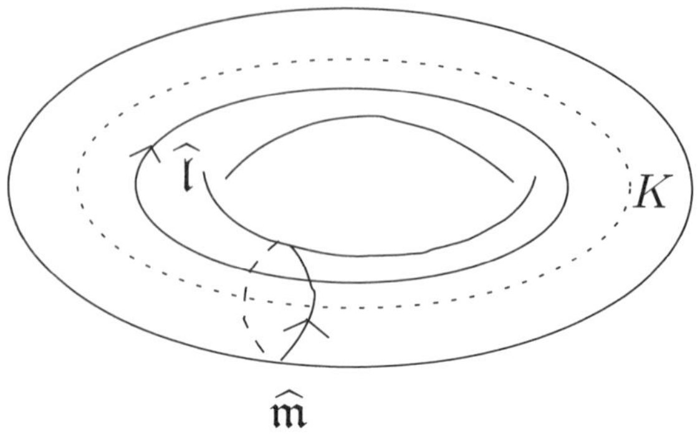

FIGURE 10. A knot K (dotted line) surrounded by a two-torus Y. On Y are drawn the meridian $\widehat{\mathfrak{m}}$, a circle whose linking number with K is 1, and the longitude $\widehat{\mathfrak{l}}$, whose linking number with K vanishes.

where it can be defined. In practice, this means that a critical point is a pair of flat connections $\mathcal{A}, \widetilde{\mathcal{A}}$ together with a lift of the pair $W(\mathcal{A})$, $W(\widetilde{\mathcal{A}})$ to $(\mathbb{C}\times\mathbb{C})/2\pi\mathbb{Z}$, where $\mathbb{Z}$ is diagonally embedded. In other words, we can shift both Chern-Simons invariants by a common multiple of 2π, but we cannot shift them independently. We can use this freedom to arbitrarily specify how we want to lift the Chern-Simons invariant of either $\mathcal{A}$ or $\widetilde{\mathcal{A}}$, but then we must consider all lifts of the other.

It is instructive to consider the usual example of a knot K in S^3. In this case, as explained in section 4.2.3, a critical point in the usual sense corresponds to an ordinary flat connection $\widetilde{\mathcal{A}}$ on S^3 plus a flat connection $\mathcal{A}$ on $S^3\backslash K$ whose monodromy around K is in a prescribed conjugacy class. As S^3 is simply connected, $\widetilde{\mathcal{A}}$ is gauge-equivalent to a trivial flat connection, and $W(\widetilde{\mathcal{A}})$ vanishes modulo 2π. It is natural to fix the complex lift of $W(\widetilde{\mathcal{A}})$ to be zero. Once we do this, we have exhausted our freedom to shift the Chern-Simons invariants, and we must consider all possible complex lifts of $W(\mathcal{A})$.

The possible critical points, then, are classified by a flat connection $\mathcal{A}$ on $S^3\backslash K$ with appropriate monodromy and with all possible choices for its Chern-Simons invariant. But this is the same set of critical points that we encounter in analytically continuing the theory on S^3 with compact gauge group H in the presence of the same knot K labeled by the same representation R. What we have just described is a nonperturbative aspect of the relation between theories with gauge groups H or $G = H_{\mathbb{C}}$ for the special case of knots in S^3. This relation was already discussed in section 2.6.

4.2.5. *Variation of the action.* To use the Morse theory formalism, we need some information about the value of the Chern-Simons function $W(\mathcal{A})$ for a classical solution. We take the gauge group to be a complex Lie group $G = H_{\mathbb{C}}$.

The following ideas prove to be useful. First of all, as sketched in fig. 10, the boundary of a small neighborhood of a knot K is a two-torus Y. On Y, we draw two circles – the meridian $\widehat{\mathfrak{m}}$, which goes the short way around Y and has linking number 1 with K, and the longitude $\widehat{\mathfrak{l}}$, which goes the long way around Y and which we choose to have linking number 0 with K.

Since the fundamental group of Y is abelian, if $\mathcal{A}$ is a flat connection, its monodromies around $\widehat{\mathfrak{m}}$ and $\widehat{\mathfrak{l}}$ (which we will denote by the same symbols) are commuting elements of G. Generically, we can conjugate them to a common maximal torus $T_{\mathbb{C}} \subset G$. According to (4.21), for a knot labeled by a representation R, $\widehat{\mathfrak{m}}$ is conjugate to $U = \exp(2\pi\mu_R/k)$. (More exactly, as explained in section 4.2.2, this is so if U is regular. In general, U is contained in the closure of the conjugacy class that contains $\widehat{\mathfrak{m}}$.)

Let us consider in detail the case $G = SL(2,\mathbb{C})$. The maximal torus of G is a copy of $\mathbb{C}^*$, consisting of diagonal matrices

$$\begin{pmatrix} \lambda & 0 \\ 0 & \lambda^{-1} \end{pmatrix}. \tag{4.34}$$

An important fact is that in a classical solution of Chern-Simons theory in the presence of a knot K, we have a little more structure than just a flat connection on the knot complement. In putting knots in the Morse theory framework, we introduced an H/T or $G/T_{\mathbb{C}}$-valued field ρ along K. In a classical solution, ρ is invariant under $\widehat{\mathfrak{m}}$ and $\widehat{\mathfrak{l}}$, and once $\widehat{\mathfrak{m}}$ and $\widehat{\mathfrak{l}}$ are given, the choice of ρ amounts to a choice of one of the two eigenvalues of $\widehat{\mathfrak{m}}$ and $\widehat{\mathfrak{l}}$, say the first eigenvalue (corresponding to λ as opposed to λ^{-1} in (4.34)). We write $\mathfrak{m}$ and $\mathfrak{l}$ for the eigenvalues of the loops $\widehat{\mathfrak{m}}$ and $\widehat{\mathfrak{l}}$ that correspond to ρ. According to (4.23), if R is the n-dimensional representation of $SU(2)$, then

$$\mathfrak{m} = \exp(i\pi(n-1)/k). \tag{4.35}$$

In section 5, we will use this formula in the semiclassical limit of large n, k with fixed ratio n/k. (In this limit, the difference between n and $n-1$ is irrelevant.)

Once K is embedded in a three-manifold – which we will take to be simply a three-sphere – $\mathfrak{l}$ and $\mathfrak{m}$ are no longer independent. They are constrained by a requirement that an $SL(2,\mathbb{C})$ flat connection on T with monodromies $\widehat{\mathfrak{l}}$ and $\widehat{\mathfrak{m}}$ should actually extend over the knot complement $S^3\backslash K$.

One simple class of $SL(2,\mathbb{C})$ flat connections are the abelian ones. These are the flat connections whose structure group reduces to $\mathbb{C}^*$, the group of diagonal matrices (4.34). They are classified very simply. Although the fundamental group of $S^3\backslash K$ can be very complicated, its first homology group is simply isomorphic to $\mathbb{Z}$, generated by $\mathfrak{m}$. This means that there is up to gauge transformation one flat $\mathbb{C}^*$-valued connection on $S^3\backslash K$ for every choice of $\mathfrak{m}$. Such a flat connection always has $\mathfrak{l} = 1$.

What about flat connections whose structure group does not reduce to $\mathbb{C}^*$? Flat connections with triangular structure group

$$\begin{pmatrix} \alpha & \beta \\ 0 & \gamma \end{pmatrix} \tag{4.36}$$

are unstable and need not be considered as critical points. What this really means is that to any strictly triangular flat connection there is a corresponding abelian one obtained by deleting the off-diagonal part of $\mathcal{A}$, and this should be considered instead. An important point which follows directly from the definition of the Chern-Simons function $W(\mathcal{A})$ is that this function is invariant under this operation:

$$\begin{pmatrix} \alpha & \beta \\ 0 & \gamma \end{pmatrix} \to \begin{pmatrix} \alpha & 0 \\ 0 & \gamma \end{pmatrix}. \tag{4.37}$$

There remain the irreducible $SL(2,\mathbb{C})$ flat connections. (A flat connection is called irreducible if the only elements of $SL(2,\mathbb{C})$ that commute with all monodromies are the central elements ± 1.) An irreducible flat connection does not exist for all values of $\mathfrak{l}, \mathfrak{m}$. Rather, for each knot K, there is a Laurent polynomial $\mathbf{A}(\mathfrak{l},\mathfrak{m})$ (a polynomial in $\mathfrak{l}$, $\mathfrak{l}^{-1}$, $\mathfrak{m}$, $\mathfrak{m}^{-1}$) such that the moduli space of irreducible $SL(2,\mathbb{C})$ flat connections on $S^3\backslash K$ is described by an equation

$$\mathbf{A}(\mathfrak{l},\mathfrak{m}) = 0 \tag{4.38}$$

which parametrizes values of $\mathfrak{l}$ and $\mathfrak{m}$ for which there exists such a connection. $\mathbf{A}(\mathfrak{l},\mathfrak{m})$, which is often called the **A** polynomial, is effectively computable for any given knot. We will make use of some examples in section 5. The **A** polynomial is invariant under the exchange

$$\mathfrak{l} \to \mathfrak{l}^{-1}, \quad \mathfrak{m} \to \mathfrak{m}^{-1}, \tag{4.39}$$

which comes from a Weyl transformation exchanging the two eigenvalues in (4.34). Such a Weyl transformation does not leave invariant the field ρ that is defined along the knot, but the definition of the **A** polynomial does not "know" about ρ. Another symmetry of the **A** polynomial is

$$\mathfrak{l} \to \mathfrak{l}, \quad \mathfrak{m} \to -\mathfrak{m}. \tag{4.40}$$

This symmetry reflects the fact that on $S^3\backslash K$, there exists a flat complex line bundle $\mathcal{L}$ with monodromies $\mathfrak{m} = -1$, $\mathfrak{l} = 1$. If $E \to S^3\backslash K$ denotes a flat complex vector bundle of rank 2 and determinant 1, corresponding to a flat $SL(2,\mathbb{C})$ connection, then the operation $E \to E \otimes \mathcal{L}$ reverses the sign of $\mathfrak{m}$ without changing $\mathfrak{l}$. It is not immediately apparent whether this operation leaves fixed the Chern-Simons function $W(\mathcal{A})$ and therefore whether it is a symmetry of Chern-Simons theory. We return to this issue in section 4.2.7.

To describe the function $W(\mathcal{A})$ once the **A** polynomial is known (see [**57**] for a mathematical treatment), we simply use the variational principle for $W(\mathcal{A})$, which follows directly from its definition:

$$\delta W(\mathcal{A}) = \frac{1}{2\pi}\int_{S^3} \operatorname{Tr}\delta\mathcal{A}\wedge\mathcal{F}. \tag{4.41}$$

On the other hand, for a flat connection on the knot complement, we have $\mathcal{F} = (2\pi\mu_R/k)\delta_K = (\ln\widehat{\mathfrak{m}})\delta_K$. In the last step, we used the fact that $\widehat{\mathfrak{m}} = \exp(2\pi\mu_R/k)$. So the right hand side of (4.41) reduces to $\operatorname{Tr}\oint_K \delta\mathcal{A}\ln\widehat{\mathfrak{m}}$. Here, once we diagonalize $\mathfrak{m}$, $\mathcal{A}$ and $\delta\mathcal{A}$ are also diagonal along K, and $\oint_K \delta\mathcal{A} = \delta\ln\widehat{\mathfrak{l}}$. So the formula becomes $\delta W = \operatorname{Tr}(\delta\ln\widehat{\mathfrak{l}})\ln\widehat{\mathfrak{m}}/2\pi$. In expressing this result in terms of the eigenvalues $\mathfrak{l}$ and $\mathfrak{m}$ of the Lie algebra elements $\widehat{\mathfrak{l}}$ and $\widehat{\mathfrak{m}}$, we get a factor of 2 from the trace:

$$\delta W(\mathcal{A}) = \frac{1}{\pi}\delta\ln\mathfrak{l}\cdot\ln\mathfrak{m}. \tag{4.42}$$

In practice, once the relation between $\mathfrak{l}$ and $\mathfrak{m}$ is known, the important results about $W(\mathcal{A})$ can usually be obtained by integrating this formula. Some slightly subtle examples will be considered in section 5. Here we will practice with some easy cases.

For an abelian flat connection, $\mathfrak{l}$ is always 1, so $0 = \delta\ln\mathfrak{l} = \delta W(\mathcal{A})$. So $W(\mathcal{A})$ is constant as a function on the moduli space of abelian flat connections on $S^3\backslash K$. It vanishes for a trivial flat connection ($\mathfrak{l} = \mathfrak{m} = 1$) so it vanishes for any abelian flat connection. For another example, consider a unitary flat $SL(2,\mathbb{C})$ connection

– in other words, one whose holonomy lies in $SU(2)$. In this case, $\mathfrak{l}$ and $\mathfrak{m}$ are both of absolute value 1, so $\ln \mathfrak{m}$ and $\ln \mathfrak{l}$ are both imaginary. Hence, δW is real on this locus, consistent with the fact that W is real for an $SU(2)$ flat connection.

Apart from the factor $\exp(ikW(\mathcal{A}))$, the integrand of the path integral in the presence of a knot also contains an additional factor $\exp(iI_R(\rho,\mathcal{A}))$, so we need to know how to evaluate this factor for a classical solution. In the usual gauge in which ρ is constant along K, this factor reduces to $\exp(-i\int_K(\mu_R,\mathcal{A}))$. This depends only on $\mathfrak{l}$ and can be understood as a character of the maximal torus in which $\mathfrak{l}$ takes values. Schematically we might write it as $\mathfrak{l}^{\mu_R}$. For the n-dimensional representation of $SU(2)$, this factor is $\mathfrak{l}^{(n-1)}$. Equivalently,

$$I_R(\rho,\mathcal{A}) = -i(n-1)\ln \mathfrak{l}. \tag{4.43}$$

4.2.6. *Branches of the logarithm.* An obvious point about this last formula is that $\ln \mathfrak{l}$ has many branches. This reflects the fact that the underlying action $I_R(\rho,\mathcal{A})$ is similarly defined only modulo 2π. There are two possible ways to proceed. If we wish to analytically continue to non-integral values of n, we should consider a critical point to be a pair consisting of a flat connection of suitable monodromy and a lift of $\ln \mathfrak{l}$ to $\mathbb{C}$ (plus a similar lift of $W(\mathcal{A})$ if we are analytically continuing in k). This amounts to replacing the loop space of H/T or $G/T_{\mathbb{C}}$ with its universal cover.

If we leave n as an integer and analytically continue only with respect to k, we need not consider the choice of a lift of $\ln \mathfrak{l}$ to be part of the definition of a critical point. Instead, the multivaluedness of the logarithm enters in the following way. Consider the quantity $\mathrm{Im}(ikW(\mathcal{A}) + iI_R(\rho,\mathcal{A}))$ which one may expect to be conserved by the flow equations. The multivaluedness of the logarithm means that this quantity is actually only conserved modulo $2\pi(n-1)\mathbb{Z}$. It is conserved in any flow provided we make suitable choices for the branches of $\ln \mathfrak{l}$, but there may be no way to choose branches of $\ln \mathfrak{l}$ that depend only on the choice of a critical point (and not on the choice of a flow line) such that $\mathrm{Im}(ikW(\mathcal{A})+iI_R(\rho,\mathcal{A}))$ is conserved in every flow.

If instead we analytically continue with respect to n, we must consider the choice of branch of $\ln \mathfrak{l}$ as part of the definition of a critical point and then the quantity $\mathrm{Im}(ikW(\mathcal{A}) + iI_R(\rho,\mathcal{A}))$ really is conserved by all flows.

See the end of section 5.2.4 for an example in which this issue becomes relevant.

4.2.7. *Tensoring with a flat line bundle.* As a further application of eqn. (4.42), we will return to the question of whether, for $H = SU(2)$, the operation $E \to E\times\mathcal{L}$, where $\mathcal{L}$ is a flat line bundle of order 2, is a symmetry of Chern-Simons theory.

Let $\Delta W(\mathcal{A})$ denote the change in $W(\mathcal{A})$ under this operation. We can differentiate $\Delta W(\mathcal{A})$ with respect to $\mathcal{A}$ by making use of (4.42). The operation $E \to E\otimes\mathcal{L}$ leaves $\mathfrak{l}$ unchanged and (as it changes the sign of $\mathfrak{m}$) shifts $\ln \mathfrak{m}$ to $\ln \mathfrak{m} - i\pi$. So $\delta \Delta W(\mathcal{A}) = (\delta \ln l/\pi)(-i\pi) = -i\delta \ln \mathfrak{l}$. Integrating, we have $\Delta W(\mathcal{A}) = -i\ln \mathfrak{l}$ up to an additive constant. In Chern-Simons theory on a general three-manifold (such as $\mathbb{RP}^3$) the constant may be nonzero.[21] But it vanishes for the special case of a knot $K\subset S^3$. This can be proved by considering the special case that $\mathcal{A}$ is a flat abelian connection (embedded in $SL(2,\mathbb{C})$) with appropriate monodromy around K. Tensoring with $\mathcal{L}$ gives another flat abelian connection $\mathcal{A}'$.

[21]It is always an integer multiple of π, though not necessarily of 2π. This can be proved by considering the Chern-Simons invariant for $SO(3,\mathbb{C}) = SL(2,\mathbb{C})/\mathbb{Z}_2$.

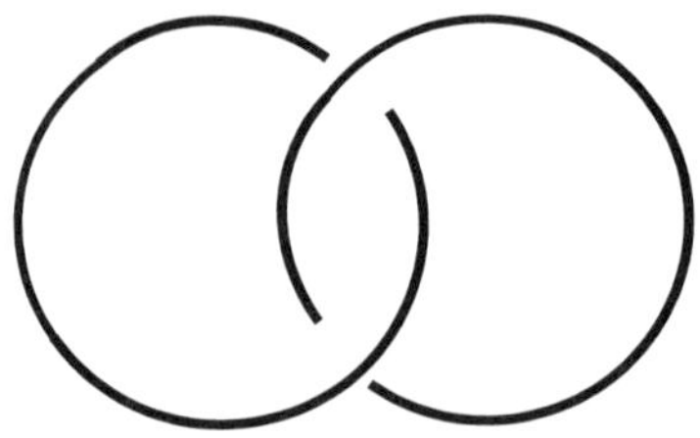

FIGURE 11. Two linked circles in S^3.

The monodromies of $\mathcal{A}$ and $\mathcal{A}'$ around the longitude and the meridian are related by $\mathfrak{m}' = -\mathfrak{m}$, $\mathfrak{l}' = \mathfrak{l} = 1$, and moreover $W(\mathcal{A}) = W(\mathcal{A}') = 0$. So for a knot in S^3, we have simply $\Delta W(\mathcal{A}) = -i \ln \mathfrak{l}$, and, making use of (4.43), we get a symmetry of $kW(\mathcal{A}) + I_R(\rho, \mathcal{A})$ if we combine $E \to E \otimes \mathcal{L}$ with $n \to n + k$. Allowing for a one-loop quantum shift described in section 2.7.1, the symmetry is actually

$$E \to E \otimes \mathcal{L}, \ n \to n + k + 2. \tag{4.44}$$

This is actually the one place in the present paper that it is important to distinguish a knot (an embedded circle in S^3) from a link (a disjoint union of embedded circles). The operation $E \to E \otimes \mathcal{L}$, where $\mathcal{L}$ has monodromy around a knot, does not affect the longitude for that knot. But in the case of two or more disjoint embedded circles with nonzero linking (fig. 11), if $\mathcal{L}$ has monodromy around one circle, then tensoring with $\mathcal{L}$ may reverse the sign of the longitude for a second circle. So (4.44) is a symmetry of Chern-Simons theory for a knot, but not necessarily for a link.

The square of the operation (4.44)

$$n \to n + 2(k + 2) \tag{4.45}$$

is a universal symmetry of Chern-Simons theory (already mentioned in eqn. (4.27)). Indeed, $\mathcal{L}^2$ is completely trivial, and tensoring with $\mathcal{L}^2$ leaves $\mathfrak{m}$ fixed.

5. Analytic continuation of the colored Jones polynomial

Here we will finally apply all this machinery to some specific examples. We will consider $SU(2)$ Chern-Simons theory at level k on S^3, with a knot K labeled by the n-dimensional representation of $SU(2)$. We denote as $J_n(q)$, where $q = \exp(2\pi i/(k+2))$, the (unnormalized) Chern-Simons path integral in the presence of this knot. $J_n(q)$ is essentially the colored Jones polynomial, though our normalization is not the most common one in the mathematical literature. (As explained at the end of section 2.5.3, the two normalizations differ by whether one chooses to divide by the Chern-Simons path integral for the unknot.)

We consider in detail two illustrative and relatively simple knots – the trefoil knot and the figure-eight knot (fig. 12). For background on these examples, see [**17**]. The trefoil knot is a basic example of a larger class of knots known as torus knots. They can be understood as orbits of a $U(1)$ action on S^3, and this leads to exceptionally simple formulas for their quantum invariants, as has been explained recently via path integrals [**34**]. We expect similar behavior for other torus knots, though we consider only the trefoil in detail. However, torus knots are atypical. A more typical knot is one whose complement in S^3 admits a complete hyperbolic

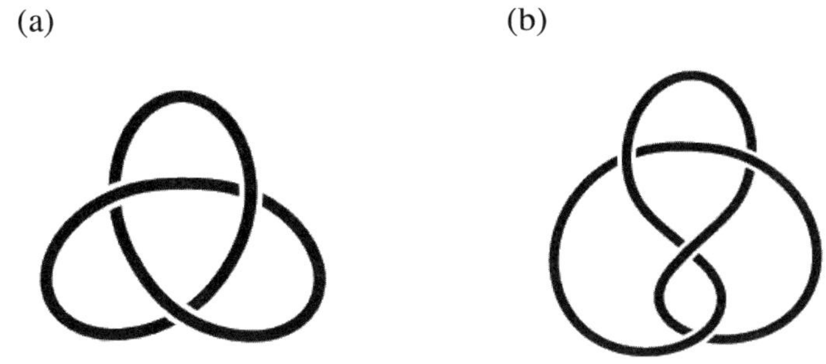

FIGURE 12. The trefoil knot (a) and the figure-eight knot (b). (From Wikipedia.)

metric.[22] The figure-eight knot (also called the 4_1 knot) is possibly the simplest example of a knot that is generic in this sense, and as a result its colored Jones polynomial has been studied extensively.

We will study the behavior of $J_n(q)$ in the limit of large n, k with fixed ratio $\gamma = n/k$. We call this the semiclassical limit. For physical values of n and k – that is, positive integers – this limit can be analyzed by standard methods. One simply sums over contributions of real saddle points of the path integral; for physical values of n and k, the expansion around such real saddle points suffices for computing $J_n(q)$ to all orders in an asymptotic expansion in $1/k$ (or in $1/n$ with γ fixed). The virtue of the Morse theory machinery of this paper is that it enables us to understand what happens in the semiclassical limit after analytic continuation away from integer values of k (and of n, if one so chooses).

5.1. Preliminaries. To apply the Morse theory machinery, we need to start at some point in the complex γ plane where we know how to express the integration cycle of Chern-Simons theory as a sum over Lefschetz thimbles. With this in hand, one can then vary γ and take Stokes phenomena into account.

If k and n are positive integers, so $\gamma > 0$, Chern-Simons theory is defined by an integral over $SU(2)$-valued connections modulo gauge transformations. In the framework of the present paper, this means that the integration cycle for $\mathcal{A} = A + i\phi$ is the real cycle $\phi = 0$, which we call $\mathcal{C}_{\mathbb{R}}$. This cycle can be expressed in terms of Lefschetz thimbles as described most precisely in eqn. (3.39):

$$\mathcal{C}_{\mathbb{R}} = \sum_{\sigma \in \Sigma_{\mathbb{R}}} \mathcal{J}_\sigma + \sum_{\sigma \in \Sigma_-} \mathfrak{n}_\sigma \mathcal{J}_\sigma, \tag{5.1}$$

Here Σ is the set of critical points – or rather critical gauge orbits, but we will speak loosely of critical points. The critical points are flat connections on $S^3 \backslash K$ with appropriate monodromy. $\Sigma_{\mathbb{R}}$ consists of the real critical points (the ones with $\phi = 0$), and Σ_- consists of complex critical points corresponding to flat connections $\mathcal{A}$ for which the real part of the exponent of the path integral is negative. This exponent, including the contribution from the knot, is

$$\mathcal{I} = ikW(\mathcal{A}) + (n-1)\ln \mathfrak{l}. \tag{5.2}$$

---

[22]The relevant metrics look near the knot like $\mathrm{d}s^2 = \mathrm{d}z^2/z^2 + z^2(\mathrm{d}x^2 + \mathrm{d}y^2)$, where the knot is located at $z = 0$, the "infrared" region in gauge/gravity duality. The "ultraviolet" region for $z \to \infty$ is cut off by the embedding in S^3.

The coefficients $\mathfrak{n}_\sigma$ must be computed by counting solutions of flow equations. Because of the difficulty of actually doing this, our results for the figure-eight knot will be incomplete.

An important topological fact about $SU(2)$ gauge theory is that for the case of a knot in S^3 and monodromy sufficiently close to the identity, the only real critical point corresponds to the abelian flat connection $\mathcal{A}_{\rm ab}$. So for sufficiently small values of $\gamma = n/k$, the contribution to $\mathcal{C}_\mathbb{R}$ from the sum over $\Sigma_\mathbb{R}$ is simply the cycle $\mathcal{J}_{\rm ab}$ associated to $\mathcal{A}_{\rm ab}$.

$\mathcal{C}_\mathbb{R}$ is initially defined as a real integration cycle in $\mathcal{U}_\mathbb{C}$, the space of complex-valued connections. To analytically continue to complex values of k, we must lift and extend $\mathcal{C}_\mathbb{R}$ to a cycle $\mathcal{C} \subset \widehat{\mathcal{U}_\mathbb{C}}$; recall that $\widehat{\mathcal{U}_\mathbb{C}}$ is a cover of $\mathcal{U}_\mathbb{C}$ where the Chern-Simons function $W(\mathcal{A})$ is well-defined as a complex-valued function. This process was described schematically in section 2.5. Concretely, analytic continuation can be made by choosing a lift to $\mathbb{C}$ of the Chern-Simons invariant $W(\mathcal{A}_\sigma)$ for each critical point that appears in (5.1) with a nonzero coefficient. In section 2.5.3, a geometrical interpretation was given for a choice in which, for some given real number α, one requires $\alpha < \mathrm{Im}\,\mathcal{I} < \alpha + 2\pi$ for all σ. The most general possible choice was described in section 2.5.2.

The choice that we will make of analytic continuation is guided by the fact that for fixed n, the colored Jones polynomial has a unique analytic continuation to a function $J_n(q)$ with no essential singularity at $k = \infty$ (or equivalently at $q = 1$). Moreover [**9**], in the semiclassical limit, $J_n(q)$ is bounded as $n \to \infty$ for sufficiently small γ. We take these facts to mean that there is a unique choice of $\mathcal{C}$ such that, in a sufficiently small neighborhood of $\gamma = 0$, the path integral has no exponentially growing contributions; it receives contributions only from critical points for which $\mathrm{Re}\,\mathcal{I}$ is zero or negative. This should hold regardless of the direction in which one approaches the point $\gamma = 0$ in the complex plane.

The Chern-Simons invariant of the abelian flat connection $\mathcal{A}_{\rm ab}$ vanishes modulo 2π. Avoiding exponential growth tells us how to lift $\mathcal{J}_{\rm ab}$ to $\widehat{\mathcal{U}_\mathbb{C}}$: we must pick the lift such that the Chern-Simons invariant $W(\mathcal{A}_{\rm ab})$ is 0 (rather than a non-zero multiple of 2π). Similarly, we pick the branch of the logarithm such that $\ln \mathfrak{l} = 0$. Henceforth, when we refer to $\mathcal{A}_{\rm ab}$ as a critical point in $\widehat{\mathcal{U}_\mathbb{C}}$ or to $\mathcal{J}_{\rm ab}$ as a cycle in $\widehat{\mathcal{U}_\mathbb{C}}$, we have in mind this lifting, which we call the canonical lifting.

For any K, the contribution of $\mathcal{A}_{\rm ab}$ to the integration cycle is always precisely $\mathcal{J}_{\rm ab}$, regardless of γ. In fact, Stokes phenomena never cause a jumping in the coefficients with which abelian critical points contribute to the integration cycle. This is explained in sections 3.3.4 and 3.4.2. Jumping definitely can occur for the coefficients of Lefschetz thimbles associated to nonabelian critical points.

For a typical knot, nonabelian critical points have nonzero $\mathcal{I}$ near $\gamma = 0$; let us assume this to be the case. At first sight, the argument of section 3.2.1 seems to show that if flat connections with nonzero $\mathcal{I}$ contribute to the path integral near $\gamma = 0$, then an essential singularity for $k \to \infty$ is unavoidable. If this were the case, we would conclude that in a neighborhood of $\gamma = 0$, the desired integration cycle $\mathcal{C}$ must precisely coincide with $\mathcal{J}_{\rm ab}$ for all knots. There seem to be two fallacies in this reasoning.

First, the argument in section 3.2.1 assumed that the flow equations conserve $\mathrm{Im}\,\mathcal{I}$. However, as discussed in section 4.2.6, in the presence of a knot, this is only so if one replaces the loop space of $G/T_\mathbb{C}$ by its universal cover, as one should do if one wishes to analytically continue $J_n(q)$ to non-integral values of n. In this case, a choice of branch of $\ln \mathfrak{l}$ is regarded as part of the data specifying a critical point. If the concept of a critical point is understood in this way, then $\mathrm{Im}\,\mathcal{I}$ is conserved in the Chern-Simons flow equations and the argument of section 3.2.1 is applicable. Otherwise, we cannot make this argument.

At best (since we note another problem shortly), the conclusion from section 3.2.1 is that if $J_n(q)$, after being analytically continued to non-integral n, has no essential singularity for $k \to \infty$ or $q \to 1$, then the integration cycle for sufficiently small γ is simply $\mathcal{C} = \mathcal{J}_{\rm ab}$.

Unfortunately, although it seems clear in the framework of the present paper that $J_n(q)$ can be analytically continued away from integer values of n, this does not seem to have been explored in the literature. The claim that $J_n(q)$ has no essential singularity at $k = \infty$ is only known in general for integer n, and we have little guidance about whether this remains so after analytic continuation in n.

The simplest example is the unknot K_0. The explicit formula[23] is

$$J_n(q; K_0) = \sqrt{\frac{2}{k+2}} \sin\left(\frac{\pi n}{k+2}\right). \tag{5.3}$$

Clearly, this function can be analytically continued in n, and the analytically continued function has no essential singularity at $k = \infty$. For the trefoil, we will see at the end of section 5.2.4 that for $n \notin \mathbb{Z}$, there is an essential singularity at infinity,[24] but the situation for more general knots is very unclear.

The second problem is that Chern-Simons theory is more complicated than the situation considered in section 3.2.1. It is assumed there that the Stokes curves are straight rays, which is not the case in Chern-Simons theory, and perhaps more important, it is assumed that there are only finitely many critical points and Stokes curves. In Chern-Simons theory, once one analytically continues away from integer k, there are infinitely many critical orbits (corresponding to different liftings of the Chern-Simons invariant of a flat connection) meeting at $\gamma = 0$, and this can play an important role. In the case of the trefoil knot, we will see that along the positive imaginary γ axis, arbitrarily close to $\gamma = 0$, an infinite number of critical points must be included to describe the integration cycle.

We do not have a general theory of how to determine the integration cycle so as to avoid exponential growth of the path integral for $k \to \infty$, so we will have to take an exploratory point of view in our examples. In our study of the trefoil, we will show that $\mathcal{J}_{\rm ab}$ is indeed the correct integration cycle if one approaches $\gamma = 0$ along any direction except the positive imaginary axis. For the figure-eight knot, we will not make an equally detailed analysis.

Once one knows the correct integration cycle near $\gamma = 0$, whether it is simply $\mathcal{J}_{\rm ab}$ or not, the appropriate integration cycle for any γ can be found in the usual way by varying γ and taking account of Stokes phenomena.

5.1.1. *Local models.* Since studying the behavior near a singularity of the moduli space of flat connections is an important technique for understanding Stokes phenomena, it may be helpful for some readers if we briefly discuss the deformation

[23]Recall that in our normalization, $J_n(q)$ is defined by a path integral rather than a ratio of path integrals; for an explanation of the relation of this to the usual normalization in the mathematical literature, see the last paragraph of section 2.5.3.

[24]I am informed by C. Beasley that this can also be verified for more general torus knots using formulas of [**34**].

theory of a flat bundle. (When we consider examples later, we do so in a more informal language, and some readers may omit the following discussion.)

Let E be a flat bundle over a manifold M, with structure group G and connection $\mathcal{A} = \mathcal{A}_0$. Let Y be the automorphism group of E, and set[25] $V = H^1(M, \mathrm{ad}(E))$, $V' = H^2(M, \mathrm{ad}(E))$. V parametrizes first order deformations of E, and V' parametrizes obstructions to such deformations. In deformation theory, it is shown that there is a Y-invariant map $\sigma : V \to V'$ such that the moduli space of deformations of E is $\sigma^{-1}(0)/Y$. (Rather than dividing by Y, it may be more convenient to think of the moduli space as the space $\sigma^{-1}(0)$ with an action of Y.) We can expand σ as a power series in a variable $v \in V$. The lowest order term is quadratic and comes from the cup product. Higher order terms come from Massey products. (For example, see [**58**]. From a physical point of view, these higher order terms arise by integrating out massive modes of the gauge field; see section 4.2 of [**59**].)

If M is a three-manifold, then V' is dual to V and the discussion simplifies. In this case, there is a Y-invariant function Φ on V, and the map σ is $\mathrm{d}\Phi$, the differential of Φ. Again Φ can be expanded as a power series in v. The leading term is cubic and comes by evaluating the classical action $\mathcal{I}(\mathcal{A})$ as a function on V. From a physical point of view, Φ is the effective action for the zero modes of the gauge field; it contains cubic couplings that can be read off from $\mathcal{I}(\mathcal{A})$, and higher order couplings (described mathematically in terms of Massey products) that arise by integrating out massive modes.

In our application, there will be an additional parameter γ, controlling the monodromy around a knot. Suppose that the flat bundle E appears for $\gamma = \gamma_0$. To include γ in the above discussion, simply set $\epsilon = \gamma - \gamma_0$. The effective action Φ can be extended from a function on V to a function on $V \times \mathbb{C}_\epsilon$, where $\mathbb{C}_\epsilon$ is a copy of the complex plane parametrized by ϵ, in such a way that flat connections, for any ϵ, correspond to points in $V \times \mathbb{C}_\epsilon$ at which the equation $\mathrm{d}_V\Phi = 0$ is obeyed. Here d_V is the exterior derivative in the V direction. In other words, flat connections for a given ϵ correspond to critical points of the restriction of Φ to the fiber of the projection $V \times \mathbb{C}_\epsilon \to \mathbb{C}_\epsilon$ that lies above that value of ϵ.

The question now arises of how many Massey products must be included to get a good model of the behavior near the flat bundle E that we started with. The answer, assuming[26] that for $\epsilon = 0$ (and hence any ϵ near 0), there are only finitely many flat bundles near E, is that one must go far enough to lift all degeneracies. If the cubic form obtained by evaluating $W(\mathcal{A})$ as a function on V is nondegenerate, this gives a complete answer for Φ; otherwise, one must carry out a power series to higher order, including enough Massey products to get a nondegenerate result.

Singularity theory gives a convenient framework for this discussion. Pick a transversal $\mathcal{U}$ to the gauge orbit of the flat bundle E. Consider $\mathcal{I}(\mathcal{A})$ as a function on $\mathcal{U}$. This function has a singularity (that is, a degenerate critical point) at the point $\mathcal{A} = \mathcal{A}_0$. Though $\mathcal{U}$ is infinite-dimensional, the singularity can be modeled in finite dimensions. In fact, after making some choices, one can embed V in $\mathcal{U}$ and write $\mathcal{U}$ as $V \times N$, where N is an infinite-dimensional space that parametrizes

[25]Here $\mathrm{ad}(E)$ is the sheaf of infinitesimal gauge transformations, leaving fixed the structure along the knot if a knot is present.

[26]This is not always true in three-manifold theory, since in general the moduli space of flat connections may have components of positive dimension. It is true in our applications involving knots. In general, one needs to include enough Massey products to lift whatever degeneracies do get lifted.

the nonzero modes of $\mathcal{A}$; we assume that $\mathcal{A} = \mathcal{A}_0$ corresponds to the product of the origin in V and a point in N that we call the origin. The decomposition $\mathcal{U} = V \times N$ can be made in a Y-invariant fashion. Singularity theory says that one can choose the decomposition so that $\mathcal{I}(\mathcal{A})$ near $\mathcal{A} = \mathcal{A}_0$ is the sum of a function on V and a function on N, such that the function on V is the effective action Φ, and the function on N has a nondegenerate critical point at the origin. The function $\Phi : V \to \mathbb{C}$ thus gives a finite-dimensional model of the singularity of the function $\mathcal{I}(\mathcal{A}) : \mathcal{U} \to \mathbb{C}$. Because of the choices that are made in identifying $\mathcal{U}$ as $V \times N$, Φ is only determined up to a diffeomorphism of V (more exactly, a Y-invariant diffeomorphism that leaves fixed the origin). Within its equivalence class, one can take Φ to be a polynomial, determined by finitely many Massey products.

The importance of this for Chern-Simons theory is as follows. The integral over Y behaves in the semiclassical limit as an infinite-dimensional Gaussian integral (with perturbative corrections). It does not contribute to Stokes phenomena near $\mathcal{A} = \mathcal{A}_0$. The interesting subtleties in the behavior of the Chern-Simons path integral near the flat bundle E in the semiclassical limit are all contained in the integral over V.

We will encounter three examples of this construction. Conceptually, the most simple example arises for the figure-eight knot at $\gamma = 1/3$, where two nonabelian flat connections meet. The flat bundle E is irreducible, and its automorphism group Y is trivial. The space V is one-dimensional, and Φ is the function $\Phi = x^3/3 - \epsilon x$ related to the Airy integral. As Φ is cubic, Massey products play no role. The other two examples involve the case that $\mathcal{A}_0 = \mathcal{A}_{\rm ab}$ is the abelian flat connection, so the group Y is $\mathbb{C}^*$. For either the trefoil or the figure-eight knot, there is a critical value of γ at which $\mathcal{A}_{\rm ab}$ meets one or more non-abelian flat connections. In each case V is two-dimensional and Y acts with weights 1 and -1. We write u, v for coordinates on V corresponding to these weights. For the trefoil knot, it turns out that the appropriate local model is $\Phi = (uv)^2/2 - \epsilon uv$. The quartic term in Φ comes from a Massey product. For the figure-eight knot, one gets instead $\Phi = (uv)^3/3 - \epsilon uv$, where now a Massey product of rather high order is involved in generating the $(uv)^3$ term.

5.1.2. *Normalization of the path integral.* Instead of simply saying, as we have just done, that the contribution to the Chern-Simons path integral of a degenerate critical point can be modeled by a corresponding finite-dimensional integral, it is possible, as we will briefly sketch, to be more precise and to describe the normalization factors. The facts we will describe could be used to sharpen the comparisons that we will describe later between the Chern-Simons path integral and computations in the mathematical literature that are based on the exact solution of Chern-Simons theory, obtained by combinatorial methods. (We will not do so in this paper and the reader might choose to omit this explanation.)

We first recall the nondegenerate case. The contribution of a nondegenerate critical point to the Chern-Simons path integral was evaluated in [**51**], compared to the exact solution (and slightly corrected[27]) in [**22**], and further refined in [**23, 27**]. In Chern-Simons theory on a three-manifold M with compact gauge group H, let $E \to M$ be an H-bundle and suppose that $H^0(M, \mathrm{ad}(E)) = H^1(M, \mathrm{ad}(E)) = 0$. The automorphism group P of E is then a finite group (in general, the Lie algebra

[27]A normalization was used in [**51**] in which η and $\log T$ were both two times smaller than usual.

of P is $H^0(M, \mathrm{ad}(E))$). Let $\mathcal{I}_E = ikW$ be the action of Chern-Simons theory, evaluated for the critical point E. The contribution of the flat bundle E to the Chern-Simons path integral is then

$$\frac{1}{\#P}\exp(\mathcal{I}_E)\exp(i\pi\eta(\mathrm{ad}(E))/4)\sqrt{T(\mathrm{ad}(E))}. \tag{5.4}$$

where $\eta(\mathrm{ad}(E))$ is the Atiyah-Patodi-Singer η-invariant of the operator $\star\mathrm{d}+\mathrm{d}\star$ acting on odd degree forms with values in $\mathrm{ad}(E)$; $T(\mathrm{ad}(E))$ is the Ray-Singer analytic torsion of this flat bundle; and $\#P$ is the order of the finite group P. The expression (5.4) arises by performing the path integral for fluctuations around the flat bundle E, in the Gaussian approximation. Higher order corrections will multiply this expression by an asymptotic series in powers of $1/k$ (the same is true for eqn. (5.5) below), as explored in [**60, 61**].

Now let us assume that E corresponds to a degenerate critical point. We continue to assume that $H^0(M, \mathrm{ad}(E)) = 0$, but we take $V = H^1(M, \mathrm{ad}(E))$ to be nonzero. We can still carry out a Gaussian integral over the nonzero modes of the gauge field (the modes that in the terminology of section 5.1.1 parametrize N). The result of integrating over the nonzero modes is still given by the formula (5.4), if properly interpreted. There is one essential point: for an acyclic flat connection, the torsion is a number, but in the general case, as explained in the original construction of analytic torsion [**62**], and used in the present context in [**23**], it is a linear form on a certain one-dimensional space – constructed as a tensor product of the determinants of the cohomology groups of the flat bundle in question. In the present situation, the only nonzero cohomology group is $V = H^1(M, \mathrm{ad}(E))$ (and its dual H^2), and $\sqrt{T(\mathrm{ad}(E))}$ is a linear form on $\det V$. It determines a translation-invariant measure on V that we will denote as $\mathrm{d}v$.

What we want to integrate over V, using this measure, is essentially $\exp(\Phi)$, where the function Φ was described in section 5.1.1. The generalization of (5.4) for $V \neq 0$ is

$$\frac{1}{\#P}\exp(i\pi\eta(\mathrm{ad}(E))/4)(k/2\pi^2)^{\dim V/2}\int_V \mathrm{d}v\exp(\Phi), \tag{5.5}$$

where we replace $\sqrt{T(\mathrm{ad}(E))}$ by the measure $\mathrm{d}v$ and use it to integrate over V. (The factor $\exp(\mathcal{I}_E)$ has disappeared as $\mathcal{I}_E$ is simply the value of Φ at $v = 0$. If $v = 0$ were a nondegenerate critical point of Φ, the factor $k^{\dim V/2}$ would be canceled upon doing the Gaussian integral over V. That this factor is precisely a power of $k/2\pi^2$ was found for $H = SU(2)$ in [**27**]. This has apparently not been established in complete generality, though it can plausibly be done using localization formulas [**32**].) The idea of (5.5) is that the integrand $\exp(i\pi\eta(\mathrm{ad}(E))/2)\,\mathrm{d}v\exp(\Phi)/\#P$ comes from the path integral over N and we complete the path integral by integrating over V.

In the general case (and in fact in two of the three examples mentioned at the end of section 5.1.1), we must consider a flat bundle E such that $H^0(M, \mathrm{ad}(E)) \neq 0$. The automorphism group P of E then has dimension $d_P = \dim H^0(M, \mathrm{ad}(E))$. Moreover, $\sqrt{T(\mathrm{ad}(E))}$ is now the ratio of a measure on V and one on the Lie algebra $\mathfrak{p}$ of P. A convenient way to proceed is to pick a metric on $\mathfrak{p}$ (such as the usual one in which short coroots have length squared two) and let $\mathrm{Vol}(P)$ be the volume of P using this metric. Once we have picked a metric and hence a measure on $\mathfrak{p}$, $\sqrt{T(\mathrm{ad}(E))}$ can be regarded just as before as a measure $\mathrm{d}v$ on V.

The generalization of (5.5) is then

$$\frac{1}{\mathrm{Vol}(P)}\exp(i\pi\eta(\mathrm{ad}(E))/4)(k/2\pi^2)^{(\dim V-\dim P)/2}\int_V dv\exp(\Phi). \tag{5.6}$$

The factor $\exp(i\pi\eta(\mathrm{ad}(E))/4)$ depends on the metric of M. The metric dependence can be removed (by subtracting a "counterterm") once a framing (or two-framing) of M is picked; this was the approach in [**51**]. A canonical two-framing exists [**63**] and is often used. The formulas specialized to this two-framing were described in detail in [**22**] for $H = SU(2)$.

5.1.3. *Orientation of the integration cycle.* As a last preliminary comment, we should perhaps discuss the fact that although the invariant $J_n(q;K_0)$ of the unknot – given in eqn. (5.3) – has no essential singularity at $k=\infty$, it does have a square root singularity there. This square root singularity is common for all knots.[28] The meaning of the square root singularity is that under analytic continuation around $k=\infty$, the orientation of the integration cycle $\mathcal{C}$ is reversed.

This reversal of orientation under analytic continuation is not an exotic phenomenon. It occurs for the simple Gaussian integral $\int_{-\infty}^{\infty} dx\,\exp(ikx^2)=\sqrt{\pi/ik}$, where the square root shows that under analytic continuation around $k=\infty$, the orientation of the integration cycle is reversed. (Concretely, this happens because if k is rotated by an angle ϕ in the complex plane, the integration cycle in the x-plane should be rotated by an angle $\phi/2$ to maintain convergence of the integral. For $\phi=2\pi$, we get back the original integration cycle with opposite orientation.) The existence of this monodromy shows that the treatment of the orientation of the integration cycle will be subtle in Chern-Simons theory, but we will phrase our discussion to minimize the role of such details.

5.2. The trefoil knot. For both of our examples, the necessary facts about the representation of the fundamental group of the knot complement are described in [**17**]. We begin with the case that our knot K is the trefoil.

The fundamental group of $S^3\backslash K$ can be described by two generators $\mathbf{x},\mathbf{y}$ with one relation $\mathbf{xyx}=\mathbf{yxy}$. $\mathbf{x}$ and $\mathbf{y}$ are both conjugate to the meridian $\widehat{\mathfrak{m}}$; for further details, see fig. 11 in [**17**]. The longitude is $\widehat{\mathfrak{l}}=\mathbf{yx}^2\mathbf{yx}^{-4}$.

For every choice of the conjugacy class of $\widehat{\mathfrak{m}}$ (apart from certain exceptional values that will be discussed), there are up to gauge transformation two $SL(2,\mathbb{C})$ flat connections on $S^3\backslash K$. The first is an abelian flat connection $\mathcal{A}_{\mathrm{ab}}$ with $\mathbf{x}=\mathbf{y}=\mathrm{diag}(\mathfrak{m},\mathfrak{m}^{-1})$. (As usual $\mathfrak{m}$ is the eigenvalue of the meridian $\widehat{\mathfrak{m}}$.) Its Chern-Simons invariant vanishes.

There is also a nonabelian flat connection $\mathcal{A}_{\mathrm{nab}}$ whose monodromies are given up to conjugation by[29]

$$\mathbf{x}=\begin{pmatrix}\mathfrak{m} & 1\\ 0 & \mathfrak{m}^{-1}\end{pmatrix},\quad \mathbf{y}=\begin{pmatrix}\mathfrak{m} & 0\\ 1-\mathfrak{m}^2-\mathfrak{m}^{-2} & \mathfrak{m}^{-1}\end{pmatrix}. \tag{5.7}$$

The eigenvalue of the longitude is

$$\mathfrak{l}=-\mathfrak{m}^{-6}. \tag{5.8}$$

According to (4.35), in $SU(2)$ Chern-Simons theory, $\mathfrak{m}$ is of modulus 1, so we will need to analyze this case carefully. If $|\mathfrak{m}|=1$, the flat connection $\mathcal{A}_{\mathrm{nab}}$ has a

[28] It cancels in ratios $J_n(q;K)/J_n(q;K_0)$, which are the functions usually considered in the mathematical literature. These ratios are Laurent polynomials in $q=\exp(2\pi i/(k+2))$.

[29] Our $\mathfrak{m}$ is called $m^{1/2}$ in [**17**].

structure group that reduces to either $SU(2)$ or $SL(2,\mathbb{R})$. This can be anticipated as follows. The complex conjugate of a representation of the fundamental group in $SL(2,\mathbb{C})$ is still a representation of the fundamental group in $SL(2,\mathbb{C})$, in general with $\mathfrak{m} \to \overline{\mathfrak{m}}$. However, if $|\mathfrak{m}| = 1$, the operation $\mathfrak{m} \to \overline{\mathfrak{m}}$ is equivalent to $\mathfrak{m} \to \mathfrak{m}^{-1}$ and does not affect the conjugacy class of the meridian, which is

$$\begin{pmatrix} \mathfrak{m} & 0 \\ 0 & \mathfrak{m}^{-1} \end{pmatrix}. \tag{5.9}$$

So for $|\mathfrak{m}| = 1$, representations of the fundamental group in $SL(2,\mathbb{C})$ occur in complex conjugate pairs or else are isomorphic to their own complex conjugates. A representation with the latter property has a structure group that reduces to a real form of $SL(2,\mathbb{C})$ – either $SU(2)$ or $SL(2,\mathbb{R})$.

For the trefoil knot, as there is only one nonabelian representation up to conjugacy, it must be equivalent to its own complex conjugate. Concretely, let us try to conjugate the representation into $SU(2)$. We set

$$C = \begin{pmatrix} 1 & -(\mathfrak{m}-\mathfrak{m}^{-1})^{-1} \\ 0 & 1 \end{pmatrix}, \tag{5.10}$$

so

$$C\mathbf{x}C^{-1} = \begin{pmatrix} \mathfrak{m} & 0 \\ 0 & \mathfrak{m}^{-1} \end{pmatrix} \tag{5.11}$$

is an element of $SU(2)$. Then

$$C\mathbf{y}C^{-1} = \begin{pmatrix} (\mathfrak{m}-\mathfrak{m}^{3})^{-1} & (\mathfrak{m}-\mathfrak{m}^{-1})^{-2} \\ 1-\mathfrak{m}^{2}-\mathfrak{m}^{-2} & (\mathfrak{m}^{-1}-\mathfrak{m}^{-3})^{-1} \end{pmatrix}. \tag{5.12}$$

If $\mathbf{x}$ and $\mathbf{y}$ can be conjugated into $SU(2)$, this can be done in a way that makes $\mathbf{x}$ diagonal. Having put $\mathbf{x}$ in the form (5.11), the only further conjugation that leaves it diagonal is by a diagonal matrix $\mathrm{diag}(\lambda,\lambda^{-1})$. Conjugation by a diagonal matrix can map a unimodular matrix of the form

$$\begin{pmatrix} a & b \\ c & \overline{a} \end{pmatrix}, \tag{5.13}$$

such as $C\mathbf{y}C^{-1}$, into $SU(2)$ if and only if $b\overline{c} < 0$. For $|\mathfrak{m}| = 1$, this is equivalent to

$$1 > \mathfrak{m}^2 + \mathfrak{m}^{-2}, \tag{5.14}$$

and that therefore is the condition that the flat connection $\mathcal{A}_{\mathrm{nab}}$ has structure group $SU(2)$. If instead

$$1 < \mathfrak{m}^2 + \mathfrak{m}^{-2}, \tag{5.15}$$

the structure group of $\mathcal{A}_{\mathrm{nab}}$ reduces to $SL(2,\mathbb{R})$.

5.2.1. *Singularities of the moduli space of flat connections.* The remaining case $1 = \mathfrak{m}^2 + \mathfrak{m}^{-2}$ requires further study. This case corresponds to a singularity of the moduli space of flat connections, where different branches meet, as we will now describe.

The condition $1 = \mathfrak{m}^2 + \mathfrak{m}^{-2}$ is equivalent to

$$\ln \mathfrak{m} = \pm i\pi/6, \ \mathrm{mod}\ \pi i\mathbb{Z}. \tag{5.16}$$

In this case, the monodromies of eqn. (5.7) are strictly upper triangular, and the structure group of a flat connection $\mathcal{A}_{\mathrm{nab}}$ with those monodromies reduces to the triangular subgroup of $SL(2,\mathbb{C})$, and not to either $SU(2)$ or $SL(2,\mathbb{R})$. In this exceptional case, there is also another flat connection $\mathcal{A}'_{\rm nab}$ that is inequivalent to $\mathcal{A}_{\rm nab}$; its monodromies are the lower triangular cousins of eqn. (5.7):

$$\mathbf{x} = \begin{pmatrix} \mathfrak{m} & 0 \\ 1 & \mathfrak{m}^{-1} \end{pmatrix}, \ \mathbf{y} = \begin{pmatrix} \mathfrak{m} & 0 \\ 0 & \mathfrak{m}^{-1} \end{pmatrix}. \tag{5.17}$$

$\mathcal{A}'_{\rm nab}$ is actually gauge equivalent to the complex conjugate of $\mathcal{A}_{\rm nab}$; indeed, eqn. (5.7), specialized to $\mathfrak{m}^2 + \mathfrak{m}^{-2} = 1$, can be transformed to (5.17) via complex conjugation composed with conjugation by the matrix $\Omega = \begin{pmatrix} 0 & 1 \\ -1 & 0 \end{pmatrix}$.

Both $\mathcal{A}_{\rm nab}$ nor $\mathcal{A}'_{\rm nab}$ represent unstable critical points in Chern-Simons theory, as their structure groups are strictly triangular. In fact, when $1 = \mathfrak{m}^2 + \mathfrak{m}^{-2}$, the only semistable critical point is the abelian one $\mathcal{A}_{\rm ab}$ with $\mathbf{x} = \mathbf{y} = {\rm diag}(\mathfrak{m}, \mathfrak{m}^{-1})$. Let us discuss the behavior of Chern-Simons theory in the vicinity of $\mathcal{A}_{\rm ab}$. We think of $\mathcal{A}_{\rm ab}$ as a diagonal flat connection with $\mathbb{C}^*$ structure group

$$\begin{pmatrix} * & 0 \\ 0 & * \end{pmatrix} \tag{5.18}$$

embedded in $SL(2, \mathbb{C})$. If $\mathfrak{m}$ is fixed (by specifying the values of n and k in Chern-Simons gauge theory), then $\mathcal{A}_{\rm ab}$ has no diagonal deformations as a flat connection. However, it does have off-diagonal deformations. There is a one-dimensional space of deformations of upper triangular form

$$\begin{pmatrix} 0 & u \\ 0 & 0 \end{pmatrix}. \tag{5.19}$$

A deformation of $\mathcal{A}_{\rm ab}$ with $u \neq 0$ gives, up to gauge transformation, the triangular flat connection $\mathcal{A}_{\rm nab}$. The value of u does not matter, as long as it is nonzero, because $\mathcal{A}_{\rm ab}$ has a $\mathbb{C}^*$ group of symmetries consisting of constant gauge transformations ${\rm diag}(\lambda, \lambda^{-1})$, and these act on u by $u \to \lambda^2 u$. Similarly (or by complex conjugation composed with the gauge transformation Ω), $\mathcal{A}_{\rm ab}$ has a one-dimensional space of lower triangular deformations

$$\begin{pmatrix} 0 & 0 \\ -v & 0 \end{pmatrix}. \tag{5.20}$$

v transforms under $\mathbb{C}^*$ by $v \to \lambda^{-2} v$, so if v is nonzero, its value does not matter. A lower triangular deformation of $\mathcal{A}_{\rm ab}$ with nonzero v gives $\mathcal{A}'_{\rm nab}$.

If we take u and v to be both nonzero, then only their product uv is $\mathbb{C}^*$-invariant. It is not possible to take both $u, v \neq 0$ without varying $\mathfrak{m}$ away from $\mathfrak{m}^2 + \mathfrak{m}^{-2} = 1$. This is evident from (5.7), according to which a two-dimensional representation of the fundamental group of $S^3 \backslash K$ that is not triangular has $\mathfrak{m}^2 + \mathfrak{m}^{-2} \neq 1$. Let $\epsilon = 1 - \mathfrak{m}^2 - \mathfrak{m}^{-2}$; ϵ is real if $|\mathfrak{m}| = 1$. In the space of possible deformations of $\mathcal{A}_{\rm ab}$ to an irreducible flat connection, we can normalize the coordinates u, v so that the relation between u, v, and ϵ is $uv = \epsilon$. This statement simply says that for a non-abelian flat connection $\mathcal{A}_{\rm nab}$, the $\mathbb{C}^*$-invariant uv has a unique value for all ϵ (since $\mathcal{A}_{\rm nab}$ is unique up to gauge transformation), with $uv = 0$ at $\epsilon = 0$ (as we know from eqn. (5.7)), and finally the zero of uv at $\epsilon = 0$ is simple (the monodromies (5.7) show that it is possible to have $u \sim 1$, $v \sim \epsilon$, and so $uv \sim \epsilon$; this should be no surprise as there is no symmetry that would force uv to have a zero of higher order).

Now let us specialize this to $SU(2)$ Chern-Simons theory. In this case, the gauge field $\mathcal{A}$ should be anti-hermitian, so we want $v = \overline{u}$. Also, the symmetry group of the abelian flat connection reduces to $U(1) = SO(2)$. If we set $u = x+iy$, $v = x - iy$, where x, y are real coordinates on the space of first order deformations of the flat $SU(2)$ connection $\mathcal{A}_{\rm ab}$, then the relation between x, y and ϵ is

$$x^2 + y^2 = \epsilon. \tag{5.21}$$

This equation has real solutions only for $\epsilon \geq 0$; this encodes the fact that $\mathcal{A}_{\rm nab}$ can be conjugated to $SU(2)$ only under that condition. Let us consider the exponent $\mathcal{I} = ikW(\mathcal{A}) + iI_R(\rho, \mathcal{A})$ of the path integral of Chern-Simons theory for small ϵ and near the abelian flat connection $\mathcal{A}_{\rm ab}$. It has critical orbits with $x = y = 0$ and with $x^2 + y^2 = \epsilon$. Moreover, $\mathcal{I}$ vanishes for the abelian flat connection $\mathcal{A}_{\rm ab}$, corresponding to $x = y = 0$, and in general is of order n in the semiclassical region of large n with fixed $\gamma = n/k$. Finally it has one nonzero critical orbit at $x^2+y^2 = \epsilon$. These facts suggest that $-i\mathcal{I}/n$ can be modeled near $x = y = \epsilon = 0$ by the simple $SO(2)$-invariant function

$$F(x, y) = c\left(\frac{1}{2}(x^2 + y^2)^2 - \epsilon(x^2 + y^2)\right) \tag{5.22}$$

with a real constant c. We will verify this shortly using (5.26), but for the moment let us discuss the implications.

Generically, in expanding around a flat connection, the Chern-Simons function is a nondegenerate Morse function in the directions transverse to the gauge orbits. When this is so, after gauge fixing, the expansion around a given critical point can be approximated for large n and k as an infinite-dimensional (oscillatory) Gaussian integral. Nondegeneracy is visible in (5.22) for $\epsilon \neq 0$. As long as ϵ does not vanish, $x = y = 0$ is a nondegenerate critical point of $F(x, y)$, and near the critical orbit $x^2 + y^2 = \epsilon$, the function $F(x, y)$ is nondegenerate in the directions transverse to the $SO(2)$ action. However, this nondegeneracy and the Gaussian nature of the Chern-Simons path integral break down at $\epsilon = 0$, where the two critical orbits meet.

The expansion of the Chern-Simons path integral around a critical point can then no longer be approximated by a Gaussian, but the Gaussian approximation fails only for the two variables x and y. To improve upon the Gaussian approximation, we need only a suitable approximation to the dependence of $\mathcal{I}$ on x and y near the singularity. Assuming that (5.22) is an appropriate approximation, we can model the integral over the variables x and y by

$$\int \mathrm{d}x\, \mathrm{d}y \exp\left(inF(x, y)\right). \tag{5.23}$$

(For the constant multiplying this integral, see (5.5).) The integral (5.23) was studied in section 3.4.2 as a basic example of an oscillatory integral with symmetry. There were two main results in that discussion.

First, in eqns. (3.59) and (3.60), we analyzed the integral by a simple scaling argument. According to (3.59), the integral is $n^{-1/2}$ times a transcendental function (essentially an error function) of $n^{1/2}\epsilon$. In the limit $n \to \infty$, $\epsilon \to 0$ with $n^{1/2}\epsilon$ fixed, the colored Jones polynomial will equal $n^{1/2}$ times this error function (times a constant described in (5.5)). The factor of $n^{1/2}$ is most obvious in the special case $\epsilon = 0$: according to (3.60), compared to the behavior with $\epsilon \neq 0$, the large n behavior of the integral gains an extra factor of $n^{1/2}$ at $\epsilon = 0$. We can immediately apply this to the invariant $J_n(k)$ at the values of n/k (such as 1/6 or 5/6) that correspond to $\epsilon = 0$ or $\mathfrak{m}^2 + \mathfrak{m}^{-2} = 1$. We expect $J_n(k)$ to grow with a factor of $n^{1/2}$ at $\epsilon = 0$, relative to its behavior for generic real ϵ:

$$\frac{J_n\big|_{\epsilon=0}}{J_n\big|_{\epsilon\neq 0}} \sim n^{1/2}, \quad n \to \infty. \tag{5.24}$$

This is a known result (Theorem 1.2 of [**16**]). We have expressed the result in terms of a ratio of values of J_n because our normalization of J_n (as a path integral rather than a ratio of path integrals) is not the most common one in the mathematical literature. With the usual normalization, the colored Jones polynomial is of order 1 for $\epsilon < 0$ and of order $n^{1/2}$ at $\epsilon = 0$.

Second, in section 3.4.2 we made a direct analysis of the Stokes phenomena in passing through $\epsilon = 0$ (or more generally in crossing Stokes lines that pass through $\epsilon = 0$). These results only depend on the singularity at $\epsilon = 0$, so they carry over immediately to Chern-Simons theory. Thus, we can understand some Stokes phenomena in $SU(2)$ Chern-Simons theory without having to solve four-dimensional flow equations. Details are discussed in section 5.2.3.

Finally let us return to (5.22), but now consider it as a model of the behavior of $SL(2,\mathbb{C})$ gauge fields near the singularity at $\epsilon = 0$. In $SL(2,\mathbb{C})$, x and y are complex and the symmetry group is $\mathbb{C}^*$. Unlike the case of $SU(2)$, where the critical orbit $x^2 + y^2 = \epsilon$ becomes equivalent to $x = y = 0$ if we set $\epsilon = 0$, in $SL(2,\mathbb{C})$ the equation $x^2 + y^2 = 0$ can be solved with $x, y \neq 0$. Modulo the $\mathbb{C}^*$ symmetry (which acts by $x \pm iy \to \lambda^{\pm 2}(x + iy)$), the locus with $x^2 + y^2 = 0$ but $(x, y) \neq (0, 0)$ consists of two points, $(x, y) = (1, \pm i)$. These two points correspond to the triangular flat connections $\mathcal{A}_{\rm nab}$, $\mathcal{A}'_{\rm nab}$. It is also possible to specialize the analysis to $SL(2,\mathbb{R})$ gauge fields – the same singularity will also play a role in $SL(2,\mathbb{R})$ Chern-Simons theory – but we leave this for the reader.

5.2.2. *The Chern-Simons function.* To understand possible Stokes phenomena and to determine the asymptotic behavior of the Chern-Simons path integral, we must evaluate the Chern-Simons function $W(\mathcal{A})$ for the flat connection $\mathcal{A}_{\rm nab}$. For this, we use (4.42), which along with (5.8) gives $\delta W(\mathcal{A}_{\rm nab}) = -(6/\pi)\delta\mathfrak{m}/\mathfrak{m} \cdot \ln \mathfrak{m} = -(3/\pi)\delta \ln^2 \mathfrak{m}$. So $W(\mathcal{A}_{\rm nab}) = -3\ln^2 \mathfrak{m}/\pi + \text{constant}$. To determine the constant, we observe that in general $W(\mathcal{A})$ vanishes mod 2π for a representation with triangular structure group, which we have in the present problem at the values of $\ln \mathfrak{m}$ given in (5.16). Thus

$$W(\mathcal{A}_{\rm nab}) = -\frac{3}{\pi}\ln^2 \mathfrak{m} - \frac{\pi}{12} - 2\pi s, \; s \in \mathbb{Z}. \tag{5.25}$$

As we are planning to analytically continue in k, we have made explicit the possibility of adding to $W(\mathcal{A})$ an integer multiple of 2π. We can now evaluate $\mathcal{I} = ikW(\mathcal{A}) + (n-1)\ln \mathfrak{l}$, using $\mathfrak{l} = -\mathfrak{m}^{-6}$ and $\ln \mathfrak{m} = i\pi(n-1)/k$ to evaluate $\ln \mathfrak{l}$. In the semiclassical limit of large n, k, we ignore the difference between $n - 1$ and n. However, we do want to take into account the multivaluedness of $\ln \mathfrak{l}$, as explained at the end of section 4.2.5. We get

$$\mathcal{I}(\mathcal{A}_{\rm nab}) = ik\left(\frac{3\pi n^2}{k^2} - \frac{\pi}{12} - 2\pi s\right) - in\left(6\pi\frac{n}{k} + 2\pi(r - 1/2)\right), \; r, s \in \mathbb{Z}. \tag{5.26}$$

Alternatively, in terms of $\gamma = n/k$, we can write

$$\mathcal{I}(\mathcal{A}_{\rm nab}) = -in\big(3\pi\gamma + (\pi/12 + 2\pi s)\gamma^{-1} + 2\pi(r - 1/2)\big). \tag{5.27}$$

We can use these formulas to justify (5.22). The relation between $\mathfrak{m}$ and γ is $\ln \mathfrak{m} = i\pi\gamma$, so $\epsilon = 1 - \mathfrak{m}^2 - \mathfrak{m}^{-2}$ vanishes precisely if

$$\gamma = \pm(1/6 + p), \ p \in \mathbb{Z}. \tag{5.28}$$

For any number y that is congruent to $\pm 1/6 \bmod \mathbb{Z}$, the nonabelian flat connection $\mathcal{A}_{\rm nab}$ with $\gamma = y$ and

$$\begin{aligned} s &= \frac{3y^2}{2} - \frac{1}{24} \\ r &= \frac{1}{2} - 3y \end{aligned} \tag{5.29}$$

is equivalent to an abelian flat connection $\mathcal{A}_{\rm ab}$ with its canonical lift (zero Chern-Simons invariant and $\ln \mathfrak{l} = 0$). The values of r and s were chosen to ensure vanishing of both contributions to $\mathcal{I}(\mathcal{A}_{\rm nab})$, namely $ikW(\mathcal{A}_{\rm nab})$ and $(n-1)\ln \mathfrak{l}$. With these values of r and s, we find for any γ

$$\mathcal{I}(\mathcal{A}_{\rm nab}) = -3\pi i n \gamma^{-1}\big(\gamma - y\big)^2. \tag{5.30}$$

So for $\gamma = y + \epsilon$, $\mathcal{I}$ is of order $n\epsilon^2$. The fact that the coefficient of ϵ^2 is nonzero is the content of (5.22). Any $SO(2)$-invariant function $F(x, y, \epsilon)$ that vanishes at $x = y = 0$, has only one critical orbit with $(x, y) \neq 0$ that approaches $(x, y) = 0$ for $\epsilon \to 0$, and is of order ϵ^2 at this orbit, can be put in the form (5.22) modulo a change of variables (an $SO(2)$-invariant redefinition of x and y that is the identity near $x = y = 0$). Possible higher order terms in F are irrelevant in the sense of singularity theory and do not affect the estimate (5.24) of the critical behavior of the Chern-Simons path integral.

5.2.3. *The integration cycle and the asymptotic behavior of the path integral.* An obvious consequence of (5.26) is that for real k and n, the nonabelian critical points have $\mathcal{I}$ imaginary. Since $\mathrm{Re}\,\mathcal{I}$ is strictly increasing or decreasing along a flow line, and vanishes on the real integration cycle $\mathcal{C}_{\mathbb{R}}$, this implies that for real n and k, there are no flows between $\mathcal{C}_{\mathbb{R}}$ and a nonabelian critical point. Hence for γ real and small, the integration cycle $\mathcal{C}_{\mathbb{R}}$ is simply the Lefschetz thimble $\mathcal{J}_{\rm ab}$ associated to the real critical point $\mathcal{A}_{\rm ab}$.

Extending this reasoning slightly, for real n and k, all critical points, abelian or nonabelian, have $\mathrm{Re}\,\mathcal{I} = 0$. So there are no non-trivial flows between them.

How then can anything nontrivial happen for real values of n, k? Something nontrivial must happen because for suitable real γ, the nonabelian flat connection $\mathcal{A}_{\rm nab}$ is $SU(2)$-valued and hence must contribute to the $SU(2)$ Chern-Simons path integral.

The answer to this question is that on the real γ axis, jumping of the Lefschetz thimbles $\mathcal{J}_\sigma$, or of the coefficients in the usual expansion $\mathcal{C} = \sum_\sigma \mathfrak{n}_\sigma \mathcal{J}_\sigma$, occurs only at values of γ where two different flat connections become coincident. At these points, which were identified in eqn. (5.28), Chern-Simons theory has jumping phenomena that can be understood from a finite-dimensional model of the appropriate singularity, as described in section 5.2.1.

Here we will analyze the integration cycle of the Chern-Simons path integral, and its asymptotic behavior, for γ close to the real axis. For a full analysis of what happens throughout the complex γ plane, one must analyze the Stokes curves; we postpone this to section 5.2.4.

We want to express the integration cycle of the Chern-Simons path integral for the trefoil knot in terms of Lefschetz thimbles associated to critical points. To this aim, we let $\mathcal{J}_{\rm ab}$ be the Lefschetz cycle associated to the abelian critical point $\mathcal{A}_{\rm ab}$ with its canonical lift. $\mathcal{J}_{\rm ab}$ is the appropriate integration cycle for sufficiently small $\gamma = n/k$. It will become clear that cycles associated to other ways of lifting the Chern-Simons invariant $W(\mathcal{A}_{\rm ab})$ will not play any role. (This is actually a general property for any knot.)

On the other hand, we do have to consider distinct cycles associated to different ways of lifting the Chern-Simons invariant $W(\mathcal{A}_{\rm nab})$ of the nonabelian critical point $\mathcal{A}_{\rm nab}$, in other words different choices of the integer s in (5.25). For each number x that is congruent to $\pm 1/6 \bmod \mathbb{Z}$, we write $\mathcal{J}_x$ for the Lefschetz thimble associated to $\mathcal{A}_{\rm nab}$, with its Chern-Simons invariant determined by the choice $s = 3x^2/2 - 1/24$ as in (5.29). These cycles differ only by the lifting of the Chern-Simons invariant, so they are equivalent when the Chern-Simons level k is an integer.

We fix the relative orientations of the cycles $\mathcal{J}_x$ by requiring that when k is an integer, the Chern-Simons path integral over $\mathcal{J}_x$ is independent of x. Momentarily, we will fix the overall orientations of the $\mathcal{J}_x$ by making a choice for $\mathcal{J}_{1/6}$.

Near $\gamma = 0$, the appropriate integration cycle of Chern-Simons theory is simply $\mathcal{J}_{\rm ab}$. Now we will explore what happens when γ departs from zero, but only near the real γ axis, so that we only need to worry about jumping due to singularities. As we increase γ along the positive real axis, nothing happens until we reach the value $\gamma = 1/6$. At this point, we meet a singularity of a type analyzed in sections 3.4.2 and 5.2.1, associated with the fact that $\mathcal{A}_{\rm ab}$ and $\mathcal{A}_{\rm nab}$ become stably equivalent[30] at $\gamma = 1/6$. The resulting Stokes phenomenon was analyzed in eqn. (3.61): $\mathcal{J}_{\rm ab}$ jumps by $\pm\mathcal{J}_{1/6}$, depending on orientation. ($\mathcal{J}_{\rm ab}$ and $\mathcal{J}_{1/6}$ correspond to $\mathcal{J}_p$ and $\mathcal{J}_{\mathcal{O}}$ in (3.61).) We pick the orientation so that for γ slightly greater than 1/6, the integration cycle of Chern-Simons theory is $\mathcal{J}_{\rm ab} + \mathcal{J}_{1/6}$.

There is a simple intuitive meaning to this result. In general, for integer values of n, k, the path integral of Chern-Simons theory is given asymptotically for large n and fixed n/k by a sum over contributions of all $SU(2)$-valued flat connections, all with weight 1. For $1/6 < n/k < 5/6$, there are two such connections, namely $\mathcal{A}_{\rm ab}$ and $\mathcal{A}_{\rm nab}$, and both must contribute with weight 1 to the integration cycle. As long as $k \in \mathbb{Z}$, the different $\mathcal{J}_x$ are equivalent, and it does not matter which of them we include. But to describe the analytic continuation away from integer k in the region just to the right of the Stokes curve that passes through $\gamma = 1/6$, the right integration cycle is $\mathcal{J}_{\rm ab} + \mathcal{J}_{1/6}$.

Increasing γ further, nothing new happens until we reach $\gamma = 5/6$. In crossing $\gamma = 5/6$, nothing happens to $\mathcal{J}_{1/6}$, but $\mathcal{J}_{\rm ab}$ will jump again, picking up $\pm\mathcal{J}_{5/6}$. We claim that (with the orientation convention chosen above) a minus sign is the right choice here, and hence that the appropriate integration cycle just beyond $\gamma = 5/6$ is $\mathcal{J}_{\rm ab} + \mathcal{J}_{1/6} - \mathcal{J}_{5/6}$.

To explain why the sign is negative, consider again the behavior for integer n, k with $5/6 < n/k < 7/6$. In this region, the only $SU(2)$-valued flat connection is the abelian one, and hence the large n, fixed n/k behavior of the path integral must be given by the contribution of $\mathcal{A}_{\rm ab}$ alone. The contribution of $\mathcal{A}_{\rm nab}$ must vanish for

[30]This means the following. $\mathcal{A}_{\rm nab}$ is triangular at $\gamma = 1/6$ and its diagonal part coincides with $\mathcal{A}_{\rm ab}$, so the gauge orbit containing $\mathcal{A}_{\rm nab}$ has $\mathcal{A}_{\rm ab}$ in its closure. Hence, to construct a sensible (Hausdorff) moduli space of flat connections, one has to consider $\mathcal{A}_{\rm nab}$ and $\mathcal{A}_{\rm ab}$ to be equivalent.

integer k. Since $\mathcal{J}_{1/6}$ and $\mathcal{J}_{5/6}$, which arise from two different liftings of $\mathcal{A}_{\rm nab}$, are equivalent for integer k, the minus sign in the integration cycle $\mathcal{J}_{\rm ab}+\mathcal{J}_{1/6}-\mathcal{J}_{5/6}$ ensures this.

Increasing γ further, nothing happens until we reach $\gamma = 7/6$. At this point, $\mathcal{J}_{\rm ab}$ jumps by $\pm\mathcal{J}_{7/6}$. In this part of the real γ axis, there is again an $SU(2)$-valued flat connection that must contribute with weight 1. So the sign is + and the integration cycle is $\mathcal{J}_{\rm ab}+\mathcal{J}_{1/6}-\mathcal{J}_{5/6}+\mathcal{J}_{7/6}$. The next jump is at $\gamma = 11/6$, where we must add $-\mathcal{J}_{11/6}$ to the integration cycle.

In general, if $x > 0$ is a number of the form $\pm(1/6+p)$, $p \in \mathbb{Z}$, then just to the right of $\gamma = x$, the appropriate integration cycle is

$$\mathcal{J}_{\rm ab}+\mathcal{J}_{1/6}-\mathcal{J}_{5/6}\pm\cdots\pm\mathcal{J}_x. \tag{5.31}$$

Similarly, for negative γ, the integration cycle is $\mathcal{J}_{\rm ab}+\mathcal{J}_{-1/6}-\mathcal{J}_{-7/6}\pm\cdots\pm\mathcal{J}_x$, for the most negative relevant value of x.

Having found the right integration cycle, it is straightforward to determine the asymptotic behavior of the Chern-Simons path integral for $n \to \infty$ with fixed γ. In general, a flat connection $\mathcal{A}_\sigma$ with $\mathfrak{n}_\sigma \neq 0$ makes a contribution to the path integral which in the semiclassical limit is of order $\exp(\mathcal{I}_\sigma)$, where $\mathcal{I}_\sigma$ is the value of the exponent of the path integral at $\mathcal{A} = \mathcal{A}_\sigma$. Unless $\mathcal{I}_\sigma = 0$, this contribution has an exponential dependence on n for large n, since $\mathcal{I}_\sigma$ is of order n in the semiclassical limit. In general, $\exp(\mathcal{I}_\sigma)$ multiplies an asymptotic series in (not necessarily integral) powers of n. In the context of an expansion around a complex critical point, this series has been studied in detail in [**18**]. The dominant contribution to the path integral, for a given γ, is the one that maximizes $\mathrm{Re}\,\mathcal{I}_\sigma$, subject to the condition $\mathfrak{n}_\sigma \neq 0$.

In the case of the trefoil knot, the abelian flat connection $\mathcal{A}_{\rm ab}$ has $\mathcal{I}_{\rm ab} = 0$. Its contribution $Z_{\rm ab}$ to the path integral is bounded (in fact, of order $1/\sqrt{n}$) for $n \to \infty$ with fixed γ. On the other hand, a non-abelian connection $\mathcal{A}_{\rm nab}$, lifted so that it meets the abelian one at $\gamma = y$, has $\mathcal{I}_y = -3\pi i n\gamma^{-1}(\gamma-y)^2$, according to (5.30). So, assuming the relevant coefficient – which we will call $\mathfrak{n}_y$ – is not zero, the asymptotic behavior of its contribution to the Chern-Simons path integral is

$$Z_y \sim \exp(-3\pi i n\gamma^{-1}(\gamma-y)^2). \tag{5.32}$$

These contributions are all oscillatory as long as γ is on the real axis. (On the real axis, they actually dominate $Z_{\rm ab}$ in absolute value by a factor of $\sqrt{n}$ for large n.) As soon as γ departs from the real axis, the Z_y grow or decay exponentially for $n \to \infty$, depending on the sign of the real part of the exponent. If $\gamma = a + ib$ with a, b real, the real part of the exponent is

$$\mathrm{Re}\,\log Z_y = 3\pi n b\left(1-\frac{y^2}{a^2+b^2}\right). \tag{5.33}$$

Near the real axis, and assuming $|a| > |y|$ (as otherwise $\mathfrak{n}_y = 0$), this is positive if and only if $b > 0$. So near the real axis, the condition for exponential growth of $J_n(q)$ is that $|a| > 1/6$ (so that one of the $\mathfrak{n}_y$ is nonzero) and $\mathrm{Im}\,\gamma > 0$. In this region, Re log Z_y is maximal for $y = \pm 1/6$, so this contribution dominates $J_n(q)$. In section 5.2.4, we will analyze the Stokes curves and extend this analysis to the rest of the complex γ plane.

What we have just obtained (except so far restricted to the region near the real γ axis) is the standard picture of the colored Jones polynomial of the trefoil knot.

For example, see eqn. (2.2) of [**7**] or Proposition 3.2 of [**16**], where this invariant, and its cousins for other torus knots, is written as the sum of contributions that correspond to our $Z_{\rm ab}$ and Z_y (with the same values of y). The starting point was an integral representation for the colored Jones polynomial of a torus knot which originally was deduced using formulas that come from braid group representations, quantum groups, and conformal field theory [**5, 30**]. This integral representation has recently been obtained directly from the Chern-Simons path integral [**34**].

5.2.4. *Stokes curves.* By a Stokes curve, we mean a curve l in the complex γ plane on which distinct critical points have the same value of the quantity $\mathrm{Im}\,\mathcal{I}$ that is conserved in the flow equation. There are many such curves. But we only really care about Stokes curves across which some jumping occurs. It turns out that there are far fewer of these.

Our basic strategy will be to show that every Stokes curve intersects the real γ axis. But we already know from section 5.2.3 what sort of jumping occurs along the real axis. It takes the form $\mathcal{J}_{\rm ab} \to \mathcal{J}_{\rm ab} \pm \mathcal{J}_y$, where $\mathcal{J}_y$ is a cycle associated to one of the lifts of $\mathcal{A}_{\rm nab}$. This will enable us to get a general description of the integration cycle. In fact, it turns out that the integration cycle always takes the form of (5.31), or its cousin with negative x, with some x that depends on γ in a way that will emerge.

A preliminary observation is that there are no flows between abelian critical points differing in the values of r, s, since all such critical points have $\mathrm{Re}\,\mathcal{I} = 0$, but $\mathrm{Re}\,\mathcal{I}$ is never constant along a non-trivial flow. So in any flow between two critical points, at least one of them is nonabelian.

To show that all Stokes curves intersect the real γ axis, we use the explicit formula for $\mathcal{I}$ and the fact that $\mathrm{Im}\,\mathcal{I}$ is conserved by the flow equations. Equivalently, the conserved quantity is $\mathrm{Im}\,\mathcal{I}/\pi n$. For a lift of $\mathcal{A}_{\rm nab}$, this can be conveniently expressed in terms of $\gamma = n/k$:

$$\mathrm{Im}\,\mathcal{I}/\pi n = -\mathrm{Re}\left(3\gamma + \left(\frac{1}{12} + 2s\right)\gamma^{-1} + 2r - 1\right). \tag{5.34}$$

If $\gamma = a + ib$ with a, b real, then

$$\mathrm{Im}\,\mathcal{I}/\pi n = -3a - \left(\frac{1}{12} + 2s\right)\frac{a}{a^2 + b^2} - 2r + 1. \tag{5.35}$$

Let us first consider Stokes curves at which there might be flows between nonabelian critical points with different values of r, s. The change in $\mathrm{Im}\,\mathcal{I}/\pi n$ in such a process is

$$\Delta(\mathrm{Im}\,\mathcal{I}/\pi n) = -2\left(\frac{a\,\Delta s}{a^2 + b^2} + \Delta r\right), \tag{5.36}$$

where Δs, Δr are the jumps in s and r. Setting this to zero, we get

$$a^2 + b^2 + \frac{a\,\Delta s}{\Delta r} = 0. \tag{5.37}$$

The case $\Delta s = 0$ is not of interest, for then vanishing of (5.36) implies $\Delta r = 0$, in which case the initial and final critical points are the same and the flow is trivial. If Δs and Δr are both nonzero, the Stokes curve is a circle that intersects the real γ axis in two points; if $\Delta r = 0$, it becomes the line $a = 0$, intersecting the real γ axis at one point. Either way, at an intersection point of one of these Stokes curves with the real axis, two nonabelian critical points with different values of r, s

remain different, and there are no flows between them, since in general there are no non-trivial flows for real γ.

In general, for two critical points p_σ, p_τ, along a Stokes curve l characterized by $\mathrm{Im}\,\mathcal{I}_\sigma = \mathrm{Im}\,\mathcal{I}_\tau$, the integer $\mathfrak{m}_{\sigma\tau}$ that counts flows between p_σ and p_τ is constant, except possibly at a point along l at which a third critical point p_ν has $\mathrm{Im}\,\mathcal{I}_\nu = \mathrm{Im}\,\mathcal{I}_\sigma = \mathrm{Im}\,\mathcal{I}_\tau$. Such a point is a triple intersection of Stokes curve (equality of any two of $\mathrm{Im}\,\mathcal{I}_\sigma, \mathrm{Im}\,\mathcal{I}_\tau$, and $\mathrm{Im}\,\mathcal{I}_\nu$ defines such a curve). At a triple intersection point there can be jumping $\mathfrak{m}_{\sigma\tau} \to \mathfrak{m}_{\sigma\tau} + \mathfrak{m}_{\sigma\nu}\mathfrak{m}_{\nu\tau}$. Eqn. (5.37) implies that there are no triple intersection points involving different lifts of $\mathcal{A}_{\mathrm{nab}}$. (Two of the curves defined by eqn. (5.37) can intersect away from the real γ axis only if they coincide identically, in which case the effects they produce can again be analyzed from what happens on the real axis.) We need not worry about the case that p_ν is a lift of $\mathcal{A}_{\mathrm{ab}}$, since in this case $\mathfrak{m}_{\nu\tau} = 0$, as we learned in section 3.3.4.

Hence the net number of flows between nonabelian critical points differing in the values of r and s is zero anywhere along one of these Stokes curves. So there are no Stokes phenomena involving flows between different lifts of $\mathcal{A}_{\mathrm{nab}}$.

The same type of argument can be used to constrain Stokes phenomena involving a flow between the abelian critical point and a nonabelian one. If we pick the canonical lift of $\mathcal{A}_{\mathrm{ab}}$, it has $\mathcal{I} = 0$ and the change in $\mathrm{Im}\,\mathcal{I}/\pi n$ is given in (5.35). Setting the right hand side of that equation to zero, we get the equation for a Stokes curve:

$$b^2 = -\frac{a(3a^2 + (2r-1)a + (2s+1/12))}{3a+2r-1}. \tag{5.38}$$

If we pick a different lift of $\mathcal{A}_{\mathrm{ab}}$, we get the same formula with shifted values of r, s. However, a byproduct of the following calculation will be to show that different lifts of $\mathcal{A}_{\mathrm{ab}}$ do not enter the integration cycle; indeed, all jumping phenomena take the form of (3.61) and do not shift the "abelian" part of the integration cycle.

Eqn. (5.38) describes many Stokes curves, labeled by the choice of r and s and a component of the part of the a axis on which the right hand side of (5.38) is nonnegative. Such a component is either a closed interval whose endpoints are zeroes of the numerator $a(3a^2 + (2r-1)a + (2s+1/12))$ or a half-open interval whose endpoints are such a zero and the zero of the denominator. Each component intersects the a axis, either in two points (the first case) or one point (the second case).

We now make the familiar argument. Since every Stokes curve l intersects the a axis, the jumping that occurs in crossing l anywhere can be deduced from what happens in crossing l where it meets that axis. If l intersects the a axis at a value of a not of the form $\pm(1/6+p)$, with $p \in \mathbb{Z}$, then l is irrelevant because no jumping can occur at such a value of a. Even if l intersects the a axis only at a point of the form $\pm(1/6+p)$, something non-trivial can happen only if r and s are of the form given in (5.29). The reason for this last statement is that as there are no non-trivial flows for real γ, all that can happen in crossing l at a point $y = \pm(1/6+p)$ is that some lift of $\mathcal{A}_{\mathrm{nab}}$ may coincide with $\mathcal{A}_{\mathrm{ab}}$ with its canonical lift. For this to happen, r and s must be related to y by (5.29).

So the relevant Stokes curves are determined by their intersection point $y = \pm(1/6+p)$ with the real axis; the curve is described by (5.38) with r, s given in

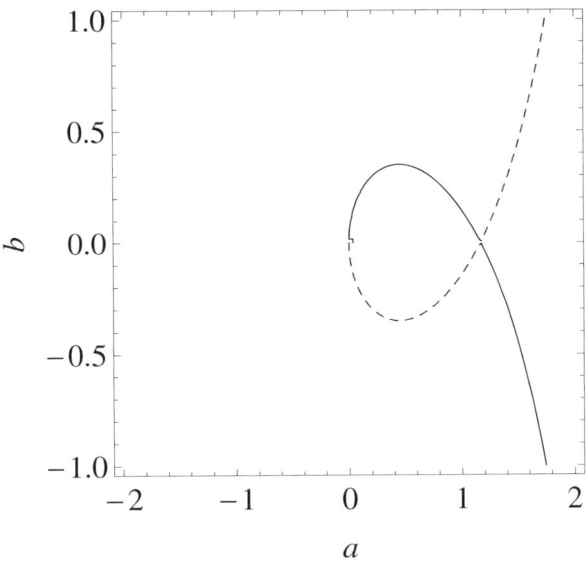

FIGURE 13. The Stokes curve (5.39) consists of both the solid and dashed curves, but just as in fig. 5, jumping occurs only across the solid part of the curve. (Only the solid parts, therefore, will be drawn in fig. 14.)

(5.29). In this special case, (5.38) simplifies to

$$b^2 = -\frac{a(a-y)^2}{a-2y}. \tag{5.39}$$

We now meet a situation described at the end of section 3.4.2. Eqn. (5.39) gives the condition that a flow is possible between $\mathcal{A}_{\rm ab}$ (with its canonical lifting) and $\mathcal{A}_{\rm nab}$ (with its chosen lifting). The flow will go from $\mathcal{A}_{\rm ab}$ to $\mathcal{A}_{\rm nab}$ if $\mathrm{Re}\,\mathcal{I}(\mathcal{A}_{\rm nab})$ is negative, and in the other direction if $\mathrm{Re}\,\mathcal{I}(\mathcal{A}_{\rm nab})$ is positive. However, we also know from the analysis in section 3.4.2 of a singularity of the type that occurs at $a = y$ (or from a more general argument in section 3.3.4) that no jumping comes from flows from $\mathcal{A}_{\rm nab}$ to $\mathcal{A}_{\rm ab}$. So just as in fig. 5, we can throw away the part of the curve (5.38) with $\mathrm{Re}\,\mathcal{I}(\mathcal{A}_{\rm nab}) > 0$. As sketched in fig. (13), the part of the curve that we keep is the part with $yb(y-a) > 0$.

Keeping only the "good" parts of these curves, they divide the complex γ plane in a way that is sketched in fig. 14. The important point here is that any point $\gamma = \gamma_0$ that is not on one of the relevant Stokes curves can be connected to the a axis in a unique way (topologically) without crossing any of those curves. The integration cycle $\mathcal{C}$ of Chern-Simons theory analytically continued to $\gamma = \gamma_0$ is given by the same formula as on the part of the a axis to which γ_0 can be connected without crossing a Stokes curve. This result for the integration cycle is in full accord with other analyses of the colored Jones polynomial of the trefoil knot. What we call the Stokes curve (5.39) – or more exactly the relevant portion of it with $yb(y-a) > 0$ – is the curve across which an extra term appears in eqn. (2.2) of [**7**].

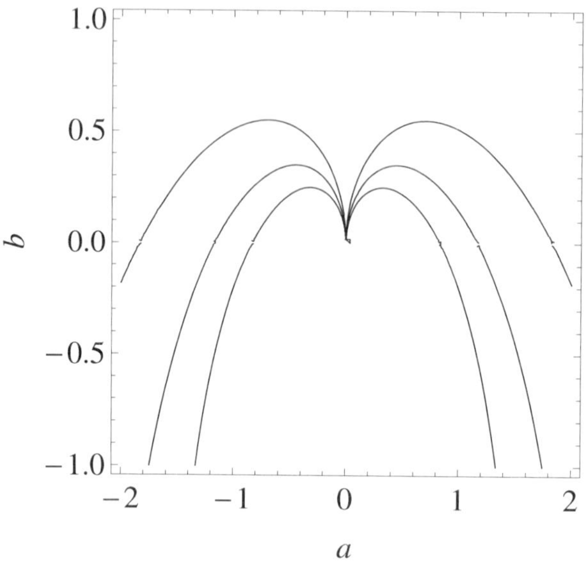

FIGURE 14. The Stokes curves across which jumping occurs for the trefoil knot divide the complex γ plane as shown here. Sketched here are the curves corresponding to $y = \pm 5/6, \pm 7/6$, and $\pm 11/6$. (All the other curves for larger or smaller values of $|y|$ are obtained from these by suitable scalings of the γ-plane.) Drawn are only the parts of each curve that actually produce jumping; these portions meet only at $\gamma = 0$, where they all join.

The analytically continued function $J_n(q)$ grows exponentially in the semiclassical region whenever $\mathrm{Re}\,\mathcal{I} > 0$ for a critical point that contributes to the path integral. This happens precisely if $\mathrm{Im}\,\gamma > 0$ and $|\gamma| > 1/6$. (Those conditions are strong enough to ensure, using eqn. (5.39), that a critical point with $y = 1/6$ or $-1/6$ contributes to the path integral, and, using eqn. (5.33), that it has $\mathrm{Re}\,\mathcal{I} > 0$.)

We are left with only one more issue, which is the behavior along the positive imaginary γ axis. As one approaches the positive imaginary γ axis from the right, one crosses all Stokes curves of positive y. As one approaches it from the left, one crosses all Stokes curves of negative y. A first consequence is that, precisely on the positive imaginary axis, the representation of the Chern-Simons integration cycle as a sum over Lefschetz thimbles is an infinite sum. A second consequence is that, unless we can consider the Lefschetz thimble $\mathcal{J}_y$ to be equivalent to $\mathcal{J}_{-y}$, this representation of the integration cycle jumps in crossing the positive imaginary γ axis. Indeed, in a complete circuit around the point $\gamma = 0$, one crosses exactly once each of the relevant Lefschetz thimbles of fig. 14. This will produce a monodromy of infinite order, and hence an essential singularity in the integral, unless $\mathcal{J}_y$ is equivalent to $\mathcal{J}_{-y}$.

Can we consider $\mathcal{J}_y$ and $\mathcal{J}_{-y}$ to be equivalent? Referring back to (5.29), we see that $y \to -y$ amounts to $s \to s$ with an integer change in r. In other words, it amounts to a change in the choice of branch of $\ln \mathfrak{m}$. The branch of $\ln \mathfrak{m}$ is irrelevant as long as $n \in \mathbb{Z}$ and we choose not to replace the loop space of $G/T_{\mathbb{C}}$ by its universal cover. This appears to be the right sense in which the integration cycle is univalent near $\gamma = 0$, but clearly, a study of more examples is needed. For $n \notin \mathbb{Z}$, the monodromy of the integration cycle is unavoidable and an essential singularity of the integral at $\gamma = 0$ should be expected.

5.3. The figure-eight knot. Here we will consider in a similar way the figure-eight knot (also known as the 4_1 knot), which we denote K. The fundamental group of $S^3\backslash K$ again has two generators $\mathbf{x}$, $\mathbf{y}$, either of them conjugate to the meridian $\widehat{\mathfrak{m}}$, with one relation. The relation is $\omega\mathbf{x} = \mathbf{y}\omega$, with $\omega = \mathbf{x}\mathbf{y}^{-1}\mathbf{x}^{-1}\mathbf{y}$.

For a generic eigenvalue $\mathfrak{m}$ of the meridian, there are up to gauge transformation three $SL(2,\mathbb{C})$ flat connections on $S^3\backslash K$. As usual, one is an abelian flat connection $\mathcal{A}_{\rm ab}$, whose monodromies are up to conjugacy $\mathbf{x} = \mathbf{y} = \mathrm{diag}(\mathfrak{m}, \mathfrak{m}^{-1})$. In addition, there are two nonabelian flat connections with monodromies

$$\mathbf{x} = \begin{pmatrix} \mathfrak{m} & 1 \\ 0 & \mathfrak{m}^{-1} \end{pmatrix}, \ \mathbf{y} = \begin{pmatrix} \mathfrak{m} & 0 \\ -d & \mathfrak{m}^{-1} \end{pmatrix}, \tag{5.40}$$

with

$$d = \frac{1}{2}\left(\mathfrak{m}^2 + \mathfrak{m}^{-2} - 3 \pm \sqrt{(\mathfrak{m}^2 + \mathfrak{m}^{-2} + 1)(\mathfrak{m}^2 + \mathfrak{m}^{-2} - 3)}\right). \tag{5.41}$$

The longitude is $\widehat{\mathfrak{l}} = \mathbf{x}\mathbf{y}^{-1}\mathbf{x}\mathbf{y}\mathbf{x}^{-2}\mathbf{y}\mathbf{x}\mathbf{y}^{-1}\mathbf{x}^{-1}$, leading to
(5.42)

$$\mathfrak{l} = \frac{1}{2}\left(\mathfrak{m}^4 - \mathfrak{m}^2 - 2 - \mathfrak{m}^{-2} + \mathfrak{m}^{-4}\right) \pm \frac{\mathfrak{m}^2 - \mathfrak{m}^{-2}}{2}\sqrt{(\mathfrak{m}^2 + \mathfrak{m}^{-2} + 1)(\mathfrak{m}^2 + \mathfrak{m}^{-2} - 3)}.$$

We denote as $\mathcal{A}_\pm$ a flat connection with the monodromies just indicated. These are the $SL(2,\mathbb{C})$ flat connections that, if $\mathfrak{m} = \pm 1$, are related to the complete hyperbolic metric on $S^3\backslash K$.

The longitude $\mathfrak{l}$ is a zero of the $\mathbf{A}$ polynomial, which is

$$\mathbf{A}(\mathfrak{l}, \mathfrak{m}) = \mathfrak{l} - \left(\mathfrak{m}^4 - \mathfrak{m}^2 - 2 - \mathfrak{m}^{-2} + \mathfrak{m}^{-4}\right) + \mathfrak{l}^{-1}. \tag{5.43}$$

Apart from the standard symmetries $\mathfrak{m} \to -\mathfrak{m}$ and $\mathfrak{m} \to \mathfrak{m}^{-1}$, this has an additional symmetry $\mathfrak{l} \to \mathfrak{l}^{-1}$, which exchanges the two branches of the square root in (5.42). This symmetry reflects the existence of a diffeomorphism τ of S^3 that maps K to itself, reversing the orientation of K but preserving the orientation of its normal bundle. (τ acts on the above presentation of the fundamental group by $\mathbf{x} \leftrightarrow \mathbf{y}$.) Such a symmetry maps $\mathfrak{m} \to \mathfrak{m}$, $\mathfrak{l} \to \mathfrak{l}^{-1}$. It reverses the orientation of S^3, so it changes the sign of the Chern-Simons function $W(\mathcal{A})$ and is not a symmetry of Chern-Simons gauge theory. Nevertheless, it will help us in understanding the behavior of the theory. A knot for which such a symmetry exists is called amphichiral. Actually, the figure-eight knot is amphichiral in the strong sense that S^3 admits another diffeomorphism $\widetilde{\tau}$ that maps K to itself (pointwise) while reversing the orientation of its normal bundle. (It acts on the presentation of the fundamental group by $\mathbf{x} \to \mathbf{y}^{-1}$, $\mathbf{y} \to \mathbf{x}^{-1}$.) Again, $\widetilde{\tau}$ reverses the orientation of S^3 and reverses the sign of the Chern-Simons invariant. It acts by $\mathfrak{m} \to \mathfrak{m}^{-1}$, $\mathfrak{l} \to \mathfrak{l}$. (We leave it to the reader to verify that the unknot is a simpler example of a knot in S^3 that admits diffeomorphisms of both of these types.) The product $\tau\widetilde{\tau}$ preserves the Chern-Simons invariant and acts by $\mathfrak{m} \to \mathfrak{m}^{-1}$, $\mathfrak{l} \to \mathfrak{l}^{-1}$.

Unlike the trefoil, where all of the interesting action occurs for $|\mathfrak{m}| = 1$, for the figure-eight knot both the real $\mathfrak{m}$ axis and the circle $|\mathfrak{m}| = 1$ are worthy of careful study. The eigenvalues of an element of $SL(2,\mathbb{R})$ are either real or of modulus 1, so in general flat connections with $SL(2,\mathbb{R})$ structure group may have $|\mathfrak{m}| = 1$ (as for the trefoil) or real $\mathfrak{m}$ (as happens for the figure-eight knot). The eigenvalues of an element of $SU(2)$ are of modulus 1, so flat connections with $SU(2)$ structure group will arise only for $|\mathfrak{m}| = 1$.

5.3.1. *Real* $\mathfrak{m}$. Suppose first that $\mathfrak{m}$ is real and that

$$\mathfrak{m}^2 + \mathfrak{m}^{-2} > 3. \tag{5.44}$$

Then by inspection, for both flat bundles $\mathcal{A}_\pm$, the matrices $\mathbf{x}$ and $\mathbf{y}$ are real, so the entire representation of the fundamental group has its image in $SL(2,\mathbb{R})$. The diffeomorphism τ exchanges $\mathcal{A}_+$ with $\mathcal{A}_-$ (since it maps $\mathfrak{l} \to \mathfrak{l}^{-1}$). It leaves $\mathcal{A}_{\mathrm{ab}}$ invariant.

The region with $\mathfrak{m}$ real and

$$\mathfrak{m}^2 + \mathfrak{m}^{-2} < 3 \tag{5.45}$$

is also easy to understand qualitatively. The flat connections $\mathcal{A}_+$ and $\mathcal{A}_-$ are still exchanged by τ. Now, however, they are no longer real (since the square roots in the monodromy formulas are imaginary) and are exchanged by complex conjugation. In fact, the region $\mathfrak{m}^2 + \mathfrak{m}^{-2} < 3$ includes the points $\mathfrak{m} = \pm 1$ where $\mathcal{A}_+$ and $\mathcal{A}_-$ are the $SL(2,\mathbb{C})$ flat connections that are related to a hyperbolic metric on $S^3\backslash K$.

As one might expect, among the most interesting points on the real $\mathfrak{m}$ axis are the points with

$$\mathfrak{m}^2 + \mathfrak{m}^{-2} = 3. \tag{5.46}$$

For such $\mathfrak{m}$, since d vanishes, the matrices $\mathbf{x}$ and $\mathbf{y}$ are both upper triangular. The flat bundles $\mathcal{A}_+$ and $\mathcal{A}_-$ become equivalent to each other. They are unstable upper triangular deformations of the abelian flat bundle $\mathcal{A}_{\mathrm{ab}}$. $\mathcal{A}_{\mathrm{ab}}$, just as in (5.18), is a diagonal flat connection

$$\begin{pmatrix} * & 0 \\ 0 & * \end{pmatrix}. \tag{5.47}$$

It has automorphism group $\mathbb{C}^*$; this symmetry group reduces to $\mathbb{R}^* \cong SO(1,1)$ if we consider $\mathcal{A}_{\mathrm{ab}}$ (for real $\mathfrak{m}$) as a flat bundle with structure group $SL(2,\mathbb{R})$. The symmetry extends to $O(1,1)$ if we include the outer automorphism of $SL(2,\mathbb{R})$ corresponding to conjugation by $\mathrm{diag}(1,-1)$ or $\mathrm{diag}(i,-i)$, and we will do so.[31] $\mathcal{A}_{\mathrm{ab}}$ has no deformations as a flat bundle with specified $\mathfrak{m}$, but as in (5.19), it has a one-dimensional space of deformations of upper triangular form

$$\begin{pmatrix} 0 & u \\ 0 & 0 \end{pmatrix}. \tag{5.48}$$

Here u is a real parameter if we consider deformations in $SL(2,\mathbb{R})$. The value of u, assuming it is not zero, does not matter modulo the action of $O(1,1)$, and the upper triangular deformation gives us the triangular flat bundle $\mathcal{A}_+$ or $\mathcal{A}_-$. Just as in (5.20), $\mathcal{A}_{\mathrm{ab}}$ has a similar lower triangular deformation, schematically

$$\begin{pmatrix} 0 & 0 \\ -v & 0 \end{pmatrix}. \tag{5.49}$$

[31] In $SL(2,\mathbb{R})$ gauge theory, one would consider two critical points that differ by this outer automorphism to be inequivalent. But in $SL(2,\mathbb{C})$ gauge theory, or analytic continuation of $SU(2)$ gauge theory, which are our focus in this paper, they are equivalent, since $\mathrm{diag}(i,-i)$ is an element of $SU(2)$.

If we take $v \neq 0$ but $u = 0$, we get a lower triangular flat bundle, and again the value of v is inessential assuming that it is nonzero.

As in our discussion of the trefoil, there is no deformation with u, v nonzero and $\mathfrak{m}^2 + \mathfrak{m}^{-2} = 3$, since a flat $SL(2, \mathbb{C})$ bundle at such a value of $\mathfrak{m}$ is triangular (or abelian). If we do turn on both u and v, the $O(1,1)$-invariant parameter is the product uv. One might expect as in the discussion of the trefoil that we would have $uv \sim \epsilon$ where $\epsilon = \mathfrak{m}^2 + \mathfrak{m}^{-2} - 3$. However, in (5.40) the upper triangular parameter is 1 and the lower triangular parameter is $-d$, so this case corresponds to $uv \sim -d$. Moreover, from (5.41), we have $d \sim \pm\sqrt{\epsilon}$ near $\epsilon = 0$. Critical points therefore are at $uv = 0$ or $\pm\sqrt{\epsilon}$. Apart from a multiplicative constant, the function with this critical behavior is

$$f(u, v) = (uv)^3/3 - \epsilon uv. \tag{5.50}$$

The oscillatory integral associated to this function was studied in section 3.4.3.

Why have we arrived at this peculiar structure, with no $(uv)^2$ term in $f(u, v)$? The answer has to do with amphichirality. As τ leaves fixed $\mathcal{A}_{\rm ab}$ and exchanges $\mathcal{A}_{\pm}$, at a point at which one of these becomes triangular and meets $\mathcal{A}_{\rm ab}$, the other must do so as well. This structure is lost if we add a $(uv)^2$ term to f. More fundamentally, τ reverses the sign of the Chern-Simons function and hence of f, which describes its critical behavior. The fact that τ exchanges $\mathcal{A}_{\pm}$ implies that if we lift the action of τ to the flat bundle with connection $\mathcal{A}_{\rm ab}$ in such a way that $\tau(u) = u$, then $\tau(v) = -v$. (By composing with the outer automorphism of $SL(2, \mathbb{R})$, one can also choose a lifting so that $\tau(u) = -u$, $\tau(v) = v$.) Thus, τ reverses the sign of uv. Since τ reverses the sign of both uv and f, f is an odd function of uv, and this accounts for the structure that we have found.

As in our analysis of the trefoil, some of the Stokes phenomena of Chern-Simons theory of the figure-eight knot can be described in terms of critical behavior of the function f. However, for brevity, the only issue that we will discuss here will be the large n behavior of the colored Jones polynomial near $\epsilon = 0$, that is, near $\mathfrak{m}^2 + \mathfrak{m}^{-2} = 3$. First we simply set $\epsilon = 0$. The parameter λ in eqn. (3.65) corresponds in Chern-Simons theory to n (or equivalently k, since the ratio n/k is fixed when we specify ϵ). So according to (3.65),

$$\frac{J_n(k)|_{\epsilon=0}}{J_n(k)|_{\epsilon<0}} \sim n^{2/3}, \quad n \to \infty. \tag{5.51}$$

Accordingly, if we normalize the colored Jones polynomial of the figure-eight knot so that it is of order 1 for $\epsilon < 0$, then it is of order $n^{2/3}$ for $\epsilon = 0$. This is a known result ([**16**], Theorem 1.1). Instead of simply setting $\epsilon = 0$, we can consider the limit $n \to \infty$, $\epsilon \to 0$ with $n^{2/3}\epsilon$ fixed. The singular contribution to the colored Jones polynomial is this limit follows from (3.64) and is an Airy function (times $n^{2/3}$ in the usual mathematical normalization), as one can see by introducing polar coordinates in the xy plane.

5.3.2. $\mathfrak{m}$ *of modulus 1.* Now let us analyze what happens for $|\mathfrak{m}| = 1$. When are the flat connections $\mathcal{A}_{\pm}$ unitary? For this to be the case, $\mathfrak{l}$ must be of modulus 1. From (5.42), the condition for this (assuming that $|\mathfrak{m}| = 1$) is

$$\mathfrak{m}^2 + \mathfrak{m}^{-2} + 1 \leq 0. \tag{5.52}$$

Conversely, if this condition is obeyed and $|\mathfrak{m}| = 1$, one can explicitly conjugate the flat bundle with monodromies (5.40) into $SU(2)$ by the same steps that we used in discussing the trefoil knot.

For $|\mathfrak{m}| = 1$ and $\mathfrak{m}^2 + \mathfrak{m}^{-2} + 1 > 0$, the bundles $\mathcal{A}_\pm$ have structure group $SL(2,\mathbb{C})$. A reduction to $SL(2,\mathbb{R})$ cannot occur because $\widehat{\mathfrak{m}}$ has eigenvalues of unit modulus while $\widehat{\mathfrak{l}}$ has real eigenvalues; commuting (and noncentral) elements of $SL(2,\mathbb{R})$ do not have this property. Also, the region $|\mathfrak{m}| = 1$, $\mathfrak{m}^2 + \mathfrak{m}^{-2} > 1$ again contains the points $\mathfrak{m} = \pm 1$ where the connections $\mathcal{A}_\pm$ are $SL(2,\mathbb{C})$ connections related to the hyperbolic metric on $S^3\backslash K$.

As usual, we have to look more closely at what happens for $\mathfrak{m}^2 + \mathfrak{m}^{-2} + 1 = 0$. It is convenient to set $\epsilon = -\mathfrak{m}^2 - \mathfrak{m}^{-2} - 1$, so that ϵ is real for $|\mathfrak{m}| = 1$ and the flat connections $\mathcal{A}_\pm$ have $SU(2)$ structure group precisely when $\epsilon \geq 0$. Unlike our previous examples, the nonabelian flat connections $\mathcal{A}_\pm$ remain irreducible at those points and as a result the abelian flat connection $\mathcal{A}_{\rm ab}$ will play no role in critical behavior near $\epsilon = 0$. All that happens when $\epsilon = 0$ is that the two critical points $\mathcal{A}_{\rm ab}$ become coincident. Moreover, near $\epsilon = 0$, the difference between the two critical points is of order $\epsilon^{1/2}$; this is evident from the formulas (5.40) for the monodromies. Finally, the critical points are real (in the natural real structure of $SU(2)$ gauge theory) if $\epsilon > 0$. These are the properties of the familiar function

$$f(x) = x^3/3 - \epsilon x \tag{5.53}$$

associated to the Airy integral. This function describes the critical behavior of the Chern-Simons function $W(\mathcal{A})$ near the points $\epsilon = 0$, and hence in studying Stokes phenomena, we will be able to borrow results from section 3.4.1.

For now, we simply describe qualitatively the behavior of the colored Jones polynomial $J_n(q)$ for real ϵ near $\epsilon = 0$. First we consider the limit $n \to \infty$ with ϵ fixed:

(1) For $\epsilon < 0$, the only real critical point is the abelian one. Its contribution to the path integral, for any knot, is of order $1/\sqrt{n}$, so we have

$$J_n(q) \sim \frac{C}{n^{1/2}}, \ \epsilon < 0 \tag{5.54}$$

with a constant C. (Actually, in a more precise description, C is the first term in an asymptotic expansion $C + C'/n + \ldots$ in integer powers of $1/n$.)

(2) For $\epsilon > 0$, the flat bundles $\mathcal{A}_\pm$ appear as real critical points in the path integral. They give oscillatory contributions, but without the factor $1/\sqrt{n}$ that appears in the contribution of $\mathcal{A}_{\rm ab}$. Including also the abelian contribution, the integral behaves as

$$J_n(\epsilon) \sim \frac{C}{n^{1/2}} + C_+ \exp(inf_+) + C_- \exp(inf_-), \ \epsilon > 0 \tag{5.55}$$

with constants $C_\pm$ (which again are the leading terms of asymptotic series). Here we have written the action $\mathcal{I}(\mathcal{A}_\pm)$ as $inf_\pm(\epsilon)$, where $f_\pm(\epsilon)$ is independent of n. We have $f_\pm(0) = 0$, since $\mathcal{A}_\pm$ coincide with $\mathcal{A}_{\rm ab}$ at $\epsilon = 0$. Also, as explained in section 5.3.3, $f_+(\epsilon) = -f_-(\epsilon)$ for real ϵ.

(3) What happens at $\epsilon = 0$? Typically the n-dependence of the integral over any one mode of the path integral is that of a Gaussian integral $\int {\rm d}x \exp(inx^2) \sim 1/n^{1/2}$. For the path integral around an irreducible flat connection with vanishing cohomology (such as $\mathcal{A}_\pm$ at $\epsilon \neq 0$), these factors, after taking into account the regularization and the volume of the gauge group, ultimately cancel in a complete

evaluation of the path integral. This is why the oscillatory terms in (5.55) do not have a power of n as a prefactor. But at $\epsilon = 0$, the integral over x, rather than being Gaussian, becomes $\int dx \exp(inx^3) \sim 1/n^{1/3} = n^{1/6}/n^{1/2}$. The denominator participates in the usual cancellation, but the numerator does not, so we expect

$$J_n(0) \sim n^{1/6}. \tag{5.56}$$

(In the usual mathematical normalization, one divides by the path integral (5.3) of the unknot, and then the colored Jones polynomial grows as $n^{2/3}$.)

So far we have described what happens for $n \to \infty$ with fixed ϵ. If instead we consider the limit $n \to \infty$ with $n^{2/3}\epsilon$ fixed, then in view of (3.55), the singular behavior of the colored Jones polynomial is an Airy function (times $n^{2/3}$ in the usual mathematical normalization).

5.3.3. *The real part of the Chern-Simons function.* For the figure-eight knot, we will study only the simplest properties of the Chern-Simons function.

First we focus on the region $|\mathfrak{m}| = 1$. We denote as $\boldsymbol{C}$ the operation of complex conjugation of a flat $SL(2,\mathbb{C})$ bundle. $\boldsymbol{C}$ maps $(\mathfrak{m}, \mathfrak{l})$ to $(\overline{\mathfrak{m}}, \overline{\mathfrak{l}})$. Let us compose this operation with the symmetry $\widetilde{\tau}$, which maps $(\mathfrak{m}, \mathfrak{l})$ to $(\mathfrak{m}^{-1}, \mathfrak{l})$. The composite acts by $(\mathfrak{m}, \mathfrak{l}) \to (\overline{\mathfrak{m}}^{-1}, \overline{\mathfrak{l}})$. If $|\mathfrak{m}| = 1$, then $\overline{\mathfrak{m}}^{-1} = \mathfrak{m}$ and the action is $(\mathfrak{m}, \mathfrak{l}) \to (\mathfrak{m}, \overline{\mathfrak{l}})$.

There are two somewhat different cases:

(1) If $\mathfrak{m}^2 + \mathfrak{m}^{-2} + 1 < 0$, the two flat bundles are $SU(2)$ bundles with complex conjugate values of $\mathfrak{l}$. The composite $\boldsymbol{C}\widetilde{\tau}$ exchanges the two values of $\mathfrak{l}$, so it exchanges the two flat bundles.

(2) If $\mathfrak{m}^2 + \mathfrak{m}^{-2} + 1 > 0$, the two flat bundles are $SL(2,\mathbb{C})$ bundles with real and unequal values of $\mathfrak{l}$. The composite $\boldsymbol{C}\widetilde{\tau}$ leaves fixed the two values of $\mathfrak{l}$, so it also leaves fixed the isomorphism class of each of these two flat bundles.

Now let us analyze the Chern-Simons invariant $W(\mathcal{A})$ in the two cases. $\boldsymbol{C}$ transforms W to its complex conjugate, while $\widetilde{\tau}$ (because it reverses the orientation of S^3) maps it to $-W$. So the composite maps W to $-\overline{W}$.

In case (1), W is real, since in general the Chern-Simons invariant of an $SU(2)$ connection is real. So W transforms simply to $-W$ under the exchange of the two flat bundles. Thus they have equal and opposite real values of W.

In case (2), W is complex. The fact that the two flat bundles are invariant under $\boldsymbol{C}\widetilde{\tau}$ tells us that their Chern-Simons functions are invariant under $W \to -\overline{W}$. This tells us nothing about $\operatorname{Im} W$, but it implies that $\operatorname{Re} W = -\operatorname{Re} W$. Before concluding that $\operatorname{Re} W = 0$, we should recall that $\operatorname{Re} W$ is only well-defined modulo 2π. So the correct conclusion is only that $\operatorname{Re} W$ vanishes mod π.

In case (2), we are left with two possible cases: $\operatorname{Re} W$ may vanish mod 2π or may equal π mod 2π. The region $\mathfrak{m}^2 + \mathfrak{m}^{-2} + 1 > 0$, $|\mathfrak{m}| = 1$, has two connected components, as $\mathfrak{m}$ may be near either 1 or -1. It turns out that $\operatorname{Re} W = 0 \bmod 2\pi$ if $\mathfrak{m}$ is near 1, and $\operatorname{Re} W = \pi \bmod 2\pi$ if $\mathfrak{m}$ is near -1.

To show this, we observe first that if $\mathfrak{m}^2 + \mathfrak{m}^{-2} + 3 = 0$, then the flat bundle is triangular, which guarantees that $W = 0 \bmod 2\pi$. Let $\mathfrak{m}_0$ be a positive root of the equation $\mathfrak{m}^2 + \mathfrak{m}^{-2} + 3 = 0$. Then $-\mathfrak{m}_0$ is a negative root. We can calculate $W(\mathcal{A})$ for $\mathfrak{m} = 1$ by integrating eqn. (4.42) along the real $\mathfrak{m}$ axis from $\mathfrak{m}_0$ to 1. We get

$$W(\mathcal{A}; \mathfrak{m} = 1) = W(\mathcal{A}; \mathfrak{m} = \mathfrak{m}_0) + (1/\pi) \int_{\mathfrak{m}_0}^{1} \ln \mathfrak{m} \, d \ln \mathfrak{l}. \tag{5.57}$$

Along the integration path, $\ln \mathfrak{m}$ is real. On the other hand, according to eqn. (5.42), $\mathfrak{l}$ is of modulus 1 for $\mathfrak{m}$ real,[32] so $\ln \mathfrak{l}$ is imaginary. Hence, the integral is imaginary. Since the integration constant $W(\mathcal{A}; \mathfrak{m} = \mathfrak{m}_0)$ vanishes, we learn that

$$\mathrm{Re}\, W(\mathcal{A}; \mathfrak{m} = 1) = 0 \bmod 2\pi. \tag{5.58}$$

Similarly we have $W(\mathcal{A}; \mathfrak{m} = -1) = W(\mathcal{A}; \mathfrak{m} = -\mathfrak{m}_0) + (1/\pi) \int_{-\mathfrak{m}_0}^{-1} \ln \mathfrak{m}\, \mathrm{d} \ln \mathfrak{l}$, and again the integration constant vanishes. But now we get a different result because the imaginary part of $\ln \mathfrak{m}$ along the negative $\mathfrak{m}$ axis is equal to π. From (5.42), we see that as $\mathfrak{m}$ varies from $-\mathfrak{m}_0$ to -1, $\mathfrak{l}$ makes a half-circuit around the unit circle from 1 to -1. So $\ln \mathfrak{l}$ changes by $i\pi$ and finally we get

$$\mathrm{Re}\, W(\mathcal{A}; \mathfrak{m} = -1) = \pi \bmod 2\pi. \tag{5.59}$$

Now let us discuss the behavior for $\mathfrak{m} = \exp(i\pi\gamma)$ as a function of real γ with $|\gamma| \leq 1$. The condition $\mathfrak{m}^2 + \mathfrak{m}^{-2} + 1 < 0$ amounts to $1/3 < |\gamma| < 2/3$. So case (2) in the above analysis corresponds to $|\gamma| < 1/3$ or $|\gamma| > 2/3$. In either of those regions, $\mathrm{Re}\, W$ can only take the values 0 or π, so it is constant by continuity.[33] Hence $\mathrm{Re}\, W = 0$ if $|\gamma| \leq 1/3$, and $\mathrm{Re}\, W = \pi$ if $|\gamma| \geq 2/3$.

For $1/3 \leq |\gamma| \leq 2/3$, there are up to isomorphism two flat $SU(2)$ connections, which we denote as $A_\pm$. At $|\gamma| = 1/3$, A_+ and A_- both coincide with the flat $SL(2, \mathbb{C})$ connection considered in the last paragraph. So in that case, $\mathrm{Re}\, W = 0 \bmod 2\pi$. It is convenient to lift the Chern-Simons invariants so that $A_\pm$ have $\mathrm{Re}\, W = 0$ at $|\gamma| = 1/3$.

Now let us follow the two $SU(2)$ flat connections A_+ and A_- continuously from $|\gamma| = 1/3$ to $|\gamma| = 2/3$. At $|\gamma| = 2/3$, we can use (5.59), along with the constancy of $\mathrm{Re}\, W$ for $|\gamma| \geq 2/3$, to deduce that $W(A_+)$ and $W(A_-)$ are each congruent to π mod 2π. Since we also showed above that $W(A_+) = -W(A_-)$, it follows that for some integer w, we have $W(A_\pm) = \pm(2w + 1)\pi$ at $|\gamma| = 2/3$.

It can be shown by integrating (4.42) that $w = 1$. (This is greatly facilitated by symmetrizing the integrand between γ and $1 - \gamma$ before attempting to integrate.) Now let us discuss what happens when we include all possible lifts of $W(A_\pm)$.

For $|\gamma| < 1/3$, critical points come in complex conjugate pairs. In passing through $|\gamma| = 1/3$, a pair of complex conjugate critical points meet and recombine into two inequivalent real critical points $A_\pm$. Actually, this occurs infinitely many times, once corresponding to each complex lift of W.

Increasing γ past $|\gamma| = 2/3$, the reverse process occurs: a pair of inequivalent real critical points recombine into a complex conjugate pair of critical points. Again this actually occurs infinitely many times.

The flow in W between $|\gamma| = 1/3$ and $|\gamma| = 2/3$ leads to a crucial detail. Actually, this detail also arose in analytic continuation of the Bessel function in section 3.5, and was illustrated in fig. 9. That figure is repeated for convenience in fig. 15, with a caption adapted to the present problem. The point is that a pair of real critical points that are "born" at $|\gamma| = 1/3$ with the same value of W no longer have the same value of W by the time they "arrive" at $|\gamma| = 2/3$. Rather they have values of W that differ by 2π. The pairs of critical points that separate at $|\gamma| = 1/3$ are not the same as the pairs that rejoin at $|\gamma| = 2/3$.

---

[32]Fundamentally, this reflects the fact that for $\mathfrak{m}$ real, the two flat bundles are invariant under the composite $\mathcal{C}\mathcal{T}$, which leaves real $\mathfrak{m}$ fixed while mapping $\mathfrak{l}$ to $\bar{\mathfrak{l}}^{-1}$.

[33]This constancy is also a direct consequence of eqn. (4.42). In the region in question, $\ln \mathfrak{m}$ is imaginary and $\ln \mathfrak{l}$ is real, so the real part of $\ln \mathfrak{m}\, \mathrm{d} \ln \mathfrak{l}$ vanishes.

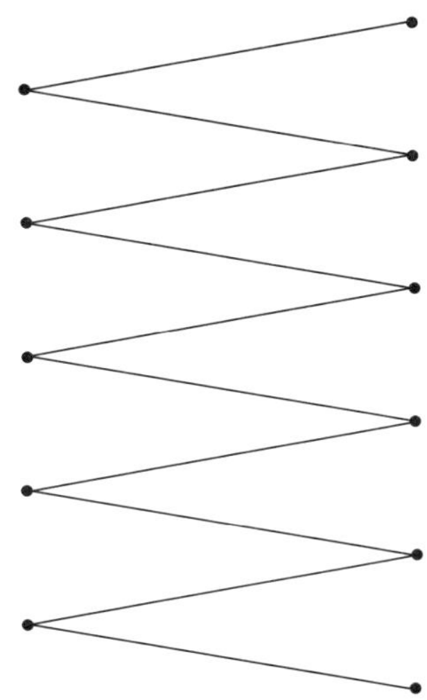

FIGURE 15. Qualitative behavior of the conserved quantity $\mathrm{Im}\, I/k$ for real critical points in the region $1/3 \leq |\gamma| \leq 2/3$. The structure is in perfect parallel with fig. 9 of section 3.5.2, except now γ is plotted on the horizontal axis. At $|\gamma| = 1/3$ (left), there are a pair of real critical points for every value of $\mathrm{Im}\, \mathcal{I}/k$ of the form $2\pi w$, $w \in \mathbb{Z}$. Increasing $|\gamma|$, $\mathrm{Im}\, \mathcal{I}/k$ increases for one of these critical points and reduces for the other. At $|\gamma| = 2/3$ (right), the critical points recombine in pairs, but now with $\mathrm{Im}\, \mathcal{I}/k$ of the form $(2w+1)\pi$, $w \in \mathbb{Z}$.

Precisely the same phenomenon led in section 3.4.2 to exponential growth of the analytically continued Bessel function for negative and non-integer values of κ. It should therefore come as no surprise that this mechanism will lead to exponential growth of the function $J_n(q)$ associated to the figure-eight knot in the region $|\gamma| > 2/3$, as long as the Chern-Simons level k is not an integer.

5.3.4. *The imaginary part of* W. We will now discuss much more briefly the simplest properties of $\mathrm{Im}\, W$ along the unit circle $|\mathfrak{m}| = 1$.

For $\mathfrak{m}^2 + \mathfrak{m}^{-2} + 1 \leq 0$, the nonabelian flat connections are $SU(2)$-valued flat connections $A_\pm$. As we have already noted, like any $SU(2)$-valued connections, these have $\mathrm{Im}\, W = 0$.

For $\mathfrak{m}^2 + \mathfrak{m}^{-2} + 1 > 1$, there are a pair of $SL(2,\mathbb{C})$ flat connections $\mathcal{A}_\pm$, exchanged by complex conjugation. Since $\mathrm{Im}\, W$ is odd under complex conjugation, we have $\mathrm{Im}\, W(\mathcal{A}_+) = -\mathrm{Im}\, W(\mathcal{A}_-)$.

The most important property of the invariant $\mathrm{Im}\, W$ is its relation to the hyperbolic volume of the knot complement $S^3\backslash K$. $S^3\backslash K$ admits a complete hyperbolic metric, unique up to diffeomorphism. Though this metric is complete, it has the property that the volume is convergent.[34] We denote this volume as V.

[34]Near K, the metric can be modeled by $\mathrm{d}s^2 = z^2(\mathrm{d}x^2 + \mathrm{d}y^2) + \mathrm{d}z^2/z^2$, where x and y parametrize a two-torus and K is at $z = 0$. Thus the region near K is the "infrared" region in the context of the AdS/CFT correspondence. This description can be used to compute the conjugacy class of $\widehat{\mathfrak{m}}$ given in eqn. (5.60) below. Both $\widehat{\mathfrak{m}}$ and $\widehat{\mathfrak{l}}$ arise in this description from shifts in x and y; such shifts correspond to unipotent elements of $PSL(2,\mathbb{C})$.

Like a hyperbolic metric on any three-manifold, the complete hyperbolic metric on $S^3\backslash K$ determines a flat $PSL(2,\mathbb{C})$ bundle on this manifold. Upon picking a spin structure on $S^3\backslash K$, the flat $PSL(2,\mathbb{C})$ bundle can be lifted to a flat $SL(2,\mathbb{C})$ bundle. The conjugacy class of the meridian for this flat bundle is

$$\widehat{\mathfrak{m}} = \pm \begin{pmatrix} 1 & 1 \\ 0 & 1 \end{pmatrix}, \tag{5.60}$$

where the sign depends on the choice of spin structure (in other words, the two flat bundles with different signs of $\mathfrak{m}$ differ by twisting by a flat line bundle of order 2 – the operation considered in section 4.2.7). The flow equations in the presence of such unipotent monodromy were discussed in section 4.2.2; for our present purposes, all we need to know is that the eigenvalue of $\widehat{\mathfrak{m}}$ is $\mathfrak{m} = \pm 1$. On the other hand, from eqn. (5.42), it follows that[35] $\mathfrak{l} = -1$, for either sign of $\mathfrak{m}$.

The relation between the hyperbolic volume and the Chern-Simons invariant is simply that, if we label $\mathcal{A}_\pm$ correctly,

$$\mathrm{Re}\,(iW(\mathcal{A}_\pm; \mathfrak{m} = \pm 1)) = \pm \frac{V}{2\pi}. \tag{5.61}$$

This means that $\mathrm{Im}\, W(\mathcal{A}_\pm; \mathfrak{m} = \pm 1) = \mp V/2\pi$. The sign may look unnatural, but it ensures that $\mathcal{A}_\pm$ make exponentially growing and decaying contributions to the path integral, respectively; this seems like the most useful mnemonic.

If we depart from $\mathfrak{m} = \pm 1$, there is no longer a complete hyperbolic metric on $S^3\backslash K$ with $\mathfrak{m}$ as the eigenvalue of the meridian. But there is an incomplete hyperbolic metric with this property (it has a conical singularity along K), and a statement along the lines of (5.61) remains valid for this incomplete metric. We will not describe details here.

We have explained the qualitative properties of $\mathrm{Re}\, W$ and $\mathrm{Im}\, W$ that will be of most concern for the limited application that we consider next. We should note, however, that (as summarized in [**17**], for example) by integrating the usual formula (4.42) and using (5.42), one can express $W(\mathcal{A})$ for flat connections on the complement of the figure-eight knot in terms of a dilogarithm function.

5.3.5. *The volume conjecture for the figure-eight knot.* The volume conjecture for knots emerged from the following observation [**2, 3**] about the analytically continued invariant $J_n(q)$ of the figure-eight knot. For γ approaching 1 from a direction away from the real axis, the large n behavior of $J_n(q)$ is, roughly speaking,

$$J_n(q) \sim \exp(nV/2\pi), \tag{5.62}$$

where V is the hyperbolic volume of the knot complement, discussed in section 5.3.4. A more precise statement will emerge below.

A partial explanation of this was given in [**6**]. It was noted that $nV/2\pi$ is the value of the exponent $\mathcal{I} = ikW(\mathcal{A})$ of the analytically-continued Chern-Simons path integral at the complex critical point $\mathcal{A}_+$. So an expansion around this critical point would give a contribution growing as $\exp(V/2\pi)$. (In this paper, and subsequent literature, generalizations of the statement to other values of γ were also made. For example, see [**3, 5, 9**].)

[35]From this we can actually give a new demonstration of one result of section 5.3.3, which was that $\mathrm{Re}\, W$ shifts by π under $\mathfrak{m} \to -\mathfrak{m}$. In section 4.2.2, we showed that the shift is $\Delta W = -i \ln \mathfrak{l}$, which here equals π mod 2π as $\mathfrak{l} = -1$.

There has always been a puzzle about this partial explanation. For physical values of k and n – positive integers – the Chern-Simons path integral is dominated in the semiclassical limit by real critical points. Contributions, if any, of complex critical points are exponentially small. Therefore, at least for positive rational values of $\gamma = n/k$, there are no exponentially growing contributions to the path integral. How then can exponential growth be observed just because γ is infinitesimally displaced from 1?

The reader who has paid close attention to sections 3.4.2 and 5.3.4 will hopefully anticipate the answer. Once we analytically continue away from integer values of k, we need to work not on the space $\mathcal{U}_{\mathbb{C}}$ of complex-valued connections modulo gauge transformations, but on a cover $\widehat{\mathcal{U}}_{\mathbb{C}}$ on which the Chern-Simons function $W(\mathcal{A})$ is $\mathbb{C}$-valued (rather than valued in $\mathbb{C}/2\pi\mathbb{Z}$). A given flat connection $\mathcal{A}$ corresponds to infinitely many critical points that differ only by the choice of a complex lift of $W(\mathcal{A})$. When the integration cycle of the analytically continued theory is expressed in terms of Lefschetz thimbles associated to critical points, it is possible for contributions from critical points that differ only in this lift to cancel for integer k, but give exponential growth as soon as k is not an integer.

Let us first express in terms of Lefschetz thimbles the integration cycle $\mathcal{C}$ of Chern-Simons theory for what proves to be the simplest case: positive integer k and n with $1/3 < \gamma < 2/3$, where as usual $\gamma = k/n$. This is easy to do because in this range there are no complex critical points, only real ones. Up to gauge transformation, there is an abelian flat connection $A_{\rm ab}$, and two nonabelian but $SU(2)$-valued flat connections $A_\pm$. As they are all $SU(2)$-valued, they all contribute to the integration cycle with a coefficient 1. This cycle is then in an obvious notation

$$\mathcal{C} = \mathcal{J}_{\rm ab} + \mathcal{J}_+ + \mathcal{J}_-. \tag{5.63}$$

As long as k and n are integers, there is also no problem in extending this answer outside the range $1/3 < \gamma < 2/3$. At either $\gamma = 1/3$ or $\gamma = 2/3$, there is an Airy-type singularity. The two real flat connections $A_\pm$ recombine to a complex conjugate pair of $SL(2,\mathbb{C})$-valued flat connections, which we will call $\mathcal{A}'_\pm$ for $\gamma < 1/3$ or $\mathcal{A}''_\pm$ for $\gamma > 2/3$. We choose the subscript so that, as in (5.61) (where $\mathfrak{m} = \pm 1$ are treated together), $\mathrm{Re}\,(iW(\mathcal{A}'_+)) > 0$ for $\gamma < 1/3$ and $\mathrm{Re}\,(iW(\mathcal{A}''_+)) > 0$ for $\gamma > 2/3$. This ensures that the integrals over the Lefschetz thimbles $\mathcal{J}'_-$ and $\mathcal{J}''_-$ associated to $\mathcal{A}'_-$ and $\mathcal{A}''$ are exponentially decaying, while the integrals over the other Lefschetz thimbles $\mathcal{J}'_+$, $\mathcal{J}''_+$ associated to $\mathcal{A}'_+$, $\mathcal{A}''_+$ are exponentially growing. When we cross the Airy-like singularities at $\gamma = 1/3$ or $2/3$, according to eqn. (3.58), the sum $\mathcal{J}_+ + \mathcal{J}_-$ recombines to a Lefschetz thimble $\mathcal{J}'_-$ or $\mathcal{J}''_-$, whose contribution to the path integral is exponentially decaying. The integration cycle thus becomes

$$\mathcal{C} = \begin{cases} \mathcal{J}_{\rm ab} + \mathcal{J}'_- & \text{for } \gamma < 1/3 \\ \mathcal{J}_{\rm ab} + \mathcal{J}''_- & \text{for } \gamma > 2/3. \end{cases} \tag{5.64}$$

Now let us continue away from integer values of k and n. Starting in the region $1/3 < \gamma < 2/3$, to make an analytic continuation, we must pick lifts of the Chern-Simons invariants $W(A_\pm)$. When we do this, we write, for example, $\mathcal{J}_{+,w}$ for the Lefschetz thimble associated to A_+ with a Chern-Simons invariant $2\pi w$; we extend this notation to the other flat connections in an obvious way. Here for $\mathcal{J}_\pm$, the value of w mod $\mathbb{Z}$ depends smoothly on γ (varying from 0 mod $\mathbb{Z}$ at $\gamma = 1/3$ to

$1/2 \bmod \mathbb{Z}$ at $\gamma = 2/3$); for $\mathcal{J}'_\pm$, w takes values in $\mathbb{Z}$; and for $\mathcal{J}''_\pm$, w takes values in $1/2 + \mathbb{Z}$.

If we do not want exponential growth for $n \to \infty$ and $\gamma < 1/3$, we must pick the lifts of $W(A_+)$ and $W(A_-)$ to be equal at $\gamma = 1/3$; otherwise, according to (3.58), in continuing past the Airy singularity, $\mathcal{J}'_+$ as well as $\mathcal{J}'_-$ will appear in the integration cycle, giving exponential growth. Similarly, if we do not want exponential growth for $\gamma > 2/3$, we need $W(A_+) = W(A_-)$ at $\gamma = 2/3$. But it is impossible to obey both of these conditions, since as was illustrated in fig. 15, if $W(A_+)$ and $W(A_-)$ are equal at $\gamma = 1/3$, then they differ by 2π at $\gamma = 2/3$, and vice-versa.

Exponential growth is therefore unavoidable, once k is not an integer, either for $\gamma < 1/3$ or for $\gamma > 2/3$. As explained in section 5.1, we wish to pick an analytic continuation that leads to no exponential growth for $\gamma < 1/3$. A further clue is that, to preserve τ and $\widetilde{\tau}$ symmetry, $J_n(q)$ should be real for real k. (Complex conjugation of the path integral reverses the sign of k, but as the figure-eight knot is amphichiral, one can compensate for this with the symmetry τ or $\widetilde{\tau}$.) This suggests that we should take the lift of $\mathcal{A}'_-$ such that $\operatorname{Re} W(\mathcal{A}'_-) = 0$. Then the integration cycle for γ just less than $1/3$ is

$$\mathcal{C} = \mathcal{J}_{\mathrm{ab}} + \mathcal{J}'_{-,0}. \tag{5.65}$$

Continuing past $\gamma = 1/3$, cycles $\mathcal{J}_{+,w}$ and $\mathcal{J}_{-,w}$ appear, initially with $w = 0$. But by the time we reach $\gamma = 2/3$, the values are $w = \pm 1/2$, as we learned in section 5.3.4. So the integration cycle as γ approaches $2/3$ from below is

$$\mathcal{C} = \mathcal{J}_{\mathrm{ab}} + \mathcal{J}_{+,1/2} + \mathcal{J}_{-,-1/2}. \tag{5.66}$$

To continue past $\gamma = 2/3$, we use eqn. (3.57) to express $\mathcal{J}_{+,1/2}$ and $\mathcal{J}_{-,-1/2}$ in terms of corresponding cycles $\mathcal{J}''_{\pm,w}$. Since w does not change in this process, the result is $\mathcal{J}_{+,1/2} \to -\mathcal{J}''_{+,1/2} + \mathcal{J}''_{-,1/2}$, and $\mathcal{J}_{-,-1/2} \to \mathcal{J}''_{+,-1/2}$. The integration cycle for γ just greater than $1/3$ is therefore

$$\mathcal{C} = \mathcal{J}_{\mathrm{ab}} - \mathcal{J}''_{+,1/2} + \mathcal{J}''_{-,1/2} + \mathcal{J}''_{+,-1/2}. \tag{5.67}$$

Near or on the real γ axis, with $\operatorname{Re}\gamma > 2/3$, exponential growth comes from the terms $\mathcal{J}''_{+,-1/2} - \mathcal{J}''_{+,1/2}$ in (5.67). Just as in the analytic continuation of the Bessel function, the relative minus sign is the reason that exponential growth is avoided when k is an integer. For γ near 1, using (5.61), the value of the path integral is

$$J_n(q) \sim \exp(nV/2\pi)\left(\exp(-i\pi k) - \exp(i\pi k)\right). \tag{5.68}$$

Here the exponential factors $\exp(\pm i\pi k)$ reflect the values $w = \pm 1/2$ for the relevant Lefschetz thimbles. Also, near $\gamma = 1$, we have written the exponentially growing factor as $\exp(nV/2\pi)$ (with n rather than k in the exponent) to put the formula in a more recognizable form for some readers.

The expression $\log J_n(q)/n$ does not have a simple large n limit if γ is on the real axis, because of the oscillatory factor $(\exp(-i\pi k) - \exp(i\pi k))$. As soon as γ is off the real axis, there is a simple limit, but it is discontinuous on the real γ axis. The reason for that is that the large n limit of $\log\left(\exp(-i\pi k) - \exp(i\pi k)\right)/n$ is $\mp i\pi k/n = \mp i\pi/\gamma$, depending on the sign of $\operatorname{Im} k$.

In the region $2/3 < \gamma < 4/3$, the flat connection $\mathcal{A}_+$, with its various possible lifts, has the maximum value of $\operatorname{Re}\mathcal{I}$, among all flat connections. This ensures that, once the cycles $\mathcal{J}''_{+,\pm 1/2}$ appear in $\mathcal{C}$ at $\gamma = 2/3$, they cannot disappear at least until $\gamma = 4/3$. At $\gamma = 4/3$, and more generally at every value of γ of the form

$\pm 1/3 \bmod \mathbb{Z}$, one meets another family of Airy-type singularities. To analyze the consequences, it may be necessary to take into account Stokes phenomena beyond what follows from the singularities. (For example, there appears to be a Stokes line crossing the real γ axis between the values 2/3 and 1; if there is a flow from $\mathcal{A}_{\rm ab}$ to $\mathcal{A}_-$ along this line, this might cause $\mathcal{J}''_{-,1/2}$ to disappear from $\mathcal{C}$.)

The normalization of $J_n(q)$ that is most often used in the mathematical literature differs from ours by dividing by a factor (namely the right hand side of eqn. (5.3)) that has no exponential growth in the semiclassical limit but does have a simple zero at $\gamma = 1$. This makes the proper statement about the semiclassical limit simpler at $\gamma = 1$, though not at other real values near $\gamma = 1$. The function $J_n(q)$, normalized as we do, vanishes (assuming that n is an integer) for $q^n = 1$ or $n = k + 2$. One may ask what that vanishing means in the context of integration over Lefschetz thimbles. The answer is that it may be hard to see the vanishing exactly, but we can certainly see it approximately. The exponentially growing contributions cancel if $n = k+2$, since after all this implies that $k = n-2$ is an integer. The remaining contribution is then dominated by $\mathcal{J}_{\rm ab}$, but this contribution vanishes at $n = k + 2$ since at this value, the dimension of the automorphism group of the abelian flat connection $A_{\rm ab}$ increases.

6. Four or five-dimensional interpretation

In section 4.1, we observed that the flow equations of three-dimensional Chern-Simons theory have a natural four-dimensional symmetry. This is a familiar fact for the flow equations of Chern-Simons theory with a compact gauge group; what is novel and perhaps surprising is that four-dimensional symmetry also emerges in flows of complex connections. Here we will make only a few simple remarks about how this fact may be interpreted. The ultimate goal of such considerations might be to give a gauge theory interpretation to Khovanov homology, which is a four-dimensional extension of three-dimensional Chern-Simons theory.

One observation is that, in the case of a finite-dimensional Kahler manifold X with a holomorphic superpotential W, the Morse theory flow equations can be interpreted as equations for supersymmetric solitons [**41**]. To interpret the Chern-Simons flow equations in this way, we need a two-dimensional theory with $\mathcal{N} = 2$ supersymmetry in which the chiral superfields are a gauge field $\mathcal{A}$ on a three-manifold M (with some gauge group H), the superpotential is the Chern-Simons function $W(\mathcal{A})$, and the gauge group is the group of maps from M to H. There is no problem in constructing such a theory. We simply start with five-dimensional maximally supersymmetric Yang-Mills theory (of course, it may be better to start with the usual ultraviolet completion of this theory in six dimensions) and compactify on M with a topological twist that preserves some supersymmetry. This gives a two-dimensional theory with $\mathcal{N} = 2$ supersymmetry and all the stated properties. In this two-dimensional theory, the Chern-Simons flow equations that we studied in section 4.1 are equations for supersymmetric solitons. This may be a useful starting point for a new perspective on three-dimensional Chern-Simons theory.

Another perhaps more direct, though mysterious, relation between three-dimensional Chern-Simons theory and a four-dimensional theory is as follows. As we observed in section 2.6, for a given three-manifold M, the possible integration cycles of Chern-Simons theory, and therefore the possible Chern-Simons path integrals, form a vector space $\mathcal{V}$. For a given choice of a direction in the complex plane for the Chern-Simons coupling parameter k, $\mathcal{V}$ has a basis corresponding to Lefschetz thimbles associated to critical points. Because of Stokes phenomena, this basis is not really natural, but $\mathcal{V}$ does have a natural lattice structure, since the condition for an element of $\mathcal{V}$ to be an integer linear combination of basis vectors is invariant under all jumping processes. We can interpret $\mathcal{V}$ as the physical "Hilbert" space of a certain topological field theory – the twisted version of $\mathcal{N} = 4$ super Yang-Mills theory that was studied in [**53**] and is related to geometric Langlands. From this point of view, one studies the flow equation on a general four-manifold X, not necessarily the product of a three-manifold and a one-manifold. If X is a four-manifold of boundary M, then by counting solutions of the flow equations on X, one would construct a vector in the physical Hilbert space of M and thus an integration cycle for Chern-Simons theory on M. Of course, for this to make sense depends on the fact that the flow equations are specializations of equations with four-dimensional symmetry.

Acknowledgments. This research was supported in part by NSF Grant Phy-0503584. The project was stimulated by lectures at the conference *Chern-Simons Gauge Theory: 20 Years After* at the Hausdorff Institute in Bonn (August, 2009). I would like to thank the organizers for their hospitality, and to acknowledge discussions during the meeting and on other occasions with J. Anderson, C. Beasley, T. Dimofte, D. Gaiotto, S. Garoufalidis, A. Givental, M. Goresky, S. Gukov, R. MacPherson, G. Moore, and H. Murakami.

References

[1] W. Li, W. Song, and A. Strominger, *Chiral gravity in three dimensions*, JHEP **0804**, 082 (2008), arXiv:0801.4566 [hep-th].

[2] R. M. Kashaev, *The hyperbolic volume of knots from the quantum dilogarithm*, Lett. Math. Phys. **39** (1997), 269–275.

[3] H. Murakami and J. Murakami, *The colored Jones polynomial and the simplicial volume of a knot*, Acta Math. **186** (2001), 85–104.

[4] H. Murakami, J. Murakami, M. Okamoto, T. Takata, and Y. Yokota, *Kashaev's conjecture and the Chern-Simons invariants of knots and links*, Experiment. Math. **11** (2002) 427-435.

[5] R. M. Kashaev and O. Tirkkonen, *A proof of the volume conjecture for torus knots*, J. Math. Sci. (NY) **115** (2003), 2033–2036, math.GT/9912210.

[6] S. Gukov, *Three-dimensional quantum gravity, Chern-Simons theory, and the A-polynomial*, Commun. Math. Phys. **255** (2005), 577–627.

[7] H. Murakami, *Asymptotic behaviors of the colored Jones polynomials of a torus knot*, Internat. J. Math. **15** (2004), 547–555.

[8] S. Garoufalidis and T. T. Q. Lê, *An analytic version of Melvin-Morton-Rozansky conjecture*, arXiv.math.GT/0503641.

[9] S. Garoufalidis and T. T. Q. Lê, *Asymptotics of the colored Jones function of a knot*, arXiv:math/0508100.

[10] J. Dubois and R. Kashaev, *On the asymptotic expansion of the colored Jones polynomial for torus knots*, Math. Ann. **339** (2007), 757–782, math.GT/0510607.

[11] S. Gukov and H. Murakami, *$SL(2,\mathbb{C})$ Chern-Simons theory and the asymptotic behavior of the colored Jones polynomial*, arXiv:math/0608324.

[12] R. van der Veen, *Proof of the volume conjecture for Whitehead chains*, arXiv:math/0611181.

[13] H. Zheng, *Proof of the volume conjecture for Whitehead doubles of a family of torus knots*, Chin. Ann. Math. Ser. B. **28** (2007), 375–388.

[14] S. Garoufalidis, *Chern-Simons theory, analytic continuation, And arithmetic*, arXiv:0711.1716.

[15] J. E. Andersen and S. K. Hansen, *Asymptotics of the Quantum Invariants of Surgeries on the Figure 8 Knot*, Journal of Knot Theory and its Ramifications **15** (2006), 1–69.
[16] K. Hikami and H. Murakami, *Colored Jones polynomials with polynomial growth*, arXiv:0711.2836.
[17] H. Murakami, *An introduction to the volume conjecture and its generalizations*, Acta Math. Vietnam **33** (2008), 209–253, arXiv:0802.0039.
[18] T. Dimofte, S. Gukov, J. Lenells, and D. Zagier, *Exact results for perturbative Chern-Simons theory with complex gauge group*, arXiv:0903.2472.
[19] R. Dijkgraaf and H. Fuji, *The volume conjecture and topological strings*, arXiv:0903.2084.
[20] D. Zagier, *Quantum modular forms*, preprint.
[21] K. Hikami and H. Murakami, *Representations and the colored Jones polynomial of a torus knot*, arXiv:1001.2680.
[22] D. Freed and R. Gompf, *Computer calculations of Witten's 3-manifold invariant*, Commun. Math. Phys. **141** (1991), 79–117.
[23] L. Jeffrey, *Chern-Simons-Witten invariants of lens spaces And torus bundles and the semi-classical approximation*, Commun. Math. Phys. **147** (1992), 563–604.
[24] D. Bar-Natan and S. Garoufalidis, *On the Melvin-Morton-Rozansky conjecture*, Invent. Math. **125** (1996), 103.
[25] L. Rozansky, *The universal R-matrix, Burau representation, and the Melvin-Morton expansion of the colored Jones polynomial*, Adv. Math. **134** (1998), 1–31.
[26] L. Rozansky, *Higher order terms in the Melvin-Morton expansion of the colored Jones polynomial*, Commun. Math. Phys. **183** (1997), 291–306.
[27] L. Rozansky, *A Large k asymptotics of Witten's invariant of Seifert manifolds*, Commun. Math. Phys. **171** (1995), 279–322, hep-th/9309099.
[28] L. Rozansky, *The trivial connection contribution to Witten's invariant and finite type invariants of rational homology spheres*, Commun. Math. Phys. **183** (1997), 23–54.
[29] J. Andersen, *The Witten invariant of finite order mapping tori, I*, preprint (ca. 1995).
[30] R. Lawrence and L. Rozansky, *Witten-Reshetikhin-Turaev invariants of Seifert manifolds*, Commun. Math. Phys. **205** (1999), 287–314.
[31] M. Mariño, *Chern-Simons theory, matrix integrals, and perturbative three-manifold invariants*, Commun. Math. Phys. **253** (2004), 25–49, hep-th/0207096.
[32] C. Beasley and E. Witten, *Non-abelian localization for Chern-Simons theory*, J. Diff. Geom. **70** (2005), 183–323, hep-th/0503126.
[33] M. Blau and G. Thompson, *Chern-Simons theory on S^1 bundles: Abelianisation and q-deformed Yang-Mills theory*, JHEP 0605:003,2006, hep-th/0601068.
[34] C. Beasley, *Localization for Wilson loops in Chern-Simons theory*, arXiv:0911.2687.
[35] V. I. Arnold, S. M. Gusein-Zade, and A. N. Varchenko, *Singularities of differentiable maps*, Vol. 2, Birkhauser, 1988.
[36] M. V. Fedorjuk, *The asymptotics of the Fourier transform of the exponential function of a polynomial*, Soviet Math. Doklady **17** (1976) 486-490.
[37] Frédéric Pham, *Vanishing Homologies And The n variable saddlepoint method*, Proc. Symp. Pure Math. **40** vol. 2 (1983), 319–333.
[38] M. V. Berry and C. J. Howls, *Hyperasymptotics*, Proc. R. Soc. **A430**, 653–668.
[39] M. V. Berry and C. J. Howls, *Hyperasymptotics for integrals with saddles*, Proc. R. Soc. **A434**, 657–675.
[40] C. J. Howls, *Hyperasymptotics for multidimensional integrals, exact remainder terms and the global connection problem*, Proc. R. Soc. London **453** (1997), 2271.
[41] S. Cecotti and C. Vafa, *On classification of $\mathcal{N} = 2$ supersymmetric theories*, Commun. Math. Phys. **158** (1993), 569, arXiv:hep-th/9211097.
[42] K. Hori, A. Iqbal and C. Vafa, *D-branes and mirror symmetry*, arXiv:hep-th/0005247.
[43] L. Fidkowski, V. Hubeny, M. Kleban, and S. Shenker, *The black hole singularity in AdS/CFT*, JHEP **0402** (2004), 014, arXiv:hep-th/0306170.
[44] J. M. Maldacena, G. W. Moore, N. Seiberg, and D. Shih, *Exact vs. semiclassical target space of the minimal string*, JHEP **0410** (2004), 020, arXiv:hep-th/0408039.
[45] D. Gaiotto, G. W. Moore and A. Neitzke, *Four-dimensional wall-crossing via three-dimensional field theory*, arXiv:0807.4723 [hep-th].
[46] P. Kronheimer and T. Mrowka, *Monopoles and Three-Manifolds*, Cambridge University Press, 2007.

[47] M. Hutchings, *Taubes's proof of the Weinstein conjecture in dimension three*, Bull. Am. Math. Soc. **47** (2010), 73.

[48] D. Bar-Natan and E. Witten, *Perturbative expansion of Chern-Simons theory with noncompact gauge group*, Commun. Math. Phys. **141** (1991), 423.

[49] D. Bar-Natan, *On the Vassiliev knot invariants*, Topology **34** (1995), 423.

[50] S. Gukov and E. Witten, *Branes and quantization*, arXiv:0809.0305 [hep-th].

[51] E. Witten, *Quantum field theory and the Jones polynomial*, Commun. Math. Phys. **121** (1989), 351.

[52] K. Corlette, *Flat G-bundles with canonical metrics*, J. Diff. Geom. **28** (1988), 361–382.

[53] A. Kapustin and E. Witten, *Electric-magnetic duality and the geometric Langlands program*, Commun. Numb. Th. Phys. **1** (2007), 1–236, arXiv:hep-th/0604151.

[54] E. Witten, *Dynamics of quantum field theory*, Lecture 7.7, in P. Deligne et. al., eds. *Quantum Fields And Strings: A Course For Mathematicians*, American Mathematical Society, 1999.

[55] S. Gukov and E. Witten, *Gauge theory, ramification, and the geometric Langlands program*, arXiv:hep-th/0612073.

[56] W. Nahm, *The construction of all self-dual multimonopoles by the ADHM method*, in *Monopoles And Quantum Field Theory*, N. S. Craigie et. al., eds., World Scientific, 1982, pp. 87–95.

[57] P. Kirk and E. Klassen, *Chern-Simons invariants of 3-manifolds decomposed along tori and the circle bundle over the representation space of* T^2, Commun. Math. Phys. **153** (1993), 521.

[58] D. Johnson and J. Millson, *Deformation spaces associated to compact hyperbolic manifolds*," in *Discrete Groups In Geometry And Analysis: Papers In Honor of G. D. Mostow*, ed. R. Howe, Birkhauser, Boston, 1987.

[59] E. Witten, *Chern-Simons gauge theory as a string theory*, Prog. Math. **133** (1995), 637–678, hep-th/9207094.

[60] S. Axelrod and I. M. Singer, *Chern-Simons perturbation theory, I*, in *Differential Geometric Methods In Theoretical Physics*, Proceedings, New York, 1991, hep-th/9110056.

[61] S. Axelrod and I. M. Singer, *Chern-Simons perturbation theory, II*, in J. Diff. Geom. **39** (1994), 173–213, hep-th/9304087.

[62] D. B. Ray and I. M. Singer, *R-torsion and the Laplacian on Riemannian manifolds*, Adv. Math. **7** (1971), 145–210.

[63] M. F. Atiyah, *On framings of three-manifolds*, *Topology* **29** (1990), 1–7.

School of Natural Sciences, Institute for Advanced Study, Einstein Drive, Princeton, NJ 08540

E-mail address: `witten@ias.edu`

Titles in This Series

50 **Jørgen E. Andersen, Hans U. Boden, Atle Hahn, and Benjamin Himpel,** Chern-Simons Gauge Theory: 20 Years After, 2011

49 **Matthias Plaue, Alan Rendall, and Mike Scherfner,** Advances in Lorentzian Geometry, 2011

48 **Lizhen Ji, Kefeng Liu, Lo Yang, and Shing-Tung Yau,** Fourth International Congress of Chinese Mathematicians, 2010

47 **Alexander Grigor'yan,** Heat Kernel and Analysis on Manifolds, 2009

46.2 **Kenji Fukaya, Yong-Geun Oh, Hiroshi Ohta, and Kaoru Ono, Editors,** Lagrangian Intersection Floer Theory, 2009

46.1 **Kenji Fukaya, Yong-Geun Oh, Hiroshi Ohta, and Kaoru Ono,** Lagrangian Intersection Floer Theory, 2009

45 **Lydia Bieri and Nina Zipser,** Extensions of the Stability Theorem of the Minkowski Space in General Relativity, 2009

44 **Eric Sharpe and Arthur Greenspoon, Editors,** Advances in String Theory, 2008

43 **Lizhen Ji, Editor,** Arithmetic Groups and Their Generalizations, 2008

42.1 **Ka-Sing Lau, Zhou-Ping Xin, and Shing-Tung Yau, Editors,** Third International Congress of Chinese Mathematicians, 2008

41 **Wen-Ching Winnie Li, Editor,** Recent Trends in Coding Theory and its Applications, 2007

40 **Ovidiu Caliin, Der-Chen Chang, and Peter Greiner, Editors,** Geometric Analysis on the Heisenberg Group and Its Generalizations, 2007

39 **Zhijie Chen, Sheng-Li Tan, Jianpan Wang, and Stephen S.-T. Yau, Editors,** Proceedings of the International Conference on Complex Geometry and Related Fields, 2007

38 **Noriko Yui, Shing-Tung Yau, and James D. Lewis, Editors,** Mirror Symmetry V, 2006

37 **Lizhen Ji, Jian-Shu Li, H. W. Xu, and Shing-Tung Yau, Editors,** Lie Groups and Automorphic Forms, 2006

36 **Chuu-Lian Terng, Editor,** Integrable Systems, Geometry, and Topology, 2006

35 **Felix Finster,** The Principle of the Fermionic Projector, 2006

34 **Ren-Hong Wong, Editor,** Computational Geometry, 2003

33 **Eric D'Hoker, Duong Phong, and Shing-Tung Yau,** Mirror Symmetry IV, 2002

32 **Xi-Ping Zhu,** Lectures on Mean Curvature Flows, 2002

31 **Kiyoshi Igusa,** Higher Franz-Reidemeister Torsion, 2002

30 **Weiman Han and Mircea Sofonea,** Quasistatic Contact Problems in Viscoelasticity and Viscoplasticity, 2002

29 **S. T. Yau and Shuxing Chen, Editors,** Geometry and Nonlinear Partial Differential Equations, 2002

28 **Valentin Afraimovich and Sze-Bi Hsu,** Lectures on Chaotic dynamical Systems, 2002

27 **M. Ram Murty,** Introduction to p-adic Analytic Number Theory, 2002

26 **Raymond Chan, Yue-Kuen Kwok, David Yao, and Qiang Zhang, Editors,** Applied Probability, 2002

25 **Donggao Deng, Daren Huang, Rong-Qing Jia, Wei Lin, and Jian Zhong Wong, Editors,** Wavelet Analysis and Applications, 2002

24 **Jane Gilman, William W. Menasco, and Xiao-Song Lin, Editors,** Knots, Braids, and Mapping Class Groups—Papers Dedicated to Joan S. Birman, 2001

23 **Cumrun Vafa and S.-T. Yau, Editors,** Winter School on Mirror Symmetry, Vector Bundles and Lagrangian Submanifolds, 2001

22 **Carlos Berenstein, Der-Chen Chang, and Jingzhi Tie,** Laguerre Calculus and Its Applications on the Heisenberg Group, 2001

21 **Jürgen Jost,** Bosonic Strings: A Mathematical Treatment, 2001

20 **Lo Yang and S.-T. Yau, Editors,** First International Congress of Chinese Mathematicians, 2001

19 **So-Chin Chen and Mei-Chi Shaw,** Partial Differential Equations in Several Complex Variables, 2001

18 **Fangyang Zheng,** Complex Differential Geometry, 2000

17 **Lei Guo and Stephen S.-T. Yau, Editors,** Lectures on Systems, Control, and Information, 2000

16 **Rudi Weikard and Gilbert Weinstein, Editors,** Differential Equations and Mathematical Physics, 2000

15 **Ling Hsiao and Zhouping Xin, Editors,** Some Current Topics on Nonlinear Conservation Laws, 2000

14 **Jun-ichi Igusa,** An Introduction to the Theory of Local Zeta Functions, 2000

13 **Vasilios Alexiades and George Siopsis, Editors,** Trends in Mathematical Physics, 1999

12 **Sheng Gong,** The Bieberbach Conjecture, 1999

11 **Shinichi Mochizuki,** Foundations of p-adic Teichmüller Theory, 1999

10 **Duong H. Phong, Luc Vinet, and Shing-Tung Yau, Editors,** Mirror Symmetry III, 1999

9 **Shing-Tung Yau, Editor,** Mirror Symmetry I, 1998

8 **Jürgen Jost, Wilfrid Kendall, Umberto Mosco, Michael Röckner, and Karl-Theodor Sturm,** New Directions in Dirichlet Forms, 1998

7 **D. A. Buell and J. T. Teitelbaum, Editors,** Computational Perspectives on Number Theory, 1998

6 **Harold Levine,** Partial Differential Equations, 1997

5 **Qi-keng Lu, Stephen S.-T. Yau, and Anatoly Libgober, Editors,** Singularities and Complex Geometry, 1997

4 **Vyjayanthi Chari and Ivan B. Penkov, Editors,** Modular Interfaces: Modular Lie Algebras, Quantum Groups, and Lie Superalgebras, 1997

3 **Xia-Xi Ding and Tai-Ping Liu, Editors,** Nonlinear Evolutionary Partial Differential Equations, 1997

2.2 **William H. Kazez, Editor,** Geometric Topology, 1997

2.1 **William H. Kazez, Editor,** Geometric Topology, 1997

1 **B. Greene and S.-T. Yau, Editors,** Mirror Symmetry II, 1997